Design for Energy
and the
Environment

Proceedings of the
Seventh International Conference
on the Foundations of
Computer-Aided Process Design

Design for Energy
and the
Environment

Proceedings of the
Seventh International Conference
on the Foundations of
Computer-Aided Process Design

Edited by

Mahmoud M. El-Halwagi
and
Andreas A. Linninger

CRC Press
Taylor & Francis Group
Boca Raton London New York

CRC Press is an imprint of the
Taylor & Francis Group, an **informa** business

CRC Press
Taylor & Francis Group
6000 Broken Sound Parkway NW, Suite 300
Boca Raton, FL 33487-2742

International Standard Book Number: 978-1-4398-0912-9 (Hardback)

Library of Congress Cataloging-in-Publication Data

International Conference on Foundations of Computer-Aided Process Design (7th : 2009
: Breckenridge, Colo.)
    Design for energy and the environment : proceedings of the Seventh International
Conference on the Foundations of Computer-Aided Process Design / editors, Mahmoud
El-Halwagi, Andreas A. Linninger.
        p. cm.
    "A CRC title."
    Includes bibliographical references.
    Papers from the conference held June 7-12, Breckenridge, Colorado.
    ISBN 978-1-4398-0912-9 (hardcover : alk. paper)
    1. Chemical processes--Data processing--Congresses. 2. Chemical process
control--Computer-aided design--Congresses. 3. Sustainable engineering--Congresses.
4. Biomass energy--Congresses. I. El Halwagi, Mahmoud M., 1962- II. Linninger,
Andreas A. III. Title.

TP155.7.I552 2009
660--dc22
                                                                    2009014910

Table of Contents

Preface

The process industries exert some of the most profound impacts on energy resources and the environment. While there are various industrial approaches to enhancing sustainability by improving energy utilization and reducing negative impact on the environment, design is the core industrial activity that addresses and reconciles the key process objectives including sustainability. This book provides state-of-the-art coverage of innovative design approaches and technological pathways that impact energy and environmental issues of new and existing processes. The book is based on selected papers from the International Conference on the Foundations of Computer-Aided Process Design (FOCAPD). The FOCAPD is the premier international conference that exclusively focuses on the fundamentals and applications of computer-aided design for the process industries. Held every five years, FOCAPD 2009, which took place in Breckenridge, Colorado from June 7–12, 2009, was the seventh in this series. The key theme of the conference was *design for energy and the environment*. The conference activities were designed to provide a forum for reviewing and assessing state-of-the-art contributions in process design, sustaining vigorous dialogue and debate on directions and opportunities for process design, and setting future trends and directions for design (particularly in areas pertaining to energy and the environment). The contributions in the book reflect these trends and directions.

Key thrust areas of FOCAPD 2009 (and the book) include:

- Energy systems design and alternative energy sources
- Design for sustainability and energy efficiency
- Robust and uncertain systems
- Design of biofuels, biological processes, and biorefineries
- Emerging technologies and processes for the environment
- Future process design education
- Multiscale and complex systems
- Process and process design
- Emerging tools and techniques in process design

We would like to acknowledge individuals and organizations that helped with the conference and the book. A great deal of appreciation goes to the authors of the book chapters that represent a fine selection of the world's leading authorities in the area of sustainable process design. All book chapters have gone through a peer review process. The reviewers and session chairs are gratefully acknowledged. We would also like to thank Ms. Robin Craven for her excellent efforts in handling the submission and the review process. CACHE Corporation and Drs. Tom Edgar and Jeff Siirola are acknowledged for the continued support of the FOCAPD and the book. We are very grateful for the outstanding job done by the Taylor & Francis team including Ms. Jill Jurgensen, Ms. Suzanne Lassandro, Ms. Allison Shatkin, and their excellent associates. Finally, we would like to express our appreciation for the financial support provided by the National Science Foundation, ASPEN Technology, Eastman Chemical Company, and Taylor & Francis Group, LLC.

Mahmoud M. El-Halwagi
and Andreas A. Linninger

Contributors

Ronald Abbasi
University of the Witwatersrand

Sheraz Abbasi
University of Illinois – Chicago

Ahmed Abdel-Wahab
Texas A&M University – Qatar

Luke Achenie
Virginia Tech University

Majid Al-Gwaiz
Saudi Aramco

Said Al-Hallaj
Illinois Institute of Technology

Mana Al-Owaidah
Saudi Aramco

Rahul Anantharaman
Norwegian University of Science
and Technology

Ryan Andress
Rensselaer Polytechnic University

Audun Aspelund
Norwegian University of Science
and Technology

Selma Atilhan
Texas A&M University

Selen Aydogan-Cremaschi
University of Tulsa

Bhavik Bakshi
Ohio State University

Michael Baldea
Praxair Technology Center

Anil Baral
Ohio State University

Massimiliano Barolo
University of Padova

Paul Barton
MIT

Sukhi Basati
University of Illinois – Chicago

Bill Batchelor
Texas A&M University

Amro El-Baz
Zagazig University

David Beneke
University of the Witwatersrand

Wayne Bequette
Rensselaer Polytechnic University

Fabrizio Bezzo
University of Padova

Abdullah Bin Mahfouz
Texas A&M University

Olav Bolland
Norwegian University of Science
and Technology

Susilpa Bommareddy
Auburn University

Vijayasekaran Boovaragavan
Tennessee Technological University

Ian Bowling
Texas A&M University

Richard Braatz
University of Illinois at
  Urbana-Champaign

Benjamin Brant
Byogy Renewables, Inc.

Joan Brennecke
Notre Dame University

Linda Broadbelt
Northwestern University

Jose Caballero
University of Alicante

Kyle Camarda
University of Kansas

Lisa Cardo
GlaxoSmithKline

Ana Carvalho
Instituto Superior Técnico

Virginie Chambost
École Polytechnique – Montréal

Chuei-Tin Chang
National Cheng Kung University

Alexandre Chapeaux
Notre Dame University

Nishanth Chemmangattuvalappil
Auburn University

Cheng-Liang Chen
National Taiwan University

Kejia Chen
University of Illinois at
  Urbana-Champaign

Irene Mei Leng Chew
University of Nottingham – Malaysia

Jun-Ki Choi
Ohio State University

Luis Cisternas
Universidad de Antofagasta
  and CICITEM

David Constable
GlaxoSmithKline

David Culver
DuPont

Louis Patrick Dansereau
École Polytechnique – Montréal

James Davis
University of California - Los
  Angeles

Prodromos Daoutidis
University of Minnesota

Urmila Diwekar
Vishwamitra Research Institute

Russell Dunn
Polymer and Chemical
  Technologies, LLC

Mario Eden
Auburn University

Thomas Edgar
University of Texas

Edi Eliezer
BioPrizM

Mahmoud El-Halwagi
Texas A&M University

Diaa El-Monayeri
Texas A&M University

Aly-Eldeen ElTayeb
Illinois Institute of Technology

Joshua Enszer
Notre Dame University

John Eslick
University of Kansas

L.T. Fan
Kansas State University

Noemi Ferrari
Politecnico di Milano

Christodoulos Floudas
Princeton University

David Follansbee
Rensselaer Polytechnic University

Dominic Chwan Yee Foo
University of Nottingham – Malaysia

Ferenc Friedler
University of Pannonia

Jeffrey Froyd
Texas A&M University

Federico Galvanin
University of Padova

Edelmira Gálvez
Universidad Catolica del Norte
    and CICITEM

Simone Gamba
Politecnico di Milano

Rafiqul Gani
Technical University of Denmark

Michael Georgiadis
Imperial College London

Krist Gernaey
Technical University of Denmark

David Glasser
University of the Witwatersrand

Charles Glover
Texas A&M University

John Gossage
Lamar University

Chrysanthos Gounaris
Princeton University

Roberto Grana
Politecnico di Milano

Johan Grievink
Delft University of Technology

Ignacio Grossmann
Carnegie Mellon University

Geoffrey Grubb
Ohio State University

Gonzalo Guillén-Gozálbez
University Rovira i Virgili

Truls Gundersen
Norwegian University of Science
    and Technology

Timothy Haab
Ohio State University

Kenneth Hall
Texas A&M University

Sean Hansrote
University of Kentucky

Andreas Harwardt
RWTH Aachen University

M. M. Faruque Hasan
National University of Singapore

Shinji Hasebe
Kyoto University

Vassily Hatzimanikatis
École Polytechnique Fédérale de
  Lausanne

Brendon Hausberger
University of Witwatersrand

Manuel Hechinger
RWTH Aachen University

Carlos Henao
University of Wisconsin – Madison

Richard Henderson
GlaxoSmithKline

Christopher Henry
Argonne National Laboratory

Diane Hildebrandt
University of Witwatersrand

Nicholas Hoffmann
University of Kansas

Simon Holland
University of Witwatersrand

Yinlun Huang
Wayne State University

Dong Hong-guang
Dalian University of Technology

Szu-Wen Hung
National Taiwan University

Konrad Hungerbühler
Swiss Federal Institute of Technology

Madhu Iyer
University of Illinois – Chicago

Peter Jansens
Delft University of Technology

Matty Janssen
École Polytechnique – Montréal

Dionicio Jantes-Jaramillo
University of Guanajuato

Robert Jernigan
Auburn University

Concepción Jiménez-González
GlaxoSmithKline

Arturo Jiménez-Gutiérrez
Instituto Tecnológico de Celaya

Sujit Jogwar
University of Minnesota

Mohd. Kamaruddin Abd. Hamid
Technical University of Denmark

Manabu Kano
Kyoto University

Iftekhar Karimi
National University of Singapore

Arunprakash Karunanithi
University of Denver

Rick Kemp
Sandia National Laboratories

Omar Khalil
Illinois Institute of Technology

Vikas Khanna
Ohio State University

Seonbyeong Kim
University of Illinois – Chicago

Antonis Kokossis
University of Surrey

Cleo Kontoravdi
Imperial College London

Sven Kossack
RWTH Aachen University

Konstantinos Kouramas
Imperial College London

Korbinian Kraemer
RWTH Aachen University

Herman Kramer
Delft University of Technology

Claudia Labrador-Darder
University of Surrey

Carl Laird
Texas A&M University

Richard Lakerveld
Delft University of Technology

Jae Lee
The City College of New York

Jui-Yuan Lee
National Taiwan University

Bao-Hong Li
Dalian Nationalities University

Kuyen Li
Lamar University

Zheng Li
Tsinghua University

Xiang Li
Lamar University

Li Li-Juan
Dalian University of Technology

David Linke
Leibniz Institute for Catalysis

Patrick Linke
Texas A&M University – Qatar

Andreas Linninger
University of Illinois – Chicago

Chaowei Liu
Lamar University

Pei Liu
Imperial College London

Zheng Liu
Wayne State University

Helen Lou
Lamar University

Eva Lovelady
Texas A&M University

Di Lu
Georgia Tech University

Angelo Lucia
University of Rhode Island

Michael Luyben
DuPont Engineering Research &
Technology

Sandro Macchietto
Imperial College London

Vladimir Mahalec
McMaster University

Y.L. Maldonado
University of San Luis Potosí

Davide Manca
Politecnico di Milano

Flavio Manenti
Politecnico di Milano

Behrang Mansoornejad
École Polytechnique – Montréal

Christos Maravelias
University of Wisconsin – Madison

Thomas Marlin
McMaster University

Wolfgang Marquardt
RWTH Aachen University

Lealon Martin
Rensselaer Polytechnic University

Henrique Matos
Instituto Superior Técnico

Samantha McLeese
University of Kansas

Pedro Medellín
University of San Luis Potosí

Anne Meyer
Technical University of Denmark

James Miller
Sandia National Laboratories

Patrick Mills
Texas A&M University

Terry Mills
Polymer and Chemical
  Technologies, LLC

Ruth Misener
Princeton University

William Mitsch
Ohio State University

Marcelo Montenegro
Universidad de Antofagasta

Daniel Montolio-Rodriguez
University of Surrey

Jeonghwa Moon
University of Illinois – Chicago

Mario Moscosa
University of San Luis Potosí

Ahmed Nadim
McMaster University

Mahmoud Bahy Noureldin
Saudi Aramco

Denny Kok Sum Ng
University of Nottingham
  – Malaysia

Andrew Odjo
University of Alicante

Seza Orcun
Purdue University

Irvin Osborne-Lee
Prairie View A&M University

Karla Ossandón
Universidad de Antofagasta

John Paccione
Rensselaer Polytechnic University

Stravos Papadokonstantakis
Swiss Federal Institute of
  Technology

Maria Papakosta
Imperial College London

Bilal Patel
University of Witwatersrand

Larissa Pchenitchnaia
Texas A&M University

Joseph Pekny
Purdue University

Laura A. Pellegrini
Politecnico di Milano

Viet Pham
Texas A&M University

Martín Picón-Núñez
University of Guanajuato

Ali Pilehvari
Texas A&M University – Kingsville

Cristina Piluso
BASF Corporation

Patricio Pinto
SKM MinMetal

Efstratios Pistikopoulos
Imperial College London

G. Pokoo-Akins
Texas A&M University

Graham Polley
University of Guanajuato

José María Ponce-Ortega
Instituto Tecnológico de Celaya

Heinz Preisig
Norwegian University of Science
and Technology

Venkatasailanathan Ramadesigan
Tennessee Technological University

Kandace Ramey
University of Kentucky

Md. Shamsuzzaman Razib
National University of Singapore

Matthew Realff
Georgia Tech University

Gintaras Reklaitis
Purdue University

Luke Richardson
University of Kentucky

Keith Riegel
DuPont Engineering Research &
Technology

Daniel Rudnick
Byogy Renewables, Inc.

Zhou Rui-Jie
Dalian University of Technology

Gerardo Ruiz
University of Illinois – Chicago

Nikolaos Sahinidis
Carnegie Mellon University

Juan Salazar
Vishwamitra Research Institute

Norman Sammons Jr.
Auburn University

Apurva Samudra
Carnegie Mellon University

Aaron Scurto
University of Kansas

Jeffrey Seay
University of Kentucky

Baraka Celestin Sempuga
University of Witwatersrand

Felipe Sepulveda
Universidad Catolica del Norte

Nilay Shah
Centre for Process Systems
Engineering

Yousef Sharifi
Virginia Tech University

Yogendra Shastri
Vishwamitra Research Institute

Luke Simoni
Notre Dame University

Gürkan Sin
Technical University of Denmark

Aditi Singh
Lamar University

Giorgio S. Soave
San Donato Milanese

Earl O. P. Solis
MIT

Charles Solvason
Auburn University

H. Dennis Spriggs
Byogy Renewables, Inc.

Mark Stadtherr
Notre Dame University

Kathrin Stephan
RWTH Aachen University

George Stephanopoulos
MIT

Paul Stuart
École Polytechnique – Montréal

Venkat Subramanian
Tennessee Technological University

Vincentius Surya Kurnia Adi
National Cheng Kung University

Brian Sweetman
University of Illinois – Chicago

Andrej Szijjarto
Swiss Federal Institute of
Technology

Raymond Tan
De La Salle University – Manila

Fouad Teymour
Illinois Institute of Technology

Osamu Tonomura
Kyoto University

Eman Tora
Texas A&M University

Robert Urban
Ohio State University

Bruce Vrana
DuPont

Michaela Vrey
University of Witwatersrand

Xiao Wu
Dalian University of Technology

Jie Xiao
Wayne State University

Qiang Xu
Lamar University

Yihui Tom Xu
DuPont Engineering Research &
Technology

Aidong Yang
University of Surrey

Xiongtao Yang
Lamar University

Prasad Yedlapalli
The City College of New York

Wei Yuan
Auburn University

Lale Yurttas
Texas A&M University

Andrea Zamboni
University of Padova

Junshe Zhang
The City College of New York

Libin Zhang
University of Illinois – Chicago

Tengyan Zhang
Kansas State University

Yu Zhu
Texas A&M University

Stephen Zitney
U.S. Department of Energy

1

"The Emerald Forest"—An Integrated Approach for Sustainable Community Development and Bio-derived Energy Generation

Fouad Teymour, Said Al-Hallaj, Aly-Eldeen ElTayeb, and Omar Khalil

Department of Chemical and Biological Engineering
Illinois Institute of Technology

CONTENTS

The welfare of humankind seems to be on a crash course with two fast approaching major potential crises: Namely, the possibility of depletion of fossil fuel reserves, which so far have single-handedly supported the complete needs of the post-industrialization economy, and the alarming increase of carbon dioxide in the atmosphere to unprecedented levels, which is predicted to lead to a catastrophic global warming phenomenon. These problems are compounded by the explosive rate of population increase in some parts of the globe and the fast depletion of worldwide usable land resources. The recent sudden decrease in oil and gas prices might lead the policy planners to relegate the plans for renewable energy research to the back burner, but we must not forget the skyrocketing rate of increase in energy costs of only a year ago. There is ample evidence that the future of the human race depends on immediate and expedient plans for sustainable development. Where energy needs are concerned, the arguments proposed here indicate the urgent need for the development of alternative fuel forms that, if not carbon-free, are at least carbon-neutral. This makes a strong case for biomass-derived fuels, which some proffer as a stop-gap measure to help us transition to the promised hydrogen economy, but which will arguably always be an essential element of the global energy economy, as detailed herein. This article is a concept paper that proposes an integrated approach, in which the

1

large-scale massive production of biomass for bio-derived fuel production is coupled with the development of sustainable communities through the reclamation of arid and/or uninhabitable regions. Many challenges exist to the successful implementation of efficient integrated developments of this type; we focus briefly on one of these challenges, namely the design of optical assemblies for optimal sunlight utilization.

The utilization of biomass as a source of fuels is a carbon-neutral process that consumes about as much carbon dioxide in growing the biomass as is eventually released during the combustion and/or gasification steps. Photosynthetic biological entities are efficient agents that primarily capture solar energy and combine it with a carbon source and some essential nutrients to increase the amount of, highly usable, biomass. To date, there have been major successes in producing bioethanol, via fermentation, from carbohydrate rich agricultural crops. Ethanol is being produced in the US and China from corn. Brazil has also developed a comprehensive ethanol fuel infrastructure based on its abundant production of sugarcane. Europe has had successes in producing biodiesel from oil-rich crops. These approaches however are heavily criticized, and arguably deserve such criticism, by their opponents for many factors: The dearth of arable land, the competition for human and cattle food supply, the need for immense land masses that represent a sizeable proportion of the earth's total, and the often necessary utilization of precious freshwater resources. Energy crops, such as switchgrass, have been proposed as alternatives that do not compete with essential food supply, still require arable land but not necessarily the premium soil quality that corn requires, and that grow more efficiently than traditional agricultural crops. These are currently being investigated and might offer a sustainable solution, but fall short of the potential detailed here of land-based aquatic biomass.

Integrated Sustainable Communities

The concept presented here integrates the production of biomass and biofuels with selected solar and wind energy technologies to develop combined sustainable living communities and green energy production facilities. An example implementation is presented in Figure 1, which shows the detailed interactions among the various renewable energy technologies used and the sustainable components. The two major elements of the system presented in Figure 1, are the "Emerald Forest" biomass production facility and the living sustainable community. Such developments involving these massive artificial forests are best located on desert lands that are in proximity of ocean waters. Many such areas are available in the United States, for example along the southeast coast, the Gulf of Mexico, the western states and parts of Nevada,

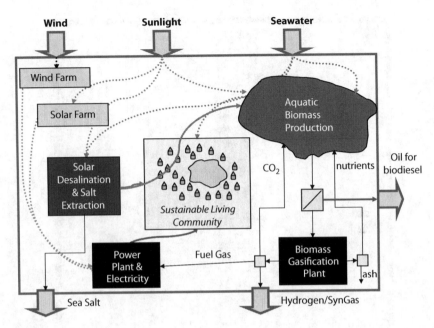

FIGURE 1
An example of sustainable community development integrated with biofuel production.

Arizona and New Mexico. The worldwide potential is orders of magnitude larger, for example the African Sahara desert can be alone sufficient for the production of a considerable portion of the world's liquid fuels need.

The forest consists of tree-like bioreactors filled with seawater and algae and continuously fed with carbon dioxide and nutrients. The tree-like design is selected for its natural efficiency in capturing sunlight. A potential design for the tree photobioreactors is presented in Figure 2 and discussed later.

The integrated development (Figure 1) is centered around a housing community with artificial lakes and ponds. These serve an important ecological function in regulating the arid desert microclimate by providing moisture through evaporation and the accompanying evaporative cooling effect. However, since excessive evaporation can lead to a non-sustainable increase in salt concentration, these ponds have to be supplemented with salinity-control technologies. A solar desalination plant is included to produce sufficient amounts of fresh water for make-up of the evaporation losses both from the communal lakes and ponds, as well as from the biomass artificial forest. Some of the energy needs of the compound derive from solar photovoltaic technology, and, if suitable to the specific location, for a windmill farm generating electrical power. However, the majority of the power is generated in a power plant that uses a portion of the fuel gas generated through gasification of the biomass produced in the artificial forest. The gasification plant receives the cellulosic portion of the produced algal biomass after the oil and carbohydrate portions are

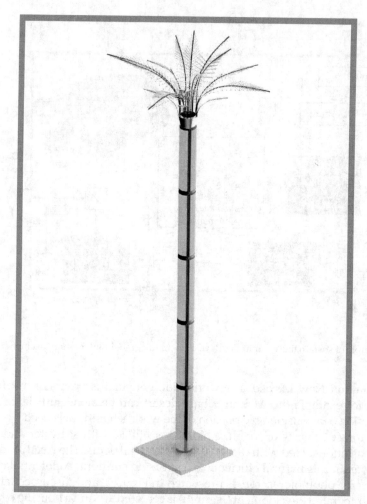

FIGURE 2
Artist rendering of one possible design for the photobioreactors.

separated. These are sent for off-site processing in biodiesel and bio-ethanol/ butanol facilities to produce liquid fuels for automotive applications. The gasification step can be optimized to produce hydrogen, or syn gas. As mentioned, a small portion of this fuel gas will be used onsite to generate the majority of the energy needed, while the larger portion is used for energy production to support the needs of nearby cities and communities.

The cornerstone of sustainability in this approach is the use of the carbon dioxide generated in the gasifier (after separation from the fuel gas) to provide a carbon-rich source for the algae nutrition and growth. The micronutrients that are also needed for algae are recycled after separation from the ash remaining after gasification.

In order for this approach to sustainability to work, the scale of production of biomass should be optimized in relation to the size and energy requirements of the living community and the available natural resources in the area. The economics have to be balanced such that the external energy production revenue exceeds the cost of internal energy generation. Calculations for typical scenarios will be presented at the conference.

Many challenges and obstacles need to be resolved for the efficient design and implementation of such integrated communities, and especially of the central tree-like bioreactors. These issues include the development of algal strains optimized for the area and for the energy need distribution, development of weather-resistant and bio-compatible polymeric materials, the development of technology for carbon dioxide separation and absorption, as well as the development of an optimized optical capture and distribution assembly. The latter is discussed in further detail below.

Design of Photobioreactors for Optimal Light Distribution

The main bottleneck in rendering closed algae production systems feasible is the limitations on the availability of light at high concentrations of biomass. Consequently, devising methods for minimizing light limitations is crucial for the future production of algae on large scales, specifically the scale required for use as a sustainable energy source and efficient medium for sequestration of carbon dioxide. Different designs have been suggested, most commonly tubular and flat-plate bioreactors. Both suffer from a main serious limitation: low ground productivity (defined as unit algae produced per unit of land). In this work, we present a tree-like photobioreactor design fitted with an optical capture assembly to maximize the collection of solar radiation, and optimize the distribution of light to dark regions inside the reactor. This design allows the use of reactors with larger diameters, and much larger capacities than the traditional algae pond production systems, hence improving the algae ground productivity.

Using optical assemblies for light collection and distribution has been proposed in the literature on multiple instances. Pulz et al. (1) experimentally studied the use of optical fibers in illuminating a plate-type photobioreactor. They concluded that optical fibers do, as expected, improve the productivity. Jin-Lan et al. (2) studied algae growth in a 40-Liter five-compartment rectangular reactor with a triangular optical guide collector system and perpendicular optical guides for light distribution. The distribution guides are plates fitted with grooves to diffuse light in the lateral direction. The compartments are connected together to allow circulation of biomass for harvesting. They used geometric optics to determine the dimensions of the optical guide that would reduce losses. The experimental data provided is however very limited and not enough to judge the efficiency of the reactor or that of the optical assembly. Cuello et al. (3) provide a short review of algae photobioreactors with optical fibers for concentration and distribution of light, focusing on the work funded by NASA and the DOE for solar-concentration and transmission using mirrors, lenses and optical fibers. They also review their own work on light distribution in flat-plate algae bioreactors. From their review, they conclude that optical-fiber based distribution systems are very promising but have not yet been sufficiently developed. They report that solar-concentrating systems have reached efficiencies higher than 45%.

The photobioreactor design exemplified in Figure 2 consists of a large vertical cylindrical reactor of up to 15 meters high, and an internal capacity of 2-3 m^3 filled with the algal suspension. The top portion of the tree sports an optical capture assembly consisting of branches and leaves built from optical waveguides and complemented with holographic concentrators and photovoltaic elements. The assembly is designed to capture sunlight and redirect it towards an internal illumination optical shaft. The light is thus dispersed into the inside of the reactor both form the internal shaft, as well as from the external transparent walls. In order to design the optical system in an optimized method, we have to explore the different methodologies for the modeling of light travel and distribution.

Modeling Light Distribution

In order to *a priori* predict the performance of new reactor designs, it is necessary to develop a reliable tool for modeling of light distribution inside the algal medium. Several models have been used in the literature, the main approaches are discussed.

Geometric Ray tracing: The simplest approach for modeling the propagation of light is ray tracing, which finds its foundation in geometric optics. Fermat's principle, the cornerstone of geometric optics, states that "Light rays follow a path that is an extremum compared to other nearby paths," which in other words means that light follows paths with the shortest optical length, i.e. the path that takes the shortest amount of time. The laws of reflection and refraction are both derived based on that same principle. The application of ray tracing generally requires the assumption of a uniform and nonabsorbent medium. Strong restrictions are thus imposed on the use of ray tracing in modeling photobioreactors, limiting their use in early stages of operation (where biomass concentration is still significantly low and the medium is for the most part non-participating i.e. no absorption and no scattering), and also for use in modeling optical assemblies that collect/distribute light. Zijffers et al (4) used ray tracing without scattering in modeling light distribution for a novel design of algae photoreactors, which includes a Fresnel lens for light collection and an optical guide for distribution. Their model was limited to the optical assembly and was used to determine the amount of radiation reaching the surface of the algal solution. Mohseni et al (5) used the Monte Carlo approach with the ray tracing method to determine light intensity distribution in a fluidized bed photoreactor for the removal of organic pollutants from water using TiO_2 catalyst. A stochastic method was first used to determine a homogeneous particle distribution in the reactor. The results were validated by comparing the measured values of effective transmittance with those predicted by the model. For the most part, the results were in good agreement with experimental data, although the transmittance was under-predicted for higher bed expansions, which the authors attribute to errors in prediction of particle distribution.

The Radiative Transport Equation (RTE) has been used since the 1960s in modeling radiative heat transfer (6), the general RTE can be derived by performing a photon balance on a fixed volume, in a manner similar to that used in deriving the general transport equation. A detailed derivation of the RTE is provided in (6). The final form of the equation for a frequency v and direction of propagation Ω is:

$$\frac{dl_v(s,\Omega)}{ds} + [k_v + \sigma_v]l_v(s,\Omega) = j_v^\theta(s,T) + \frac{\sigma_v}{4\pi} \int_{\Omega^t=4\pi} P(\Omega' \to \Omega)I_v(s,\Omega)\,d\Omega'$$

where $I_{\Omega,s}(s,\Omega)$ is the amount of radiation reaching point s in direction of propagation Ω; k_v and σ_v are the absorption and scattering coefficients respectively; j_v^s is the emission term, usually modeled using the Planck's blackbody radiation model making it a strong function of temperature; and $P(\Omega \to \Omega')$ is the phase function, which determines the probability of in-scattering of radiation from all solid angles Ω' into a volume defined by solid angle Ω.

The first term on the left hand side of the equation represents the amount of radiation per unit volume, whereas the second term represents the extinction by absorption and out-scattering. The first term on the right hand side represents the radiation emitted, and the last term represents the radiation entering the control volume (which is defined by a solid angle) by scattering from other control volumes. It is important to note that this equation is written for a certain waveband, v, for which the absorption, scattering and emission are defined. The equation may therefore be solved for different frequencies of interest separately, or solved for an average frequency by averaging the different properties that are functions of frequency.

Several assumptions are made in the derivation of this equation (6), most notably:

1. Scattering is coherent, i.e. the frequency of radiation is not changed after scattering.

2. Single scattering: the energy scattered is not re-scattered again.

3. The incident radiation at any point, G, can then be found by integrating the radiation intensity on all solid angles:

$$G_v(x,y,z) = \int_\Omega I_v(s,\Omega)\,d\Omega = \int_{\theta=0}^{\pi/2}\int_{\phi=0}^{2\pi} I_v(x,y,z,\theta,\phi)\sin\theta\,d\phi\,d\theta$$

The integro-differential RTE equation has no analytical solution except for highly idealized cases. Thus the discrete ordinates (DO) method is used to solve it by discretizing the terms in the equation spatially and directionally (i.e. angularly). The main advantage of the DO model is that it solves the complete RTE, with no assumptions leading to inherent errors. The main source of error is thus associated with the discretization and can be minimized by using a fine spatial and angular grid.

The original DO model does not conserve radiant energy at the surfaces in complex geometries (7), and therefore a conservative variant of the DO model, the finite volume method, is usually used in commercial CFD packages. This allows relatively simple integration of fluid dynamics with radiation modeling in the same environment.

There has been a lot of work on modeling light in photoreactors using the DO model, especially reactors for removal of pollutants in the presence of a photocatalyst, usually TiO_2. Sgalari et al (8) used the DO method to model the distribution of light in an annular photoreactor, and used results from Monte Carlo simulations to validate their results. For photo removal of a pollutant, Adesina et al (9) used the DO model with the granular Eulerian-Eulerian model to determine light intensity distribution with scattering due to the presence of photocatalyst. Trujillo et al (10-11) used the DO model in conjunction with Eulerian-Eulerian approach to describe the liquid-gas mixture to model the light intensity distribution in photocatalytic reactor. In both cases, the flow problem was solved first to obtain local solid volume fractions/gas holdups (since catalyst particles and bubbles are the main scatterers in these problem). The data are then used to determine the absorption and scattering coefficients, and hence predict distribution of light in the reactor and the local amounts of pollutant removed. Pareek et al (12) used the DO to model the same annular photoreactor with TiO_2 catalyst, and included a distributed light source using the commercial CFD package FLUENT (Ansys, USA). They assessed the effects of wall reflectivity, catalyst loading and phase function parameter.

In our work, the DO implementation in FLUENT is used to solve the RTE for several candidate designs for the optical capture and distribution assembly, the internal light shaft, and the algal suspension. Results are discussed in the presentation.

References

1. Pulz, O., Gerbsch, N., Buchholz, R., 1995, Light energy supply in plate-type and light diffusing optical fiber bioreactors. *Journal of Applied Phycology*, Vol. 7, pp. 145–149.
2. Xia Jin-lan, Levert J. M, Benjelloun F., Glavie P., Lhoir P. 4, 2002, Design of a novel photobioreactor for culture of microalgae. *Wuhan University Journal of Natural Sciences*, Vol. 7, pp. 486–492.
3. Ono, E., Cuello, J.L., 2004, Design Parameters of solar concentrating systems for CO2-mitigating algal bioreactors. *Energy*, Vol. 29, pp. 1651–1657.
4. Jan-Willem F. Zijffers*, Sina Salim, Marcel Janssen, Johannes Tramper, René H. Wijffels. 2008, Capturing sunlight into a photobioreactor: Ray tracing simulations of the propagation of light from capture to distribution into the reactor. *Chemical Engineering Journal*, (In press).
5. Mohseni, M., Gustavo E. Imoberdorf, Fariborz Taghipour, Mehrdad Keshmiri, Predictive radiation field modeling for fluidized bed photocatalytic reactors. 2008, *Chemical Engineering Science*, Vol. 63, pp. 4228–4238.
6. Chandrasekhar, S., Radiative Transfer. New York : Dover Publishers, 1960.
7. J. Pruvosta, J.-F. Cornet, J. Legrand. 2008, Hydrodynamics influence on light conversion in photobioreactors: An energetically consistent analysis. *Chemical Engineering Science*, Vol. 63, pp. 3679–3694.
8. Vishnu K. Pareek, Adesoji A. Adesina, 2004, Light intensity distribution in a photocatalytic reactor using finite volume. *AIChE Journal*, Vol. 50, pp. 1273–1288.
9. G. Sgalari, G. Camera-Roda, and F. Santarelli. 5, 1998, Discrete ordinate method in the analysis of radiative transfer in photocatalytically reacting media. *Int. Comm. Heat Mass Transfer*, Vol. 25, pp. 651–660.
10. Francisco J. Trujillo, Tomasz Safinski, Adesoji A. Adesina. 2007, CFD analysis of the radiation distribution in a new immobilized catalyst bubble column externally illuminated photoreactor. *Journal of Solar Energy Engineering*, Vol. 129, pp. 27–36.
11. Francisco J. Trujillo, Ivy A-L Lee, Chen-Han Hsu, Tomasz Safinski, Aesoji A. Adesina. 2008, Hydrodynamically-enhanced light intensity distribution in an externally-irradiated novel aerated photoreactor: CFD simulation and experimental studies. *International Journal of Chemical Reactor Engineering*, Vol. 6.
12. V.K. Pareek, S.J. Coxb, M.P. Brungsb, B. Youngc, A.A. Adesina. 2003, Computational fluid dynamic (CFD) simulation of a pilot-scale annularbubble column photocatalytic reactor. *Chemical Engineering Science*, Vol. 58, pp. 859–865.

2

Dynamic Optimisation and Multi-Parametric Model Predictive Control of Hydrogen Desorption in Metal-Hydride Bed Storages

**M. Papakosta[1], K.I. Kouramas[1], C. Kontoravdi[1],
M.C. Georgiadis[2] and E.N. Pistikopoulos[1*]**

[1]*Centre for Process Systems Engineering, Department of Chemical Engineering,
Imperial College London, London SW7 2AZ, UK*
[2]*Department of Engineering Informatics and Telecommunications, University
of Western Macedonia, Karamanli and Lygeris Street, 50100 Kozani, Greece*

CONTENTS

ABSTRACT This paper presents a systematic approach for the optimal design and evaluation of a novel model-based control method for the control of hydrogen desorption in metal-hydride bed storages. An existing dynamic model for a metal-hydride storage tank, developed in Kikkinides et al. (2006) and Georgiadis et al. (2009), is used for performing optimization and control studies for the hydrogen desorption in metal-hydride beds. Two types of controllers are presented, a Proportional Integral and a Multi-Parametric MPC, and their performances are evaluated with simulations in nominal and perturbed operating conditions.

KEYWORDS *Hydrogen storage, metal-hydride beds, dynamic optimization, multi-parametric MPC*

* Corresponding author, email: e.pistikopoulos@imperial.ac.uk

Introduction

The potential of hydrogen as a promising alternative, environmentally friendly fuel carrier, that can be produced by a variety of renewable energy sources with nearly zero emissions of pollutants and greenhouse gases (Dhaou et al., 2006), has led to a significant research effort in hydrogen technologies. However, the worldwide conversion from fossil fuels to hydrogen and the commercialization of hydrogen especially for mobile applications (such as transportation) still has to overcome several barriers in terms of storage. Technologies that have been developed for hydrogen storage, such as compression and liquefaction, are commercially impractical due to the heavy gas tanks involved and high energy requirements. The storage of hydrogen in metal hydrides has been proposed as an alternative technology to overcome these issues (Jemni and Nasrallah, 1995).

A hydrogen storage alloy is capable of adsorbing and releasing hydrogen, with the adsorption and release rates being controlled by adjusting the temperature or pressure. Several studies and analysis have been performed on modeling the dynamic behavior of hydrogen storage in metal-hydride beds. In Askri et al. (2003) a 2D mathematical model was presented for a metal-hydride bed to examine the effects of heat and flow transfer while Dermican et al. (2005), Man et al. (2004) and Muhittin et al. (2005) investigated various bed geometries and their effect on the adsorption process. Bhouri et al. (2000), performed preliminary computational studies of hydrogen desorption. Kikkinides et al. (2006) and Georgiadis et al. (2009) presented a systematic approach for the optimal design, simulation and control of an enhanced tubular metal-hydride bed with additional heat exchangers. It was shown that by appropriately controlling the thermal dynamics of the bed, the adsorption rate of hydrogen in a metal-hydride can be controlled.

The study of advanced process control methods, such as Model Predictive Control, has been limited for hydrogen storages based on metal-hydride beds. In Georgiadis et al. (2008) an advanced Multi-Parametric Model-Based Predictive Controller (mp-MPC) was developed for the first time for the adsorption of hydrogen in metal-hydride beds. This paper offers a similar design and evaluation strategy of mp-MPC control for the desorption of hydrogen in metal-hydride beds. Based on the previous work of Kikkinides et al. (2006) and Georgiadis et al. (2009), a mathematical model for desorption of hydrogen in metal-hydride beds is presented and will be used as the base for performing optimization and control design studies. Dynamic simulation of the bed and dynamic optimization studies are performed where the control objective for the adsorption process, is to ensure minimum time desorption of hydrogen in the bed. Simulations and optimization results are then used to obtain reduced order input-output models via model identification, suitable for applying mp-MPC design studies. Finally the evaluation of the developed controller is performed by comparing its performance to that of a conventional PID controller.

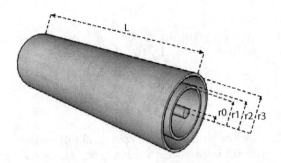

FIGURE 1
Enhanced metal-hydride bed reactor with heat exchangers (Georgiadis et al., 2009).

Mathematical Model of the Metal-Hydride Bed

This study focuses on an enhanced metal-hydride tubular reactor, shown in Figure 1. The reactor is filled with a metal hydride alloy, namely LaNi$_5$ and is equipped with a cooling/heating system that controls the desorption process. The heating/cooling system consists of a concentric tube of diameter r_0, a concentric annular ring of thickness $\delta r = r_2 - r_1$ and a jacket filled with a heating medium surrounding the reactor. Heating is provided to the system to enhance the endothermic hydrogen desorption process. A 2D mathematical model for the desorption process and the heat exchangers was developed in Kikkinides et al. (2006) and is presented here briefly. This mathematical model is based on the following assumptions: 1) at the beginning of the process the hydride bed is filled with hydrogen and the bed temperature coincides with the inlet temperature, 2) the ideal gas law rules in the gas phase, 3) axial and radial dispersion are included in the mass balances for interstitial fluid, 4) the axial and radial pressure drops in the bed depend linearly on the fluid velocity through Darsy's law ot Blake-Cozeny equation.

The model of the metal hydride reactor with the heat exchangers comprises of a set of integral, partial differential and algebraic equations (IPDAE) which are described next.

1. *Mass balance of hydrogen in the gas phase (bulk of the reactor)*

$$\frac{\partial \hat{\rho}}{\partial \tau} + \frac{1}{\varepsilon} \nabla \cdot (\hat{\rho}\hat{u}) - \left(\frac{1}{Pe_m}\right) \cdot \nabla^2 \hat{\rho} + W_0 \cdot \frac{\partial \hat{\rho}_s}{\partial \tau} = 0 \qquad (1)$$

where τ is the dimensionless time, $\tau = L / u_0$ (L is the bed length and u_0 is a reference velocity), $\hat{u} = \left(\hat{u}_z, \hat{u}_r\right)$ is the superficial gas velocity vector, $\hat{\rho} = \rho/\rho_e$ is the density of hydrogen in the gas phase, ρ_e being the density at the equilibrium, and $\hat{\rho}_s$ is the solid (metal) density of the alloy. Pe_m and W_0 are dimensionless parameters defined below and ε is the void fraction of the bed.

2. *Pressure droop equations*

The steady state momentum balance of laminar gas flow through a packet bed can be expressed by the Darcy's law

$$\hat{u}_z = -K_z \frac{\partial \hat{P}}{\partial z}, \quad \hat{u}_r = -K_r \frac{\partial \hat{P}}{\partial r} \tag{2}$$

where K_z, K_r are the dimensionless permeability values of the bed and P_0 is the equilibrium pressure, variables $z = z^*/L$, $r = r^*/R$ are the dimensionless axial and radial distance, normalized by the total length L and bed radius R. It is assumed that in Eq. (2) the transient and inertial terms are negligible (Yang et al., 1993).

3. *Energy balance equation*

$$(\hat{\rho} + \hat{\rho}_s \cdot Le) \cdot \frac{\partial \theta}{\partial \tau} + \left(\frac{1}{\varepsilon}\right) \cdot \hat{u} \cdot \nabla \theta - \left(\frac{1}{Pe_t}\right) \cdot \nabla^2 \theta$$

$$- W_0 \cdot \frac{\partial \hat{\rho}_s}{\partial \tau} \left[\beta + \left(1 - \frac{C_{ps}}{C_{pg}}\right) \theta \right] = 0 \tag{3}$$

where $\theta = T/T_0$ is the gas temperature, T_c being the temperature of the equilibrium state and θ_c is the inlet temperature of the heating medium that is constant in time and space. *Le* is the dimensionless Lewis number.

4. *Mass balance of the metal hydride*

$$(1-\varepsilon) \cdot \frac{\partial \hat{\rho}_s}{\partial \tau} = C_a \cdot t_{res} \cdot \exp(-\varepsilon_d/\theta) \cdot \ln\left(\frac{\hat{P}}{\hat{P}_{eq}}\right) \cdot \left(\hat{\rho}_{sat} - \hat{\rho}_s\right) \tag{4}$$

where $\hat{P}_{eq}$ is the dimensionless equilibrium pressure, $\hat{\rho}_{sat}$ is the saturated bed density and ε_d is the dimensionless activation energy of the reaction kinetics process.

5. *Definition of equilibrium pressure*

$$P_{eq} = f(H/M) \cdot \exp\left[\frac{\Delta H}{R_g} \cdot \left(\frac{1}{T_0 \theta} - \frac{1}{T_{ref}}\right)\right] \tag{5}$$

Where the function f (H/M) is the equilibrium pressure at the reference temperature T_{ref} and is usually obtained by fitting experimental data to a polynomial of nth degree (see more in Kikkinides et al., 2006).

The equations and variables of the model, presented so far, are in dimensionless form. Hence several dimensionless parameters are introduced which are given below

$$W_0 = \left(\frac{1-\varepsilon}{\varepsilon}\right)\frac{\rho_{s0}}{\rho_0}, \quad Le = \frac{W_0 \cdot C_{ps}}{C_{pg}}, \quad \beta = \frac{-\Delta H}{C_{pg} \cdot MW \cdot T_0}$$

$$K_z = \frac{K_0 \cdot P_0}{\mu \cdot u_0 \cdot L}, \quad K_r = \frac{K_0 \cdot P_0}{\mu \cdot u_0 \cdot R}, \quad \varepsilon_d = \frac{E_d}{R_g T_0}, \quad t_{res} = \frac{L}{u_0}$$

$$\lambda_e = \varepsilon\lambda_g + (1-ge)\lambda_g, \quad Pe_{t,z} = \frac{\varepsilon \cdot C_{H_2O} \cdot u_0 \cdot C_{pg} \cdot L}{\lambda_e}$$

$$Pe_{t,r} = \frac{\varepsilon \cdot C_{H_2O} \cdot u_0 \cdot C_{pg} \cdot R}{\lambda_e}, \quad Pe_{m,z} = \frac{\varepsilon u_0 L}{D_z}$$

$$Pe_{m,r} = \frac{\varepsilon u_0 R}{D_r}, \quad Bi_z = \left(\frac{hL}{\lambda e}\right), \quad Bi_r = \left(\frac{hR}{\lambda e}\right)$$

6. *Modeling the heat exchange process*

An energy balance for each heat exchanger inside the reactor has also to be introduced

$$\frac{\partial \theta_{f1}}{\partial \tau} - \frac{1}{Pe_{h1}} \cdot \frac{\partial^2 \theta_{f1}}{\partial z^2} + \hat{u}_{f1} \frac{\partial \theta_{f1}}{\partial z} + B_{iM} \cdot (\theta_{f1} - \theta(z, r_0)) = 0$$

$$\frac{\partial \theta_{f2}}{\partial \tau} - \left(\frac{1}{Pe_{h2}}\right) \cdot \frac{\partial^2 \theta_{f2}}{\partial z^2} + \hat{u}_{f2} \frac{\partial \theta_{f2}}{\partial z}$$

$$+ B_{iM} \cdot (\theta_{f2} - \theta(z, r_1)) + B_{iM} \cdot (\theta_{f2} - \theta(z, r_2)) = 0$$

(6)

7. *Boundary and initials conditions*

The necessary boundary and initials conditions to complete the IPDAEs of the model are given next

$$\hat{\rho} = 1, \quad \theta = 1, \quad \hat{P} = 1 \tag{7}$$

$$\frac{\partial \hat{\rho}}{\partial z} = 0, \quad -\frac{\partial \theta}{\partial z} = Bi_z \cdot (\theta - \theta_c), \quad \hat{u}_z = \hat{u}_r = 0 \tag{8}$$

$$\frac{\partial \hat{\rho}}{\partial r} = 0, \quad \frac{\partial \theta}{\partial r} = Bi_r (\theta - \theta_c), \quad \frac{\partial \hat{u}_z}{\partial r} = \frac{\partial \hat{u}_r}{\partial r} = 0 \tag{9}$$

$$\frac{\partial \hat{\rho}}{\partial r} = 0, \quad -\frac{\partial \theta}{\partial z} = Bi_z (\theta - \theta_c), \quad \hat{u}_z = \hat{u}_r = 0 \tag{10}$$

Equation (7) are the Dirichlet boundary conditions at the tank inlet ($z = 0$), Eq. (8) are the boundary conditions at the tank outlet ($z = 0$), Eq. (9) are the boundary conditions at the tank centre ($r = 0$) and Eq. (10) are the boundary conditions at the tank wall ($r = 1$). Furthermore, the use of the concentric annular ring heat exchanger introduces two extra boundary conditions

$$\frac{\partial \hat{\rho}}{\partial r} = 0, \quad -\frac{\partial \theta}{\partial r} = Bi_r(\theta - \theta_c), \quad \frac{\partial \hat{u}_z}{\partial r} = \frac{\partial \hat{u}_r}{\partial r} = 0 \tag{11}$$

$$\frac{\partial \hat{\rho}}{\partial r} = 0, \quad -\frac{\partial \theta}{\partial r} = Bi_r(\theta - \theta_c), \quad \frac{\partial \hat{u}_z}{\partial r} = \frac{\partial \hat{u}_r}{\partial r} = 0 \tag{12}$$

where Eq. (11) are the boundary conditions at $r = r_1$ and Eq. (12) are the boundary conditions at $r = r_2$.

Optimal Design and Control of Hydrogen Desorption

The purpose of this work is to offer a preliminary study for the use of advanced model-based control methods for metal-hydride bed hydrogen storages. The control objective for this study is to ensure the minimum-time hydrogen release from the metal-hydride bed reactor (similar objectives were set for the problem of hydrogen adsorption in Georgiadis et al., 2009). More control objectives and scenarios can be considered however will be left for future work. In order to achieve the control objective, certain parameters (such as the volumetric flow of the heating medium, its inlet temperature etc.) have to be optimally decided while certain operating constraints, which we will present later, have to be satisfied.

In Papakosta (2008), a number of dynamic simulation studies showed that the rate of hydrogen release follows the same profile with the rate of bed temperature increase $T_{(z=1)}$ at the exit of the reactor ($z = 1$). Furthermore, it was also shown that by increasing the heating medium flowrate U_f the rate of hydrogen release decreases. Therefore the bed temperature $T_{(z=1)}$ and heating medium flowrate U_f will be considered as the controlled and manipulated variables for the control design studies that follow.

During the desorption process a number of path constraints on the heating medium flowrate, heat exchanger pressure drop and bed temperature have to be satisfied to ensure a safe and economic operation (Papakosta, 2008).

$$U_f \leq 3.87 \text{ m/sec}, \, T_{max} \leq 1.02 \, (296K), \, \Delta p_f < 0.01 \text{atm},$$

$$\Delta p_f = f \frac{\rho_f U_f^2}{2D_t} L, \, f = 0.184 Re_{Dt}^{-0.2}, \, D_t = 2r_0 R$$

The desorption is considered to be completed when at least 99% of the total volume of hydrogen in the reactor has been released which poses an additional constraint $H_2(\%) > 0.99$ at the end of the desorption time.

A dynamic optimization problem can then be solved to obtain the optimal profiles of the manipulating and controlled variable for which the minimum time hydrogen release objective is achieved. The mathematical formulation of the dynamic optimization is

$$\min_{u_f, T, \dot{P}} t_s$$

subject to:

- Detailed model equations (1)–(12)
- Constraints on U_f, T, Δp_f and $H_2(\%)$

where t_s is the time horizon of the desorption process. The optimization problem was solved by using an implementation of the control vector parameterization (CVP) method in gPROMS modeling system (Process System Enterprise Ltd, 2007). The centered finite difference (CFD) method was used to discretize the spatial and radial domain of the model using 30 elements. The time horizon is assumed to be divided in 34 time intervals.

The results of the optimization are presented in Figure 2 and 3 where the optimal time profiles of the temperature $T_{(z=1)}$ and hydrogen mass are shown. Figure 3 shows that the time needed for 99% of the hydrogen mass to be released is 120 sec. The main idea is to maintain the exit bed temperature $T_{(z=1)}$ close to the optimal profile in Figure 2, in order to achieve the minimum release time of hydrogen. To this end, feedback control methods have to be

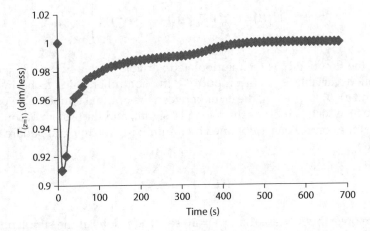

FIGURE 2
Optimal time profile of $T_{(z=1)}$.

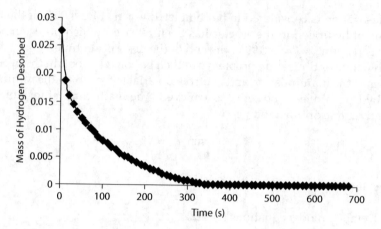

FIGURE 3
Optimal time profile of hydrogen mass.

employed especially in the presence of fluctuations of the process operating conditions from the optimal ones. In this work we employ two feedback control methods the Proportional Integral (PI) control and multi-parametric MPC (mp-MPC) which will be discussed next.

Proportional Integral Control

A proportional integral (PI) controller is initially designed for tracking the optimal temperature time profile (Figure 3)

$$U_f(t) = U_{f0} + K_C \left(e(t) + \frac{1}{\tau_i} \int_0^t e(\tau)d\tau \right)$$ (13)

where the flowrate U_f is the manipulated variable (system input), $T_{(z=1)}$ is the controlled variable (system output), K_C the controller gain, τ_i the integral time and $e = T_{(z=1),ref} - T_{(z=1)}$ the error between the optimal temperature profile (Figure 3), which is used as the reference signal, and the system output. The gains of the controller are obtained by minimizing the Integral Square Error criterion

$$ISE = \int_0^t e^2(\tau)d\tau$$ (14)

and are given by $K_C = -0.01$ and $\tau_i = 10$ sec. The results of the simulations of the PI controller will be used for comparison evaluations with the mp-MPC controller that is discussed next.

Multi-Parametric Model Predictive Control

A multi-parametric Model Predictive Controller is designed next for the desorption process in the metal-hydride reactor. In model predictive control (MPC) an open-loop optimal control problem is solved at regular intervals (sampling instants), given the current process measurements, to obtain a sequence of the current and future control actions up to a certain time horizon (in a receding horizon control fashion), based on the future predictions of the outputs and/or states obtained by using a mathematical representation of the controlled system. Only the first input of the control sequence is applied to the system and the procedure is repeated at the next time instant when the new data are available. Being an online constrained optimization method, MPC not only provides the maximum output of the controlled process but also takes into account the various physical and operational constraints of the system.

The benefits of MPC have long been recognized form the viewpoint of cost and efficiency of operations. Nevertheless, its applications maybe restricted due to increased online computational requirements related to the constrained optimization. In order to overcome this drawback, mp-MPC was developed (see Pistikopoulos et al., 2002 and Pistikopoulos et al., 2007b) which avoids the need for repetitive online optimization. In mp-MPC the online optimization problem is solved off-line with multi-parametric programming techniques (Pistikopoulos et al, 2007a) to obtain the objective function and the control actions as functions of the measured state/outputs (parameters of the process) and the regions in the state/output space where these functions are valid i.e. as a complete map of the parameters. The control is then applied by means of simple function evaluations instead of demanding online optimization computations.

Three steps are followed to design the mp-MPC controller. First, a reduced-order input-output model is obtained with model identification. Then, the MPC optimization problem is formulated. Finally, the MPC problem is solved with multi-parametric programming.

Model Identification

An ARX input-output model is obtained with model identification from the data of the simulations of the desorption process. The data are derived with a sampling time of 1 sec. The mathematical representation of the ARX is the following

$$A(q)T_{(z=1)}(t) = B(q)U_f(t) + e(t) \tag{15}$$

where $A(q) = 1 - 2.695q^{-1} + 2.467q^{-2}$, $B(q) = -0.342q^{-1} + 0.7139q^{-2} + 1.055q^{-3} - 0.2121q^{-4} - 0.7749q^{-5} - 0.4399q^{-6} - 1.61 \cdot 10^{-7}q^{-7}$ and q^{-1} is the unit delay operator

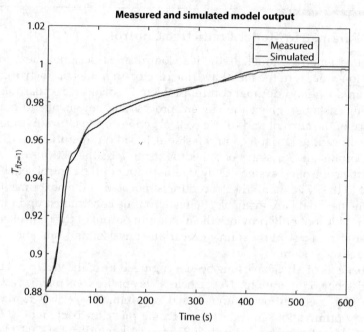

FIGURE 4
Process and ARX model comparison.

which is defined as $q^{-k}V(t) = V(t-k)$. The comparison between the simulations of the actual process and the ARX model are shown in Figure 4. It is obvious that the ARX model approximates closely the behavior of the actual process with a small approximation error, as it can be seen from the differences of the simulation lines in Figure 4.

Model Predictive Control Formulation and mp-MPC

The following MPC formulation is considered for the hydrogen desorption process in the metal-hydride reactor

$$\min_{u_{t+1},...,u_{t+5}} J = \sum_{i=1}^{N_y} Q(y_i - y_{ref,i})^2 + \rho \sum_{j=1}^{N_u} R(u_j - u_{j-1})^2 \qquad (16)$$

$$\text{s.t.} \quad A(q)y_i = B(q)u_i$$

$$0 \le u_i \le 129, \quad 1 \le y_i \le 1.2$$

$$y_i = T_{(z=1)}(t+i), \quad i = 1,....N_y$$

$$u_j = U_f(t+j), \quad j = 1,...,N_u$$

$$Q = 100, \quad \rho = 0.001, \quad R = 1$$

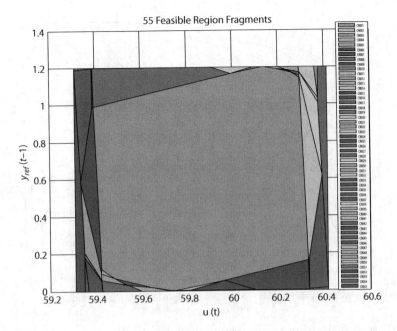

FIGURE 5
Critical regions of mp-MPC in the $u_t - y_{ref}$ sub-space.

where u is the manipulated variable U_f, y is the controlled variable $T_{(z=1)}$, y_{ref} is the optimal temperature profile and $N_y = N_u = 5$. The above optimization involves five optimization variables $u_{t+1}, \ldots, u_{t+5}$ and six parameters $x = [u_t, u_{t-1}, y_t, y_{t-1}, y_{t-2}, y_{ref}]$ which represent past input and output measurements and the reference signal. The objective function is set to minimize the quadratic norm of the error between $T_{(z=1)}$ and its optimal profile while the constraints on u and y are also introduced. The optimization problem (15) is thus a multi-parametric Quadratic Programming (mp-QP) problem and can be solved with the standard parametric programming techniques (Pistikopoulos et al., 2007a). For this problem the Parametric Optimization Software was used (ParOS Ltd, 2003) to solve (16) and obtain the mp-MPC controller.

The solution of the MPC (16) as an mp-QP (Pistikopoulos et al., 2002) is a functional map of parameters consisting of 523 critical regions with an equal number of corresponding control laws. An illustration of the critical regions is shown of the solution is given in Figure 5 where a projection of the critical regions on the $u_t - y_{ref}$ sub-space is given. Each of the critical regions is described by a number of linear inequalities $A_i x \leq b_i$ and its corresponding control is piecewise linear $U_f = K_i x + c_i$, where i is the index of solutions. The online implementation of the controller is as following

$$\text{If } x \text{ in } A_i x \leq b_i \text{ Then } U_f = K_i x + c_i$$

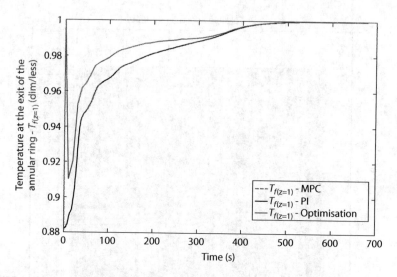

FIGURE 6
Temperature profiles of the mp-MPC and PI controllers for nominal conditions.

In Figure 6 the simulation results of the mp-MPC and PI implementation are compared for nominal operating conditions (same conditions as in dynamic optimization), where $T_f = 350K$. It can be observed that both controllers demonstrate almost similar behavior. In the first few seconds the temperature diverges from the optimal temperature but after 350 sec converge to the optimal profile. In Figure 7 one can see that the hydrogen release

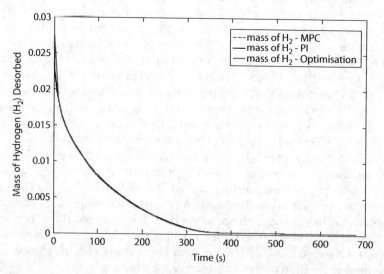

FIGURE 7
Hydrogen mass desorbed at nominal conditions.

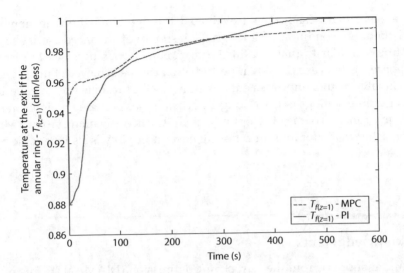

FIGURE 8
Temperature profiles for mp-MPC and PI controllers — perturbation of $T_f = 0.86T_{f,nom}$.

rate is maintained close to the optimal one while in the few seconds a small deviation is noticeable. Both controllers achieve a 99% hydrogen release in 680sec. In Figure 8 the simulations of both mp-MPC and PI are shown when a perturbation is applied on the system, by decreasing the heating medium temperature to $T_f = 0.86T_{f,nom}$ of its nominal value. A noticeable difference between the responses of the two controllers occurs. The PI controller shows a similar performance with the previous one under nominal conditions. The mp-MPC on the other hand demonstrates a faster response in the first 300 secs as it is able to predict the changes in the system due to the disturbance and maintain a smoother response. The difference in the responses of the two controllers after 400 sec is also very small (less than 1% of the response of the PI).

Conclusions

This work presents a systematic approach for the design and evaluation of advanced multi-parametric MPC controllers for the desorption of hydrogen in metal-hydride beds. Based on a detailed model of a metal-hydride bed for hydrogen storage, dynamic optimization and control studies were performed to obtain the optimal temperature profile that can guarantee hydrogen release in minimum-time (an evaluation criterion previously used for hydrogen adsorption control studies). A PI and mp-MPC controller have been then designed and evaluated through simulations in nominal and operating conditions. Although the response of advanced model-based control is similar to that of a PI during nominal operations, an improved response for the case of system perturbation can be achieved. Future work will involve further improvements on the mp-MPC's response by considering better tuning settings (such as longer horizon times or higher weights in the objective function of the MPC optimization). The investigation of robust MPC approaches that offer immunization against uncertainty is also the subject of future work.

Acknowledgments

Financial support from the European Commission (DIAMANTE ToK project, Contract No: MTKI-CT-2005-IAP-029544) and EPSRC (EP/E047017/1) are gratefully acknowledged.

References

Askri, F., Jemni, A., Nasrallah, S.B. (2003). Study of two-dimensional and dynamic heat and mass transfer in a metal-hydrogen reactor. *Int. J. of Hyd. En.*, 28, 537.

Bhouri, M., Askri, F., Jemni, A., Nasrallah, S.B. (2000). Computational analysis of H2 desorption in LaNI5 reactor. *Laboratoire d'Etudes des Systemes Thermiques ET Energetiques, Ecole Nationale d'Ingenieurs de Monastir.*

Demircan, A., Demiralp, M., Kaplan, Y., Matb, M.D., Veziroglu, T.N. (2005). Experimental and theoretical analysis of hydrogen adsorption in LaNi5-H2 reactors. *Int. J. of Hyd. En.*, 30, 1437.

Dhaou, H., Mellouli, S., Askri, F., Jemni, A., Nasrallah, S.B. (2006). Experimental and numerical study of discharged process of metal-hydrogen tank. *Int. J. of Hydr. En.*, 32, 1922.

Georgiadis, M.C., Kikkinides, E.S., Makridis, S.S., Kouramas, K.I., Pistikopoulos, E.N. (2009). Design and optimization of advanced materials and processes for efficient hydrogen storage. *Comput. Chem. Eng.* to appear.

Jemni, A., Nasrallah, S.B. (1995). Study of two-dimensional heat and mass transfer during adsorption in a metal-hydrogen reactor. *Int. J. of Hyd. En.*, 20, 43.

Man, Y.H., In, K.K., Ha, D.S., Sikyong, S., Dong, Y.L. (2005). A numerical study of thermo-fluid phenomena in metal-hydride beds in the hydriding process. *Int. J. of Heat and Mass Transf.*, 47, 2901

Mat, M.D., Kaplan, Y., Aldas, K. (2002). Investigation of three-dimensional heat and mass transfer in a metal-hydride reactor. *Int. J. of Hyd. En.*, 26, 973.

Muhittin, B.O., Ercan, A. (2005). Numerical analysis of hydrogen adsorption in a P/M metal bed. *Powder Technology*, 160, 141.

Parametric Optimization Solutions Ltd (2007), *Parametric Optimization (POP) Software*, UK

Papakosta, M. (2008). Dynamic optimization and multi-parametric model predictive control of hydrogen desorption in metal hydride bed. *MSc Thesis*, Imperial College London, UK.

Pistikopoulos, E.N., Dua, V., Bozinis, N.A., Bemporad, A., Morari, M. (2002). On-line optimization via off-line parametric optimization tools. *Comput. Chem. Eng.*, 26, 175.

Pistikopoulos, E.N., Georgiadis, M.C., Dua, V. (2007a). Multi-parametric Programming. *Volume 1 in Process Systems Engineering series*, Wiley-VCH, Weinheim.

Pistikopoulos, E.N., Georgiadis, M.C., Dua, V. (2007b). Multi-parametric Model-Based Control. *Volume 2 in Process Systems Engineering series*, Wiley-VCH, Weinheim.

Process Systems Enterprise Ltd (2006). *gPROMS Introductory User Guide*, UK

3

Embedding Sustainability into Process Development: GlaxoSmithKline's Experiences

Concepción Jiménez-González[1*], David J. C. Constable[2], Richard K. Henderson[3], Rebecca DeLeeuwe,[3] and Lisa Cardo[2]

[1]*GlaxoSmithKline CEHSS, 5 Moore Drive, RTP, NC, 27709*
[2]*GlaxoSmithKline CEHSS, 1 Franklin Plaza, Philadelphia, PA, 19101*
[3]*GlaxoSmithKline CEHSS, Ware, UK*

CONTENTS

ABSTRACT GlaxoSmithKline (GSK) aspires to be a sustainable company and has developed a series of tools and processes to apply Green Chemistry and Engineering principles to new product development. These tools and processes are used routinely to design-out hazardous chemicals, and bring products to market more cost effectively as we believe more sustainable practices will enable GSK to produce medicines with greater efficiency in mass and energy utilization.

The Eco-Design toolkit is currently composed of five modules: Green Chemistry/Technology Guide; Materials Selection Guides; FLASC™ (Fast Lifecycle Assessment for Synthetic Chemistry); Green Packaging Guide; and a Chemicals Legislation Guide. In addition, GSK has developed an EHS Milestone

---

[*] To whom all correspondence should be addressed

Aligned Process to ensure EHS and Sustainability are integrated into each phase of new product development. These tools and processes are designed to help scientists and engineers consider potential EHS and sustainability impacts from the manufacture of the raw materials through to the ultimate fate of products and wastes in the environment. In this paper, we will describe at a high-level some of these tools and processes. Examples of practical applications and some of their benefits to the company will also be covered.

KEYWORDS   *Sustainability, pharmaceuticals, new process, green chemistry, new process development*

Introduction

GlaxoSmithKline (GSK) embraces sustainability as a driver for achieving competitive advantage as a leading Pharmaceutical company. To facilitate this move towards more sustainable practices, GSK has developed the Eco-Design toolkit which enables scientists and engineers to identify process improvements and address Environment, Health and Safety (EHS) issues early in the product development process.

In addition, an Environment, Health and Safety Milestone Aligned Process (EHS MAP) was developed to ensure EHS and Sustainability aspects are considered at each phase of new product development. These tools and processes enable consideration of EHS impacts from the manufacture of the raw materials through to the ultimate fate of products and wastes in the environment. Implementing practices from the Toolkit will enable GSK to produce medicines with fewer EHS impacts throughout their life cycle.

The Eco-Design Toolkit

The Eco-Design Toolkit was developed using state-of-the-art scientific advancements and standards in Green Chemistry, Green Technology and Life Cycle Assessment to assist chemists and engineers in designing-out hazardous chemicals while bringing products to market more cost effectively (Constable *et. al.*, 2005).

The toolkit is accessible through the GSK intranet and is currently composed of five modules: Green Chemistry/Technology Guide, Material Selection Guides, FLASC™ (Fast Lifecycle Assessment for Synthetic Chemistry); Green Packaging Guide; and a Chemicals Legislation Guide (CLG). The combination of these tools helps to ensure that all EHS impacts are considered across a range of disciplines and phases of development. For example, using FLASC™ ensures that the life cycle impacts of chemicals used in syntheses while the CLG is used to identify chemicals of concern for each pilot plant campaign from the earliest 1 kg makes to multi-kilo production.

The Eco-Design Toolkit continues to be updated to integrate scientific advances. Currently, about 50 additional solvents have been added to the solvent selection guide and there is ongoing work to identify replacements for chemicals of concern (e.g. dichloromethane, N,N-dimethyl formamide) using computer modeling methodologies. We are also working to identify solvents that can be derived from renewable sources. Each module of the Eco-Design toolkit is described in the paragraphs below.

Green Chemistry and Technology Guide

The Green Chemistry/Technology Guide (Figure 1) offers guidance to GlaxoSmithKline scientists and engineers on how the application of Green Chemistry concepts can enable more efficient use of resources, reduce environment, health and safety impacts, reduce the use of toxic materials, and minimize costs. It includes:

- a ranking and summary of the most used chemistries and 'best-in-class' examples from well-developed GlaxoSmithKline chemical synthetic processes (Curzons *et. al.*, 2001; Constable *et. al.*, 2002)
- an EHS ranking and summary of common technology alternatives for chemical processing to improve the use of resources and minimize life cycle impacts (Jiménez-González *et. al.*, 2001; Jiménez-González *et. al.*, 2002).
- guidance on materials, process alternatives, synthetic route strategies and metrics for evaluating chemistries, technologies and processes
- a ranking and review of issues encountered during process design and development

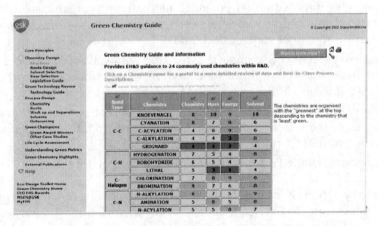

FIGURE 1
Screenshot of the green chemistry guide.

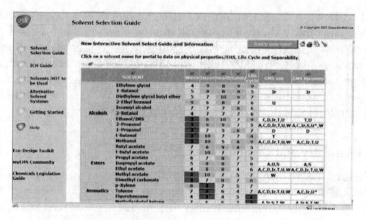

FIGURE 2
Screenshot of the solvent selection guide.

Material Selection Guides: Solvents and Bases

Given the important role that solvents have in the environmental footprint of pharmaceuticals (Jiménez-González *et. al.*, 2004; Constable *et. al.*, 2007), one of the first tools developed was the Solvent Selection Guide (Curzons *et. al.*, 1999). The Solvent Selection Guide (Figure 2) contains information on a wide range of solvents used within GSK and identifies solvents that should be avoided. It allows for the selection solvents with fewer EHS impacts and:

- compares and ranks solvents according to waste profile (treatability), environmental impact (eco-toxicity), safety profile (e.g. flamability, explosivity) and health impact (toxicity)

- provides information on potential alternatives to organic solvents, such as solvent-less reactions, water and aqueous systems, carbon dioxide and other supercritical fluids.

- assesses the life cycle impacts associated with solvent manufacture to stress solvent minimization and recycling (Jiménez-González *et. al.*, 2005)

- provides information on boiling point and azeotrope formation to assist in the selection of separable co-solvents to minimize energy required for separation and treatment.

- provides detailed information on physical properties.

The Base Selection Guide builds on the success of the solvent selection guide (Constable *et. al.*, 2005). Requested by R&D chemists, the base selection guide contains information on a wide range of chemical bases used within GlaxoSmithKline R&D and manufacturing operations. It

- compares and ranks 42 bases routinely used by R&D chemists according to their environmental waste profile, environmental impact, safety profile and health impact
- provides detailed information on physical properties and EHS issues

As part of a continuing vision to develop additional guidance on materials, a prototype materials selection guide was developed for acids, catalysts and reducing agents.

FLASC™—Fast Life Cycle Assessment for Synthetic Chemistry

FLASC™ allows chemists to perform streamlined environmental life cycle assessments of processes and measure green metrics such as mass efficiency (Curzons *et. al,* 2007). The FLASC™ tool development was based on a full LCA of an Active Pharmaceutical Ingredient or API (Jiménez-González *et. al.,* 2004). It assesses eight different environmental life cycle impact categories associated with materials used in synthetic routes: Mass Intensity, Cummulative Energy Requirements, Global Warming Potential, Photochemical Ozone Creation Potential, Acidification, Eutrophication, Total Organic Carbon (pre-treatment) and oil and natural gas depletion for raw materials manufacture.

FLASC helps scientists and managers to rapidly identify the greenest option, from a materials use perspective by:

- comparing and benchmarking GSK synthetic processes with a color coded score (Figure 3)
- estimating the environmental life cycle impacts of the materials used

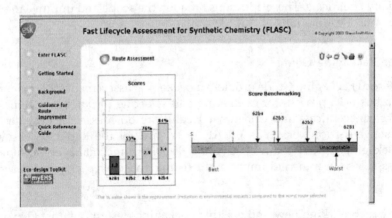

FIGURE 3
Example of an output of FLASC™.

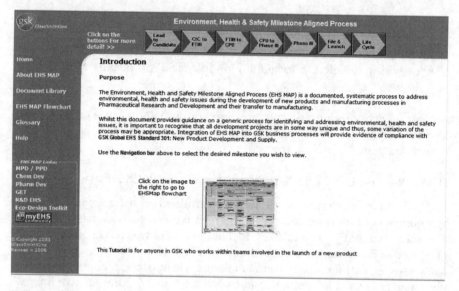

FIGURE 4
Screenshot of EHS MAP.

- identifying the materials with the biggest contribution to those life cycle impacts
- providing guidance on how to reduce those impacts.

FLASC can also be used to track synthetic route or manufacturing process improvements. Additional life cycle inventory data for materials of interest to the Pharmaceutical Industry are currently being obtained for the FLASC database. These materials include key raw materials, catalysts and complex specialty chemicals. These additions broaden the scope and improve the life cycle impact estimations FLASC delivers.

Green Packaging Guide

The Green Packaging Guide provides a packaging assessment tool, guidance and a business process for evaluating and selecting packaging options for the Pharmaceuticals and Consumer Healthcare businesses. It provides an interactive section known as WRAP — Wizard for the Rapid Assessment of Packaging. WRAP allows packaging designers and managers to rapidly assess the environmental impacts of existing and new packaging designs, including:

- benchmarking new and existing packaging designs against Glaxo-SmithKline's existing product portfolio,
- a best-in-class example in each packaging category,

- green packaging guides for nutritional healthcare products and con-sumer healthcare products,
- a score with a simple color-coded report that clearly shows if the packaging associated with a product is better or worse than the appropriate benchmark.

WRAP also allows more detailed analysis of the underlying issues around packaging and enables users to compare alternative packaging designs through scenario analysis.

Chemical Legislation Guide

The CLG is the Eco-Design toolkit's newest addition. While most of our tools focus on best practice and issue identification, the CLG was developed in anticipation of chemicals legislation (e.g. Homeland Security lists, CA Proposition 65, REACH) in various parts of the world to phase out high hazard substances from routine use (chemicals of concern). This provides R&D and manufacturing scientists and engineers with a user friendly tool that considers hazard, volume and phase of use to deliver targeted guidance about a variety of chemicals. Because the tool is a spreadsheet it is easy to update and requires no special training to use. Consequently, uptake of the tool has been very rapid and implementation was almost immediate.

Environment, Health and Safety Milestone Aligned Process. EHS MAP

EHS MAP is a documented, systematic process to address EHS and Sustainability aspects during the development of new products and manufacturing processes in Pharmaceutical Research and Development and their transfer to Manufacturing facilities.

The EHS MAP Process is integral to GSK business processes and aligned with new product development and supply milestones; identifying responsibilities, accountabilities, deliverables and communications. To embed Sustainability principles into new product development, EHS MAP enables the identification of activities that can improve process efficiencies, eliminate waste and facilitate the new product development and supply process.

The outputs of the EHS MAP process are a better understanding and appreciation of EHS and Sustainability throughout a products life cycle; the implementation of best practices throughout the company; and finally having people across the company engaged and committed to making

EHS and Sustainability an integral part of new product development and supply.

Examples of Applications and Benefits

The Eco-Design toolkit and EHS MAP are routinely used in GlaxoSmithKline, primarily to support new product development. Their utilization has rendered many tangible results that enable GSK to drive towards more sustainable practices. Some examples of success of this operational sustainability are described below:

1. Green chemistry metrics have been adopted by and integrated into R&D Chemical Development evaluations of pilot plant campaigns. These metrics are used by chemists and engineers to eliminate toxic materials and to reduce material consumption and waste.

2. Life cycle impacts and chemicals of concern (hazardous chemicals) are identified in pilot plant campaigns and plans are set to improve the environmental profile as the process progresses through the development milestones. For instance, during 2008, the mass percent of chemicals of concern decreased 9-fold (Figure 5) and estimated average life cycle impacts were reduced 4-fold as compounds moved to the last stage of development (Figure 6).

3. More than 500 life cycle assessments of synthetic routes have been performed using FLASC in R&D and Primary Manufacturing.

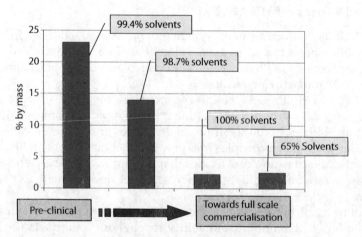

FIGURE 5
Materials of concern in the 2008 portfolio.

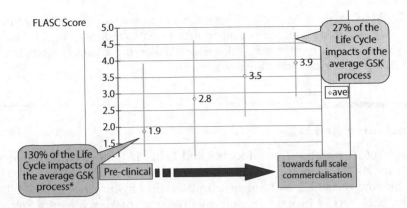

FIGURE 6

Life cycle Impacts on the 2008 development portfolio. The average performance of the bench-mark routes (1990–2000) was assigned a FLASC score of 2.3.

4. A target has been set to double the average mass efficiency of processes for new products introduced between 2006 and 2010. There is already progress in this area, such as the manufacturing route for a development compound; which compared to the previous route it eliminated several hazardous chemicals, halved estimated life cycle impacts, and doubled the mass efficiency (Table 1). This work won the 2nd place in the Green Chemistry category of GlaxoSmithKline CEO's EHS Excellence Awards in 2005.

5. The solvent selection guide is widely used in GSK facilities throughout the world and it has been integrated into other solvent selection tools in R&D. To promote a spirit of collaboration, the solvent guide has been shared within collaborative consortiums.

6. The work to identify replacement for chemicals of concern (e.g. dichloromethane, N,N-dimethyl formamide) using computer modeling methodologies developed to identify solvents that would provide the same process performance with a smaller impact (Gani *et. al*, 2005, Gani *et. al.*, 2006) has been verified in initial laboratory experiments and proved that an alternative solvent not only provided a better environmental profile, but improved process performance.

TABLE 1

Comparison of Selected Green Metrics for Two Routes of a Development Compound

Route	No. of Steps	Mass Intensity (kg/kg product)	Mass Efficiency (%)	Solvent Intensity (kg/kg product)
F (old)	6	162	0.6 %	99
G (new)	3	61	1.6 %	36

7. WRAP is used to perform assessments of environmental benefits derived from reductions and improvements in packaging. A new version is currently being developed following requests from the Consumer Healthcare business.

Concluding Remarks

GlaxoSmithKline's Eco-Design toolkit and EHS MAP process are a unique-in-their-class set of tools to design safer, greener, more efficient processes by minimizing their EHS and life cycle impacts and by making sustainability an integral part of process and product design. Tools such as the Solvent Guide, FLASC or the Green Technology Guide have been first in their class and have been recognized for their innovation and for leading the green design of pharmaceuticals.

Although there are still challenges in the quest for sustainability that need to be addressed, the achievements to date are an excellent sample of GSK's vision for Operational Sustainability and its commitment to strive towards more sustainable business practices.

Acknowledgments

The authors wish to acknowledge the efforts of the many colleagues who have significantly contributed to the development and implementation of the Eco-Design Toolkit and EHS MAP: James R. Hagan, Alan D. Curzons, Virginia Cunningham, Robert Hannah, Ailsa Duncan, John Hayler, Steve Binks, Teresa Oliveira, Tom Roper, Luisa Freitas Dos Santos, Giuseppe Lo Biundo, Ugo Cocchini, Jim McCann, Benjamin Holladay, Nga Vo, in particular, and more generally, the Sustainable Processing Team and the Corporate Environment, Health, Safety and Sustainability department at GlaxoSmithKline.

References

Constable DJC, Curzons AD and Cunningham VL (2002). Metrics to green chemistry—which are the best? Green Chem., 4, 521 – 527

Constable DJC, Curzons A, Duncan A, Jiménez-González C, Cunningham VL. (2005) The GSK Approach to Sustainable Development and The GSK Approach to Metrics for Sustainability. Chapter 8.7 and Section 6.1.2 in "Transforming Sustainability Strategy into Action: The Chemical Industry" 568pp, Beth Beloff Eds. Wiley and Sons, Oct 2005

Constable DJC, Jiménez-González C, Henderson RK. (2007) Perspective on Solvent Use in the Pharmaceutical Industry. Organic Process Research and Development, 11, 133–137

Curzons AD, Constable DJC, Cunningham VL. (1999). Clean Products and Processes, 1, 82-90. Solvent Selection Guide: A Guide to the Integration of Environmental, Health and Safety Criteria into the Selection of Solvents.

Curzons A, Constable DJC, Mortimer D, Cunningham VL (2001). So you think your process is green, how do you know?—Using principles of sustainability to determine what is green–a corporate perspective. Green Chem., 3, 1 - 6,

Curzons A, Jiménez-González C, Duncan A, Constable D, Cunningham V. (2007). Fast Life-cycle Assessment of Synthetic Chemistry, FLASCTM Tool. Int J LCA 12(4)272–280

Gani R, Jiménez- González C and Constable DJC. (2005). Method for Selection of Solvents for Promotion of Organic Reactions. Computers and Chemical Eng. 29(7)1661–1676

Gani R, Jiménez-González C, ten Kate A, Crafts PA, Jones M, Powell L, Atherton J, Cordiner J. (2006). A modern Approach to Solvent Selection. Chemical Engineering 113(3) 30–43.

Jiménez-González C, Curzons A, Constable D, Overcash M, Cunningham V. (2001). How do you select the 'greenest' technology? Development of Guidance for the Pharmaceutical Industry. Clean Products and Processes, 3: 35–41

Jiménez-González C, Constable DJC, Curzons AD and Cunningham VL. (2002). Developing GSK's Green Technology Guidance: Methodology for Case-Scenario Comparison of Technologies. Clean technology and Environmental Policy, 4:44–53.

Jiménez-González C, Curzons AD, Constable DJC, Cunningham VL. (2004) Cradle-to-Gate Life Cycle Inventory and Assessment of Pharmaceutical Compounds: a Case-Study. Int J LCA 9(2) 114–121.

Jiménez-González C, Curzons AD, Constable DJC, Cunningham VL. (2005). Expanding GSK's Solvent Selection Guide — Application of Life Cycle Assessment to Enhance Solvent Selections. Journal of Clean Technology and Environmental Policy, 7:42–50

4

Green Process Design, Green Energy, and Sustainability: A Systems Analysis Perspective

Urmila M. Diwekar* and Yogendra N. Shastri

Center for Uncertain Systems: Tools for Optimization & Management
Viswamitra Research Institute
Clarendon Hills, IL 60514

CONTENTS

ABSTRACT This paper presents a systems analysis perspective that extends the traditional process design framework to green process design, green energy and industrial ecology leading to sustainability. For green process design this involves starting the design decisions as early as chemical and material selection stage on one end, and managing and planning decisions at the other end. However, uncertainties and multiple and conflicting objectives are inherent in such a design process. Uncertainties increase further in industrial ecology. The concept of overall sustainability goes beyond industrial ecology and brings in time dependent nature of the ecosystem and multi-disciplinary decision making. Optimal control methods and theories from financial literature can be useful in handling the time dependent uncertainties in this problem. Decision making at various stages starting from green process design, green energy, to industrial ecology, and sustainability

---

* To whom all correspondence should be addressed

is illustrated for the mercury cycling. Power plant sector is a major source of mercury pollution. In order to circumvent the persistent, bioaccumulative effect of mercury, one has to take decisions at various levels of the cycle starting with greener power systems, industrial symbiosis through trading, and controlling the toxic methyl mercury formation in water bodies and accumulation in aquatic biota.

KEYWORDS   *Green process design, green energy, sustainability*

Introduction

Chemical process simulation tools and models allow engineers to design, simulate and optimize a process. Steady state simulators like PRO-II and ASPEN Plus are well known in this area and are extensively used for simulation of continuous processes. In recent years, chemical process industries have become aware of the importance of waste reduction and environmental consciousness, which demands an effort extending far beyond the capability of existing process simulation to model processes with environmental control options. For tracking trace components, non-equilibrium based models are implemented. Packages like Waste Reduction Algorithm (WAR) (EPA, 2002) provide data related to various environmental impacts like toxicity and exposure. Designing green processes with "process integration" which takes into consideration the entire process is now possible with the new tools. However, there is still a long way to attain the goal of sustainability. Unlike traditional design where engineers are looking for low cost options, environmental considerations include objectives like the long term and short term environmental impacts. Green process design and green energy involves not only extending the design framework to include process integration, environmental control technologies, starting as early as the material selection stage, and going beyond just green energy, green processing, and green management, but to look at industrial sector level management through industrial ecology as shown in Figure 1. In industrial ecology, this decision making changes from the small scale of a single unit operation or industrial production plant to the larger scales of integrated industrial park, community, firm or sector. Uncertainties increase as one goes from traditional process design to green design and to industrial ecology. The concept of overall sustainability goes beyond industrial ecology and brings in time dependent nature of ecosystem. Decisions regarding regulations and human interactions with ecosystem come in picture. It involves dealing with various time scales and time dependent uncertainties. This work presents a systems analysis approach to various steps involved from green process design to sustainability.

Mercury has been recognized as a global threat to our ecosystem, and is fast becoming a major concern to the environmentalist and policy makers.

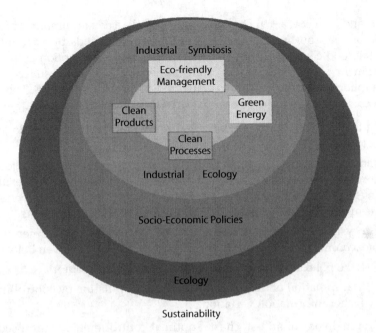

FIGURE 1
Green design to industrial ecology to sustainability.

Mercury is a major pollutant from power plants. The task of mercury pollution management is arduous due to complex environmental cycling of mercury compounds. Successful handling of the issues calls for a sustainability based approach. This work presents the systems analysis approach to sustainability with the case study of mercury.

Mercury Cycle

Mercury can cycle in the environment in all media as part of both natural and anthropogenic activities (USEPA, 2000). Majority of mercury is emitted in air in elemental or inorganic form, mainly by coal fired power plants, waste incinerators, industrial and domestic utility boilers, and chloro-alkali plants. However, most of the mercury in air is deposited into various water bodies such as lakes, rivers and oceans through processes of dry and wet deposition. In addition, the water bodies are enriched in mercury due to direct industrial waste water discharge, storm water runoffs, and agricultural runoffs. Once present in water, mercury is highly dangerous not only to the aquatic communities but also to humans through direct and indirect effects. Methylation of inorganic mercury leads to the formation of methyl mercury which accumulates up the aquatic food chains, so that organisms in higher trophic levels have higher mercury concentrations

(Jensen and Jarnelov, 1969; Desimone et al., 1973). The consumption of these aquatic animals by humans and wild animals further aids bioaccumulation along the food chain. As a result, contaminated fish consumption is the most predominant path of human exposure to mercury. This has resulted in fish consumption advisories at various water bodies throughout the US. The work proposes sustainable management strategies at various levels of mercury cycle.

- Industry level management: Environmental control technologies selection and design.

- Industrial sector (inter-industry) level management: Symbiosis through trading combined with industry level management resulting in mixed integer nonlinear programming (MINLP) and stochastic mixed integer nonlinear programming (SMINLP) problems.

- Ecosystem level management: Effective control strategies of mercury bioaccumulation in water bodies. These strategies are given below.

  - Lake pH control to manage methyl mercury formation.

  - Manipulation of the regimes of species population by controlling Fisher information variation.

  Optimal control and stochastic optimal control methods are used for these strategies.

The following section presents the algorithmic framework for this work.

Algorithmic Framework

The algorithmic framework is shown in Figure 2. The optimization framework is used for green process design and industrial ecology, while stochastic optimal control is used for time dependent decisions under uncertainty.

Level 1: is the inner most level and corresponds to models for processes. For ecological level management, at this level optimal control and stochastic optimal control problems are formulated. Optimal control problems in engineering have received considerable attention in the literature. In general, solutions to these problems involve finding the time-dependent profiles of the decision (control) variables so as to optimize a particular performance index. The dynamic nature of the decision variables makes these problems much more difficult to solve compared to normal optimization where the decision variables are scalar. In general mathematical methods to solve these problems involve calculus of variations, the maximum principle and the dynamic programming technique. Nonlinear Programming (NLP) techniques can also be used to solve this problem provided all the system of differential equations is converted to nonlinear algebraic equations. For details of these methods, please see Diwekar (2008). In the maximum principle, the objective function is reformulated as a linear function in terms of final values of state variables and the values of a vector of constants resulting in ordinary

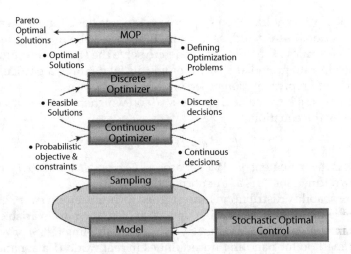

FIGURE 2
Algorithmic framework.

differential algebraic equation that are easier to solve as compared to calculus of variations or dynamic programming. However, this maximum principle formulation needs to include additional variables and additional equations. We use the maximum principle and the stochastic maximum principle formulation (with sampling) (Rico-Ramirez and Diwekar, 2003) along with NLP optimization technique to obtain the control profiles.

Level 2: Sampling loop: It is a common practice to use probability distribution functions like normal, lognormal, uniform distributions, to model uncertainties as stated above. However, these distributions are used for scalar parameter uncertainties. Modeling dynamic or time-dependent uncertainties is a difficult task. Recently, Diwekar (2003, 2008) presented basic concepts for modeling time dependent uncertainties. These concepts are derived from the financial and economics literature where time dependent uncertainties dominate. The following paragraphs present these concepts briefly.

A Wiener process can be used as a building block to model an extremely broad range of variables that vary continuously and stochastically through time. A Wiener process has three important properties:

1. It satisfies the Markov property. The probability distribution for all future values of the process depends only on its current value.

2. It has independent increments. The probability distribution for the change in the process over any time interval is independent of any other time interval (non-overlapping).

3. Changes in the process over any finite interval of time are normally distributed, with a variance that increases linearly with the time interval.

Stochastic processes like Weiner processes do not have time derivatives in the conventional sense and, as a result, they cannot be manipulated using the ordinary rules of calculus as needed to solve the stochastic optimal control problems. Ito provided a way around this by defining a particular kind of uncertainty representation based on the Wiener process.

An Ito process is a stochastic process $x(t)$ on which its increment dx is represented by the equation:

$$dx = a(x,t)\, dt + b(x,t) \tag{1}$$

where dz is the increment of a Wiener process ($dz = \varepsilon\sqrt{dt}$), and $a(x,t)$ and $b(x,t)$ are known functions. ε is a unit normal distribution.

Once probability distributions are assigned to the uncertain parameters, the next step is to perform a sampling operation from the multi-variable uncertain parameter domain. Hammersley Sequence Sampling (HSS) provides an efficient method for handling uncertainties in real world (Kalagnanam and Diwekar, 1997; Diwekar, 2008) and is used here.

Level 3: Continuous optimizer: This step involves continuous decisions like design and operating conditions for a process. Derivative based quasi-Newton methods are used for this step.

Level 4: Discrete optimizer: This involves dealing with discrete decisions such as various point sources and environmental control options. Decomposition strategies are used for MINLP problems.

Level 5: Multi-Objective Programming (MOP): This represents the outermost loop in Figure 2. There are a large array of analytical techniques to solve this MOP problem; however, the MOP methods are generally divided into two basic types: preference-based methods and generating methods. Preference-based methods like goal programming attempt to quantify the decision-maker's preference, and with this information, the solution that best satisfies the decision-maker's preference is then identified. Generating methods, such as the weighting method and the constraint method, have been developed to find the exact Pareto set or an approximation of it. A new variant of constraint method that MInimizes the Number of Single Objective Optimization Problems (MINSOOP) (Fu and Diwekar, 2004) to be solved which is based on the HSS method can be used for this framework. HSS method can be combined with the weighting method also.

Industrial Level and Industrial Sector Level Mercury Management

In the wake of increasingly stringent discharge regulations on mercury, efficient management at the individual level is not sufficient. Innovative methods are required that will analyze the problem from industrial sector level achieving simultaneous economic and ecological sustainability.

TABLE 1

Data for the Various Environmental Control Technologies.

Process	Mercury Reduction Capability (ng/lit)	Capital Requirement ($/1000 gallons)
Activated carbon adsorption (A)	3.0	1.5
Coagulation and Filtration (B)	2.0	1.0
Ion exchange (C)	1.0	0.6

Industrial Level Management: Environmental Control Technologies

Three environmental control technologies are considered for this problem and they are available to all industries for implementation. These include: coagulation and filtration, activated carbon adsorption and ion exchange process. The capital requirement and reduction capability of any process is expected to be nonlinearly related to the capacity of the treatment plant and the form and concentration of the waste to be treated, amongst many other factors. The total plant cost is reported as a function of the waste volume (USDOI, 2001). Since waste volumes encountered in this case study are mostly greater that 1 MGD, asymptotic values reported in USDOI (2001) are used. The treatment efficiencies depend on the waste composition and concentration. In general though, a more efficient treatment is likely to be more expensive. This criterion, along with data given in USEPA (1997a), is used to decide the treatment efficiencies. Table 1 gives the technology data. The nonlinear cost functions are reported in USEPA (1997b). The models are not reproduced here for the sake of brevity and interested readers are referred to the mentioned reference.

Industrial Sector Level Management: Pollutant Trading

Pollutant trading is a market based strategy to economically achieve environmental resource management. The goal is to attain the same or better environmental performance with respect to pollution management at a lower overall cost for the industrial sector. The concept is attributed to Crocker (1966), Dales (1968), and Mongomery (1972).

Various aspects of watershed based trading are extensively discussed in USEPA (1996) and USEPA (2003) and hence not reproduced here. To summarize the aspects relevant for this work, the state or federal authority proposes a regulation such as Total Maximum Daily Load (TMDL) which establishes the loading capacity of a defined watershed, identifies reductions or other remedial activities needed to achieve water quality standards, identifies sources, and recommends waste load allocation for point (and nonpoint) sources. To comply with the regulation, a point source (industry) in the watershed may need to reduce its discharge level. It has two options to accomplish this: (1) the point source can implement an environmental control technology,

(2) the point source can trade a particular amount of pollutant to another point source in the watershed that is able to reduces its discharge more than that specified by the regulation.

Trading optimization problem formulation: The basic optimization model assumes that all information is deterministically known. Under such as assumption, the model is formulated as follows.

$$\text{Minimize} \quad \sum_{i=1}^{N}\sum_{j=1}^{M} f_j(\phi_j, D_i).b_{ij} \tag{2}$$

$$t_{ii} = 0 \qquad \forall i = 1,...,N \tag{3}$$

$$red_i \leq \sum_{j=1}^{M} q_j.D_i.b_{ij} + \sum_{k=1}^{N} t_{ik} - r\sum_{k=1}^{N} t_{ki} \tag{4}$$

$$P_i \geq \sum_{j=1}^{M} b_{ij}.f_j(\phi_j, D_i) + F\left(\sum_{k=1}^{N} t_{ik} - \sum_{k=1}^{N} t_{ki}\right) \tag{5}$$

where there are N point sources (PS) disposing pollutant containing waste water to a common water body or watershed. D_i is discharge quantity of polluted water from PS_i [volume/year]. red_i is desired pollutant quantity reduction in discharge of PS_i [mass/year]. P_i is treatment cost incurred by PS_i. There are M technologies available for waste reduction. f_j is the linear or nonlinear cost function for technology j [$]. ϕ_j is set of design variables for technology j. q_j is pollution reduction possible from technology j implementation [mass/volume]. r is the trading ratio and F is the transaction cost [$/mass]. b_{ij} are the binary variables representing point source technology correlation. The variable is 1 when PS_i installs technology j. t_{ik} [mass/year] is the amount of pollutant traded by PS_i with PS_k. All parameters are on annual basis. The objective function gives the sum of technology cost for all point sources.

The proposed model is applied to the Savannah River watershed. There are 29 major polluters in this watershed including a power plant.

The optimization model presented in the previous section assumes that all data is deterministically known. However, there are various possible sources of uncertainty in this framework. For example, the Mercury Study Report to Congress (USEPA, 1997b) states that uncertainty in point estimates of anthropogenic mercury emissions ranges from medium (25%) to high (50%). This results in a stochastic optimization (stochastic programming) problem.

Health care cost: The bioaccumulative nature of mercury and its slow dynamics make the long term effects of mercury exposure important. Hence, it is essential to account for such effects while quantifying health care costs. Majority of mercury accumulates in the food chain as methyl mercury. Therefore, quantification of health care costs based on methyl mercury concentration is most

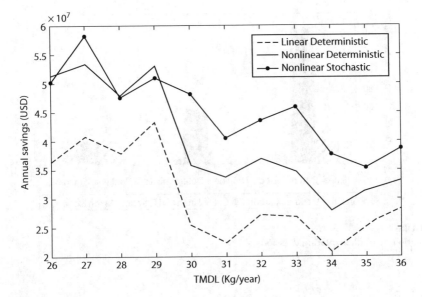

FIGURE 3
Effect of nonlinearity and uncertainty on annual savings due to trading.

appropriate. Health care cost is assumed to be a function of fish consumption, safe concentration if fish, and LC50 value for mercury. Addition of health care cost in the formulation results in a multi-objective optimization problem.

Results and discussions: Fig. 3 plots the annual saving due to trading implementation for the considered TMDL range (26 Kg/year to 36 Kg/year) for three different models. It is observed that approximate linear models underestimate the annual savings. Inclusion of uncertainty in the analysis predicts even higher savings for most TMDL values. It should be noted here that trends in savings do not necessarily reflect the trends in overall cost. Thus, although nonlinear stochastic model leads to higher savings than nonlinear deterministic model, the total cost with trading for nonlinear stochastic model is not necessarily lower than the total cost with trading for nonlinear deterministic model. This is because the savings for a particular model are calculated over the technology option for the same model setting.

Fig. 4 shows the implications of nonlinearity and uncertainty inclusion on technology selection for trading option. The figure shows the number of times each technology is implemented over the complete TMDL range. It can be seen that there are definite implications on technology selection. With linear technology models, various small industries implement technologies along with large industries. However, for nonlinear model, large industries implement most of the technologies and smaller industries satisfy the regulations by trading with these large industries. Due to space limitations, multi-objective optimization results are not presented here.

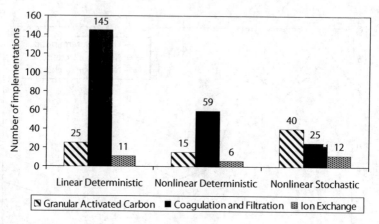

FIGURE 4
Technology implementation decisions.

Ecological Level Management

Mercury and its compounds exist in different segments of the water body such as water column, sediment (active and passive), and biota (fish). Mercury can undergo various transformations in a water body such as oxidation, reduction, volatilization, methylation and demethylation. All these transformations are simultaneously observed in a given water body. The relative concentration of each chemical form depends on the extent of various reactions, which can differ for different water bodies. Of the various chemical forms of mercury, methyl mercury (MeHg) is considered to be the most dangerous due to its bioaccumulative potential. As a result, the concentration of methyl mercury in large aquatic animals (such as predatory fishes) is many times more than the water column or sediment concentration. This work explores two strategies for ecosystem mercury management: (1) the time dependent liming strategy of lakes and rivers to control water pH and (2) controlling nutrient flow to manipulate eating habits of organisms.

Liming and pH Control

Methylation of mercury to MeHg is a key step in the bioaccumulation of mercury in aquatic food chains (Sorensen et al., 1990). The exact mechanism of the methylation reaction is however not well understood. Studies have also been carried out to understand the effect of physical and chemical conditions such as pH, dissolve oxygen, dissolved organic carbon (DOC), temperature, salinity etc., on methylation (Winfrey & Rudd, 1990; Driscoll et al., 1995).

These studies have shown a strong correlation between acidic conditions (low pH values) and high mercury bioaccumulation in fish.

Although lake liming for pH control has been relatively successful in Scandinavian countries, there are various issues related to liming that need further in-depth research. These are:

(1) Liming accuracy: Currently, most of the liming decisions (liming dosage) are based on rule of thumb. The amount of lime to be added is decided using parameters such a lake volume, current lake pH, targeted pH, water salinity etc. (Hakanson & Boulion, 2002). These are mostly static decisions and do not take into account the dynamic nature of the natural system. It is obvious that such heuristics based decisions do not lead to accurate liming results.

(2) Cost of liming: Liming entails considerable costs. Hence, it is essential that the liming operation is optimized to reduce expenses. Even though the liming technique is the major factor deciding the expenses, efficient implementation of the selected technique can reduce expenses. Previous work in this area includes Hakanson (2003a) and Riely & Rockland (1988).

(3) Presence of uncertainty: Liming operation has to deal with presence of various kinds of uncertainties, such as lack of information on the exact pH of the lake, seasonal variations in lake pH, and topological effects of liming. Moreover, the spatial and temporal effects of liming on lake biota are subjective. In order to make liming implementable, one needs to incorporate these uncertainties in the analysis. Due to these issues, lake liming has not been a widespread practice in North America.

To make liming more accurate, an effective approach is to use time dependent liming where liming decisions (amount of lime to be added) change with time based on the current lake conditions. The reliability of liming can be further improved if these dynamic liming decisions are based on a systematic approach rather than heuristics.

Basic liming model: The basic lake liming model is presented in Ottosson & Hakanson (1997) and further discussed in Hakanson & Boulion (2002) and Hakanson (2003b). It is a mixed model consisting of both statistical regression and dynamic interactions. An empirical model is used to predict the initial pH (mean annual pH). The model also includes a regression that predicts natural pH. In addition to these empirical sub-models, the lake liming model consists of dynamic (time dependent) interactions. It is a compartmental model with three different compartments, namely, water, active sediment and passive sediment. Accordingly, the three model variables are: lime in water, lime in active sediment and lime in passive sediment. Four continuous flows of lime connect the three compartments: sedimentation to active

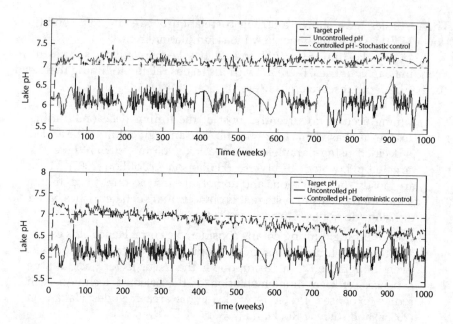

FIGURE 5
Liming pH control, deterministic and stochastic.

sediments, internal loading from active sediments to water, outflow from the lake water and transport from active to passive sediments. In addition, two flows give the inflow of lime from the liming, one to the lake water and one directly to the active sediments.

Natural pH of a lake varies seasonally and hence constitutes an uncertain parameter. In this work, mean reverting Ito process is used to model fractional variation in pH owing to its success in modeling various time dependent stochastic parameters (Diwekar, 2008; Shastri & Diwekar, 2006a,b).

Optimal control problems require establishing an index of performance for the system and designing the course of action so as to optimize the performance index. The goal in lake liming operation is to maintain the pH value at some desired level or within a desired range. Since cost of liming is also a concern, this converts the problem into multi-objective optimal control problem.

Results and discussions: Figure 5 presents the result of deterministic and stochastic optimal control problems indicating that the targeted pH is effectively achieved. The plots also show that the stochastic optimal control leads to better pH control. Due to space limitations multi-objective optimal control results are not presented here.

Manipulation of Regime

It has been illustrated that a major portion of mercury found in the tissues of various aquatic organisms enters through food (ingestion). As a consequence, the eating habits of these organisms are expected to have a significant impact on the mercury intake by these organisms. The eating habits depend to quite an extent on the various species populations and their pattern of fluctuations at a given time in the water body. In ecological literature, these different patterns are referred to as regimes. A regime, therefore, if maintained for sufficient duration, is expected to affect the steady state mercury bioaccumulation levels in different species. As a result, manipulation of the regimes of these species populations presents a tool to control mercury bioaccumulation levels. This work performs an optimal control analysis to achieve regime shifts in a predator-prey model (Wang et al., 1998; Monson et al., 1998). The predator-prey model and the bioaccumulation model are inter-related by correlating the food intake of any particular species with the mercury intake for the bioaccumulation model. Changes in the dynamics of the Canale's model change the instantaneous food intake for the predators and super-predators (due to changing predation rates). This affects the total mercury that is taken by these species through food. Hence, any regime shift in the predator-prey model, which affects the predation rates, affects the mercury intake by the species. If the particular regime is maintained for a sufficient duration, the steady state mercury concentration in these species can alter. This is the basic foundation for the proposed work.

Regime change and optimal control: Optimal control theory presents an option to derive time dependent management strategies that can effectively achieve regime shifts in food chain models. Past work by the authors has illustrated the success of this approach (Shastri and Diwekar, 2006a, 06b). That work uses Fisher information based sustainability hypothesis, proposed by Cabezas and Fath (2002), to formulate time dependent objective functions for the control problem. A similar approach has been used in this work. The regime shift is to be achieved by minimizing the variation of the time averaged Fisher information around the constant Fisher information of the targeted regime. Canale's model exhibits various regimes such as cyclic low frequency, cyclic high frequency, stationary, and chaotic (Gregnani al., 1998). The idea proposed in this work is to achieve regime shift from a regime leading to high mercury bioaccumulation to a regime resulting in low mercury bioaccumulation. The control variables to achieve the regime shift are: nutrient inflow rate and nutrient input concentration.

Results and discussion: Simulations for the integrated model (Canale's model and the bioaccumulation model) illustrate that there is a strong correlation between the regime and steady state mercury bioaccumulation in predator and super-predators. Hence, the objective of causing a regime change in justified. Figure 6 shows the regime shift achieved by control of nutrient flow.

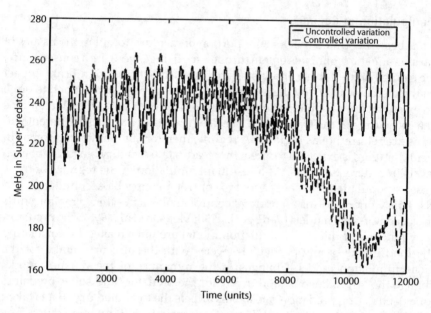

FIGURE 6
Controlling nutrient flow.

References

Boer M.P., Kooi B.W. and Kooijman S.A.L.M. (1998), Food chain dynamics in the chemostat, *Mathematical Biosciences*, 150, 43–62.

Burtraw, D., Evans, D., Krupnick, A., Palmer, K., Toth, R. (2005). Economics of pollution trading for SO2 and NOx. *Annual Review and Environment and Resources*, 30, 253–289.

Crocker, T. (1966). *The economics of air pollution*, chap. The structuring of air pollution control systems. New York: W.W. Norton.

Dales, J. (1968). *Pollution, property and prices*. Toronto: University of Toronto Press.

DeSimone, R., Penley, M., Charbonneau, L., Smith, G., Wood, J., Hill, H., Pratt, J., Ridsdale, S., Williams R. (1973). The kinetics and mechanism of cobalamine-dependent methyl and ethyl transfer tomercuric ion. *Biochimica et Biophysica Acta*, 304, 851–863.

Diwekar, U. (2003). *Introduction to Applied Optimization*. Kluwer Academic Publishers, Dordrecht.

Diwekar U. (2008). *Introduction to Applied Optimization*, 2nd edition, Springer, New York.

Driscoll, C., Blette, V., Yan, C., Schofield, C., Munson,R., Holsapple, J. (1995). The role of dissolved organic carbon in the chemistry and bioavailability of mercury in remote Adirondack lakes. *Water, Air and Soild Pollution*, 80, 499–508.

EPA. (2002), *Waste Reduction Algorithm Graphical Interface, Version 1.0*, ORD, USEPA, November.

Fath B. and H. Cabezas. (2002). Towards a theory of sustainable systems, *Fluid Phase Equilibria*, 2, 194-7.

Gragnani, A., De Feo, O. and Rinaldi, S. (1998). Food chains in the chemostat: Relationships between mean yield and complex dynamics. *Bulletin of Mathematical Biology*, 60 (4), 703–719.

Fu Y. and U. Diwekar, (2004). An efficient sampling approach to multi-objective optimization, (2004), *Annals of Operations Research*, 132, 109-120.

Hakanson, L. (2003). A general management model to optimize lake liming operations. Lakes & Reservoirs: *Research and Management*, 8, 105–140.

Hakanson, L., Boulion, V. (2002). *The lake foodweb*. Backhuys Publishers, Leiden.

Henrikson, L., Brodin, Y. (1995). *Liming of acidified surface waters*. Springer, Berlin.

Jensen, S., Jernelov, A. (1969). Biological methylation of mercury in aquatic organisms. *Nature*, 223, 753–754.

Kalagnanam J. and U. Diwekar.(1997). An efficient sampling technique for off-line quality control, *Technometrics*, 39(3), 308–19.

Landrum, P.F., Lee II, H. and Lydy, M.J. (1992). Toxicokinetics in aquatic systems: Model comparisons and use in hazard assessment, *Environmental Toxicology and Chemistry*, 11,1709–1725, 1992.

Monson B.A. and Brezonik P.L. (1998). Seasonal patterns of mercury species in water and plankton from softwater lakes in Northeastern Minnesota, *Biogeochemistry*, 40, 147–162.

Montgomery D.(1972). Markets in licences and efficient pollution control programs, *J. Economic Theory*, 5, 395-418.

Ottosson, F., Hakanson, L. (1997). Presentation and analysis of a model simulating the pH response of lake liming. *Ecological Modelling*, 105, 89–111.

Rico-Ramirez, V., Diwekar, U. (2004). Stochastic maximum principle for optimal control under uncertainty. *Computers and Chemical Engineering*, 28, 2845–2845.

Riely, P., Rockland, D. (1988). Evaluation of liming operations though benefit-cost analysis. *Water, Air and Soil Pollution*, 41, 293–228.

Shastri, Y., Diwekar, U. (2006a). Sustainable ecosystem management using optimal control theory: Part 1 (Deterministic systems). *Journal of Theoretical Biology*, 241, 506–521.

Shastri, Y., Diwekar, U. (2006b). Sustainable ecosystem management using optimal control theory: Part 2 (Stochastic systems). *Journal of Theoretical Biology*, 241, 522–532.

Sorensen, J., Glass, G., Schmidt, K., Huber, J., Rapp, G. (1990). Airborne mercury deposition and watershed characteristics in relation to mercury concentrations in water, sediments, plankton and fish of eighty northern minnesota lakes. *Environmental Science and Technology*, 24, 1716–1727.

USDOI (2001). Total plant costs: For contaminant fact sheets. *Technical report, U.S. Department of Interior*, Bureau of Reclamation, Water treatment engineering and research group, Denver CO 80225.41

USEPA (1996). Draft Framework for Watershed-Based Trading. *Tech. rep., EPA 800-R-96-001*. Washington, DC: United States Environmental Protection Agency, Office of Water.

USEPA (1997a). Capsule report: Aqueous mercury treatment. *Technical report: EPA/625/R-97/004*, United States Environmental Protection Agency, Office of Research and Development, Washington DC 20460.

USEPA (1997b). Mercury study report to congress. *Report to congress: EPA-452/R-97-003*, United States Environmental Protection Agency.

USEPA (2000a). Guidelines for preparing economic analysis. *Technical report: EPA 240-R-00-003*, United States Environmental Protection Agency.

USEPA (2000b). Mercury research strategy. *Technical report: EPA/600/R-00/073*, United States Environmental Protection Agency, Office of Research and Development, Washington DC 20460.

USEPA (2001a). Total Maximum Daily Load (TMDL) for total mercury in fish tissue residue in the middle and lower Savannah river watershed. *Report, United States Environmental Protection Agency, Region 4.*

USEPA (2001b). *Watershed Characterization System User's Manual.*, U.S. Environmental Protection Agency, Region 4, Atlanta, Georgia.

USEPA (2003). Water *Quality Trading Policy. Tech. rep.*, Office of Water, Washington, DC: United States Environmental Protection Agency.

Wang, W-X., Stupakoff I., Gagnon C. and Fisher N.S. (1998). Bioavailability of Inorganic and Methylmercury to a Marine Deposit-Feeding Polychaete, *Environmental Science and Technology*, 32, 2564-2571.

Winfrey M. R., Rudd J. W. M. (1990). Environmental factors affecting the formation of methylmercury in low pH lakes. *Environmental Toxicology and Chemistry*, 9, 173–174.

5

Scope for the Application of Mathematical Programming Techniques in the Synthesis and Planning of Sustainable Processes

Ignacio E. Grossmann[1*] and Gonzalo Guillén-Gosálbez[2]

[1]*Department of Chemical Engineering, Carnegie Mellon University, Pittsburgh, US*
[2]*Department of Chemical Engineering, University Rovira i Virgili, Tarragona, Spain*

CONTENTS

ABSTRACT Sustainability has recently emerged as a key issue in process systems engineering (PSE). Mathematical programming techniques offer a general modeling framework for including environmental concerns in the synthesis and planning of chemical processes. In this paper, we review major contributions in process synthesis and supply chain management, highlighting the major optimization approaches that are available, including the handling of uncertainty and the multi-objective optimization of economic and environmental objectives. Finally, we discuss challenges and opportunities identified in the area.

---

* Corresponding author: grossmann@cmu.edu

KEYWORDS *Sustainability, uncertainty, process synthesis, supply chain management*

Introduction: Sustainability in PSE

In the past, the methods devised in PSE to assist in the optimization of chemical processes have traditionally concentrated on maximizing an economic criterion. However, in the recent past there has been an increasing awareness of the importance of incorporating environmental aspects in the decision-making process. As a result, the scope of the analysis carried out in PSE is being enlarged with the aim to guide practitioners towards the adoption of more sustainable alternatives.

Including environmental issues in the synthesis and planning of chemical processes poses significant challenges that have not yet been fully solved, and hence merit further attention. One major critical issue is how to systematize the search for alternatives leading to reductions in environmental impact. Furthermore, aside from anticipating the effect of uncertainties, which are quite pronounced in this area, there is the issue on how to cope with competing economic and environmental objectives. Hence, there is a clear need to develop sophisticated optimization and decision-support tools to help in exploring and analyzing diverse process alternatives under uncertainty, and so as to yield optimal trade-offs between environmental performance and profit maximization. These methods should be employed to improve the environmental performance at different hierarchical levels, covering both single-site and multi-site industrial applications.

The aim of this paper is to summarize major contributions made in these fields, paying special attention to those based on mathematical programming. We center our discussion on two specific areas of PSE that can potentially help to identify and establish for environmental improvements: process synthesis and supply chain management (SCM).

Single-site Level: Process Synthesis

Process synthesis deals with the selection of the topology of a process in order to convert a set of inputs into a desired set of outputs (Rudd et al., 1973). Commonly the objective is to find designs that minimize cost or maximize profit. However, objectives such as maximizing efficiency or minimum usage of a resource (e.g. energy or freshwater) can also be considered. The area of process synthesis was especially active between the 70s and early 90s, in large part due to the increase in the cost of energy (Nishida et al., 1981). The area has addressed a number of major subproblems such as the synthesis of heat exchanger networks, distillation sequences, reactor networks, steam and power plants, mass exchange networks and process water networks, including total process flowsheets.

The area of process synthesis is particularly relevant to sustainability for two major reasons. First, process synthesis can help to identify the most efficient and/or economical process. This means that instead of using old technology to assess the environmental impact or energy use of a process in a life cycle analysis as is commonly done in many policy studies, one can rely on state-of-the-art process technology. A second reason is that process synthesis addresses subproblems such as pollution prevention (El-Halwagi, 1997), minimization of energy use (Linnhoff, 1993) and freshwater consumption (Wang and Smith, 1994) that lie at the heart of the environmental performance of a process. Unfortunately, the introduction of these considerations at the early stages of the process development increases the complexity of the design task, which is further complicated by the need to account for different conflictive criteria in the decision-making as well as various sources of uncertainty brought about by several problem parameters (costs, prices, demand, etc.). The development of efficient modeling and solution strategies capable of dealing with these issues constitutes a major challenge in PSE.

Multi-site Level: SCM and EWO

Supply Chain Management (SCM) is a relatively new discipline that aims to integrate manufacturing plants with their suppliers and customers in an efficient manner (Shapiro, 2001). In the context of PSE, the optimal integration of the operations of supply, manufacturing and distribution activities is the main goal of the emerging area known as Enterprise-wide optimization (EWO), which as opposed to SCM, places more emphasis on the manufacturing stage (Grossmann, 2005).

The major goal in the design and planning of sustainable SCs is to reduce the environmental impact of a process over its entire life cycle. This implies expanding the boundaries of the analysis typically performed in process synthesis in order to embrace a wider range of logistic activities. Note that besides the challenges posed by the standard economic optimization of these systems, which have been already discussed in Grossmann (2005), there are some additional issues associated with the inclusion of environmental aspects in SCM/EWO that deserve further attention. The first critical point is how to measure the environmental impact of a process/ product through all the stages of its life so this can be explicitly included as an additional criterion to be optimized. As pointed out in the literature (Freeman and Harten, 1992), the lack of accepted metrics to support objective environmental assessments still represents a major limitation in the area. The second aspect, which is strongly linked to the previous one, is how to incorporate these concerns in a modeling framework and effectively solve the resulting formulations by devising efficient algorithms and computer architectures. Note that the consideration of environmental design objectives further complicates the optimization problems arising in SCM and EWO, which are *per se* quite complex. Finally, as in the previous case, the problem is affected by different sources of uncertainty (inventory of emissions, damage model, waste generated, etc.) that can greatly impact the conclusions and recommendations made at the end of the environmental analysis (Geisler et al., 2005). Therefore, another major issue is the development of novel and meaningful stochastic methods capable of effectively anticipating the effect of these variations. All these aspects are expected to be the focus of future research.

General Techniques

In this section we summarize the main methodologies and techniques that can be used to reduce the environmental impact in process synthesis and SCM/EWO.

Process Synthesis

Major approaches to synthesizing process flowsheets that are cost effective, energy efficient and with potentially low environmental impact, include the use of heuristics, the development of physical insights (commonly based on thermodynamics), and the optimization of superstructures of alternatives. Major contributions in the first two approaches have been hierarchical decomposition (Douglas, 1988), and pinch analysis (Linnhoff, 1993) that has proved to be very successful in industrial applications.

The more recent trend has been to combine some of these concepts with the mathematical programming approach (see Grossmann et al., 1999), which consists of three major steps. The first is the development of a representation of alternatives from which the optimum solution is to be selected. The second is the formulation of a mathematical program that generally involves discrete and continuous variables for the selection of the configuration and operating levels, respectively. The third is the solution of the optimization model (commonly a mixed-integer nonlinear programming, MINLP, or a generalized disjunctive programming, GDP, model) from which the optimal solution is determined.

While superstructures can be developed in a systematic way for subsytems (e.g. see Yee and Grossmann (1990) for heat exchanger networks), their development for general process flowsheets is more complex. Here two approaches that have emerged are the axiomatic approach by Friedler et al. (1993), and the State-Task and State-Equipment Networks by Yeomans and Grossmann (1999). As for the problem formulation it is important to note that synthesis models can be formulated at three major levels of detail: a) Aggregated models that are high level and concentrate on major features like energy flows (e.g. LP transshipment model for HEN by Papoulias and Grosmann, 1983; NLP heat and mass exchanger by Papalexandri and Pistikopoulos, 1996); b) Short-cut models that involve cost optimization (investment and operating costs), but in which the performance of the units is predicted with relatively simple nonlinear models (e.g. MINLP heat exchanger networks by Yee and Grossmann, 1990; MINLP process flowsheets by Kocis and Grossmann, 1987); c) Rigorous models that rely on detailed superstructures and involve rigorous and complex models for predicting the performance of the units (e.g. MINLP synthesis of distillation sequences, Smith and Pantelides, 1995; and GDP models, Grossmann et al., 2005).

At this point there are still very few papers that have reported the use of process synthesis techniques with the explicit incorporation of sustainability issues (eg. Steffens et al., 1999; Halasz et al., 2005). Some of them have applied optimization techniques to the molecular design of solvents and the synthesis of the associated separation processes (Pistikopoulos and Stefanis, 1998; Hostrup et al., 1999), whereas an increasing number of publications are addressing the synthesis of biofuels plants (e.g. Agrawal et al., 2007; Karuppiah et al., 2008).

SCM/EWO

The combination of environmental management and SCM into a single framework has recently led to a new discipline known as Green Supply Chain Management (GrSCM). An exhaustive review on the area of GrSCM can be found in the work of Srivastava (2007). According to the author, there are two main types of approaches in GrSCM: empirical studies and mathematical modeling. Within the latter group, we can find a variety of tools and techniques, such as mathematical programming (LP, NLP, MILP, MINLP and dynamic programming), Markov chains, Petri Nets, input-output models, game theory, fuzzy logic, data envelopment analysis (DEA), descriptive statistics and simulation. These methods have been applied to the two main areas of GrSCM: green design and green operations. The former one involves the environmentally conscious design of products and processes, whereas the second one deals with green manufacturing and remanufacturing, reverse logistics, network design and waste management. Both areas share the same holistic approach in which the key issue is to take into account the complete life cycle of the product/process under study. This global perspective avoids technological alternatives that decrease the impact locally at the expense of increasing the environmental burdens in other stages of the life cycle of the product.

The application of these techniques to the design of sustainable processes has followed two different approaches. The first one, which has been the most common approach, has focused on including them as additional constraints to be satisfied by the optimization model. As pointed out by Cano-Ruíz and McRae (1998), the environmental issues should be regarded as new design objectives and not merely as constraints on operations. This second approach is better suited to account for environmental concerns at the design stage, since it can lead to the discovery of novel alternatives that simultaneously improve the economic and environmental performance of the process (Hugo and Pistikopoulos, 2005). As mentioned before, such consideration leads to more complex problems that require specific multi-objective optimization methods, some of which can be used in conjunction with the mathematical tools previously described.

Optimization using mathematical programming is probably the most widely used approach in SCM. General literature reviews have been made by Thomas and Griffin (1996) and Maloni and Benton (1997), whereas a more specific work devoted to process industries can be found in Grossmann (2005).

From the modeling point of view, the preferred tool has been mixed-integer linear programming (MILP). This choice has been motivated by the fact that these formulations tend to be represented at a high level, and hence apply fairly simple representations of capacity that avoid nonlinearities and allow them to be easily adapted to a wide range of industrial scenarios. Such simplification can sometimes lead to approximate solutions, the accuracy of which may vary depending on the specific application. In contrast to SCM, EWO focuses more on process industries and often includes more realistic capacity models involving nonlinear equations.

In the aforementioned MILP formulations, continuous variables are used to represent materials flows and purchases and sales of products, whereas binary variables are employed to model tactical and/or strategic decisions associated with the network configuration, such as selection of technologies and establishment of facilities and transportation links. These models have been traditionally solved via branch and bound techniques, which in some cases have been applied in conjunction with other strategies such as Lagrangean (Graves, 1982), Benders (Spengler et al., 1997) and bi-level decomposition methods (Iyer and Grossmann, 1998). In the area of GrSCM, mathematical programming has been employed in green manufacturing, remanufacturing, reverse logistics, network design and waste management (for a detailed review see Srivastava, 2007). In contrast, in EWO the use of mathematical programming in the design and planning of sustainable SCs has been rather limited and it is still waiting for further research.

Mathematical Programming Techniques

A discussed in the previous sections, mathematical programming has been widely applied in process synthesis and SCM. The next sections summarize the major methods, including the handling of uncertainty and multiobjective optimization.

Mixed-integer Optimization

Developing the full range of models for EWO often involves MILP techniques for which efficient software such as CPLEX and XPRESS are available for solving fairly large-scale problems. However, EWO problems may require that nonlinear process models be developed for representing manufacturing and inventories in the planning and scheduling of production facilities. This gives rise to mixed-integer nonlinear programming (MINLP) problems since they involve discrete variables to model assignment and sequencing decisions, and continuous variables to model flows, amounts to be produced and operating conditions (e.g. temperatures, yields). While MINLP optimization is still largely a rather specialized capability, it has been receiving increasing attention by the OR community in the last few years. Furthermore, modeling systems like GAMS now offer multiple methods for solving these problems (e.g. DICOPT, SBB, a-ECP, Bonmin, BARON). A recent review on MINLP methods can be found in Grossmann (2002). Major methods include branch and bound, outer-approximation, Generalized Benders decomposition, extended cutting planes, and LP/NLP based branch and bound. While these methods have proved to be effective, they are still largely limited to moderate-sized problems (few hundreds of 0–1 variables, several thousands of continuous variables and constraints). In addition there are several difficulties that must be faced in solving these problems. For instance in NLP subproblems with fixed values of the binary variables, a significant number of equations and variables are often set to zero as they become redundant when units "disappear." This in turn often leads to singularities and poor numerical performance. There is also the possibility of getting trapped in suboptimal solutions when nonconvex functions are involved. Finally, there is the added complication when the number of 0–1 variables is large, which is quite common in planning and scheduling problems.

To circumvent some of these difficulties, the modeling and optimization of Generalized Disjunctive Programs (GDP) seems to hold good promise for process synthesis and EWO problems. The GDP problem is expressed in terms of Boolean and continuous variables that are involved in constraints in the form of equations, disjunctions and logic propositions (Raman and Grossmann, 1994). This has the effect of greatly simplifying the modeling of discrete/continuous problems. Furthermore, the logic-based outer approximation for nonlinear GDP problems (Turkay and Grossmann, 1996) has the important feature of generating NLP subproblems where redundant equations and constraints of non-existing units are not included, which improves the robustness of the optimization. The only software available for solving GDP models is LOGMIP (Vecchietti and Grossmann, 1999). Another major challenge is obtaining the global optimum solution. Here a number of global optimization algorithms (Floudas, 2000; Sahinidis, 1996; Tawarmalani and Sahinidis, 2002) have emerged, mostly through spatial branch-and-bound schemes. BARON has become the major global optimization solver in modeling systems such as GAMS. For GDP, there are still few global optimization techniques (e.g. Lee and Grossmann, 2001).

Multi-objective Optimization

As mentioned before, multi-objective optimization (MOO) is well suited to incorporate environmental concerns in the optimization of sustainable processes, since it allows to treat them as decision-making objectives. The use of these methods requires translating such environmental aspects into suitable environmental performance indicators that should be optimized in conjunction with the traditional economic-based criteria. There are three main types of MOO approaches: (1) those based in the transformation of the problem into a single-objective one (see Ehrgott, 2000), (2) the Non-Pareto approaches, which use search operators based in the objectives to be optimized and (3) Pareto approaches, which directly apply the concept of dominance (see Deb, 2008). Whereas the first approach can be easily applied in conjunction with standard exact algorithms (i.e., branch and bound), the second and third ones are better suited to work with meta-heuristics. Note that any of the traditional exact methods employed in process synthesis and SCM (LP, MILP, MINLP, GDP and global optimization) can be coupled with single-objective MOO approaches, such as aggregation methods, the epsilon constraint method, goal programming and goal attainment. As pointed by Hugo and Pistikopoulos (2005), many of these methods have been employed to account for environmental concerns in process design problems that focus on single-site scenarios. However, their application to the multi-objective optimization of entire supply chain networks has been rather limited.

Uncertainty

All the problems mentioned in the previous sections are further complicated by different sources of uncertainty that can be encountered in practice. Many of the existing methods in process synthesis and SCM/EWO assume nominal values for the uncertain parameters and do not consider their variability. However, this simplification may lead to solutions that perform well in the most likely scenario but exhibit poor performance under other circumstances. Special techniques able to assess process alternatives under uncertainty can avoid this situation and guarantee a good performance for any possible outcome of the uncertain parameters. We next review the two major proactive methods that account for uncertainty considerations in the decision-making process, focusing on their applications in process synthesis and SCM/EWO.

Stochastic Programming with Recourse

In stochastic programming (Birge and Louveaux, 2000; Sahinidis, 2004), mathematical programs are defined with a set of uncertain parameters, which are normally described by discrete distributions. These in turn give rise to scenarios that correspond to a particular realization of each of the uncertain parameters. Furthermore, stochastic programs are solved over a number of stages. The fundamental idea behind stochastic programming is the concept of recourse. Recourse is the ability to take corrective action after a realization of a scenario has taken place. Between each stage, some uncertainty is resolved, and the decision maker must choose an action that optimizes the current objective plus the expectation of the future objectives. The most common stochastic programs are two-stage models in which typically stage-1 decisions involve selection of topology or design variables in a synthesis problem, or planning decisions for the first month in SCM problem. Stage-2 decisions involve variables that can be adjusted according to the realization of the scenarios (e.g. recycles in a flowsheet, production levels in a SCM problem). Two-stage programming problems may be solved in a number of methods including decomposition methods (Ruszczyński, 2003) and sampling-based methods (Linderoth, et al., 2006).

When the second-stage (or recourse) problem is a linear program these problems are straightforward to solve, but the more general case is where the recourse is a MILP or a MINLP. Such problems are extremely difficult to solve since the expected recourse function is discontinuous and nonconvex (Sahinidis, 2004). It should be noted that the more general stochastic program corresponds to the multistage model. In this case, decision variables and constraints are divided into groups of corresponding temporal stages. At each stage some of the uncertain quantities become known. In each stage one group of decisions needs to be fixed based on what is currently known, along with trying to compensate for what remains uncertain. The model essentially becomes a nested formulation. Although these problems are difficult to solve, there is extensive potential for applications. The strength of stochastic programming is that it is one of the very few technologies for optimization under uncertainty that allows models to capture recourse.

Planning in the chemical process industry has used stochastic programming for a number of applications (Liu and Sahinidis, 1996; Clay and Grossmann, 1997). The scheduling of batch plants under demand uncertainty using stochastic programming has only recently emerged as an area of active research. Engell et al. (2004) use a scenario decomposition method for the scheduling of a multi-product batch plant by two-stage stochastic integer programming. Balasubramanian and Grossmann (2004) present an approach for approximating multistage stochastic programming to the scheduling of multiproduct batch plants under demand uncertainty. More recent applications of stochastic programming to supply chain optimization can be found in Pistikopoulos et al. (2007).

Robust Optimization and Probabilistic Programming

Robust optimization, which was first introduced by Ben-Tal and Nemirovski (1998), seeks to determine a robust feasible/optimal solution to an uncertain problem. This means that the optimal solution should provide the best possible value of the original objective function and also be guaranteed to remain feasible in the range of the uncertainty set considered for a predefined probability level. A major difference between robust optimization and stochastic programming with recourse is the explicit consideration of feasibility issues. In robust optimization, the solution must ensure that a set of constraints will be satisfied with a certain probability when the uncertainty is realized. Instead, stochastic optimization either assumes complete recourse, that is, every scenario is supposed to be feasible, or allows infeasibilities at a certain penalty. Furthermore, robust optimization cannot handle recourse variables. Thus, it can be considered as a particular category of single-stage here-and-now problems, where the uncertain parameters are enclosed in an inequality constraint subject to a probability or reliability level. Because of this simplification (absence of recourse actions), robust optimization usually leads to lower computational burdens. However, the difficulty in solving these problems still lies in the computation of the probability and its derivatives of satisfying inequality constraints (Li et al., 2008).

The concept of robust optimization is somehow linked to chance-constrained programming. Chance-constrained programming (Charnes and Cooper, 1959), also known as probabilistic programming, considers the uncertainty by introducing a probabilistic level of constraint satisfaction. This method is useful to deal with inequality constraints the satisfaction of which is highly desirable, but not absolutely essential. In practice, chance-constrained programming provides the mathematical framework that allows to deal with the probabilistic constraints employed in robust optimization.

Robust optimization has been widely applied in optimization under uncertainty (see Uryasev, 2000). Specifically, many previous works have focused on efficiently solve this type of problems (see Nemirovski and Shapiro, 2006). However, the application of robust optimization in PSE has been rather limited and usually restricted to operational/tactical problems, such as the scheduling of batch plants under uncertainty (Petkov and Maranas, 1997; Janak et al., 2007). Gupta et al. (2000) also used a chance-constrained approach in conjunction with a two-stage stochastic programming model to analyze the tradeoffs between demand satisfaction and production costs for a mid-term supply chain planning problem. Extensions of these strategies to deal with strategic problems as well as different sources of uncertainty are still waiting for further research.

EXAMPLES

In this section, we present four examples that illustrate the challenges cited in this article on problems encountered in the synthesis and planning of sustainable chemical processes. The first example addresses the energy optimization of a bioethanol plant. The second problem deals with the synthesis of a process network with uncertain yields. The third example illustrates the use of multi-objective optimization coupled with life cycle assessment (LCA) to address a SCM problem. Finally, the fourth problem deals with the design of sustainable petrochemical SCs and extends the framework presented in example 3 to account for different sources of uncertainty that affect the environmental impact calculations.

EXAMPLE 1. SYNTHESIS OF BIOFUELS

Karuppiah et al. (2008) considered the energy optimization of the "dry-grind" process for the corn-based bio-ethanol plant. In such plants, fuel ethanol is produced using corn-kernels as the feedstock. Fuel grade ethanol has to be 100% pure before it can be blended with gasoline to be used in automobiles. However, conventional distillation columns produce an azeotropic mixture of ethanol and water (95% ethanol – 5% water), which has to be purified further for making fuel ethanol. The main challenge in the way of producing fuel ethanol commercially is that the process is very energy intensive and requires large amounts of steam and electricity for use in the rectifiers to get an azeotropic mixture of ethanol and water and requires the use of expensive molecular sieves to get 100% pure ethanol.

Karuppiah et al. (2008) developed a simplified model to predict the performance of the bio-ethanol flowsheet that includes grinding, scarification, fermentation, centrifugation and drying operations (see Fig. 1). A superstructure was also postulated in which some of the major alternatives include separation by molecular sieves and or corn grits, and different ways to accomplish the drying for the dried grains solids, the cattle feed by-product. The objective was to optimize the structure, determining the connections in the network and the flow in each stream in the network, such that the energy requirement of the overall plant is minimized while trying to maximize the yields.

The optimization without heat integration (MINLP model) led to a decrease of the manufacturing cost from $1.61/gal (base case) to $1.57. In the next step heat integration was considered in the optimization, which further reduced the cost to $1.51/gal. However, it became clear that the scope of heat integration is limited by the relatively low temperature in the fermentor. In order to improve the potential for heat integration the authors considered multi-effect distillation in the "beer" column and in the azeotropic column as

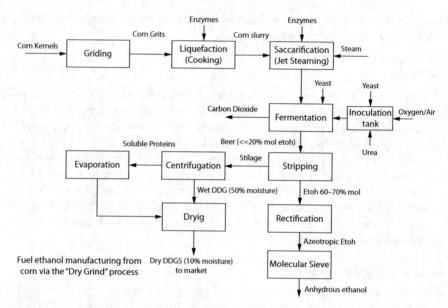

FIGURE 1
Flowsheet of dry-grind process for bioethanol.

alternatives for the optimization (see Fig. 2). This finally, led to a 65% savings in steam consumption and cost reduction down to $1.43/gal! This example then illustrates on the one hand the potential for cost reduction in biofuel plants, and on the other hand the potential pitfall when policy researchers (e.g. Pimentel, 1991) perform life cycle analyses without accounting for the fact that the cost and efficiency of the manufacturing technology can be substantially improved as was the case in this example.

EXAMPLE 2. STOCHASTIC PROGRAMMING OF PROCESS NETWORK

We consider as a second example the problem studied by Tarhan and Grossmann (2008), a multi-period synthesis of process networks under *gradual uncertainty reduction* in the process yields and with possible investments in pilot plants for reducing uncertainties. It is assumed that a process network is given with availabilities of raw materials, intermediates and demands for final products over T time periods. The problem is to determine in each time period t whether the capacity of specific processes should be expanded or not (including new or existing processes), whether specific processes should be operated or not, and whether pilot plants for reducing uncertainties in new processes should be installed or not. In addition, other decisions include selecting the actual expansion capacities of the processes, the flowrates in the network, and the amount of purchase and sales of final products. The objective is to select these decisions to maximize the expected net present value.

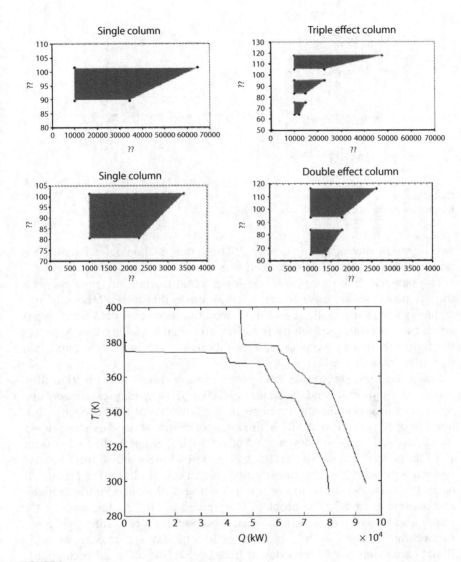

FIGURE 2
Profiles for multieffect columns for beer and azeotropic column, and T-Q curves for optimized process.

Figure 3 shows an example of a process network that can be used to produce a given product. Currently, the production of A takes place only in Process III which consumes an intermediate product B that is purchased. If needed, the final product A can also be purchased so as to maintain its inventory. The demand for the final product, which is assumed to be known, must be satisfied for all periods over the given time horizon. Two new technologies (Process I and Process II) are considered for producing the intermediate from

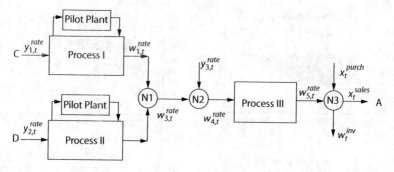

FIGURE 3
Example of process network with uncertain yields.

two different raw materials C and D. These new technologies have uncertainty in the yields which is reduced over time.

The scenario tree representations for gradual uncertainty resolution for the two processes are given in Fig. 4. It is assumed that at step 2 the only realizable yields are the highest and the lowest of all possible values. At step 3, when the uncertainty is totally revealed, all possible yields are realized. In the figure there are two possible realizations for yields in step 2 and four possible yields in step 3.

Using the multistage stochastic programming model and the algorithm proposed by Tarhan and Grossmann (2008), the capacity expansion and operation decisions for the problem in Fig. 3 were optimized over a time horizon of 10 years. Process III is already operational with an existing capacity of 3 tons/day and a known yield of 70%. All possible realizations of the yield for process I at step 3 are 69, 73, 77 and 81% where only 69 and 81% are realizable in step 2 of the uncertainty resolution. Similarly for process II, 60 and 90% are two realizations in step 2 with 60, 70, 80 and 90% as possible realizations at step 3. The problem was solved within 2% tolerance of the upper and lower bounds with the proposed method. The solution proposes expanding Process I up to a capacity of 10 tons/day and making an additional expansion of 4.93 tons/day at time period 3 if the yield turns out to be 69%. If the yield for Process I was found to be 81% then an expansion of 2.98 tons/day is made at the time period 4. This solution did not involve the

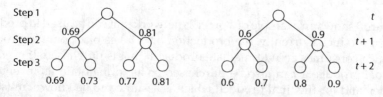

FIGURE 4
Gradual resolution of uncertainty in yields of procesees 1 and 2.

use of a pilot plant, and yielded an expected net present value of $8,050,500. The proposed branch and bound algorithm based on Lagrangean relaxation required about 5,000 secs of CPU-time. This example shows the potential for applying stochastic programming approaches to new processes that have uncertain yields, a problem of relevance to biofuels.

EXAMPLE 3. DESIGN OF HYDROGEN SCs FOR VEHICLE USE

This example deals with the optimal design of a hydrogen SC for vehicle use in UK taking into account economic and environmental concerns. The problem, which was first proposed by Almansoori and Shah (2006), considers different technologies for production, storage and transportation of hydrogen to be established in a set of geographical regions distributed all over the country (see Fig. 5). The goal is to determine the optimal network configuration in terms of its economic and environmental performance.

The problem can be formulated as a bi-criterion MILP that seeks to minimize the total cost of the network and its environmental impact. In this formulation, integer variables indicate the number of plants and storage facilities to be opened in a specific region (i.e., grid), whereas binary variables are employed to denote the existence of transportation links connecting the SC entities. The importance of climate change in the transition towards a hydrogen energy system motivated the selection of the damage caused by

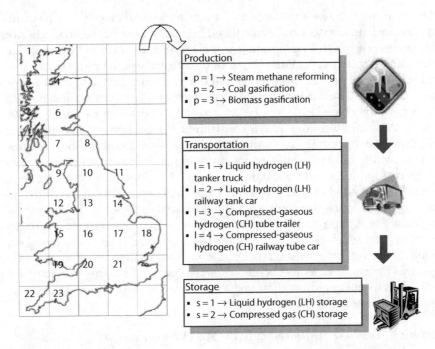

FIGURE 5
Superstructure of example 3.

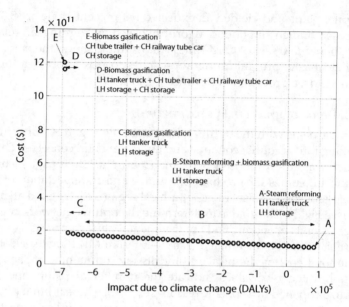

FIGURE 6
Pareto set of example 3.

the green house gases emissions as the environmental objective to be mini-mized. Such an impact can be calculated by making use of the Eco-indicator 99 framework, which incorporates the most recent advances in LCA method-ology and hence covers all the stages of the life cycle of the process.

The difficulty of this approach is that the size of the problem can become very large as the number of periods increases. For instance, a problem with 10 periods involves 21,160 binary variables, 1,880 discrete variables, 25,396 continuous variables and 70,976 constraints. To circumvent this problem, Guillén-Gosálbez et al. (2008) developed a bi-level decomposition scheme that proved to be approximately one order of magnitude faster than the full space method for small optimality tolerances (i.e., less than 1%). The Pareto set calculated (see Fig. 6) showed that important reductions in the contribu-tion to global warming can be achieved by replacing steam reforming by biomass gasification and also by establishing more decentralized hydrogen networks in which the transportation tasks are minimized. On the other hand, the results also revealed that replacing liquid hydrogen by compressed gaseous hydrogen is not a good choice, since such option leads to a marginal reduction in the environmental impact at the expense of a large increase in the total cost of the network.

Example 4. Design of sustainable chemical SCs under uncertainty

This example addresses the optimal design of petrochemical SCs tak-ing into account economic and environmental concerns and considering

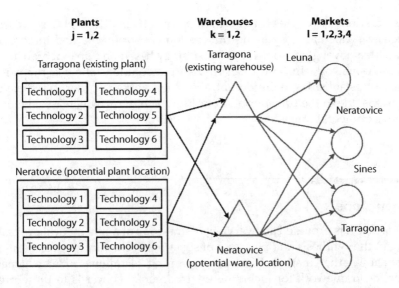

FIGURE 7
Superstructure of example 4.

different sources of uncertainty affecting the environmental assessment of the process. We consider a superstructure based on a three-echelon SC (production-storage-market) with different available production technologies for plants, potential locations for SC entities and transportations links (see Fig. 7). The goal is to maximize the NPV of the SC and minimize its environmental impact. As in the previous example, the latter performance indicator is calculated over the entire life cycle of the process by using the Eco-indicator 99 approach, which in this case includes not only the damage due to global warming, but also the remaining impacts in the human health, ecosystem quality and resources depletion. The example also accounts for the uncertainty of the life cycle inventory of emissions and feedstock requirements associated with the network operation. Guillén-Gosálbez and Grossmann (2009) proposed a novel MINLP formulation to deal with this problem in which the environmental performance of the network under uncertainty was measured via probabilistic constraints that were converted into standard deterministic inequalities by applying concepts from chance-constrained programming. Specifically, the example in Fig. 7 led to a bi-criterion MINLP involving 78 binary variables, 1,837 continuous variables and 1,963 constraints. The authors solved this problem by a novel decomposition technique based on parametric programming that allowed the calculation of the complete Pareto set in 7 iterations after 68.30 CPU seconds.

This modeling framework was later expanded in scope to deal with the uncertainty of the parameters of the damage model, and also to allow for the simultaneous control of different damage categories included in the Eco-indicator 99 (Guillén-Gosálbez and Grossmann, 2008). These new considerations gave rise to a bi-criterion nonconvex MINLP, the solution of which was calculated by using the epsilon constraint method in conjunction with a novel global optimization strategy based on a spatial branch and bound framework. In both cases, the Pareto solutions showed the convenience of establishing more decentralized networks in order to reduce the emissions due to the transportation tasks, which in turn decreases the overall environmental impact.

Conclusions

This article has provided an overview of the scope of mathematical programming in the synthesis and planning of sustainable chemical processes. It has been shown that mathematical programming techniques offer a general modeling framework for including environmental concerns in these problems. It was also shown that process synthesis and supply chain management are two key areas in PSE that lend themselves very well to addressing sustainability issues together with the common economic targets. We highlighted the major optimization approaches that are available, including the handling of uncertainty and the multi-objective optimization.

Some of the major challenges have been highlighted throughout the paper and several examples presented to illustrate the nature of the applications and the problems that are faced. While perhaps obvious, it is clear that the area of sustainability offers a great opportunity to renew the interest in process synthesis since it appears that many of the new biofuel plants have not had the benefit of being subjected to more systematic and thorough optimizations as their petrochemical counterparts. As we also discussed, this can lead to flawed analyses when comparing energy content or life cycle analysis of competing energy technologies. We should note, however, as was illustrated in the bioethanol example, that it will not be sufficient to simply apply the known synthesis techniques to these new processes. Major reasons include having to deal with exothermic reactions that take place at lower temperatures and separations of highly diluted systems. Furthermore, a major challenge not encountered in conventional process synthesis is that many of the biofuel plants are rather small and therefore cannot benefit of the economies of scale. This would seem to imply that process intensification could hold the promise of making these processes economically viable. This area has been virtually unexplored. An interesting possibility might involve developing superstructures that contemplate alternatives for process intensification.

In the area of supply chain management it is clear that progress has been made in terms of incorporating models for environmental impact within a multiobjective optimization framework. However, the greatest challenge still lies in properly accounting for the uncertainties associated with the parameters of these models (e.g. emissions, potential harm, etc.). An interesting possibility would be to characterize the various sources of uncertainty and establish what would be the more meaningful stochastic programming strategies to anticipate their effect.

Finally, although one can in principle formulate the associated optimization problems discussed above, it is clear that models are often very large, defeating current computational capabilities. Hence, developing effective solution approaches and algorithms continues to be a very real need.

Acknowledgments

Ignacio Grossmann would like to acknowledge support from the Center for Advanced Process Decision-making at Carnegie Mellon University. Gonzalo Guillén-Gosálbez wishes to acknowledge support from the Spanish Ministry of Education and Science (research project DPI2008-04099).

References

Agrawal, R., Singh, N.R, Ribeiro, F.H., Delgass, W. N. (2007). Sustainable fuel for the transportation sector. *PNAS*, 104 (12) , 4828.

Almansoori, A., Shah, N. (2006). Design and operation of a future hydrogen supply chain: snapshot model. *Chem. Eng. Res. Des.*, 84(A6), 423.

Balasubramanian, J., Grossmann, I.E. (2004). Approximation to Multistage Stochastic Optimization in Multiperiod Batch Plant Scheduling under Demand Uncertainty. *Ind. Eng. Chem. Res.*, 43, 3695.

Ben-Tal, A., Nemirovski, A. (1998). Robust convex optimization. *Math. Oper. Res.*, 23(4), 769.

Birge, J.R. and F. Louveaux. (2000). *Introduction to Stochastic Programming*. Springer.

Cano-Ruiz, J.A., McRae, G.J. (1998). Environmentally conscious chemical process design. *Annu. Rev. Energy Env.*, 23, 499.

Charnes, A., Cooper, W.W. (1959). Chance-constrained programming. *Manage. Sci.*, 6(1), 73.

Clay, R. L., Grossmann, I. E. (1997). A disaggregation algorithm for the optimization of stochastic planning models, *Comput. Chem. Eng.* 21, 751.

Deb, K. (2008). *Multi-objective Optimization Using Evolutionary Algorithms*. John Wiley & Sons Inc.

Douglas, J.M. (1988). *Conceptual Design of Chemical Processes*. McGraw-Hill.

Ehrgott, M. (2000). *Multicriteria optimization*. Springer.

El-Halwagi, MM. (1997). *Pollution Prevention Through Process Integration: Systematic Design Tools*. San Diego, CA: Academic.

Engell, S., Markert, A., Sand, G., Schultz, R. (2004). Aggregated scheduling of a multi-product batch plant by two-stage stochastic integer programming. *Optimization and Engineering*, 5, 335–359.

Floudas, C. A. (2000). *Deterministic Global Optimization: Theory, Methods and Applications.* Dordrecht, The Netherlands: Kluwer Academic Publishers.

Freeman, H., Harten, T., Springer, J., Randall, P., Curran, M.A., Stone, K. (1992). Industrial pollution prevention: a critical review. *J. Air Waste Manage. Assoc.*, 42(5), 618.

Friedler, F., Tarjan, K., Huang, Y.W., Fan, L.T. (1993). Graph-theoretic Approach to Process Synthesis: Polynomial Algorithm for Maximal Structure Generation. *Comput. Chem. Eng.*, 17, 929.

Geisler, G., Hellweg, S., Hungerbuhler, K. (2005). Uncertainty analysis in life cycle assessment (LCA): Case study on plant-protection products and implications for decision making. *Int. J. Life Cycle Assess.*, 10(3), 184.

Graves, S.C. (1982). Using Lagrangean techniques to solve hierarchical production planning problems. *Manage. Sci.*, 28(3), 260.

Grossmann, I.E. (2002). Review of Nonlinear Mixed-Integer and Disjunctive Programming Techniques. *Optimiz. Eng.*, 3, 227.

Grossmann, I.E. (2005). Enterprise-wide optimization: A new frontier in process systems engineering. *AIChE J.*, 51(7), 1846.

Grossmann, I.E., Aguirre, P.A., Barttfeld, M. (2005). Optimal Synthesis of Complex Distillation Columns Using Rigorous Models. *Comput. Chem. Eng.*, 29, 1203.

Grossmann, I.E., Caballero, J.A., Yeomans, H. (1999). Mathematical Programming Approaches for the Synthesis of Chemical Process Systems. *Korean J. Chem. Eng.*, 16, 407.

Guillén-Gosálbez, G., Grossmann, I.E. (2008). A global optimization strategy for the environmentally conscious design of chemical supply chains under uncertainty. Submitted to *Comput. Chem. Eng.*

Guillén-Gosálbez, G., Grossmann, I.E. (2009). Optimal design and planning of sustainable chemical supply chains under uncertainty. *AIChE J.*, 55(1), 99.

Guillén-Gosálbez, G., Mele, F., Grossmann, I.E. (2008). A bi-criterion A bi-criterion optimization approach for the design and planning of hydrogen supply chains for vehicle use. Submitted to *AIChE J.*

Gupta, A., Maranas, C. D., McDonald, C. M. (2000). Midterm supply chain planning under demand uncertainty: Customer demand satisfaction and inventory management. *Comput. Chem. Eng.*, 24, 2613.

Halasz, L., Povoden, G., Narodoslawsky, M. (2005). Sustainable processes synthesis for renewable resources. *Resour., Conserv. Recycl.*, 44, 293.

Hostrup, M., Harper, P.M., Gani, R. (1999). Design of environmentally benign processes: integration of solvent design and separation process synthesis. *Comput. Chem. Eng.*, 23, 1395.

Hugo, A., Pistikopoulos, E.N. (2005). Environmentally conscious long-range planning and design of supply chain networks. *J. Clean. Prod.*, 13(15), 1471.

Iyer, R.R., Grossmann, I.E. (1998). A bilevel decomposition algorithm for long-range planning of process networks. *Ind. Eng. Chem. Res.*, 37(2), 474.

Janak, S.L., Lin, X.X., Floudas, C.A. (2007). A new robust optimization approach for scheduling under uncertainty - II. Uncertainty with known probability distribution. *Comput. Chem. Eng.*, 31(3), 171.

Karuppiah, R., Grossmann, I.E. (2008). Energy optimization for the design of corn-based ethanol plants. *AIChE J.*, 54(6), 1499.

Kocis, G.R., Grossmann, I.E. (1987). Relaxation Strategy for the Structural Optimization of Process Flow Sheets. *Ind. Eng. Chem. Res.*, 26, 1869.

Lee, S., Grossmann, I.E. (2001). A Global Optimization Algorithm for Nonconvex Generalized Disjunctive Programming and Applications to Process Systems. *Comput. Chem. Eng.*, 25, 1675.

Li, P., Arellano-Garcia, H., Wozny, G. (2008). Chance constrained programming approach to process optimization under uncertainty. *Comput. Chem. Eng.*, 32, 25.

Linderoth, J., Shapiro, A., Wright, S. (2006). The Empirical Behavior of Sampling Methods for Stochastic Programming. *Annals of Oper. Res.*, v142, 215.

Linnhoff, B. (1993). Pinch Analysis - A state of the art overview. *Trans. IchemE.*, 71(A), 503.

Liu, M. L., Sahinidis, N. V. (1996). Optimization in process planning under uncertainty, *Ind. Eng. Chem. Res.*, 35, 4154.

Maloni, M.J., Benton, W.C. (1997). Supply chain partnerships: Opportunities for operations research. *Eur. J. Oper. Res.*, 101(3), 419.

Nemirovski, A., Shapiro, A. (2006). Convex approximations of chance constrained programming. *SIAM J. Optimiz.*, 17, 969.

Nishida, N., Stephanopoulos, G., Westerberg, A.W. (1981). Review of Process Synthesis. *AIChE J.*, 27, 321.

Papalexandri, K.P., Pistikopoulos, E.N. (1996). Generalized Modular Representation Framework of Process Synthesis. *AIChE J.*, 42, 1010.

Papoulias, S.A., Grossmann, I.E. (1983). A structural optimization approach in process synthesis. II: Heat recovery networks. *Comput. Chem. Eng.*, 7(6), 707.

Petkov, S. B., Maranas, C. (1997). Multiperiod planning and scheduling of multiproduct batch plants under demand uncertainty. *Ind. Eng. Chem. Res.*, 36, 4864.

Pimentel, D. (1991). Ethanol fuels: energy security, economics, and the environment. *J. Agr. Environ. Ethic.*, 4, 1.

Pistikopoulos, E.N., Georgiadis, M., Papageorgiou, L. (2007). *Supply-chain Optimization, Vol.4.* Wiley.

Pistikopoulos, E.N., Stefanis, S.K. (1998). Optimal solvent design for environmental impact minimization. *Comput. Chem. Eng.*, 22(6), 717.

Raman, R., Grossmann, I.E. (1994). Modeling and computational techniques for logic-based integer programming. *Comput. Chem. Eng.*, 18(7), 563.

Rudd, D. F., Powers G. J., Siirola, J. J. (1973). *Process Synthesis.* Prentice-Hall, Englewood Cliffs, NJ.

Ruszczyński, A. (2003). *Decomposition Methods*, in A. Ruszczyński and A. Shapiro (eds.), Stochastic Programming, Handbooks in Operations Research and Management Science Vol. 10, Elsevier.

Sahinidis, N.V. (1996). BARON: A General Purpose Global Optimization Software Package. *J. Global Optim.*, 8, 201.

Sahinidis, N.V. (2004). Optimization under Uncertainty: State of the Art and Opportunities. *Comput. Chem. Eng.* 28, 971.

Shapiro, J. F. (2001). *Modeling the Supply Chain.* Duxbury, Pacific Grove.

Smith, E.M.B., Pantelides C.C. (1995). Design of Reaction/Separation Networks Using Detailed Models. *Comput. Chem. Eng.*, S231, 19, s83.

Spengler, T., Puchert, H., Penkuhn, T., Rentz, O. (1997). Environmental integrated introduction and recycling management. *Eur. J. Oper. Res.*, 97(2), 308.

Srivastava, S.K. (2007). Green supply-chain management: A state-of-the-art literature review. *Int. J. Manage. Rev.*, 9(1), 53–80.

Steffens, M.A., Fraga, E.S., Bogle, I.D.L. (1999). Multicriteria process synthesis for generating sustainable and economic bioprocesses. *Comput. Chem. Eng.*, 23, 1455.

Tarhan, B., Grossmann, I.E. (2008). A multi-stage stochastic programming approach with strategies for uncertainty reduction in the synthesis of process networks with uncertain yields. *Comput. Chem. Eng.*, 32(4–5), 766.

Tawarmalani, M., Sahinidis, N.V. (2002). *Convexification and global optimization in continuous and mixed-integer nonlinear programming: theory, algorithms, software and applications*. Springer.

Thomas, D.J., Griffin, P.M. (1996). Coordinated supply chain management. *Eur. J. Oper. Res.*, 94, 1.

Turkay, M. and I.E. Grossmann. (1996). Logic-Based MINLP Algorithms For the Optimal Synthesis Of Process Networks. *Comput. Chem. Eng.*, 20, 959.

Uryasev, S. (2000). *Probabilistic constrained optimization: Methodology and applications*. Dordrecht: Kluwer Academic Publishers.

Vecchietti, A. and I.E. Grossmann. (1999). "LOGMIP: A Disjunctive 0-1 Nonlinear Optimizer for Process Systems Models, Computers and Chemical Engineering 23, 555–565.

Wang, Y. P., Smith, R. (1994). Wastewater minimization. *Chem. Eng. Sci.*, 49, 981.

Yee, T.F., Grossmann, I.E. (1990). Simultaneous Optimization Models for Heat Integration, II: Heat Exchanger Network Synthesis. *Comput. Chem. Eng.*, 14, 1165.

Yeomans, H., Grossmann, I.E. (1999). A Systematic Modeling Framework of Superstructure Optimization in Process Synthesis. *Comput. Chem. Eng.*, 23, 709.

6

Rigorous Propagation of Imprecise Probabilities in Process Models

Joshua A. Enszer and Mark A. Stadtherr*

University of Notre Dame
Notre Dame, IN 46556

CONTENTS

ABSTRACT   Models of process dynamics often involve uncertain parameters, inputs and/or initial states. Even if probability distributions for the uncertainties are available, they too may be imprecise. An approach for rigorously and tightly bounding the effects of such uncertainty in process models is described here, and it is shown how this can be extended to determine rigorous bounds on the probabilities of achieving desired outcomes.

KEYWORDS   *Process dynamics, ordinary differential equations, initial value problems, probability bounds analysis*

---

* To whom all correspondence should be addressed

Introduction

The process models used in analysis and design, whether static or dynamic, often involve uncertainties in parameters, inputs, and/or initial states. Determining how these uncertainties propagate through a model to affect its outputs, and doing so rigorously, can be a very challenging problem, especially for nonlinear dynamic systems. The problem is further complicated by the fact that the probability distributions describing the uncertainties may not be known precisely, if they are known at all. If there is no known probability distribution for an uncertain quantity, but only bounds, then the uncertainty can be modeled using an interval. If some knowledge of the probability distribution is available, but it is imprecise, then this can be modeled using a probability box (p-box), which provides upper and lower bounds on the cumulative probability distribution function for the uncertain quantity. We will concentrate here on the latter case, in which uncertain quantities in the process model are characterized by imprecise probabilities represented by p-boxes. Furthermore, we will focus on the difficult case of a nonlinear dynamic model, i.e., a nonlinear system of ordinary differential equations (ODEs) for which an initial value problem (IVP) must be solved.

One common approach for dealing with this problem is Monte Carlo analysis. However, in this approach, it is not possible to investigate the complete space of uncertain quantities in a finite number of simulations, and thus Monte Carlo analysis may fail to capture all possible system behaviors, especially in the case of nonlinear systems. We will describe here an approach for rigorously and tightly bounding the effects of uncertainty in process models, and show how this can be extended to determine rigorous bounds on the probabilities that desired outcomes are achieved. This approach is enabled by the use of Taylor models to represent the solution of IVPs with uncertain parameters and/or initial states, as described recently by Lin and Stadtherr (2007b).

The rest of this paper is organized as follows. In the next section, we provide some background on the approaches used here for representing uncertainties, as well as on the use of Taylor models. This is followed by a formal problem statement and then a description of the solution procedure used. Finally a demonstrative example is provided, with comparison to results obtained from Monte Carlo simulation.

Background

There are several ways to treat numerical uncertainty in mathematical models. In this section, we provide background on the specific approaches used here, namely intervals and p-boxes. We also provide background on the use of Taylor models.

Intervals

A real interval X is the set of real numbers between and inclusive of its lower bound $\underline{X}$ and upper bound $\overline{X}$; that is, $X = [\underline{X}, \overline{X}] = \{x \in \Re \mid \underline{X} \le x \le \overline{X}\}$. Thus, an interval can be used to represent an uncertain quantity for which no information is available other than its lower and upper bounds. Intervals are also used to represent computational uncertainties due to machine rounding. That is, a real number that is not exactly machine representable is bounded by an interval determined from the real number's nearest floating-point representations. An interval vector $X = (X_1, X_2, ..., X_n)^T$ has n real interval components and can be thought of as an n-dimensional rectangle or box. Interval matrices are similarly defined.

Basic arithmetic operations are defined on intervals according to

$$X \operatorname{op} Y = \{x \operatorname{op} y \mid x \in X, y \in Y\} \tag{1}$$

for $\operatorname{op} \in \{+, -, \times, \div\}$ and, in the case of division, $0 \notin Y$, though division in the case of Y containing zero is allowed in extensions of interval arithmetic (e.g., Hansen and Walster, 2004). Commutativity and associativity hold for addition and multiplication, but these operations are only subdistributive. Interval versions of the elementary functions can also be defined.

For a real function $f(x)$, an interval extension $F(X)$ encloses the range of $f(x)$ for all $x \in X$. When $f(x)$ can be written as a series of arithmetic operations and elementary functions, substituting X into $f(x)$ and evaluating using interval arithmetic gives the "natural" interval extension. However, computing the interval extension in this manner often results in overestimation of the function range due to the "dependency" problem. This issue may arise if there are multiple occurrences of the same variable in the function, since in computing the natural interval extension each such occurrence is treated as being independent, though clearly this is not the case. Another source of overestimation that may arise in the use of interval methods is the "wrapping" effect. This occurs when an interval is used to enclose (wrap) a set of results that is not an interval. If overestimation due to either of these issues is propagated from step to step in an integration procedure for ODEs, it can quickly lead to the loss of a meaningful enclosure. One approach for addressing these issues is the use of Taylor models, as discussed later in this section.

Several good introductions to interval analysis, as well as interval arithmetic and other aspects of computing with intervals, are available (e.g., Hansen and Walster, 2004; Jaulin et al., 2001; Kearfott, 1996; Neumaier, 1990). Implementations of interval arithmetic and elementary functions are also readily available, and recent compilers from Sun Microsystems directly support interval arithmetic and an interval data type.

Probability Boxes (P-boxes)

If more is known about an uncertain quantity than simply its upper and lower bounds, then this can be represented in a number of ways. We will assume here that some information, not necessarily precise, is known about the probability distribution of the uncertainty. In other situations, the use of fuzzy numbers (e.g., Dubois and Prade, 1978), clouds (e.g., Neumaier, 2004), and other representations of uncertain knowledge may be appropriate, depending on the type of information that is available.

For some quantity (variable or parameter) x, the cumulative distribution function (CDF) $F(z)$ gives the probability that $x \leq z$. In practice, knowledge of the probability distribution describing an uncertainty is often itself uncertain. To deal with imprecise probability distributions, we use probability boxes (p-boxes) (e.g., Ferson, 2002; Ferson et al., 2004). A p-box, as defined below, is a way to bound probability distributions, in much the same way that an interval is used to bound a real number. Furthermore, arithmetic operations with p-boxes can be performed, again in much the same way as done with intervals. Computations with p-boxes allow for more information about the uncertainty of a quantity to be utilized in modeling and analysis.

Formally, a p-box is the set of all CDFs enclosed by two bounding CDFs $F(z)$ and $G(z)$; that is,

$$(F, G) = \{H(z) \mid F(z) \geq H(z) \geq G(z)\} \tag{2}$$

Less formally, a p-box can be thought of as a set of interval bounds on a cumulative distribution function, and thus, in practice, computation with p-boxes and intervals are analogous (Ferson, 2002). The bounding functions $F(z)$ and $G(z)$ are decomposed into interval-mass pairs, and interval arithmetic is then applied. Therefore, computation with p-boxes involves the same issues of dependency and (especially) wrapping that occur in computations with intervals. For a p-box represented as n interval-mass pairs, a single arithmetic operation with another independent p-box provides a result with n^2 interval-mass pairs, and a p-box with n interval-mass pairs must then be used to condense (wrap) this result.

A p-box may be constructed from any available information about an uncertain quantity, including, but not limited to, any combination of its maximum, minimum, mean, median, or standard deviation. An interval is a special case of a p-box where only the maximum and minimum are known. P-boxes may also be created by assuming a particular form of probability distribution for the bounding functions $F(z)$ and $G(z)$. Some sample p-boxes are shown in Figure 1. Figure 1(a) shows a p-box representation of a standard interval. Figure 1(b) gives the p-box for a variable whose minimum, median, and maximum are known. Figure 1(c) is a p-box created from two different uniform distributions as bounding functions. It is important to note that the true probability distribution simply lies between the bounding functions and does not necessarily take the same form as a bounding function; that is, a distribution within a p-box bounded by uniform distributions is not necessarily also uniform.

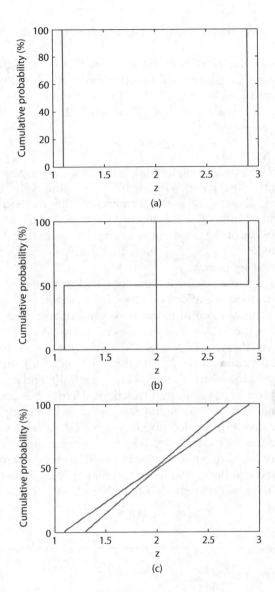

FIGURE 1
Examples of p-boxes for given knowledge of uncertainty. See text for discussion.

Taylor Models

In order to alleviate the overestimation problems that occur in interval computations, Makino and Berz (1996, 1999) have described a remainder differential algebra (RDA) approach for bounding function ranges. In this method, a function is represented using a model consisting of a real-valued

Taylor polynomial and an interval remainder bound. Such a model is called a Taylor model.

One way of forming a Taylor model of a function is through direct use of the Taylor theorem. Consider a real function $f(x)$ that is $(q + 1)$ times partially differentiable on X and let $x_0 \in X$. The Taylor theorem states that for each $x \in X$, there exists a real ζ with $0 < \zeta < 1$ such that

$$f(x) = p_f(x - x_0) + r_f(x - x_0, \zeta) \tag{3}$$

where p_f is a q-th order polynomial (truncated Taylor series) in $(x - x_0)$, and r_f is a remainder term, which can be quantitatively bounded over $0 < \zeta < 1$ and $x \in X$ using interval arithmetic or other methods to obtain an interval remainder bound R_f. A q-th order Taylor model $T_f = p_f + R_f$ for $f(x)$ over X then consists of the polynomial p_f and the interval remainder bound R_f and is denoted by $T_f = (p_f, R_f)$. Note that $f \in T_f$ for $x \in X$; therefore, T_f encloses the range of f over X.

In practice, it is more useful to compute Taylor models of functions by performing arithmetic operations on other Taylor models. Arithmetic operations, including addition, multiplication, reciprocal, and intrinsic functions, can be done using the RDA operations described by Makino and Berz (1996; 1999; 2003). Using these, it is possible to start with simple functions such as the constant function $f(x) = k$, for which $T_f = (k, [0, 0])$, and the identity function $f(x_i) = x_i$, for which $T_f = (x_{i0} + (x_i - x_{i0}), [0, 0])$, and then to compute Taylor models for more complicated functions. Hence, it is possible to compute a Taylor model for any function representable in a computer environment by simple operator overloading through RDA operations. It has been shown that the Taylor model often yields sharper bounds for modest to complicated functional dependencies compared to other rigorous bounding methods (Makino and Berz, 1996, 1999; Neumaier, 2003). A discussion of the uses and limitations of Taylor models has been given by Neumaier (2003).

Problem Statement

Consider the parametric, autonomous IVP

$$\mathbf{y}'(t) = f(\mathbf{y}, \theta), \ \mathbf{y}(t_0) = \mathbf{y}_0 \in Y_0, \theta \in \Theta \tag{4}$$

over the time interval $t \in [t_0, t_m]$ where $t_m > t_0$. Here, $\mathbf{y}$ is the n-dimensional vector of state variables whose initial value is $\mathbf{y}_0$, and θ is a p-dimensional vector of time-invariant parameters. The uncertainties in the initial states and parameters are enclosed in the interval vectors Y_0 and Θ, respectively. Further, additional probabilistic information is available for at least one component of $\mathbf{y}_0$ or Θ, and this information is expressed in the form of a p-box.

We also assume that f is $(k-1)$ times continuously differentiable with respect to y and $(q+1)$ times continuously differentiable with respect to θ. Here, k is the order of the truncation error in the interval Taylor series (ITS) used in the solution procedure (see below), and q is the order of the Taylor model used to represent dependence on initial values and parameters. We also assume that f is representable by a finite number of standard functions.

Our goal is twofold: 1) to obtain verified (e.g., mathematically and computationally guaranteed) enclosures of the state variables y at specified times t_k of interest from t_0 to t_m, and 2) to obtain a probability distribution, in the form of a p-box, for the values of y within these enclosures.

Solution Procedure

In this section, we summarize the solution methods used to achieve the two goals stated above.

Enclosing the State Variables

Interval methods (also called validated methods or verified methods) for ODEs provide a natural approach for computing the desired enclosure of the state variables in the problem stated above. Traditional interval methods usually consist of two processes applied at each integration step. In the first process, existence and uniqueness of the solution are proved using the Picard-Lindelöf operator and the Banach fixed point theorem, and a rough enclosure of the solution is computed. In the second process, a tighter enclosure of the solution is computed. In general, both processes are realized by applying interval Taylor series (ITS) expansions with respect to time, and using automatic differentiation to obtain the Taylor coefficients. An excellent review of the traditional interval methods has been given by Nedialkov et al. (1999), and more recent work has been reviewed by Neher et al. (2007). For addressing this problem, there are various packages available, including AWA (Lohner, 1992), VNODE (Nedialkov, 1999; Nedialkov et al., 2001) COSY VI (Berz & Makino, 1998) and ValEncIA-IVP (Rauh et al., 2006). In this study, we will use a new validating solver VSPODE (Lin & Stadtherr, 2007b) for parametric ODEs, which is capable of determining guaranteed bounds on the solutions of dynamic systems with interval-valued initial states and parameters, and which offers significant performance improvements over the popular VNODE package. The method makes use, in a novel way, of the Taylor model approach (Makino & Berz, 1996, 1999, 2003) to deal with the dependency and wrapping problems on the uncertain quantities (parameters and initial values). We will summarize here the basic ideas of the method used.

As in traditional interval methods, each integration step in VSPODE consists of two phases, as noted above. Assume that at time t_j there is a known enclosure (computed in the previous time step) Y_j of $y_j = y(t_j)$. In the first phase of the next time step, a step size $h_j = t_{j+1} - t_j$ and coarse enclosure $\tilde{Y}_j$ are determined such that a unique solution $y(t) \in \tilde{Y}_j$ is guaranteed to exist for all $t \in [t_j, t_{j+1}]$, all $y_j \in Y_j$ and all $\theta \in \Theta$. This is achieved using a high-order (k) ITS with respect to time and the traditional approach using the Picard-Lindelöf operator and the Banach fixed-point theorem. In the second phase of the method, a tighter enclosure $Y_{j+1} \subseteq \tilde{Y}_j$ is computed, such that $y_{j+1} \in Y_{j+1}$ for all $y_0 \in Y_0$ and all $\theta \in \Theta$. This is done by using an ITS approach to compute $T_{y_{j+1}}(y_0, \theta)$, a Taylor model of y_{j+1} in terms of the initial values y_0 and parameters θ. To compute this Taylor model, we begin by representing the interval initial states and parameters by the Taylor models (identity functions) T_{y_0} and T_θ, respectively. Then, we can determine Taylor models $T_{f^{[i]}}$ of the Taylor series coefficients $f^{[i]}(y_0, \theta)$ by using RDA operations to compute $T_{f^{[i]}} = f^{[i]}(T_{y_0}, T_\theta)$. Using an ITS for y_{j+1} with coefficients given by $T_{f^{[i]}}$, and using the mean-value theorem, one can obtain $T_{y_{j+1}}(y_0, \theta)$, the desired Taylor model of y_{j+1} in terms of y_0 and θ. To control the wrapping effect, the state enclosures are propagated using a new type of Taylor model consisting of a polynomial and a *parallelepiped* (as opposed to an interval) remainder bound. Complete details of the computation of $T_{y_{j+1}}$ using VSPODE are given by Lin and Stadtherr (2007b), who also describe a procedure for efficient bounding of $T_{y_{j+1}}(y_0, \theta)$ over $y_0 \in Y_0$ and $\theta \in \Theta$ to obtain the final state enclosure Y_{j+1}.

Probability Distribution of State Variables

Using the method summarized above, we can obtain, for the specified time of interest t_k, a Taylor model $T_{y_k}(y_0, \theta)$, that gives the state variables $y_k = y(t_k)$ as a polynomial function $p_{y_k}(y_0, \theta)$ of the initial states $y_0 \in Y_0$ and the parameters $\theta \in \Theta$, plus a small remainder bound. If probability distributions (p-boxes) are available for y_0 and for θ, then these can be substituted directly into $T_{y_k}(y_0, \theta)$, and a p-box giving bounds on the probability distribution for y_k can be computed using p-box operations.

Straightforward application of p-box operations to evaluate the Taylor model $T_{y_k}(y_0, \theta)$ may lead to significant overestimation of bounds on the true probability distribution of the state variables, due to the dependency problem and the wrapping effect. One method to obtain a much tighter enclosure is subinterval reconstitution (SIR). In this procedure, when p-box operations are done using the Taylor model, the intervals of the decomposed p-box are further partitioned into subintervals, which are then projected through the Taylor model separately. The final bounds are then reconstituted using the union of the subinterval results (Ferson and Hajagos, 2004). Results obtained from use of SIR will, in general, also overestimate the bounds somewhat, but if a reasonably large number of subintervals are used, the p-box bounds can become quite good. P-box operations and evaluation of the Taylor model, including optional use of SIR, can be performed using the risk analysis software RAMAS Risk Calc (Ferson, 2002). We also employ our own skeletal Matlab implementation of p-box arithmetic and SIR.

Example: Bioreactor Process

We consider here the microbial growth of a single biomass feeding on a substrate in a bioreactor. The process is described by the ODE system

$$X' = (\mu - \alpha D)X \tag{5}$$

$$S' = D(S_f - S) - k\mu X, \tag{6}$$

where X and S represent the biomass and substrate concentrations, respectively, α is the process heterogeneity parameter, D is the dilution rate, S_f is the concentration of substrate in the influent, k is the yield coefficient, and the growth rate μ of biomass follows Monod reaction kinetics (Bastin and Douchain, 1990; Bequette, 2003). For Monod kinetics,

$$\mu = \frac{\mu_{max} S}{K_S + S}, \tag{7}$$

where μ_{max} is the maximum growth rate and K_s is the saturation parameter. The initial states ($t = 0$) are X_0 and S_0.

For this example, we will consider two quantities to be uncertain, namely the maximum growth rate parameter μ_{max} and the initial biomass concentration X_0. Two different cases for these uncertain quantities will be considered. The other parameters, including the initial substrate concentration, are taken to be fixed at the values shown in Table 1 (Lin and Stadtherr, 2007a).

For all p-box operations, the p-boxes used were discretized into 100 interval-mass pairs (each interval corresponding to a single percentile). In applying

TABLE 1

Fixed Quantities in Bioreactor Example.

Parameter	Value	Units
S_0	0.8	g S/L
α	0.5	
D	0.36	day^{-1}
S_f	5.7	g S/L
k	10.53	g S/g X
K_s	7.0	g S/L

VSPODE, the order of the interval Taylor series used was $k = 17$, while the order of the Taylor model used was $q = 5$. In the integration procedure, a constant step size of $h = 0.2$ was used, though this step size is automatically reduced if needed. The Taylor model remainder bounds were obtained using a QR-factorization process (Lin and Stadtherr, 2007b). In the SIR procedure for computation with p-boxes, each of the 100 interval-mass pairs of the p-box was bisected. All problems were solved on an Intel Pentium 4 3.2 GHz machine running Red Hat Linux.

Case 1: $X_0 \in [0.794, 0.864]$ and $\mu_{max} \in [1.15, 1.25]$

In this first case, there is an uncertainty of about ±4.2% (relative to the mean) in both μ_{max} and X_0. We shall assume that the uncertainty in both these quantities can be described by p-boxes bounded by uniform distributions, as shown in Figure 2. We further assume that these two uncertainties are independent from one another (i.e., there is no correlation between them).

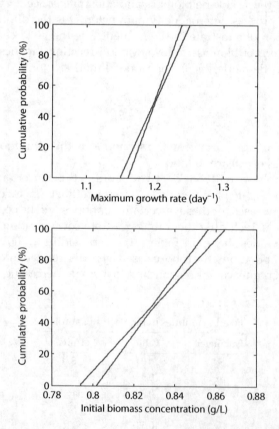

FIGURE 2
P-box representation of uncertainties for case 1 in bioreactor example.

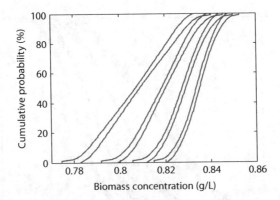

FIGURE 3

Case 1 results from Taylor model method. P-box bounds for biomass concentration $X(t)$ at (left to right) $t = 2.5, 5, 7.5$ and 10 days.

Using the procedure described above, we computed p-box representations for the probability distribution of the state variables at four different times, $t = 2.5, 5, 7.5$, and 10 days. These results are plotted in Figure 3 for the biomass concentration $X(t)$, showing (from left to right) the p-box results as time increases. For example, these results show that the probability that $X \leq 0.82$ is bounded by the interval $[72.1, 78.3]\%$ at $t = 2.5$ days, by $[49.4, 54.8]\%$ at $t = 5$ days, by $[14.9, 20.7]\%$ at $t = 7.5$ days, and by $[0, 3.5]\%$ at $t = 10$ days. These bounds are mathematically and computationally rigorous. The computational expense of obtaining these results was quite small. Use of VSPODE to determine the Taylor model for the state variables at $t = 10$ days required 0.444 seconds. Once the Taylor model was obtained, the p-box operations (with SIR) needed to get the final results for $t = 10$ days required 188.2 seconds using Matlab.

As a basis for comparison, we also determined probability bounds for $X(t)$ at the same four points in time using Monte Carlo simulation. To do this first requires sampling the space of the probability distributions for μ_{max} and X_0. We used 100 samples, each a uniform distribution chosen randomly from the p-boxes for μ_{max} and X_0. For each of these 100 distributions, we then ran (using Matlab with ode45) 50,000 simulations to obtain a probability distribution for X. Combining the results for each of the 100 input distributions, we obtain Figure 4, again showing the results for increasing time from left to right. The results (Figure 3) obtained using the Taylor model approach described here are clearly consistent with the MC results. It is important to note: 1) Probability bounds obtained from MC analysis are not rigorous, but those obtained from the Taylor model analysis are. For the number of trials done here, which is relatively large to ensure meaningful results, the computational expense was quite large, about 9 hours (vs. about 3 minutes for the more rigorous Taylor model approach). 2) The probability bounds from MC become quite narrow at the median, less so than obtained from the Taylor

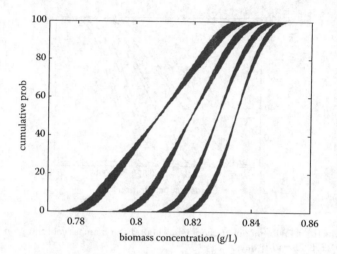

FIGURE 4

Case 1 results from Monte Carlo analysis. Probability distributions for biomass concentration $X(t)$ at (left to right) $t = 2.5, 5, 7.5$ and 10 days.

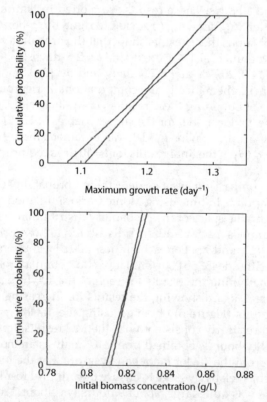

FIGURE 5

P-box representation of uncertainties for Case 2 in bioreactor example.

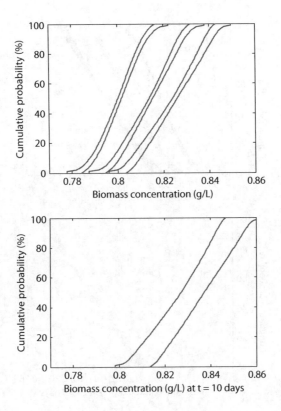

FIGURE 6
Case 2 results from Taylor model method. P-box bounds for biomass concentration $X(t)$ at (top, left to right) $t = 2.5, 5, 7.5$ and (bottom) 10 days.

model analysis. This reflects the use of only uniform distributions in the MC analysis. A p-box with uniform bounds also contains non-uniform distributions, and this is accounted for in results of Figure 3.

Case 2: $X_0 \in [0.81, 0.83]$ and $\mu_{max} \in [1.08, 1.32]$

In this case, we consider a larger degree of uncertainty in μ_{max}, now $\pm 10\%$, than in the previous case, and a smaller degree of uncertainty in X_0, now about $\pm 1.2\%$. Again we assume probability distributions contained in p-boxes with uniform bounds, as shown in Figure 5.

Figure 6 shows the p-box enclosures for the reactor biomass concentration obtained from the Taylor model approach with p-box arithmetic, and Figure 7 shows the probability distributions obtained using Monte Carlo analysis. Though the Taylor model results are quite good for $t = 2.5, 5$ and 7.5 days, they are not as good for $t = 10$ days, the p-box for which is noticeably wider than

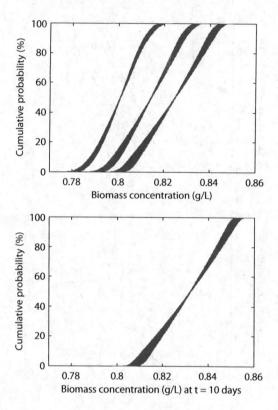

FIGURE 7

Case 2 results from Monte Carlo analysis. Probability distributions for biomass concentration $X(t)$ at (top, left to right) $t = 2.5, 5, 7.5$ and (bottom) 10 days.

the other p-boxes, and in comparison to the MC results. This is occurring in part because of growth, as time increases, in the width of the remainder bound term in the Taylor model, and is a reflection of the relatively large range of μ_{max} considered. Better results could be obtained by bisecting μ_{max} and employing a SIR-like procedure on the Taylor model level (i.e., a different Taylor model for each μ_{max} subinterval).

Concluding Remarks

The parametric nonlinear ODEs that arise in process models for design and analysis often include uncertainty in parameters and initial states, and the distribution of this uncertainty is often not known precisely. We have presented here a new approach, based on Taylor models and probability boxes (p-boxes) for propagating such imprecise probability distributions into the state variable trajectories, enabling the computation of rigorous bounds on the probabilities that desired outcomes can be achieved. In comparison to Monte Carlo analysis, this new approach provides not only guaranteed bounds, but also a reasonable computational cost. Though we can demonstrate this for a variety of process models, we have focused here on the case of a bioreactor process, computing bounds on the probability distribution of the biomass trajectory.

Acknowledgments

This work was supported in part by the U. S. Department of Energy grant DE-FG36-08GO88020-A0, and by a Lilly Foundation graduate fellowship (JAE). Computational work was performed in part using facilities at the Notre Dame Center for Research Computing.

References

Bastin, G., Douchain, D. (1990). *On-line Estimation and Adaptive Control of Bioreactors.* Elsevier, New York, NY.

Bequette, B. W. (2003). *Process Control: Modeling, Design, and Simulation.* Prentice-Hall, Upper Saddle River, NJ.

Berz, M., Makino, K. (1998). Verified integration of ODEs and flows using differential algebraic methods on high-order Taylor models. *Reliable Computing, 4*, 361.

Dubois, D., Prade, H. (1978). Operations on fuzzy numbers. *International Journal of Systems Science, 9*, 613–626.

Ferson, S. (2002). *RAMAS Risk Calc 4.0: Risk Assessment with Uncertain Numbers.* Lewis Press, Boca Raton, FL.

Ferson, S., Hajagos, J. G. (2004). Arithmetic with uncertain numbers: Rigorous and (often) best possible answers. *Reliability Engineering and System Safety, 85*, 135–152.

Ferson, S., Nelson, R. B., Hajagos, J., Berleant, D. J., Zhang, J., Tucker, W. T., Ginzburg, L. R., Oberkampf, W. L. (2004). *Dependence in probabilistic modeling, Dempster-Shafer theory, and probability bounds analysis.* Technical Report, Sandia National Laboratories.

Hansen, E. R., Walster, G. W. (2004). *Global Optimization Using Interval Analysis.* Marcel Dekker, New York, NY.

Jaulin, L., Kieffer, M., Didrit, O., Walter, É. (2001). *Applied Interval Analysis.* Springer-Verlag, London, UK.

Kearfott, R. B. (1996). *Rigorous Global Search: Continuous Problems.* Kluwer, Dordrecht, The Netherlands.

Lin, Y., Stadtherr, M. A. (2007a). Guaranteed state and parameter estimation for nonlinear continuous-time systems with bounded-error measurements. *Industrial & Engineering Chemistry Research, 46*, 7198.

Lin, Y., Stadtherr, M. A. (2007b). Validated solutions of initial value problems for parametric ODEs. *Applied Numerical Mathematics, 57*, 1145.

Lohner, R. J. (1992). Computations of guaranteed enclosures for the solutions of ordinary initial and boundary value problems. In: Cash, J., Gladwell, I. (Eds.), *Computational Ordinary Differential Equations.* Clarendon Press, Oxford, UK, 425.

Makino, K., Berz, M. (1996). Remainder differential algebras and their applications. In: Berz, M., Bishof, C., Corliss, G., Griewank, A. (Eds.), *Computational Differentiation: Techniques, Applications, and Tools. SIAM, Philadelphia*, 63.

Makino, K., Berz, M. (1999). Efficient control of the dependency problem based on Taylor model methods. *Reliable Computing, 5*, 3.

Makino, K., Berz, M. (2003). Taylor models and other validated functional inclusion methods. *International Journal of Pure and Applied Mathematics, 4*, 379.

Nedialkov, N. S. (1999). *Computing rigorous bounds on the solution of an initial value problems for an ordinary differential equation.* Ph.D. Thesis, University of Toronto, Toronto, Canada.

Nedialkov, N. S., Jackson, K. R., Corliss, G. F. (1999). Validated solutions of initial value problems for ordinary differential equations. *Applied Mathematics and Computation, 105*, 21.

Nedialkov, N. S., Jackson, K. R., Pryce, J. D. (2001). An effective high-order interval method for validating existence and uniqueness of the solution of an IVP for an ODE. *Reliable Computing, 7*, 449.

Neher, M., Jackson, K. R., Nedialkov, N. S. (2007). On Taylor model based integration of ODEs. *SIAM Journal on Numerical Analysis, 45*, 236.

Neumaier, A. (1990). *Interval Methods for Systems of Equations.* Cambridge University Press, Cambridge, UK.

Neumaier, A. (2003). Taylor forms—Use and limits. *Reliable Computing, 9*, 43.

Neumaier, A. (2004). Clouds, Fuzzy Sets, and Probability Intervals. *Reliable Computing, 10*, 249.

Rauh, A., Hofer, E. P., Auer, E. (2006). VALENCIA-IVP: A Comparison with Other Initial Value Solvers. In: *Proceedings 12th GAMM—IMACS International Symposium on Scientific Computing, Computer Arithmetic, and Validated Numerics (SCAN 2006)*, Duisburg, Germany, 36.

7

Are Mechanistic Cellulose-hydrolysis Models Reliable for Use in Biofuel Process Design?— Identifiability and Sensitivity Analysis

Gürkan Sin[*], Anne S. Meyer and Krist V. Gernaey

Department of Chemical and Biochemical Engineering,
Technical University of Denmark
Building 229, DK-2800 Kgs. Lyngby, Denmark

CONTENTS

ABSTRACT An in-depth statistical analysis of a dynamic enzymatic cellulose hydrolysis model, the NREL model, is presented. The uncertainty of model parameters was analysed using confidence intervals (CI) of parameter estimates from non-linear regression. The 95% confidence of the estimated parameters showed that some of them had unacceptably large bands (e.g. 120 times the mean estimates). This simply means the true parameter value can be found anywhere between these large intervals, hence *statistically unidentifiable*. Such results mean that the model is over-parameterized w.r.t. experimental data. To better understand the underlying reasons, sensitivity and collinearity analysis of the model structure was performed. The sensitivity

---

[*] To whom all correspondence should be addressed

analysis showed that among 26 model parameters, only few (ca 10) were significant while the rest had negligible impact on the model predictions. The collinearity analysis showed that significant correlations existed between the parameters, hence explaining the source of the large confidence intervals. To obtain unique estimates with acceptable confidence bands, therefore, one has to use an identifiable subset. In this case, many potentially identifiable subsets were found, however only six parameters (out of 26) could be *uniquely* identified. Finally, it should be emphasized that other dynamic hydrolysis models are likely to have severe identifiability issues, which need to be overcome before one can reliably use them for engineering purposes such as biofuel process design.

KEYWORDS *Cellulose hydrolysis, design, dynamic modeling, identifiability, sensitivity, collinearity*

Introduction

Biofuel production from lignocellulosic biomass is a complex process. Physical pre-treatment of biomass and subsequent hydrolysis of cellulose to simple sugars are the chief challenges of bioprocess development and feasibility. Development and transfer of these processes from proof-of-concept to industrial scale are mainly done on an empirical basis, and typically rely on experiences from conventional one-pot conversion processes. This approach is rather inefficient and costly in terms of time and resource investments.

This study introduces a model based simulation framework for biofuel process design (see Figure 1). The hypothesis is that simulations with a dynamic mechanistic model will facilitate rational process development and boost innovative designs. Central to this design approach is the availability of reliable models of all the reactions and unit processes involved in the biofuel production, foremost the enzymatic hydrolysis.

The proposed model-based design framework is realized in two phases, which are complimentary to each other and iterative in nature. In the first phase of the procedure, one aims at developing a reliable model for describing the kinetics of multiple enzyme actions using a systematic identification procedure (see Figure 1). Once a reliable model is obtained, one then moves to the next phase in which the kinetic hydrolysis model is used in developing integrated process models (all unit operations in the bio-ethanol production process). These integrated process models are used to generate innovative configurations for upscaling the targeted enzyme catalyzed process using integrated process models. The enzyme kinetic model identified in phase 1 forms a key part of these integrated process models, which is the focus in this paper.

The description of the kinetics of multiple enzymes on insoluble cellulosic material is on its own a complex process, which has lured interest

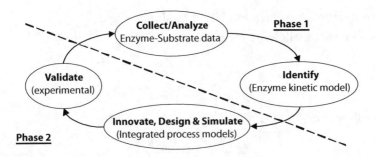

FIGURE 1
A model-based simulation framework for design of new enzyme processes.

of researchers over two decades. An up-to-date review of research activities in this field is given by Zhang and Lynd (2004). As a result, a number of dynamic models for hydrolysis of cellulose have been proposed in the past (see e.g. Okazaki and Moo-Young, 1978; Wald et al., 1984; Gan et al., 2003; Zhang and Lynd, 2004; Andersen, 2007). To the best of our knowledge, none of these models have been independently validated nor verified. This remains as a weak point in their credibility hence applicability for design purposes. To shed light on this vital issue, this paper performs an in-depth identifiability analysis of a dynamic cellulose hydrolysis model, namely the one developed by NREL (Kadam et al., 2004). For the identifiability analysis, parameter uncertainty (confidence intervals) and sensitivity and collinearity analysis are used.

Methods

Dynamic Cellulose Hydrolysis Model of NREL

The NREL model was developed by Kadam et al (2004) to describe enzymatic hydrolysis of lignocellulosic biomass (e.g. corn stover) with the purpose of process optimization. The model is based on a number of assumptions among others (1) both amorphous and crystalline cellulose regions are considered as lumped substrate (2) a multi-step reaction scheme is assumed for the enzymatic action of cellulases and (3) no enzyme deactivation/decay. The kinetic expressions for the different steps of the hydrolysis are described as follows:

(1) enzyme adsorption follows Langmuir kinetics:

$$E_{Bi} = \frac{E_{\max i} K_{adi} E_{Fi} S}{1 + K_{adi} E_{Fi}} \tag{1}$$

(2) competitive (product) inhibition by simple sugars

Cellulose to cellobiose reaction:

$$r_1 = \frac{k_1 E_{B1} R_S S}{1 + \frac{G_2}{K_{IG2}} + \frac{G}{K_{IG}} + \frac{X}{K_{IX}}} \tag{2}$$

Cellulose to glucose:

$$r_2 = \frac{k_2 (E_{B1} + E_{B2}) R_S S}{1 + \frac{G_2}{K_{2IG2}} + \frac{G}{K_{2IG}} + \frac{X}{K_{2IX}}} \tag{3}$$

Cellobiose to glucose:

$$r_3 = \frac{k_3 E_{F2} R_S G_2}{K_{3M}\left(1 + \frac{G}{K_{3IG}} + \frac{X}{K_{3IX}}\right) + G_2} \tag{4}$$

(3) Arrhenius temperature effect on the rate

$$k_{iT2} = k_{iT1} \exp\left(-\frac{E_a}{R}\left(\frac{1}{T_1} - \frac{1}{T_2}\right)\right) \tag{5}$$

(4) non-linear substrate (cellulose) reactivity, R_S,

$$R_S = \alpha \frac{S}{S_0} \tag{6}$$

For further details about the reaction mechanism and mass balances, the reader is referred to Kadam et al. (2004).

The model has 8 variables and contains a total number of 26 parameters, which were estimated from dedicated experiments. The experimental data include typical hydrolysis progress curves as well as inhibition experiments, in which typically glucose and cellobiose are measured.

Parameter Uncertainty and Correlation

To estimate the confidence interval of the parameters, a linear approximation of the covariance matrix of parameter estimators, COV(θ), was performed (Seber and Wild, 1989):

$$\text{COV}(\theta) = \frac{J(\theta)}{N-p}\left(\left(\frac{\partial y}{\partial \theta}\right)^T Q^{-1}\left(\frac{\partial y}{\partial \theta}\right)\right) \tag{7}$$

In Eq.(1), $J(\theta)$ is the minimum sum of squared errors obtained from the least-squares parameter estimation method. $\frac{\partial y}{\partial \theta}$ is the sensitivity matrix corresponding to sensitivity of model variables, y, to parameters, θ. $\mathbf{Q}$ is the covariance matrix of measurement errors, while N is the total number of measurements and p is the number of estimated parameters. The confidence interval of the parameters, θ, at α significance level is given as:

$$\theta_{1-\alpha} = \theta \pm \sqrt{diag(\text{COV}(\theta))} \cdot t(N - p, \alpha/2) \tag{8}$$

In Eq.(2), $t(N - p, \alpha/2)$ is the t-distribution value corresponding to the $\alpha/2$ percentile with $N - p$ degrees of freedom, and diag represents the diagonal elements of $\text{COV}(\theta)$.

The correlation matrix between two parameters, $COR(\theta_i, \theta_j)$, is given as

$$COR(\theta_i, \theta_j) = \frac{COV(\theta_i, \theta_j)}{\sqrt{\sigma_{\theta_i}^2 \sigma_{\theta_j}^2}} \, .$$

Identifiability Analysis

The identifiability methodology given in Table 1 is largely based on the analysis of the sensitivity of the model variables to the parameters and is adopted from Brun et al. (2002). The sensitivity analysis is then used to screen for parameter significance ranking by calculating a measure called δ^{msqr} (see

TABLE 1

Identifiability Methodology

Steps	Description
Absolute sensitivity[1]	$\mathbf{S_a} = \{s_{a,ij}\}$ *where* $s_{a,ij} = \dfrac{\partial y_i}{\partial \theta_j}$
Non-dimensional sensitivity[1]	$\mathbf{S_{nd}} = \{s_{nd,ij}\}$ *where* $s_{nd,ij} = \dfrac{\partial y_i}{\partial \theta_j} \cdot \dfrac{\theta_j}{sc_i}$
Sensitivity measure, δ^{msqr}	$\delta_j^{msqr} = \sqrt{\dfrac{1}{N} \sum_{i=1}^{N} (s_{nd,ij})^2}$
Normalized sensitivity[1]	$\mathbf{S_{norm}} = \{s_{norm,ij}\}$ *where* $s_{norm,ij} = \dfrac{s_{nd,ij}}{\|s_{nd,ij}\|}$
Collinearity index[2], γ_K	$\gamma_K = \dfrac{1}{\sqrt{\min \lambda_K}}$ *where* $\lambda_K = eigen\left(\mathbf{S}_{norm,K}{}^T \mathbf{S}_{norm,K}\right)$

[1] y_i stands for the ith model variable, (e.g. cellulose y_1 & glucose y_2) and θ_j stands for the jth parameter, and sc_i is the scaling factor for the variable y_i.

[2] K stands for the index of the parameter subset, which is a combinatorial function of the parameter vector, q.

Table 1), and for analyzing the near-linear dependency between parameters by a measure called the collinearity index, γ_K.

If the sensitivity functions of two parameters are orthogonal (meaning independent), the γ_K is equal to unity, otherwise it approaches infinity (meaning the parameters are linearly dependent). This information is used to find an identifiable parameter subset, for which a threshold value for γ_K between 10–15 is typically used. Any parameter combination with higher γ_K (than the threshold) is deemed unidentifiable (Brun et al., 2002; Sin and Vanrolleghem, 2007).

The model implementation, simulation and the abovementioned statistical methods are performed in Matlab (The Mathworks, Natick, Massachusetts). The model equations (ODE and algebraic) were solved using a stiff-solver (ode15s with integration accuracy set to 1.0E-07), while parameter estimation was performed using the lsqnonlin algorithm available in Matlab.

Results

Model Fits, Parameter Uncertainty and Correlation

The model fits to the experimental data of Kadam et al (2004) are shown in Figure 2.

Visually speaking the goodness of the model fits is quite remarkable considering that the experimental data include enzymatic hydrolysis of cornstover solids for a range of initial conditions. These fits were obtained from identification of 12 kinetic parameters involved in these multi-step reactions using non-linear regression.

The mean values of the estimated parameters as well as the uncertainty of the parameter estimates represented as 95% confidence interval (CI) are shown in Table 2. All the parameters are found to have a too large CI, e.g. some parameters had upper and lower confidence bands as much as 120 times the mean parameter estimate (see for K_{2IG2}). This indicates simply that the true value of the parameter estimate is located in a very wide confidence band (e.g. as much as 120 times the mean value for K_{2IG2}). Statistically speaking this means the parameter in question is *unidentifiable*. It is important to note that despite variation of the true value of the parameter estimate within this large confidence band, the model fit to the data may still remain unchanged (it doesn't compromise the model fit quality). Put another way, these estimates are not unique but conditional on each other, which indicates that one should refrain from attaching physical meaning to these parameter values. The proper way of interpreting this result is *to consider the estimated parameters as one convenient set of values that provides a good fit to the data.*

The underlying reason of these huge variations is simply the strong correlation between the parameters (the correlation matrix (12x12) could not be shown due to page format/size limitation). But effectively, the enzymatic rate

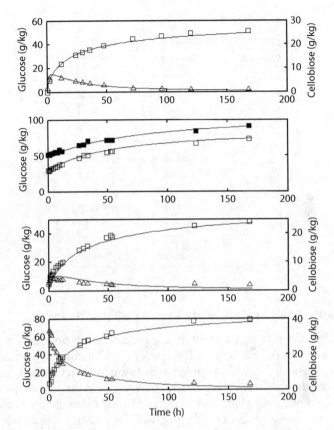

FIGURE 2
Model fits to experimental data of Kadam et al (2004): hydrolysis of cellulose with 10% w/w corn-stover solids (A), A+initial glucose 30 & 50 g/kg (B), A+ initial xylose 40 g/kg (C), and A+ initial cellobiose 30 g/kg (D). Square and triangle symbols refer to glucose and cellobiose respectively.

(k) and the inhibition coefficients are correlated to a high extent, with correlation coefficients that are around 1.0. This means that any change in one parameter could be compensated by a change in the other one, hence making it extremely difficult (if not impossible) to find a unique estimate for these correlated parameters. This correlation stems from the model structure itself (Okazaki and Moo-Young, 1978; Gan et al., 2003; Kadam et al., 2004).

There are several ways to reduce the correlation and thereby bring the parameter uncertainty to acceptable levels: (1) modify the model structure, (2) increase the information content of experimental data by a proper design of experiments, and (3) search for a parameter subset that can be reliably estimated from the given data. In this paper, we perform an identifiability study to address point (3).

TABLE 2

Parameter Estimation Uncertainty

Parameter	Estimate	95% CI
k_1	19.16	±824.00
K_{1IG2}	0.07	±3.28
K_{1IG}	0.03	±1.21
K_{1IX}	0.16	±6.78
k_2	5.74	±139.23
K_{2IG2}	91.14	±11178.58
K_{2IG}	0.07	±1.61
K_{2IX}	0.93	±22.93
k_3	98.38	±1761.89
K_{3M}	104.86	±2317.94
K_{3IG}	0.77	±3.62
K_{3IX}	326.62	±14196.32

Identifiability Analysis: Sensitivity Versus Collinearity Index

The identifiability analysis largely depends on the interpretation of the sensitivity function of parameters with respect to model outputs — usually corresponding to the measured variables such as glucose. In Figure 3, sensitivity functions of some selected parameters are shown. From the shape of these functions, one observes that there is quite a linear dependency between the effects of these parameters on the glucose. Particularly worth mentioning is the parameter k_1 (the enzymatic rate constant of the first step) versus K_{IG} (inhibition coefficient of glucose on the enzymatic step that converts cellulose to cellobiose). This outcome is actually quite known for Miche;alis-Menten type kinetics, hence not surprising (Holmberg, 1982). Moreover this explains the huge correlation coefficients mentioned above, which were around 1.0.

The identifiability procedure is performed in two steps. First, the parameters are ranked based on the sensitivity measure δ^{msqr} as shown in Figure 4. The higher the magnitude of δ^{msqr}, the more significant a parameter is. One observes that the 10 highest ranking parameters make up about 95% of the total effect on the model outputs (sensitivity). The remaining 16 parameters contribute to only 5% of the total parameter effect. It is noteworthy that reaction conditions (temperature (T), enzyme loading (EL)) as well as biomass characteristics such as the composition of lignocellulosic biomass (fraction of cellulose, CP) and the substrate reactivity (alfa) were found the most significant. Added to the significant list were also enzyme adsorption kinetics, e.g. maximum adsorbed concentration of enzyme (E_1max) versus adsorption equilibrium (K_1ad).

Enzyme kinetics (k_1, k_2 and k_3) and inhibition coefficients (K_{1IG}, K_{3M} and K_{3IG}) were found significant albeit to a lower extent. The impact of this sensitivity ranking on parameter estimation uncertainty is that only those

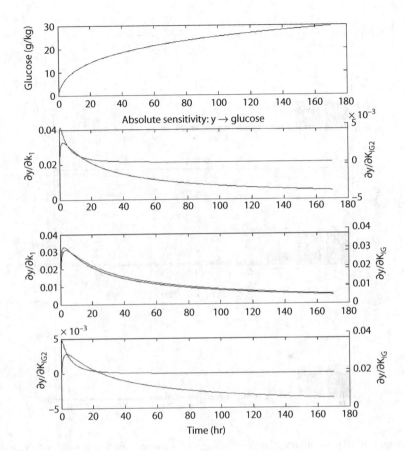

FIGURE 3
Sensitivity of glucose to three selected parameters k_1, K_{IG2}, versus K_{IG}.

parameters with significant uncertainty can be expected to be identified from a given data/experimental condition.

As a second part of the identifiability analysis, one now measures an identifiability index, the so-called collinearity index. The collinearity index is measured for all possible parameter combinations starting from size 2 up to the maximum size, which is 26. The sensitivity measure above indicated that some of the parameters had negligible impact on the outputs, and those insensitive parameters were therefore excluded from the identifiability analysis.

The results of the collinearity index for a number of combinations of parameter subsets (from 2 up to 19) are shown in Figure 5. In total about 500,000 parameter combinations exist, while the identifiability results showed that 1024 parameters combinations are qualified as potentially identifiable subsets (meaning that they have identifiability index lower than the threshold, i.e. 15). This corresponds to 0.25% of the total possible parameter subsets, hence only a tiny fraction is in principle identifiable from the given data.

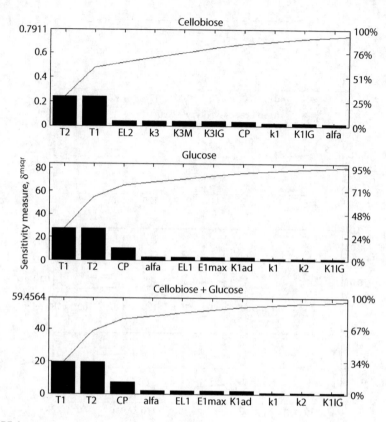

FIGURE 4

Parameter significance ranking based on the sensitivity measure δ^{msqr}: only the first 95% of the cumulative distribution is shown.

Further, the size of identifiable subsets ranges from 2 up to 6. Essentially what these results mean is the following: (1) many identifiable parameter combinations exist (in this case 1024), not only one specific subset; (2) there is (only) a maximum number of parameters that can be identified uniquely, in this case six. These results are typical for over-parameterized numerical models, which are subjected to limited data (Brun et al., 2002; Ruano et al., 2007; Sin and Vanrolleghem, 2007).

This identifiability analysis (sensitivity plus collinearity) essentially diagnosed the issue, which is causing the large confidence intervals (parameter uncertainty) reported in Table 2. Due to high correlation between the model parameters, the identifiability of that particular parameter subset (shown in Table 2) is rather poor (e.g. its identifiability measure (the collinearity index) is around 1400, which is way above the minimum threshold to be qualified as identifiable (i.e. 15)). This already indicates severe identifiability problems as revealed during parameter estimation mentioned above.

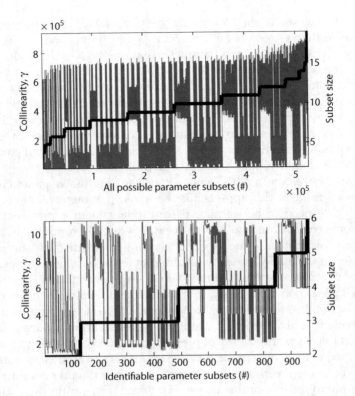

FIGURE 5
Collinearity analysis for all possible combinations of parameter subsets (top) potentially identifiable subsets (bottom).

To overcome this issue, one should instead use an identifiable subset. Unique parameter estimate means that the parameter shall have an acceptably low correlation (to any of the other parameters, e.g. with a correlation coefficient lower than 0.5) and a low confidence interval. One parameter subset providing these conditions is k_1, K_{2IG}, k_3, K_{1IX}, K_{2IX} and K_{1IG2}. (One has to mention a caveat here that there are 45 other parameter combinations with size 6 that meet these constraints. Hence the abovementioned parameter subset is just one of them).

Implications to Future Research in Modeling Cellulose Hydrolysis and Some Caveats

It is likely that the identifiability issues shall be valid for other cellulose hydrolysis models since most of these models (defined as functional by Zhang and Lynd, 2004) employ a similar model structure based on modification of Michaelis-Menten type kinetics/inhibition combined with adsorption kinetics. Therefore, when identifying these models, i.e. when reporting parameter estimates, one should exercise extra caution. Indeed, as shown here, any attempt to estimate a parameter that is not *UNIQUELY* identifiable from a given data set is bound to lead to arbitrary values having far too large confidence intervals. A good model fit to the data does not imply that the parameters providing that fit are unique estimates, and thus reliable. On the contrary, many parameter combinations can provide a satisfactory fit to the data. What is challenging is to ascertain an acceptable confidence level on the obtained parameter estimates. Hence when interpreting reported parameter values in the literature, one should bear in mind this caveat.

One way of overcoming this issue is to use an identifiable parameter subset. The downside of this approach is that those parameters that are found unidentifiable need to be estimated from independent experiments. This may become costly. A second alternative is to work on improving the model structure. This can be done by incorporating more mechanistic/process knowledge in the model, possibly resulting in fewer parameters to estimate.

A third alternative to deal with model structure uncertainty is to evaluate the model prediction uncertainty. To this end established methodologies exist such as the Monte Carlo technique. Model prediction uncertainty quantifies to what extent input parameter uncertainty of the model is propagated to the output uncertainty, e.g. uncertainty of prediction of glucose. In this way, one can decide whether the model prediction uncertainty is still feasible to work with, e.g. for process engineering purposes such as process design or optimization.

Returning to the title of this paper – are mechanistic cellulose-hydrolysis models reliable for use in biofuel process design?, the right answer probably is that the models are not yet reliable! Based on the analysis reported in this paper one may expect that a significant uncertainty exists in the parameter values reported in the literature for this class of models. Hence more research (as outlined above) needs to be done before such models can be deemed reliable for design purposes. Considering the fact that there is an enormous activity w.r.t. development of biofuel processes, this type of research needs to be conducted as soon as possible. If not, design of biofuel production plants is likely to remain based on steady-state and empirical models.

Conclusions

An exhaustive identifiability analysis has been performed for a dynamic cellulose hydrolysis model in view of finding whether the model is reliable enough for biofuel process design. As a case study the dynamic NREL model was used. The following was concluded:

- Parameter estimation showed that the confidence intervals are too large (i.e. several times the mean estimated values), meaning that a huge parameter uncertainty exists.

- Sensitivity analysis revealed that only few parameters (10 out of 26) were found to be significant. The rest had negligible impact on the predictions of the model.
- Collinearity analysis showed that (i) many parameter combinations can be uniquely estimated, whereas (ii) the maximum number of parameters that are uniquely identifiable is low. In this case, only 6 parameters out of 26 were identifiable based on the presented available data.
- Combining sensitivity plus collinearity analysis, one concludes that the model is over-parameterized w.r.t. available data.

Based on this analysis, one expects that the identifiability issue is also valid for other dynamic cellulose hydrolysis models, since essentially they are all based on extension/modifications of Michaelis-Menten type kinetics). Hence caution should be exercised when interpreting these reported parameter values.

Last but not least, it should be emphasized that more research needs to be done to overcome these identifiability issues. Future research should have focus on issues such as prediction uncertainty analysis or model structure simplification. Only by performing this type of research can these models be deemed reliable for process design purposes.

Acknowledgments

This work is funded by a research grant from the Danish Research Council for Technology and Production Sciences (FTP project # 274-07-0339).

References

Andersen N. (2007) Enzymatic hydrolysis of cellulose – experimental and modelling studies. PhD thesis. BioCentrum, Technical University of Denmark, Lyngby, Denmark.

Brun R., Kuhni M., Siegrist H., Gujer W. and Reichert P. (2002) Practical identifiability of ASM2d parameters – systematic selection and tuning of parameter subsets. *Wat. Res. 36(16), 4113-4127.*

Gan Q., Allen S.J. and Taylor G. (2003) Kinetic dynamics in heterogeneous enzymatic hydrolysis of cellulose: an overview, an experimental study and mathematical modeling. *Proc. Biochem. 38,1003–1018.*

Helton J.C. and Davis F.J. (2003) Latin hypercube sampling and the propagation of uncertainty in analyses of complex systems. *Reliab Engng Syst Saf 81, 23–69.*

Holmberg A. On the practical identifiability of microbial growth models incorporating Michaelis-Menten type nonlinearities. Math. Biosci., 1982, 62,23–43.

Okazaki M. and Moo-Young M. (1978) Kinetics of enzymatic-hydrolysis of cellulose - analytical description of a mechanistic model. *Biotech. Bioeng. 20, 637–663.*

Kadam K.L.,. Rydholm E.C and. McMillan J.D (2004) Development and validation of a kinetic model for enzymatic saccharification of lignocellulosic biomass. *Biotechnol. Prog., 20, 698–705.*

Ruano M.V., Ribes J., De Pauw D.J.W. and Sin G. (2007) Parameter subset selection for the dynamic calibration of activated sludge models (ASMs): experience versus systems analysis. *Wat. Sci. Technol. 56(8), 107–115.*

Sin G. and Vanrolleghem P.A. (2007) Extensions to modeling aerobic carbon degradation using combined respirometric–titrimetric measurements in view of activated sludge model calibration. *Wat. Res. 41, 3345–3358.*

Seber G. and Wild C. (1989) Nonlinear regression. New York, Wiley.

Wald S., Wilke C.R. and Blanch H.W. (1984) Kinetics of the Enzymatic Hydrolysis of Cellulose. *Biotech. Bioeng. 26, 221–230.*

Zhang Y.-H.P. and Lynd L.R. (2004) Toward an Aggregated Understanding of Enzymatic Hydrolysis of Cellulose: Noncomplexed Cellulase Systems. *Biotech. Bioeng. 88, 797–824.*

8

Future System Challenges in the Design of Renewable Bio-energy Systems and the Synthesis of Sustainable Biorefineries

Antonis C. Kokossis[1]* and Aidong Yang[2]

[1]*National Technical University of Athens, Greece*
[2]*University of Surrey, UK*

CONTENTS

ABSTRACT The chemical industry experiences a steady growth in the use of renewables induced by the gradual depletion of oil, uncertainties in energy supplies and a commanding requirement to reduce GHG emissions and save the planet. Renewables introduce an impressive range of options with biorefining at the centre of attention as an emerging industrial concept, uniquely attached to chemical engineering and aiming to transform plant-derived biomass into a variety of products including transport fuels, platform chemicals, polymers, and specialty chemicals. In competing with conventional processes, biorefineries should match maximum efficiencies with better design and process integration. The paper highlights the pivotal role of systems technology to foster innovation, preview options, and support high-throughput computational experimentation, arguing that systems-enabled platforms could function as powerful environments to generate ideas for integrated designs and offer tremendous services to the complex

---

* To whom all correspondence should be addressed.

and large problems produced by the numerous portfolios of feedstocks, unknown portfolios of products, multiple chemistries, and multiple processing paths. Complexities certainly exceed capabilities of previous methodologies but established achievements and experience with similar problems are excellent starting points for future contributions. Besides a general discussion, the paper outlines opportunities for innovation in design, concept-level synthesis, process integration, and the development of supply chains.

KEYWORDS   *Renewables, biorefineries, process synthesis, process integration, optimization*

Introduction

The first decade of the 21st century is witnessing a drive towards sustainable manufacturing, GHG reductions and the increasing use of renewable resources. The chemical industry continues to improve its energy efficiency and emissions, but remains highly dependent on oil and gas (80% of its feedstock and energy according to Eurostat) with oil reserves to probably last for only 40 years and natural gas for 60 years (BP, 2008). Fossil-oil transportation fuels are major products of the chemical industry but continue to contribute heavily to GHG emissions consuming nearly 25% of the total net primary energy and 70% of the energy provided by petroleum and NGPL (EIA, 1999). In EU alone, 90% of the increase of CO2 emissions between 1990 and 2010 will be attributable to transport (BRAC, 2006). Biomass offers a promising alternative to the needs of modern society, now providing only 13% of the world energy needs (IEA Statistics 2005) and with most of its annual production left underexploited (220 billion tones per year).

Motivated by the need for sustainable solutions and general uncertainties in oil prices and energy supplies, the industry has been experiencing a steady growth in the production of biofuels (Demirbas, 2006; Huber et al, 2006) that is now developing into the emerging concept of biorefining (Fernando et al., 2006; Clark, 2007). Biorefining bears striking analogies with fossil-oil refining, fractionating biomass into a family of products to include transport fuels, platform chemicals, polymers, and specialty chemicals with yields and distributions that vary widely on the chemical and physical nature of the feedstock. With the production concentrated on biofuels, oil refining remains more viable as every drop of oil is used to make commodities (fuels, bulk chemicals) and specialty chemicals. Following the example of the petroleum industry, sustainable multiple product biorefineries should target a greater proportion of biomass, producing multiple streams of high volume/low value as well as low volume/high value molecules (Audsley and Annetts, 2003). Looking into the near future, the expected annual growth rate for fermentation products is 5% (compared to 2-3% for the overall chemical production) with McKinsey (Riese, 2006) predicting that by 2010 biobased products will account for 10% of the chemical industry (96€ billion).

From a systems perspective, the design and the synthesis of biorefineries are open and complex problems as biorefineries, to compete with conventional processes, should achieve maximum efficiencies with better design and process integration. Dimian (2007) discusses opportunities for computer-aided process engineering and renewable raw materials, particularly with regard to conceptual process design. Klatt and Marquardt (2008) highlighted the processing of renewable feedstocks as one of the emerging application domains in process systems engineering. First generation biofuels proved that using plant material does not necessarily improve sustainability. Accordingly, the development of biorefineries has to follow a holistic approach, with new challenges to account for the wide range of feedstocks and a need to formulate local and regional patterns of solutions. Unlike the case of fossil oil and gas as they are produced and processed on a global basis, the sustainable utilization of biomass requires a production close to, and closely integrated with, their source. Biorefineries will then probably take the form of regional developments designed to best exploit resources and regional forms of renewable energy, following paradigms of Industrial Symbiosis (Chertow, 2004).

The remaining sections review developments in the first and the second generation of biofuels, systems challenges in the optimization of their supply chains, and a position that the total systems approach could seize a chance for major and lasting contributions. In each section, the paper highlights opportunities to apply or develop systems methods, arguing that the systems technology is the natural approach to produce solutions for the complex problems posed by biorefineries.

Early Developments in the Production of Biofuels

Early efforts in the production of biofuels revived, quite separately, established biochemical and thermochemical routes (cf. Figure 1). The biochemical paths addressed transportation fuels and included fermentation and extraction paths based on edible agricultural products. As fermentation requires sugars, sugar canes have been primal choices leading to Brazil's gasohol programme. In USA, a long history of corn-based bioethanol technology, previously unable to compete with cheaper petroleum prices, has experienced a revival, especially in the grain growing states of Midwest. Using plant oils and transesterification with methanol, Europe shifted towards the production of biodiesel (fatty acid methyl ester) bringing substance, with over 400 outlets selling biodiesel in Germany, to an old claim by Rudolf Diesel that his *'engines can be fed with vegetable oils helping the agriculture of the countries*

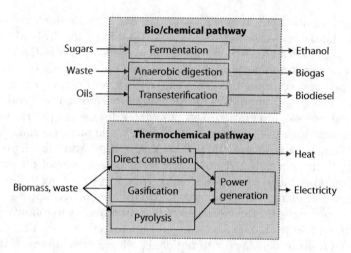

FIGURE 1
Early developments of biofuels.

that use it'. At a limited scale, also using a biochemical path, biogas and LFG gases, essentially byproducts of waste management, increasingly contributed to the generation of power and heat. However, a wider production of heat and electricity is available using thermochemical technology to produce gaseous fuels (and charcoal) and run gas turbines and engines, using a large variety of biomass and waste. Such technologies tread well-known paths of combustion, gasification (Fisher-Tropsch), and pyrolysis, also bringing to the fore possibilities to integrate gasification with pyrolysis, and chemical routes with thermochemical products (synthesis gas, methanol, hydrogen, diesel, bio-oil).

From a systems perspective, biorefineries represent ordered combinations of feedstocks, processing pathways, processing technologies, and products. At least in reference to the biochemical transformations, early developments actually addressed a limited number of biofuel products (mainly ethanol and biodiesel), well-known processing pathways, a limited number of processing technologies, and a variety of different feedstocks.

Still, systems technology had on offer powerful methods to systematize improvements, upgrade flowsheets, target the performance of processing units, achieve better flowsheet integration, and assess the impact of seasonality and variability of the feedstocks. Superstructure methods could have supported, for instance, a systematic screening of technological options for the pre-treatment stages or a rigorous assessment of performance limits for reactors and separators. Multi-period optimization models could have systematically assessed the variability in raw materials and compositions, whereas Pinch Analysis could have determined quick targets for energy and water use, setting incentives for better integration. Instead, literature evidence indicates a predominant use of general flowsheeting technology for techno-economic and case-by-case analysis, using Aspen Plus for instance, as in the work of NREL on bioethanol production (Wooley et al, 1999) and that of Haas et al. (2006) for biodiesel. In reference to thermochemical processes, systems technology tools are similarly restricted to general flowsheeting and CFD modelling, as in the work of pyrolysis reactors (Bridgwater et al., 1999; Wurzenberger et al., 2007; Blasi, 2008) with an emphasis on simulation and validation of experimental results. Even though consistently praised as important, the integration between different processing routes is unnoticeable in the literature with processing paths on Figure 1 essentially studied in isolation. Systematic approaches certainly missed a chance for better contributions due to a number of reasons, not excluding a lack of awareness or confidence by the practice groups for the advanced systems methods available.

Second Generation Fuels and Multi-product Biorefineries

While the production of first generation biofuels continues to increase, its sustainability and viability remain uncertain and questionable. The viability depends on world prices for sugar, grains and oil, all of which have varied widely and rapidly over the years. Major concerns relate to the sustainability of production, as feedstocks compete with food supplies whereas processing of low-density biomass yields excessive requirements for land. Focused on sugars and starch, the production neglects the energy-rich cellulosic and lignocellulosic content that, although more difficult to process, accounts for a larger portion of the biomass available.

Second generation (2G) biorefineries advocate a whole crop approach, featuring additional processing paths for residues and leading to complex portfolios of multiple products (biofuels and specialty chemicals), drawing direct analogies to conventional refineries (Fernando et al., 2006; Clark, 2007). The range of suitable feedstocks is much larger whereas the additional processing paths may separately follow either the biochemical or the thermochemical conversions discussed earlier. Biochemical conversions, as illustrated in Figure 2, make repeated use of enzyme technology to produce, not only C6 and C5 sugars required for the cellulosic ethanol, but numerous intermediate fractions in the form of carbohydrates, proteins, and phenolics. Several intermediates in the form of oligomers and monomers account for potential building blocks in the production of specialty chemicals, organic acids and polyols, natural 'health food' components (phytosterols, folates, phytates), prebiotics, and additives. Engineering yields and efficiencies are subject to continuous improvements using better enzyme technologies and catalytic processes. Lignocellulosic biomass, still difficult to break with enzyme technology, is processed thermochemically for either heat (or power), or to produce chemicals (methanol, synthetic diesel, bio-oil). 2G biodiesel combines gasification with syngas and FT synthesis to also produce LPG, naphtha, jet

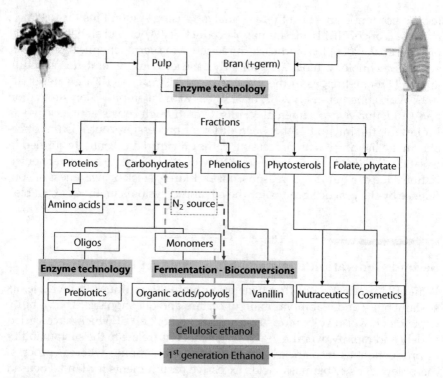

FIGURE 2
Biochemical conversions in 2G biorefineries (courtesy of INRA).

oil and lubricants. Hydrogenation of plant oils to 'green diesel' yields fuels with superior energy and GHG properties over conventional biodiesel.

In comparison with first generation biofuels, 2G biorefineries involve an extensive number of possible products, numerous processing paths, and extensive options for processing technologies. Figure 3 illustrates the case in the processing of lignocellulosics. As the design and decision-making problems are complex and large, essentially spanning all the known scales of process development, systems engineering technologies emerges with unique capabilities to systematize the analysis and contribute with the development of innovative solutions.

Major challenges relate to technologies in process synthesis, process integration, retrofitting, and general process modelling.

(i) Synthesis and Process Integration

Given an abundance of degrees of freedom for the processing paths and products, synthesis and process integration are natural technologies to coordinate a concept-based level analysis (strategic decisions), ahead of detailed evaluations and flowsheeting studies. The analysis, required for the holistic

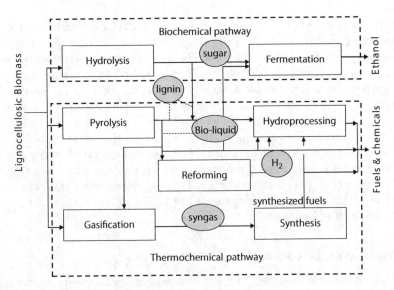

FIGURE 3
Processing routes in lignocellulosics (adapted from Bridgwater, 2007).

evaluation of the techno-economic trade-offs between the different objectives, would yield robust portfolios of products to match market and process uncertainties or variable feedstocks. Compact representations, probably in aggregate form, are required to combine production paths (reaction and fractionation) and processing steps (processing technologies and units) in the biochemical and thermochemical paths, deploying degrees of freedom over pathways (selection of enzyme and enzyme cocktails), chemical building blocks (cellulose, hemicellulose, lingins) and processing stages, letting optimization to scope for the fractionation and the refinement paths required. The development of effective and economical synthesis representations is a major challenge, as exhaustive superstructures are highly unlikely to bear fruitful results in such large and complex problems. Concerning the integration of pathways, one should probably differentiate separate methods with a focus either on the feedstocks (Gulati et al., 1996) or the processing technologies. Thermodynamics, targeting and shortcut methods may bear useful insights to explore and simplify each one of these problems.

(ii) 'Retrofitting' the Petrochemical Refineries

The integration of biomass processing is an attractive way to upgrade conventional refineries and represents a modern systems version of a retrofit problem. Examples include the production of 'green biodiesel', the NexBTL process, and the catalytic cracking of pyrolytic lignin (CGR, 2008). Green biodiesel (Petrobras/H-BIO, UOP with ENI) is produced widely with the hydrogenation of plant oils (or animal fat) using hydrogen available at the refinery. Fortum Oil Oy uses the proprietary NExBTL process to produce an isoparaffinic fuel (not FAME) still compatible with existing diesel engines (capacities range from 170-800 kT/yr). The hydrotreating and catalytic cracking of pyrolytic lignin (UOP, 2005) produces gasoline and aromatics, reforming further the water-soluble phase of bio-oil to produce hydrogen. The systematic development of such integrated scenarios could use a systems approach to differentiate between available feedstocks (biomass and fossil), processing routes (biomass, petrochemical refinery), and available chemicals. This comparator could produce scenarios for integration badly needed in reviewing the numerous options available in practice. Starting in 2009, a DEFRA project is specifically committed to deploy a systems platform to this purpose with a focus on developments of the Teesside industrial complex at the North East of England.

(iii) General Process Modelling and Flowsheeting

Separate efforts have addressed the improvement of process units and especially reactors, as in the work of cellulosic hydrolysis (Zhang and Lynd, 2004), and the integration of fermentation with hydrolysis. Using alternative reactors and better integration, and in reference to a base case stirred-tank reactor, NREL reported a two-fold increase in the concentration of sugars, a ballpark figure for the expected benefits in the yields. Instead of case-by-case analysis, systems tools are able to, systematically and rigorously, target efficiencies, setting the scope for integration (El-Halwagi, 1997, 2006; El-Halwagi, et al., 2008) and enabling the evaluation of fractionation, enzyme technologies and reactive-separation schemes (Mehta & Kokossis, 1997, 1998; Linke & Kokossis, 2003). Flowsheeting and simulation studies certainly dominate the literature, as in the work of Cardona and Sanchez (2006) that used Aspen Plus to evaluate process configurations in the production of ethanol from lignocellulosic biomass. Configurations experimented with different arrangements for hydrolysis, fermentation and purification, and several scenarios for water recycle. Using alternative simulators, Pfeffer et al. (2007) evaluated the energy demand over several options to utilize tillage. Piccolo and Bezzo (2008) simulated the production of lignocellulosic ethanol comparing fermentation routes with enzymatic hydrolysis over gasification. Gutierrez et al. (2009) simulated scenarios to integrate biodiesel production from palm oil with ethanol produced from lignocellulosic residues of the plant. In comparison to flowsheeting analysis, synthesis and process integration applications are rare but exist. Duret et al. (2005) applied energy integration to study the gasification of wood in the production of synthesis gas. Sanchez et al. (2006) provide a synthesis study for the production of lignocellulosic ethanol. More recently, Gassner and Marechal (2008a, b) optimized wood gasification processes enabling the possible integration with electrolysis and producing promising flowsheets for validation, setting the type of screening expected from the synthesis and process integration methods.

Challenges in the Design of Supply Chains

The selection of processing technologies depends not only on the production costs and on the value of the biorefinery products, but also on the supply chains required and the business models available. The development of viable business models is a challenging and difficult problem with several dimensions. Biomass resource systems have to match the quality characteristics of existing distribution systems but, due to the bulky nature of biomass, road transportation affects carbon and energy balances and is expensive in relation to the product value.

Several literature studies demonstrate optimization models that combine processing and transportation costs in the supply chains of renewables-based manufacturing, using mathematical programming to illustrate techno-economic trade-offs between production and logistics. In the production of bioethanol from sugar cane and sweet sorghum, Nguyen and Prince (1996) optimized transportation costs and the scale of the production plant. Using a variety of lignocellulosic feedstocks, Kaylen et al. (2000) optimized the location and the capacity of feedstocks to maximize bioethanol and furfural production. Recent studies evaluate product portfolios, as in the work of Sammons et al. (2007) that optimized the allocation of biorefinery products accounting for sales revenues, processing and feedstock costs. The demonstration included the production of syngas from chicken litter with syngas allocated between hydrogen and electricity stations. In reference to wheat bran and wheat-based biorefineries, Sadhukhan et al. (2008) reported a comparable value analysis approach that models marginal contributions from processing routes and products and compares different manufacturing scenarios.

Instead of the centralized facilities addressed by these studies, the consensus is that the future supply chains would probably feature decentralised networks to reduce costs and improve efficiencies in the energy value per transport unit. Several studies (e.g. Ranta, 2005; Shi et al., 2008) investigate the logistic issues of a distributed system that utilize biomass, yet assuming rather simple usage scenarios and with little consideration on the biomass processing aspects. Second generation developments suggest the progression towards new platforms to produce fuels and power but also a clear potential in the integration of processing paths. To exploit the regional availability of feedstocks, decentralised centres would have to connect to the network adopting processing technologies as appropriate to their region.

A systems representation to handle a distributed network could use nodes for each processing centre, letting degrees of freedom for the capacity of each node and its processing technologies. Nodes could feature a different potential based on the local resources available, and their proximity to relevant industrial sites (i.e. petrochemical complexes), power stations, or urban and municipal sites (i.e. rendering opportunities to exploit waste streams).

As a special case, such representation could handle distributed biomass-to-liquid (BTL) platforms that produce second-generation biodiesel. The processing

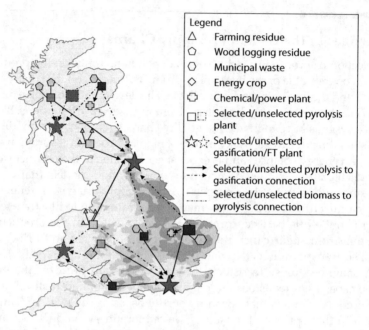

FIGURE 4
Illustrative BTL links and supply chains.

technologies include gasification and pyrolysis. Gasification can process a wide range of feedstocks (wood waste, agricultural waste, energy crops, and animal waste) but favors large capacities and, given the volumes needed, the cost of logistics is often prohibitive to run the gasification centrally. Bio-oil can provide the link, as pyrolysis units can be localised and the bio-oil is possible to move into bulk ships to a centralised facility. Using a network model, the location and the scale of the gasification and the pyrolysis units will be degrees of freedom to search with optimization, determining the best capacities and the best sites to use. Figure 4 illustrates a possible scenario produced by the solution of this hypothetical problem. In more complex cases, degrees of freedom could incorporate the optimal selection of processing technologies (e.g. those open to thermochemical as well as other processing pathways). Graph theory, optimization technology and general methods in operations research are welcome to tackle the complex and large problems, contributing with insightful recommendations to the development of business models.

A Total Systems Approach to the Emerging Developments

While 2G technology makes steady progress through several projects worldwide, there is already discussion about Phase III biorefineries (cf. Figure 5) that avoid any competition with food commodities. Unlike previous

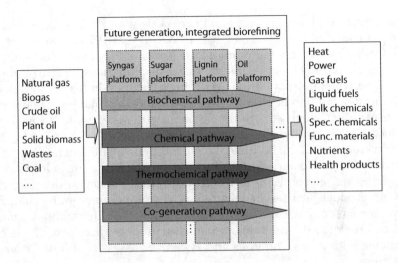

FIGURE 5
Integrated platforms-biorefineries.

generations, there is no pre-occupation in the production of particular fuels (or even fuels at all) with the problem base expanding further in reference to feedstocks, fractionation paths and processes. Concerning feedstocks, the expanding base includes energy crops such as short rotating coppices (in EU-25 about 142 Mtoe, EEA report, 2006), embracing mineral raw materials (gas, crude oil, coal and lignite), waste bioproducts (biogas, municipal and agricultural waste), and aquatic biomass. Indeed, in the form of algae structures, the latter source accounts for the most promising biomass at the time of writing (Dismukes et al., 2008; Raja, et al., 2008). Requiring considerably less land use than terrestrial biomass, algal species can grow at mild conditions, offering much higher (solar) energy yields in comparison with terrestrial plants and the possibility to clean industrial CO_2 emissions from flue gases and/or industrial and municipal wastewater. From a synthesis and design perspective, available processing options include closed culture systems (e.g. photosynthetic bioreactors or PBRs) and open ponds for massive algal production (cf. Borowitzka, 1999; Merchuk et al., 2007; Ugwu et al., 2008). Systems approaches to target the performance, design and optimize PBRs could rely on superstructure methods, drawing apparent analogies between temperature profiles and light utilization. The biomass produced from the culture of algae could follow the 2G pathways discussed previously but the design and the optimization of the supply chains would certainly need an alternative approach.

Rather than scoping for products and pathways, developments are important to further scope for common chemicals platforms suitable to share energy supplies and products between biorenfineries. As Phase III biorefineries assume no specific feedstocks and products, the synthesis problems remain open to all energy fuels and chemicals with choices for platforms among sugars, oils, syngas, and lignin. NREL and PNNL (Werpy et al., 2004, Holladay et al., 2007) produced an impressive list of potential building blocks, secondary chemicals, intermediates and products. The list demonstrates overwhelming similarities to the petrochemical industry, underlying the importance of the chemical engineering standpoint in the developments. Corma et al (2007) reviewed the development in this area, with a focus on the products derived from heterogeneous catalysis. Hermann and Patel (2007) and Haveren et al. (2008) discussed the substitution of mineral feedstocks by renewable biomass components, offering suggestions that include ethylene, propylene, and glycol. With few exceptions, the production of intermediates and products features bewildering opportunities, presenting challenging problems for synthesis. Taking into account available techno-economic data and regional distributions of biomass, process systems methods can contribute with a more rigorous and systematic assessment of the suitable platforms.

The wide range of data and the multiple though parallel dimensions of analysis (e.g. the potential in processing different algal strains, the economic viability of particular platforms) set a colossal undertaking to the techno-economic studies, rendering flowsheeting too basic and a primitive option to choose unless in coordination and combination with other advanced methods. Such methods could link farming and renewables production and, subsequently, biomass analysis with biomass processing, stimulating demand for arable farming whilst reducing manufacturing impact on the environment. Process synthesis and conceptual programming can assess overall routings (global view), and the scope for integration between adjacent sites, processes and supply chains. Flowsheeting and optimisation can evaluate design parameters of the supply chain (best technology), performance of unit operations (efficiency), operational aspects (reliability, flexibility, seasonality), and the cost efficiency of each venture.

The potential to coordinate different systems tools suggests the significance of an environment for high-throughput testing of ideas and processing technologies, gaining additional value in further collaboration with initiatives addressing strategic objectives (carbon footprints, LCA), and others focusing on the detailed analysis of processes (experimental investigation, detailed modeling, CFD modeling). One would wish that developments produce an environment open to modelling or simulation software (probably using CAPE-OPEN standards) offering opportunities to experiment with alternative solution technologies, algorithms and decomposition techniques.

In the light of challenges posed by the biorefineries on one hand, and the merits and the capacities of the systems tools on the other, a Total Systems Approach is justified, particularly with respect to process synthesis and design. The approach could support a holistic and structured way to understand and formulate problems in biorefineries, enabling a variety of solution methods and engaging links between different models. As illustrated

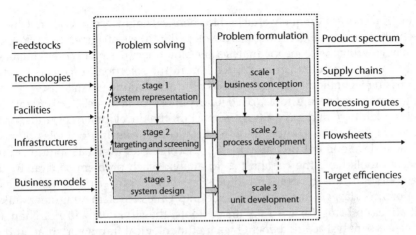

FIGURE 6
A total systems approach to biorefineries.

in Figure 6, the approach could stratify problems formulated over different levels (scales), echoing similar multi-level and multi-scale paradigms in the literature (Marquardt et al., 1999; Pantelides, 2001; Grossmann, 2004; Ingram, Cameron & Hangos, 2004). At the scale of a preliminary strategic investigation (or business conception), issues would address the regional planning and the evaluation of processing paths in combination with feedstock portfolios and logistics. Process development would target at the synthesis of flowsheets, the definition of products and the integration of units, whereas lower scales would develop unit parameters and operating conditions.

To solve each problem the approach could employ a multi-stage procedure with screening and targeting stages (Linke and Kokossis, 2003) to precede the development of resourceful, economical, and inexpensive representations ready for optimization. Early stages could rely on concise and aggregate mathematical models, setting performance targets and reducing the space to search with the more detailed models. The above methodological framework, with possible adaptations, may be applied as a whole or in part by regional planning agencies and/or manufacturing companies with vested interest in the area.

Conclusions

As renewable systems have to match maximum efficiencies to compete with conventional processes, they could invariably benefit by the systematic use of optimization and modeling. However, in no other system is the potential of systems engineering more pronounced as in the design of biorefineries, an industrial concept uniquely attached to chemical engineering. Within its relatively short history, biorefining has developed into several generations. Rather than narrowing the scope for analysis, the biorefinery design becomes increasingly complex, as the degrees of freedom in the selection of feedstocks, the processing technologies and its various products continue to increase.

Systems engineering tools, particularly those in synthesis, optimization, and modelling could have a huge impact on the development of units, processes, supply chains, and the biorefinery concept itself. Principles of sustainable development probably preclude scenarios for global solutions but systems tools have the capabilities to analyze each problem separately. In the course of first generation biofuels, systems engineering demonstrated its potential largely through flowsheeting. Unless systems engineers take initiatives to bring forward advanced tools, the situation will not change. Advanced analysis could benefit by a methodological framework that, in the form of a systems platform, could support high-throughput testing, analysis, and computational experimentation using a total systems approach to combine multi-scale formulations with multi-stage problem solving, and building capabilities to tackle new problems with new methods, proving the power of systems engineering to support novelty and innovation.

References

Audsley, E., Annetts, J. E. (2003). Modelling the value of a rural biorefinery--part I: the model description, Agricultural Systems, 76 (1), 39–59.

Batterham, R. J. (2003). Ten years of sustainability: where do we go from here, Chemical Engineering Science, 58 (11), 2167–2179.

Blasi, C. D. (2008). Modeling chemical and physical processes of wood and biomass pyrolysis. Progress in Energy and Combustion Science 34, 47–90.

Borowitzka , M. A. (1999). Commercial production of microalgae: ponds, tanks, tubes and fermenters. Journal of Biotechnology 70, 313–321.

BP (2008), BP Statistical Review of World Energy, June 2008 (available online) www. bp.com

BRAC (2006). Biofuels in the European Union – A Vision for 2030 and Beyond. The Biofuels Research Advisory Council, 14 March, 2006.

Bridgwater, T. (2007). Renewable transport fuels from biomass. Presented on International Biofuels Opportunities, Royal Society (UK), 23 & 24 April 2007.

Bridgwater, T., A.V., Meier, D., Radlein, D. (1999). An overview of fast pyrolysis of biomass. Organic Geochemistry 30, 1479–1493.

Cardona, C.A., Sanchez O.J. (2006). Energy consumption analysis of integrated flowsheets for production of fuel ethanol from lignocellulosic biomass. Energy 31, 2447–2459.

Cardona C.A., Sanchez, O. J. (2007). Fuel ethanol production: Process design trends and integration opportunities. Bioresource Technology 98, 2415–2457.

Carlos A. Cardona, Óscar J. Sánchez (2007). Fuel ethanol production: Process design trends and integration opportunities. Bioresource Technology, 98 (12), 2415–2457

Chertow, M. R. (2004). Industrial Symbiosis, In: Cutler J. Cleveland, Editor(s)-in-Chief, Encyclopedia of Energy, Elsevier, New York, 2004, Pages 407–415.

CGR (2008). The integration of biofuels inside the refinery gate: implementation, logistics, and strategies. Study proposal, The Catalyst Group Resources, Inc., March 2008. Online available at http://www.catalystgrp.com/IntegrationOfBiofuels.html, accessed December 2008.

Clark, J. H. (2007). Green chemistry for the second generation biorefinery – sustainable chemical manufacturing based on biomass. J Chem Technol Biotechnol 82:603–609.

Corma, A., Iborra, S., & Velty, A. (2007). Chemical routes for the transformation of biomass into chemicals. Chem. Rev., 107, 2411–2502.

Demirbas A. (2007). Progress and recent trends in biofuels. Progress in Energy and Combustion Science 33, 1–18.

Dimian, A. C. (2007). Renewable raw materials: chance and challenge for computer-aided process engineering. Computer Aided Chemical Engineering, 24, 309–318.

Dismukes, G.C., Carrieri, D., Bennette, N., Ananyev, G.M., Posewitz, M.C. (2008). Aquatic phototrophs: efficient alternatives to land-based crops for biofuels. Current Opinion in Biotechnology 2008, 19:235–240.

EIA (1999). Annual Energy Review, Energy Information Administration, 1999.

El-Halwagi, M. M.(1997). Pollution Prevention through Process Integration: Systematic Design Tools", Academic Press, San Diego.

El-Halwagi, M. M. (2006). "Process Integration", Elsevier, Amsterdam.

El-Halwagi, M., Dustin Harell, H. Dennis Spriggs (2008). Targeting cogeneration and waste utilization through process integration. Applied Energy, In Press.

Fernando S., Adhikari, S., Chandrapal, C., Murali, N. (2006). Biorefineries: Current Status, Challenges, and Future Direction. Energy & Fuels, 20, 1727–1737.

Gassner, M, Marechal, F. (2008). Methodology for the optimal thermo-economic, multi-objective design of thermochemical fuel production from biomass, Computers & Chemical Engineering, In Press.

Grossmann, I. E. (2004). Challenges in the new millennium: product discovery and design, enterprise and supply chain optimization, global life cycle assessment, Computers & Chemical Engineering, 29, 29–39.

Gulati, M., Kohlman, K., Ladish, M.R., Hespell, R., Bothast, R.J. (1996). Assessment of ethanol production options for corn products. Bioresource Technology 5, 253–264.

Gutierrez, L. F., Sanchez, O. J., Cardona, C. A. (2009). Process integration possibilities for biodiesel production from palm oil using ethanol obtained from lignocellulosic residues of oil palm industry, Bioresource Technology 100, 1227–1237.

Haas, M.J., Mcaloon, A.J., Yee, W.C., Foglia, T.A. 2006. A process model to estimate biodiesel production costs. Bioresource Technology. 97:671–678.

Haveren, J. v., Scott, E. L., Sanders, J. (2008). Bulk chemicals from biomass. Biofuels, Bioprod. Bioref. 2, 41–57.

Hermann, B.G., Patel, M.(2007). Today's and Tomorrow's Bio-Based Bulk Chemicals From White Biotechnology: A Techno-Economic Analysis. Applied Biochemistry and Biotechnology 136, 361-388.

Holladay, J.E., Bozell, J.J., White., J.F., Johnson, D. (2007). Top Value-Added Chemicals from Biomass, Volume II: Results of Screening for Potential Candidates from Biorefinery Lignin, PNNL, October 2007.

Huber, G. W., Iborra, S., & Corma, A. (2006). Synthesis of transportation fuels from biomass: Chemistry, catalysts, and engineering. Chem. Rev., 106(9), 4044–4098.

Ingram, G. D., Cameron, I. T., Hangos, K. M. (2004). Classification and analysis of integrating frameworks in multiscale modelling, Chemical Engineering Science 59, 2171–2187.

Kaylen, M., van Dyne, D. L., Choi, Y.-S., Blasé, M. (2000). Economic feasibility of producing ethanol from lignocellulosic feedstocks. Bioresource Technology, 72 (1), 19–32.

Kempener, R. Beck, J., Petrie, J. (2007). Multi-scale modelling of bio-energy networks: a complex systems approach. European Congress of Chemical Engineering – 6, Copenhagen 16-21 September 2007.

Klatt K.-U., Marquardt, W. (2008). Perspectives for process systems engineering—Personal views from academia and industry. Computers and Chemical Engineering, in press.

Linke, P., and A.C. Kokossis (2003). Attainable designs for reaction and separation processes from a superstructure-based approach. *AIChE Journal* 49(6), 1451–1470.

Marquardt, W., von Wedel, L., & Bayer, B. (2000). Perspectives on lifecycle process modeling. In M. F. Malone & J. A. Trainham (Eds.), Proceedings of the 5th international conference foundations of computer-aided process design, FOCAPD 1999, AIChE Symp. Ser. No. 323, Vol. 26 (pp. 192–214).

Merchuk, J. C. , Garcia-Camacho, F., Molina-Grima, E. (2007). Photobioreactor Design and Fluid Dynamics. Chem. Biochem. Eng. Q. 21 (4) 345–355.

Mehta, V. L., Kokossis, A. (1997). Development of novel multiphase reactors using a systematic design framework, Computers & Chemical Engineering, 21, S325–S330.

Mehta, V. L., Kokossis, A. (1998). New generation tools for multiphase reaction systems: A validated systematic methodology for novelty and design automation, Computers & Chemical Engineering, 22, S119-S126

Nguyen, M. H., Prince, R. G. H. (1996). A simple rule for bioenergy conversion plant size optimisation: Bioethanol from sugar cane and sweet sorghum. Biomass and Bioenergy, 10(5-6), 361-365.

Pantelides, C. C. (2001). New challenges and opportunities for process modelling, In: Rafiqul Gani and Sten Bay Jorgensen, Editor(s), Computer Aided Chemical Engineering, Elsevier, Volume 9, Pages 15–26.

Pfeffer, M., Wukovits, W., Beckmann, G., Friedl, A. (2007). Analysis and decrease of the energy demand of bioethanol-production by process integration. Applied Thermal Engineering 27, 2657–2664.

Piccolo, C., Bezzo, F. (2008). A techno-economic comparison between two technologies for bioethanol production from lignocellulose. Biomass and Bioenergy, In Press.

Ranta, T. (2005). Logging residues from regeneration fellings for biofuel production: GIS-based availability analysis in Finland. Biomass and Bioenergy 28, 171–182.

Raja, R., Hemaiswarya, S., Kumar, N. A., Sridhar, S., Rengasamy, R. (2008). A Perspective on the Biotechnological Potential of Microalgae. Critical Reviews in Microbiology, 34:77–88, 2008.

Riese J. (2006). Industrial Biotechnology -- Turning Potential into Profits. Plenary presentation, the third annual World Congress on Industrial Biotechnology and Bioprocessing, Toronto, July 11–14, 2006.

Sadhukhan, J., Mustafa, M.A., Misailidis, N., Mateos-Salvador, F., Du, C., Campbell, G.M. (2008). Value analysis tool for feasibility studies of biorefineries integrated with value added production. Chemical Engineering Science 63, 503-519.

Sammons, N., Eden, M., Cullinan, H., Perine, L., Connor, E. (2007). A flexible framework for optimal biorefinery product allocation. Environmental Progress, 26, 349–354.

Sendich, E. D., Dale, B. E., Kim, S. (2008). Comparison of crop and animal simulation options for integration with the biorefinery, Biomass and Bioenergy, 32, 1162–1174.

Shi, X., Elmoreb, A., Li, X., Gorenced, N. J., Jin, H., Zhang, X., Wang, F. (2008). Using spatial information technologies to select sites for biomass power plants: A case study in Guangdong Province, China. Biomass and Bioenergy, 32, 35–43.

Sokhansanj, S., Kumar, A., Turhollow, A. F. (2006). Development and implementation of integrated biomass supply analysis and logistics model (IBSAL). Biomass and Bioenergy 30, 838–847.

Ugwu, C.U., Aoyagi, H., Uchiyama, H.(2008). Photobioreactors for mass cultivation of algae. Bioresource Technology 99, 4021–4028.

UOP (2005). Opportunities for biorenewables in oil refineries. Final technical Report to US Department of Energy, December 2005.

Werpy T., Petersen, G., Aden, A., Bozell, J., Holladay, J., White, J., Manheim, A. (2004). Top Value Added Chemicals From Biomass, Volume I: Results of Screening for Potential Candidates from Sugars and Synthesis Gas. NREL and PNNL, August 2004.

Wooley, R., Ruth, M., Glassner, D., Sheejan, J. (1999). Process design and costing of bioethanol technology: a tool for determining the status and direction of research and development. Biotechnology Progress 15, 794–803.

Wurzenberger, J. C., Wallner, S., Raupenstrauch, H., Khinast, J. G. (2002). Thermal Conversion of Biomass: Comprehensive Reactor and Particle Modeling. AIChE Journal, 48 (10), 2398-2411.

Zhang, Y.-H., Lynd, L. R. (2004). Toward an Aggregated Understanding of Enzymatic Hydrolysis of Cellulose: Noncomplexed Cellulase Systems, Biotechnology and Bioengineering, 88 (7), 797–824.

9

Bioethanol Production System in Northern Italy: Supply Chain Design and Cost Optimisation

A. Zamboni[1], N. Shah[2] and F. Bezzo[1]

[1]DIPIC—*Dipartimento di Principi e Impianti di Ingegneria Chimica*
Università di Padova, via Marzolo 9, I-35131, Padova, Italy
[2]CPSE—*Centre for Process Systems Engineering*
Imperial College London, South Kensington Campus, SW7 2AZ London, UK

CONTENTS

ABSTRACT Italy has recently set as mandatory the minimum blending fraction of renewable fuels within conventional ones at 3% by energetic content for 2009 and 5.75% for 2010 in order to comply with the EU Commission guidelines regarding biofuels. As a consequence, decision makers need to cope with several issues mainly related to the integration of an alternative fuel within the conventional supply system. This work proposes a quantitative tool for the strategic design and the economic optimisation of biofuel supply networks with a particular view to the oncoming Italian corn-based ethanol production. Over a short-term time horizon, the entire supply chain as well as the secondary distribution logistic network has been optimised under a cost minimisation

---

[1] To whom all correspondence should be addressed: andrea.zamboni@unipd.it

criterion. The modelling framework has been based on a Mixed Integer Linear Programming (MILP) approach. Ethanol production via dry grind processes has been taken into account. The economic performance has been assessed by means of Supply Chain Analysis (SCA) techniques, focussing on biomass cultivation site locations, ethanol production capacity assignment and facilities location as well as transport system optimisation.

KEYWORDS *Ethanol supply chain optimisation, strategic design, capacity allocation*

Introduction

Over the last few years, important issues regarding fossil fuel depletion and energy supply security have led to a worldwide debate centred on the future routes of energy policy. The EU Commission (EC, 2003) has been driving the EU Members to a general effort in reaching a sustainable solution for the global energy supply question. Liquid biofuels have been identified as the best energy carrier to obtain a partial substitution of fossil energy in the transport sector (EC, 2007). Italy has complied with EU guidelines by setting the minimum blending fraction of biofuels within the conventional ones at 3% by energetic content for 2009 and 5.75% for 2010 (Legge Finanziaria 2008, L. 24.12.2007 n° 244).

Currently, bioethanol is widely considered the most appropriate solution for a short-term gasoline replacement (Cardona and Sànchez, 2007). However, although in other countries first generation ethanol production is a well-established industrial practice, some doubts still persist on whether ethanol production from starchy biomass brings any effective economic and environmental benefits (Granda et al., 2007). Romero-Hernandez et al. (2008) also outlines that both the economic profitability and especially the environmental sustainability of bioethanol production strongly depend on the production system geographical and social conditions.

Moreover, even though ethanol can be blended with gasoline without significant difficulties in using the existing distributing infrastructure (Bernard and Prieur, 2007), the transition from an oil-based fuel system to a biomass-based one represents a complex strategic design problem. However, at the moment there has not been any well-measured attempt to develop an appropriate strategic policy for the emerging Italian bioethanol system so as to ease the optimised integration within the present fuel supply infrastructure. Decision makers should be provided with specific optimisation tools capable of assessing the economic interactions along the entire fuel supply network so as to adopt the best policies for a sustainable transition.

The "supply chains of the future", among which the crops for non-food utilisation are included, should be designed before they develop organically in order to support national and international policy as well as strategic decisions in industry (Shah, 2005). Supply chain network design is defined as a strategic decision process regarding where to locate new facilities (production, storage and logistics), what supplier to use for each facility (sourcing decision) as well as allocation decisions. In tackling such high-level decision problems, analytical modelling has been recognised as the best optimisation option especially in the early stage of unknown structures design (Beamon, 1998). In particular, as stated by Kallrath (2000), mixed integer programming (MIP) represents one of the most suitable tools in determining the optimal solutions of complex supply chain design problems.

To the best of our knowledge, little attention has been directed so far towards using optimisation models to design a first generation bioethanol supply system although in other industrial sectors (including the design of novel biomass energy systems) this is a common practice. Indeed, there is a lot of work on the economic performance of bioethanol production from corn; however, the primary objective has never been the assessment of the entire supply chain in order to support the cost-optimal strategic design of future biofuels systems. It is our belief that, especially in those countries where first generation production is not yet established, the design of such a system before it develops organically is of the utmost importance, even more if first generation ethanol is viewed as a preliminary step towards a second generation biofuels supply system.

This paper proposes a spatially explicit modelling framework developed for the strategic design of biofuels supply chains with a particular view to the forthcoming Italian corn-based ethanol production. A Mixed Integer Linear Programming (MILP) model has been formulated for the optimisation of the entire supply network considering cost minimisation. Northern Italy has been considered as geographical benchmark and the dry grind process has been taken as the standard technology for ethanol production. The modelling framework has been applied to design the bioethanol supply chain in order to steer strategic decisions including biomass cultivation site locations, production facilities allocation and capacity assignment as well as transport system definition. Two different demand scenarios derived from Italian policy are taken into account.

The economic performance has been defined by means of Supply Chain Analysis (SCA) techniques; the plant capacity allocation problem has been assessed by taking into account the scale factors affecting the production costs according to a purpose-designed financial model (Franceschin et al., 2008).

Bioethanol Supply Chain

The economic assessment of a production system by means of SCA techniques requires a rigorous preliminary work focused on the characterisation of the supply chain components as well as of the logistic nodes in terms of both infrastructure availability and operating costs.

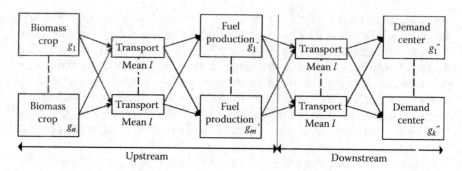

FIGURE 1
Bioethanol supply chain superstructure.

Figure 1 depicts the general structure of the supply chain of the ethanol production system. A biomass-based fuel supply chain can be divided into two main substructures: the first one is concerned with the fuel upstream production network and involves biomass cultivations, biomass delivery and fuel production sites; the latter is related to the downstream product distribution to the demand centres.

Biomass Cultivation

Considering the first generation production technology as the best solution over a short-term horizon, corn has been identified as the most convenient biomass for ethanol production in Italy. Spatially specific data regarding yield and land availability for corn crops have been collected from Governmental institution web sites (ISTAT, 2007; APAT, 2000). Corn production costs have been derived from actual data (CRPV, 2006): fixed costs have been separated from yield dependent ones, so as to create the grid dependent set of parameters reported in Figure 2, where every diamond represents the production cost as a function of the crop yield. The approach adopted has been validated by comparing the obtained results with the actual production costs collected from regional databases and through industrial information.

Since the Italian corn cultivation data are not classified in terms of the final utilization (either food or industrial purposes), a maximum biomass utilisation quota should be assumed in order to avoid the potential risk of a conflict between "biomass-for-food" and "biomass-for-fuel". This quota has been set equal to the estimation reported by the United States Department of Agriculture for corn production (USDA, 2005), in which the corn amount deployed for industrial purposes has been envisaged to reach an asymptotic threshold of about 14% of the overall domestic production. The assumption seems quite reasonable considering that the Italian region under investigation

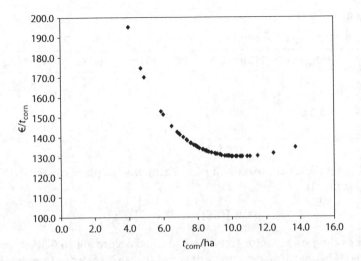

FIGURE 2
Unit production costs for corn cultivation.

present some similarities to the American corn belt with concern to soil conditions, corn yield and farming practices.

Transport System

The Northern Italy distribution infrastructure includes a full-scale range of transport options available for industrial purposes. The whole set has been considered so that trucks, rail, barges and ships have been included as the possible delivery means. Transport costs have been gathered from the literature (Buxton, 2008) and then validated comparing them with actual data from confidential information. The availability of each transport option has been characterised through the definition of feasibility constraints on transport means suitability. A tortuosity factor has also been introduced in order to take account that the actual product transport route is not linear: this is a multiplication factor to be applied to the local linear distance between network nodes and has been specifically defined according to the different nature of transport options. For instance, the delivery distance might be extremely different if covered by truck rather than by ship although the linear distance between nodes has been set as a single independent parameter.

Ethanol Production

Dry-grind technology has been considered in characterising the production facilities. Ethanol production costs are sensitive to plant capacity as there is an economy of scale effect on capital and accordingly on operating costs.

TABLE 1

Plant Capacity Parameters

p	PCap kt/y	PCap^{max} kt/y	PCap^{min} kt/y	PCC M€	UPC €/kg
1	110	120	80	70	0.160
2	150	160	140	91	0.154
3	200	210	190	115	0.151
4	250	260	240	139	0.149

Capital costs (*PCC*) are usually a power function of plant capacity (*PCap*) as shown by Eq. (1):

$$PCC \ [M€] = a \cdot PCap^r \ [kt/y] \tag{1}$$

where *r* is the power factor (set equal to 0.836 according to Gallagher et al., 2005) and *a* is an estimated constant equal to 1.132. In order to meet the linearity need imposed by the model formulation, *PCap* has been discretised into four intervals (*p*) and the central value of each interval has been taken as reference point for estimating *PCC*. Once the capital costs have been defined, unit production costs (*UPC*) for each different interval have been estimated by using the financial model developed by Franceschin et al. (2008): both biomass costs and capital investment depreciation charges have been deducted from ethanol production costs because they will be considered in the overall supply chain operating costs assessment.

Table 1 summarises the model parameters related to each plant capacity interval.

Demand Centres

A recent Governmental report (INDIS, 2007) outlining the Italian fuel-for-transport supply chain raises some important issues regarding gasoline distribution: downstream products (gasoline, diesel and LPG) are distributed to internal depots located in the neighbourhoods of the main transport nodes (rail stations or highways) and then delivered to filling stations mainly by road tankers; besides, the strong instability of blended mixtures forces the addition of ethanol to gasoline straight before the final distribution. Therefore, internal depots have to be assumed as the actual demand centres for bioethanol. Data about provincial gasoline demand perspectives for 2009 and 2010 as well as internal depot locations and maximum distribution capacity have been collected from Governmental web sites (MSE, 2007).

Given the demand-driven nature of the optimisation problem considered, the demand assignment to blending centres must be solved as a secondary distribution problem before optimising the ethanol supply chain. Accordingly, the overall problem has been decomposed into two sub-problems:

- the blended fuel secondary distribution optimisation problem, carried out to define the ethanol demand by allocating final demand to blending nodes over the two time scenarios;
- the bioethanol supply chain optimisation problem, implemented to design the entire fuel system over the two demand scenarios prospected.

Secondary Distribution

The blended fuel demand allocation is a typical case of terminal assignment to drop zones. Northern Italy has been divided into grids (drop zones, g) characterised by a homogeneous blended fuel demand (DEM_g). Internal depots (t) need to be assigned the drop zones they are going to serve. DEM_g values have been extrapolated from local gasoline demand, while the maximum terminal throughput allowed (THR^{max}_t) has been derived from the internal depot capacities. The actual throughput ($DEMT_t$) is assigned to each terminal by implementing the optimisation model. Once $DEMT_t$ is known, the ethanol demand (D^T_g) can be easily derived by fixing the blending percentage that characterise each demand scenario.

The mathematical formulation has been based on the MILP modelling approach commonly applied in the optimisation of fuel distribution systems (Kong, 2002) and solved in GAMS (Rosenthal, 2006). A graphical representation of the optimal distribution network ensuing from the model implementation is illustrated in Figure 3, while model parameters and optimisation results are reported in Table 2.

Supply Chain Optimisation

Once the ethanol demand has been defined, the entire bioethanol supply system has been optimised by means of a spatially explicit modelling framework. The core of the model is based on the MILP approach adopted in optimising the strategic design and the operation of multi-echelon supply networks for renewable fuels (Almansoori, 2006).

As mentioned above, strategic decisions in designing a biofuel production network deal with the geographical location of biomass cultivation sites, logistic definition of transport system and location as well as with the capacity assignment of production facilities.

The key variables are:

- amount of corn production for ethanol in each cell;
- corn distribution processes from crop fields to production facilities;

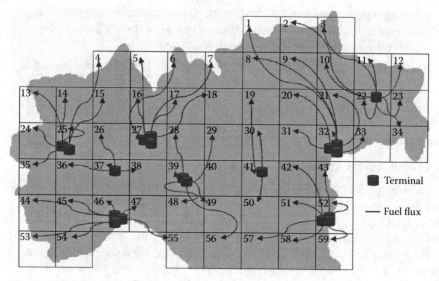

FIGURE 3
Blended fuel distribution system.

- location and capacity of production facilities;
- ethanol distribution processes from production facilities to blending terminals.

The model formulation has to be a mathematical representation of the most general supply chain network embracing all the possible configurations and interactions between the chain echelons as shown in Figure 1. Thus, the geographical region has been divided into grid squares of equal size (g) each one corresponding to the drop zone previously defined in the secondary distribution problem, and representing a potential location for each supply chain logistic node. Within this general superstructure, the optimisation algorithm searches for the best configuration excluding the undesired options, according to the optimisation criteria defined as well as to the logical constraints imposed.

Equation (2) represents the objective function adopted as optimisation criteria:

$$\min TDC = \frac{FCC}{a} CCF + FOC + TC \qquad (2)$$

where:

TDC	supply network daily operating costs (€/d)
FCC	facilities capital costs (€)
a	network operating period (340 d/y)
CCF	capital charge factor (1/3 y⁻¹)

TABLE 2

Secondary Distribution Results

2009 (3%)

G	DEMT, t/y	D^T_g t/y
22	470,910	22,133
25	630,248	29,622
27	1,342,313	63,089
32	634,624	29,827
37	311,355	14,634
39	670,830	31,529
41	446,185	20,971
46	468,319	22,011
52	684,060	32,151
Tot.	5,658,844	265,966

2010 (5.75%)

g	DEMT, t/y	D^T_g t/y
22	420,318	37,829
25	640,965	57,687
27	1,211,709	109,054
32	566,437	50,979
37	199,468	17,952
39	660,324	59,429
41	489,716	44,074
46	342,839	30,856
52	519,098	46,719
Tot.	5,050,873	454,579

FOC   facilities operating costs (sum of biomass cultivation, *bcost*, and etha-
       nol production costs, *ecost*; €/d)

TC   transport costs (€/d)

All the cost variables depend on design variables related to product demand (D^T_{ig}), production (P^T_{ig}) and mass flows between grids $(Q_{ilgg'})$. The representation of the supply chain features is captured by the definition of logical constraints. Equation (3) and Eq. (4) are demonstrative examples:

$$P^T_{ig} = D^T_{ig} + \sum_{l,g'} (Q_{ilgg'} - Q_{ilg'g}) \quad \forall i, g \tag{3}$$

$$Q^{min}_{il} X_{ilgg'} \leq Q_{ilgg'} \leq Q^{max}_{il} X_{ilgg'} \quad \forall i, l, g, g' \tag{4}$$

where i is the product (ethanol or corn), l the transport option and $X_{ilgg'}$ is the binary decision variable that is assigned the value 1 if the transportation of product i by means l is allowed from g to g', or 0 otherwise.

Case Study

The MILP optimisation framework has been solved to assess two different demand scenarios derived from the ethanol market penetration imposed by the current Governmental policy:

A   3% penetration by energy content for 2009;
B   5.75% penetration by energy content for 2010.

For each scenario two case studies have been formulated:

1   optimisation without plant location and capacity constraints;
2   optimisation by fixing plant locations and capacities according to the Italian Industry plans.

The industrial plans have been laid out according to the present status of the future production system: at the moment a production plant is under construction in Porto Marghera and another one is going to be set up in Porto Viro; two other plants have been scheduled but still are waiting for endorsement, and they should be located in Tortona and Trieste. Detailed characteristics of the production plants listed are provided in Table 3.

As mentioned, the aim of the proposed framework is to outline the structure of the best possible ethanol supply chain and to compare it with the industrial situation as prospected, in order to provide the decision makers with a detailed assessment which they can base their decisions on. Figure 4 and Figure 5 show the graphical representation of the best optimum in the two demand scenarios as resulting from the optimisation solution. In terms of transport system, it appears that truck delivery is the preferred transport option for high density products (e.g. ethanol) whereas rail is chosen when large amounts of product need to be transported (corn from crop fields to production plants or ethanol from plants to truck distribution nodes). With respect to plant locations and capacity assignment, they are optimised finding the best trade off between production costs (asking for high capacity centralised plants) and transportation costs (reduced by a more distributed system). As a consequence, the unconstrained optimisation within scenario A

TABLE 3

Italian Industry Plans

Location	Grid g	Size p	Nominal Capacity t/y
Porto Marghera	32	1	110,000
Porto Viro	43	2	160,000
Tortona	37	2	160,000
Trieste	34	1	100,000

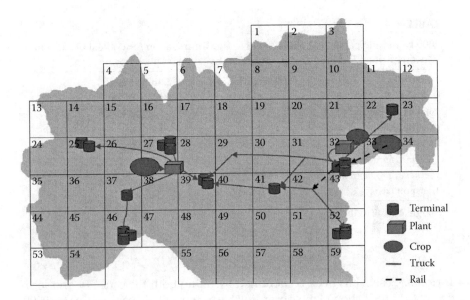

FIGURE 4
Scenario A-case 1.

(Figure 4) sets the location of two production plants in grid 32 and 27. The first one matches exactly with both the location and the capacity planned for the plant under construction in Porto Marghera (110,000 t/y). The second one is set at a capacity of 160,000 t/y and is placed on grid 27 (quite close to

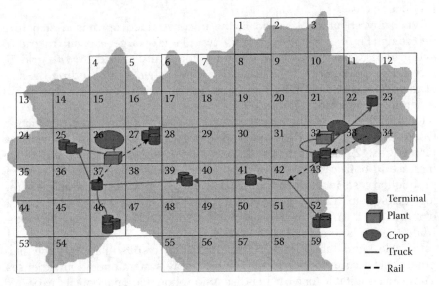

FIGURE 5
Scenario B-case 1.

TABLE 4

2009 Demand Scenario (A): Results of Cost Minimisation for (1) Best Optimum (Unconstrained Plant Location) and (2) Industrial Perspectives

A. 2009 Demand Scenario	1. Unconstrained €/d	2. Industrial Perspective €/d	Δ perc
Total daily costs	647,909	671,815	3.7%
Facilities capital costs	157,827	157,827	0.0%
Facilities operating costs	437,869	442,894	1.1%
Biomass production costs	315,428	320,355	1.6%
Ethanol production costs	122,440	122,539	0.1%
Transport costs	52,213	71,094	36.2%
Marginal costs (€/10^6 kJ$_{etoh}$)	24.585	25.492	3.7%

Tortona). Case 2 has been constrained by fixing both the capacity and location of the production plants according to the most likely perspective for 2009, i.e. the operation of the two plants already approved (grid 32 and grid 43). As a consequence, the resulting transport logistics are suboptimal with respect to Case 1. This is clear from the data summarised in Table 4: in Case 2 the total operating costs increase by about 4%; in particular, that is a direct result of the transportation costs that are 36% higher than in Case 1.

The non optimised plant location also entails a slight worsening of the biomass production site locations that account for a 1.1% increase in the biomass production costs.

When investigating Scenario B, the unconstrained optimised structure depicted in Figure 5 proposes to locate two plants of larger capacity in grid 26 and grid 32 (Case 1). However, given the contiguity of grid 26 with grid 27, it could be assumed that the target is accomplished by expanding the size of the plant scheduled in Scenario A (Case 1), without affecting the overall economic performance significantly.

On the other hand, the actual industrial plan for bioethanol production would outline a more distributed system based on four smaller capacity plants (Case 2). As reported in Table 5, this obviously causes a non-trivial increase in both capital and operating costs (corresponding to an ethanol production costs increase of about 5%) in addition to a clear worsening of the logistics system (transport costs rise about 44%). This shows a global increase of more than 8% in the overall supply chain operating costs when the industrial situation is compared with the best optimum.

In Table 4 and Table 5 are also reported some results relating to the marginal costs accounting for to the entire supply system operation expenses. They can be suitable for a further discussion about national policies expected to support the bioethanol industry. If Case 2 within demand scenario B is

TABLE 5

2010 Demand Scenario (B): Results of Cost Minimisation for (1) Best Optimum (Unconstrained Plant Location) and (2) Industrial Perspectives

B. 2010 Demand Scenario	1. Unconstrained €/d	2. Industrial Perspective €/d	Δ perc
Total daily costs	1,161,039	1,256,640	8.2%
Facilities capital costs	272,522	315,655	15.8%
Facilities operating costs	802,545	817,141	1.8%
Biomass production costs	586,734	590,404	0.6%
Ethanol production costs	215,812	226,737	5.1%
Transport costs	85,972	123,844	44.1%
Marginal costs (€/10^6 kJ$_{etoh}$)	23.793	25.735	8.2%

taken as an example, the marginal cost for operating the entire supply chain would be equal to 25.7 €/10^6 KJ$_{etoh}$[1]. In this situation the breakeven with gasoline production costs occurs when the oil price is about 100 $/bbl. Therefore, as long as the oil price stays below this threshold, the ethanol industry would need some kind of government intervention to enable the penetration of the alternative fuel within the conventional market. The Italian Government may be called upon to pass an appropriate regulation in order to impose on the oil companies the minimum ethanol content in their gasoline blend. However, this kind of policy would entail an increase in the fuel price (at least for an oil price below the breakeven value). Alternatively, a different form of subsidy might be the reductions on renewable fuel taxation (Salomon et al., 2007). The current regulations set the inland duties for biofuels at 13.5 €/10^6 kJ against the quota of 17.7 €/10^6 kJ applied to gasoline. This has the obvious outcome to partially fill the gap between gasoline and ethanol production costs (it has the effect to lower the breakeven point down to 97 €/bbl) but with the social side effect to require the use of financial resources which otherwise would be assigned to other sectors.

Finally, the modelling framework has been used to assess a further issue that relates to the sustainability of bioethanol production processes. According to the sustainability factor previously mentioned, the maximum ethanol production obtainable by using domestic corn only has been determined and is equal to 548,900 t/y, about 100,000 t/y more than the actual need for 2010. This excess production could be used for reaching a higher market penetration (corresponding to 6.5% in energy content).

---

[1] In evaluating the operating costs a quote for DDGS has been detracted considering an allocation factor deduced from literature data (Hammerschlag, 2006)

Conclusions

A spatially explicit modelling framework for the strategic design of biofuels supply networks has been developed. The aim of the study is to build a general modelling tool that may be helpful to steer an economic conscious design for biofuels supply chains. The bioethanol production system of Northern Italy has been chosen as a case study in order to demonstrate the model capabilities. Two different scenarios derived from ethanol demand perspectives have been assessed. The analysis has shown that in meeting Government requirement for 2009 (Scenario A) the best solution is to establish ethanol plants in Porto Marghera (grid 32) and in the industrial area of Milan (grid 27) with a production capacity respectively of 120,000 t/y and 150,000 t/y. For the 2010 perspective (Scenario B) the optimal supply network configuration provides for a capacity increase of the plant in grid 32 up to 240,000 t/y, and the construction of a similar capacity plant in grid 26. This solution would allow about an 8% saving on the total daily operating costs when compared to the likely oncoming scenario. The modelling tool can be used to provide consistent results in order to drive political decisions about energy policies for the future biofuels industry. For instance, the representation of production costs in terms of costs per unit of service provided by a fuel can be a consistent indicator to assess the actual fuel performance.

This research work should be considered as a starting point for a future more comprehensive framework. Future work will focus on:

1 considering the option of importing biomass from external suppliers

2 introducing profit performance indicators in order to develop a financial risk management tool capable of dealing with the uncertainties related to fuel demand and biomass prices

3 coupling environmental criteria with the economic ones in order to implement a multi-objective optimisation framework for investigating the benefits (if any) of biofuels on global warming.

Acknowledgement

A.Z. gratefully acknowledges the financial support of the University of Padova under Progetto di Ateneo 2007 (cod. CPDA071843): "Bioethanol from lignocellulosic biomass: process and equipment development".

References

Almansoori A. (2006). Design and Operation of a Future Hydrogen Supply Chain. *PhD Thesis*, Imperial College London.

APAT database: www.clc2000.sinanet.apat.it

Beamon B. (1998). Supply Chain Design and Analysis: Models and Methods. *International Journal of Production Economics, 55*, 281.

Bernard F., Prieur A. (2007). Biofuel Market and Carbon Modelling to Analyse French Biofuel Policy. *Energy Policy, 35*, 5991.

Buxton L. (2008). All Aboard. *Biofuel International, 1*.

Cardona C. A., Sànchez O. J. (2007). Fuel Ethanol Production: Process Design Trends and Integration Opportunities. *Bioresource Technology, 98*, 2415.

CRPV database: www.crpv.it

EC (2003). EC Directive 2003/30/EN of the European Parliament and of the Council of 8 May 2003 on the Promotion of the Use of Biofuels or Other Renewable Fuels for Transport. *Official Journal of the European Parliament*, Brussels.

EC (2007). Presidency Conclusions OR.EN. 7224/07 of the European Parliament and of the Council, Brussels.

Franceschin G., Zamboni A., Bezzo F., Bertucco A. (2008). Ethanol from Corn: a Technical and Economical Assessment Based on Different Scenarios. *Chem. Eng. Research and Design, 86*, 488.

Gallagher P.W., Brubaker H., Shapouri H. (2005). Plant Size: Capital Cost Relationships in the Dry Mill Ethanol Industry. *Biomass Bioenergy, 28*, 565.

Granda C. B., Zhu L., Holtzapple M. T. (2007). Sustainable Liquid Biofuels and their Environmental Impact. *Environmental Progress, 26*, 233.

Hammerschlag, R. (2006). Ethanol's Energy Return on Investment: a Survey of the Literature 1990–present. *Environ. Sci. Technol., 40*, 1744.

INDIS. (2007). *Istituto Nazionale Distribuzione e Servizi, La filiera dei carburanti per autotrazione-Un'analisi*, INDIS: Roma, IT.

ISTAT database: www.istat.it

Kallrath J. (2000). Mixed Integer Optimisation in the Chemical Process Industry: Experience, Potential and Future Perspectives. *Trans. IchemE-Part A, 78*, 809.

Kong M. T. (2002). Downstream Oil Products Supply Chain Optimisation. *PhD Thesis*, University of London.

MSE database: dgerm.sviluppoeconomico.gov.it

Romero-Hernández O., Salas-Porras E. D., Rode M. I. (2008). Panorama of the Social, Environmental, and Economic Conditions of Corn Ethanol in Mexico. *Instituto Tecnológico Autónomo de México (ITAM)*. Mexico City, Mexico.

Rosenthal R. E. (2006).GAMS-A Users' Guide (Version 22.5). *GAMS Development Corporation*. Washington, DC.

Solomon, B. D., Barnes, J. R. and Halvorsen, K. E. (2007). Grain and Cellulosic Ethanol: History, Economics, and Energy Policy. *Biomass Bioenergy, 31*, 416.

Shah N. (2005). Process Industry Supply Chains: Advanced and Challenges. *Computers and Chemical Engineering, 29*, 1225.

USDA. (2005). *USDA, Feed Grains Baseline*, USDA: Washington, USA.

10

Discovery of Novel Routes for the Production of Fuels and Chemicals

Linda J. Broadbelt[*1], Christopher S. Henry[2] and Vassily Hatzimanikatis[*3]

[1]*Department of Chemical and Biological Engineering, Northwestern University Evanston, IL 60208*
[2]*Argonne National Laboratory, Argonne, IL 60439*
[3]*École Polytechnique Fédérale de Lausanne, Lausanne, Switzerland*

CONTENTS

ABSTRACT   We have developed a computational framework that offers a new paradigm for the discovery of known and novel compounds and reactions in biochemical systems. The approach is built on the concept of generalized enzyme function and utilizes the methods of automated network generation based on graph theory. The number of pathways and compounds that can be generated can be staggering. Thus, we have begun to implement methods to identify the most promising novel pathways that can be implemented practically, including pathway length, degree of novelty, and thermodynamic landscape. The guiding principles are applicable to a wide range of different targets for biochemical production, including fuels, chemicals and pharmaceuticals.

KEYWORDS   *Computer generated reaction networks, generalized enzyme function, novel biotransformation, gibbs free energy of reaction, thermodynamic analysis*

---

[*] To whom all correspondence should be addressed

Introduction

Carbon-based compounds are marked for a transition. Fuels and chemicals derived from biomass are viewed as a companion, or even as a successor, to compounds derived from petroleum. However, the present portfolio is far from balanced (Ragauskas *et al.* 2006). At the beginning of the 20th century, numerous industrial materials were made from agricultural crops, but by the late 1960s, many were displaced by petroleum derivatives (van Wyk 2001). Growing energy consumption will place heavy pressure on petroleum resources, and it is generally accepted that relying on finite petroleum reserves is not a satisfactory policy for the long term (Ragauskas *et al.* 2006). In addition, biochemical processes offer potential advantages over typical petrochemical production routes, usually requiring lower temperature and pressure and using renewable resources as raw materials. A transition from petroleum-based feedstocks and conventional catalytic processes to renewable feedstocks and biochemical processes is an important component of any strategic initiative for long-term production of fuels and chemicals (Figure 1).

How do we begin to move in the direction of biochemical conversion of renewable resources? There are numerous possible compounds as targets, yet their production using bioprocessing may not have been demonstrated yet. Even if a compound is currently produced using biochemical reactions, there are opportunities for process improvement via the use of alternative routes. To navigate among the vast number of possible combinations of chemicals and processes for producing them, we are developing a computational

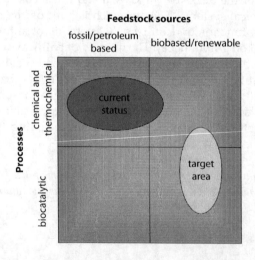

FIGURE 1
A transition to sustainable processes would involve migration from petroleum-based feedstocks and chemical processes toward renewable feedstocks and biocatalytic processes.

discovery platform. It is not sufficient to simply search databases of known compounds and reactions. The diversity of chemistry and biochemistry suggests that there are pathways that have not been discovered yet, and it is even possible that novel compounds have yet to be synthesized. Thus, the computational discovery platform that we are developing:

- Generates biochemical reaction pathways consisting of known and novel reactions to synthesize both known and novel compounds
- Screens and ranks the reaction pathways based on their feasibility for implementation

The reaction pathways are evaluated based on whether they meet one of the following criteria:

- Present a novel but superior route to a compound that is already produced biochemically
- Offer a novel route to produce a compound biochemically that is typically produced via traditional organic synthesis
- Propose a synthesis route to a novel compound

Methodology

The core of the computational framework is the concept of generalized enzyme reaction that we recently introduced, which is based on the Enzyme Commission (EC) classification system, for a systematic formulation of reaction rules governing known enzyme chemistry (Tipton and Boyce 2000). The Enzyme Commission established a classification scheme that involves a four-tiered hierarchical classification: EC *i.j.k.l*. The first three numbers classify enzymes according to the chemical rules of the reaction they catalyze, and the fourth number corresponds to the participating substrates and products. Our generalized enzyme reactions are defined at the *i.j.k* level and provide a set of operators that represent known enzyme chemistry in a general way. The application of these operators to a set of substrates and their progeny using automated pathway generation not only reproduces known compounds and reactions but also creates novel compounds and novel biochemical routes to both known and novel compounds.

The generalized enzyme reactions are implemented via the computer using automated network generation based on graph theory. A graph is composed of a finite set of edges and vertices. To represent a molecule, each vertex of the graph corresponds to an atom and each edge corresponds to a bond between different atoms. The graphs can be stored and manipulated within the computer using several different representations. One representation that illustrates the basic concept clearly is the bond-electron matrix (BEM) that cannot only describe molecular connectivity but also store the formal electronic state for every atom in

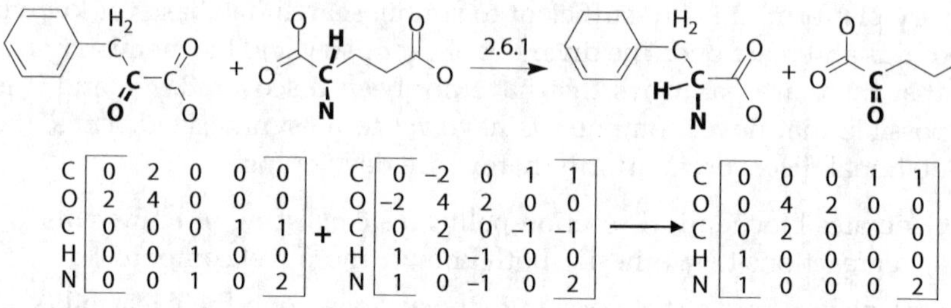

$$
\begin{array}{c|ccccc}
 & C & O & C & H & N \\
\hline
C & 0 & 2 & 0 & 0 & 0 \\
O & 2 & 4 & 0 & 0 & 0 \\
C & 0 & 0 & 0 & 1 & 1 \\
H & 0 & 0 & 1 & 0 & 0 \\
N & 0 & 0 & 1 & 0 & 2
\end{array}
+
\begin{array}{c|ccccc}
 & C & O & C & H & N \\
\hline
C & 0 & -2 & 0 & 1 & 1 \\
O & -2 & 4 & 2 & 0 & 0 \\
C & 0 & 2 & 0 & -1 & -1 \\
H & 1 & 0 & -1 & 0 & 0 \\
N & 1 & 0 & -1 & 0 & 2
\end{array}
\longrightarrow
\begin{array}{c|ccccc}
 & C & O & C & H & N \\
\hline
C & 0 & 0 & 0 & 1 & 1 \\
O & 0 & 4 & 2 & 0 & 0 \\
C & 0 & 2 & 0 & 0 & 0 \\
H & 1 & 0 & 0 & 0 & 0 \\
N & 1 & 0 & 0 & 0 & 2
\end{array}
$$

FIGURE 2

Example of the implementation of enzyme-catalyzed reactions through addition of bond-electron matrices. The generalized reaction operator for the EC 2.6.1 class is shown.

the molecule. The diagonal elements, ii, of the BE matrix denote the non-bonded valence electrons of atom i; the non-diagonal elements, ij, give the connectivity via bonding between different atoms and the bond order between atoms i and j. The power of the BEM approach is that enzyme-catalyzed reactions can be represented using similar notation. Figure 2 represents the enzyme-catalyzed reaction EC 2.6.1 based on its generalized reaction transformation. The sub-matrices for the reactant and product(s) were formed first based on those atoms whose electronic environment and/or bonding will change as a result of reaction. By subtracting the reactant matrix from the product matrix, the reaction operator was obtained, which then can be used to act on other substrates undergoing the EC 2.6.1 reaction. When the matrix representing the reaction is added to the BEM for the substrate, the BEM formed specifies the products of the reaction.

The automated pathway generation algorithm operates in an iterative manner. A set of molecules is given as input and this set of molecules and their descendents are evaluated to see if they have the appropriate functionality to undergo the specified reaction families, or EC classes. The reactions are next implemented through matrix addition, forming new molecules. An iteration count is maintained as new molecules are created, keeping track of the generation number of each species, which is the number of steps required to create a given product from the original reactant(s). A maximum generation number can be specified, and thus the generation number can be used to determine if a given molecule is allowed to react in the next generation. If the generation number is above the specified maximum, the given molecule is a terminus in the reaction network. Imposing a maximum generation number provides crucial control of the growth of the networks generated.

Results

To date, we have curated 33 of the ~250 third-level ($i.j.k$) enzyme classes. We have focused heavily on those EC classes that are central to aromatic amino acid synthesis and bioremediation of aromatics, and a total of 86 operators have been created. This number is higher than the number of EC classes that have been curated because operators have been formulated in both the forward and reverse directions where appropriate, and we did find that many EC classes do not have one single chemical pattern that characterizes them. In addition, we detected cases of enzyme misclassification, where an enzyme assigned to a given *i.j.k* class did not catalyze the same transformation as the other enzymes in the same EC class. Interestingly, these 86 operators covered nearly half of the diverse biochemical reactions contained in both the KEGG and the iJR904 metabolic model of metabolism (Reed *et al.* 2003), even though only a small portion of the KEGG was curated. This suggests that our approach could provide an alternative organizational scheme for enzyme chemistry that may possibly prove to be more concise.

A methodology has been established to evaluate the pathways for their thermodynamic feasibility. Experimental thermodynamic data is most desirable if available. However, the large majority of the compounds and reactions will have thermodynamic parameters that are not known from experiment, particularly those that are novel. Thus, we have implemented an estimation method based on group additivity to calculate the Gibbs free energy of reaction in aqueous systems (Jankowski 2006). Inspired by the classic group additivity approach of Benson (Benson 1968), Mavrovouniotis (Mavrovouniotis 1990; Mavrovouniotis 1991) initially developed a group contribution method for the rapid calculation of accurate estimates of ΔG_f and ΔG_r for a wide variety of biological reactions and compounds at a pH of 7 and room temperature. As illustrated in Figure 3, the properties of a molecule are estimated by summing the contributions from its constituent groups. The Gibbs free energy of reaction is estimated by tallying the contributions of the groups that are created or destroyed during reaction. While the method is powerful, the group contribution method of Mavrovouniotis has some limitations. The

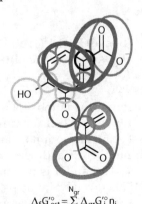

$$\Delta_f G'^o_{est} = \sum_{i=0}^{N_{gr}} \Delta_{gr} G'^o_i n_i$$

Group	$\Delta_{gr}G'^o$	n	SE_{gr}
[−COO⁻]	−83.2	2	0.1
[−OH]	−41.4	1	0.1
[−CH=]$_{ring}$	8.17	3	0.3
[−O−]	−22.9	1	0.4
[>C=]	15.9	1	0.4
[>CH−]$_{ring}$	4.65	2	0.2
[>CH=]$_{ring}$	11.7	1	0.4
[=CH₂]	6.73	1	0.3
OCCC	−1.52	2	0.2
CCCC	−4.75	1	0.4

Chorismate: −170.4 ± 1.4 kcal/mol

FIGURE 3
The free energy of formation of chorismate is calculated based on the contributions from its constituent groups.

method is incapable of estimating properties for molecules involving some nitrogen, halogen and sulfur substructures. Additionally, ΔG_f estimations calculated using the method different significantly from literature values for many phosphorylated compounds (Alberty 1998; Alberty 2006), and ΔG_r estimations differ significantly from experimentally observed ΔG_r values for reactions involving the formation of thioester bonds or the formation of conjugated double bonds. In addition, the confidence intervals are given very generally, rather than specific to each group. We recently addressed these limitations by utilizing a larger and more current training set of ΔG_r and ΔG_f data including new tables of thermodynamic data found in the NIST database by Goldberg and Tewari (see all references by Goldberg et al.) and the work of Alberty (Alberty 1998; Alberty 2006), Thauer (Thauer 1998) and Dolfing and coworkers (Dolfing and Harrison 1992; Holmes *et al.* 1993; Huang *et al.* 1996). The expanded group contribution method can estimate 93% of the reactions and 79% of the compounds in KEGG. While this represents a significant increase compared to the method of Mavrovouniotis, there are still groups that are found in KEGG that are missing, such as NO_2 groups and carbon atoms in a variety of fused ring environments. In addition, it can be expected that the pathway generation algorithms will create other molecules that possess additional unknown groups. Thus, current efforts are focused on estimating properties of molecules to fill in these gaps.

As a case study, we have applied the computational framework to explore novel biosynthetic routes for the production of 3-hydroxypropanoate (3HP) from pyruvate. Among the pathways to 3HP generated by the framework were all of the known pathways for the production of 3HP. In our analysis to date, we can report that many novel pathways were discovered, two of which involve fewer steps than any of the known pathways to 3HP. The overall reactions of all of the pathways discovered for which 3HP is the only product that is not a cofactor had the following form:

$$\text{(a) NADH + (a-c-d) H+ + (b) glu + (b+c+d) H}_2\text{O + (c+d)}$$
$$\text{ATP + pyruvate} \leftrightarrow \text{(a) NAD+ + (b) akg + (b) NH}_3 \text{ + (c+d)}$$
$$\text{phosphate + (c+d) ADP + 3HP}$$

where a is the number of oxidation reactions minus the number of reduction reactions involved in the pathway, b is the number of 2.6.1.a (amino transferase) reactions minus the number of 2.6.-1.a (reverse amino transferase) reactions involved in the pathway, c is the number of additions and subsequent removals of CoA involved in the pathway, and d is the number of carboxylations and subsequent decarboxylations involved in the pathway.

Thermodynamics-based Metabolic Flux Analysis (TMFA) has been utilized to determine the maximum yield for the production of 3HP in *E. coli* growing anaerobically on glucose for each of the pathways discovered. The thermodynamic constraints of TMFA also allow for the maximum achievable intracellular activity of 3HP to be determined for each of the pathways.

Two of the novel four-step pathways discovered matched the yield of the implemented pathway. In one of these pathways, (pyruvate → α-alanine → β-alanine → propenoate → 3HP) propenoate is utilized as the third intermediate instead of 3-oxopropanoate, and the maximum achievable 3HP activity is slightly lower than in the implemented pathway. The other novel pathway discovered involves the same intermediate compounds and has the same maximum achievable 3HP activity as the implemented pathway. This pathway differs from the implemented pathway in that both of the 2.6.1 reactions are replaced with 1.4.1 reactions. The implemented pathway has no clear advantages or disadvantages over this novel pathway. These results demonstrate the ability of the computational framework to produce numerous promising alternative pathways for the biosynthesis of 3HP despite the extensive studies already performed using conventional methods.

Acknowledgments

The work is supported by the US Department of Energy, Genomes to Life Program, the DuPont Young Professor's grant (VH), and an NSF IGERT Complex Systems Fellowship (CSH).

References

Alberty, R. A. (1998). "Calculation of standard transformed entropies of formation of biochemical reactants and group contributions at specified pH." *Journal of Physical Chemistry A* **102**(44): 8460–8466.

Alberty, R. A. (2006). "Standard molar entropies, standard entropies of formation, and standard transformed entropies of formation in the thermodynamics of enzyme-catalyzed reactions." *Journal of Chemical Thermodynamics* **38**(4): 396–404.

Benson, S. W. (1968). Thermochemical kinetics: *Methods for the estimation of thermochemical data and rate parameters*. New York, John Wiley and Sons, Inc.

Dolfing, J. and B. K. Harrison (1992). "Gibbs free-energy of formation of halogenated aromatic-compounds and their potential role as electron-acceptors in anaerobic environments." *Environmental Science & Technology* **26**(11): 2213–2218.

Goldberg, R. N. and Y. B. Tewari (1994a). "Thermodynamics of enzyme-catalyzed reactions. 2. Transferases." *Journal of Physical and Chemical Reference Data* **23**(4): 547–617.

Goldberg, R. N. and Y. B. Tewari (1994b). "Thermodynamics of enzyme-catalyzed reactions. 3. Hydrolases." Journal of Physical and Chemical Reference Data **23**(6): 1035–1103.

Goldberg, R. N. and Y. B. Tewari (1995a). "Thermodynamics of enzyme-catalyzed reactions. 4. Lyases." *Journal of Physical and Chemical Reference Data* **24**(5): 1669–1698.

Goldberg, R. N., Y. B. Tewari, D. Bell, K. Fazio and E. Anderson (1993a). "Thermodynamics of enzyme-catalyzed reaction: Part 1. Oxidoreductases." *J. Phys. Chem. Ref. Data* **22**: 515–579.

Goldberg, R. N., Y. B. Tewari and T. N. Bhat (2004). "Thermodynamics of enzyme-catalyzed reactions - a database for quantitative biochemistry." *Bioinformatics* **20**(16): 2874–2877.

Holmes, D. A., B. K. Harrison and J. Dolfing (1993). "Estimation of Gibbs free-energies of formation for polychlorinated-biphenyls." *Environmental Science & Technology* **27**(4): 725–731.

Huang, C. L., B. K. Harrison, J. Madura and J. Dolfing (1996). "Gibbs free energies of formation of pcdds: Evaluation of estimation methods and application for predicting dehalogenation pathways." *Environmental Toxicology and Chemistry* **15**(6): 824–836.

Jankowski, M. D. (2006). On the feasibility of novel biochemical transformations: A thermodynamics and constraints- based study, Evanston, Northwestern University. PhD thesis.

Mavrovouniotis, M. L. (1990). "Group contributions for estimating standard Gibbs energies of formation of biochemical compounds in aqueous solution." *Biotechnol. Bioeng.* **36**: 1070-1082.

Mavrovouniotis, M. L. (1991). "Estimation of standard Gibbs energy changes of biotransformations." *Journal of Biological Chemistry* **266**(22): 14440–14445.

Ragauskas, A. J., C. K. Williams, B. H. Davison, G. Britovsek, J. Cairney, C. A. Eckert, W. J. Frederick, J. P. Hallett, D. J. Leak, C. L. Liotta, J. R. Mielenz, R. Murphy, R. Templer and T. Tschaplinski (2006). "The path forward for biofuels and biomaterials." *Science* **311**(5760): 484–489.

Reed, J. L., T. D. Vo, C. H. Schilling and B. O. Palsson (2003). "An expanded genome-scale model of Escherichia coli k-12 (ijr904 gsm/gpr)." *Genome Biology* **4**(9).

Thauer, R. K. (1998). "Biochemistry of methanogenesis: A tribute to Marjory Stephenson." *Microbiology-Sgm* **144**: 2377–2406.

Tipton, K. and S. Boyce (2000). "History of the enzyme nomenclature system." *Bioinformatics* **16**(1): 34–40.

van Wyk, J. P. H. (2001). "Biotechnology and the utilization of biowaste as a resource for bioproduct development." *Trends in Biotechnology* **19**(5): 172–177.

11

Smart Process Manufacturing— A Vision of the Future

James F. Davis[1] and Thomas F. Edgar[2]
[1]*University of California, Los Angeles*
Los Angeles, CA 90095
[2]*University of Texas, Austin, TX 78712*

CONTENTS

ABSTRACT Smart process manufacturing (SPM) refers to a design and operational paradigm involving the integration of measurement and actuation; environment, safety, and health (ES&H) protection; regulatory control; real-time optimization and monitoring; and planning and scheduling. This integrated approach provides the basis for a strong predictive mode of operation with a much swifter incident-response capability. Smart process manufacturing is the enterprise-wide application of "smart" technologies, tools, and systems coupled with knowledge-enabled personnel to plan, design, build, operate, maintain, and manage process manufacturing facilities in full concert with the business and manufacturing missions of the enterprise. This paper provides a roadmap for the chemical processing industry, and for the corporations, universities and consortia who endorse the concept of smart process manufacturing to move from today's operations to a future desired state. That future state is defined as a set of research thrusts laid out in pathways, or lanes, that lead to a realization of the vision for SPM.

KEYWORDS *Smart manufacturing, process systems, computing*

Introduction

There are a number of forces driving the U.S. chemical industry in the 21st century, including shareholder return, globalization, efficient use of capital, faster product development, minimizing environmental impact, new ways to realize economic value, improved and more efficient use of research, workforce transitions and efficient use of people. As the chemical industry responds to these forces and tries to redefine and achieve these goals, it is investigating the expanded use and application of new computational technologies employed in areas such as modeling, computational chemistry, design, control, instrumentation, and operations. The key technology driver over the past 20 years has been the continuing advances in digital computing. The 100-fold increase in computer speed each decade, the concomitant reductions in the cost of storing data, the significantly increased capability of high fidelity multiscale models, the advances in networking and connectivity and the expansive potential of integration regardless of geographical location have tremendously increased the scope of computer applications in chemistry and chemical engineering. The new paradigm that encompasses the process systems engineering (PSE) topics mentioned above is *smart process manufacturing*, which will require the development of computational research environments to support such activities, also known as *cyberinfrastructure*.

Smart process manufacturing (SPM) refers to a design and operational paradigm involving the integration of measurement and actuation, safety and environmental protection, regulatory control, high fidelity modeling, real-time optimization and monitoring, and planning and scheduling. SPM is the enterprise-wide application of advanced technologies, tools, and systems, coupled with knowledge-enabled personnel, to plan, design, build, operate, maintain, and manage process manufacturing facilities. This framework provides the basis for a strong predictive and preventive mode of operation with a substantially more rapid incident-response capability, ensuring safe and health-conscious operations. Driving towards zero incidents and zero-emissions in the smart manufacturing paradigm recognizes that energy usage, energy production, safety, and environmental impact are inextricably linked.

What are the attributes of SPM? SPM systems have intelligent actions and possess a learning process to decide appropriate actions including the ability to actually implement that intelligent action. SPM systems are self-aware and proactive. They can adapt to new situations or perturbations (abnormal situations), and, by using feedback, smart systems can adapt or evolve for better performance. People are essential to SPM but play a more significant role than before. SPM systems integrate human expertise in new ways that allow for much more strategic, real-time decision making. In SPM pertinent information related to system performance is available, accessible, timely and appropriate and is understandable to the various parties or functions that need the information. Plant assets are integrated and self-aware (via sensors) of their state. Assets may be many things: plant, equipment, knowledge, experience, models, or data properties. Field devices have CPU and sensors, recognize their condition, and publish that information so they or other devices can take appropriate actions. A typical device receives, processes, translates, and publishes or shares information.

Specific process status variables can be related to larger business issues (profitability, zero incidents, zero emissions, etc.), i.e., beyond specific process and cross-industry impacts. Sustainable smart manufacturing includes reuse, with a life-cycle view of products and processes and a minimum environmental footprint (energy, water, emissions). Human resources (people) have to be knowledgeable, trained, empowered, connected (via cyber initiatives), and able to adapt/improve the system's performance.

A research roadmap for Smart Process Manufacturing is discussed in this white paper. Coming out of a 2006 NSF Workshop (Davis, 2006, Christofides et al., 2007), the Smart Process Manufacturing Engineering Virtual Organization (EVO) from industry and academia was formed in 2007, with the following steering committee:

- Jim Davis — UCLA (co-chair)
- Tom Edgar — University of Texas-Austin (co-chair)
- Jay Boisseau — Texas Advanced Computing Center
- Jerry Gipson — Dow Chemical Corporation
- Ignacio Grossmann – Carnegie Mellon University
- Peggy Hewitt — Honeywell Process Solutions
- Ric Jackson — FIATECH
- Jim Porter — DuPont
- Rex Reklaitis — Purdue University
- Allan Snavely — San Diego Supercomputer Center
- Bruce Strupp — CH2M Hill
- Jorge Vanegas — Texas A&M University

Primary input for this report was gathered at a workshop conducted April 21-22, 2008, at the NSF headquarters in Arlington, VA. The workshop was sponsored by the National Science Foundation under a grant to the CACHE Corporation on behalf of the Smart Process Manufacturing Engineering Virtual Organization (SPM-EVO).

The objective of the Smart Process Manufacturing workshop was to provide much-needed focused collaboration through facilitated assessment and creation of the base foundation for a roadmap. This white paper is a key step in the road-map development process. The roadmapping methodology was provided by IMTI, Inc. (Integrated Manufacturing Technology Initiative) for this workshop to generate, capture, and provide content for the roadmap, which contains:

1. A current state assessment for the three pillars (Technology Management, Systems and Facilities Management, and Enterprise Management), identifying technical and intellectual barriers and deficiencies, state of practice, and best practices and emerging research.

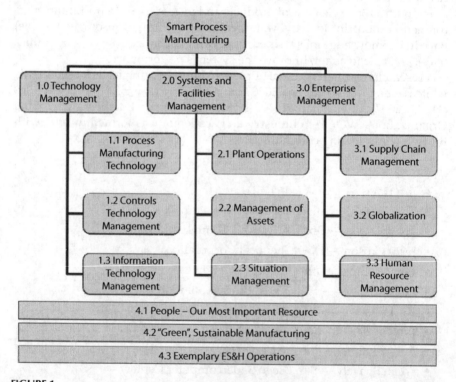

FIGURE 1

The smart process manufacturing functional model provides a hierarchical, logical frame-work for analysis of technology research and development requirements.

2. A future state vision for each pillar, articulating high-level goals and objectives for research and development (R&D) focus.

3. A framework of identified and prioritized issues and solutions within each pillar.

4. A review and prioritization by workshop participants of key solutions to be pursued across the scope of Smart Process Manufacturing.

For more details on the research roadmap, see http://www.oit.ucla.edu/nsf%2 Devo%2D2008/ and Edgar and Davis (2008).

Figure 1 depicts a functional framework for SPM that has been developed by PSE researchers from industry and academia.

There are a number of key elements in smart process manufacturing as shown in Figure 1. Each element is discussed below in specific sections of this white paper

Technology Management

Technology management addresses the determination of all the technological resources required to sustain, protect, and improve operations of the manufacturing enterprise, and includes the sub areas of process manufacturing technology, control technology management, and information technology management. Typically, many types of technology are involved, and they act as enablers for both local and cross-enterprise/global issues. Key components are specialized processes and systems, tools, control systems, modeling capabilities, and the people whose knowledge and practical expertise make the processing successful.

Increased use of technology has led to mountains of data and associated data management problems. However, the relationships of data to underlying processes and material properties is not well understood; the data are not integrated between systems and so important data are not fully exploited. The challenge is to understand relevant technologies and their associated data well enough to target acquiring and processing only the needed data. There is, however, increasing use of models and efforts to grow their scope, interoperability, and applicability to production operations—not just design. Technology investment decisions tend to be made with too short-term a view, expecting a payoff too quickly. Operations decisions tend to be highly reactive instead of proactive. A frustrating daily concern is that process engineers have to spend too much time worrying about cybersecurity and other IT details—software complexity and maintenance, version control, etc. A widely discussed concern is that current engineering education at the B.S. level does not prepare students well for the manufacturing-specific tasks they will face in actual production facilities. Needed topics frequently mentioned include optimization of process systems, data rectification, and other manufacturing skill sets such as power technology.

In the future, smart plants will be developed, designed, and operated using molecularly-informed engineering and will operate in a robust fashion. Models and all associated knowledge will be maintained and enriched as part of the plant's routine operation. Models will be developed to the level of detail needed, including multi-scale modeling to achieve the higher fidelity needed for some functions. Business goals will be directly translated into technology plans. This will enable flexibility of operations, robustness, agility, and the ability to change product streams quickly (especially in batch) or to make grade changes (for continuous processing).

Process Manufacturing Technology

Process manufacturing technology encompasses the capability to transform raw materials into useful products. This includes transformation of materials, intensification of materials, molecular transformation, and other processing and packaging. Companies still tend to design their products and processes based on specific customer needs and the lowest cost. There is increasing realization of the need to optimize design and production over the entire product life-cycle, with greater consideration of larger business and sustainability objectives. Even when the organization has ongoing research and development efforts, the results of those efforts are often not integrated with production engineering even within the same company. Process design is faced with challenges of increasing complexity in integrating multiple objectives and technologies, yielding multidimensional optimization problems. There is growing movement to sustainability, reuse, and approaches like pinch technology in facing current process design problems.

Currently there is an excessive amount of sensor-generated data, which needs to be sorted out and processed to be useful. Understanding the relationships embedded in the data would be facilitated by the availability of high fidelity models that have been validated on training data sets. When such high fidelity models are used, it is typically to analyze specific events rather than running in the background on a daily or hourly basis. Often there are multiple models that are used for a given asset, but they are not integrated into a single model covering the entire process. Models can increase the amount of molecular information to describe the applicable phenomena.

Enhancing the reusability of products is a complex problem for which there is not currently a large base of experience. Retrofit of existing plants is challenging when introducing new technology. Process intensification especially to reduce energy costs and capital requirements is becoming more important. Product design is based on product needs optimized from a customer perspective, but this needs to expand to have a product life-cycle perspective as well as a global rather than regional focus (involving multiple objectives beyond cost).

In the future smart manufacturing will be in common practice, using molecularly informed engineering of sufficient accuracy to enable fully customized products and sustainable processes. Processes will be designed to maximize the number of productive transformations and reduce the number of harmful transformations and corrective steps, undesired byproducts and waste. A single, universally recognized set of criteria for evaluating risks and potential mitigations will be applied from the process inception through the process life-cycle, from design into manufacturing. Knowledge- and data-driven models based on first principles will be used for design of every process, and will identify deviations from optimal operation. Plants will not only use the best models, but they will have appropriate instrumentation to assist in the process. Existing plants will be able to apply cutting-edge technology to cost-effectively maintain and refit operations. Furthermore, the long-term implications and resulting economic value of adding technology to achieve zero incidents and zero emissions processing will be understood and exploited.

Control Technology Management

With *controls technology management*, a sense, analyze and control environment ensures that all critical parameters operate within control limits, providing needed information and communications, and respond properly to abnormal conditions. Further, the analysis of existing conditions can predict and proactively respond to a changing environment. Digital control systems are now in wide use, generally based on PC-level machines and Microsoft architectures. Smart instruments are more common, but the challenge remains to integrate models built for specific functions and make wider use of them, for example for safety and environmental purposes. There is pressure to allow more input variability while production requirements mandate less output variability. Consequently, control systems need better ability to quickly determine the varying physical properties of feedstock and respond appropriately. In seeking this capability, the challenge is understanding, modeling and successfully moving operations from bench scale to production scale control systems.

Current control models need better physical property models and more of a molecular basis, especially to deal with increased variability of the process inputs (e.g., feedstocks). Models are often built by different people at different times for different purposes, with few similarities and a lack of consistency. Integration of model and controller information is performed at the user level but not at the more general (higher) levels. Often the data to be collected are not well-matched with the hardware specified for the process. There is also a need to better integrate real-time process control systems with fault detection, diagnosis, and management. The treatment of data from multiple sensor sources is not performed adequately. There needs to be an ability to aggregate information from multiple distributed sources and get the right data to the right place. New sensor technology that is cheaper and simpler is needed to measure more properties on-line. In order to provide adequate support, there is a concern that the B.S. chemical engineering education does not cover key topics from process systems engineering such as optimization and data rectification.

In the future plant automation advances will enable a transformative shift in the efficiency and sustainability of both existing and future facilities. Smart plants will be robust, proactive and able to predict impending trouble and take proper corrective action before serious consequences occur, using monitoring, controls, and actuator and sensor networks. Plants will align their operational objectives with overall business goals in real time, based on market needs as well as environmental, health and safety considerations. Sensors for real-time physical, chemical and composition property measurement will be smaller, simpler, and cheaper. Different types of models (statistical, first principles, business, supply chain, physical properties, etc.) will be combined and integrated in a hierarchical architecture to appropriate level of detail to realize intelligent operation toward business objectives. The resulting models and monitoring and control algorithms will be understood at a level that will provide universal availability and uptime (continuously productive operation) and will bring about safer operation and better product quality.

Information Technology Management

Information technology management ensures that the right information is available, at the right time, and in the right format. This includes the provision of a secure environment that protects the interests of the company and the individuals and is compliant with all relevant regulations. There is a general movement from proprietary to open systems, and growth in mobile computing and wireless communication. However, current technology capabilities should be made available to design and production engineers and operators in forms that can easily be used, automatically maintaining required security, while not requiring extensive training or deep understanding of either the underlying technology or the application software complexities. In addition, better tools are needed for understanding and managing the value of data streams.

Current off the shelf software (largely Microsoft) is not security-oriented and is vulnerable to bugs and viruses. Cybersecurity is important but directs resources from other areas. Application software is usable for areas like optimization without the user having a deep understanding of the inner workings, but there needs to be expertise to deal with unusual problems. The number of control loops expected to be managed by a single operation is growing. The management of data could be improved, starting with discovery and moving to commercialization. Data mining techniques are being used in a variety of scientific and engineering fields and could be applicable to process manufacturing. Knowledge management techniques are needed to avoid having islands of data that are not sufficiently integrated. Visualization of such multidimensional data is also a challenge. As computations grow in intensity, there will need to be utilization of supercomputing capability (multi-core parallel processing) in manufacturing plants, especially for multiphase and multiscale modeling.

The goal is to make the cyber infrastructure functional, foolproof, ubiquitous, secure and invisible, where process engineers do not have to worry about IT details. Domain knowledge will be available and trustworthy to all who need it, with seamless integration of different software tools.

Systems and Facility Management

Systems and facility management encompasses the oversight and assurance that the assets of the company and the enterprise are available to execute all needed functions within the defined operating envelope. Subareas include plant operations, asset management, and situation management.

One of the largest barriers to continuous improvement is the need to maintain current production and not change anything. There needs to be more of a balance between keeping the plant at constant operations vs. improving the profitability. The application of computing power to analyzing system performance has been limited to personal computers. Applying advanced technologies such as model predictive control and real-time optimization has been successful for many continuous petrochemical plants but is not applied as much to batch processes and more specialized products. There are concerns about the cost of supporting and maintaining such technologies. An awareness of the value of computing technology does not uniformly exist in process manufacturing, and there needs to be more meaningful metrics for valuing different technologies.

In the future systems and facilities will be managed for optimum availability and performance. Continuous sensing and control, at the appropriate level based on the need, will assure safe, optimized performance with the realization of zero incidents and zero emissions. Models will accurately predict plant performance and will support automated and computer-assisted decision-making for assurance of operation within the envelope for safety, environmental responsibility, and cost optimization.

Plant Operations

Plant operations include the execution of processes through multiple molecular, chemical, and physical transformations to produce the desired products. Predictive operation requires access to information and data but sometimes there is insufficient information about equipment. Many manufacturing execution systems are operated manually and use different systems that are not integrated, and data from different systems/vendors cannot be integrated due to interoperability issues. Standard definitions of data structures do not exist and existing standards leave too much room for interpretation. The skill requirement of operators is changing and requires additional training and improved human-machine interfaces. The characterization of the complete production process and corresponding models in incomplete and often does not use science-based information. Workflow is often not coordinated within a plant between organizations inside the plant. Wireless sensing has the potential to have significant economic impact but is still not considered ready for critical sensing and control activities.

Fundamental understanding and pervasive models enable skilled people to optimize manufacturing processes for maximum profit in a safe, socially responsible, sustainable environment that automatically adapts to uncertainty and faults. Ubiquitous use of sensors of all types (not just process sensors) can enrich status models and optimize facility operation, integrating environmental issues and market conditions into business decisions. The outcome of the models, sensors, and associated knowledge will be complete understanding of the system at any point in time, yielding meaningful metrics based around long-term technology and performance improvement instead of short-term financial gain. The realization of open, integrated systems for managing multiple facilities will remove the barriers to multi-plant operations and global companies. As a result, human and automated systems will have a better ability to manage the facility, and predict and analyze different options.

It is desired that process plants be flexible, scalable, tolerant of uncertainty, and adaptive to change with dynamic response. Models should accurately represent plant operations, including the automatic capture and management of knowledge. This will enable all information needed for optimized operation to be "pushed" to the right place, at the right time to assure pro-active production and recommended decision alternatives. The information will be presented in a self-guided and intuitive form that will make the right result transparent.

Management of Assets

Management of assets involves the optimization of asset procurement, retirement, liquidation, and all other aspects to ensure the appropriate assets are available to meet corporate objectives. In a typical plant there are conflicting objectives and key performance indictors (KPIs). Service factors conflict with profitability and energy KPIs conflict with profitability. Most assets do not have adequate monitoring due to lagging practice and capability limitations, also cost of hardware and software for management of assets. Models and data are also assets that are not typically viewed as such, and there are no systems for creating, managing, and valuing models as assets. Data collection regarding plant assets is often not focused; it is not clear what the data will be used for and how will it be used. Sometimes the cost of additional sensors is an impediment to collecting important information. There is a lack of prediction, coherent decision making frameworks for maintaining and assuring reliability of assets. There is also the lack of a systematic approach to capture the experience and knowledge of the workforce as well as training the workforce to operate in a model-rich environment.

In the SPM environment, assets will have the capacity to self-evaluate their individual and integrated roles and consider sustainability metrics. Assets, many of them augmented with "smart" capabilities, can be managed according to their contribution and risk to assure the best total value, including profit, safety, sustainability, and social responsibility, employing an optimal degree of automation. In this content assets include people, systems, facilities and equipment, and knowledge—all of the elements that make the plant work.

In the future optimized processes will be based on open, integrated and flexible models wherein all operations are conducted within acceptable limits (time, quality, environmental, etc.). The result will be a plant that operates within all regulations and guidelines and is adaptable to change, enabling profitability with minimum human effort. Smart assets integrated with intelligent models of the process facility will be able to assess their operating environment and configure themselves for optimized operation. Devices, equipment and other assets will have self-assessment capabilities and predictive capabilities that allow detection of performance trends and requirement changes and "intelligent" coordinated adjustments to optimize total value and plant operation.

Situation Management

Situation management is the detection, situational awareness, evaluation and understanding of existing conditions compared to the normal envelope for conduct of operations. In the event of deviations from the normal conduct of operations, the assessment includes the evaluation of options and the determination of best response. Abnormal situations are not handled uniformly from one plant to another; sometimes abnormal situations are detected but it is not possible to respond to it or the information needed for the response is not available. Often the response is based on knowledge and experience rather than on a systematic analysis. Training for abnormal responses is difficult because the situations are hard to simulate, and the data from normal operations are not relevant to abnormal situations. The loss of process operating knowledge and skills as the workforce ages limits the ability to diagnose and respond. Pressure to reduce costs and increase the scope of the operator due to reduced staffing are potential problems. Sometimes the process and systems are so complex that evaluation of event response automation is not easy. The current culture in operations is predominately reactive vs. proactive for situation prevention and response, and quick response is not possible.

The goal is for reliable and trustworthy systems that quickly diagnose and resolve all abnormal situations with no compromise of safety and for prognostic systems to predict failure so reliably that failure rarely occurs, and resolution of pending situations is managed without interruption to critical processes. Even when faced with unknown and undocumented events, assimilation and analysis of relevant accumulated operational data and recognition or patterns and behavior (including emergent behavior) can permit identification of both potential situations and best responses. If multiple sites are involved, collaborative technology between associated facilities will enable effective decision-making and rapid marshalling of appropriate resources.

Enterprise Management

Enterprise management takes an integrated view of all enterprise activities, from process loops (perhaps in many facilities within single company) up through strategic direction setting. It includes multiple plants working together, interaction between companies plus the global view, and integration and optimization of processing and business functions. Subareas include supply chain management, globalization, and human resource management. Business planning determines the right business/product mix and how products and technologies should evolve. Achieving smart manufacturing capabilities in the process manufacturing industry will require a much higher level of integration, flexibility, and adaptability in enterprise management than exists today. Although many business functions and many manufacturing functions have been integrated in large, complex software systems (ERP/ERM, PDM, etc.), there is still a lack of interoperability of business and technical functions across the enterprise. Furthermore, there is no direct connection between the manufacturing production goals and the larger business objectives of the enterprise. As the enterprise extends its relationship with supplier partners, there is a lack of consistent understanding of quality, safety and environmental standards across the supply chain, especially on a global basis. In addition, as companies become more closely integrated with their supply network and global partners, there is still a reluctance to share information freely and risk losing the company's intellectual property.

Another challenge to enterprise management is lost expertise as the industry's aging workforce has the potential of creating a loss of critical knowledge by attrition. Companies find it difficult to acquire needed skill sets and hard to attract skilled workers for some jobs, such as facility management. University curricula are not producing sufficient numbers of graduates with the needed practical and technical skills, so there is an ongoing dialogue on how U.S. universities should address this need.

In the future enterprise management tools, cost-effective technologies and cross-industry standard practices will enable U.S. manufacturing companies to successfully collaborate and compete in the global economy. Beneficial technologies and productivity improvements will spread across global industry via the development of industry-wide standards. Interoperability of systems will enable use of the complete set of plant and enterprise data in decision making. ES&H concerns will be effectively addressed by collective cross-industry practices on a global basis, enabled by the better tools and technologies that comply with industry standards. Cybersecurity technology will automatically provide protection of intellectual property, and liability and trust issues will be resolved to the extent that cross-industry relationships are encouraged. The cost-effective, flexible and adaptive operation that results from these improved technologies will ensure U.S. competitiveness in the global market.

Supply Chain Management

Supply chain management is the coordination and management of the supply base to ensure that the right components come together at the right place, at the right time, and in such a way that a useful product results all of the time. This includes the satisfaction of all business, technical, cultural, and regulatory obligations for every element of the supply network.

Although the network of suppliers has expanded greatly, reaching many hundreds of relationships for most large process manufacturing companies, there are still major challenges on achieving smoothly automated, just-in-time provision of needed resources. The different supply partners use different terminology and business systems (e.g., SAP, Oracle) and it is difficult to align and integrate the multiple database structures (even when nominally using the same major tools) so that automated planning and procurement can occur. The usual practice today is to develop custom models around the specific partners' business environments to get information flowing between manufacturing operations and the supply chain, and make that part of standard operations. There are a number of supply chain management tools to aid the process, but most tools are too locally focused and suboptimized, failing to deal with the complications as supplier relationships grow from cross-town to global in nature. It is difficult to address and understand uncertainty in the whole supply and demand chain, and to associate risk with the uncertainty. Furthermore, supplier and customer collaboration should broaden beyond basic business concerns to address energy and environmental issues such as carbon footprint of their products and processes.

In the future the supply chain will be a cyber-enable ecosystem that is fully integrated, optimized, flexible, agile, cost-effective, reliable and sustainable. It will ensure the right resources reach the right people when needed. The supply chain will be managed at multiple levels (strategic, tactical, operational, and practical) through the use of large-scale, multi-level (time and space) optimization models, where uncertainty is treated through adaptable, self-learning, self-correcting, self-maintaining robust solutions. It will employ fully-defined metrics to support continuous improvement in sustainability, safety, security, environment, and health across the enterprise. Partners in the supply chain will openly collaborate, optimally allocate their tasks, and share their information and systems to meet common goals on sustainability, safety, environment, health as well as profitable operation. With these improvements, the partners will progress to become a supply network or ecosystem that supports safe, optimized operations and mass-customization of product to suit the customer's changing needs.

Globalization

Globalization is concerned with decisions concerning location of operations, satisfaction of requirements for operation in the global community, balance of operations to assure that corporate goals are not compromised, and cultural issues such as language and traditions. Globalization is the accepted normal mode of doing business, but forming global teams and partnerships is still a logistics challenge and funding issue. Going global makes everything more complicated, like dealing with time differences and resources in geographically remote locations, and determining how to work together and how to allocate tasks. The technology tools used need to be more globally adaptive, accommodate different languages and different toolsets at different locations, and able to integrate into different IT infrastuctures.

In most companies, value knowledge is not formally inventoried and certainly not universally shared. There is great concern over loss of developed intellectual property due to off-shore partnerships and facilities, and a variety of mitigating measure are being tried. When setting up off-shore facilities, the issue of portability of knowledge arises, determining how to build new capacity including the needed human perspectives, and whether to develop knowledge locally or import people. Most companies are moving to use of local expertise as much as possible. An additional globalization issue concerns the different interpretations on safety, health, economic biases, and their effect on plant management, the product life cycle. Global, virtual teaming and knowledge management deliver continued improvement in enterprise value creation and human resource renewal. Industry and academia are aligned on competency, capability, and the need for international experience. Smart manufacturing offers an attractive career path for chemical engineers and other process manufacturing specialists.

The smart process manufacturing company of the future will be a cyber-enabled enterprise that focuses on value-creation, innovation, and sustainability across the enterprise, and competes effectively and efficiently using a diverse and borderless workforce, both globally and locally. Work practices will be anchored on achieving high levels of ES&H as well as producing quality, enhanced decision-making, and reduced liability, as a result of accessibility and connectivity to both global and local information and knowledge, experience, and wisdom. Compliance with industry-wide standards will yield beneficial technologies and tools and productivity improvements that will spread across the global industry. Securely sharing information and collaborating between partners, the U.S. companies and workforce will be able to successfully compete in the global market. There is a growing cross-industry perspective, including environmental concerns, as well as investment and R&D and product portfolio management.

Human Resource Management

Human Resource Management addresses the assurance of ready and sustained staffing to meet the needs today and at any point in the future. This includes trends and staffing projections, education, training and life-long learning, work rules, and compensation. The major human resource issues facing the process manufacturing industry are the graying workforce and loss of their technical knowledge and experience with attrition, and the lack of sufficiently qualified workers to replace them. The speed of change in industry's technology and operations has not been kept up with in academia; process manufacturing is still viewed as a fairly low-tech industry and not that attractive to many students. A number of studies have pointed out the issues and problems, but it has been difficult to translate that into actions that will assure an adequate supply of qualified graduates.

Internationally there is much greater competitiveness (in number and quality) in engineering and science graduates; fewer students go into engineering in the U.S. than in other countries, so fewer U.S. B.S. graduates possessing less technical knowledge are available to be hired by the U.S. process manufacturing industry. Furthermore, domestic students at both undergraduate and graduate levels typically lack language skills and international exposure needed for globalized operations. Fewer tenure-tracked faculty are teaching some needed practical topics like control and design, and universities are de-emphasizing these courses, which compounds the problems and leads to lack of faculty trained to teach control and design. There is growing dialogue between industry and academia on how to fit needed skills and technical expertise (e.g. in bio- or nano-technologies) into engineering curricula which are already deemed too large to fit into the typical four-year university time-frame. Meanwhile, in response to the lack of continuing education programs available in manufacturing, many organizations have internal "universities" for continuing education. Many companies are also making expanded use of available cooperative education and co-op programs.

In an SPM environment, people are critical in setting strategies and assuring that those strategies are carried out. Graduates with needed specialties will regularly flow from the educational "pipeline", and continuing education programs will assure continued refresh of skills and knowledge for successful competition in the global economy ("K-to-Gray" education. A significant aspect of valuing knowledgeable workers is the capture and reuse of their knowledge and experience, in order to improve the overall plant performance.

Green Sustainable Manufacturing

Green, Sustainable Manufacturing in an SPM facility make environmentally sound practices become automatic, which are part of the business drivers that guide all operations. The benefits of improved environmental performance on manufacturing operations can extend beyond improving the quality of the environment and the need to take a life cycle perspective.

Exemplary Environmental, Safety, and Health Operations

Exemplary ES&H Operations beyond ES&H regulatory compliance is a prominent business driver to "zero-incident" operations, where the goal is to have no negative impact on personnel, surrounding communities, or the environment in general. Smart plants proactively prevent environmental, health, and safety problems while they at the same time seize opportunities to optimize operational and financial performance and monitor environmental conditions, and any aberrations are immediately noted and mitigated.

Conclusions

The research agenda described in this paper is quite ambitious, and the first steps at addressing how this grand challenge problem can be attacked are being taken by the EVO on Smart Process Manufacturing under sponsorship by the National Science Foundation. Some specific aspects of the roadmap can be researched under the NSF CDI (CyberDiscovery and Innovation) Program but other more applied projects could be pursued by an industry consortium similar to FIATECH (www.fiatech.org). Faculty and industrial practitioners are invited to participate in the EVO and provide input to the EVO's planning by sending an email to jdavis@oit.ucla.edu.

References

Christofides, P.D., J.F. Davis, N.H. El-Farra, D. Clark, K.R.D. Harris, and J.N. Gipson, "Smart Plant Operations: Vision, Progress and Challenges," *AIChE J.* Vol. 53, 2734-2741 (2007).

Davis, J.F., "NSF Workshop on Cyberinfrastructure in Chemical and Biological Systems: Impact and Directions", September 25-26, Final Report, February (2007).

Davis, J.F. and T.F. Edgar, "NSF Roadmap Development Workshop: Zero-Incident, Zero-Emission Smart Manufacturing", April 21 and 22, 2008, Draft Report, May (2008).

12

Toward Sustainability by Designing Networks of Technological-Ecological Systems

Bhavik R. Bakshi[1]*, Robert A. Urban[1], Anil Baral[1], Geoffrey F. Grubb[1] and William J. Mitsch[2]

[1]*Department of Chemical and Biomolecular Engineering*
[2]*School of Environment and Natural Resources*
The Ohio State University
Columbus, Ohio 43210, USA

CONTENTS

ABSTRACT Sustainability of human activities is entirely dependent on the availability of ecosystem goods and services such as carbon sequestration, mineral and fossil resources, sunlight, biogeochemical cycles, soil formation, pollination, etc. However, most existing methods in sustainable engineering and industrial ecology ignore this crucial role played by nature. The use of

---

* To whom all correspondence should be addressed

such a narrow boundary can lead to perverse and misleading results. This paper introduces the idea of designing networks of technological systems along with their supporting ecological systems. Such technological-ecological networks (TEco-Nets), can emulate ecosystems by having closed material loops and minimum exergy loss, leading to truly self-sustaining systems. Methods for designing TEco-Nets could be developed by extending existing process synthesis and design approaches to include ecological models. Such an approach would integrate industrial ecology with ecological engineering and require collaboration between engineers and ecologists. It presents many new challenges and opportunities for process systems engineering to contribute to the sustainability of engineered systems. The idea of TEco-Nets is illustrated via several practical case studies, with focus on a typical residential system and the life cycle of corn ethanol.

KEYWORDS *Sustainability, design, ecosystems, life cycle, networks*

Introduction

The strongest motivation for sustainable engineering is provided by recent studies such as the Millennium Ecosystem Assessment (MA, 2005), which quantify the huge impact of anthropogenic activities on the very goods and services that are essential for all activities on earth. These include provisioning, supporting, regulating and cultural goods and services such as air, minerals, genetic diversity, sunlight, carbon sequestration, biogeochemical cycles, pollination services, pest regulation etc., and are also referred to as Natural Capital (NC). Examples of such deterioration include ozone depletion, climate change due to exceeding the sequestration ability of the carbon cycle, loss of pollinators such as bees, eu-trophication due to disruption of the nitrogen cycle, depletion of biological diversity and abiotic resources, etc. In response to such studies, many researchers, non-governmental groups, corporations, and governments are realizing the urgent need for transforming human activities to make them sustainable.

This poses some of the most formidable challenges to most disciplines, including engineering.

Challenges of Sustainability

Process Systems Engineering can and must play a significant role in meeting the challenges of achieving sustain-ability, but requires an expansion of the traditional PSE boundary beyond the process and enterprise to include the life cycle and associated economic and ecological systems (Bakshi and Fiksel, 2003; Sikdar, 2003). Life cycle assessment (LCA) represents a broad class of methods that consider this larger boundary, and includes methods for assessing the impact of emissions (Bare and Gloria, 2006), the reliance on fossil and other resources (Spreng, 1988), and the transformation of exergy

(Ukidwe and Bakshi, 2007). These methods have been combined with traditional process design by treating the life cycle aspects as design objectives along with the traditional economic objectives (Azapagic et al., 2006; Pistikopoulos, 1999; Fu et al., 2000). Furthermore, many efforts are directed toward developing new products and processes that are likely to have a smaller life cycle environmental impact. Examples include products based on nanotechnology such as solar cells and water purification devices, fuels based on biomass, green chemistry and environmentally benign manufacturing systems. Also, many corporations are actively reducing the life cycle environmental impact or footprint of their activities.

These efforts are certainly encouraging, but unfortunately, in many, if not most cases, there is little reason to believe that their success will lead to greater sustainability. This is because technology alone cannot lead to sustainability since it involves aspects beyond engineering and technology, which must be taken into account to prevent unpleasant and unexpected surprises. For example, over the decades, despite increasingly efficient technologies, total consumption of energy has continued to increase. This is due to factors such as the economic rebound effect and consumerism. Thus, accounting for socio-economic aspects should be a part of sustainable engineering. However, even in the ideal situation where socio-economic and other non-technological and non-scientific effects are accounted for, existing efforts need *not* lead to sustainability if they ignore the role of ecosystems. Excluding this pivotal role from the engineering analysis and design boundary could lead to decisions and designs that may even accelerate the deterioration of NC, which would certainly not be sustainable. Another shortcoming of existing LCA and related methods is that they do not consider the carrying capacity of ecosystems for providing the resources used in the life cycle or for absorbing the impact of emissions. Again, this could lead to perverse decisions.

The second law of thermodynamics also indicates that no technological solution, as practiced currently, can lead to sustainability. This is because the second law implies that creating order in a system must result in even greater disorder in the surroundings. This disorder usually manifests itself as environmental impact. Since virtually all technological activities aim to create order in the form of manufactured goods and services, environmental impact is inevitable. This implies that no single technology, product or process can be claimed to be sustainable. In fact, it also implies that no individual technology by itself, that is available now or will be developed in the future, can lead to sustainability. This poses a severe dilemma for engineering research and technology development.

Learning from Ecosystems

The only systems that have sustained themselves for millennia are ecological systems. Despite the limits imposed by the second law, ecosystems are able to sustain themselves by building a strongly connected network where all materials are recycled and the only waste is that of low quality heat. Such networking means that individual ecological processes do not have to be particularly efficient, as long as there are other processes that can take advantage of the waste. Consequently, it has often been suggested that technological systems should learn from and emulate ecosystems. This has been the source of ideas and approaches such as Industrial Ecology (Graedel and Allenby, 2003), Ecologically Balanced Industrial Complexes (Nemerow, 1995), Biomimicry, and Ecological Engineering (Mitsch and Jørgensen, 2004). The first two approaches focus on developing networks of industrial systems where "waste = food". Thus, the waste from one process should be used as a resource in other processes. Many efforts have focused on developing such industrial ecosystems (Chertow, 2007). However, a crucial shortcoming of such approaches is that they ignore the role of ecosystems, and in general, there is no consideration of ecosystem ecology in industrial ecology (Tilley, 2003). The approach of ecological engineering has been developed mainly by ecologists, and aims to engineer ecosystems to provide goods and services essential for human activities (Mitsch and Jørgensen, 2004). The underlying belief being that since ecosystems are self-sustaining, they can be better for supporting human activities than traditional technological alternatives. For example, instead of conventional methods for treating waste water, the ecological engineering solution would be to allow a wetland to self-organize for treating the waste. Similarly, for dealing with degraded soil, ecological engineering suggests building the soil ecosystem such that it can make soil and enhance its quality, instead of the traditional approach of trucking in new soil. There is little doubt about the sustainability of such approaches, but little attention has been directed toward considering such engineered ecosystems in industrial design. The idea of "ecological design" (der Ryn and Cowan, 1996) has been popular in some fields such as architecture, but it is quite narrow in scope and ad hoc due to the lack of systematic methods.

The main contribution of this paper is in introducing the idea of designing Technological-Ecological Networks (TEco-Nets) as essential components in sustainable systems. Such an approach explicitly accounts for the role of ecosystems in human activities and has the potential to overcome the most significant shortcoming of existing approaches in sustainable engineering. It merges industrial ecology with ecological engineering by introducing ecosystem ecology into industrial ecology and connecting ecological engineering with the design of industrial processes. It focuses on the design of TEco-Nets that contain closed loops for materials, and are extremely efficient. Furthermore, considering TEco-Nets can automatically account for the limits posed by the carrying capacity of supporting ecosystems and ensure that engineering activities take place within ecological constraints (Schulze, 1996; Realff, 2006). This work is related to the approach of Ecologically-Based LCA (Eco-LCA) (Zhang et al., 2008), which focuses on analyzing economic goods and services while accounting for the role of ecosystems. This work focuses only on analysis, while the current paper considers the synthesis or design problem.

The rest of this paper is organized as follows. The next section provides examples of some typical TEco-Nets to provide insight into the nature of the design problem. This is followed by a discussion of the challenges in designing TEco-Nets and possible approaches based on existing methods in process design. This includes the nature of ecological models, use of methods such as exergy analysis and impact assessment for identifying design opportunities, and the use of process synthesis methods including hierarchical design and optimization. The proposed ideas and approaches are illustrated via case studies of TEco-Nets for the home and corn ethanol systems. The paper ends with a discussion of the challenges and the urgent need for more research.

Examples of TEco-Nets

This section sheds more light on the types of ecological goods and services that technological systems rely on by considering specific such networks. Technological-Ecological Networks are always present for all technological and industrial activities, but are only rarely considered explicitly or quantitatively. The examples in this section are not meant to be comprehensive or complete, but are qualitative illustrations to demonstrate the concept of a TEco-Net and to identify the nature of the design challenges.

Coal Burning Power Plant TEco-Net

The major waste flows from a coal burning power plant include carbon dioxide and other air emissions, and ash. The resources used include the coal, air, and the large quantity of water needed for any thermo-electric process. These are shown in Figure 1. Approaches such as industrial symbiosis suggest integration of this power plant with other processes that can utilize the waste. Thus, the SO_2 and warm water could be used in making cement, and the ash for making concrete. While this technological network does reduce the impact ofthese industries, several emissions and resource use still cause negative environmental impact due to the presence of open resource loops. These include the ecological resources of coal, limestone, and water, and ecological services of carbon sequestration, oxygen production, and the water cycle. Consequently, a TEco-Net for this system, as shown in Figure 1, includes the ecological processes of forest ecosystems for absorbing the CO_2 and producing O_2, and the river and wetland ecosystems to clean the water for reuse. Replacing nonrenewable resources of coal and limestone is more challenging, but could be achieved via technologies that enable recycling of such resources or by ecological services that make reduced carbon from its oxidized form. This is illustrated in this example via using the CO_2 for growing algae or using the forest biomass as fuel instead of coal. Considering such a network brings the role of supporting ecosystems to the forefront and

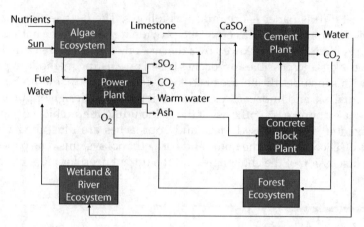

FIGURE 1
Technological-ecological network for coal burning power plant.

can indicate the extent and type of ecosystems needed to sustain the technological processes. Selecting the appropriate technological and ecological alternatives, while considering economic aspects may be approached as a design problem. For some emissions such as greenhouse gases, and other ecosystem services such as those from wetlands, the alternative of buying offsets or participating in cap and trade systems may also be available.

Residential TEco-Net

A typical suburban North American home also represents a TEco-Net, as shown in Figure 2. A residential system is defined here as the house, its surrounding property including lawn and trees (the ecosystems), its inhabitants, and the cars, appliances, and other machinery (the technological system) within the home. The flows with a dotted line indicate that in this diagram these flows do not form a loop to synergize these flows with the rest of the network. These flows indicate opportunities for improvement in that modifying them can cause them to mesh with the residential network or even a network outside of the system boundary. For example, if the solid waste was diverted from the landfill to a municipal solid wasted (MSW) electricity generation plant, it could be then looped back into the residential system. Technological changes such as the use of more insulation, more efficient heating, etc. can also reduce the emissions and need to be considered in solving the TEco-Net design problem.

Corn Ethanol TEco-Net

The life cycle impact of biofuels may be reduced by designing a TEco-Net. Such a partial network for corn ethanol is shown in Figure 3. Emissions such

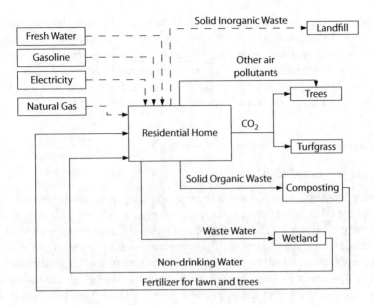

FIGURE 2
Technological-ecological network for typical residential system.

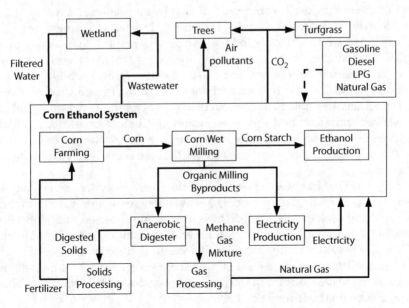

FIGURE 3
Potential technological-ecological network for the corn ethanol life cycle.

as nitrogen and phosphorus runoff could be treated via engineered wetland ecosystems. Technological or ecological alternatives also exist for closing the loop on organic wastes, such as integration with cattle farming, anaerobic digestion, no-till farming, etc. Current corn ethanol production also relies on fossil fuels throughout the life cycle. These could be replaced by bio-fuels. The effect of such alternatives is quantified in the case study section.

Designing TEco-Nets

Designing technological-ecological networks requires methods to identify opportunities for improving existing systems, models or data representing relevant technological and ecological systems, and methods for obtaining potential solutions. Each step represents an area of on-going research, and presents many challenges where the experience with traditional process design is likely to help. This section discusses some relevant issues and techniques. A unique challenge in the design of TEco-Nets is that the entire network with all ecosystem goods and services and emissions can be large and difficult to model in its entirety. Such challenges are also encountered in LCA, but are beyond the scope of this article.

Identifying Improvement Opportunities

Environmental impact is commonly caused by emissions and resource use. For identifying improvement opportunities, impact assessment methods such as those used commonly in LCA, can be very helpful. To identify opportunities for reducing resource consumption, exergy analysis has been popular in engineering, and may also be used for analyzing TEco-Nets. Design attention could focus on process changes for reducing emissions and/or including new technological or ecological processes in the network that can absorb the emissions and create a closed-loop for resource use. Similarly, resource use may be reduced by modifying existing technological systems and/or including systems that can recover the resource from byproducts and emissions.

Since modeling has been a popular activity in engineering and ecology for many decades, many useful models are available for a large variety of technological and ecological systems. These models may be used to design TEco-Nets, but it is important to understand and account for some fundamental differences between technological and ecological systems. After discussing some such differences, this section describes typical ecological models that are used in the case studies in the next section followed by some ideas about using design methods for TEco-Nets.

Usually, technological systems are designed by humans and behave in a relatively deterministic or predictable manner since most properly designed technological systems will give the desired performance or output right away. Thus, a pump will provide the expected head, and a car, the designed mileage. Also, technological systems usually work best for the designed task but may not provide many other goods or services. In contrast, ecological systems are usually not as well understood as technological systems, but can design themselves, given some basic resources. For example, given basic biological resources and a "hole in the ground", a wetland can design itself and become capable of purifying water (Mitsch and Jorgensen, 2004). In addition, most ecosystems will provide a variety of other ecosystem goods and services such as wetlands providing biodiversity enhancement, flood regulation, and carbon sequestration, that may or may not be needed for supporting the selected technological systems. Ecological systems also seem to exhibit more complex dynamics and nonlinearities than technological systems, and are usually more resilient to disturbances. Fortunately, for design purposes, particularly at early stages of process synthesis, it is common to use relatively crude technological models such as those representing steady-state operation. A similar approach may also be used with ecological models, that is, ecosystems may be treated as process unit operations for TEco-Net synthesis, as illustrated in the case studies.

Models exist for many common ecosystems such as trees (i-Tree), turfgrass (CENTURY, 2008), and wetlands (Restrepo et al., 1998). The CENTURY model (CENTURY, 2008) is a dynamic model capable of predicting the behavior of turfgrass systems under a variety of conditions, such as temperature, precipitation amount, and perturbations such as natural disasters. This model has been used to predict carbon sequestration in turfgrass systems under a variety of management schemes (Qianetal., 2003). Trees provide services such as air pollution removal and carbon sequestration as well as energy benefits by providing shade to buildings. These have also been modeled extensively for individual trees, urban forests, and other forest ecosystems (Nowak et al., 2006; USFS). As mentioned earlier, wetlands can be used to filter wastewater and remove phosphorus and nitrogen minerals as well as other contaminants (Verhoeven and Meuleman, 1999).

Design Approach

Methods for process design, particularly at early stages rely on steady state models of process units. In this work, ecological systems are included in the integrated design problem by treating them like another process unit. The assumption of steady state operation is quite reasonable for industrial processes at this stage of design since their behavior, as compared to that of ecological systems, is quite predictable and may be reached in a relatively short period of time. However, as mentioned above, ecosystems are constantly evolving, with many systems taking a very long time to mature, certainly much longer than the life of a typical manufacturing plant. This fundamental difference between technological and ecological systems may be dealt with by using ecosystem models over multiple time periods.

The TEco-Net design problem may be solved for a single technological system or the entire life cycle. As discussed in the introduction, ideally, the resulting TEco-Net should have closed loops for materials use, with minimum exergy losses. Achieving such a closed network may not be possible for many technological systems in the short run due their very large reliance on nonrenewable resources. In such cases, the TEco-Net design approach can identify the main improvement opportunities for transforming the system and relevant network toward becoming self-sustaining. A typical TEco-Net design problem is expected to involve trade-offs between changing the technological processes by making them more resource efficient, adding new technological processes that have a symbiotic relationship with the main system, modifying ecological processes that provide goods and services, adding new ecological processes to the network, or relying on economic mechanisms to offset emissions and resource use. Thus, the TEco-Net design problem includes the challenges of designing conventional processes, networks of processes and ecosystems, and ecological systems, making it significantly more challenging that traditional engineering design. In addition, market mechanisms such as carbon trading and ecological offsets also need to be included in the TEco-Net design problem. This design problem requires consideration of multiple objectives to account for economic and ecological considerations.

Despite these new challenges of TEco-Net design, many traditional engineering design methods could be used for solving this problem, and present exciting opportunities for research at the interface of engineering, ecology, and economics. Popular approaches for process design include hierarchical methods (Douglas, 1988), process integration (El-Halwagi, 2006), and mathematical programming (Biegler et al., 1997). These and other approaches could be adopted for solving the proposed design problem, and willbe subjects of future publications.

Case Studies

This section describes the preliminary results of two case studies. These studies demonstrate the benefits of considering integrated technological-ecological networks and suggest possible designs. They do not solve the design problem in a systematic manner by any of the methods mentioned in the previous section. This is the topic of on-going work, and will be included in future publications.

Residential TEco-Net

This example considers carbon sources and sinks in a typical North American suburban residence. Only the direct emissions of CO_2 are considered, which means that indirect emissions such as those in generating electricity from coal are not included. Typical sources are as follows.

- *Home Energy Use.* Direct carbon emissions from the home itself are created by fuel consumption for central heating, water heating, cooking, and other appliances but not necessarily all or even any of them, as a house could, in theory, use electricity for all energy needs. This is an issue that will need to be addressed for complete specification of the system, but for this base case, it is assumed that the house uses natural gas for central heating, water heating, cooking, and other typical natural gas using processes. Natural gas consumption is taken as the average for a Midwestern household (EIA) and emissions factors for natural gas (GREET) are used to calculate the carbon emissions from natural gas consumption. As shown in Table 1, this emission is 1, 573 kg C/year.
- *Automobile Use.* Based on average gasoline consumption for a Midwestern household (EIA) and emissions factors for gasoline (GREET), 2,863 kg C/year is emitted. These emissions are over a larger area than that of the house, but in this example, are allocated to the residence.
- *Lawn Care.* Although lawn care can have diverse impacts on the environment due to fertilizer runoff and water use, only carbon emissions from lawn mower operation are considered here. An average walk-behind mower uses 1.0 gallons of gasoline per acre (Sahu). It is assumed that the lawn is mowed for 9 months per year at a rate of 4 mows per month, which equates to 36 mows per year. Using emissions factors from the GREET model, carbon emissions for lawn mowing can be calculated to be 216 kg C/(ha-yr).

TABLE 1

Carbon Sources and Sinks in Residential TEco-Net

Residential System Unit	Value
Natural Gas Combustion	1573 kg/yr (EIA)
Gasoline Combustion (Automobile)	2863 kg/yr (EIA)
Gasoline Combustion (Lawn Mower)	$216 \dfrac{kg}{ha \cdot yr}$ (Sahu)
Tree Sequestration	$N \cdot C \cdot D \cdot \Delta(sfw); \Delta(sfw) = e^A \cdot [(dbh^\circ + r \cdot t)^B - (dbh^\circ + r \cdot (t-1))^B]$ 1b/yr (USFS)
Lawn Sequestration	$\begin{cases} 620 \text{ kg/(ha} - \text{yr)} & \text{if } t \le 50 \text{ yr,} \\ 350 \text{ kg/(ha} - \text{yr)} & \text{if } t > 50 \text{ yr.} \end{cases}$ (Singh, 2007)
Carbon Credits	\$22/1000 lbs CO_2 (TP)

The two main sinks in a residential system and their contribution to carbon sequestration are as follows.

- *Trees.* Models for calculating carbon sequestration in individual trees are available, and the equation used in this work is shown in Table 1. For this study, willow and maple trees are considered to be present in the residential landscape. The rate of sequestration varies overtime as the tree grows and matures, and models for individual trees, urban forests and more natural forest ecosystems are available. For the equations used here, $\Delta(sfw)$ is the change in dry storage weight of the tree from the previous year. $dbh°$ is the initial diameter at breast height (dbh) of the tree in inches. r is the growth rate, assumed to be 0.3 in/yr. The parameters C and D refer to the percent dry mass and percentage of carbon in the dry mass of the tree, respectively. The parameters A and   are empirically determined parameters. N refers to the number of trees on the property (USFS).

- *Turf and Soil.* Sophisticated models exist for modeling turfgrass systems, which can determine the effect of many different variables on the ability of the turfgrass to sequester carbon. The preliminary results in this work use a highly simplified model for the base case scenario. From a study by Qian et al. (2003), it was found that under a clippings returned management scenario and a fertilization rate of 150 kg/(ha-yr), carbon is sequestered at a rate of 620 kg/(ha-yr) for the first 50 years after establishment and a rate of 350 kg/(ha-yr) from 50 years to 100 years.

Figure 4 shows the net carbon "footprint" of the residential TEco-Net. The discontinuity in this figure is due to the piecewise nature of the lawn sequestration model used. This calculation assumes a lawn size of 0.2 acres (0.08 ha) and that there are four trees on the property. The figure indicates that the carbon emitted from the technological subsystem cannot be canceled by the carbon sequestered by the ecological subsystem. Adding in the upstream effects will probably make the net carbon footprint increase even more since many quantities are neglected in this base case.

Figure 5 shows the individual contributions to the net carbon footprint at the year at which maximum carbon is sequestered in the turfgrass and trees. This calculation also assumes the trees are sugar maples. This figure shows the sources on the positive y-axis and the sinks on the negative y-axis. It is clear from this figure that the sequestration of carbon by trees and lawn are dwarfed by emissions from fossil fuel combustion. The emissions from gasoline consumption for lawn maintenance is, however, canceled by the sequestration of carbon through the lawn and trees. This calculation assumes a fixed lawn size and number of trees; these variables can be optimized to determine the lawn area and number of trees required to sequester the

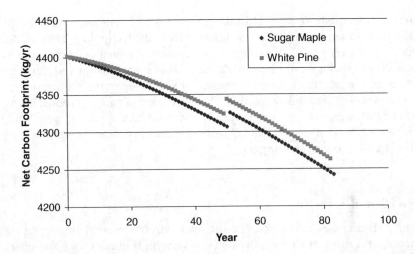

FIGURE 4
Net carbon footprint.

amount of carbon emitted from the technological subsystem. This is shown in equation 1, where N is the number of trees and M is the lawn area, assuming that the maximum tree/area ratio is 20 $\frac{\text{trees}}{\text{ha}}$.

$$N \cdot 25 \left(\frac{\text{kg CO2}}{\text{tree}} \right) + M \cdot 610 \left(\frac{\text{kg CO2}}{\text{ha turf}} \right)$$

$$= M \cdot 216 + 4436 \text{ (kg CO2)}$$

(1)

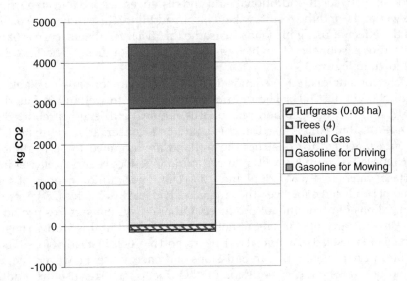

FIGURE 5
Individual carbon contribution.

Solving this equation, it would take 5 hectares of lawn area and 100 sugar maple trees to sequester the carbon emitted from the technological subsystem. This may not be practically feasible in a residential system, but it nonetheless indicates that it would take much more than average lawn and treescapes to counter the average emissions created from the operation of a residential system. Another possibility would be the purchase of carbon credits. Using a commercial carbon offset company (TP), it would cost approximately $200 to purchase carbon offsets for the net carbon emissions created over a one year period in the preceding example.

Corn Ethanol TEco-Net

Several studies have assessed the life cycle environmental impact of corn ethanol, and most of them show its relatively high impact and low energy return on investment (Hill et al., 2006). Nevertheless, with depletion of fossil fuels and concerns about climate change, biomass based fuels and products are likely to become increasingly attractive. Therefore, this case study takes a constructive approach and focuses on ways of enhancing the sustainability of corn ethanol by designing a TEco-Net. Detailed data underlying this study are available in Baral and Bakshi (2008). One of the large environmental impacts of corn ethanol is on the nitrogen cycle due to run-off of fertilizers and pesticides. These pollute water bodies and create a hypoxic or "dead zone" in the Gulf of Mexico. This study considers the use of engineered wetlands to reduce this impact, and provides insight into the trade-off between the additional land and resources needed for establishing the wetlands, and the resulting benefits. In addition, this study also considers the effect of using biosolids instead of artificial fertilizers, corn ethanol instead of gasoline, and soybean biodiesel instead of diesel. The TEco-Net is shown in Figure 3.

Data for engineered wetlands to treat farm run off are available in Millhollon et al. (2004). The ratio of cropland area to wetland is found to be 42, and the nitrogen and phosphorus removal efficiency is reported to be 40–90%, but 50% is used in this study as a conservative estimate. The resources needed to construct the wetland are assumed to be negligible. With these assumptions, wetlands are found to reduce N and P releases into surface waters by 9.6 MT of N and 0.2 MT of P per million gallons of corn ethanol produced. However, they increase land use by 62 acres (2%) per million gallons of corn ethanol produced. Other ecosystem services provided by wetlands are substantial but not yet incorporated in this study. The use of biosolids instead of artificial fertilizers, and biofuels instead of fossil fuels result in further reduction in emissions and energy use. However, the use of biofuels does increase emission of N_2O due to fertilizer use for producing biomass. Using biofuels instead of fossil fuels for full fuel substitution

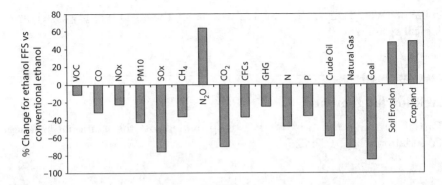

FIGURE 6
Percentage change in emissions and resource use for wetlands with full fuel substitution versus conventional corn ethanol. Note that the increase in land area and soil erosion is mainly due to the use of biofuels instead of fossil fuels. Increase in land use due to wetlands is only 2%.

also increases soil erosion and land use. The percentage change in the flow of selected resources and emissions for the full fuel substitution with wetlands scenario versus the conventional corn ethanol scenario is shown in Figure 6.

This study demonstrates the potential benefits of developing a TEco-Net for corn ethanol. Designing the TEco-Net requires consideration of other ecological and technological processes for dealing with products such as organic wastes and other pollutants, and is the subject of on-going research. These include alternatives for handling byproducts such as distillers dried grain with solubles (DDGS), and different agricultural practices.

Conclusions

This article suggests that for transforming engineering activities toward greater sustainability it is essential to design integrated networks of technological-ecological systems. Such networks can explicitly consider the essential role that ecosystems play in supporting and sustaining all activities, and ensure their availability. The role of this natural capital is being ignored in most existing sustainable engineering and related methods. Modeling the TEco-Net can also help in identifying specific types of natural capital that may be most stressed by the selected technological processes, and ways of reducing this stress may be designed. The TEco-Net design problem can be very complicated due to the use of models across disciplines and the highly interconnected nature of of these networks. As discussed in this article, process synthesis and design methods are promising starting points for developing the necessary approach. The proposed ideas present exciting opportunities for greater synergy between engineering and ecology and for

collaborative research that can truly enhance the sustainability of technological systems.

Acknowledgements

Partial funding for this work was provided by the National Science
Foundation (CBET-0829026).

References

Azapagic, A., Millington, A., and Collett, A. (2006). A methodology for integrating
    sustainability considerations into process design. *Chemical Engineering Research
    & Design*, 84(A6):439–452.
Bakshi, B. R. and Fiksel, J. (2003). The quest for sustainability: Challenges for process
    systems engineering. *AIChE Journal*, 49(6):1350–1358.
Baral, A. and Bakshi, B. R. (2008). Hybrid LCA of cornethanol, gasoline, soybean
    biodiesel, and diesel: Accounting for natural capital and net energy analysis.
    Technical report, The Ohio State University.
Bare, J. C. and Gloria, T. P. (2006). Critical analysis of the mathematical relationships and
    comprehensiveness of life cycle impact assessment approaches. *Environmental
    Science & Technology*, 40(4):1104–1113.
Biegler, L. T., Grossmann, I. E., and Westerberg, A. W. (1997). *Systematic Methods of
    Chemical Process Design*. Prentice Hall.
CENTURY (accessed November 11, 2008). CENTURY Soil Organic Matter Model,
    Version 5. www.nrel. colostate.edu/projects/century5/.
Chertow, M. R. (2007). "uncovering" industrial symbiosis. *J. Industrial Ecology*, 11(1):
    11–30.
der Ryn, S. V. and Cowan, S. (1996). *Ecological Design*. Island Press, Washthington, DC.
Douglas, J. M. (1988). *Conceptual Design of Chemical Processes*. McGraw-Hill.
EIA. Official Energy Statistics from the U.S. Government. http://www.eia. doe . gov/,
    accessed July 23, 2008.
El-Halwagi, M. M. (2006). *Process Integration*. Academic Press.
Fu, Y., Diwekar, U. M., Young, D., and Cabezas, H. (2000). Process design for the
    environment: A multi-objective framework under uncertainty. *Clean Products
    and Processes*, 2:92–107.
Graedel, T. E. and Allenby, B. R. (2003). *Industrial ecology*. Prentice Hall.
GREET. The Greenhouse Gases, Regulated Emissions, and Energy Use in Transportation
    (GREET) Model, argonne national laboratory. www.transportation.
    anl. gov/modeling_s imulation/GREET, accessed July 23, 2008.
Hill, J., Nelson, E., Tilman, D., Polasky, S., and Tiffany, D. (2006). Environmental,
    economic, and energetic costs and benefits of biodiesel and ethanol biofuels.
    Proceedings of the National Academy of Sciences, 103(30):11206–11210.

i-Tree. i-Tree: Tools for assessing and managing community forests. http: / /www. itreetools.org.

MA (2005). *2005 Millennium Ecosystem Assessment (MEA) , Ecosystems and Human Wellbeing: Synthesis,*. Island Press. www.maweb.org, Accessed Jan 12, 2008.

Millhollon, E. P., Raab, J. L., Anderson, R., and Lis-cano, J. (2004). Constructed wetlands for improving water quality of agricultural runoff in northwest louisiana. www.usawaterquality.org/conferences/2004/posters/mill-hollonLA.pdf.

Mitsch, W. J. and Jørgensen, S. E. (2004). *Ecological Engineering and Ecosystem Restoration.* John Wiley & Sons, Inc.

Nemerow, N. L. (1995). *Zero pollution for industry.* Wiley-Interscience.

Nowak, D., Crane, D., and Stevens, J. (2006). Air pollution removal by urban trees and shrubs in the United States. *Urban Forestry & Urban Greening,* 4(3-4):115–123.

Pistikopoulos, E. N. (1999). Design and operations of sustainable and environmentally benign processes. *Computers and Chemical Engineering,* 23:1363.

Qian, Y., Bandaranayake, W., Parton, W., Mecham, B., Hari-vandi, M., and Mosier, A. (2003). Long-Term Effects of Clipping and Nitrogen Management in Turfgrass on Soil Organic Carbon and Nitrogen Dynamics The CENTURY Model Simulation. *Journal of Environmental Quality,* 32(5):1694–1700.

Realff, M. J. (2006). Environmentally benign design and manufacturing (ebdm): Future directions presentation. In
The NSF Symposium on Environmentally Benign Design and Manufacturing for Sustainable Economic Competitiveness, St. Louis, MO.

Restrepo, J., Montoya, A., and Obeysekera, J. (1998). A
Wetland Simulation Module for the MODFLOW Ground Water Model. *Ground Water,* 36(5):764–770.

Sahu, R. Technical Assessment of the Carbon Sequestration Potential of Managed Turfgrass in the United States. Technical report. Prepared on behalf of the Outdoor Power Equipment Institute (OPEI).

Schulze, P., editor (1996). *Engineering Within Ecological Constraints.* National Academies Press.

Sikdar, S. K. (2003). Sustainable development and sustain-ability metrics. *AIChE Journal,* 49(8):1928–1932.

Singh, M. (2007). *Soil Organic Carbon Pools in Turfgrass Systems ofOhio.* The Ohio State University.

Spreng, D. T. (1988). *Net-EnergyAnalysis.* Praeger.

Tilley, D. R. (2003). Industrial ecology and ecological engineering: Opportunities for symbiosis. *Journal ofIndus-trial Ecology,* 7(2):13–32.

TP. www.terrapass.com. accessed Jan4, 2009.

Ukidwe, N. U. and Bakshi, B. R. (2007). Industrial and
ecological cumulative exergy consumption of the united states via the 1997 input-output benchmark model. *Energy,* 32(9):1560–1592.

USFS. The Urban Forest Effects (UFORE) Model,
United States Forest Service. www.fs.fed.us/ne/ syracuse/Tools/tools.htm, accessed December 2, 2008.

Verhoeven, J. and Meuleman, A. (1999). Wetlands for wastewater treatment: Opportunities and limitations. *Ecological Engineering,* 12(1-2):5–12.

Zhang, Y., Baral, A., and Bakshi, B. R. (2008). Toward accounting for natural capital in life cycle assessment. Technical report, The Ohio State University.

13

Advanced Co-Simulation for Computer-Aided Process Design and Optimization of Fossil Energy Systems with Carbon Capture

Stephen E. Zitney

Collaboratory for Process & Dynamic Systems Research
U.S. Department of Energy, National Energy Technology Laboratory
Morgantown, WV 26507-0880

CONTENTS

ABSTRACT   In this paper, we describe recent progress toward developing an Advanced Process Engineering Co-Simulator (APECS) for use in computer-aided design and optimization of fossil energy systems with carbon capture. The APECS system combines process simulation with multiphysics-based equipment simulations, such as those based on computational fluid dynamics. These co-simulation capabilities enable design engineers to optimize overall process performance with respect to complex thermal and fluid flow phenomena arising in key plant equipment items. This paper also highlights ongoing co-simulation R&D activities in areas such as reduced order modeling, knowledge management, stochastic analysis and optimization, and virtual plant co-simulation. Continued progress in co-simulation technology—through improved integration, solution, deployment, and analysis—will have profound positive impacts on the design and optimization of high-efficiency, near-zero emission fossil energy systems.

KEYWORDS   *Process simulation, computational fluid dynamics, co-simulation, virtual engineering, fossil energy*

Introduction

Meeting the increasing demand for clean and affordable energy is arguably the most important challenge facing the world today. Fossil fuels currently provide nearly 75 percent of the world's energy supply and contribute more than 85 percent to the U.S. energy market (DOE/EIA, 2008a,b). Projections indicate that fossil fuels, especially coal, will continue to be an important part of the energy portfolio in the U.S. Coal is widely available at sufficiently low prices and currently generates more than half of the nation's electric power (MIT, 2007). With an estimated 250-year supply at current U.S. consumption levels, coal is a potential source of energy not only for meeting future electric power needs, but also for the polygeneration of chemicals, liquid fuels, substitute natural gas, and hydrogen (DOE/FE/NETL, 2008).

The use of coal shifts the challenge from one of energy supply to that of preventing adverse impact on the environment. While the fossil energy industry continues to improve the environmental performance of conventional pulverized coal combustion power plants, advanced technologies such as coal gasification, under development by the Department of Energy (DOE), offer the potential to produce significantly lower quantities of air pollutants such as fine particulates, SO_x, NO_x, and mercury (DOE/FE, 2006). An even greater challenge for coal in the future is the reduction of greenhouse gas emissions. Future power plants based on advanced technologies will have the ability to capture and sequester most of the potential carbon dioxide emissions (DOE/FE/NETL, 2007).

Developing environmentally friendly and affordable technologies for using coal in future power plants is a major research focus at the DOE's National Energy Technology Laboratory (NETL, 2007a). NETL is collaborating closely with industry to address the grand challenge of reducing the time, cost, and technical risk of designing the next-generation of advanced energy systems with carbon capture and storage. Such systems include coal-fed oxy-combustion plants, integrated gasification combined cycle (IGCC) plants, chemical looping systems, and polygeneration plants. These emerging systems must be designed with unprecedented efficiency and near-zero emissions, while optimizing profitably amid cost fluctuations for raw materials, finished products, and energy. To help achieve the aggressive goals for energy and the environment, NETL and its research partners in industry and academia are relying increasingly on the use of sophisticated computer-aided process design and optimization tools, which have become major enabling technologies for complex process systems engineering and analysis (Zitney, 2007a).

One advanced enabling technology of particular interest is *co-simulation* which involves the integration of multiple component simulation tools for the purpose of overall system design and optimization. The independently simulated components are described using domain-specific models with appropriate fidelity. Furthermore, co-simulation enables the reuse and combination of already existing and validated component simulations without re-entering model data. Such co-simulation technology has been used to address complex integrated applications in a wide variety of engineering fields including fluid-structure interaction (Sett, 2006), automotive engine design (Flowmaster, 2007), petroleum reservoir characterization (Mata-Lima, 2006), continuous/discrete systems (Nicolescu et al., 2006), field-circuit design (Zhou et al., 2005), and building design and evaluation (Wang and Wong, 2009). Recent research efforts in the area of chemical process and energy process co-simulation have been described by Bezzo et al. (2000), Aumiller et al. (2002), Mota et al. (2004), and Scheffknecht et al. (2007).

In this paper, we describe a process/equipment co-simulation software framework and its application to advanced fossil energy systems with carbon capture. The co-simulation framework, known as *APECS (Advanced Process Engineering Co-Simulator)*, combines steady-state process simulation with one or more equipment items simulated by a different simulation tool running simultaneously and exchanging information (e.g., Zitney and Syamlal, 2002). APECS provides capabilities to design and optimize overall plant performance with respect to complex thermal and fluid flow phenomena in equipment items described by high-fidelity computational fluid dynamics (CFD) models (Sloan et al., 2005; Zitney et al., 2007c).

We will also highlight ongoing co-simulation R&D in the following areas: reduced-order modeling (ROM) based on CFD results; data, model, and knowledge management; stochastic analysis and optimization; and virtual engineering concepts. Continued R&D aimed at these and other advanced process design and optimization technologies will accelerate the development of fossil energy systems with desired environmental outcomes for the U.S. and the world.

Advanced Process Engineering Co-Simulator

Developed since the year 2000 by NETL, ANSYS, and other research partners, the Advanced Process Engineering Co-Simulator (APECS) is an innovative software tool that provides process/equipment co-simulation capabilities for model-based decision support in process design and optimization (e.g., Zitney et al., 2006; Schowalter et al., 2007).

As shown in Figure 1, the APECS software is built on the process industry *CAPE-OPEN (CO)* standard which provides plug-and-play model interoperability (e.g., Syamlal et al., 2004; Zitney, 2005, 2006; Swensen et al., 2007). Using APECS, design engineers can integrate, solve, and analyze co-simulations combining CO-compliant equipment simulations with CO-compliant process simulations (e.g., Aspen Plus®, COCO, gPROMS®, HYSYS®, and PRO/II®) (Zitney, 2007b, c). APECS provides easy-to-use configuration wizards for use

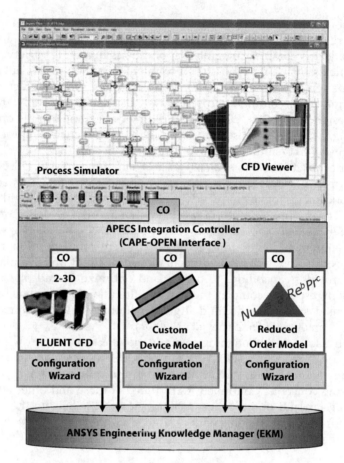

FIGURE 1
APECS software architecture.

with a wide variety of CO-compliant equipment models including computational fluid dynamics (CFD) models, custom engineering models (CEMs), and reduced-order models (ROMs).

CFD models provide a detailed and accurate representation of complex thermal and fluid flow phenomena occurring within a wide variety of power plant equipment items, such as combustors, gasifiers, syngas coolers, carbon capture devices, steam and gas turbines, heat recovery steam generators, cooling towers, and fuel cells. The commercial CFD code, FLUENT®, is available for use in APECS as a CO-compliant unit operation model (Zitney and Syamlal, 2002). APECS also provides a FLUENT-specific configuration wizard for enabling the user to specify boundary conditions and CFD model parameters to make available in the process simulator as CAPE-OPEN ports and parameters. Examples of common equipment parameters include the current and voltage for a fuel cell (Zitney et al., 2004), or the angle of injection for the coal slurry in a gasifier (Zitney, 2007d). The CAPE-OPEN COM/ CORBA bridge implementation in APECS allows process simulations running under the Windows operating system to use CFD-based equipment simulations running locally or remotely and in serial or parallel under a different operating system such as Linux (Osawe, 2005).

CEMs are typically engineering models that calculate mass and energy balances, phase and chemical equilibrium, and reaction kinetics. Custom models can also be in-house computer codes based on empirical data obtained from many years of experience in designing and operating certain equipment items. Such models are typically very fast and accurate within a narrow parameter range. A pulverized coal boiler design code from ALSTOM Power was made available as a CO-compliant unit operation model by using the CO-wrapper template for CEMs provided in APECS (Sloan et al., 2002).

ROMs are a class of equipment models that can be based on pre-computed CFD solutions over a range of parameter values, but are much faster than CFD models. The APECS system currently provides for automatically generating and using CO-compliant ROMs based on (piecewise) multiple linear regression and artificial neural networks (ANN). Syamlal and Osawe (2004) demonstrated the use of the linear regression ROM to predict performance of a continuously stirred tank reactor (CSTR) with respect to impeller speed. The fast ROM accurately approximated the high-fidelity CFD results for a two-dimensional CSTR model in the co-simulation and optimization of a reaction-separation-recycle flowsheet (Zitney and Syamlal, 2002). The ANN-based ROM in APECS is built on the well-known logistic activation function and a hybrid of simulated annealing and conjugate gradient training algorithms. Osawe et al. (2006) described the ANN implementation, training, and application to a CFD model of heat recovery steam generator (HRSG) used in APECS co-simulations of a natural gas-fired combined cycle power plant and a coal-fired integrated gasification combined cycle system with carbon capture and hydrogen production.

In recent research, Lang et al. (2007) proposed a ROM strategy for APECS based on principal component analysis (PCA). In a gas turbine combustor case study, a significant CPU time saving is observed with reasonable accuracy when comparing the PCA-based ROM to a rigorous CFD model. The errors in the combustor outlet temperatures are less than 3.0% for all 45 random test points in the input space. By using a neural network as the non-linear regression tool, the PCA-based ROM is well-matched at the designed points. However, it is important to overcome the overfitting during training of the neural network and to reduce interpolation errors at unknown points. The issue of trading off overfitting and underfitting; that is, improving the performance of interpolation at points away from the designed ones, is an area of ongoing research.

Future ROM research for APECS will also focus on network-of-zones or multizonal approaches (e.g., Bezzo et al., 2004; Chilka and Orsino, 2007), as well as equation-driven approaches such as proper orthogonal decomposition

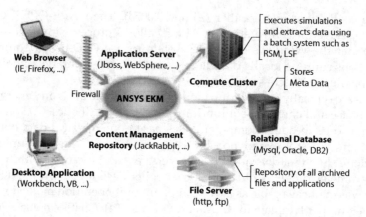

FIGURE 2
ANSYS engineering knowledge manager™ (EKM™) plug-and-play environment.

(e.g., Cizmas et al., 2003). Additional improvements in efficient space-filling experimental designs are also needed to minimize the number of CFD simulations required for ROt2M development. In addition, a systematic procedure is needed for the use of ROMs in advanced process/equipment optimization applications. With the development of error bounds, optimization strategies can be applied that remain in a trust region for a given ROM. With the evolution of the optimization, the ROM can be updated while the trust region is adjusted so that a high-fidelity optimization can still be performed.

In APECS, the CO-compliant CFD models, CEMs, and ROMs are stored in the *ANSYS® Engineering Knowledge Manager™ (EKM™)*. With EKM, APECS users can manage CO-compliant equipment simulation data/model archiving and retrieval processes through the use of a single server-based, enterprise-wide repository (Widmann et al., 2008). As shown in Figure 2, EKM is a design and simulation framework, which is aimed at hosting all simulation data, processes, and tools while maintaining a tight connection between them.

EKM is based on modern open standards which can accommodate a wide range of enterprise environments. More details on EKM software architecture, functionality, capabilities, and features can be found in the EKM brochure and technical specifications sheet (ANSYS Inc., 2007, 2008).

APECS offers the co-simulation user the ability to define flexible solution strategies consisting of a combination of one or more models stored in EKM. For example, one common solution strategy is to use a fast ROM for the initial flowsheet iterations and use a high-fidelity CFD model for the final iterations. In this way, a process engineer can customize solution strategies from a hierarchy of models, thereby achieving the desired trade-off between speed and accuracy. If a parallelized solver is available for a given equipment model (e.g., FLUENT), improved performance can be achieved by using multiple processes that may be executed on the same computer, or on different computers in a network (Zitney, 2004). An APECS user can specify the number of processors to be used, message passing protocol, and hosts file containing the list of computers on which to run the parallel job. By providing solutions on both ends of the performance spectrum, including parallel execution of the CFD models on high-performance computers and the use of fast ROMs based on CFD results, APECS addresses the performance issue that equipment simulations based on high-fidelity CFD models require much more computational time than process simulations based on simplified models, especially for cases in which one or more CFD models are embedded in the iterative flowsheet solution process.

The APECS system provides a variety of analysis tools for optimizing overall plant performance with respect to mixing and fluid flow behavior. For process optimization under uncertainty and in the face of multiple, and sometimes conflicting, objectives, APECS offers stochastic modeling and multi-objective optimization capabilities (Subramanyan et al., 2005; Shastri et al., 2007). Shastri et al. (2008) demonstrated the usefulness of these CO-compliant capabilities on an integrated gasification combined cycle with a single-stage, entrained-flow gasifier and carbon capture.

For post-processing and visualization, APECS provides a CFD viewer to display, within the process simulator, the results of a CFD simulation conducted as a part of a co-simulation. Typical CFD results include 2D contours of velocity, temperature, pressure, and species mass fractions for a specified surface in the equipment item. The ParaView scientific visualization tool (Squillacote, 2008) is available in APECS for viewing 3D CFD results.

Together with researchers from Iowa State University and Ames Laboratory, NETL is working toward the integration of the APECS system with the immersive and interactive virtual engineering software, VE-Suite (McCorkle et al., 2007; Zitney et al, 2007). The VE-Suite open-source library of tools provides a 3D virtual plant walkthrough environment for running APECS co-simulations and analyzing the coupled process simulation and CFD equipment results (see Figure 3). This integration represents a necessary step in the development and deployment of virtual power plant co-simulations.

FIGURE 3
APECS/VE-Suite integration.

APECS Applications

In the power industry, design engineers are employing APECS to design and optimize commercial-scale power plants, including conventional pulverized coal-fired (PC), oxy-fired PC systems, natural gas-fired combined cycles (NGCC), and advanced gasification-based systems. Oxy-fired and gasification-based systems are adaptable for use with technologies for capture and sequestration of carbon emissions from fossil fuels.

ALSTOM Power Inc., through its U.S. Power Plant Laboratories in Windsor, Connecticut, have developed APECS co-simulations for a conventional 30 MWe PC steam plant for municipal electricity generation and an advanced 250 MW, NGCC power plant. In the PC co-simulation shown in Figure 4, an Aspen Plus process design specification is used to adjust a FLUENT CFD model parameter for the boiler damper position (bypass resistance) to maintain a specified steam temperature over a range of loads, from the load at the maximum continuous rating to a control load, below which the boiler cannot sustain the required turbine inlet temperatures (Sloan et al., 2004, 2007).

For the NGCC co-simulation shown in Figure 5, ALSTOM is using an Aspen Plus process design specification to manipulate designated control parameters for the FLUENT CFD model of the heat recovery steam generator (HRSG) so that a specified superheat steam temperature is maintained for various load points over the range from 100% to 50% gas turbine load (Sloan et al., 2005).

ALSTOM extended the FLUENT CFD model of the HRSG for use in an APECS co-simulation of an IGCC power and hydrogen co-production plant with carbon capture (Sloan and Fiveland, 2005). The physical model port concept in APECS allows a coolant stream from the Aspen Plus cycle to connect

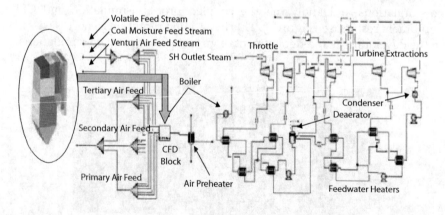

FIGURE 4

APECS process/CFD Co-Simulation application for an ALSTOM conventional steam plant (30MWe) with 3D CFD boiler.

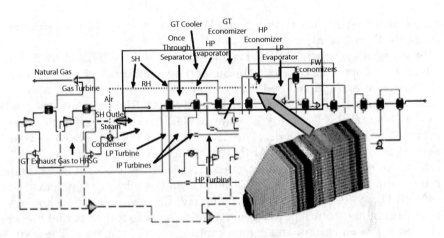

FIGURE 5
APECS application for an ALSTOM NGCC (250MWe) with 3D CFD HRSG.

directly with one of several tube banks configured by the pseudo 1-D heat exchanger model in FLUENT (see Figure 6). In that manner, a CFD heat exchanger model that computes both the hot gas-side and the coolant-side of the heat exchanger can be linked directly with the corresponding streams on the flowsheet. An efficient methodology was implemented to solve the heat exchanger blocks associated with the HRSG island on the flowsheet and couple those results interactively with the HRSG CFD simulation. By using the CO-compliant COM-CORBA bridge implementation in APECS, the HRSG demonstration case was successfully run over a local area network

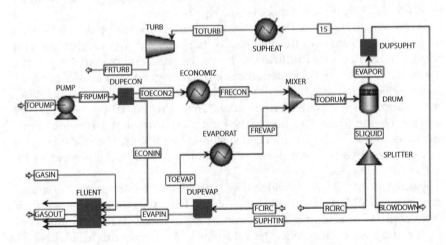

FIGURE 6
HRSG portion of IGCC power and hydrogen co-production plant with FLUENT CFD model of HRSG.

(LAN) with Aspen Plus running on a Windows PC and FLUENT running on a LINUX platform.

In an ongoing project, ALSTOM is collaborating with NETL and ANSYS to advance and apply APECS to chemical looping combustion (CLC) and oxy-fired circulating fluidized bed (CFB) applications which are well suited to carbon capture (DOE/FE, 2008; Andrus et al., 2007). The project is designed to demonstrate the feasibility of using APECS to develop efficient, time-averaged, reduced-order models (ROMs) for high-fidelity CFD models of transient, dense multiphase flow. At the completion of this project, the APECS toolkit will be capable of combining dense-phase, gas-solids riser ROMs with process simulations of commercial-scale CLC and oxy-fired CFB systems.

At NETL, system analysts are applying APECS co-simulation technology to emerging and future power generation systems that incorporate higher efficiency, low emissions, and carbon capture and sequestration. These systems range from multi-kilowatt fuel cell applications to commercial-scale power plants generating many hundreds of megawatts of electricity.

Through DOE's 10-year multimillion dollar Solid State Energy Conversion Alliance (SECA) program (Surdoval, 2008), one ready-market opportunity for APECS is the co-simulation of fuel cell-based auxiliary power units (APU) for transportation applications. With fuel-to-electricity conversion efficiencies approaching 50%, fuel cell APUs can dramatically reduce diesel fuel consumption, cost, and pollutant emissions for the idling of heavy-duty truck engines. Using APECS, Zitney et al. (2004) showed that the overall performance of solid oxide fuel cell (SOFC) auxiliary power units (APUs) modeled using Aspen Plus can be optimized with respect to the local fluid flow, heat and mass transfer, electrochemical reactions, current transport, and potential field in the SOFC simulated using detailed, three-dimensional, steady-state FLUENT CFD models (see Figure 7).

As shown in Figure 8, the APECS co-simulations are performed over a range of fuel cell currents to generate a voltage-current curve and analyze the effect of current on fuel utilization, power density, and overall system efficiency. The fuel cell APU system considered here generated 4.3 kW of power and yielded a maximum fuel-to-electricity conversion efficiency of 45.4% at a current of 18 amperes (Zitney et al., 2004). Process/CFD co-simulations provide a better understanding of the fluid mechanics that drive overall performance and efficiency of fuel cell systems. In addition, the analysis of the fuel cell using CFD is not done in isolation but within the context of the whole APU process.

Syamlal et al. (2003) considered a natural gas based proton exchange membrane (PEM) fuel cell power system. The Aspen Plus process flow-sheet consists of a reformer, shift converter, fuel cell, anode exhaust combustor and heat exchangers. The reformer is modeled with a FLUENT 3D CFD model and is heated with hot gases from the anode exhaust burner. The APECS co-simulation enables the CFD analyst to easily account for the effect of the recycled hot gases in the CFD model. The CFD model predicts

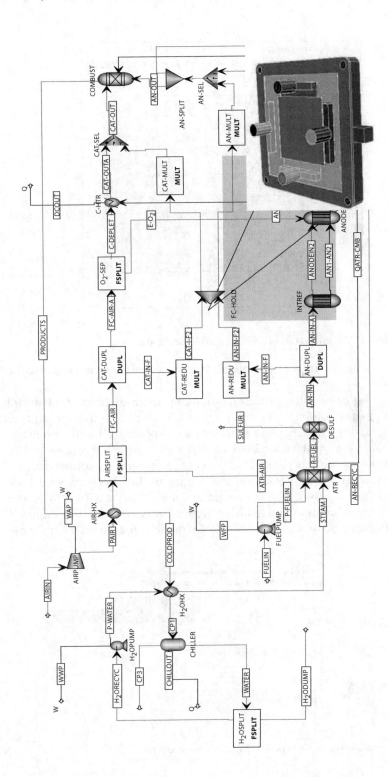

FIGURE 7
APECS process/CFD optimization of a SECA fuel cell APU application.

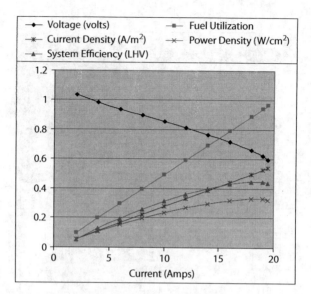

FIGURE 8
APECS process/CFD optimization of a SECA fuel cell APU application.

conversions that account for the limitations imposed by the heat transfer to the catalyst bed.

NETL is also developing process/equipment co-simulations of advanced coal-fired, gasification-based electricity plants with carbon dioxide capture and storage Using APECS, plant performance is optimized with respect to coupled fluid flow, heat and mass transfer, and chemical reactions in key equipment items such as gasifiers, syngas coolers, carbon capture devices, steam and gas turbines, HRSGs, and cooling towers. In a recent paper, Zitney et al. (2006) describe an APECS co-simulation of a 250-MWe IGCC power and hydrogen co-production plant with carbon capture. As shown in Figure 9, the co-simulation couples FLUENT CFD models for an entrained-flow

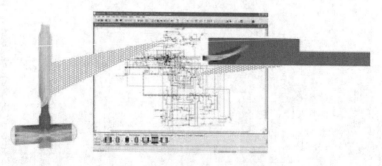

FIGURE 9
APECS co-simulation of an IGCC power/hydrogen co-production.

gasifier and a gas turbine combustor with the Aspen Plus steady-state process simulation of the co-production plant.

Zitney et al. (2006) analyze gasifier performance where fluid dynamics strongly affect synthesis gas quality and carbon conversion, and gas turbine performance where the blending of air and fuel is at the heart of power generation, efficiency, and environmental compliance. APECS automatically executes the gasifier and turbine combustor CFD models as needed to converge the tail gas recycle loop and a design specification on the gas turbine inlet temperature. The design specification is met by manipulating the synthesis gas split between power production and hydrogen production. For the process design case, the APECS results show that the target for turbine inlet temperature is met when 43% of the syngas is sent to the gas turbine combustor and the remainder goes to the pressure swing adsorption unit for hydrogen production. The net equivalent power output from the plant is 243.8 MW, corresponding to an HHV thermal efficiency of 53%. These results illustrate how APECS is helping NETL system analysts to better understand and analyze the complex thermal and fluid flow phenomena that impact overall power plant performance.

Under the auspices of a U.S.-U.K. Memorandum of Understanding for Fossil Energy Research and Technology Development, NETL participated in a three-year collaboration on power plant co-simulation with a project team supported by the Department for Business, Enterprise & Regulatory Reform in the United Kingdom (NETL, 2007b). In the U.K. Virtual Plant Demonstration Model (VPDM) program, ALSTOM Power Ltd, Engineous Software Ltd, ANSYS/Fluent Europe Ltd, RWE npower plc, Doosan Babcock Energy Ltd, Process Systems Enterprise Ltd (PSE), and the University of Ulster leveraged APECS and the CAPE-OPEN software standard to integrate high-fidelity FLUENT CFD equipment models into overall power plant models developed with PSE's gPROMS® simulator Zitney, 2007d). The APECS technology was demonstrated by co-simulating a conventional coal-fired power station (Patel and Wang, 2006; Wang and Patel, 2006). Using APECS, the VPDM team coupled a detailed CFD furnace simulation with a steady-state process simulation of the power plant. The co-simulation enabled U.K. process engineers to analyze and optimize overall performance of the power station with respect to the thermal and fluid dynamics behavior occurring in the furnace.

Conclusions

The fossil energy industry operates some of the most complex and expensive plants in the world, spending millions of dollars annually in plant innovation, design, operation, and management. This vital industry also faces the enormous challenge of designing next-generation plants to operate with increased efficiency and near-zero emissions, while ensuring profitability amid changes in environmental regulations and fluctuations in the cost of raw materials, finished products, and energy. Continued research and development on advanced co-simulation technology will help this industry to address the grand challenge of designing next-generation plants with higher efficiencies and improved environmental performance. Future APECS co-simulation R&D will focus on carbon management by integrating power plant simulations with CO_2 pipeline simulation and reservoir simulation for carbon storage.

References

Andrus, H.E., Beal, C., Brautsch, A. (2007). Alstom's Chemical Looping Coal-Fired Power Plant Development Program. *Proc. of the 32nd International Technical Conference on Coal Utilization & Fuel Systems*, June 10–15, Clearwater, FL.

ANSYS Inc. (2007). ANSYS Engineering Knowledge Manager. Canonsburg, PA.

ANSYS Inc. (2008). ANSYS EKM 1.2 Software Technical Specifications. Canonsburg, PA.

Aumiller, D.L., Tomlinson, E.T., Weaver, W.L. (2002). An Integrated RELAP5-3D and Multiphase CFD Code System Utilizing a Semi-Implicit Coupling Technique. *Nuclear Engineering and Design, 216*, 77–87.

Bezzo, F., Macchietto S., Pantelides, C.C. (2000). A General Framework for the Integration of Computational Fluid Dynamics and Process Simulation. *Computers and Chemical Engineering, 24*, 653-658.

Bezzo, F., Macchietto S., Pantelides, C.C. (2004). A General Methodology for Hybrid Multizonal/CFD Models: Part I. Theoretical Framework. *Computers and Chemical Engineering, 28*, 501–511.

Chilka, A., Orsino, S. (2007). Multi-zonal Model for Accurate and Faster Combustion Simulations. *Proc. of the 32nd International Technical Conference on Coal Utilization & Fuel Systems*, June 10-15, Clearwater, FL.

Cizmas, P.G., Palacios, A., O'Brien, T.O., Syamlal, M. (2003). Proper-Orthogonal Decomposition of Spatio-temporal Patterns in Fluidized Beds. *Chemical Engineering Science, 58*, 4417–4427.

Department of Energy, Energy Information Administration (2008a). *International Energy Outlook 2008*. DOE/EIA-0484.

Department of Energy, Energy Information Administration (2008b). *Annual Energy Outlook 2008*. DOE/EIA-0383.

Department of Energy, Office of Fossil Energy (2008). *Project Fact Sheet: Process/ Equipment Co-Simulation of Oxy-Combustion and Chemical Looping Combustion*. DE-NT0005395, DOE/FE.

Department of Energy, Office of Fossil Energy (2006). *Office of Clean Coal: Strategic Plan*. DOE/FE.

Department of Energy, Office of Fossil Energy, National Energy Technology Laboratory (2007). *Carbon Sequestration Technology Roadmap and Program Plan: Ensuring the Future of Fossil Energy Systems through the Successful Deployment of Carbon Capture and Storage Technologies*. DOE/FE/NETL.

Department of Energy, Office of Fossil Energy, National Energy Technology Laboratory (2008). *Hydrogen from Coal Program: Research, Development, and Demonstration Plan for the Period 2008 through 2016*. DOE/FE/NETL.

Flowmaster Ltd. (2007). BMW Motoren, Steyr couples 3D CFD with 1D Flowmaster. www.flowmaster.com.

Lang, Y., Biegler, L.T., Munteanu, S., Madsen, J.I., Zitney, S.E. (2007). Advanced Process Engineering Co-Simulation using CFD-based Reduced Order Models. Presented at the *AIChE 2007 Annual Meeting*, 4th Annual U.S. CAPE-OPEN Meeting, November 4–9, Salt Lake City, UT.

Massachusetts Institute of Technology (MIT) (2007). *The Future of Coal: Options for a Carbon-Constrained World*. MIT, Cambridge, MA.

Mata-Lima, H. (2006). Reservoir Characterization with Iterative Direct Sequential Co-Simulation: Integrating Fluid Dynamic Data into Stochastic Model. *Journal of Petroleum Science and Engineering*, 62(3–4), 59–72.

McCorkle, D., Yang, C., Jordan, T., Swensen, D., Zitney, S.E., Bryden, M. (2007). Towards the Integration of APECS with VE-Suite to Create a Comprehensive Virtual Engineering Environment. *Proc. of the 32nd International Technical Conference on Coal Utilization & Fuel Systems*, June 10-15, Clearwater, FL.

Mota, J.P.B., Esteves, I.A.A.C., Rostam-Abadi, M. (2004). Dynamic modeling of an adsorption storage tank using a hybrid approach combining computational fluid dynamics and process simulation. *Computers and Chemical Engineering*, 28, 2421–2431.

National Energy Technology Laboratory (2007a). *2007 NETL Accomplishments*.

National Energy Technology Laboratory (2007b). *U.S-U.K. Collaboration on Virtual Plant Simulation*. NETL Fact Sheet, October.

Nicolescu, G., Bouchhima, F., Gheorghe, L. (2006). Codis — A Framework for Continuous/Discrete Systems Co-Simulation. *Proc. of the 2nd IFAC Conference on Analysis and Design of Hybrid Systems*, June 7–9, Alghero, Italy, 274–275.

Osawe, M.O. (2005). Fluent CAPE-OPEN COM/CORBA Bridge and CO-Compliant Unit Operation. Presented at the 2nd Annual U.S. CAPE-OPEN Meeting, May 25–26, Morgantown, WV.

Osawe, M.O., Sloan, D.G., Fiveland, W.A., Madsen, J.I. (2006). Fast Co-Simulation of Advanced Power Plants Using Neural Network Component Models. *Proc. of the AIChE 2006 Annual Meeting*, 3rd Annual U.S. CAPE-OPEN Meeting, November 12–17, San Francisco, CA.

Patel, V.C., Wang, M. (2006). The Development of a Virtual Plant Demonstration Model, *18th International Conference on System Engineering*, Coventry, UK.

Scheffknecht, G., Hetzer, J., Dieter, H., Leiser, S., Schnell, U. (2007). Coupled Simulation of the Combustion and Steam Generation Process in Large Utility Boilers. *Proc. of the 32nd International Technical Conference on Coal Utilization & Fuel Systems*, June 10–15, Clearwater, FL.

Schowalter, D.G., Madsen, J.I., Collins, R.L. (2007). Power Plant Simulation for the 21st Century. *Proc. of the 32nd International Technical Conference on Coal Utilization & Fuel Systems*, June 10–15, Clearwater, FL.

Sett, S. (2006) FSI Controls Flow Rate. *FLUENT News*, Spring, pp. 11–12.

Shastri, Y., Diwekar, U., Zitney, S.E. (2007). CAPE-OPEN Compliant Stochastic Modeling Capability. Presented at the *AIChE 2007 Annual Meeting*, 4th Annual U.S. CAPE-OPEN Meeting, November 4–9, Salt Lake City, UT.

Shastri, Y., Diwekar, U., Zitney, S.E. (2008). Uncertainty Analysis of an IGCC System with Single-Stage Entrained-Flow Gasifier. Presented at *AIChE 2008 Annual Meeting*, November 16–21, Philadelphia, PA.

Sloan, D.G., Fiveland, W. (2005). Development of Technologies and Analytical Capabilities for Vision 21 Energy Plants: Description of Demonstration Case 3 Deliverable. DOE/NETL Topical Report.

Sloan, D.G., Fiveland, W., Zitney, S.E., Osawe, M.O. (2007). Plant Design: Integrating plant and equipment models. *Power Magazine*, 151(8).

Sloan, D.G., Fiveland, W., Zitney, S.E., Syamlal, M. (2002). Software Integration for Power Plant Simulations. *Proc. of the 27th International Technical Conference on Coal Utilization & Fuel Systems*, March 4–7, Clearwater, FL.

Sloan, D.G., Fiveland, W., Zitney, S.E., Syamlal, M. (2004). Power Plant Simulations Using Process Analysis Software Linked to Advanced Modules. *Proc. of the 29th International Technical Conference on Coal Utilization & Fuel Systems*, April 18–22, Clearwater, FL.

Sloan, D.G., Fiveland, W., Zitney, S.E., Syamlal, M. (2005). Demonstrations of Coupled Cycle Analyses and CFD Simulations over a LAN. *Proc. of the 30th International Technical Conference on Coal Utilization & Fuel Systems*, April 17–21, Clearwater, FL.

Squillacote, A.H. (2008). The ParaView Guide. *Kitware, Inc.*, Clifton Park, NY.

Subramanyan, K., Xu, W., and Diwekar, U., (2005). Stochastic Modeling and Multi-objective Optimization for APECS System. DOE/NETL Final Report, December.

Surdoval, W.A. (2008). DOE's SECA Program: Progress & Plans. *Proc. of the 25th Annual International Pittsburgh Coal Conference*, September 29–October 2, Pittsburgh, PA (2008).

Swensen, D.A., Zitney, S.E., Bockelie, M. (2007). Development of Cape-Open Unit Operations for Advanced Power Systems Modeling. Presented at the *AIChE 2007 Annual Meeting*, 4th Annual U.S. CAPE-OPEN Meeting, November 4–9, Salt Lake City, UT.

Syamlal, M., Madsen, J.I., Rogers, W.A., Zitney, S.E. (2003). Application of an Integrated Process Simulation and CFD Environment to Model Fuel Cell Systems. Presented at the *AIChE 2003 Spring Meeting*, March 30 – April 3, New Orleans, LA.

Syamlal, M., Osawe, M.O. (2004). Reduced Order Models for CFD Models Integrated with Process Simulation. *Proc. of the AIChE 2004 Annual Meeting*, November 7–12, Austin, TX.

Syamlal, M., Zitney, S.E., Osawe, M.O. (2004). Using CAPE-OPEN Interfaces to Integrate Process Simulation and CFD. *CAPE-OPEN Update Newsletter*, Volume 7.

Wang, M. and Patel, V.C. (2006). VPDM: Enabling Framework for Secure Distributed Whole Power Plant Modelling, *UK Automatic Control International Conference (UKACC)*, Glasgow, UK.

Wang, L., Wong, N.H. (2009). Coupled Simulations for Naturally Ventilated Rooms between Building Simulation (BS) and Computational Fluid Dynamics (CFD) for Better Prediction of Indoor Thermal Environment. *Building and Environment*. 44(1), 95–112.

Widmann, J., Munteanu, S., Madsen, J.I., Zitney, S.E. (2008). Coupling Computational Fluid Dynamics with Process Modeling for Improved Plant Design. Presented at *2008 AspenTech User Conference*, April 7-11, Houston, TX.

Zhou, P., Lin, D. Fu, W.N., Ionescu, B., Cendes, Z.J. (2005). A General Co-Simulation Approach for Coupled Field–Circuit Problems. *Proc. of COMPUMAG Conference*, Shenyang, China, June.

Zitney, S.E. (2004) High-Performance Process Simulation of Advanced Power Generation Systems. Presented at *SMART TechTrends 2004*, August 3–6, Pittsburgh, PA.

Zitney, S.E. (2005). CAPE-OPEN Integration for Advanced Process Engineering Co-Simulation. Presented at the *2nd Annual U.S. CAPE-OPEN Meeting*, May 25–26, Morgantown, WV.

Zitney, S.E. (2006). CAPE-OPEN Integration for CFD and Process Co-Simulation. *Proc. of the AIChE 2006 Annual Meeting*, 3rd Annual U.S. CAPE-OPEN Meeting, November 12–17, San Francisco, CA.

Zitney, S.E. (2007a). Computational Research Challenges and Opportunities for the Optimization of Fossil Energy Power Generation Systems. *Proc. of the 32nd International Technical Conference on Coal Utilization & Fuel Systems*, June 10–15, Clearwater, FL.

Zitney, S.E. (2007b). Use of CAPE-OPEN Standard in US-UK Collaboration on Virtual Plant Simulation. Presented at the *AIChE 2007 Annual Meeting*, 4th Annual U.S. CAPE-OPEN Meeting, November 4–9, Salt Lake City, UT.

Zitney, S.E. (2007c). Using Process/CFD Co-Simulation for the Design and Analysis of Advanced Energy Systems. Presented at 2007 Aspen Engineering Suite (AES) User Group Meeting, April 10-11, Houston, TX.

Zitney, S.E. (2007d). Virtual Engineering for Power Plant Design. Presented at *Virtual Environments, Virtual Worlds, and Applications*, November 1, Ames, IA.

Zitney, S.E., McCorkle, D., Yang, C., Jordan, T., Swensen, D., Bryden, M. (2007). Towards the Integration of APECS and VE-Suite for Virtual Power Plant Co-Simulation. Presented at *Virtual Engineering 2007*, May 1–2, Ames, IA.

Zitney, S.E., Osawe, M.O., Collins, L., Ferguson, E., Sloan, D.G., Fiveland, W.A., Madsen, J.M. (2006). Advanced Process Co-Simulation of the FutureGen Power Plant. *Proc. of the 31st International Technical Conference on Coal Utilization & Fuel Systems*, May 21–25, Clearwater, FL.

Zitney, S.E., Prinkey, M.T., Shahnam, M., Rogers, W.A. (2004). Coupled CFD and Process Simulation of a Fuel Cell Auxiliary Power Unit. Presented at the *2nd International Conference on Fuel Cell Science, Engineering and Technology*, June 14–16, Rochester, NY.

Zitney, S.E., Syamlal, M. (2002). Integrated Process Simulation and CFD for Improved Process Engineering. *Proc. of the European Symposium on Computer Aided Process Engineering –12, ESCAPE-12*, J. Grievink and J. van Schijndel, Eds., May 26–29, The Hague, The Netherlands, 2002, pp. 397–402.

14

A MINLP-Based Revamp Strategy for Improving the Operational Flexibility of Water Networks

Bao-Hong Li[1] and Chuei-Tin Chang[*,2]

[1]*Department of Chemical Engineering, Dalian Nationalities University*
Dalian 116600, People's Republic of China
[2]*Department of Chemical Engineering, National Cheng Kung University*
Tainan, Taiwan 70101, Republic of China

CONTENTS

ABSTRACT   In designing and operating any water network, a typical issue that must be addressed is concerned with the uncertain process conditions. A systematic procedure is proposed in this paper to enhance their operational flexibility. Specifically, two design options have been investigated, i.e., (1) relaxation of the allowed maximum freshwater consumption rate in the nominal design and (2) installation of auxiliary pipelines and/or elimination of existing ones. The flexibility index model proposed by Swaney and Grossmann (1985) has been modified and formulated in a generalized format to evaluate the impacts of introducing various modifications into a given network. Since this model is a Mixed Integer NonLinear Program (MINLP),

---

[*] To whom all correspondence should be addressed

it is in general very difficult to reach the global optima in the iterative solution processes. A simple and effective initialization procedure has also been devised in this work to facilitate successful convergence. Finally, the effectiveness of the proposed approach is demonstrated with an example in this paper.

KEYWORDS *Water network, flexibility index, mathematical programming model, initialization strategy*

Introduction

In recent years, the important issues of freshwater conservation and wastewater reduction have drawn increasing attention in the process industries (Dunn and Wenzel, 2001). Process integration techniques have often been adopted to realize the reuse, regeneration-reuse and regeneration-recycling of process and utility waters in chemical plants (Wang and Smith, 1994). In implementing these design procedures, it is usually assumed that the process data are fixed and well-defined. However, the actual operating conditions of the freshwater supplier and the basic processing units in a water network, e.g., the mass loads of water users, the removal ratios of treatment operations, and the upper limits of contaminant concentrations at the inlet and outlet of each unit, may fluctuate over time. Such fluctuations could lead to deterioration in the water qualities of effluents and even operation disruption if the water network design is not flexible enough to cope with these uncertain disturbances (Tan et al, 2007).

It is thus widely recognized that there is a need to develop systematic techniques to improve the operational flexibility of one or more given network obtained on the basis of economic criteria only. To this end, two design options have been thoroughly studied in the present work, i.e., (1) relaxation of the upper limit of freshwater supply rate and (2) installation of auxiliary pipelines and/or elimination of original ones. The flexibility index model proposed by Swaney and Grossmann (1985) has been modified and formulated in a generalized format to evaluate the impacts of introducing various combinations of these additional features into an existing network. The uncertain disturbances considered are those in the freshwater quality, the mass loads of water-using units, the removal ratios of wastewater treatment units, and the maximum inlet and outlet concentrations of these two types of units. The control variables used for compensating disturbances are assumed to be the freshwater consumption rate and the flow ratios associated with the outward branches connected to every splitter in the network.

Since the flexibility index model is a complex mixed integer nonlinear program (MINLP), the global optimum cannot always be obtained in the iterative solution process. A good initial guess is often needed to facilitate the search process. To satisfy this need, a simple and effective initialization procedure has also been developed in this work to systematically solve the MINLP models within GAMS environment (Brooke et al, 2005).

The remaining paper is organized as follows. A concise problem statement is first provided in the next section. Next the detailed formulation of the generalized flexibility index model is given. The implementation procedures for incorporating the aforementioned design options in a given network are then presented. To demonstrate the effectiveness of the proposed approach, the numerical results obtained in a simple example are described. Finally, conclusions are outlined in the last section.

Problem Statement

As mentioned previously, the primary objective of this work is to develop a set of systematic methods to assess and then enhance the operational resiliency of one or more given nominal water-network design. A nominal design should include specifications of the freshwater consumption rate, the effluent flow rates, all unit throughputs, the network configuration and the water flow rate in every branch, and the contaminant concentrations at the sinks and the inlets and outlets of all water-using and wastewater-treatment units.

Generally speaking, the flexibility level in a given process is dependent upon the maximum range of variation in each uncertain parameter that the plant can tolerate. The so-called *flexibility index* $\delta (\geq 0)$ is a measure of the largest size of feasible operation region in the space of uncertain parameters θ. More specifically, this parameter space Θ can be expressed as

$$\Theta(\delta) = \{\theta | \theta^N - \delta \Delta \theta^- \leq \theta \leq \theta^N + \delta \Delta \theta^+\} \tag{1}$$

where, θ^N is a vector of parameter values from which the nominal water-network design is obtained, and $\Delta \theta^+$ and $\Delta \theta^-$ denote the expected deviations of uncertain parameters from their nominal values in the positive and negative directions respectively. In this study, the uncertain parameters are assumed to be: the freshwater quality, the mass loads of water-using units, the removal ratios of wastewater-treatment units, and the maximum inlet and outlet concentrations of these two types of units.

The flexibility index model developed by Swaney and Grossmann (1985) has been adopted to determine a quantitative measure for use as the selection criterion of additional features to be introduced into the nominal design.

It is our intention to answer the following two important questions with this model:

(1) Is the given nominal design flexible enough?
(2) If not, can it be improved by relaxing the upper limit of freshwater supply rate and/or by modifying the network structure of the given design?

Generalized Flexibility Index Model

Since it is very tedious and inefficient to construct different versions of the flexibility index model for various candidate network configurations and then carry out the needed optimization computations, a generalized model has been formulated and used in this work as a design tool for all possible alternative structures.

Superstructure

In order to develop the general model, it is necessary to first build a superstructure in which all possible flow connections are embedded. The superstructure presented here is essentially a modified version of that suggested by Chang and Li (2005). In its original form, a distinct label is assigned to each of the given water-using unit, wastewater-treatment unit, water source and sink, i.e., U, T, W and D respectively. This scheme can be represented by Figure 1, in which the symbols S and M denote the splitting and mixing node respectively.

Complete Model Formulation

Due to length limit of this paper, the complete model formulation is not provided here and interested reader can find it elsewhere (Chang et al.,

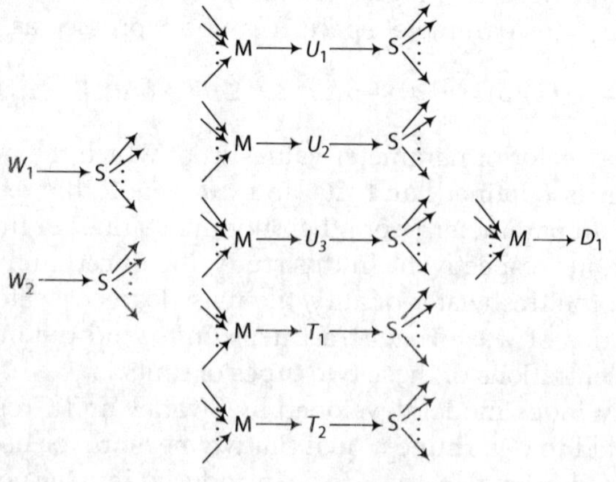

FIGURE 1
Superstructure of water network.

2009). Notice that the resulting GAMS code for any given problem should be the same for all possible revamp designs. It is only necessary to change the parameter values in each optimization run.

Initialization Procedure

In this work, the GAMS modules DICOPT and BARON have been used to solve the aforementioned flexibility index models. Since the proposed model is written in a general code, only those constraints to define the existing network must be revised and then introduced in every optimization run. In addition, it has been well recognized that the starting point usually exerts a profound influence on the convergence process. A reliable initialization procedure has thus been developed to facilitate effective solution. The detailed procedure is not given here because of the same reason as before and interested reader can refer to Chang et al. (2009). Instead, the main steps are summarized below:

Step 1: Impose the additional structure constraints according to the given network configuration.

Step 2: Set the initial guesses of the flow rates in the existing branches to be their nominal flow rates in the given network.

Step 3: Compute the initial guesses of throughputs in water-using units and wastewater-treatment units by substituting the initial guesses of branch flow rates obtained in Step 2 into flow balance equations.

Step 4: Set the initial values of the flexibility index and all uncertain multipliers to be 1.

Step 5: Determine the initial values of the inlet concentrations of processing units and sinks according to their nominal values in the given network design.

Step 6: Estimate the initial guesses of slack variables and initialize the binary variable according to the definition of active constraint.

Step 7: Generate the initial guesses of Lagrange multipliers for the inequality constraints by two steps.

Step 8: The upper bound of flexibility index δ is estimated.

A Simple Example

Let us consider the revamp designs of an existing water network shown in Figure 2, which consists of a single water source and two water-using units. The design specifications and nominal operating conditions of the

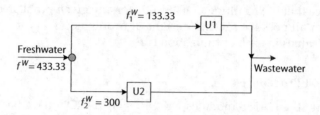

FIGURE 2
Nominal water network.

water-using units in the original design are provided in Tables 1 and 2 respectively. Notice that only one key contaminant is adopted in the network design. The maximum allowable freshwater supply rate is 433.33 tonne/hr (which is the minimum freshwater usage when the opportunities of wastewater reuse is ignored) and the nominal contaminant concentration in freshwater is 20 ppm. Notice that a splitter is located at the source and it is marked by a small circle in Figure 2. The split ratios of its two branch streams can be adjusted to compensate external disturbances during operation. It is assumed in this example that the maximum inlet and outlet contaminant concentrations of unit 1, i.e., C_1^{in} and C_1^{out}, and the maximum contaminant concentration at the outlet of unit 2, i.e., C_2^{out}, may vary with the ambient temperature. The corresponding uncertain multipliers are referred to as θ_1, θ_2 and θ_3 respectively. It is further assumed that

$$\Delta\theta_1^- = \Delta\theta_2^- = \Delta\theta_3^- = 0.04$$

$$\Delta\theta_1^+ = \Delta\theta_2^+ = \Delta\theta_3^+ = 0.05$$

which mean that the lower and upper bounds of above three uncertain parameters are 96% and 105% of their nominal values respectively.

Notice that $\theta_1^N = \theta_2^N = \theta_3^N = 1$ because they are the multipliers of their nominal values. A total of 19 variables (including 3 binary variables which are corresponding to three maximum inlet or outlet concentration constraints) are needed to formulate the corresponding flexibility index model. This model was easily solved with GAMS modules (Version 22.4) on a HP Compaq DC7700 convertible minitower. The default NLP solver in GAMS is CONOPT3 and MIP solver is CPLEX, while DICOPT and BARON are both adopted to solve MINLP for comparison and validation purposes. The

TABLE 1

Design Specifications of Water-using Units

Units	C^{in} (ppm)	C^{out} (ppm)	Mass Load (Kg/hr)	Limiting F. (ton/hr)
u1	70	170	20	200
u2	20	120	30	300

TABLE 2

Nominal Operating Conditions of Water-using Units

Units	c^{in} (ppm)	c^{out} (ppm)	Mass Load (Kg/hr)	Flowrate (ton/hr)
u1	20	170	20	133.33
u2	20	120	30	300

optimization computation converged within 1 second and nearly no initialization steps were required for this simple problem. It can be found from the optimal solution that the flexibility index in this case is *zero*, which is clearly undesirable. The following steps were then taken to evaluate the benefits of introducing additional design changes:

Attach Auxiliary Pipeline(s)

An additional pipeline from $u2$ to $u1$ was first added to the original network (see Figure 3), while the upper limit of freshwater usage was still kept unchanged. As a result, the total number of variables in the corresponding model becomes 20. It was found that the flexibility index could be increased to 1.6026 with the aforementioned auxiliary pipeline and the corresponding optimum solution is presented in Table 3. Since $\delta > 1$, it is clear that the improved network design is operable or has enough flexibility to counteract all possible disturbances by adjusting the control variables. From the above optimum solution, it can also be observed that the most constrained point in design is located at where every uncertain multiplier reaches its lower bound, i.e., 0.936. This finding is of course consistent with our intuitive prediction.

Other possible auxiliary pipelines, such as those from $u1$ to $u2$ and from water source to sink, were excluded from consideration. Notice that the former design change obviously violates the optimality conditions of water network design (Savelski and Bagajewicz, 2000). More specifically, the outlet contaminant concentration of unit $u1$ should reach its upper bound (when the total freshwater usage of the network is to be minimized) which is larger than that of unit $u2$, and the water reuse stream from $u1$ to $u2$ can

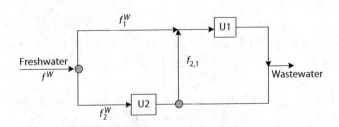

FIGURE 3
Improved water network.

TABLE 3

Obtained Optimum Solution of Water-using Units

Units	c^{in} (ppm)	c^{out} (ppm)	Mass Load (Kg/hr)	Flowrate (ton/hr)
u1	65.51	159.10	20	213.70
u2	20	112.31	30	325.00

only increase the total freshwater usage of the network. On the other hand, since the freshwater cannot be directed to the water-treatment units with a source-to-sink pipeline and no upper limit is imposed upon the contaminant concentration in effluent, it is clearly unnecessary to consider the latter structural modification.

Relax Upper Bound of Freshwater Usage

Based on the improved configuration obtained in previous step, the generalized flexibility index model was also solved repeatedly for different levels of maximum freshwater supply rate. From the corresponding results presented in Figure 4, it is obvious that the flexibility index of this water network increases almost linearly with the supply limit of freshwater. Furthermore, 420 tonne/hr is the minimum freshwater capacity for ensuring adequate operational flexibility under the influences of anticipated uncertain disturbances.

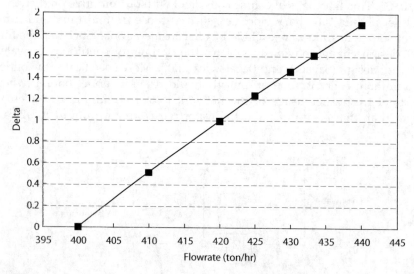

FIGURE 4

Impact of increasing freshwater supply capacity on flexibility index.

Conclusions

A MINLP-based design procedure is developed in this work for assessing and improving the operational flexibility of existing water networks under uncertain disturbances. Specifically, a generalized flexibility index model and its initialization procedure are proposed for evaluating the benefits of introducing additional design options, i.e, relaxation of the upper limit of freshwater supply rate and incorporation of structural modifications. It has been shown in the case studies that this approach is feasible and efficient.

Acknowledgments

Financial support provided by National Natural Science Foundation of China under Grant NO. 20806015 is gratefully acknowledged by the first author.

References

Brooke, A., Kendrik, D., Meeraus, A., Ramam, R. (2005). GAMS: A User Guide. GAMS Development Corp.: Washington, DC,

Chang, C. T., Li, B. H. (2005). Improved optimization strategies for generating practical water-usage and -treatment network structures. *Ind. Eng. Chem. Res.*, *44*, 3607.

Chang, C. T., Li, B. H., Liou, C. W. (2009). Development of a generalized MINLP model for assessing and improving the operational flexibility of water network designs. *Ind. Eng. Chem. Res.* 48, 3496.

Dunn, D. F., Wenzel, H. (2001). Process integration design methods for water conservation and wastewater reduction in industry. Part I: Dessign for single contaminant. *Clean. Prod. Processes*, *3*, 307.

Savelski, M. J., Bagajewicz, M. J. (2000). On the Optimality Conditions of Water Utilization Systems in Process Plants with Single Contaminants. *Chem. Eng. Sci.*, *55*, 5035.

Swaney, R. E., Grossmann, I. E. (1985). An Index for Operational Flexibility in Chemical Process Design Part I: Formulation and Theory. *AIChE J.*, *31*, 621.

Tan, R. R., Foo, D. C. Y., Manan, Z. A. (2007). Assessing the sensitivity of water networks to noisy mass loads using Monte Carlo simulation. *Comput. Chem. Eng.*, *31*, 1355.

Wang, Y. P.; Smith, R. (1994). Wastewater minimization. *Chem. Eng. Sci.*, 49, 981.

15

A Simultaneous Approach for Batch Water-allocation Network Design

Li-Juan Li[1], Hong-Guang Dong[1*], Rui-Jie Zhou[2] and Wu Xiao[1]

[1]*School of Chemical Engineering, Dalian University of Technology, Dalian, 116012, PRC*
[2]*Department of Economics, Dalian University of Technology, Dalian, 116024, PRC*

CONTENTS

ABSTRACT The mathematical technique presented in this work deals with one step design of batch water-allocation network, where batch production schedules, water-reuse subsystems, and wastewater treatment subsystems are all taken into account. The proposed model formulation is believed to be superior to the available ones in the following aspects. In the first place, a continuous based time model is introduced and superstructures incorporating State-Task Network (STN), State-Equipment Network (SEN) and Gantt chart are adopted to capture all basic information of batch production and water-allocation network. Specifically, our superstructure is able to generate a class of optimum network structures which can never be captured by all previous frameworks. Then, costs of junctions and pipelines are added to evaluate and reduce the complexity of the network configuration. Finally, a hybrid optimization strategy integrating DICOPT and GA is developed for the resulting mixed-integer nonlinear programming (MINLP) model and one example is presented to demonstrate the advantages of the proposed approach.

KEYWORDS *Batch water-allocation network, superstructures, network complexity*

---

* To whom all correspondence should be addressed

Introduction

In the past, the tasks of optimizing batch schedules, water-reuse and waste-water treatment subsystems were performed individually. Cheng and Chang (2007) first developed an effective procedure to incorporate these three components into a single comprehensive model. However, in their study, the discrete-time model embedded in their model is rigid and unrealistic. In addition, it is very difficult to incorporate all possible network configurations and the relationship between operations and equipments, which is a key feature of time-variant batch process, were not illustrated in their superstructure. Finally, no effective algorithm was adopted and even good local optimum cannot be guaranteed. In this paper, to overcome the aforementioned shortcomings, we have developed a simultaneous optimization model characterized by the following features. First, a continuous-time based time model is introduced and two novel superstructures are designed for the ambiguity-free representation of batch production and water-allocation network. Specifically, these superstructures are designed not only to incorporate all possible elements (states, tasks, equipments and time) in one general conceptual framework, but also designed to capture all the structural characteristics of the integrated water-allocation network. Then, a hybrid optimization method combining DICOPT and GA is addressed to guarantee the global optimum with a much higher probability. Finally, costs of splitters and mixers are first considered in our work, and their impacts on the network configurations are explored.

Superstructures

In terms of the General framework for process synthesis, the basic elements were identified as state, task and equipment. The relationship that each of these elements can have, leads to two representation approaches: the STN (Kondili et al., 1993), and the SEN (Smith, 1996).

Before introducing our superstructure for batch water-allocation network, the concept of multistage mixing/splitting will be discussed at first. Suppose Unit 1, Unit 2 and Unit 3 are three water users operating in the same time interval and their inlet flow conditions are assumed as follows (see Figure 1 left). When multistage mixing/splitting options are introduced, the inlet flow configuration can be revamped by allowing stream to mix and split on a sequential basis, as shown in the right part of Figure 1. Such flow policy, which has never been revealed by previous works, is termed as multistage mixing/splitting.

Our design specifications for our superstructures are presented as follows (see Figure 2 and 3). In Figure 2, firstly, each of the material state has been divided into two parts: material states to be consumed (on the top row) and

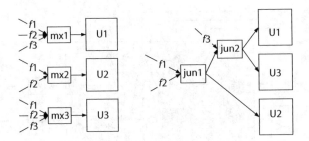

FIGURE 1
Illustration of multistage mixing/splitting.

produced (at the bottom row). Then, each equipment is illustrated as a block with many sub-blocks which correspond to operations performed at certain event points. Finally, a reference axis of time is set and the starting and ending points of all operations can be clearly recognized.

The improved state-space superstructure (Bagajewicz and Manousiouthakis, 1992) is viewed as three inter-connected blocks (see Figure 3). The OP block has been divided into two parts: units in water network and junctions used to mix and/or split all streams in the system. Like Figure 2, each unit is illustrated as a block with many sub-blocks which correspond to operations performed at certain event points. As for junctions, the optimal number and the corresponding assignment of units to junctions are not fixed a priori, thus being subjected to optimization process. It is obvious that in this framework, the same junction can be shared by different units in the whole time horizon and multi-level mixing among streams can also be realized. Another marked feature of our superstructures is that the concept of

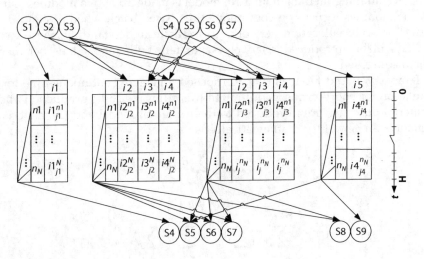

FIGURE 2
Superstructure for batch production.

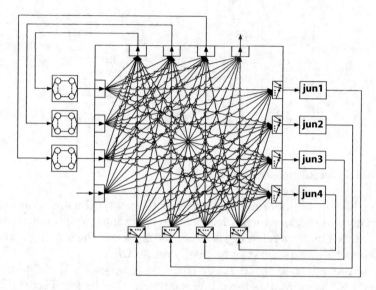

FIGURE 3
Superstructure for water-allocation network.

STN and SEN has also been incorporated. Detailed explanations cannot be provided due to the length of this paper.

Mathematical Model

The overall integrated mathematical model is made up of two modules. One of the modules focuses on continuous-time based batch schedules and the other on water-allocation network. To facilitate understanding, these two parts will be introduced relatively separately with interactions between them addressed.

Ierapetritou and Floudas (1998) proposed a general continuous-time formulation for short-term scheduling. In this study, periodic production and material purchase are introduced to their original model and the material balance can be written mathematically as:

$$ST_{s,n} = ST_s^{in} - \sum_{i \in I_s} \rho_{s,i}^c \sum_{j \in J_i} B_{i,j,n} + r_{s,n} - d_{s,n}$$

$$\forall s \in S, n \in N(n = n_1) \tag{1}$$

$$ST_{s,n} = ST_{s,n-1} - \sum_{i \in I_s} \rho_{s,i}^c \sum_{j \in J_i} B_{i,j,n} + \sum_{i \in I_s} \rho_{s,i}^p \sum_{j \in J_i} B_{i,j,n} + r_{s,n} - d_{s,n}$$

$$\forall s \in S, n \in N(n > n_1) \tag{2}$$

$$ST_{s,N} = ST_s^{in} \qquad \forall s \in S \tag{3}$$

As other constraints are essentially the same as the original study, they are omitted altogether due to the length of the paper.

On the other hand, the mathematical model for water-allocation network mainly involves the following constraints:

(1) Mass and flow balances of operations

$$f_{u,t}^{in} = \sum_{u' \in U} fs_{u',u,t} \qquad \forall u \in U, t \in T \tag{4}$$

$$f_{u,t}^{in} \cdot c_{u,k,t}^{in} = \sum_{u' \in U} fs_{u',u,t} \cdot cs_{u',k,t} + \sum_{jun \in JUN} fs_{jun,u,t}$$

$$\forall u \in U, k \in K, t \in T \tag{5}$$

$$f_{u,t}^{out} = \sum_{u' \in U} fs_{u,u',t} + \sum_{jun \in JUN} fs_{u,jun,t} \qquad \forall u \in U, t \in T \tag{6}$$

$$c_{u,k,t}^{out} = cs_{u,k,t} \qquad \forall u \in U, k \in K, t \in T \tag{7}$$

(2) Connection relationship between operations and equipments

$$ne(e,e') \leq \sum_{u_e \in U_e} \sum_{u_{e'} \in U_{e'}} \sum_{t \in T} nfs(u_e, u_{e'}, t) \leq N^{\max} \cdot ne(e,e')$$

$$\forall e \in E, e' \in E \tag{8}$$

$$ne(eq, jun) \leq \sum_{u_e \in U_e} \sum_{t \in T} nfs(u_e, jun, t)$$

$$+ \sum_{t \in T} nfs(jun', jun, t) \leq N^{\max} \cdot ne(eq, jun)$$

$$\forall e \in E, eq \in E \cup JUN, jun' \in JUN, jun \in JUN, t \in T \tag{9}$$

$$ne(jun, eq) \leq \sum_{u_e \in U_e} \sum_{t \in T} nfs(jun, u_e, t)$$

$$+ \sum_{t \in T} nfs(jun, jun', t) \leq N^{\max} \cdot ne(jun, eq)$$

$$\forall e \in E, eq \in E \cup JUN, jun' \in JUN, jun \in JUN, t \in T \tag{10}$$

where Eq.(8) is used to identify the connection between units; while Eqs.(9) and (10) are enforced to specify the mixing and/or splitting function of each junction.

(3) Logic constraints

$$\frac{f_{u,t}^{in}}{F_{u,t}^{in,max}} \le \sum_{jun\in JUN} nfs_{jun,u,t} + \sum_{u'\in U} nfs_{u',u,t} < \frac{f_{u,t}^{in}}{F_{u,t}^{in,max}} + 1$$

$$\forall u \in U, t \in T \tag{11}$$

$$\frac{f_{u,t}^{out}}{F_{u,t}^{out,max}} \le \sum_{u'\in U} nfs_{u,u',t} + \sum_{jun\in JUN} nfs_{u,jun,t} < \frac{f_{u,t}^{out}}{F_{u,t}^{out,max}} + 1$$

$$\forall u \in U, t \in T \tag{12}$$

$$N^{min} \cdot w(jun) \le \sum_{jun\in JUN} nfs_{jun,u,t} + \sum_{jun'\in JUN} nfs_{jun,jun',t}$$

$$+ \sum_{jun\in JUN} nfs_{u,jun,t} + \sum_{jun'\in JUN} nfs_{jun',jun,t} < N^{max} \cdot w(jun)$$

$$\forall jun \in JUN, t \in T \tag{13}$$

where Eqs.(11) and (12) enforce that only one stream is allowed to enter and/or discharge from each unit in certain time interval; Eq.(13) is imposed to specify if certain junction is selected in the optimal network configuration.

The criterion used in this part of study is the minimizing of cost, which can be expressed as follows:

$$Obj = (total\ income - purchasing\ cost)$$

$$- (cost\ of\ fresh\ water + cost\ of\ water\ treatment$$

$$+ cost\ of\ buffer\ tanks + cost\ of\ junctions) \tag{14}$$

Interactive Solution Strategy

The proposed solution procedure can be broadly divided into two stages. The first stage is designed to get a set of feasible solutions while providing the candidate search region for stage two; and the second stage is focused on refined search. In stage two, GA is introduced to provide an evolutionary region of initial values which facilitate DICOPT solver to identify the true global optimal. Since both parallel and adaptive techniques are incorporated in GA, the searching efficiency of our algorithm can be greatly improved.

Application Example

The example presented below was originally solved simultaneously based on discrete-time representation and the modified MUS (mixer-unit-splitter) superstructure (Papalexaddri and Pistikopoulos, 1994) by Cheng and Chang (2007). All the data used can be found in the original work.

On the basis of a time of horizon of 4 hr, the most appropriate schedules, as well as network configurations for both cases are presented in Figures 4 to 7. In Case I, although the schedule scheme (see Figure 4) is almost the same as the former study, the cost of network is found to be 310.32 units, which represent a 18.1% reduction due to the improvement of network configuration. It can be observed that the network in Figure 5 is assembled with 4 junctions and 24 pipelines. In Case II, when network complexity is taken into account, the optimal production schedule (see Figure 6) is slightly different from the former case. This is due to the fact that trade-offs between schedule and syntheses are properly balanced. On the other hand, Figure 7

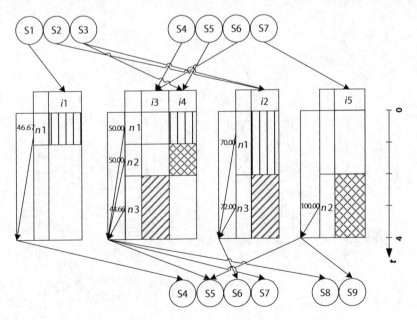

FIGURE 4
Optimal schedule for Case I.

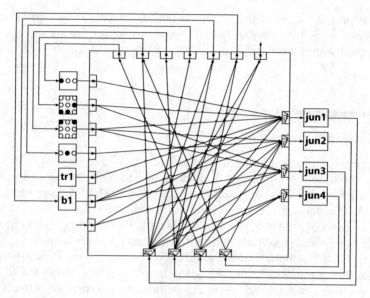

FIGURE 5
Optimal network configuration for Case I.

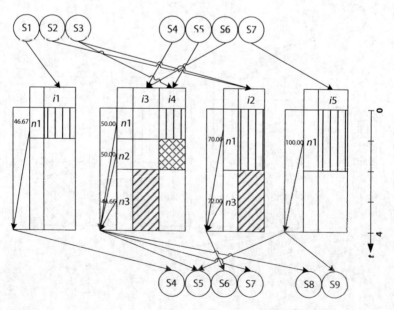

FIGURE 6
Optimal schedule for Case II.

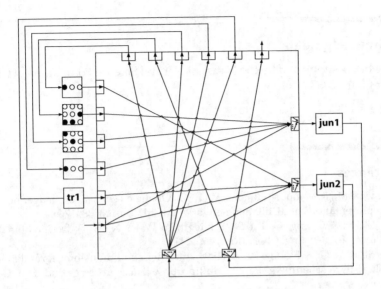

FIGURE 7
Optimal network configuration for Case II.

shows the topology of the real network system. It is apparent that the complexity of this network is significantly reduced and only 2 junctions and 13 pipelines are needed. It is worthy of note, that the stream from jun1 to jun2 in Figure 7 is a kind of multi-level mixing and such common situation in process industries can only be obtained with our proposed superstructure. It should also be mentioned that buffer tank is not selected in the optimal network configuration in Case II and this can explain that buffer tanks are not necessary in some batch process if optimal operating procedures and policies are adopted.

Conclusion

Two novel superstructures have been modified in this study to formulate a MINLP model for batch water-allocation system design. The advantage of this approach is that not only all possible water flow options, but batch concept as well, can be easily incorporated. In addition, the complexity of network can always be reduced by introducing the cost of junctions, and example provided proves that it is also an effective way for network designs and complexity evaluation. Finally, by interactively applying our proposed solution techniques, the global optimum can almost always be identified in all case studies presented in this paper.

Acknowledgments

This work is supported by the National Natural Science Foundation of China under Grant 20876020.

References

Bagajewicz, M., & Manousiouthakis, V. (1992). On the mass/heat exchanger network representations of distillation networks. AIChE Journal, 38, 1769.

Cheng, K. F., & Chang, C. T. (2007). Integrated Water Network Designs for Batch Processes. Ind. Eng. Chem. Res., 46, 1241.

Ierapetritou, M. G., & Floudas, C. A. (1998). Effective continuous-time formulation for short-term scheduling. Part 1. Multipurpose batch processes. Ind. Eng. Chem. Res., 37, 4341.

Kondili, E., Pantelides, C. C., & Sargent, R. W. H. (1993). A general algorithm for short-term scheduling of batch operations—I. MILP formulation. Comput. Chem. Eng., 17, 211.

Papalexaddri, K.P., & Pistikopoulos, E.N. (1994). A multi-period MINLP model for the synthesis of flexible heat and mass exchange network. Comput. Chem.Eng., 18, 1125.

Smith, E. M. (1996). On the optimal design of continuous processes, Ph.D. Dissertation, under supervision of C. Pantelides. Imperial College of Science, Technology and Medicine, London, UK.

16

Process Integration and System Analysis for Seawater Cooling in Industrial Facilities

Abdullah Bin Mahfouz[1], Mahmoud M. El-Halwagi[1], Bill Batchelor[1], Selma Atilhan[2], Patrick Linke[2] and Ahmed Abdel-Wahab[2]

[1]*Texas A&M University, College Station, TX 77843, USA*
[2]*Texas A&M University-Qatar, Doha, Qatar*

CONTENTS

ABSTRACT Using seawater in cooling systems is a common practice in many parts of the world where there is a shortage of freshwater. Biofouling is one of the major problems associated with the usage of seawater in cooling systems. Microfouling is caused by the activities of microorganisms, such as bacteria and algae, creating a very thin layer sticks to the inside surface of the heat exchangers. In some instances 250 micrometer thickness of fouling film would reduce 50% of the heat exchanger heat transfer coefficient. On the other hand, macrofouling is the blockage of marine relatively large organisms, such as oysters, mussels, clams, and barnacles. Therefore, a biocide is typically added to eliminate or at least reduce microfouling by intermitted dosages and macrofouling by continuous dosages. The objective of this work is to develop a systematic approach to the optimal design and integration of seawater cooling system. Specifically, the paper will address the following tasks:

1. Identification of the reaction pathways for the biocide from the mixing basin to the discharge points

2. Kinetic modeling of the biocide and byproducts throughout the process

3. Process integration for the reduction of biocide usage and discharge.

KEYWORDS   *Biocide, seawater cooling, seawater chemistry, energy integration, process integration*

Introduction

The use of seawater in industrial cooling is a common practice in many parts of the world that have limited fresh-water resources. One of the primary operational problems of using seawater in cooling is biofouling. Because of the biological activities of micro-organisms in seawater, biofilms are formed. These biofilms tend to stick to heat-exchange surfaces, thereby significantly reducing heat-transfer coefficients For instance, the heat-transfer coefficient may be reduced by 50% when a 250 μm thick biofilm is formed (Goodman 1987). In some cases, excessive bio-fouling can lead to plugging of heat exchangers. Therefore, biocides must be used in order to avoid or minimize the effect of fouling. Controlling microbial growth is usually achieved by using an oxidizing agent such as chlorine in an easy-to-disperse form such as hypochlorous acid, hypochlorite ion or gaseous form such as chlorine gas or chlorine dioxide. Intermitted chlorine dosage of 2–5 mg/L for 10 minutes a day to prevent microufouling and continuous dosage of 0.5 mg/L during the week two to four of breeding season to prevent blockage of macrofouling. At continuous biocide dosage animals like oysters and mussels close their shells tightly for weeks if needed but might die due to asphyxiation from extended continuous dosage. Chlorine forms are most widely used because of cost and effectiveness factors. Chlorine is a non selective oxidant (reacts with organics and non organics) and deactivates microbes as well. Also, chlorine reacts with natural organic matter (NOM) leading to the formation of numerous byproducts (Ben Waren 2006). While there are other means to prevent biofouling such as periodic cleaning with sponge balls, tube heating and drying, and antifouling paint, nonetheless, chlorine dosing is most widely used and cost effective method.

Because of the strong interaction between the process cooling demand, operating conditions, and biocide needs and performance, it is important to develop an integrated approach to optimizing biocide usage and discharge by understanding the key process factors and seawater chemistry aspects and reconciling them in an effective manner. The objective of this paper is to develop a systematic approach to the optimization of biocide usage and discharge by integrating seawater chemistry and process performance issues. These include modeling the mechanism and kinetics of the biocide, relating the biocide kinetics to process conditions, and reducing biocide usage by lower the cooling needs of the process via heat integration. The usage and discharge of seawater is linked to the process requirements including cooling duties. So, any reduction in cooling duties will have a direct impact to the usage and discharge of seawater along with usage and discharge of biocide.

Problem Statement

The problem to be addressed in the paper can be formally stated as follows:

Given a process with certain cooling duties which are satisfied by using seawater for cooling, a biocide (e.g., chlorine) is used to prevent or decrease the formation of biofilms. The process intake of seawater is referred to as $F_{S.W.}^{Intake}$ and the load of added biocide is designated by $L_{Biocide}^{Intake}$. Currently, the process discharges a flowrate, $F_{S.W.}^{Discharge}$, of used seawater and a biocide concentration of $C_{Biocide}^{Discharge}$ leading to a discharge load of biocide being $L_{Biocide}^{Discharge} = F_{S.W.}^{Discharge} * C_{Biocide}^{Discharge}$. Because of environmental regulations, it is desired to reduce the load of discharged biocide to $L_{Biocide}^{Regulated}$ or by controlling the concentration of discharged biocide $C_{Biocide}^{Discharge}$. It is desired to

- Develop a systematic procedure for understanding the chemical and kinetic aspects of biocide usage
- Relate the formation of byproducts to the characteristics of the seawater, the biocide, and the process
- Develop optimal policies for process modification and biocide dosing that will optimize biocide usage and discharge

Approach

Figure 1 provides a summary of the approach. First, heat integration is carried out using graphical thermal pinch analysis, algebraic techniques, or optimization formulations (e.g., (Smith 2005; El-Halwagi 2006; Kemp 2007)). The objective of this step is to minimize cooling and heating utilities of the process. Biocide dosage is proportional to the cooling duty. As such, reducing the cooling utility of the process leads to reducing the usage and discharge of the biocide.

Next, it is necessary to develop a mechanistic model for the reaction pathways involving the biocide and the various species in the seawater. Let us first start with some of the overall reactions involved when chlorine is added as a gas to seawater. First, it will dissolve and hydrolize rapidly and completely to $HOCl$ (hypochlorous) acid with reaction rate constant of 5×10^{14}.:

$$Cl_2 + H_2O \leftrightarrow HOCl + HCl \tag{1}$$

Hypochlorous acid is the most germicidal species, but it is a weak acid that will dissociate to hydrogen and hypochlorite ions with pKa of 7.5 at 30°C :

$$HOCl \leftrightarrow H^+ + OCl^- \tag{2}$$

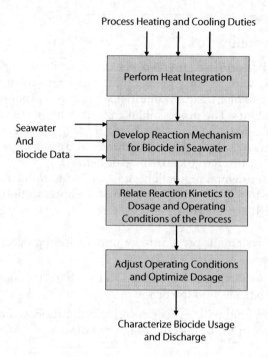

FIGURE 1
Proposed approach for optimizing biocide usage and discharge.

where OCl^- is the hypochlorite ion. In terms of disinfection effectiveness, hypochlorous acid is much stronger (almost two orders of magnitude) than the hypochlorite ion. Since, the hydrogen ion appears on the right side, this reaction is pH-dependant. Hypochlorous acid will reach its maximum concentration at pH ranges between 4 and 6 (Hostgaard-Jensen, Klitgaard et al. 1977). However, the effectiveness of a chemical as a disinfectant may not be the same as its effectiveness in removing biofilms. Controlling biofilm is achieved by weakening the polysaccharide matrix of microbial cells. There is experimental evidence that shows that chlorination is more effective in causing biofilm detachment at pH values greater than pH 8, where OCl^- concentration is more dominant than $HOCl$. (Characklis W. G. 1979)

Usually, seawater contains organic and non-organic species. Of particular importance are ammonia and bromide species. Ammonia, as well as other reactive nitrogenous compounds, will be chlorinated to yield monochloramine (NH_2Cl), dichloramine ($NHCl_2$), and trichloramine (NCl_3) by replacing the hydrogen atom of the ammonia molecule with a chlorine atom while maintaining its positive charge according to the following reactions:

$$HOCl + NH_3 \rightarrow NH_2Cl \text{ (monochloramine)} + H_2O \qquad (3)$$
$$NH_2Cl + HOCl \rightarrow NHCl_2 \text{ (dichloramine)} + H_2O \qquad (4)$$
$$NHCl_2 + HOCl \rightarrow NCl_3 \text{ (trichloramine)} + H_2O \qquad (5)$$

Those reactions depend on pH, temperature, contact time, but mainly on chlorine to ammonia ratio. All the free chlorine (hypochlorous acid) will be converted to monochloramine at pH 7-8 (fastest conversion is at pH 8.3) when there is 1:1 molar ratio of chlorine to ammonia (5:1 by wt) or less. Then, within the same range of pH, dichloramine is produced at a molar ratio of 2:1 of chlorine to ammonia (10:1 by wt). This reaction is relatively slow, so it may take an hour. Also, within the same range of pH, trichloramine will be produced at a molar ratio of 3:1 of chlorine to ammonia (15:1 by wt) and at equal molar ratios but at pH 5 or less. The two reactions producing di- and tri-chloramine are known as the breakpoint reactions where the chloramines are reduced suddenly to the lowest level. The significance of breakpoint reaction is that chlorine reaches its highest concentration and germicidal efficiency (at 1:1 molar ratio of chlorine to ammonia) just before reaching this point. Also, at the breakpoint monochloramine and dichloramine react together (which reduces chlorine residuals) to produce nitrogen gas, nitrate, and trichloramine.

Dichloramine decomposes to an intermediate reactive product (NOH) which consumes mono-, di-chloramine, and hypochlorous acid producing nitrogen gas and nitrate. Also, excessive chlorine will form trichloramine.

$$NHCl_2 + H_2O \rightarrow NOH + 2\,H^+ + 2\,Cl^- \tag{6}$$

$$NOH + NH_2Cl \rightarrow N_2 + H_2O + H^+ + Cl^- \tag{7}$$

$$NOH + NHCl_2 \rightarrow N_2 + HOCl + H^+ + Cl^- \tag{8}$$

$$NOH + 2\,HOCl \rightarrow NO_3^- + 3\,H^+ + 2\,Cl^- \tag{9}$$

$$NCl_3 + H_2O \rightarrow NHCl_2 + HOCl \tag{10}$$

The reaction of chlorine into these forms steer it away from the disinfection function and render the biocide less effective. Consequently, it is important to understand such side reactions.

Hypochlorous acid rapidly reacstwith bromide producing hpobromous acid, which also can be produced from the reaction of bromide with monochloramine.as follows:

$$HOCl + Br^- \Leftrightarrow HOBr + Cl^- \tag{11}$$

$$NH_2Cl + Br^- + H_2O \rightarrow HOBr + Cl^- + NH_3 \tag{12}$$

where HOBr is hypobromous acid. Additionally, the hypochlorite ion may undergo a slow reaction with the bromide ion as follows:

$$OCl^- + Br^- \Leftrightarrow OBr^- + Cl^- \tag{13}$$

where OBr^- is the hypobromite ion. Bromide in seawater may also react directly with added chlorine to give bromine and chloride:

$$Cl_2 + 2Br^- \Leftrightarrow Br_2 + 2Cl^- \tag{14}$$

It is worth noting that the presence of ammonia and other nitrogenous compounds in the seawater will react with HOBr to yield monobromamine (NH_2Br), dibromamine ($NHBr_2$), and tribromamine (NBr_3).

$$HOBr + NH_3 \rightarrow NH_2Br \text{ (monobromamine)} + H_2O \qquad (15)$$

$$HOBr + NH_2Br \rightarrow NHBr_2 \text{ (dibromamine)} + H_2O \qquad (16)$$

$$HOBr + NHBr_2 \rightarrow NBr_3 \text{ (tribromamine)} + H_2O \qquad (17)$$

The bromine breakpoint happens when the dibromamines are produced rapidly, leading to the formation of nitrogen gas:

$$NHBr_2 + H_2O \rightarrow NOH + 2H^+ + 2\ Br^- \qquad (18)$$

$$NOH + NHBr_2 \rightarrow N_2 + HOBr + H^+ + Br^- \qquad (19)$$

It is also important to consider the effect of bromide which naturally exists in seawater at (50–70 mg/l). This is in stoichiometric excess of chlorine dosage as well as ammonia concentration don't exceed 2 to 3 mg/L. The relative amount produced of bromine species to ammonia species is propotional to bromide concentration over ammonia concentration if we assume both reactions are rapid and simultaneous.

In order to understand the various species interactions and reaction pathways, we have constructed the reaction mechanism shown in Fig. 2. On these diagrams, starting species and intermediate and final byproducts are represented in boxes. Arrows correspond to reaction steps. Boxes on the arrows represent reactive species that contribute to that reaction.

Next, it is necessary to develop a kinetic model of the process. The consumption of chlorine is due to the reaction with organic and non organic compounds, biofilm, and corrosion. The chlorine decay in a kinetics in a batch cooling system (X in 2003) may be described as follows:

$$\frac{dC}{dt} = -\frac{k_W C_W}{r_h} - \frac{W}{r_h} - k_b C \qquad (20)$$

Based on examining numerous experimental results for the kinetics of seawater chlorination, decay kinetics have been correlated to several factors including temperature, pH, contact time, but mainly on the ratio of chlorine dosage to ammonia (Haag 1981). The reduction or decay of chlorine in seawater is due to reactions with organic and non-organic compounds in seawater. The chlorine decay occurs in three stages starting with a very fast rate during the first the 2 minutes due to reaction with inorganic reducing agents. The second phase is slower and usually does not last more than two hours. It involves reactions mainly with organic compounds that started in the first phase. Then, chlorine decays continuously via a very slow rate.

Of particular importance is the dependence on residence time and temperature. Given the specific path of seawater inside the process, the following model is developed to account for chlorine decay throughout the process. The seawater goes through a number of pipes and units. The process is discretized

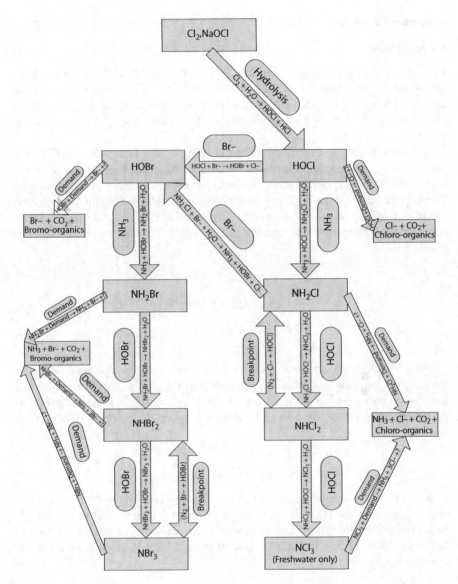

FIGURE 2
Proposed reaction mechanism for biocide reactions.

into a number of segments. Each segment, i, represents a portion of the seawater pipeline or a heat exchanger (e.g., cooler, coil in a hot unit, cooling jacket of a unit, etc.). Consider the N key species represented in Fig. 2a and refer to their concentrations in segment i as: $C_{i,1}, C_{i,2}, ..., C_{i,j}, ...C_{i,N}$. For the jth species, in the ith segment, the following kinetic expression may be written

$$C_{i+1,j} = \psi_j(C_{i,1}, C_{i,2}, ..., C_{i,j}, ...C_{i,N}, T_i, pH_i) \tag{21}$$

Case Study

Consider a urea process described by (Bin Mahfouz 2006). The current usage of seawater cooling utility is 89.2 MM Btu/hr and employs a continuous dosage of 0.75 mg/L. For maintaining appropriate biocide effect, the residual chlorine should be kept at levels higher than 0.05 mg/L throughout the system.

By carrying out heat integration, the cooling utility is reduced to 77.1 MM Btu/hr which corresponds to about 14% reduction in the biocide usage. Next, kinetic modeling is used to track the biocide and its key reaction by products. The experimental data of (Ben Waren 2006) of residual chlorine at different chlorine dosages at 25°C are used to develop the following dynamic decay functions for the three continuous dosages (1.0, 5.0, and 10 mg/L, respectively) as:

$$C_1 = 0.42e^{(-0.0015*t)} \tag{22}$$

$$C_2 = 3.80e^{(-0.0004*t)} \tag{23}$$

$$C_3 = 8.55e^{(-0.0003*t)} \tag{24}$$

where C is chlorine concentration in mg/L and t is the residence time in minutes. A regression analysis was carried out with the experimental data for a residence time of 20 minutes for the case study. A linear expression was derived to relate the residual chlorine (C_{out}) to the chlorine dosage (C_{in}) through:

$$C_{out} = 0.9C_{in} - 0.52 \tag{25}$$

In order to maintain 0.05 mg/l concentration of chlorine residual at the discharge we need a dosage concentration of not more than 0.63 mg/L which provides 19% reduction in biocide usage from the current practice which is 0.78 mg/L (in addition to the 14% reduction in biocide load already achieved via heat integration).

Acknowledgments

The authors would like to acknowledge support from the King Abdullah University of Science and Technology (KAUST), the Saudi Ministry of Higher Education, and Qatar National Research Fund (QNRF).

References

Ben Waren, A. H., Cynthia Joll, and Robert Kagi (2006). "A new method for calculation of the chlorine demand of natural and treated waters." *Water research* 40: 8.

Bin Mahfouz, A. S., El-Halwagi, Mahmoud M. and Abdel-Wahab, Ahmed (2006). "Process Integration Techniques for Optimizing Seawater Cooling Systems and Biocide Discharge." *Clean Technologies and Environmental Policy* 8(3): 203–215.

Characklis W. G., T., M. G., Chang Lin-Chiang (1979). *Oxidation and Disinfection of Microbial Films*, Ann Arbor Science.

El-Halwagi, M. M. (2006). *Process Integration*, Academic Press.

Goodman, P. D. (1987). "Effect of chlorine on materials for sea water cooling systems: A review of chemical reactions." *British Corrosion Journal* 22(1): 56–62.

Haag, W. R., Lietzke, M.H. (1981). "A Kinetics Model for Predicting The Concentrations of Active Halogens Species in Chlorinated Saline Cooling Waters." *Oak Ridge National Laboratory* W-7405(eng-26).

Hostgaard-Jensen, P., J. Klitgaard, et al. (1977). Chlorine decay in cooling water and discharge into seawater. *J. Water Pollut. Control Fed.*; Vol/Issue: 49:8. United States: Pages: 534a, 1832–1841.

Kemp, I. (2007). *Pinch Analysis and Process Integration: A User Guide on Process Integration for the Efficient Use of Energy*, Butterworth-Heinemann/Elsevier.

Smith, R. (2005). *Chemical Process Design and Integration*. New York Wiley.

Xin, L., Da-ming, GU., Jing-yao, QI., Ukita, M., Hong-bin, ZHAO (2003). "Modeling of residual chlorine in water distribution system." *Jounal of Environmental Science (China)* 15(1): 136–144.

17

Significance of Dead-state-based Thermodynamics in Designing a Sustainable Process

Tengyan Zhang* and L. T. Fan

Department of Chemical Engineering, Kansas State University
Manhattan, KS 66506, USA

CONTENTS

ABSTRACT The standard state of thermodynamics is regarded to be in thermal and mechanical equilibrium with the environment having the temperature and pressure of 298.15 K and 1 atm, respectively. In contrast, the dead state, or extended standard state, is defined such that it is not only in thermal and mechanical equilibrium with the environment but also in chemical equilibrium with the environment. Consequently, the enthalpy of formation and free-energy of formation of the substance at the dead state are invariably non-negative. As such, the energy and available energy balances can be executed in a straightforward manner around any given system (process), or arbitrary segment of the system. This gives rise to the energy loss from and the available energy consumption, i.e., exergy dissipation, of the process or segment of the process. Thus, dead-state-based thermodynamics renders it possible to carry out the multi-scale thermodynamic analysis of a process necessary for the assessment of its sustainability on the uniformly consistent platform at any desired level of details. This is not the case with the standard-state-based

---

* To whom all correspondence should be addressed. Tel.: +1-785-532-4327. Fax: +1-785-532-7372. E-mail address: tengyan@ksu.edu

thermodynamics: The standard enthalpy of formation and free-energy of formation can be positive or negative depending on the substances.

KEYWORDS *Exergy, sustainability, process design, dead state, thermodynamics*

Introduction

Thermodynamic constraints on any system or process in terms of entropy increase manifest themselves in exergy dissipation by the process, which can be evaluated by resorting to the available energy balance. The available energy balance results from the combination of the first and second laws of thermodynamics (Keenan, 1951; Hatsopoulos and Keenan, 1965; Fan and Shieh, 1980; Petit and Gaggioli, 1980; Szargut et al., 1988). In reality, however, it also implicitly embodies the mass conservation law: The available energy balance entails the detailed accounting of each material species involved in the process. This renders it possible to determine the transformation of its inherent available energy pertinent to its chemical changes, which is termed chemical exergy. Note that all the thermodynamic properties are measured relative to the same reference state that is specifically defined.

Reference States

In principle, any arbitrary state may serve as a datum level to evaluate thermodynamic state properties of any material species. This base is generally referred to as the reference state, where the values of one or more state properties of matter are deemed to be zero. Nevertheless, while the values of the state functions vary according to the reference states designated, the first and second laws of thermodynamics remain valid (Hatsopoulos and Keenan, 1965; Fan and Shieh, 1980).

Standard State

In the conventional treatment of classical thermodynamics, the standard state is defined in terms of the prevailing environmental temperature, T^0, the prevailing environmental pressure, P^0, and all the pure elements (see, e.g., Keenan, 1951; Hatsopoulos and Keenan, 1965; Szargut et al., 1988). For convenience, T^0 and P^0 are specified, respectively, as

$$T^0 = 298.15 \ K \qquad P^0 = 1 \ atm$$

By definition, the concentration of any pure element i at the standard state, x_i^0, is an unity, i.e.,

$$x_i^0 = 1; \quad i = 1, 2, \ldots k \quad (k = number\ of\ elements)$$

The values of state properties, such as enthalpy and free energy, of various substances at the standard state relative to this reference state are available in most treaties on classical thermodynamics (see, e.g., Lide, 2008); for instance, enthalpy and free energy are called the standard enthalpy of formation and the standard free energy of formation, respectively. While the values of these thermodynamic functions for any element are zero, those for most, but not all, of the compounds are negative. This renders it impossible to consistently evaluate the thermodynamic efficiency straightforwardly, especially when chemical reactions are involved in the process of concern. Thus, defining an appropriate reference state is of paramount importance for the thermodynamic analysis (Hatsopoulos and Keenan, 1965; Szargut et al., 1988). In fact, from a practical viewpoint, it would often be desirable to shift the reference state from that mentioned above to the so-called dead state.

Dead State

The dead state is a natural extension of the notion of the standard state; therefore, it can be regarded as the extended standard state. The rationale behind such an extension is that the pure elements seldom prevail under the environmental conditions. Frequently, they are chemically active and they react spontaneously with the substances in the surrounding environments, thus releasing the chemical form of latent energy or available energy. In fact, some elements even explosively react in the surrounding environments. For instance, with the slightest disturbance, hydrogen H_2 and dust of carbon C may explosively react with oxygen O_2 in the air, thereby yielding water H_2O (l) and carbon dioxide CO_2, respectively. The foregoing arguments imply that it is indeed untenable to recognize or identify all the pure elements as totally neutral and stable substances under the environmental conditions; hence, not all the pure elements are appropriate to serve as the references for measuring various thermodynamic functions or thermal properties of other materials. To follow the paradigm of specifying the environmental temperature and pressure as the components of the standard state, another set of reference substances, each corresponding to a single element, must be specified. This has given rise to the notion of the datum level materials. These materials in conjunction with the environmental temperature and pressure constitute the dead state (Keenan, 1951; Hatsopoulos and Keenan, 1985; Szargut et al., 1988); analogous to the standard state, the environmental temperature and pressure are usually taken to be 298.15 K and 1 atm, respectively.

Formally, the datum level materials are defined to be the compounds or elements that are not only thermodynamically stable but also exist in abundance in the environment; hence, they are regarded as void of energy or available energy at the dead or extended standard state. In other words, each element has a natural tendency to be part of its datum level material. For instance, the aforementioned H_2O (l) and CO_2 are the datum level materials for H and C, respectively (Denbigh, 1956; Hatsopoulos and Keenan, 1965; Fan and Shieh, 1980; Fan et al., 1983; Szargut et al., 1988). According to Petit and Gaggioli (1980), the dead state is the state that each constituent of the substance is in complete stable equilibrium with the components in the environment. Note that subscript 0 stands for the dead state or extended standard state, instead of superscript 0 for the standard state; thus,

$$T_0 = 298.15 \ K \qquad P_0 = 1 \ atm$$

x_{i0} = *environmental composition of the corresponding datum level material*
$i = 1, 2, \ldots, k$ (k = *number of elements*)

In general, enthalpy and exergy are measured relative to the dead state or extended standard state (Keenan, 1951; Denbigh, 1956; Hatsopoulos and Keenan, 1965; Fan and Shieh; 1980; Szargut et al., 1988; Fan et al., 2006). The enthalpy of formation and free-energy of formation of the substance at the dead state are invariably non-negative. As such, the energy and available energy balances can be executed in a straightforward manner around any given system (process), or arbitrary segment of the system. This gives rise to the energy loss from and the available energy consumption, i.e., exergy dissipation, of the process or segment of the process. Thus, dead-state-based thermodynamics renders it possible to carry out the multi-scale thermodynamic analysis of a process necessary for the assessment of its sustainability on the uniformly consistent platform at any desired level of details. This is not the case with the standard-state-based thermodynamics: The standard enthalpy of formation and free-energy of formation of a substance are evaluated relative to the pure elements contained in it. Moreover, under environmental conditions, many of these elements are neither the most stable nor abundant among all the substances comprising the elements. The standard enthalpy of formation and free-energy of formation, therefore, can be positive or negative depending on the substances.

Note that the system's energy (enthalpy) and available energy (exergy) depend on the extent of its deviation from the dead state. In other words, any deviation (the physical, thermal and/or chemical deviations) of the system's state from the dead state, induced by physical, thermal and/or chemical processes, results in the system's energy and available energy. Thus, $\bar{\beta}$, $\bar{\gamma}$, and $\bar{\varepsilon}$ can be estimated from the following equations (Denbigh, 1956; Hatsopoulos and Keenan, 1965; Fan and Shieh, 1980; Yantovskii, 1994).

$$\bar{\beta} = \bar{\beta}_0 + \bar{\beta}_T + \bar{\beta}_P$$

$$\begin{bmatrix} \text{partial molar} \\ \text{enthapy} \\ \text{at T and P} \end{bmatrix} \quad \begin{bmatrix} \text{partial molar} \\ \text{chemical} \\ \text{enthapy} \end{bmatrix} \begin{bmatrix} \text{partial molar} \\ \text{thermal} \\ \text{enthapy} \end{bmatrix} \quad \begin{bmatrix} \text{partial molar} \\ \text{pressure} \\ \text{enthapy} \end{bmatrix}$$

$$= \bar{\beta}_0 + \int_{T_0}^{T} \bar{c}_p dT + \int_{P_0}^{P} \left[\bar{V} - T \left(\frac{\partial \bar{V}}{\partial T} \right)_P \right] dP \tag{1}$$

$$\begin{bmatrix} \text{partial molar} \\ \text{chemical} \\ \text{enthapy} \end{bmatrix} \quad \begin{bmatrix} \text{temperature} \\ \text{effect} \end{bmatrix} \begin{bmatrix} \text{pressure} \\ \text{effect} \end{bmatrix}$$

$$\bar{\gamma} = \bar{\gamma}_0 + \bar{\gamma}_T + \bar{\gamma}_P$$

$$\begin{bmatrix} \text{partial molar} \\ \text{entropy} \\ \text{at T and P} \end{bmatrix} \quad \begin{bmatrix} \text{partial molar} \\ \text{chemical} \\ \text{entropy} \end{bmatrix} \begin{bmatrix} \text{partial molar} \\ \text{thermal} \\ \text{entropy} \end{bmatrix} \quad \begin{bmatrix} \text{partial molar} \\ \text{pressure} \\ \text{entropy} \end{bmatrix}$$

$$= \bar{\gamma}_0 + \int_{T_0}^{T} \frac{\bar{c}_p}{T} dT + \left(-\int_{P_0}^{P} \left(\frac{\partial \bar{V}}{\partial T} \right)_P dP \right) \tag{2}$$

$$\begin{bmatrix} \text{partial molar} \\ \text{chemical} \\ \text{entropy} \end{bmatrix} \quad \begin{bmatrix} \text{temperature} \\ \text{effect} \end{bmatrix} \begin{bmatrix} \text{pressure} \\ \text{effect} \end{bmatrix}$$

$$\bar{\varepsilon} = \bar{\varepsilon}_0 + \bar{\varepsilon}_T + \bar{\varepsilon}_P$$

$$\begin{bmatrix} \text{partial molar} \\ \text{exergy} \\ \text{at T and P} \end{bmatrix} \quad \begin{bmatrix} \text{partial molar} \\ \text{chemical} \\ \text{exergy} \end{bmatrix} \begin{bmatrix} \text{partial molar} \\ \text{thermal} \\ \text{exergy} \end{bmatrix} \quad \begin{bmatrix} \text{partial molar} \\ \text{pressure} \\ \text{exergy} \end{bmatrix}$$

$$= \bar{\varepsilon}_0 + \int_{T_0}^{T} \bar{c}_p \left(1 - \frac{T_o}{T} \right) dT + \int_{P_0}^{P} \left[\bar{V} - (T - T_o) \left(\frac{\partial \bar{V}}{\partial T} \right)_P \right] dP \tag{3}$$

$$\begin{bmatrix} \text{partial molar} \\ \text{chemical} \\ \text{exergy} \end{bmatrix} \quad \begin{bmatrix} \text{temperature} \\ \text{effect} \end{bmatrix} \begin{bmatrix} \text{pressure} \\ \text{effect} \end{bmatrix}$$

The values of $\bar{\beta}_0$, $\bar{\gamma}_0$, and $\bar{\varepsilon}_0$ of many material species in the above expressions can be found in various sources (Fan and Shieh, 1980; Yantovskii, 1994).

The evaluation of the energetical quantities, $\bar{\beta}_0$ and $\bar{\varepsilon}_0$, demands the detailed accounting of the flow of each material species participating in the process, the outcome of which is the inherent coupling of the mass flow and the energy flow. Naturally, the specific chemical enthalpy, the specific chemical entropy, and specific chemical exergy (or the specific chemical availability) are represented by β_0, γ_0, and ε_0, respectively.

Multi-scale Thermodynamic Analysis of a Process

The thermodynamic first-law conservation (process) efficiency, $(\eta_1)_p$, of a process system is defined as (see, e.g., Keenan, 1951; Denbigh, 1956; Hatsopoulos and Keenan, 1965; Fan and Shieh, 1980; Petit and Gaggioli, 1980; Szargut et al., 1988; Fan et al., 2006)

$$(\eta_1)_p \equiv \frac{[\text{energy transfer of the desired kind achieved by the system}]}{[\text{energy input into the system}]} \tag{4}$$

Moreover, the corresponding thermodynamic second-law conservation efficiency, $(\eta_2)_p$, is defined as

$$(\eta_2)_p \equiv \frac{[\text{available energy transfer of the desired kind achieved by the system}]}{[\text{available energy input into the system}]}$$

$$\tag{5}$$

The thermodynamic efficiency may assume many forms and expressions based on the above two expressions. Nevertheless, the significance of each form or expression is best understood in the light of the mass, energy, and available energy balances (Fan and Shieh, 1980; Fan et al., 2006). Thus, it is of primary importance in the thermodynamic analysis of any system to establish the "legitimate" mass, energy and available energy balances and "rigorous" evaluation of various loss and dissipation terms in the balance expressions.

The thermodynamic analysis of a process can be carried out in various scales by expanding or contracting the boundary enclosing the process. As illustrated in Figure 1, the system of concern can be a reactor only, as indicated by the dotted-line; a process plant comprising the reactor and all the ancillary facilities necessary for maintaining its operation, as indicated by the solid line; or the entire system including not only the process plant but also all the facilities and infrastructures required to maintain the plant operation, as indicated by the dashed line.

The available energy balance around a process or system at any scale can generally be expressed as (Fan and Shieh, 1980; Szargut et al., 1988)

$$E_{xf} + E_{xi} = E_{xp_m} + E_{xp_e} + E_{xw} + E_{xd} \tag{6}$$

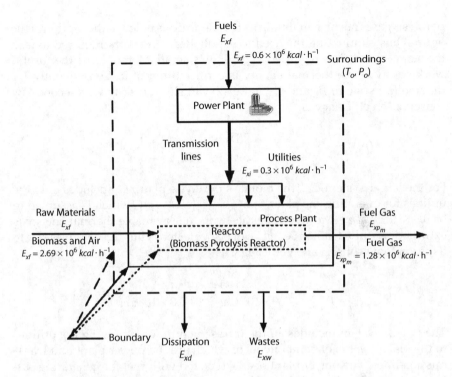

FIGURE 1

Multi-scale evaluation of thermodynamic efficiencies: Biomass pyrolysis system where biomass is fed into a biomass pyrolysis reactor, i.e., pyrolizer, to generate pyrolysis gas as product and char as waste; a portion of the pyrolysis gas is burnt in the pyrolizer to maintain its temperature; and the remainder is usable for the burner and boiler.

where E_{xf} is the exergy of the feeds including raw materials and/or fuels; E_{xi}, the exergy of the work and available thermal energy (heat) supplied to the system; E_{xp_m}, the exergy of products; E_{xp_e}, the work and available thermal energy usefully extracted out of the system; E_{xw}, the exergy of the wastes; and E_{xd}, the dissipated exergy. In Eq. (6), the sum, $(E_{xf} + E_{xi})$, in the left-hand side represents the input of exergy to the system, and the sum, $(E_{xp_m} + E_{xp_e})$, in the right-hand side represents the output of useful exergy from the system. Hence, according to the definition of $(\eta_2)_p$ expressed by Eq. (5), we have

$$(\eta_2)_p = \frac{E_{xp_m} + E_{xp_e}}{E_{xf} + E_{xi}} \tag{7}$$

Obviously, $0 \le (\eta_2)_p < 1$, and the larger the $(\eta_2)_p$, the more sustainable the process. Thus, $(\eta_2)_p$ might also be regarded as a sustainability index.

The reactor in Figure 1 is a biomass pyrolizer for which the pertinent data are available (Ishimi et al., 1983). For this pyrolizer, a portion of the pyrolysis gas is burnt to maintain its temperature as noted in the caption of the figure, thus eliminating the need to supply heat; and work required to feed the biomass and air is regarded as negligibly small. Moreover, neither useful work nor available thermal energy is extracted out of it. Consequently, for the smallest system, E_{xi} and E_{xp_e} in Eq. (7) vanish; accordingly, its second-law conservation efficiency is

$$(\eta_2)_p = \frac{E_{xp_m} + E_{xp_e}}{E_{xf} + E_{xi}} = \frac{1.28 \times 10^6 \; kcal \cdot h^{-1} + 0}{2.69 \times 10^6 \; kcal \cdot h^{-1} + 0} = 47.58\%$$

For the larger system, i.e., the biomass pyrolysis plant, E_{xi} includes only the utilities for operating the plant, $\left[\left| W_{in} \right| + \left| Q_{in} \right| \left(1 - T_0/T_{in} \right) \right]$, which is assumed to be 0.3×10^6 $kcal \cdot h^{-1}$. Moreover, neither work nor available thermal energy is extracted usefully from it, i.e., E_{xp_e} can be neglected. Thus, Eq. (7) gives rise to its second-law conservation efficiency as

$$(\eta_2)_p = \frac{E_{xp_m} + E_{xp_e}}{E_{xf} + E_{xi}} = \frac{1.28 \times 10^6 \; kcal \cdot h^{-1} + 0}{2.69 \times 10^6 \; kcal \cdot h^{-1} + 0.3 \times 10^6 kcal \cdot h} = 42.81\%$$

The largest system includes all the infrastructure for supplying the utilities to the plant, the major components of which are the power plant and transmission lines. The power plant needs to be fed with fuel for generating sufficient power for the utilities and that for compensating the transmission loss. The exergy content of fuel, which is part of E_{xf} of this system, therefore, is assumed to be 0.6×10^6 $kcal \cdot h^{-1}$. Obviously, neither work nor available thermal energy is extracted usefully out of the system. Thus, Eq. (7) yields its second-law conservation efficiency as

$$(\eta_2)_p = \frac{E_{xp_m} + E_{xp_e}}{E_{xf} + E_{xi}} = \frac{1.28 \times 10^6 \; kcal \cdot h^{-1} + 0}{(2.69 \times 10^6 + 0.6 \times 10^6) \; kcal \cdot h^{-1} + 0} = 38.91\%$$

where coal is considered to be the fuel for the power plant. If coal is replaced by photons in the form of solar radiation, which is regarded as free (Ewing and Pratt, 2005), the corresponding second-law conservation efficiency is increased to

$$(\eta_2)_p = \frac{E_{xp_m} + E_{xp_e}}{E_{xf} + E_{xi}} = \frac{1.28 \times 10^6 \; kcal \cdot h^{-1} + 0}{(2.69 \times 10^6 + 0) \; kcal \cdot h^{-1} + 0} = 47.58\%$$

Note that the largest system symbolizes the notion of ecological economists; quoting Nadeau (2006), "...low-entropy matter-energy in a closed system is always transformed into high-entropy matter-energy."

The thermodynamic analysis at each scale, as illustrated with the biomass pyrolysis process, has its own utility. For example, the data generated at the smallest scale will be needed by a developer or designer of the pyrolizer, and the information obtained at the largest scale will be useful by those involved in corporate-level decision making or ecologists.

Conclusions

The underlying rationale is elucidated for defining the notions of the dead state and concomitant datum-level materials. These notions render it possible to execute the consistent and straightforward evaluation of the thermodynamic efficiency necessary for the thermodynamic analysis of any given process at various scales and a desired level of details.

References

Denbigh, K. G., (1956). The Second-Law Efficiency of Chemical Processes. *Chem. Eng. Sci.*, 6, 1.

Ewing, R. A., Pratt, D. (2005). Got Sun? Go Solar: Get Free Renewable Energy to Power Your Grid-Tied Home. *PixyJack Press, LLC*. Masonville, CO.

Fan, L.T., Shieh, J. H. (1980). Thermodynamically Based Analysis and Synthesis of Chemical Process Systems. Energy, 5, 955.

Fan, L. T., Zhang, T., Schlup, J. R. (2006). Energy Consumption Versus Energy Requirement. *Chemical Engineering Education*, 40, 132.

Hatsopoulos, G. N., Keenan, J. H. (1965). Principles of General Thermodynamics. *Wiley*. New York.

Ishimi, T., Shieh, J. H., Fan, L.T. (1983). Thermodynamic Analysis of a Biomass Pyrolysis Process in Wood and Agricultural Residues, pp. 439-465, E.J. Soltes, ed., *Academic Press*. New York.

Keenan, J. H. (1951). Availability and Irreversibility in Thermodynamics. *Br. J. Appl. Phys.*, 2, 183.

Lide, David R., ed. (2008).CRC Handbook of Chemistry and Physics, 89th Edition. *CRC Press, Inc.*, Boca Raton, Florida.

Nadeau, Robert L. (2006). Chapter 7. A Green Thumb on The Invisible Hand: Environmental Economics and Ecological Economics, in THE ENVIORNMENTAL ENDGAME: Mainstream Economics, Ecological Disaster and Human Survival, *Rutgers University Press*. New Brunswick, New Jersey, and Londo.

Petit, P. J., Gaggioli, R. A. (1980). Second Law Procedures for Evaluating Processes in Thermodynamics: Second Law Analysis, ACS Symposium Series 122, Ed. R.A. Gaggioli. *ACS*. Washington, D. C.

Szargut, J., Morris, D. R., Steward, F. R. (1988). Exergy Analysis of Thermal, Chemical, and Metallurgical Processes. *Hemisphere Publishing Corporation*. New York.

Yantovskii, E. I. (1994). Energy and Exergy Currents. *NOVA Science Publishers*. U.S.A.

18

Analysis and Generation of Sustainable Alternatives: Continuous and Batch Processes Using Sustainpro Package

A. Carvalho[1,2]**, H.A. Matos**[1] **and R. Gani**[2]

[1]*CPQ- Dep of Chem. & Bio. Eng, Ins. Sup. Técnico,*
Av Rovisco Pais, 1049-001 Lisboa, Portugal
[2]*CAPEC- Dep of Chem Eng, Tech Univ of Denmark, DK-2800 Lyngby, Denmark*

CONTENTS

ABSTRACT The objective of this paper is to present a new methodology that is able to generate, screen and then identify sustainable alternatives in chemical process operating in continuous or batch operational models by locating the operational, environmental, economical, and safety related bottlenecks inherent to the process. The methodology is able to analyze a wide range of processes that operate in continuous mode, in semi-continuous and/or in batch mode. The main steps of the methodology are described, highlighting the main differences between continuous and batch processes. The software, SustainPro, has been further developed to include the latest features of the indicator-based methodology. Through a case study involving a batch-process, the important features of the methodology and the corresponding software are illustrated together with the obtained sustainable process design.

KEYWORDS *Sustainability, process design alternatives, indicators*

Introduction

The concerns about issues such as the global warming, greenhouse effect, water/land acidification, etc., are growing everyday. In order to ensure that the future generations will be allowed to live in a sustainable world, the industrial activities related to chemicals-based products need to be improved. In this way, it is necessary to develop systematic methods and tools, which enable the generation of more sustainable alternatives and also improve the ability to adapt to the future needs.

Many methodologies have been presented in order to improve/retrofit chemical processes operating in continuous mode. Lange, (2002) has proposed a methodology that directly relates the process design alternatives to improvements in sustainability of the processes. This author claims that the optimal solution is a trade-off between the different performance criteria, (sustainability in a process). A mmethodology for the synthesis of the entire process by incorporating operating units with enhanced performance has been proposed by, Liu *et al.*, (2006).

Regarding the batch operations, several methodologies for retrofit design have been proposed. For example, Halim and Srinivasan (2008) proposed an intelligent simulation–optimization framework for identifying comprehensive sustainable alternatives for batch processes. Simon *et al.* (2008), on the other hand, presented an indicator, heuristics, and process model based decision support framework for retrofitting chemical batch processes. Their framework considered the identification of improvement opportunities in a batch plant by considering first the product market situation.

Recently, Carvalho *et al.*, (2009) presented an extended version of their systematic indicator-based methodology for sustainable design. This new version is generic (i.e., it can handle a wider range of problems with the same models and algorithm), is able to generate, screen and then identify sustainable alternatives in chemical processes operating in continuous or batch modes. This was achieved through the use of a new set of indicators, and, sustainability and safety parameters, which allow the location of the process bottlenecks across the process.

The objective of this paper is to give a general overview of the new combined methodology together with the new version of SustainPro. A batch process involving the production of insulin is used as a case study to illustrate the new features of the methodology and the software.

Methodology

The main steps of the methodology are briefly described below. For continuous processes the main features of the methodology can be found in Carvalho *et al.* (2008) and the extended methodology for batch processes is

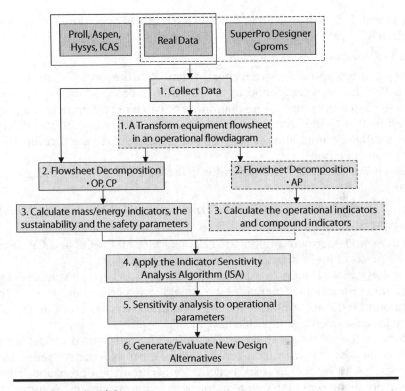

FIGURE 1
Flow-diagram of the indicator-based methodology.

described in Carvalho *et al.* (2009). Figure 1 presents the flow-diagram of the indicator-based methodology.

Step 1: Collect data

In this step, the process data needed to apply the methodology is collected from different sources (plant and/or model generated). In continuous processes, steady state data related to the mass and the energy balance is needed. In the batch processes the information required is the time of each operation, the equipment volume, the initial and the final mass for each compound in each operation and the energy used in each step. The purchase and sale prices for each chemical are needed for both cases. All these data can be collected from real plant and/or generated through model-based simulations.

Step 1A: Transform equipment flowsheet in operational flow-diagram

The batch processes will be treated as "continuous" processes in terms of the material and energy (data) flow from operation to operation. Thus, the equipment based-flowsheet is transformed to an operation based flow-diagram (Carvalho *et al.* 2009)

Step 2: Flowsheet decomposition

Flowsheet decomposition is performed to identify all the open- (OP) and closed-paths (CP) for each compound in the continuous process flowsheet as well as in the batch operations flow-diagram. For batch operation flow-diagram, a path related to the accumulation of mass and energy is introduced. This path is called accumulation-path (AP) and corresponds to the accumulation in a given operation.

SustainPro generates automatically a list of all the open- and closed-paths.

Step 3: Calculate the indicators, the sustainability and the safety parameters

The mass and energy indicators are calculated for all the OP and CP determined in step 2 for both batch and continuous processes.

Two sets of indicators for batch processes have been developed: The Operation Indicators (compare the operation performance) and the Compound Indicators (indicate for each operation which compound is most likely to cause operational problems).

Through the values obtained for the full-set of indicators it is possible to identify the locations within the process where the mass/energy "paths" face "barriers" with respect to costs, benefits, or accumulation problems. These critical points present high potential for process improvements. A summary of the indicators application is shown in figure 2.

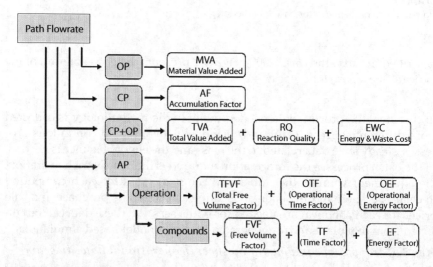

FIGURE 2
Overview of indicators application.

The analysis of the process impact is determined through the calculation of a set of sustainability metrics. The safety of the process is measured by the use of a set of safety indices.

Step 4: Indicator Sensitivity Analysis (ISA) algorithm

In this step the target indicators are determined using the ISA algorithm (Carvalho *et al.* 2008). To apply this algorithm the indicators having the highest potential for improvements are identified first. Then an objective function such as gross-profit or process total cost is specified. A sensitivity analysis is then performed to determine the indicators that allow the best improvements in the objective function. The target values for the indicators are also specified in this step.

Step 5: Operational Sensitivity Analysis

A sensitivity analysis with respect to the operational (parameters) variables, which influence the target indicators, is performed. The analysis identifies the operational variables that need to be changed to improve the process in the desired direction.

Step 6: Generation of new design alternatives

New alternatives are generated using a systematic analysis where a collection of synthesis algorithms are used.

SustainPro

Based on the work-flow, data-flow and calculations of the methodology described above, a new version of SustainPro has been developed using EXCEL as the software environment

The inputs for SustainPro are the mass and the energy balance data (continuous or batch mode) as well as the prices of the chemicals present in the process. SustainPro is able to read the mass and the energy balance data from an EXCEL-file, thereby making the transfer of data generated through process simulators or from real plants simple and easy. SustainPro follows all the steps of the extended methodology, giving as output, suggestions for new alternatives. The needed data on chemical properties is retrieved tools such as the CAPEC data-base, property prediction package (ProPred) and the WAR Algorithm, which is available in ICAS. Figure 3 shows the interconnection between SustainPro and the other tools.

Case Studies

This methodology has been applied to processes operating in batch mode (Laundry Case Study and Insulin Production) and continuous mode (MTBE, Ammonia and VCM production). In this paper, the results obtained for the

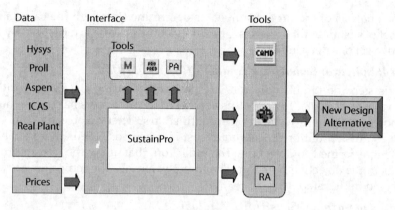

FIGURE 3
SustainPro interface.

sustainable design of an Insulin Production process operating in the batch-mode are highlighted in detail. Table 1 summarizes the main results for the above mentioned case studies.

Insulin Production Case Study (BatchProcess)

The insulin process, Petrides, *et al.* (1995), is divided into four sections: Fermentation, Primary Recovery, Reactions and Final Purification.

TABLE 1

Summary for Different Analyzed Case Studies

Case Study	New Alternative	Improvements
Continuous Processes		
MTBE	Reduce Water in Recycle	1) Target Indicator: 20%
		2) Energy Metric: 3%
		3) Water Metric: 4%
		4) Profit: 1,6%
Ammonia	Recycle Water	1) Target Indicator: 100%
		3) Water Metric: 26%
		4) Profit: 82%
VCM	Reduce Raw Material Recycle	1) Target Indicator: 31%
		2) Energy Metric: 2%
		3) Water Metric: 2,8%
		4) Profit: 0,34%
Batch Processes		
Laundry	Recycle Water	1) Cont. Target Indi.:100%
		2) Batch Target Indi.:27%
		3) Water Metric: 20%
		4) Cost: 36%

Step 1: Collect the steady state data

The required detailed process data for the insulin synthesis plant was taken from a simulation available on SuperPro Designer software package. The prices and costs were taken from Petrides, *et al.* (1995), where the insulin production simulation is described in detail.

Step 1.A: Transform equipment flowsheet in an operational flow-diagram

The equipment flowsheet consists of 31 units. Taking into account the sequence of operations, the operational flow-diagram is determined (92 operations, 169 streams and 38 compounds).

Step 2: Flowsheet decomposition

For this case study the operational flow-diagram decomposition generated 418 closed-paths, 1022 open-paths and 3344 accumulation-paths.

Step 3: Calculate the indicators, the sustainability and the safety metrics

For the entire set of flow-paths, the full-set of indicators were calculated and only the most sensitive indicators were selected for further study. The sustainability metrics and the safety index were also calculated.

Step 4: Indicator Sensitivity Analysis (ISA) algorithm

To apply the ISA algorithm the indicators for section 3 were selected as possible target indicators. After applying the ISA algorithm it is seen that from the selected indicators, the MVA (Material Value Added) indicator related to OP591 for Formic acid is the most sensitive. Consequently, this indicator is considered the target indicator to achieve process improvements. For batch indicators, the most sensitive indicator in section 1 is the TF of ammonia in the fermentation operation (V-102R).

Step 5: Process sensitivity Analysis

From a sensitivity analysis of the operational parameters influencing the target indicator (MVA - OP591) it was found that the most significant operational parameter is the flowrate of OP591.

Through analysis of the operational parameters that influence the batch target indicator (TF), it was possible to verify that the ammonia (NH_3) concentration is the most significant parameter in order to reduce the time of the reaction.

Step 6: Generation of new design alternatives

To reduce the OP flowrate, the recycle of formic acid coming from this OP needs to be considered. To recycle formic acid, a separation operation needs to be inserted in order to purify/recover this compound. Applying the process separation algorithm of Jaksland and Gani (1996), a set of feasible separation techniques for the recovery of formic acid were identified and pervaporation was selected as the separation operation, because it involves lower operational costs when compared with the other separation techniques and it does not need solvents. In the literature, Nakatani *et al.* (1994) found that membranes such as aromatic imide polymer asymmetric are available to purify/recover formic acid from water. Using this information the process was simulated again in order to validate the new design alternative. To reduce the fermentation time the concentration of ammonia needs to be increased. The concentration was increased by 2% and 0.2% of fermentation time reduction was achieved. This is not a significant improvement. This fact indicates that the fermentation process is already optimized and further improvement is not possible. Also, the fermentation operation has additional constraints (such as enzyme inhibition) that cannot be violated without changing the enzyme. For the new sustainable design alternative, which consists of recycling of formic acid, the following improvements were achieved: the profit increased by 1.98%, the water and the energy metrics per value added improved by 2%. The material metrics improved by 2% and 4%, respectively, per kg of final product and per value added. Finally, the environmental impact output was improved by 31.7%. The rest of the performance criteria parameters have remained constant. The target indicator was improved by 99.9% (Initial Value MVA = –17340$/y). These results show that a more sustainable design alternative is presented.

Other Case Studies

In Table 1 the most significant results for other case studies are presented.

Conclusions

The development of a systematic and generic indicator-based methodology for continuous and batch process analysis and for generating sustainable processes improvements has been presented and highlighted. With the development of corresponding SustainPro software, the application of the methodology to study different chemical processes has become more efficient. The software generates new design alternatives and also performs sustainable evaluation of any proposed design alternative. As shown through the insulin case study, the size of the problem is not an issue since the objective is to locate the most sensitive indicators and their related variables, which are usually not that many. It can be concluded that this methodology has a wide range of applicability (see Table 1). Current and future work will further develop the methodology to incorporate more features to handle various types of batch and continuous processes as well as improve the usability of SustainPro.

Acknowledgments

The authors gratefully acknowledge financial support from Fundação para a Ciência e a Tecnologia (under Grant No. SFRH/BD/24470/2005).

References

Carvalho, A., Matos, H. A., Gani. R., 2008, Design of Sustainable Chemical Processes: Systematic Retrofit Analysis Generation and evaluation of alternatives, *Process Safety and Environmental Protection*, 86: 328–346

Carvalho, A., Matos, H. A., Gani. R., 2009, Computers and Chemical Engineering, submited to publication

Halim, I. and Srinivasan, R., 2008, Designing sustainable alternatives for batch operations using an intelligent simulation–optimization framework, *Chemical engineering research and design*, 86: 809–822

Jaksland, C., Gani, R. and Lien, K., 1996, An integrated approach to process/product design and synthesis based on properties-process relationship, *Comput.Chem. Eng*, 20: 151–156.

Lange J.P., 2002, Sustainable development: efficiency and recycling in chemical manufacturing, *Green Chem*, 4(6): 546–550.

Liu, J.; Fan L. T.; Seib, P.; Friedler, F.; Bertok, B., 2006, Holistic Approach to Process Retrofitting: Application to Downstream Process for Biochemical Production of Organics, *Ind. Eng. Chem. Res.*, 45: 4200–4207.

Nakatani, M., Sumiyama, Y., Kusuki, Y., 1994, EP0391699: Pervaporation method of selectively separating water from an organic material aqueous solution through aromatic imide polymer asymmetric membrane

Petrides, D., Sapidou, E., Calandranis, J., 1995, Computer-aided process analysis and economic evaluation for biosynthetic human insulin production—a case study. *Biotechnology and Bioengineering*, 48 (5), 529–541.

Simon L. L., Osterwalder N., Fischer U., Hungerbuhler K., Systematic Retrofit Method for Chemical Batch Processes Using Indicators, Heuristics, and Process Models, *Ind. Eng. Chem. Res.*, 47: 66–80

19

Process Synthesis Using Integrated Energy Networks

M. M. Faruque Hasan and I. A. Karimi*

Department of Chemical & Biomolecular Engineering,
National University of Singapore, 4 Engineering Drive 4, Singapore 117576

CONTENTS

ABSTRACT Despite recent works to integrate energy in various unit and network levels, most literature does not rigorously perform energy integration during flow sheeting. Moreover, most design procedures are heavily dependent on heuristics. Given a number of streams with their initial and final temperatures, pressures and states, we rigorously optimize a process flow diagram for minimum energy requirement. We achieve this through the optimal sequencing of operations, such as cooling, heating, compression, expansion, and pumping. The challenges, however, arise mainly due to the nonlinear relations between various process variables, such as temperature and pressure, temperature and enthalpy, etc. In this work, we apply a cubic correlation for boiling point and pressure, and use binary variable to select the stream state. Using a stage-wise superstructure representation, we propose a mixed integer nonlinear program (MINLP) for optimally sequencing different energy consuming operations. From the solution, we can further integrate energy by developing heat exchanger and compressor networks for the process. Finally, we present a case study to demonstrate the applicability of our approach.

---

* To whom all correspondence should be addressed

KEYWORDS   *Energy networks, process synthesis, design, optimal PFD, phase change*

Introduction

Energy is an immediate concern for chemical industries. High energy prices, tightening regulations on CO_2 emissions, intense competition in an increasingly global market, etc. underline the importance of efficient use of energy. The industrial sector is the largest consumer of energy in the USA, which consumed nearly 32% of the world's energy in 2007 (EIA/Monthly Energy Review, May 2008). As much as 40% of the total operational cost of a chemical plant is attributable to energy. This is why energy integration in chemical industries has been a major concern over the years.

Designing energy efficient processes is crucial and the inevitable first step for reducing energy consumption in a chemical plant. Process synthesis combines different processing and auxiliary units in a systematic way to achieve desired product(s) at specified conditions from raw material(s). It involves gathering information, representation of alternatives, assessment of preliminary designs, and generating and searching among alternatives (Biegler et al., 1997). Despite recent works to integrate energy in various unit and network levels, most synthesis literature does not consider it in the conceptual design phase or during flow sheeting, but treat as a separate and to-do-later entity in the form of, for example, design and operation of networks for heat exchangers (Furman and Sahinidis, 2002), fuel gas (Hasan et al., 2008a), reactors (Floudas, 1995), separation (Floudas, 1995), mass exchange (El-Halwagi et al., 2003) and utility (Bruno et al., 1998). Moreover, most design procedures are heavily dependent on heuristics. Examples of such heuristics include, using few equipment as possible, compressing before heating, etc.

Heuristics often results in sub-optimal flow sheet in terms of total energy consumption. However, in spite of the extensive literature on process synthesis, no rigorous and automated methodologies are available for designing a process in the most energy efficient manner at the first place. To the best of our knowledge, a rigorous methodology for dealing with optimal sequencing of various process operations during the development of process flow diagram (PFD) is missing.

Synthesis of energy efficient and cost effective PFD poses several challenges. The first concerns the nonlinear temperature-enthalpy (T-H) curves. A typical T-H curve for a multi-component mixture may have multiple zones (Hasan et al., 2009a). Moreover, it varies nonlinearly with pressure. Second, although phase changes abound in industries such as petrochemicals, gas processing, cryogenic, and chemical, where operations such as distillation, stripping, refrigeration, etc. are common, it is not rigorously modeled in process synthesis literature. Hasan et al. (2009b) incorporated multi-component phase change in HENS. However, they did not consider synthesis of PFD. As we show in the motivating example presented in the next section, we can effectively sequence operations or tasks in a PFD and use phase change to reduce total utility and power consumption of a process significantly.

In this work, we present a mixed integer nonlinear program (MINLP) for optimal sequencing of operations, such as cooling, heating, compression, pumping, etc. for minimum utility and fixed cost. Finally, we show the utility of our method using a case study.

Motivating Example

Let us consider a synthesis problem where natural gas or NG (methane 63%, ethane 22.84%, propane 4.26%, n-butane 1%, and nitrogen 8.9%) with a flow of 1 kgmol/s is required to be pressurized from a state (S1) of 160 kPa at 30°C to a state (S2) of 4200 kPa at 30°C. Both S1 and S2 represent gas phase. We develop the first alternative (A1) by applying three heuristics, namely compress before cool, avoid phase change, and use minimum units. A compressor having 75% adiabatic efficiency is used to increase the pressure from 160 kPa to 4200 kPa, which requires about 14.83 MW of power. However, after compression, the temperature is also increased to 320°C. Therefore, NG is cooled to 30°C before it finally reaches S2. The cooling utility is 15.96 MW. In total, A1 requires 30.79 MW of energy, one compressor and one cooler.

However, the total theoretical change in energy content from S1 to S2 for NG is only 1.13 MW (using ASPEN HYSYS with Peng-Robinson as the fluid package). Therefore, A1 is far away from the optimal design in terms of energy usage.

Let us consider another alternative (A2), which involves three units, namely a heat exchanger, a pump, and a cooler in a sequence as mentioned. Instead of being compressed, NG is first liquefied from 30°C to –177°C. the liquid is then pumped from 160 kPa to 4200 kPa using a pump with 75% adiabatic efficiency. Pumping requires only 0.21 MW of utility and increases the temperature by 1°C. This liquefied and pressurized NG is then heated to change the phase from liquid to gas again and further superheated to 56°C in a heat exchanger which is integrated with the liquefaction. Therefore, liquefaction and evaporation are completely energy integrated and no external utility is required. Such integration saves 18.59 MW of utility for liquefaction and the same for evaporation. Finally, superheated natural gas is cooled to 30°C using a cooler that uses only 1.35 MW of cooling utility. In total, A2 requires an external energy of 1.56 MW, which is close to the theoretical minimum energy requirement for the process. In other words, although phase change is present and one more unit is used in A2, it is a much better alternative than A1 in terms of energy usage. Although the process has a net gain of energy, A2 is non-intuitive and involves rejection of heat and phase change.

The above motivating example leverages on the fact that optimal sequencing of heating, cooling, compression, etc. is crucial. Moreover, phase change plays a role in energy integration. Most importantly, it is now shown that the optimal PFD might require heat integration, even before the stage when traditional HENS is applied. This would lead to the rigorous optimization of the flow sheet at the first place.

We now present the sequencing problem for process synthesis.

Problem Statement

Let I ($i = 1, ..., I$) process streams with flow rate F_i are given. Let T_i^{IN} and P_i^{IN} be the initial temperature and pressure for stream i respectively. Develop a process network that transfers I streams to a final state with temperature and pressure of T_i^{OUT} and P_i^{OUT} respectively for minimum operating and fixed cost or a suitable design objective.

We assume the following.

(1) Ideal gas behavior

(2) Isothermal phase change

(3) Linear T-H relations

(4) No pressure drops in heaters and coolers, and no change in temperatures in pumps.

We now present the MINLP formulation to develop a conceptual design (CD) or block diagram for the aforementioned problem.

MINLP Formulation

Let J ($j = 1, ..., J$) operations or tasks can be performed. We use a stage-wise superstructure representation (Figure 1) to represent all plausible alternatives. K ($k = 1, ..., K$) stages are present in the superstructure. At each stage k, stream i can either undergo heating ($j = 1$), cooling ($j = 2$), compression ($j = 3$), expansion ($j = 4$), or pumping ($j = 5$).

To select task j to be conducted at stage k, we define a binary variable x_{ijk} as follows.

$$x_{ijk} = \begin{cases} 1 & \text{if stream } i \text{ undergoes task } j \text{ in stage } k \\ 0 & \text{otherwise} \end{cases}$$

To ensure that at most one task can be performed at each stage, we impose,

$$\sum_j x_{ijk} \leq 1 \tag{1}$$

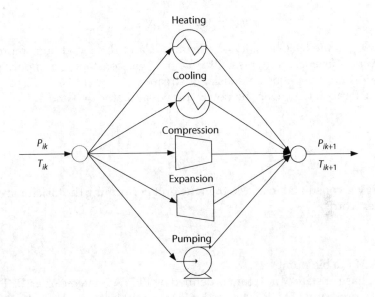

FIGURE 1
Stage *k* of the superstructure.

Let P_{ik} and T_{ik} be the pressure and temperature respectively when stream *i* enters stage *k*. Therefore, $P_{i1} = P_i^{IN}$, $T_{i1} = T_i^{IN}$, $P_{iK+1} = P_i^{OUT}$, and $T_{iK+1} = T_i^{OUT}$. Let ΔP_{ijk} and ΔT_{ijk} be the change in pressure and temperature respectively when stream *i* undergoes task *j* at stage *k*. Therefore,

$$P_{ik+1} = P_{ik} + \sum_j \Delta P_{ijk} \tag{2}$$

$$T_{ik+1} = T_{ik} + \sum_j \Delta T_{ijk} \tag{3}$$

Note that $\Delta P_{i1k} = 0$, $\Delta P_{i2k} = 0$, $\Delta T_{i5k} = 0$.

Let Q_{ijk} be the energy consumed during task *j* when x_{ijk} is one. Energy balance for each unit yields,

$$\textit{heater:} \ Q_{i1k} = F_i C p_i \Delta T_{i1k} \tag{4}$$

$$\textit{cooler:} \ Q_{i2k} = -F_i C p_i \Delta T_{i2k} \tag{5}$$

$$\textit{compressor:} \ Q_{i3k} = F_i \alpha_i \Delta T_{i3k} \tag{6}$$

$$\textit{turbine:} \ Q_{i4k} = -F_i \beta_i \Delta T_{i4k} \tag{7}$$

$$pump: \ Q_{i5k} = F_i \gamma_i \Delta P_{i5k} \tag{8}$$

where, C_{pi} is the heat capacity of stream i, α_i and β_i depend on compressor or turbine efficiency, compressibility factor Z, gas constant R, and polytropic constant, and γ_i depends on density of stream i.

The T-P relation is given the polytropic equation for $j = 3$ and 4,

$$\frac{\Delta T_{ijk}}{T_{ik}} = \frac{\left(P_{ik} + \Delta P_{ijk}\right)^{\frac{\delta_i - 1}{\delta_i}}}{\left(P_{ik}\right)^{\frac{\delta_i - 1}{\delta_i}}} - 1 \tag{9}$$

If energy is consumed in a task, then the unit performing the task must exist. In other words,

$$Q_{ijk} \leq M x_{ijk} \tag{10}$$

where, M is a big number.

The phase or state of a stream is defined by BPT. As stated earlier, BPT varies with pressure. Let BPT_{ik} be the boiling point temperature of stream i at pressure P_{ik}. We use the following correlation to model the relation between BPT and pressure.

$$BPT_{ik} = BPT_i^{IN} \pm \left[A_i \Delta PR_{ik} + B_i \Delta PR_{ik}^2 \right] \tag{11}$$

where, A_i, and B_i are fitted parameters, BPT_i^{IN} is the BPT of stream i at P_i^{IN}, and $\Delta PR_{ik} = P_{ik} - P_i^{IN}$.

To select the state of stream i, we define binary variable y_{ik} as follows.

$$y_{ik} = \begin{cases} 1 & \text{if stream } i \text{ is in gas phase when it enters stage } k \\ 0 & \text{otherwise} \end{cases}$$

From the definition of y_{ik}, we require,

$$T_{ik} \geq BPT_{ik} - M(1 - y_{ik}) \tag{12}$$

$$T_{ik} \leq BPT_{ik} + M y_{ik} \tag{13}$$

Eqs. 12–13 ensure that $T_{ik} \geq BPT_{ik}$ when y_{ik} is one and $T_{ik} \leq BPT_{ik}$ when y_{ik} is zero.

Since no liquid can be present in compressors and turbines, we must ensure that compressors and turbines are used only for gases. Therefore,

$$y_{ik} \geq x_{i3k} \tag{14}$$

$$y_{ik+1} \geq x_{i3k} \tag{15}$$

$$y_{ik} \geq x_{i4k} \tag{16}$$

$$y_{ik+1} \geq x_{i4k} \tag{17}$$

Similarly, to ensure that pumps are only used for liquids, we have,

$$1 - y_{ik+1} \geq x_{i5k} \tag{18}$$

$$1 - y_{ik+1} \geq x_{i5k} \tag{19}$$

Objective Function

The objective here is to minimize total energy consumption and the fixed cost of the units.

$$\text{Min} \sum_{i}\sum_{i}\sum_{k} UC_{ij}Q_{ijk} + \sum_{i}\sum_{i}\sum_{k} FC_{ij}x_{ijk} \tag{18}$$

where, UC_{ij} and FC_{ij} are the unit cost of power and fixed cost for stream i to perform task j.

Once CD is finalized, we can apply the generalized HENS (GHENS) technique (Hasan et al., 2009c) for the synthesis of optimal heat exchanger networks for further integration of energy, if required.

Case Study

The case study involves two streams with initial and final states given in Table 1.

The computing platform used for the case study is an AMD Athlon™ 64 × 2 Dual Core Processor 6000 + 3.00 GHz, 3.00 GB of RAM using BARON v.7.5 (MINLP solver) in GAMS 22.2. The model has 217 continuous and 72 binary variables, 281 constraints, and 919 nonzero elements. The model is

TABLE 1

Data for the Case Study

	Pressure kPa		Temperature (K)	
	Initial	Final	Initial	Final
stream 1	160	4200	303	350
stream 2	200	500	340	400

solved within 312 CPU s. The total network cost is $ 3.63×10^6. We then apply GHENS to the PFD generated by the above model. The final network involves two compressors and a heater and a cooler. The final process involves heat integration in both conceptual level and network level. The result shows improvement in energy requirement compared to the alternatives generated using various heuristics.

Conclusions

We have presented an MINLP model for optimal synthesis of PFD and to integrate energy and minimize cost at the first place. Using a motivating example and a case study, we have shown that integrating energy in the initial design phase can yield significant savings in energy. Further work is needed to exploit the full potential of such an approach.

Acknowledgements

The authors would like to acknowledge the financial support for this work in the form of a Research Scholarship from the National University of Singapore.

References

Biegler, L. T., Grossmann, I. E., Westerberg, A. W. (1997). Systematic Methods of Chemical Process Design. *Prentice Hall PTR*. Upper Saddle River, N.J.

Bruno, J. C., Fernandez, F., Castells, F., Grossmann, I. E. (1998). MINLP Model for Optimal Synthesis and Operation of Utility Systems. *Chemical Engineering Research and Design*, 76, 246–258.

El-Halwagi, M. M.; Gabrien, F.; Harell, D. (2003). Rigorous Graphical Targeting for Resource Conservation via Material Recycle/Reuse Networks. *Ind. Eng. Chem. Res.* 42, 4319–4328.

Energy Information Administration/Monthly Energy Review May 2008. Web Page: http://www.eia.doe.gov/emeu/mer/consump.html.

Floudas, C. A. (1995). Nonlinear and mixed-integer optimization: fundamentals and applications. *Oxford University Press*. New York.

Furman, K. C, Sahinidis, N. V. (2002). A Critical Review and Annotated Bibliography for HE Network Synthesis in the 20th Century. *Ind. Eng. Chem. Res.* 40:2335–2370.

Hasan, M. M. F., Karimi, I. A., Alfadala, H. E., Grootjans, H. (2009a). Operational Modeling of Multi-Stream Heat Exchangers with Phase Changes. *AIChE Journal*, 55(1): 150-171.

Hasan, M. M. F., Karimi, I. A. (2009b). Synthesis of Heat Exchanger Networks with Multi-component Phase Changes. Submitted to *AIChE Journal.*

Hasan M. M. F., Karimi IA, Alfadala HE. (2009c). Synthesis of Heat Exchanger Networks Involving Phase Changes. In: *Proc. 1st Annual Gas Processing Symposium.* 185–192.

20

Optimal Synthesis of Compressor Networks

Md. Shamsuzzaman Razib, M. M. Faruque Hasan and I. A. Karimi*

Department of Chemical & Biomolecular Engineering,
National University of Singapore, 4 Engineering Drive 4, Singapore 117576

CONTENTS

ABSTRACT Energy production/usage and its impact on environment are global concerns. The cleanest energy is the one that we never use, thus energy efficiency and conservation are extremely critical in mitigating the impact on environment and climate. Process industries are major users of energy and plants are increasingly keen on increasing efficiency. In many chemical plants such as LNG, refineries, air enrichment, ammonia, etc, compression is a major consumer of energy. In such plants, some process streams may need compression, while others may need expansion. Optimal integration of these streams can yield major savings in energy. Even when streams need compression only, e.g. for natural gas transmission, optimal design and operation of compression networks is essential. In other words, a synthesis problem analogous to that for a heat exchange network can be defined for compression as well. While a lot of work exists on heat exchanger network synthesis, there is virtually no work on compressor network synthesis. The objective of this paper is to minimize the total power costs by reducing energy consumption demanded by compressors. We employ the concept of superstructure and formulate the entire problem as a mixed-integer nonlinear program (MINLP). We propose a methodology for solving this difficult problem and demonstrate the benefits of such an optimized network.

KEYWORDS *Compressor networks, energy integration, process synthesis, MINLP*

---

* To whom all correspondence should be addressed

Introduction

High energy price due to restriction of sources of energy is a global alert. Besides this, conversion of energy also causes dreadful impact on environment emitting CO_2. Chemical plants are considered as main users of energy. Its high requirement coupled with its increasing price trigger chemical plants to have an efficient use of energy.

Because of process requirements in chemical plants i.e. refineries, petrochemical, natural gas (NG) and liquefied natural gas (LNG) production and synthesis plants, some process streams may need compression, and or expansion. Compressors, valves and turbines are used for compression and expansion. About 40% of energy is consumed by compressors in these industries. Hence, optimal integration of these streams can be seen as an opportunity to yield major savings in energy e.g by designing integration of compressor networks.

Del Nogal et al. (2005) presented an MILP model for driver and power plant selection using the idea of superstructure for multistage compressors. In a follow-up paper (Del Nogal et al., 2006), they used multistage compressors with intercoolers to design a multi-component refrigerant system. However, their focus was compressor stage arrangement rather than network design.

In almost all chemical process industries, low pressure (LP) streams need to be pressurized, while high pressure (HP) streams need to be depressurized to meet the condition of stream requirements for different purposes in the plant. However, to the best of our knowledge, no systematic approach is available in the literature for the synthesis of a network that utilizes the pressure energy and integrates energy from HP to LP streams. In this work, we consider a compressor network synthesis problem, where streams at different pressures need either compression or expansion, and energy integration involves extracting energy from high-pressure streams and utilizing it for low-pressure streams. The network may involve compressors, turbines (gas or steam), single-shaft turbo-compressors, valves, exchangers, etc. and steam at different pressure levels. It will involve realistic work with a goal to minimize energy consumption and/or costs.

In this paper, we employ the concept of superstructure and formulate the entire problem as a mixed-integer nonlinear program (MINLP). We also demnostrate the benefits of such an optimized network using a case study.

Problem Statement

Let I ($i = 1,..., I$) HP and J ($j = 1,..., J$) LP streams are present in a chemical process. HP streams need to be depressurized from P_{io}^{in} to P_{io}^{out}, and LP streams need to be pressurized from P_{jo}^{in} to P_{jo}^{out}. These streams also have different stream properties as they come from different sections of the plant. Valves and turbines can be used for depressurization, while compressors can be used for pressurization. The intergation is done by developing a turbo-compressor (coupling of turbines and compressors in a single shaft) network. The streams which do not need to change pressure are not included in this network. When HP streams go through valves to reduce pressure removed energy cannot be utilized from the system. Hence, we lose energy when we intend to use valve. On the Contrary, if we use turbine to reduce pressure we can extract energy from the streams and utilize this energy either to operate compressor or generate electricity that reduce utility energy consumption to drive compressors as well as reduce operating cost to drive compressors. However, fixed cost for valve is lower than turbine. Therefore, there is a trade-off of using valves and turbines for depressurization purpose. The LP streams needed to be pressurized can either go through the compressors driven by those turbines which extracted energy from HP streams or through the compressors driven by external utility energy sources. As compressors efficiency are rapidly influenced by rise of temperature, cooler is required after certain increase of temperature due to compression. Hence, cooler is added between two compression stages to mollify compression efficiency.

Compression pressure ratio, inter-stage temperature, number of stage, efficiency and energy requirement is variables of this turbo-compressor network those affect compression operations. Inlet and outlet streams conditions i.e. pressure, temperature, flow rate, composition should be defined. Also compression path exponent, compressibility and heat capacity for a certain gas streams should be defined.

We assume the following.

1. All streams are in gas phase.
2. Adiabatic (isentropic) compression/expansion.
3. Constant compression path exponent, compressibility and heat capacity.
4. No pressure drops in the inter-stage coolers.
5. Cooling is implemented after each compression.

Model Formulation

We use a stage-wise superstructure with K ($k = 1, \ldots, K$) stages that allows almost all possibilities for integrating pressure energy from HP to LP streams. Figure 1 shows the superstructure for the turbo-compressor network with two HP and two LP streams.

In each stage of the superstructure, HP stream supplies energy to a turbine and LP stream uses energy through a compressor. The HP streams

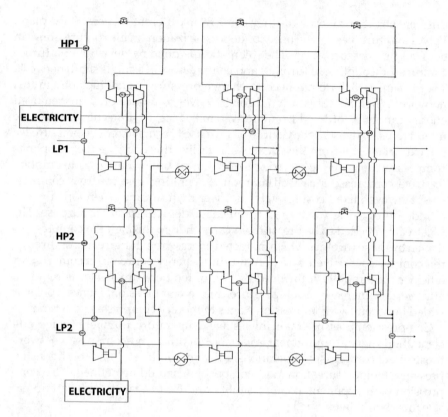

FIGURE 1
Superstructure of the compressor network.

have options to go through either valve and/or turbine. If HP streams go through turbines, they will release energy and the energy will be sent to the central utility energy source. On the other hand, if HP streams go through the valves, no energy will be utilized (loss of energy). Meanwhile, the LP streams have also options to go through the compressors driven by either turbines extracted energy from HP or external utility energy source. In the superstructure only one compressor is shown in one shaft which can be more than one depending on stream flow. After compression the streams have the option to go through inter-stage cooler between each stage. The streams will pass through cooler if temperature of the streams becomes high due to compression as it affects compression efficiency.

Let s and m denote splitting of HP and LP streams respectively. The HP has two options to flow. If $s = 1$, then it goes through valve, and if $s = 2$, it goes through turbine.

We now define two binary variables for the structural decision of the network as follows.

$$y_{i,s,k} = \begin{cases} 1 & \text{if stream } i \text{ passes through unit } s \text{ in stage } k \\ 0 & \text{otherwise} \end{cases}$$

$$z_{j,m,k} = \begin{cases} 1 & \text{if stream } j \text{ passes through unit } m \text{ in stage } k \\ 0 & \text{otherwise} \end{cases}$$

Let $F_{i,s,k}$ ($F_{j,m,k}$) be the flow of stream i (j) that passes through unit s in stage k. To model the stream splitting, we require,

$$F_{io} = \sum_s F_{i,s,k} \tag{1}$$

$$F_{jo} = \sum_m F_{j,m,k} \tag{2}$$

Where, F_{io} and F_{jo} are the initial flow of stream i and j respectively. To ensure a minimum flow exists if $y_{i,s,k}$ is one, we require,

$$UV_{min}y_{i,1,k} \leq F_{i,1,k} \leq UV_{max}y_{i,1,k} \tag{3}$$

$$UT_{min}y_{i,2,k} \leq F_{i,2,k} \leq UT_{max}y_{i,2,k} \tag{4}$$

Similar constraints are required for $F_{j,m,k}$. Now, compression ratio at each stage is similar and can be expressed as the following where ER represents expansion ratio and CR represents compression ratio. If there will be no compression ER and CR will be one. Besides this, CR_{min} and ER_{max} are always one. Let $P_{i,k}$ be the pressure of i stream at stage k. Note that $P_{i,k-1}$ is P_{io}^{in} for $k = 1$, and $P_{i,K} = P_{io}^{out}$. Pressure change at different stage can be represented as

$$P_{i,k} = ER_{i,k}P_{i,k-1} \tag{5}$$

As minimum expansion ratio is required to conduct a feasible operation in turbine, maximum and minimum expansion duty at one stage is introduced.

$$ER_{min} \leq ER_{i,k} \leq ER_{max} \tag{6}$$

When stream does not flow through a unit, ER must be one. Hence,

$$ER_{i,k} \geq 1 - y_{i,2,k} \tag{7}$$

Similarly constraints are used for $CR_{j,k}$. Note that $P_{j,m,k-1}$ is P_{jo}^{in} for $k = 1$, and $P_{j,K} = P_{jo}^{out}$. Therefore,

$$P_{j,m,k} = CR_{j,m,k}P_{j,m,k-1} \tag{8}$$

Compression ratios of all splitting at same stage must be equal as after compression same type of streams mix and go through next compressor. Therefore,

$$P_{j,m-1,k} = P_{j,m,k} \tag{9}$$

When stream does not flow through a unit, CR must be one. Hence,

$$CR_{j,m,k} \le 1 + Mz_{j,m,k} \tag{10}$$

Where, M is a big number. Furthermore, CR has also maximum and minimum values and CR must be one if stream does not flow through a unit. Now, temperature and energy equations are introduced to calculate temperature at each stage as well as energy. Let $Q_{j,m,k}$ be the cooling duty in the interstage cooler. Therefore,

$$Q_{j,m,k} = F_{jo}^{in} C_p \left(T_{j,m,k} - T_{jo}^{in} \right) \tag{11}$$

Let $T_{i,k}$ and $T_{j,m,k}$ represent temperature of high and LP streams after expansion and compression respectively. Therefore,

$$T_{i,k} = T_{i,k-1} \exp\left[\frac{(\gamma - 1)}{\gamma} \log(ER_{i,k}) \right] \tag{12}$$

$$T_{j,m,k} = T_{jo}^{in} \exp\left[\frac{(\gamma - 1)}{\gamma} \log(CR_{j,m,k}) \right] \tag{13}$$

Where, T_{jo}^{in} is the suction temperature of compressor as cooler always cool streams to it initial temperature, and $T_{i,k-1}$ is T_{io}^{in} for $k = 1$.

Let $W_{i,s,k}$ be the energy extracted by turbine while $W_{j,m,k}$ is energy required by compressors. If HP passes through valves, such that s is one, no energy is extracted. Hence,

$$W_{i,1,k} = 0 \tag{14}$$

$$W_{i,2,k} = 0.0853 F_{i,2,k} T_{i,k} \left[1 - ER_{i,k}^{z(\gamma-1)/\gamma} \right] \tag{15}$$

$$W_{j,m,k} = 0.0853 F_{j,m,k} T_{jo}^{in} \left[CR_{j,m,k}^{z(\gamma-1)/\gamma} - 1 \right] \tag{16}$$

Compressors are driven by either utility energy source or turbines those extract energy from HP. When m is one, compressors are driven by external energy source while for other values of m it uses HP stream energy. Let $We_{i,k}$ be the electricity produced from turbine using HP. Therefore,

$$W_{i,2,k} = \sum_j W_{j,m,k} + We_{i,k} \quad m > 1 \tag{17}$$

The objective is to reduce utility requirements to drive compressors as well as minimize cost of a turbo-compressor network. Let *FCC*, *FCV*, and *FCT* be the fixed cost of compressors, valves, and turbines respectively, and *CC*, *CH* and *CT* are the price of unit energy for compressors, heat exchanger and electricity respectively. Therefore, the objective function is as follows.

$$\text{Min } FCC \sum_j \sum_m \sum_k z_{j,m,k} + FCV \sum_i \sum_k y_{i,1,k}$$

$$+ FCT \sum_i \sum_k y_{i,2,k} + CC \sum_j \sum_k W_{j,1,k}$$

$$+ CH \sum_j \sum_m \sum_k Q_{j,m,k} - CT \sum_i \sum_k We_{i,k} \tag{18}$$

This completes the MINLP formulation.

Case Study

In our study, we have two HP (HP1-2) and LP (LP1-2) streams. Data for these streams are provided in Table 1.

The model was solved on a Dell Optiplex GX280 with Pentium IV HT 3.20 GHz, 2 GB RAM using BARON v.7.5 in GAMS 22.2. The minimum cost found

TABLE 1

Stream Data for the Case Study

Strea	Pressure (Psi)		Flow Rate (MMSCFD)
	Initial	Final	
Hp1	500	200	400
Hp2	500	20	200
Lp1	14.7	152	300
Lp2	29.4	200	180

from GAMS is US$5202K while without utilizing the energy from HP stream cost around 8.34% increase in power cost. The result justifies the feasibility and usefulness of the model in process field.

Conclusion

In this paper, we present a cost minimization strategy of turbo-compressor network in a process industry. This is an introductory approach to design a turbo-compressor network. Compression and expansion are the main decision variables in this project. The objective of the model is to minimize the total power cost for process plants. In this model compressibility, heat capacity and compression path exponent are assumed to be constant. Furthermore, we need to consider pressure and temperature affect on compressibility, heat capacity and compression path exponent. Besides this, the streams are cooled after each compression. A decision variable can also be introduced to make a decision for cooling duty. Finally, the model can be generalized for any type of process industry where vapor streams are ready for compression and expansion.

References

Brown, Royce, N. Compressors Selection and Sizing, Page 38, Third Edition, *Gulf Publishing*, 2005.

Bruno, J.C., Fernandez, F., Castells, F. and Grossmann, I.E., A Rigorous MINLP Model for the Optimal Synthesis and Operation of Utility Plants. *Trans IChemE*, Vol. 76, Part A, March 1998.

Cumpsty, N.A, Compressor Aerodynamics, Page 39, First Edition, Longman Scientific and Technical, 1989.

Del Nogal, F. L., Townsend, D.W. and Perry, S.J., Synthesis of power systems for LNG plants, *GPA Europe Spring Meeting*, Bournemouth, UK, 2003.

Del Nogal, F.L., Kim, J., Perry, S.J. and Smith, R. (2005). Systematic driver and power plant selection for power-demanding industrial processes, Energy Efficiency for a Sustainable Economy, *AIChE Spring National Meeting*, Atlanta, USA

Del Nogal, F.L., Kim, J., Perry, S.J. and Smith, R. Integrated Approach for the Design of Refrigeration and Power, 6th Topical Conference on Natural Gas Utilization, *AIChE Spring National Meeting*, Florida, USA, 2006.

Edgar. Thomas F., Himmelblau, David.M, Lasdon, Leon. S, Optimization of Chemical Processes, Page 472, 2nd Edition, *McGraw Hill*, 2001.

Hasan, M. M. F., Karimi, I. A, Alfadala, H. E. Optimizing Compressor Operations in an LNG Plant. Proceedings of the *1st Annual Gas Processing Symposium*. Editors H.Alfadala G.V, Rex Reklaitis and M.M El-Halwagi. 2009 Elsevier B.V.

Heckel, B.G. and Davis, F.W. Starter Motor Sizing for Large Gas Turbine (Single Shaft) Driven LNG Strings. 3rd Doha Conference on Natural Gas, Doha, Qatar, March 1999.

Manninen, J. and Zhu, X. X. Optimal Gas Turbine Integration to the Process Industries. *Ind. Eng. Chem.Res.*, Vol. 38, pp. 4317–4329, 1999.

21

Developing Sustainable Processes for Natural Gas Utilization by Using Novel Process Synthesis Methods and Tools

Truls Gundersen[*] and Audun Aspelund

Department of Energy and Process Engineering
Norwegian University of Science and Technology
Kolbjoern Hejes vei 1.B, NO-7491 Trondheim, Norway

CONTENTS

ABSTRACT A Liquefied Energy Chain (LEC) for utilizing stranded natural gas with CO_2 capture and offshore storage for EOR has been developed. The LEC is a novel energy- and cost-effective transport chain consisting of an offshore section, a combined carrier and a receiving terminal that is integrated with an Oxy-fuel power plant and an Air Separation Unit (ASU). The LEC concept elegantly and efficiently solves the CCS problem for the power plant, and it is resource efficient in the sense that natural gas that otherwise is reinjected into the oil field is utilized for power production. The concept is energy efficient for a number of reasons. First, the offshore process is self-supported with power and hot/cold utilities. Secondly, nitrogen which is a valuable byproduct from the ASU is utilized as a carrier of cold exergy (a refrigerant) and not wasted. Benchmarking has shown that the LEC is 2-3 times more energy efficient than conventional transport chains for stranded natural gas with CCS. As a crucial part of the work, design methods and tools have been developed for subambient processes by combining and expanding Process Synthesis methods such as Heuristic Rules, Pinch and Exergy

---

[*] To whom all correspondence should be addressed

273

Analyses and Optimization. The single most important contribution of this work is the inclusion of pressure as a vital design variable.

KEYWORDS *Process synthesis, pinch analysis, exergy analysis, optimization, subambient, LNG, CCS, EOR*

Introduction

There are multiple objectives behind the work described in this paper, such as improved resource utilization, reduced environmental effects and improved design methodologies for subambient processes. Stranded natural gas is utilized for onshore power production through a Liquefied Energy Chain (LEC). The CO_2 from the power plant is captured and stored offshore for Enhanced Oil Recovery (EOR), thus replacing the stranded natural gas that otherwise would have been reinjected. The LEC provides an energy and exergy efficient solution to the CCS problem as well as the utilization of stranded gas.

When developing this new concept; it became obvious that existing methodologies had severe limitations in tackling subambient design situations. Even though Pinch Analysis has been extensively used by operating as well as engineering & contracting companies to optimize the use of energy in the process industry, only temperature is used as a quality parameter for the streams. In subambient applications, stream pressure and phase are important design and decision variables that can and should be used to develop superior designs.

This important issue can be illustrated by two simple examples. First, a pressurized cold stream can be expanded to provide more cooling at a lower temperature. Secondly, the required work to increase the pressure of a stream can be drastically reduced if this is done in liquid phase (pump) rather than in gas phase (compressor).

In subambient process design, there is a lot of focus on exergy; and Exergy Analysis has the inherent capability of including all stream properties (temperature, pressure and composition); however, this methodology has its focus on the equipment units, rather than the flowsheet level. In addition, there is no strong link between exergy and cost; in fact, there is often a conflict between reducing exergy losses and cost, especially above ambient conditions. Below ambient, exergy losses translate directly into added work in the refrigeration compressors, and this is the reason for the extensive use of exergy in such processes.

Optimization is another emerging Process Synthesis tool that has been used to design subambient processes. A typical and very challenging problem is the optimization of composition for mixed refrigerants. Structural design problems have also been the subject of optimization, and both Deterministic methods (Mathematical Programming) and various Stochastic algorithms have been used.

A New Problem Definition

As indicated in the previous section, Process Design and Integration of sub-ambient processes require a new and expanded problem definition compared to the traditional heat recovery problem of Pinch Analysis. The following problem describes the design objective of this work:

> "Given a set of process streams with a supply state (temperature, pressure and the resulting phase) and a target state, as well as utilities for heating, cooling and power; design a system of heat exchangers, expanders and compressors in such a way that the irreversibilities (or other objective functions) are minimized".

Since the new problem definition goes beyond heat exchangers and temperatures into pressure and phase as well as compressors and expanders, the design task is vastly more complex than Heat Exchanger Network Synthesis. The most challenging issue is the fact that the "path" from supply state to target state is not fixed and can take any possible shape, including elements of heating, cooling, compression, expansion and phase change.

The Liquefied Energy Chain (LEC)

The novel transport chain for stranded natural gas referred to as the Liquefied Energy Chain (LEC) is shown in Fig. 1. It includes an offshore section, a combined gas carrier, and an onshore integrated receiving terminal. Due to the combined gas carrier and the fact that cold exergy is recovered and reused in the offshore and onshore sections, the transport chain is both energy and cost efficient.

In the offshore section, natural gas is liquefied by exchanging heat with liquid CO_2 and liquid nitrogen that are supplied by the ship. After utilization as a carrier of cold exergy, CO_2 is compressed to reservoir conditions and injected for storage as well as EOR. Nitrogen is simply vented to the atmosphere after use.

Natural gas in the form of LNG is transported to the onshore section in the same combined gas carrier. Its cold exergy is recovered by liquefaction of CO_2 and nitrogen. The evaporated LNG is used as fuel in an Oxy-combustion power plant. An Air Separation Unit (ASU) supplies oxygen to the power plant and nitrogen to the LEC.

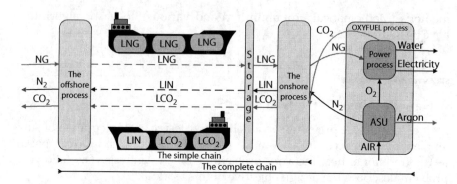

FIGURE 1
The liquefied energy chain (LEC).

In the onshore part of the chain, a storage system provides the required flexibility and offers continuous operation, while the use of two ships (combined carriers) solves the logistic problems offshore. It should also be mentioned that emptying and filling the tanks on the ship involves change of grade. The number of tanks has been subject to optimization and the logistics related to the change of grade has been studied rigorously. Finally, the transport pressures affect the efficiency of the offshore and onshore sections as well as the ship utilization factor, and are important optimization variables for the entire chain.

As indicated in Fig. 1, the "simple" chain is basically a transport chain for LNG and CO_2, while the "complete" chain is a (Liquefied) Energy Chain from well to grid. Details about the overall concept, the offshore and onshore sections, the combined gas carrier as well as benchmarking and sensitivity analyses are given in a 4 paper series by Aspelund et al. (2008a-d).

Important Features of the LEC

The Liquefied Energy Chain outlined in Fig. 1 has a number of favorable features related to resource utilization, environmental friendliness, energy efficiency, and process simplicity. Due to space limitation, these characteristics can only be mentioned in a bulleted list with little room for explanations and details. The most important properties of the LEC concept are:

- An elegant and cost effective solution to the CCS problem for natural gas based power plants.
- Stranded natural gas that normally is reinjected in the reservoir is utilized for power production.

- The offshore section is self-supported with power, hot and cold utilities and can operate with little rotating equipment and no flammable refrigerants.

- While tight integration normally means complex processes that are difficult to operate, the offshore section is very simple despite its heavy integration.

- The losses of natural gas to cover energy (power) requirements in the LEC are roughly one third of the losses in a conventional transport chain for stranded natural gas with CCS.

- The complete LEC has an exergy efficiency of 46.4%, which is considerably better than traditional LNG chains with CCS (42.0%) and similar to pipeline transport with CCS (46.4%). For long distances and moderate volumes, pipeline transport cannot compete from an economic point of view.

New Design Methodologies and Tools

In parallel with the development of the LEC concept, our group has extended existing synthesis methodologies and developed a few new design tools. Graphical tools and representations of Pinch Analysis have been combined with exergy calculations while thermodynamic insight has been matched with process and engineering knowledge in the formulation of a set of Heuristic rules. Finally, both Deterministic and Stochastic Optimization have been used; the former to design individual chain processes by simultaneously addressing temperature and pressure, and the latter to address optimization issues at the chain level.

The following sections give a brief description of some of these achievements. It should be emphasized that when considering subambient processes such as LNG, the use of rigorous Process Simulation and advanced Thermodynamic Packages is an indispensable part of the design exercise. The extremely low temperature driving forces in the heat exchangers ask for high level of accuracy in the calculations, and the correct identification of phase situation for multicomponent mixtures is critical.

The ExPAnD Procedure for Subambient Design

The first step towards a systematic and general design methodology for subambient processes is the Extended Pinch Analysis and Design (ExPAnD) procedure developed in parallel with the offshore section of the LEC and aiming at solving the expanded problem definition given above. Details of

ExPAnD have been presented by Aspelund et al. (2007a) and only highlights
are listed here:

- The use of expansion (as well as compression) of a cold (or hot) stream
  can be used to manipulate the Composite Curves in order to obtain
  tighter integration and reduced exergy losses. The options (sequence
  of heating and expansion) have been illustrated by using the well
  known concept in Process Synthesis referred to as Attainable Region.
- A set of 10 Heuristic Rules related to compression and expansion as
  well as phase changes are proposed to guide the design in order to
  obtain Composite Curves with close to parallel behavior.
- By decomposing thermomechanical exergy into temperature and
  pressure based components, designs can be improved by switching
  between the two components using expansion and compression.

The "proof of the pudding" as far as the ExPAnD methodology is concerned was its use in the development of the offshore section of the LEC concept. The final and thermodynamically optimized design is shown in Fig. 2.

A more recent development of ExPAnD has resulted in new insight regarding Appropriate Placement of Compressors and Expanders in heat recovery
problems. Compressors add heat to the system and should be placed just
above the Pinch. Expanders add cooling to the system and should be placed
just below the Pinch. This deviation from proven design practice (compress
at low temperature and expand at high temperature) has been demonstrated
by Gundersen et al. (2009) using Thermodynamic arguments.

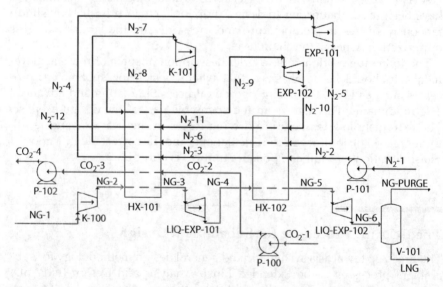

FIGURE 2
The offshore natural gas liquefaction part of the LEC.

A Mathematical Programming Approach

While the insight behind ExPAnD is powerful and can be used to design close to optimal processes from a thermodynamic point of view, the multiple trade-offs in process design calls for an automated approach that can handle both continuous and discrete design decisions.

When using Deterministic Optimization such as Mathematical Programming, three distinct stages are involved in the formulation and solution; superstructure development (representing multiple flowsheets), modeling (objective function and constraints) and the numerical solution phase. Interestingly, these three stages are closely connected, and the success in using Mathematical Programming is closely related to the ability to develop efficient superstructures that can be modeled in a way that reduces the computational efforts in the solution phase.

While Fig. 2 shows the result of a manual/interactive design effort, Fig. 3 shows an example of a superstructure that can and has been used to derive and optimize the flowsheet in Fig. 2 in an automatic way. As indicated, the

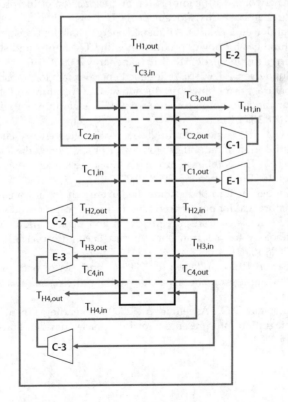

FIGURE 3
Pinch and pressure operator.

superstructure allows both heating and cooling as well as compression and expansion for both hot and cold streams. The logical sequence of these actions is based on the new insight mentioned above regarding Appropriate Placement of compressors and expanders. The details about this superstructure (which has been referred to as a combined Pinch and Pressure Operator) as well as the mathematical model and the optimization results can be found in Aspelund et al. (2009). A preliminary version of the work was presented by Aspelund et al. (2007b).

References

Aspelund A., Berstad D.O., Gundersen T. (2007a). An extended Pinch Analysis and design procedure utilizing pressure based exergy for subambient cooling. *J. Appl. Thermal Eng.*, vol. 27, no. 16, pp. 2633–2649.

Aspelund A., Barton P.I., Gundersen T. (2007b). Minimal irreversibilities for heat exchanger networks with compression and expansion of the process streams. AIChE Mtg., Salt Lake City, November 2007.

Aspelund A., Gundersen T. (2008a). A liquefied energy chain for transport and utilization of natural gas for power production with CO_2 capture and storage–Part 1. *Applied Energy* (in press), doi: 10.1016/j.apenergy.2008.10.010.

Aspelund A., Gundersen T. (2008b). A liquefied energy chain for transport and utilization of natural gas for power production with CO_2 capture and storage–Part 2: The offshore and onshore processes. *Applied Energy* (in press), doi: 10.1016/j.apenergy.2008.10.022.

Aspelund A., Tveit S.P., Gundersen T. (2008c). A liquefied energy chain for transport and utilization of natural gas for power production with CO_2 capture and storage–Part 3: The combined carrier and onshore storage. *Applied Energy* (in press),z doi: 10.1016/j.apenergy.2008.10.023.

Aspelund A., Gundersen T. (2008d). A liquefied energy chain for transport and utilization of natural gas for power production with CO_2 capture and storage–Part 4: Sensitivity analysis of transport pressures and benchmarking with conventional technology for gas transport. *Applied Energy* (in press), doi: 10.1016/j.apenergy.2008.10.021.

Aspelund A., Barton P.I., Gundersen T. (2009). Synthesis of heat exchanger networks with compression and expansion of the process streams. Manuscript in preparation.

Gundersen T., Berstad D.O., Aspelund A. (2009). Extending Pinch Analysis and Process Integration into pressure and fluid phase considerations. PRES'2009, Rome, May 2009.

22

Synthesis of Property-Based Recycle and Reuse Mass Exchange Networks

José M. Ponce-Ortega[1,2], **Mahmoud M. El-Halwagi**[3] and **Arturo Jiménez-Gutiérrez**[1*]

[1] *Instituto Tecnológico de Celaya, Celaya, Gto., México*
[2] *Universidad Michoacana de San Nicolás de Hidalgo, Morelia, Mich., México*
[3] *Texas A&M University, College Station, TX, USA*

CONTENTS

ABSTRACT   A mathematical programming model for the synthesis of mass exchange networks is presented. The approach considers simultaneously direct recycle-reuse networks together with wastewater treatment processes to provide an optimal structure constrained by a given set of environmental regulations. The model combines both mass and property integration; the latter incorporates in-plant property constraints as well as properties impacting the environment, such as toxicity, theoretical oxygen demand, *pH*, color, and odor. A disjunctive programming formulation is developed to optimize the recycle/reuse of process streams to units and the performance of wastewater treatment units. The formulation gives rise to a mixed-integer nonlinear programming model, which is used to minimize the total annual cost of the network (cost of fresh sources plus the annualized costs for piping and property interceptors). A numerical application that shows the advantages of this integrated approach with respect to a sequential optimization strategy is included.

KEYWORDS   *Property interception networks, mass exchange networks, environmental constraints, optimization*

---

* To whom all correspondence should be addressed

Introduction

Mass integration is a recognized tool to provide significant economic and environmental benefits (El- Halwagi, 2006). In the area of sustainable design, mass integration can be used to lower the consumption of fresh sources and to reduce the waste materials discharged to the environment. Minimizing fresh water usage and wastewater discharge through recycle/reuse strategies is a particular application of the mass integration tools, and the paper by Bagajewicz (2000) provided a review for the procedures reported to design water networks. Although previous works provide major contributions to the synthesis of water recycle/reuse networks, they have a couple of limitations. First, their focus has been given to in-plant recycle strategies with constraints limited to process units and without considering the effects of the resulting terminal wastewater streams and the environmental regulations. To satisfy the environmental regulations, the wastewater stream needs to be treated prior to be discharged to the environment; this situation may increase the total cost associated to the recycle-reuse mass integration process. For example, when the solution that minimizes fresh sources consumption yields a high wastewater treatment cost, then such a solution may not correspond to the overall optimal solution. In this case, it may be preferable to use more fresh sources and lower the wastewater treatment cost. Therefore, it is important to consider simultaneously the optimization of the process integration together with the wastewater treatment process to take into account the trade offs between both factors. Second, process characterization and recycle constraints have been based on compositions. It is worth noting that there are many industrial cases when such constraints should be based on properties. Examples include applications when the performance of the process units is affected by the properties of its feed and the cases when environmental regulations impose limits for specific properties of waste streams such as toxicity, theoretical oxygen demand (ThOD), pH, temperature, color, odor, viscosity and density. These properties are difficult to quantify as a function of composition because of the many components present in the process streams.

This paper presents a mathematical programming model to simultaneously optimize the direct recycle networks together with the wastewater treatment process in order to satisfy a set of process and environmental constraints. An optimization formulation is developed based on disjunctive programming. The model is formulated to consider constrained properties such as toxicity, ThOD, pH, color, odor, and any other property that may cause pollution to the environment, in addition to the mass and composition constraints for hazardous compounds in the waste stream. The problem is formulated with proper constraints as an MINLP model that minimizes the total annual cost of the system, which includes the cost for the fresh sources, the piping cost for the process integration and the wastewater treatment cost.

Model Formulation

The problem addressed is defined as follows. Given is a set of process and fresh sources with known flowrates, compositions and properties. Also given is a set of sinks (process units) with constraints for the inlet flowrates and allowed compositions and properties. In addition, there is a set of constraints given by the environmental regulations for the waste streams discharged to the environment. The problem then consists of finding the optimal mass and property integration network that includes a direct recycle strategy that meets the environmental regulations for the waste streams and minimizes the total annual cost of the overall process. Figure 1 shows the schematic representation of the problem addressed in this paper.

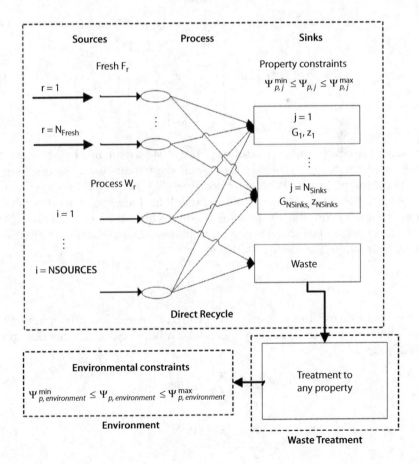

FIGURE 1
Superstructure representation.

The process modeling constraints are the following,

$$W_i = \sum_{j \in NSI} w_{i,j} + waste_i, i \in NSO \tag{1}$$

$$F_r = \sum_{j \in NSI} f_{r,j}, r \in NF \tag{2}$$

$$WASTE = \sum_{i \in NSO} waste_i \tag{3}$$

$$G_j = \sum_{i \in NSO} w_{i,j} + \sum_{r \in NF} f_{r,j}, j \in NSI \tag{4}$$

$$z_{j,c}^{in} G_j = \sum_{i \in NSO} [z_{i,c} \, w_{i,j}] + \sum_{r \in NF} [z_{r,c} \, f_{r,j}], j \in NSI, c \in NC \tag{5}$$

$$\psi_p \left(p_{j,p}^{in} \right) G_j = \sum_{i \in NSO} \left[\psi_p (p_{i,p}) w_{i,j} \right] + \sum_{r \in NF} \left[\psi_p (p_{r,p}) f_{r,j} \right], j \in NSI, p \in NP \tag{6}$$

$$p_{p,j}^{min} \leq p_{p,j}^{in} \leq p_{p,j}^{max}, j \in NSI, p \in NP \tag{7}$$

To satisfy the environmental regulation, the waste stream needs to be treated prior to be discharged. The environmental regulations limit the discharge of hazardous materials. Whenever possible, they are based on component constraints, but for properties that are difficult to characterize because of the large number of components in some waste streams, they are formulated as property constraints. First, it is necessary to determine the ratio contribution for each stream to the waste as follows,

$$Ratio_i = {}^{waste_i} \Big/ \sum_{i \in NSO} waste_i \tag{8}$$

In this fashion, for any property, the property integration principle yields the following (for the property operators mixing rules see Shelley and El-Halwagi, 2000),

$$f_{Pr}(Pr_{mean}) = \sum_{i \in NSO} f_{Pr}(Ratio_i \, Pr_i) \tag{9}$$

$$Wn^{Pr} = Wn \, Pr_{mean} \tag{10}$$

$$Wn^{Pr\,Reg} = Wn \, Pr_{Reg} \tag{11}$$

The property treatment units are modeled according to the following disjunction,

$$
\begin{bmatrix}
Y^{\text{Pr}} \\
Wn^{\text{Pr}} \geq Wn^{\text{PrReg}} \\
Cost^{\text{Pr}} = (Wn^{\text{Pr}} - Wn^{\text{PrReg}})Costu^{\text{Pr}}H_Y
\end{bmatrix}
\vee
\begin{bmatrix}
\neg Y^{\text{Pr}} \\
Wn^{\text{Pr}} \leq Wn^{\text{PrReg}} \\
Cost^{\text{Pr}} = 0
\end{bmatrix}
$$

The previous disjunction is modeled through a set of linear relationships using the convex-hull formulation for any property that needs to be treated in the waste stream.

The objective function consists of the minimization of the total annual cost associated with both the mass integration and the waste treatment processes as follows,

$$
TAC = \sum_{r \in FR} Cost_r^{Fr} F_r\, H_Y + \sum_{\substack{i \in NSO \\ j \in NSI}} pip_{i,j}\, w_{i,j}
$$
$$
+ \sum_{\substack{r \in NF \\ j \in NSI}} pip_{r,j}\, f_{r,j} + \sum_{P \in NP} Cost_P^{\text{Pr}}
$$

(12)

Results

The production of phenol from cumene yielding acetone as subproduct is taken as example problem. The sources and sinks data are shown in tables 1 and 2. Two fresh sources are available with impurities of 0 and 0.012 mass fraction of phenol, and with unit costs of 6×10^{-4}/lb and 4×10^{-4}/lb, respectively.

The environmental regulations for this example are as follows (Hortua, 2007). The maximum concentration to be discharged to the environment for acetone and phenol is 0.005. The toxicity must be zero. The theoretical oxygen demand of the waste stream must be lower than 75 mg O_2/l, and the *pH* must be between 5.5 and 9.0.

TABLE 1

Sources Data for the Example

i	W [lb/hr]	z_i, Phenol	z_i, Acetone	ThOD [gO_2/l]	pH	ρ[lb/l]
1	8,083	0.016	0.000	0.187	5.4	2.205
2	3,900	0.024	0.010	48.850	5.1	2.205
3	3,279	0.22	0.028	92.100	4.8	2.205

TABLE 2

Sinks Data for the Example

j	G_j [lb/hr]	z_j^{max}
1	6,000	0.013
2	4,400	0.013
3	2,490	0.1

The unit costs for the recovery processes are 0.065 $/lb and $0.033 $/lb for phenol and acetone. The unit cost for aeration is 0.006$/lb of air diffused, and the unit costs for the H_2SO_4 and NaOH used for the neutralization process are $46/l and $31/l.

Figure 2 shows the optimal solution for the case when environmental and process constraints are considered simultaneously. In addition, Figure 3 shows the solution for the process optimization without considering the

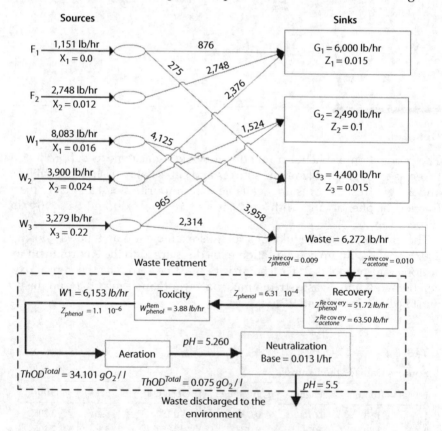

FIGURE 2
Optimal integrated solution for the example (Sol. A).

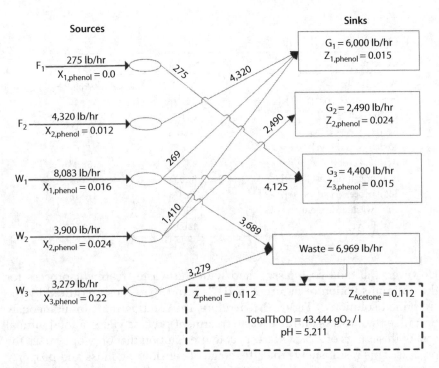

FIGURE 3
Integration without environmental constraints (Sol. B).

environmental constraints. Using the DICOPT solver (Brooke et al., 2006), the CPU times required for the solutions of Figures 2 and 3 in a Pentium IV at 2.8 GHz were 0.52 s and 0.23 s, respectively.

The total waste is reduced in the solution that incorporates the waste treatment process by 11.1% with respect to the solution that considered only the process constraints. In addition, in the optimal solution shown in Figure 2, the total consumption of fresh source 1 (with an impurity concentration of phenol of 0) is increased by 76.1% with respect to the solution of Figure 3; this yields a decrease in the concentration of phenol and acetone in the waste stream by 22.2% and 30.0%, respectively. The theoretical oxygen demand in the waste stream for the optimal solution that considers the waste treatment process is reduced by 27.4% with respect to the simplified solution of Figure 3. The pH in the optimal solution is increased by 0.93% with respect to the simplified solution.

Table 3 shows a summary of the costs associated with both solutions obtained here. For comparison purposes, the costs of the waste treatment processes that would be needed for the solution of Figure 3 were calculated after the optimization procedure. Notice in Table 3 that the sum of the costs associated with the process (fresh sources and piping costs) in the optimal solution of Figure 2 is 8% higher than those required for the solution of Figure 3.

TABLE 3

Comparison of Results

Concept	Sol. A	Sol. B
Waste [lb/hr]	6,273	6,969
Fresh sources cost [$/yr]	14,318	15,146
Piping cost [$/yr]	31,128	26,669
Recovery cost [$/yr]	43,661	61,859
Toxicity cost [$/yr]	5,098	7,226
Aeration cost [$/yr]	43,363	161,083
Neutralization cost [$/yr]	3,237	4,567
Total process costs [$/yr]	45,446	41,816
Total waste treatment costs [$/yr]	95,360	134,734
TAC [$/yr]	140,807	176,550

However, the total costs associated with the waste treatment process for the optimal solution of Figure 2 is 41.3% lower than those required by the simplified solution of Figure 3. Therefore, the solution that simultaneously optimize the process and the waste treatment process yields a total annual cost with savings of 25% with respect to the solution that only optimizes the process. This result shows the advantage of considering mass and property integration simultaneously, and the importance of accounting for the trade off between the direct recycle of fresh resources and the treatment of waste streams while satisfying process constraints and environmental regulations. A sequential solution in which the mass exchange network is optimized without including the waste treatment cost can lead to an inferior design strategy, as shown in the example presented here.

Conclusions

A new framework has been introduced to integrate the design of in-plant recycle networks with the end-of-pipe waste treatment facilities; the model has been formulated to include property-based constraints in addition to the composition constraints commonly taken into account in these formulations. The model is based on disjunctive programming, and considers the technologies that are commonly used to treat wastewater streams so that they can meet environmental regulations. The model has been formulated with linear relationships within each one of the disjunctions, which aids its numerical solution. In addition to standard mass balance and composition constraints, the model incorporates property-based models, property mixing rules, and environmental constraints for properties such as toxicity, theoretical oxygen demand, pH, color, and odor, and provides a general framework to include any other property that may cause pollution. The methodology has shown how property integration can be used for cases in which a given property is difficult to estimate when there are many compounds in a waste stream. The formulation results in an MINLP model, which has been used to minimize the total annual cost of the system. The solution of the proposed model shows that the simultaneous consideration of mass and property integration and the tradeoff between in-plant recycle/reuse with wastewater treatment sections can yield important economic savings with respect to the implementation of a typical sequential solution. The model could be readily applied to any other set of environmental constraints to study the effect of different environmental regulations.

References

Bagajewicz M. (2000). A review of recent design procedure for water networks in refineries and process plants. *Comput. Chem. Eng.,. 24,*:2093.

Brooke, A., Kendrick, D., Meeraus, A. A., Raman, R. (2006). GAMS- A User's Guide. *The Scientific Press*. San Francisco, CA.

El-Halwagi M. M. (2006). Process integration. *Academic Press*. New York.

Hortua, A. C. (2007). Chemical process optimization and pollution prevention via mass and property integration. M.S. Thesis, Texas A&M University.

Shelley M. D., El-Halwagi, M. M. (2000). Componentless design of recovery and allocation systems: A functionality-based clustering approach. *Comput. Chem. Eng.,. 24,* 2081.

23

Point-Based Standard Optimization with Life Cycle Assessment for Process Design

Di Lu and Matthew J. Realff[*]
School of Chemical & Biomolecular Engineering
Georgia Institute of Technology - Atlanta
Atlanta, GA 30332

CONTENTS

ABSTRACT   There has been a proliferation of building and product sustainability standards based on the success of the LEED™ program from the Green Building Council for new commercial building construction. These standards have several common features. 1) They are point-based, in which points are earned for undertaking various activities or using certain materials. 2) The points are aggregated to achieve an overall score. 3) The score is compared to a threshold that determines the rating of a building or a product. The question is whether the standards actually promote products that are better from a life cycle perspective, or whether they are biased towards certain activities based on a perception that they are inherently better than others (recycling or bio-based materials for example). In this work we use optimization methods coupled with life cycle inventory information to explore this question. A carpet standard is used to compare the life cycle optimization against an optimization to earn the maximum number of points in the standard. Our work is extended to show how sustainability assessment standards can be redesigned to make them congruent with life cycle measures so that there is less opportunity for distortions in product design to

---

[*]   To whom all correspondence should be addressed

maximize performance against the standard and have a less than optimal life cycle impact.

KEYWORDS *Process design, life cycle assessment (LCA), point-based standard*

Introduction

Developing sustainable environmental policies and strategies in government and sustainable environmental processes in industry has evolved towards a more quantitative approach. One essential component is a life cycle inventory (LCI), which serves as input to a number of activities such as process development, design and synthesis, and environmental assessments. LCI optimization evaluates environmental burdens associated with all aspects of a production process in an effort to minimize those burdens while satisfying operational constraints, with objective functions that reflect environmental life cycle considerations. LCI optimization shares a similar objective with *point-based standards*, which aim to minimize the environmental impact by maximizing awarded points. Point-based standards have several common features. First, points are earned for undertaking various activities or using certain materials. Second, the points are aggregated to achieve an overall score. Third, the score is compared to a threshold that determines the rating. An important component in awarding some of the points is the measure of environmental performance within a table having different threshold values. Leadership in Energy and Environmental Design (LEED™) and NSF/ANSI 140-2007 (2007) are the major two point-based standards developed with substantial life cycle assessment metrics.

An assumption in point-based standards is that points earning from different activities or categories are equal in value. For instance, one system earning *N1* points from Category I and *N2* points from Category II is evaluated as the same as another system earning *N2* points from Category I and *N1* points from Category II. However, the two systems could possibly have quite different environmental impacts. This happens because the points were assigned to activities or categories without understanding how they related to underlying changes in life cycle inventories. This occurs because at the time the standards were developed, such information was not available to the stakeholder groups. In time, the standard matures, and the allocation can be changed. The lack of congruency between life cycle impacts and points creates potential opportunities for production design distortions that maximize the performance against the standard, but have a less than optimal life cycle impact.

The studies of Azapagic and Clift (1999) and (Stefanis et al., 1997) are seminal contributions to the LCA optimization field. Lu and Realff (2008) developed a mathematical programming framework that combines LCI and optimization together in a straightforward way. The framework first systematically generates all possible alternatives to be analyzed. Then it evaluates all generated alternatives from an environmental perspective and selects the best or the best combination by optimization. In this work, we use optimization methods, coupled with LCI information, to explore how sustainability assessment standards are related to life cycle measures and optimization. The carpet standard, NSF 140-2007, is used as a case study to compare life cycle optimization with optimization to earn the maximum number of points in the standard.

Methodology

Point-Based Standards

Generally the points in a point-based standard can be classified into two categories, check-off-points and threshold-points. "Check-off-points" are earned when a process or product complies with some predefined rule. One check-off-point example from NSF/ANSI 140-2007 (2007) is as follows,

> "A manufacturer shall receive one point for identifying material composition for components present at 1% (10 parts per thousand) or greater of the incoming raw materials, including materials identified as persistent, bio accumulative, and toxic (PBT) as found in Annex B."

"Threshold-points" are earned according to a pre-defined threshold-point table which specifies the points that the process or product earns when it exceeds a given threshold. Table 1 is one threshold-point example from NSF/ANSI 140-2007, which shows the thresholds and their corresponding points a carpet product can earn when reducing energy consumption. For example, if the product saves more than 75% energy, it will be awarded 12 points. The "threshold-point" scheme encourages standard users earn more points by achieving higher level compliance, and which is hypothesized will eventually minimize product environmental impacts.

Point-Based Standard Optimization

In this paper, we will focus on optimizing point-based standards mainly based on threshold-point tables. More specially, for point-based standards

TABLE 1

Points Awarded for Energy Reduction

Percent Energy Reduction	Points Awarded
≥ 1%	2
......	......
≥ 75%	12

in which the stakeholders weigh different environmental impacts (e.g., NSF/ANSI 140-2007), we will re-evaluate the standards by coupling LCA-based mathematical programming techniques, developed in our previous work, to mixed integer representations of the standard. We propose a new model to optimize point-based standards with LCA analysis. The major challenges in developing such a model are; how to represent different threshold-point tables, and how to combine LCA optimization with the point-based standard. As a case study, we will optimize NSF/ANSI 140-2007 with the new model and compare the results of maximizing the awarded point in NSF/ANSI 140-2007 with the results of minimizing environmental impact directly from life cycle measures to see whether these two models are consistent with stakeholders' environmental values.

Generally, different manufacturers have different production lines or pathways to produce the same product or deliver the same functional unit. When producing the same amount of final product, different pathways consume different raw materials and energy, while generating different amounts of wastes and emissions. Therefore, it is valuable to evaluate the performance of each possible pathway from the environmental perspective and choose the pathways that are more energy-efficient, consume fewer raw materials, and release less waste and emissions. In previous work, we developed a mathematical programming model, and related techniques, to automatically generate all possible pathways and select pathways based on different environmental objectives. Our goal was to identify the optimal production alternatives according to available processes and different objectives. Our proposed two-phase synthesis optimization framework breaks the complex problem into two relatively simple sub-problems and solves them separately using LCI information collected independently. In this study, we proposed a new point-based life cycle inventory optimization model as follows,

$$Max \sum_k N_k * X_k + \sum_b C_b * Y_b + \sum_L U_L + \sum_L S_L \tag{1}$$

$$\sum_i P_i = D \tag{2}$$

$$M_j \geq \sum_i a_{(i,j)} * P_i \tag{3}$$

$$\frac{\sum_{kp} P_{kp}}{D} \geq \sum_k T_k * X_k \tag{4}$$

$$\sum_k X_k \leq 1 \tag{5}$$

$$\frac{\left(D*Benchmark - \sum_i P_i*E_i\right)}{D*Benchmark} \geq \sum_b Q_b * Y_b \tag{6}$$

$$\sum_b Y_b \le 1 \tag{7}$$

$$Z * (1 - U_L) + (\sum_{cat} R_{cat, L} - UU) \ge 0 \tag{8}$$

$$Z * U_L + \left(UU - 1 - \sum_{cat} R_{cat, L} \right) \ge 0 \tag{9}$$

$$Z * (1 - S_L) + (\sum_{cat} R_{cat, L} - SS) \ge 0 \tag{10}$$

$$Z * S_L + (SS - 1 - \sum_{cat} R_{cat, L}) \ge 0 \tag{11}$$

$$\frac{D * Average_{cat} - \sum_i \left(P_i * \sum_f Emission_{(i,f)} * Cat_f \right)}{D * Average_{cat}} \ge W_L * R_{cat, L} \tag{12}$$

Nomenclature

Indices:

k	List of thresholds in the standard table for recycled contents.
b	List of thresholds in the standard table for energy reduction.
L	List of thresholds in the standard table for cross categories of emissions.
i	Alternative pathways of manufacturing the product.
f	Chemicals involved in emissions.
j	Raw materials.
cat	Emission categories.
kp	Pathways of manufacturing the product with recycle contents.

Decision variables:

P_i	The amount of product manufactured through pathway (i).
X_k	Binary variable (0,1): if threshold (k) is crossed, $X_k = 1$, otherwise, $X_k = 0$.
Y_b	Binary variable (0,1): if threshold (b) is crossed, $Y_b = 1$, otherwise, $Y_b = 0$.
S_L	Binary variable (0,1): if SS categories exceed threshold (L), $S_L = 1$, otherwise, $S_L = 0$.
U_L	Binary variable (0,1): if UU categories exceed threshold (L), $U_L = 1$, otherwise, $U_L = 0$.
$R_{(cat, L)}$	Binary variable (0,1): if SS categories exceed threshold (L), $R_{(cat, L)} = 1$, otherwise, $R_{(cat, L)} = 0$.

Parameters:

D	Production demand.
M_j	The amount of available raw materials (j).

$a_{(i,j)}$	Coefficient of raw material (j) to manufacturing one functional unit of product for pathway (i).
T_k	Thresholds value in recycled contents table.
N_k	Awarded points in recycled contents table.
Q_b	Thresholds value in energy reduction table.
C_b	Awarded points in energy reduction table.
E_i	The amount of the energy consumption for each pathway (i).
Z	A big number (10 for the NSF/ANSI 140-2007 cross categories table).
Cat_f	The environmental potential (cat) of chemical (f).
uu	The lower number of crossed categories.
ss	The higher number of crossed categories.
W_L	Thresholds value in cross categories table for emissions.
Benchmark	Benchmark value of the energy consumption.
$Average_{cat}$	Benchmark value of the environmental impact for emission category (cat).
$Emission_{(i,f)}$	The amount of emissions of chemical (f) for pathway (i).

The above presented approach applies mixed-integer mathematical modeling techniques to the optimization of sustainable production standards with LCI information. In our model, all environmental burdens are expressed as a function of the continuous decision variable $E(i)$ and Emission (i,f). The binary variables X_k, Y_b, $R_{(cat,L)}$, S_L, and U_L denote whether the corresponding threshold is crossed or not, and appear linearly in the objective function and also in the constraints. The inequalities, Eq. (3) to (12), include raw material limits, which are also linear inequalities. Generally, the representation of emissions constraints may lead to very complex models. Our framework helps to avoid this situation. Instead, specific LCI databases, which contain the inventory of emissions of a wide range of chemical processes, are used to establish the overall emissions for a pathway and this lumped value used in further calculations.

The objective function of the optimization model is to maximize the sum of awarded points in terms of LCI calculations from three perspectives. The first one, which is denoted by $\sum_k N_k * X_k$, is the total points awarded by using recycled content. The second part $\sum_b C_b * Y_b$ represents the total points awarded by saving energy consumption. The last part $\sum_L U_L + \sum_L S_L$ represents the total points awarded by reducing emissions of environmental impact. In addition, the constraints are divided into four sub groups. Equations (2) and (3) are the basic material balances as mentioned in optimization model. Equations (4) and (5) are incorporated with the table of recycled contents, while Equations (6) and (7) represent the table of energy efficiency. Equations (8) to (12) deal with Table 2, which evaluate the system from a different perspective. Among those constraints, constraints (4), (6), and (12) link LCI calculations with the point-based standard. The model can be solved using the

TABLE 2

Life Cycle Points Awarded

Percent Reduction	Across Four Impact Categories	Across Eight Impact Categories
≥ 10%	1pt	2pts
......	......	......
≥ 75%	1pt	2pts

general algebraic modeling system (GAMS) combined with a mixed integer solver such as CPLEX.

Case Study

In this case study, a carpet production system was analyzed in terms of life cycle optimization and standard optimization. The evaluated standard of carpet is NSF 140-2007(2007), which is a point-based standard designed for carpet products that provides benchmarks for sustainable carpet improvement and innovation and would help consumers identify certified carpets with lower environmental impacts. The environmental impact is measured through the Tool for the Reduction and Assessment of Chemical and Other Environmental Impact (TRACI) method (Bare et al., 2006), which is a reasonable reflection of the current state of the art in LCA methodology and application. In our case study, eight categories including global warming, acidification, eutrophication, photochemical smog, human health, fossil fuel depletion, ecological toxicity, and solid waste were used to measure the product's life cycle impact. Table 2 is a special point-based table, which represents the life cycle impact categories. If more than four and less than eight impact categories are crossed at each range indicated in Table 2, one point will be awarded accordingly. In addition, another one point will be awarded if all eight impact categories are crossed at each range.

Two sets of optimization programming experiments were conducted in this case study. One is based on life cycle measures with the objective to minimize energy consumption directly from life cycle inventory information. The other is focused on maximizing the awarded points from the standard point-based tables, which has three sources: using recycled or bio-based contents, reducing energy consumption, and reducing emissions to the environment. Energy use is categorized into two types: electricity and fuel. Table 3 shows the results of the case study: optimizing the LCI will save more energy but gain less points in the standard. This suggests that the allocation of points is not consistent with an environmental goal of minimizing energy use.

TABLE 3

Case Study Results Summary

		Points	Energy KJ/Kg Carpet)
Optimize	Maximize electricity points	30	4.47E + 03
Standard	Maximize fuel points	32	7.52E + 03
Optimize	Minimize electricity	26	4.44E + 03
LCI	Minimize fuel	23	6.53E + 03

Conclusion

In this work, we use optimization methods coupled with life cycle inventory information to propose a method to synthesize products earning the maximum number of points in a standard. A number of alternatives of manufacturing the product are examined according to the point-based standard in our model. Our proposed method is intended to guide the decision-makers towards the adoption of a sustainable production design from a standard-based perspective, consequently leading to a reduction of the overall environmental impact. In the future we intend to study reallocating the awarded points of the point-based standard to ensure congruency of life cycle impacts and points.

References

Azapagic, A., Clift, R. (1999). The application of life cycle assessment to process optimisation. *Comput. Chem. Eng., 10, 1509.*

Bare, J.C., Gloria, T.P., Norris, G. (2006). Development of the Method and U.S. Normalization Database for Life Cycle Impact Assessment and Sustainability Metrics. *Env. Sci. and Tec.,* 40, 16.

Lu, D., Realff, M. (2008). A mathematical programming tool for LCI synthesis: A case study for a carpet product. Env. Sci. Tec., (in preparation)

NSF/ANSI 140-2007. (2007). Sustainable carpet assessment. *NSF international.* Ann Arbor, MI.

Stefanis, S. K., Livingston, A. G., Pistikopoulos, E. N. (1997). Environmental impact considerations in the optimal design and scheduling of batch processes. *Comput. Chem. Eng., 21, 1073.*

U.S. Green Building Council. LEED standard. http://tinyurl.com/57y3p6

24

Process Synthesis and Design of Sustainable Systems: A Case Study in Long-term Distant Space Exploration

Selen Aydogan-Cremaschi[*1], **Seza Orcun**[2], **Joseph F. Pekny**[2] **and Gintaras V. Reklaitis**[2]

[1]*The University of Tulsa*
Tulsa, OK 74104
[2]*Purdue University*
West Lafayette, IN 47907

CONTENTS

ABSTRACT This paper introduces a novel algorithm utilizing simulation-based optimization approach to provide insight to early-stage design decisions for systems with high recycle rates and uncertain process performances. The algorithm combines a stochastic discrete-event simulation with a deterministic mathematical programming approach to generate multiple, unique realizations of controlled evolution of a system. The realizations are then analyzed to determine necessary technologies, their retrofit plan, and necessary inventory levels. The process synthesis and design optimization of a life-support-system for survival and well-being of the crew during a space mission is discussed as an application.

---

[*] To whom all correspondence should be addressed.

KEYWORDS   *Process synthesis, simulation-based optimization, retrofitting, sustainable systems*

Introduction

A life-support system is a set of physical, chemical and biological processes that provide the basic elements necessary to support human life, such as oxygen, potable water, and food to the crew for short and long-term space-flights. The input materials for such a system are the wastes generated by the crewmembers. Ongoing research is being conducted to identify, develop and test the life support system technologies that would fulfill these needs. As a result, there are numerous technologies that can be used for the same tasks. However, synthesizing the integrated life support system from these candidate technologies with an appropriate deployment schedule is not a trivial problem. Moreover, such a system is subject to a myriad of uncertainties because most of the technologies involved are still under development and their performance in the space environment is unknown. As a result high levels of uncertainties exist in the estimates of the model parameters, such as recovery rates or process efficiencies. Due to the high recycle rates within the system (as will be the case for sustainable systems), the uncertainties are amplified and propagated, resulting in a complex decision problem under uncertainty.

The problem of selecting the optimum set of technologies for a life-support system is similar to the process synthesis problem faced in chemical engineering. One of the process synthesis approaches is superstructure optimization. The deterministic problem has been studied (e.g., Yeomans and Grossmann, 1999) and the developed methodologies have been applied to obtain lower cost flowsheet designs (e.g., Lee et al., 2003). Dua and Pistikopoulos (1998) summarized the mixed integer optimization approaches for process synthesis and material design problems under uncertainty. The authors obtained solution profiles as a function of uncertainty in the product demand. However, what the inventory levels should be and whether and how the process synthesis will change depending on the reliabilities of the processes was not addressed. A practical solution to open-ended flowsheet synthesis problems under uncertainty was also presented by Chakraborty and Linninger (2003). Their methodology incorporated variations in expected waste loads into plant-wide waste-management policies. The authors concluded that simplifying assumptions such as linearized cost models, and uncertainty in the waste loads with invariant composition are satisfactory to model uncertainty for multipurpose manufacturing sites. Chakraborty et al. (2003) extended their work to long-term site-management strategies under uncertainty. An industrial case study was used for validating their methodology. The multi-period, mixed-integer programming framework was successfully used to model plant inventory and equipment investments. However, how reliabilities of the processes affect the design and inventory levels, and how sharp demand changes might affect the design over time (retrofitting) are open for further research in this field.

In this paper, a computational architecture, simulation-based optimization (SIMOPT) that combines combinatorial optimization with a simulation environment, is proposed to solve simultaneous process synthesis and retrofitting problem under uncertainty for life support system for manned space missions. In the following sections, the problem is defined and the computational approach is described in detail. A case study is presented to demonstrate the application of the proposed framework. Finally, last section provides conclusions and future directions.

Problem Definition

Given a set of technologies, crew requirements and demands, crew wastes, and the mission specifications; the objective is to determine a technology list, its deployment schedule, main life support elements' supply amounts (inventory levels), and the reliability of the mission (99% confidence) that would minimize the mission cost (Equivalent System Mass–ESM; Levri et al., 2003). Here, we refer to the reliability of the mission as the time system is able to support crew life.

Computational Approach

The framework (Figure 1) includes a technology selection model, a simulation module and a data analysis module. Given crew member requirements and loads, and available technologies and their properties, the technology selection module determines the optimum technology list and its deployment schedule that would minimize the mission cost. The simulation module is used to predict the behavior of the integrated system (technologies selected by the optimization module) under uncertainties such as technology performance variations (in recovery rates or process efficiencies) and random events e.g. technology malfunctions. Trigger events are caused by technology malfunctions. In case of a trigger event, the simulation is stopped and the state of the system and the information gathered up to that time is fed to the optimization to revise the technology list and its deployment schedule. The simulation is resumed with the new list and schedule. This loop (inner loop in Figure 1) is continued till the end of the simulation time (exploration duration analogous to plant life), resulting in one timeline. In the outer loop, the data analysis module uses the information gathered from

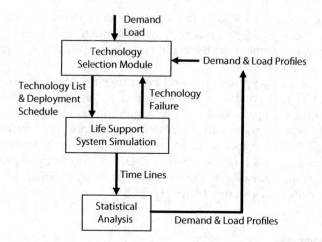

FIGURE 1
SIMOPT framework for life support system synthesis.

a statistically significant number of timelines to update the demand and load profiles in the optimization module. At the termination of SIMOPT framework, a technology list, its deployment schedule, main life support elements' supply amounts (inventory levels), the cost of the mission, and its reliability are determined.

Technology Selection Module

Figure 2 depicts the superstructure of the life support system, which involves all possible tasks and states. The system boundary is the outside of the shaded area. The system interacts with the crew and atmosphere through crew demands and wastes. The MILP formulation of the superstructure can be found in Figure 3. In order to preserve the linearity of the model, the following assumptions are made: The processing capacity of any technology can only be increased by a fixed predetermined small amount. The cost (ESM) attributes of the technologies change linearly with its capacity. These are acceptable early-stage design considerations for space exploration.

Life Support System Simulation

In the simulation, all life-support-system tasks and their interactions with the crew, atmosphere, and buffers, and all states are modeled using process models on a daily basis. There are seven tasks, i.e., nitrogen supply, carbon dioxide removal, carbon dioxide reduction, oxygen supply/generation, waste water recovery, food preparation, and solid-waste processing. The buffers, analogous to intermediate storages, are used for storing components such as oxygen, carbon dioxide, nitrogen, food, potable water, grey water, solid waste, side products, and waste.

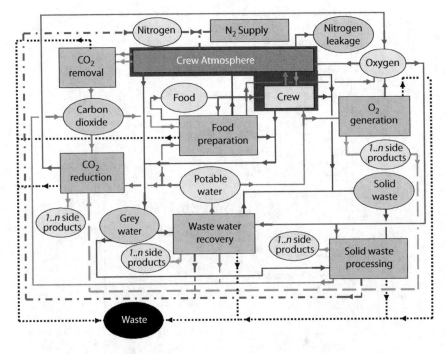

FIGURE 2
Life support system superstructure.

For each task, the simulation contains a list of available technology options that should be specified at the beginning of the simulation. Each technology option contains its processing models, efficiency distribution and a reliability distribution. The efficiency distribution for each technology is estimated as uniform distribution between lower and upper bounds determined by the data available in literature. The reliability is defined as the possibility of malfunction and modeled using exponential distribution with a mean of expected time to failure for each technology (determined by the data available in literature). Once a technology fails, it is not used in the system any more.

There are three random variables related to the crew module: diet-consumption efficiency, crew-member-activity schedule, and crew-member-water-usage rates. The activity schedules are modeled in the form of 8-hour shifts and assumed to change between very light physical activity to heavy physical activity. The activity schedule changes are represented by a triangular distribution, with mode being the crew usual activity schedule, maximum corresponding to eight hours of heavy physical activity and minimum as eight hours of very light physical activity. Diet consumption efficiency is modeled as a uniform distribution according to the values given in Hanford (2004). Crew member water-usage rates define the total amount of water consumed by the crew members for drinking, cleaning and hygiene purposes.

Model

Objective Function

$$\text{Min TC}$$

Subject to

Cost Function

$$TC = \sum_m \sum_p C_{m,p,p} ESM\_NC_m$$

$$+ \sum_m \sum_t \sum_p C_{m,t,p} CT_m + \sum_k \sum_t S_{k,t}$$

Crew Demands Requirements

$$\sum_m \sum_p C_{m,t,p} G_{k,m} \geq D_{k,t} \qquad \forall t, k \in D$$

Accumulation of Wastes

$$LA_{k,t} = L_{k,t} + LA_{k,t-1} - \sum_m \sum_p C_{m,t-1,p} G_{k,m} \qquad \forall t, k \in L$$

Supply/Accumulation

$$\sum_m \sum_p C_{m,t,p} G_{k,m} = A_{k,t} - S_{k,t} \qquad \forall t, k \notin D \wedge k \notin L$$

Level of sustainability

$$\sum_k \sum_{m \in Supply} \sum_{p \leq t} Y_{m,t,p} G_{k,m} \geq (1 - Cl_p) \sum_{k \in D} D_{k,t} \qquad \forall p$$

Reliability:

$$C_{m,t+1,p} = C_{m,t,p} \qquad \forall m, t \in [p, p + Rl_m)$$

$$C_{m,t,p} = 0 \qquad \forall m, t \geq p + Rl_m$$

Nomenclature

TC: Total Cost

$C_{m,t,p}$: Capacity of technology m installed at time p and in operation at time t

$ESM\text{-}NC_m$: Equivalent System Mass without crew time and supply mass for technology m with capacity CS

CT_m: Crew Time to support technology m with capacity CS

$S_{k,t}$: Amount of material k supplied at time t.

$D_{k,t}$: Amount of material k demanded by the crew at time t

$G_{k,m}$: Amount of material k produced/consumed by technology m with capacity CS.

$LA_{k,t}$: Amount of material k available for processing at time t

D: Demand vector

L: Load vector

$L_{k,t}$: Amount of waste k generated by crew members at time t.

$A_{k,t}$: Amount of material k accumulated at time t.

Cl_p: Closure level at time p

Rl_m: Mean failure time of technology m.

FIGURE 3

MILP model formulation of LSS superstructure.

They are modeled as uniform and triangular distributions according to the values given in Hanford (2004).

More detailed information about the simulation can be found elsewhere (Aydogan, 2006).

Data-Analysis Module

The data-analysis module gathers the necessary data from each timeline to construct an average of each demand and load every time inner loop is traversed. At the end of each inner loop completion, demand and load vectors are calculated. After a randomly high number of inner loop repetitions, the initial estimate of each demand and load is compared to the calculated demand and load profile. If the mean of the inner loop repetitions is not equal to the initial estimate for any demand or load (with $\alpha = 0.0005$), the initial estimates of demand and load are updated with the new means.

Case Study: Evolution of Mars Base

The mission period is 13 years with 6 trips between Earth and Mars that are 26 months apart. The analysis of this application focuses mainly on the phases in which the crew lives on the Mars surface and ignores the crew requirements during the transit time between Mars and the Earth. A crew of six is supported on the Mars surface for a period of 13 years with six launches between the Earth and Mars to supply the basic life-support needs for the crew members. The data used to populate the simulation and optimization models, i.e., initial crew demand and waste amounts, infrastructure costs, technology list and their data, can be found at Aydogan (2006).

The SIMOPT framework for life-support-system synthesis and design for the Early Mars mission converged in 20,000 timelines. The computing time required on a Solaris Workstation with UltraSPARK II Processors, 4 GM RAM, and 4*400-Mhz processor speed was about 81.6 hours. The mean ESM for the life-support system for this mission is about 315 tons (± 0.5 tons at 99% confidence level) for a crew of six members for a duration of 4680 days. Figure 4 shows the frequency plot of ESM that is attainable with 20,000 timelines. From 30 technology options considered, 12 of them are selected for launch, with the following schedule $(1 + 2 + 1 + 3 + 3 + 2)$. During the mission, 167 tons of waste is accumulated with 67% as grey water, 26% as carbon dioxide and rest as solid waste. The total main life-support-system supply amount is about 170 tons, with 51% as potable water, 28.6% as oxygen, 20% as food, and rest as nitrogen and nitric acid. Based on the assumptions of the system examined and with the given supply and storage amounts, the system reliability is calculated as 4410 days with 99% confidence.

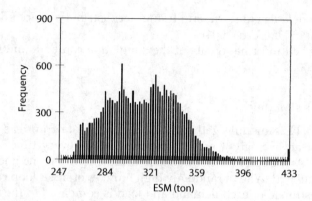

FIGURE 4
ESM for mars mission.

Conclusion and Future Directions

This paper summarizes the development and proof of concept of a novel process synthesis method that incorporates: 1) uncertainties in the product demands and raw material availabilities, 2) changes in process efficiencies, 3) reliabilities of the processes. The method describes an algorithm for synthesis of chemical flow sheets with retrofitting option and determination of inventory levels.

The reliabilities of technologies are assumed to be independent of each other in the current framework setting, due to lack of data. A straightforward extension for this framework can be the accounting for dependencies among technology reliabilities. Then, the reliability data generated from a statistically significant number of timelines can be used to modify the technology-selection model. In the proposed framework, the technology failure mode is not considered. Once a technology fails, it is assumed that it cannot be repaired and re-used by the system. The addition of different failure modes, such as repairable failures, and complete failures is another potential improvement avenue.

Acknowledgments

Financial support from NASA Specialized Center of Research and Training in Advanced Life Support under grant NAG5-12686 is greatly acknowledged.

References

Aydogan, S. (2006). A Simulation-Based Optimization Approach to Model and Design Life Support Systems for Manned Space Missions. *PhD. Dissertation*. Purdue University, West Lafayette, IN.

Chakraborty, A., Linninger, A. A. (2003). Plant-Wide Waste Management. 2. Decision Making under Uncertainty. *Ind. Eng. Chem. Res.*, *42*, 357.

Chakraborty, A., Colbery, R. D., Linninger, A. A. (2003). Plant-Wide Waste Management. 3. Long-Term Operation and Investment Planning Under Uncertainty. *Ind. Eng. Chem. Res.*, *42*, 4772.

Dua, V., Pistikopoulos, E. N. (1998). Optimization Techniques for Process Synthesis and Material Design under Uncertainty. *Trans IChemE*, *76*, 408.

Lee, S., Logsdon, J. F., Foral, M. J. Grossmann, I. E. (2003). Superstructure Optimization of the Olefin Separation Process. *In Proceedings of 13th European Symposium on Computer Aided Process Engineering*, Finland, 191.

Levri, J. A., Drysdale, A. E. et al. (2003). *Advanced Life Support Equivalent System Mass Guidelines Document*, NASA Ames Research Center. Moffett Field, CA.

Yeomans, H., and Grossmann, I.E. (1999). A Systematic Modeling Framework of Superstructure Optimization in Process Synthesis. *Comput. Chem. Eng*, *23*, 709.

25

Equilibrium-Staged Separation of CO_2 and H_2 Using Hydrate Formation and Dissociation

Jae Lee[*], Prasad Yedlapalli and Junshe Zhang

Department of Chemical Engineering, The City College of New York, New York, NY 10031

CONTENTS

ABSTRACT   This work addresses a conceptual design for separating carbon dioxide (CO_2) and hydrogen (H_2) in a multi-stage gas hydrate formation and dissociation process. The van der Waals and Platteeuw statistical thermodynamic model modified by John and Holder is employed to determine equilibrium conditions. The experimental data obtained from a high-pressure differential scanning calorimeter (DSC) were used to tune the reference chemical potential difference of water phase. Using this parameter in the John-Holder model, we generate a vapor-hydrate binary phase diagram of CO_2 + H_2. Based on this diagram, we propose a flowsheet for the separation of CO_2 from the pre-combustion stream.

KEYWORDS   *Separation, conceptual design, caron dioxide, hydrogen, gas hydrate*

---

[*] To whom all correspondence should be addressed

Introduction

One of global issues is climate change. There may be many factors for contributing to the climate changes but a primary reason is due to the increased concentration of greenhouse gases (GHGs) in the atmosphere. CO_2 is the most abundant species of GHGs, causing about 60% of the greenhouse effect. The CO_2 concentration in the atmosphere has increased from 280 ppm at the pre-industry level to currently 350 ppm as a result of anthropogenic CO_2 emission (Gupta et al., 2003). The current dependence of fossil fuels on energy consumption is around 90%. This dependence may become smaller as renewable energy technologies advance but still it's unavoidable to use fossil fuels. The sustainable way to continuously use fossil fuels is to develop an energy-efficient method for capturing and sequestering CO_2.

One of promising methods is to employ gas hydrate formation and dissociation to capture CO_2 from process streams. Gas hydrates are one group of clathrates that is composed of host and guest molecules (Holder et al., 1988, Sloan and Koh, 2008). In clathrate hydrates, water is the host molecule and small gas or organic molecules, such as hydrogen (H_2), methane, ethane, propane, CO_2, and cyclopentane (CP), are guest molecules. Water molecules form a cage by hydrogen bonds where one guest generally is enclathrated.

This study presents a new conceptual design for the separation of CO_2 from the pre-combustion stream (CO_2 and H_2 are the predominant components) using gas hydrate formation and dissociation. The concentration of CO_2 in the pre-combustion stream after gasification is around 20 to 40 vol % (Rand and Dell, 2008). There is enough thermodynamic driving force between CO_2 hydrate (2.9 MPa at 280 K) and H_2 hydrate (300 MPa at 280 K) formation (Sloan and Koh, 2008, Mao and Mao, 2004). However, a primary obstacle to implement a hydrate-based CO_2 separation process from the pre-combustion stream is a very low hydrate formation rate. Our recent study reported the fast kinetics of CO_2 hydrate formation with a small amount of CP (Zhang and Lee, 2008a). Based on this promise of fast formation kinetics of CO_2 hydrates, we proposed a multi-stage equilibrium separation process of CO_2 and H_2 using clathrate hydrate formation and dissociation. First, we present a statistical thermodynamic model to predict equilibrium conditions of multi-component gas hydrates. The reference chemical potential difference of water is then adjusted according to the measured equilibrium data of mixed CO_2 + H_2 + CP hydrates. The equilibrium data are obtained from a high-pressure differential scanning calorimeter (DSC). Finally, we use this thermodynamic model to generate a hydrate-vapor phase diagram that is used to generate a conceptual design of hydrate-based CO_2 separation from the pre-combustion stream.

Thermodynamic Model

At equilibrium, the chemical potentials of water in the hydrate and in the other coexisting phases are equal as written in equation 1.

$$\mu^L_w (T,P) = \mu^H_w (T,P) \tag{1}$$

where $\mu^L_w (T,P)$ is the chemical potential of water in the aqueous phase and $\mu^H_w (T,P)$ is the chemical potential of water in hydrates at the given temperature and pressure (T & P).

Using the chemical potential of empty hydrate lattice, μ^β, at the reference state (273.15 K and zero atmospheric pressure), equation 1 becomes

$$\Delta\mu^L_w = \Delta\mu^H_w \tag{2}$$

where $\Delta\mu^L_w = \mu^\beta - \mu^L_w$ and $\Delta\mu^H_w = \mu^\beta - \mu^H_w$.

The statistical thermodynamic model for the hydrate phase as derived by van der Waals and Platteeuw (1959) is

$$\frac{\Delta\mu^H_w}{RT} = -\sum_{i=1}^{2} v_i ln\left(1 - \sum \theta_{ij}\right) \tag{3}$$

where v_i is the number of i-type cavities per water molecule and θ_{ij} is the fractional occupancy of i-type cavities with j-type molecules. It is expressed as

$$\theta_{ij} = \frac{C_{ij}f_j}{\left(1 + \sum_j C_{ij}f_j\right)} \tag{4}$$

where f_j is the fugacity of guest molecules. Peng and Robinson equation of state (Peng and Robinson, 1976) is used to calculate the fugacity of gas in this paper and C_{ij} is the Langmuir constant. The smooth cell Langmuir constant can be calculated as

$$C_{ij} = \frac{1}{kT} \int_0^R 4\pi r^2 \exp\left(-\frac{W_1(r) + W_2(r) + W_3(r)}{kT}\right) dr \tag{5}$$

where, $W_1(r)$, $W_2(r)$, and $W_3(r)$ are the smooth cell potentials of the first, second, and third shells and R is the radius of the hydrate cavity (John and Holder, 1982). These smooth cell potentials are used to calculate the Langmuir constants of hydrogen and CP. The Langmuir constant for CO$_2$ is determined by an empirical correlation that is proposed by Munck et al. (1998). We assume here the single occupancy for hydrogen or CO$_2$ in the small cavity while a single CP molecule is engaged in the large cavity.

The chemical potential of water in the aqueous phase at equilibrium with the hydrate phase is given as (Holder et al., 1988)

$$\frac{\Delta \mu_W^L}{RT_F} = \frac{\Delta \mu_W^0}{RT_0} - \int_{T_0}^{T_F} \frac{\Delta h_W'}{RT^2} dT + \int_0^P \frac{\Delta V_w'}{RT_F} dP - \ln \gamma_w X_w \qquad (6)$$

The first term on the right hand side is the reference chemical potential difference between the pure water and empty hydrate lattice, which is experimentally determined based on the John and Holder's three-shell model. The second term gives the temperature dependence of enthalpy at constant pressure and the proper parameters with formula are available in Holder et al. (1988). The third term accounts for the change in chemical potential difference due to pressure. The fourth term gives the activity change of water.

Experimental Determination of HVLL Equilibrium

The equilibrium of quaternary system of $CO_2 + H_2 + CP + H_2O$ containing hydrate-vapor-liquid-liquid (HVLL) is determined using a high-pressure micro-differential scanning calorimeter (μ-DSC). The high-pressure μ-DSC is less time-consuming to measure phase equilibria of clathrate hydrates compared with the classical PVT technique (Dalmazzone et al., 2002, Le Parlouër et al., 2004). The principle behind the measurement can be referred to Dalmazzone et al. (2002) and the detailed setup for the μ-DSC can be found in our previous work (Zhang and Lee, 2008b).

A total volume of 25 μL liquid (10 μL deionized water and 15 μL CP) was charged to a measuring cell, followed by purging the cell with the $CO_2 + H_2$ gas mixture twice. Then, the measuring cell was pressurized between 1.5 MPa and 7.3 MPa. The reference cell is empty to have the difference of heat flow between the reference and the measuring cell. The cells were cooled down to 238.2 K and then heated up to 274.3 K. It keeps this temperature for 1 hour. The cells were heated up again to 303 K at a rate of 1 K min^{-1}. The measurement at each data point was repeated more than three times and the average value was reported in this study.

Figure 1 in the previous page shows the experimental data (symbols) of the equilibrium temperatures and pressures at different CO_2 molar fractions in vapor gas phase. We ignore the partial pressures of CP and water in the vapor phase because they are very small compared to that of CO_2 and H_2. As the CO_2 mole fraction increases in vapor phase (or the H_2 fraction decreases), the equilibrium pressure of mixed $CO_2 + H_2 + CP$ hydrates decreases. This follows the same trend as the case without CP (Sugahara et al., 2005). The dissociation temperature of $CO_2 + H_2$ binary hydrates without CP is about

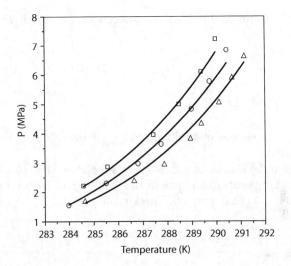

FIGURE 1

P-T diagrams for HLLV of H$_2$ + CO$_2$ + CP + H$_2$O mixtures. CO$_2$ mole compositions in vapor phase = 0.26, 0.30, 0.4 for symbols in squares, circles, and triangles.

276 K at a CO$_2$ mole fraction of 0.26 and a pressure of 7.23 MPa (Sugahara et al., 2005) while the dissociation temperature at the same pressure and CO$_2$ composition is about 290K. The presence of CP acting as a thermodynamic promoter in hydrate formation can facilitate the CO$_2$ separation at moderate operating conditions.

Thermodynamic Calculation Procedure

The calculation procedure for determining equilibrium temperatures, pressures, and compositions is as follows:

1) For given temperatures and vapor compositions, the hydrate phase chemical potential difference ($\Delta\mu_H$) is calculated using equations 2 to 5 with an initial guess of pressure.

2) We set $\Delta\mu_H$ equal to the right side of equation 6 and solve for pressure (P). Then, this new pressure is used to update $\Delta\mu_H$. This iteration is continued until we satisfy equation 2 with a converged pressure.

To determine the optimal value of water reference chemical difference ($\Delta\mu^{\circ}_w$) in equation (6), the percent of average absolute deviation (AAD) between the experimental dissociation temperatures and calculated ones in equation (2) is minimized with respect to its normal range between 900 and 1,400 J/mol.

$$\text{ADD \%} = \frac{\sum_{1}^{n} \left(\left(T_{i,\text{exp.}} - T_{i,\text{cal.}} \right) \times 100 \Big/ T_{i,\text{exp.}} \right)}{n} \tag{2}$$

where $T_{i,exp.}$ is the experimental values, $T_{i,cal.}$ is the calculated ones, n is the total data points.

The minimum AAD is obtained as 4% when $\Delta\mu^{o}_{w} = 1,060$ J/mol. The calculated values of temperature and pressure are shown in Figure 1 with respect to experimental data that were obtained in the μ-DSC.

Conceptual Design: Multi-stage CO_2 + H_2 Separation

This section shows the application of the previous thermodynamic calculation procedure to a hydrate-aided CO_2 separation from a CO_2 + H_2 mixture with CP. Here, we determine the binary vapor-hydrate phase diagram by excluding CP and water. Figure 2 shows this diagram at a constant temperature of 282 K. If we assume a molar composition of the stream of pre-combustion as 40% CO_2 and 60% H_2, we need to have only three equilibrium stages to produce the 99% CO_2 stream as shown in Figure 2.

Upon entering the first reactor, the feed stream with 40% of CO_2 (60% of H_2) is split into vapor and hydrate phases. The bottom stream of the first reactor containing clathrate hydrates dissociates into vapor by reducing the pressure. The vapor stream at the lower pressure is then separated into vapor and hydrate phases in a second reactor. The hydrate stream of the second reactor becomes a vapor stream by reducing pressure further and it enters the third reactor. CO_2 can be enriched to 99% in the hydrate phase after three stages of formation/dissociation. We can put more reactors by connecting the vapor stream of the first reactor to further purify the hydrogen.

Conclusions

We have demonstrated a conceptual design of a hydrate-based separation of CO_2 from the pre-combustion stream by coupling the μ-DSC experimental measurements and a statistical thermodynamic procedure. The high-pressure μ-DSC allows us to quickly obtain the equilibrium data with a very small volume of samples (tens of micro-liters). The John-Holder model with an

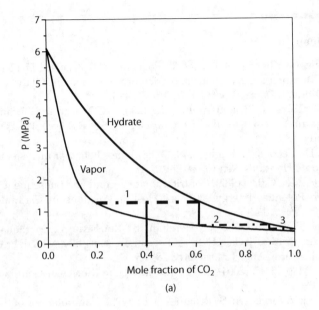

(a)

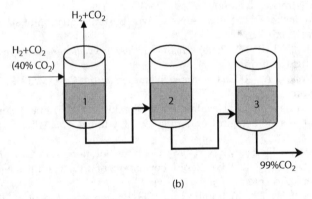

(b)

FIGURE 2

HVLL equilibrium of $H_2 + CO_2 + CP + H_2O$. (a) pressure-vapor/hydrate composition diagram for $H_2 + CO_2$ hydrates with a water and CP free basis. (b)conceptual flow diagram corresponding to the phase diagram.

optimized value of the reference chemical potential difference can predict the equilibrium conditions within an average absolute deviation (AAD) of 4%. This model is used to generate a $CO_2 + H_2$ binary diagram of hydrate and vapor phases, which aids to generate a separation process to produce pure CO_2 in the form of clathrate hydrates. The actual implementation may require an additional piece of equipment to dissociate clathrate hydrates in each reactor bottom. In the real operation, the flow assurance of hydrate slurries will be one of important issues to develop a continuous separation process.

References

Dalmazzone, D., Kharrat. M, Lachet, V., Fouconnier, B., Clausse, D. (2002). DSC and PVT Measurements Methane and Trichlorofuoromethane Hydrate Dissociation Equilibria. *J. Therm. Anal. Calorimetry. 70*, 493.

Gupta, M., Colye, I., Thambimuthu, K. (2003). CO_2 Capture Technologies and Opportunities in Canada. *1st Canadian CC&S Roadmap Workshop*, Calgary, Canada, 18.

Holder, G. D., Zetts, S. P., Pradgan, N. (1988) Phase Behavior in Systems containing Clathrate Hydrates. *Rev. Chem. Eng. 5*, 1.

John, V.T., Holder, G. D. (1982). Contribution of Second and Subsequent Water Shells to the Potential Energy of Guest-Host Interactions in Clathrate Hydrates. *J. Phys. Chem., 86*, 455.

Le Parlouër, P., Dalmazzone, C., Herzhaft, B., Rousseau, L., Mathonat, C. (2004). Characterization of Gas Hydrates Formation Using a New High Pressure micro-DSC. J. Therm. Anal. Calorimetry. *78*,165.

Mao, W. L., Mao, H. K. (2004). Hydrogen Storage in Molecular Compounds. *PNAS, 101*, 708.

Munck, J., Skjold-Jorgensen, S., Rasmussen, P. (1998). Computations of the Formation of Gas Hydrates. *Chem. Eng. Sci., 43*, 2661.

Peng. D.Y., Robinson, D. B. (1976). A New Two-constant Equation of state. *Ind. Eng. Chem. Fundam, 15*, 59.

Rand, D. A. J., Dell, R. M. (2008). *Hydrogen Energy Challenges and Prospects*. RSC Publishing: Cambridge.

Sloan, E. D., Koh, C. A. (2008). *Clathrate Hydrate of Natural Gases*, 3rd ed., CRC Press, Boca Raton.

Sugahara, T., Murayama, S., Hashimoto, S., Ohgaki, K (2005) Phase Equilibria for $H_2 + CO_2 + H_2O$ System Containing Gas Hydrates. *Fluid Phase Equilib., 233*, 190.

van der Waals, J. H., Platteeuw, J. C. (1959). Clathrate Solutions. *Adv. Chem. Phys., 2*, 1.

Zhang, J. S., Lee, J. W. (2008a). Rapid Formation of CO_2 Hydrates under Static Conditions. Accepted to *Ind. Eng. Chem. Res.*

Zhang J. S., Lee, J. W. (2008b). Equilibrium of Hydrogen + Cyclopentane and Carbon dioxide + Cyclopentane Binary Hydrates. *J. Chem. Eng. Data*, in press, web-release: 15-Jul-2008, DOI: 10.1021/je800219k.

26

Optimal Design and Operation of a Circulating Fluidized Bed Reactor for Water Polishing Featuring Minimum Utility Cost

David M. Follansbee, John D. Paccione and Lealon L. Martin[*]
Rensselaer Polytechnic Institute
Troy, NY 12180-3590

CONTENTS

ABSTRACT   In this paper we present a circulating fluidized bed model for a continuous advanced oxidation process consisting of an adsorptive moving packed bed, a UV reactor, and a riser for hydraulic transport of photoactive particles. The model is employed as part of an optimization-based strategy to identify optimal equipment design parameters, flexible operating conditions, and to aid in the selection of composite, photoactive particles with tailored system properties. The particles are to be circulated as a "working solid" for the decomposition of chemical contaminants in the presence of ultra-violet (UV) light in aqueous media. Based on our model, we formulate a nonlinear mathematical program with a minimum utility cost objective. Globally optimal solutions are obtained through direct search with interval

---

[*] To whom all correspondence should be addressed

analysis; and these solutions can provide insight into optimal design and operation of the system. Our methodology is demonstrated for the design of a composite titanium-dioxide (photocatalytic)-activated carbon (adsorbent) material immobilized on numerous substrates. The particles are studied and evaluated based on their performance in the aforementioned system for the degradation of benzene in a contaminant stream. A sensitivity analysis is performed to identify qualitative trends implicit in the proposed mathematical model. Furthermore, we study the effects of small perturbations in the optimal particle design parameters on overall system design parameters, such as annular bed height.

KEYWORDS   *Photocatalysis, organic removal, activated carbon*

Introduction

Circulating fluidized bed (CFB) systems have been used in a wide range of industrial applications (synthesis of fine chemicals, petrochemicals, and petroleum refining) for fluid-particle phase reactions and separations (Grace et al. 1997, Dudukovic et al. 2002). A Typical CFB system is comprised of a moving packed bed with a riser section for hydraulic or pneumatic transport of solid particles. The multiphase hydrodynamics that govern these systems (gas/solid and liquid/solid contacting) have been studied for decades by a significant number of researchers (Grbavcic et al. 1991, Grbavcic et al. 1992, Khan and Richardson 1989, Littman et al. 1993). CFBs offer several attractive features, among which is the ability to independently control the solid recirculation rate and the fluid flowrate through the moving packed bed.

It has been shown that the efficient decomposition of chemical contaminants from water can be performed with a CFB-type system (Follansbee et al. 2008), shown in Figure 1, and is modeled here as three individual processes connected by particle circulation:

(i) Physical adsorption of organic contaminant onto particles in the annular bed.

(ii) Transport of the contaminant-laden particles to a photoreactor.

(iii) Regeneration of spent particles through UV exposure in the photoreactor.

In this paper we present a NLP formulation to study the operation and design of a CFB system for the advanced oxidation of aqueous organics at the minimum utility cost. The NLP formulation is also used for the selection of particles with optimal bulk catalyst properties. A case study is then

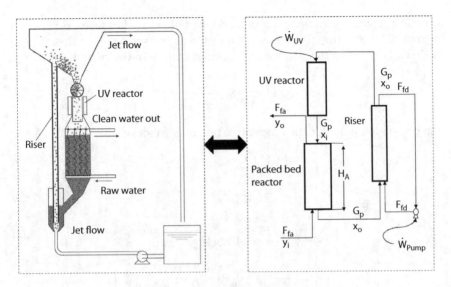

FIGURE 1
Process schematic and diagram.

performed on the oxidation of benzene in water with a TiO_2/activated carbon composite immobilized on a solid oxide support to demonstrate the strength of the proposed methodology and to discuss the operation and the design of the CFB system.

System Modeling

Particle Construction

The photoactive particles are modeled using three parameters (i) the mass fraction of adsorbent to substrate, θ_g (kg_{ads}/kg_{sub}) (ii) the mass fraction of adsorbent to photocatalyst, θ_{load} (kg_{ads}/kg_{photo}) and (iii) the (normalized) amount of photocatalyst, θ_{cat} (ratio of total mass of photocatalyst in composite particle to the maximum loading (mass) of photocatalyst on substrate). These parameters describe the adsorptive and photocatalytic capacity of the particle.

Contaminant Adsorption

Adsorption of the contaminant onto the surface of the adsorbent depends primarily on the overall mass transfer of the contaminant from the rich (liquid) phase to the lean (solid surface) phase and phase equilibrium. The annular bed is modeled as a single-component, continuous, counter-current mass exchanger. A counter current configuration is employed to maximize interfacial contact area between rich and lean phases (Manousiouthakis and Martin 2004). A Langmuir adsorption isotherm is used to relate the concentration adsorbed on the surface to the concentration in the rich phase.

$$x = x_{max}\theta_g \frac{K_A y}{1 + K_A y} \tag{1}$$

The complete model for the adsorptive bed is as follows:

$$M = Q_{f_A}\Delta y = Q_{f_A}(y_i - y_o) \tag{2}$$

$$Q_{f_A}\Delta y = G_p \Delta x \tag{3}$$

$$H = \frac{M}{A_a K_l a \Delta y_{lm}} \tag{4}$$

$$\Delta y_{lm} = \frac{\left(y_i - \frac{x_o}{(x_{max}\theta_g - x_i)K_A}\right) - \left(y_o - \frac{x_i}{(x_{max}\theta_g - x_i)K_A}\right)}{\ln\left(\dfrac{y_i - \frac{x_o}{(x_{max}\theta_g - x_i)K_A}}{y_o - \frac{x_i}{(x_{max}\theta_g - x_i)K_A}}\right)} \tag{5}$$

Equations (2) and (3) are component (contaminant) mass balances within the annular bed, while Eq. (4) and Eq. (5) are effective bed height and concentration driving force, respectively.

Particle Transport

A one-dimensional, two-fluid model with individual fluid and particle phase momentum equations describes the hydrodynamics of multiphase, vertical transport of particles. To minimize the complexity of the transport line model the following assumptions are made:

1. Non-uniformities in fluid/particle flow patterns near the wall are negligible.
2. Particle acceleration in the transport line is negligible for this fluid/particle system.

Based on these assumptions the following equations are used to describe the behavior of particle mass flowrate (G_p), fluid volumetric flowrate (Q_{fl}), voidage within the transport line (ε_T), and the particle and interstitial fluid velocities within the transport line (u and v).

$$Q_{f_T} = u\varepsilon_T A_T \tag{6}$$

$$G_p = v(1 - \varepsilon_T)\rho_p A_T \tag{7}$$

$$u_{sl} = u - v = \left[\frac{\varepsilon_T(1 - \varepsilon_T)(\rho_p - \rho_f)g - (1 - \varepsilon_T)F_f + \varepsilon_T F_p}{\beta_T} \right]^{1/2} \tag{8}$$

The dominant frictional, gravitational, and inertial effects are sufficiently captured through relations between the pressure gradient and voidage in the transport line, and the interphase drag between the fluid and the particles. For a more detailed discussion of Eq. (6)–Eq. (8) the reader is referred to Follansbee et al. (2008).

Photocatalytic Activity and Chemical Decomposition

Contaminant degradation occurs via photon-driven chemical decomposition at the catalyst surface and the following assumptions are made in the UV reactor model:

1. No overall mass transfer limitations of contaminant from the bulk liquid to the particle surface.
2. First order kinetics (above a minimum intensity of UV throughout the reactor)
3. Plug flow with a uniform particle velocity
4. No wall effects or particle-particle interactions.

The final assumption is valid, since the particle density throughout the UV chamber is in the lean phase. The UV reactor design model is then:

$$x_i = x_o \exp\left[-\hat{k} \frac{z}{v_t} \right] \tag{9}$$

where $\hat{k}$ is the apparent reaction rate — accounting for contaminant diffusion mass transfer limitations on the particle surface. $\hat{k}$ has been shown to have a logarithmic dependence on θ_{load} (Torimoto et al. 1997). As θ_{load} increases, θ_{cat} decreases along with $\hat{k}$. This is consistent with the fact that only a fraction of the photon flux (or intensity) can reach available active sites, according to the Beer-Lambert law. The decay is additive and it is attributed to the absorption of light by fluid medium and the extinction of light by the particles themselves:

$$I_o = I \exp\left[\alpha \frac{D_{UV}}{2} \right] \tag{10}$$

$$\alpha = w(1 - \varepsilon_{UV}) + \alpha_o \tag{11}$$

$$G_p = v_t(1 - \varepsilon_{UV})\rho_p A_{UV} \tag{12}$$

$$\dot{W}_{UV} = I_o \pi D_p z(1 - \varepsilon_{UV}) \tag{13}$$

Equation (12) is the particle flow rate given as a function of terminal velocity, reactor voidage, and cross-sectional area. Eq. (13) is the power required to ensure that a minimum intensity (I_o) is maintained at the centerline of the chamber (validates first order kinetics).

Mathematical Formulation

We now pose the following problem statement:

Given adsorptive mass transfer rates, degradation rates for a contaminant species, specific adsorbent and photo-oxidative materials with physical property information, a steady-state model quantifying overall input-state-output relationships, influent flow rate and concentration (Q_{fA} and y_i), and the target effluent concentration (y_o); determine optimal design parameters and operating conditions that satisfy target outlet conditions and minimizes utility cost; and identify the optimal loadings for the circulating photoactive particles.

The proposed mathematical model employs both empirically and theoretically determined equations. These equations outline the conservation of mass, energy, and momentum throughout the integrated system. However, constraints must be applied to ensure that the mathematical model will be consistent with physically relevant hydrodynamic phenomena:

$$G_p < (1 - \varepsilon_{min})A_a \rho_p v_{a_{max}} \tag{14}$$

$$\varepsilon_{vc} < \varepsilon_T < 1 \tag{15}$$

$$u \geq 1.5 v_t \tag{16}$$

$$\rho_p > \rho_f \tag{17}$$

$$y > \frac{x}{K_A(x_{max}\theta_g - x)} \tag{18}$$

The function ζ is defined to represent system utility cost as follows:

$$\zeta = \gamma_1(\dot{W}_{UV} + \dot{W}_{pump}) + \gamma_2 H + \gamma_3 G_p\left(\frac{L}{v} + \frac{z}{v_t}\right) + \gamma_4 \theta_{load} \tag{19}$$

Having defined the feasible region based on hydrodynamic transportation of particles, adsorption and degradation of contaminant, and particle construction; and having defined a general cost objective function the NLP for the CFB system is as follows:

$$\min_{Q_{f_T}, G_p} \zeta \qquad\qquad (P1)$$

$$\text{s.t. } (1)-(18)$$

As written this problem has two degrees of freedom, which are chosen to be the jet flowrate (Q_{ft}) and particle mass flowrate (G_p). A method for solving (P1) globally has been presented by Chen et al. (2006) and Follansbee et al. (2008). The solution strategy employs interval analysis to identify acceptable ranges (intervals) for all variables and subsequently optimized within this range. In this work we search directly over the acceptable ranges for all parameters.

Results and Discussion

The algorithm was coded into Matlab and executed on a 2.2 GHz Intel Core 2 Duo processor with a CPU time of 46.0934 seconds.

CFB Operation

The utility cost of the system is indeed sensitive to perturbations in operating conditions. This sensitivity is shown in Figure 1 for an annular flow of 1 GPM with inlet and outlet concentrations of 100 and 10 ppm, respectively.

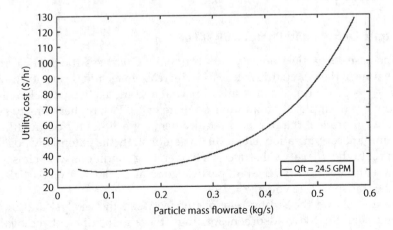

FIGURE 2
Effect of G_p on utility cost (silica particle, $\theta_{load} = 0.2$).

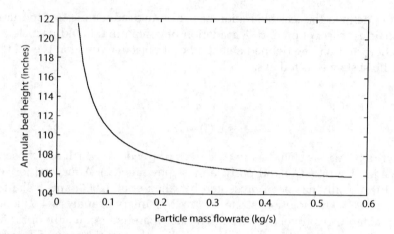

FIGURE 3
Effect of G_p on annular bed height (silica particle, $\theta_{load} = 0.2$).

Sensitivity is largely due to dramatic increases in the operating cost for the UV chamber with small changes in particle mass flow rate. The UV chamber becomes more densely populated, as particle mass flow rate increases, and consequently light decays much more rapidly through the system. Therefore, to ensure that the intensity at the centerline of the UV chamber is maintained, the wall intensity must be increase. However, there is an equivalent increase in the particle regeneration rate, resulting in a higher adsorptive driving force within the annular bed. An increased driving force lowers the effective bed height, as shown in Figure 3, and has a positive impact on cost minimization with respect to particle utility cost and (although not considered here) capital costs.

Substrate Density and Photocatalyst Loading

An increase in particle density proportionally decreases the particle residence time in the UV chamber, due to an increase in terminal velocity. Lower particle residence times can result in partial and ineffective particle regeneration. For example, at an annular flowrate of 1 GPM, higher density substrates cannot meet the imposed requirements of a 10-fold drop in effluent contaminant concentration. However, to demonstrate the potentially positive effects of higher density substrates, CFB operating conditions were chosen as follows: annular flow = 0.5 GPM, particle circulation rate = 3800 particles/s, and the transport line voidage = 0.99.

Figure 4 shows the strong relationship between the catalyst and annular bed height. At low adsorbent loadings, there is insufficient removal of contaminant in the bed. As the loading increases, the annular bed height decreases to an optimal loading between 25% and 30%.

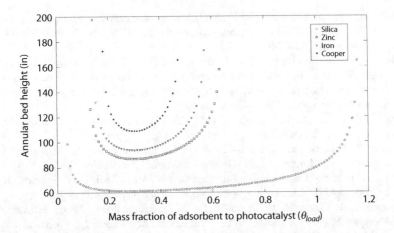

FIGURE 4
Effect of θ_{laod} on annular bed height ($G_{fa} = 0.5$ GPM, $\varepsilon_T = 0.99$, 3800 particles/s).

Conclusion

We have formulated a mathematical model to study the operation and design of a circulating fluidized bed system for the advanced oxidation of aqueous organics at the minimum utility cost. A case study was performed on the oxidation of benzene in water with a TiO_2/activated carbon composite immobilized on a solid oxide support. It was determined that the required adsorption bed height has a strong dependence on the catalytic properties of the particle and a lesser dependence on the actual system operating conditions. We also determined that the minimum utility cost of the system is highly sensitive to changes in particle mass flowrate due to sharp increases in particle transport and UV delivery costs.

Acknowledgments

Financial support was provided by Army Research Office W911NF-06-1-0260 and greatly appreciated.

References

Chen, K. I., Winnick, J., Manousiouthakis, V. I. (2006) Global optimization of a simple mathematical model for a proton exchange membrane fuel cell. *Computers & Chemical Engineering, 30*, 1226–1234.

Dudukovic, M., Larachi, F., Mills, P. (2002) Multiphase catalytic reactors: A perspective on current knowledge and future trends. *Catalysis Reviews-Science and Engineering*, 44, 123–246.

Follansbee, D. M., Paccione, J. D., Martin, L. L. (2008) Globally optimal design and operation of a continuous photocatalytic advanced oxidation process featuring moving bed adsorption and draft-tube transport. *Industrial & Engineering Chemistry Research*, 47, 3591–3600.

Grace, J. R., Avidan, A. A., Knowlton, T. M. (1997) *Circulating Fluidized Beds*, Blackie Acedemic & Professional.

Grbavcic, Z. B.,Garic, R. V., Hadzismajlovic, D. E., Jovanovic, S., Vukovic, D. V., Littman, H., Morgan, M. H. (1991). Variational Model For Prediction Of The Fluid Particle Interphase Drag Coefficient And Particulate Expansion Of Fluidized And Sedimenting Beds. *Powder Technology*, 68, 199–211.

Grbavcic, Z. B., Garic, R. V., Vukovic, D. V., Hadzismajlovic, D. E., Littman, H., Morgan, M. H., Jovanovic, S. D. Hydrodynamic Modeling Of Vertical Liquid Solids Flow. *Powder Technology* **1992**, 72, 183–191.

Khan, A. R., Richardson, J. F. (1989). Fluid-particle interactions and flow characteristics of fluidized-beds and settling suspensions of spherical-particles. *Chemical Engineering Communications*, 78, 111–130.

Littman, H., Morgan, M. H., Paccione, J. D., Jovanovic, S. D., Grbavcic, Z. B. (1993) Modeling and measurement of the effective drag coefficient in decelerating and non-accelerating turbulent gas-solids dilute phase flow of large particles in a vertical transport pipe. *Powder Technology*, 77, 267–283.

Manousiouthakis, V., Martin, L. L. (2004) A minimum area (MA) targeting scheme for single component MEN and HEN synthesis. *Computers & Chemical Engineering*, 28, 1237–1247.

Torimoto, T., Okawa, Y., Takeda, N., Yoneyama, H. (1997). Effect of activated carbon content in TiO2-loaded activated carbon on photodegradation behaviors of dichloromethane. *Journal of Photochemistry and Photobiology A-Chemistry*, 103, 153–157.

27

Energy Integration of an IGCC Plant for Combined Hydrogen and Electricity Production—Methodology and Tools Integration

Rahul Anantharaman[*], Olav Bolland and Truls Gundersen

Department of Energy & Process Engineering,
Norwegian University of Science and Technology
Kolbjoern Hejes vei 1B, NO-7491, Trondheim, Norway

CONTENTS

ABSTRACT Carbon dioxide capture is associated with significant energy penalties that affect the economic viability of such plants. This paper focuses on the energy integration of combined hydrogen and electricity production of an IGCC unit to increase its energy efficiency. Two energy integration methods are used in succession, with the ELCC forming the screening tool for heat integration in the Sequential Framework. Different modeling and optimization tools are integrated to evaluate the performance of the IGCC plant. Results show that the methodology designs efficient processes with flexibility in being able to choose from different heat integrated schemes based on qualitative and quantitative parameters.

---

[*] To whom all correspondence should be addressed

KEYWORDS *Energy integration, heat exchanger network synthesis, IGCC, CO_2 capture*

Introduction

IPCC estimates that the economic potential of Carbon Capture and Storage (CCS) could be between 10% and 55% of the total carbon mitigation effort until the year 2100. Major research is in the field of CO_2 capture from power plants, particularly coal based power generation. Pre-combustion capture of carbon dioxide by decarbonizing the fuel enables polygeneration—large scale electricity generation combined with the production of hydrogen and synthetic fuels. In this context, Integrated Gasification Combined Cycle (IGCC) is one of the important concepts for coal based power plants with CO_2 capture. Carbon dioxide capture is associated with significant energy penalties that reduce the economic viability of such plants.

This paper presents a framework combining different methodologies and tools for energy integration of combined hydrogen and electricity production of an IGCC unit to improve its energy efficiency. The subsequent sections provide a brief description of the methodologies, tools, process and results from energy integration.

Methodologies for Energy Integration

Two methodologies for energy integration employed in this paper are briefly described below.

Energy Level Composite Curves

Energy Level Composite Curves (ELCC)—a synergy of Exergy Analysis and the Composite Curves of Pinch Analysis—is a novel method for energy integration that incorporates pressure and composition changes in the process in addition to temperature.

Energy levels at target and supply conditions are evaluated as:

$$\Omega = \frac{(H - H_0) - T_0(S - S_0)}{H - H_0} \qquad (1)$$

Streams with increasing energy levels are energy sinks while those with decreasing energy levels are energy sources. Energy sources at higher energy levels can be potentially integrated with energy sinks at lower energy levels.

The ELCC are energy levels—enthalpy curves constructed by plotting energy levels of process streams against cumulative values of enthalpy differences.

It functions as a screening tool or idea generator, giving physical insight for energy integration between streams on the energy source curve and the energy sink curve. Details of the methodology are provided in Anantharaman et al. (2006).

Sequential Framework for Heat Exchanger Network Synthesis

As a compromise between Pinch Analysis and simultaneous MINLP models, a sequential and iterative framework has been in development in our group with the main objective of finding near optimal heat exchanger networks for industrial size problems.

The subtasks of the design process are solved sequentially using Math Programming. Briefly, these steps involve: establishing the minimum energy consumption (LP), determining the minimum number of units (MILP), finding sets of matches and corresponding heat load distributions (HLDs) for minimum or a given number of units (MILP), and network generation and optimization (NLP) as shown in Figure 1.

The Sequential Framework is based on the recognition that the selection of HLDs impacts both the quantitative (network cost) and the qualitative aspects such as network complexity, operability and controllability. The Vertical MILP model for selection of matches and the subsequent NLP model for generating and optimizing the network form the *core engine* of the framework. The loops in the framework simulate the three-way trade-off between energy consumption (E), heat transfer area (A) and how this area is split up into a number of heat transfer units (U). Loops 1 and 2 can be thought of as the *area* loops, loop 3 as the *unit* loop and finally loop 4 as the *energy* loop.

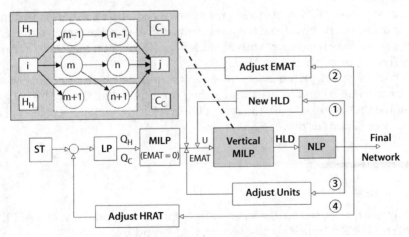

FIGURE 1
Sequential framework for HENS with the vertical transportation mode.

Significant user interaction is built into the framework in the form of iterative loops to enable the designer to explore and evaluate the most promising networks with respect to Total Annual Cost (TAC), network complexity (number of units, splits, etc.), operability and controllability.

Details of the Sequential Framework can be found in Anantharaman and Gundersen (2006).

IGCC Process Description

The IGCC process is modeled to produce 400 MW of electricity and 50 MW of hydrogen. The main characteristics of the process are:

- Oxygen blown Shell gasification process—48 bar
- Syngas recirculation
- Two stage sour shift conversion of the syngas
- Separate capture of H_2S and CO_2 using Selexol process
- GE9FB gas turbine burning H_2 rich fuel
- High pressure cryogenic Air Separation Unit (ASU)

Douglas Premium hard coal is taken as the fuel for the process. The Acid Gas Removal (AGR) system utilizes a separate removal of H_2S and CO_2 based on the Selexol process. The Sulphur recovery unit is a Claus plant using oxygen for combustion and steam from the flue gas is condensed. This enables CO_2 and traces of H_2S to be recycled to the system with no flue gas emissions from the Claus Unit to the atmosphere. The CO_2 product (98% purity) is compressed to 110 bar.

The ASU takes 50% of its total air feed as bleed air from the gas turbine. Some air has to be bled from the gas turbine to balance the large volumetric flow rate of the H_2 rich fuel with N_2 added as diluent to reduce NOx formation. The O_2 purity from the ASU is 95%.

A part of the H_2 rich gas from the AGR is purified by a Pressure Swing Adsorption (PSA) unit to 99.9% and is then compressed to 70 bar. The tail gas from the PSA is mixed with the rest of the H_2 rich stream and sent as fuel to the power island for power generation.

Tools

The main tools used in the energy integration methodology are ASPEN HYSYS, GT PRO from Thermoflow Inc, GAMS and Excel.

Aspen HYSYS is used to model the gasification island including CO_2 and H_2 purification and compression. A CAPE-OPEN entrained flow type gasifier model was developed for use in HYSYS. The ASU is modeled as a black box while the PSA is modeled as a simple separator with 85% H_2 yield.

GT PRO is used to model the combined cycle gas turbine that forms the power island. A steam cycle with three pressure levels with reheat is selected. Stream information between HYSYS and GT PRO is shared using an Excel add-in developed in-house.

The ELCC and energy targets associated with it are generated using an Excel add-in HYSYS XLink, also developed in-house.

The sub-problems in the Sequential Framework are modeled in GAMS. An Excel add-in developed for the Sequential Framework, SeqHENS, forms the interface between stream data in HYSYS and GAMS as well as for user interaction.

Energy Integration

The plant can be broken down into the following process areas:

- 1a: Gasifier island including ASU and shift reactors
- 1b: AGR unit and CO_2 compression
- 1c: H_2 purification and compression
- 2: Power island

The ELCC for process areas 1a, 1b and 1c show that there is not much scope for integration between these processes. The PSA tail gas is compressed to 25 bar to mix with H_2 rich gas from AGR to the gas turbine. The resulting exergy loss can be minimized if the tail gas is used as fuel at atmospheric conditions, e.g. duct burning in the Heat Recovery Steam Generator (HRSG). This alternate scheme has not been evaluated. The ELCC is expected to be useful when integrating the ASU with the gasifier island and AGR.

The Sequential Framework is used for heat integration of the process. The energy targeting LP model is modified to handle steam generation. Steam raised in process areas 1a, 1b and 1c have to be superheated in the HRSG due to material limitations. Three different integration schemes are considered based on the level of integration between the four process areas.

- Scheme 1: Reboilers in the AGR unit are integrated in the process **directly**.
- Scheme 2: Reboilers in the AGR unit are integrated in the process **indirectly** with LP steam for reboilers extracted from the steam turbine. Process streams act as economizers for BFW used for process steam generation.

- Scheme 3: Reboilers in the AGR unit are integrated in the process **indirectly** with LP steam for reboilers generated in the process.

Assumptions, Utility and Cost Data

A HRAT of 20°C is taken for gas/gas heat exchange and 15°C for a gas/liquid heat exchange. The 3 pressure levels chosen for the steam system are: 144, 54 and 3.5 bar respectively for HP, IP and LP steam.

Cost of electricity is taken to be 63 €/MWh. Steam utility costs are calculated form:

- 1 kg/s of sat HP steam ≈ 0.92 MWe
- 1 kg/s of sat IP steam ≈ 0.68 MWe
- 1 kg/s of sat LP steam ≈ 0.42 MWe

The heat exchanger cost law is taken to be 10,000 € + (800 €)*(Area)$^{0.8}$.

Results and Discussion

The results for the three different integration schemes are presented in Table 1. The Composite Curves for process areas 1a, 1b and 1c for integration schemes 1, 2 and 3 are shown in Figure 2, 3 and 4 respectively. The HRSG (process area 2) is not shown as part of this figure for clarity.

The efficiency presented in Table 1 is the plant net electric efficiency and does not include hydrogen production. The gas turbine power output, 287.5 MWe, is constant for all schemes. The plant equivalent efficiency including hydrogen production for all the three schemes is approximately 36.6% including CO_2 capture and compression.

The results indicate that the different integration schemes do not have a direct impact on the efficiency of the plant. The integration chosen will thus

TABLE 1

Process Metrics for the 3 Integration Schemes

	Steam Production (kg/s)			Steam extraction from ST (kg/s)		ST Power Output (MW)	No. HX Units	Efficiency (%)
Scheme	HP 144 bar	IP 54 bar	LP 3.5 bar	IP 41 bar	LP 7 bar			
1	157.5	27.9	17.4	59.5		206.6	34	33.60
2	160.2	29.0	23.3	59.5	12.0	202.2	33	33.58
3	155.6	24.6	21.4	59.5		202.2	36	33.58

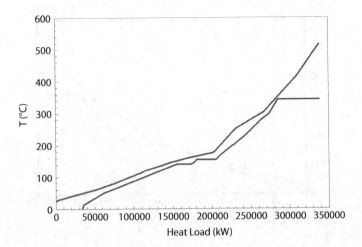

FIGURE 2
Composite curves for integration scheme 1.

depend on other parameters such as the Total Annualized Cost, operability etc. It must be pointed out that the Sequential Framework produces many heat exchanger networks with comparable TAC for each of the integration schemes.

The optimization scheme for HENS does not include carbon tax/credit. This could be included in the SuperTargeting step for identifying the "optimum" ΔT_{min}.

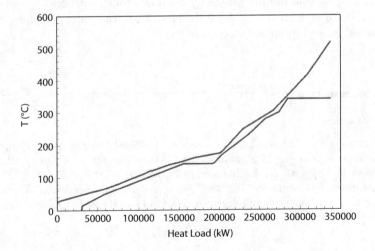

FIGURE 3
Composite curves for integration scheme 2.

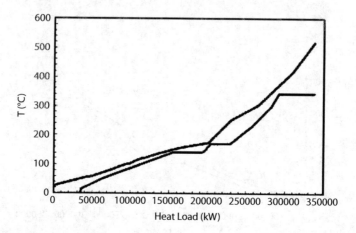

FIGURE 4

Composite curves for integration scheme 3.

Conclusions

Energy integration of an IGCC plant for combined hydrogen and electricity production is presented. Two energy integration methods are used in succession with the ELCC forming the screening tool for heat integration in the Sequential Framework. Different modeling and optimization tools are integrated to evaluate the performance of the IGCC plant.

Results show that the methodology designs efficient processes with flexibility in being able to choose from different heat integrated schemes based on qualitative and quantitative parameters.

References

Anantharaman, R., Abbas, O. S., Gundersen. T. (2006). Energy Level Composite Curves—a new graphical methodology for the integration of energy intensive processes. *Applied Thermal Engineering, 26, 1378.*

Anantharaman, R., Gundersen, T. (2006). Developments in the Sequential Framework for Heat Exchanger Network Synthesis of industrial sized problems. *Computer Aided Chemical Engineering, 21A, 725.*

28

The Application of a Task-based Design Approach to Solution Crystallization

R. Lakerveld*, H.J.M. Kramer, P.J. Jansens and J. Grievink
Delft University of Technology
Leeghwaterstraat 44, Delft, The Netherlands

CONTENTS

ABSTRACT   A new task-based approach is applied to design a solution crystallization process unit. Task-based design involves the conceptual built-up of a process (unit) from functional building blocks called tasks, which represent fundamental physical events. The motivation for developing this approach is to get a better control over the physical events governing crystalline product quality. To deliver a proof of concept, two lines of research are followed. First of all, several small scale experiments are designed to demonstrate practical feasibility of the approach. The new lab-scale equipment allows for isolation and manipulation of individual crystallization tasks. Secondly, a model based on the experimentally tested tasks is developed for a crystallizer design and used for dynamic optimization in batch mode of product quality with minimum energy consumption. The results show that a task based crystallizer is capable of keeping the state variables very closely to ideal values. This is the direct result of the ability to control the rates at which individual crystallization tasks are executed as well as the material flows between those tasks.

KEYWORDS   *Crystallization, task-based design, energy minimization, crystal product quality control*

---

* To whom all correspondence should be addressed

Introduction

Crystallization is one of the oldest and economically most important separation and product formation technologies. The design of crystallization processes still poses many challenges despite its wide application. The selection of crystallization equipment is traditionally done from a limited number of state-of-art crystallizers followed by optimization of that particular type of equipment. Present industrial crystallizers harbor many physical phenomena such as primary nucleation, crystal growth, generation of supersaturation, and attrition. Optimization of each of these individual physical phenomena is not well possible as in present industrial crystallizers these phenomena are strongly entangled.

This lack of control over physical phenomena causes crystallizers to exhibit complex process behavior and limits flexibility for both design and operation. The physical phenomena are however of key importance as in the end they determine the properties of the crystalline product and also the efficiency in terms of energy consumption. Therefore there is a need for a design approach which considers the important phenomena as starting point for design rather than the equipment itself. This contribution discusses the application of such a design methodology, which is a task based design (TBD)(Menon, 2007).

In TBD an attempt is made to conceptually construct the crystallization process from fundamental building blocks called physical tasks (similar to Kondili et al. 1993). In principle the TBD approach for solution crystallization can be embedded in existing generalized frameworks for the representation of superstructures and optimization models in process synthesis such as derived by Yeomans & Grossmann (1999). For crystallization processes it is however difficult to specify physical devices that will execute a single crystallization task as current crystallizers facilitate many of those tasks and the control of individual tasks is not well possible. Therefore, the first part of this paper summarizes the key results of small scale experiments with newly built dedicated equipment showing that it is possible to isolate and optimize single crystallization tasks. The results, of which detailed descriptions are described elsewhere (Lakerveld et al. 2008[a,b,c]), demonstrate the practical feasibility of the approach showing the isolation and optimization of the crystallization tasks growth, supersaturation generation and nucleation. In the second part of this paper a model based on the experimentally tested tasks is developed for a task based crystallizer design. It is used in dynamic optimization of a batch case study tailored towards energy minimization with constraints on product quality. The results show that simply by changing the operational policy, a tight control over product quality is possible and simultaneously minimizing energy demand.

Experimental Isolation of Single Tasks

The control over a single crystallization task means in practice that the operating conditions in a physical device must be tailored such, that the resulting driving forces enable a certain independent task to be dominant. The following experimental targets are considered:

1. For separation of the task crystal growth and attrition, minimize shear stress acting on crystals as growth and attrition are competing at high shear stress.
2. For separation of the task crystal growth and primary nucleation, keep supersaturation below the primary nucleation threshold at any location.
3. To generate new crystals, evaluate the use of ultrasound to induce and control the task nucleation at low supersaturation and low shear stress.

To isolate the task Crystal Growth a physical environment is needed which can minimize attrition as attrition increases crystal surface area, which competes for growth with the existing crystals. The environment that was selected consisted of a bubble column in which supersaturation was created by simultaneous cooling and evaporation of the solvent by sparging air (Lakerveld et al., 2008a). The crystals were kept in suspension by the upward velocity of the bubbles, eliminating the need for a stirrer or a circulation pump. In this way, attrition caused by collisions between crystals and moving mechanical parts was absent. Seeded batch experiments in a 3-l bubble column showed that an initial seed population could grow without an increase in number of crystals. Similar experiments in agitated crystallizers (Hojjati et al. 2005, Lakerveld et al. 2007) showed a clear increase in number of crystals due to attrition. It demonstrated that the concept is very promising to isolate the task Crystal Growth by effectively suppressing nucleation.

The second objective of the experimental work is related to tight control of supersaturation at any location in a future crystallizer to prevent spontaneous nucleation bursts and to maximize crystal growth. Membranes can selectively remove the solvent and offer an interesting opportunity to control supersaturation gradients in new crystallizer designs (Azouy et al, 1986; Curcio et al., 2001; Tun et al. 2005). Potentially they can also contribute to reduced energy consumption as an expensive vapor liquid equilibrium can be avoided. An experimental setup was constructed and tested for both an NH_4SO_4 water system and an adipic acid water system to assess the potential application of reverse osmosis membranes for crystallization processes

(Lakerveld et al. 2008[b]). It showed practical feasibility of the concept and the membrane is used in the next section for design.

Ultrasound is an interesting tool to induce nucleation at low supersaturation in a controlled way (see for example Li et al., 2003; Virone et al., 2006). The applicability of ultrasound was illustrated in a number of experiments in which an ultrasound generator was placed inside a supersaturated solution of an NH_4SO_4 water system. The supersaturation was kept low to avoid spontaneous nucleation. The objective of the experiments was to relate the power input and insonation time to the number of nuclei produced. It was concluded that nuclei could be generated at low supersaturation. Moreover, the nucleation rate could be optimized, by changing the power input and insonation time (Lakerveld et al. 2008[c]).

Dynamic Optimization of a Task-based Crystallizer

The starting point of the development of a TBD for solution crystallization, was the development of new process units that could isolate single tasks as described in the previous section. Those new unit operations provide possible building blocks for a task superstructure, which can be optimized resulting in an optimal task structure. In this section, the experiments summarized in previous section are combined in a model, which represents a possible outcome of such a task superstructure optimization, and optimized. The aim is to illustrate flexibility for process design resulting in improved product quality with reduced energy demand.

A drawing of the modeled system is given in Figure 1. It contains one compartment with a gas, a liquid and a solid phase (labeled GLS), which behaves as the bubble column. This compartment is used to grow crystals with negligible attrition. Ultrasound can be applied within the compartment to produce nuclei at low supersaturation with the experimentally determined rate. From this compartment a clear solution is taken to a second compartment (labeled ML) operating at higher temperature, in which the task supersaturation generation is realized. A reverse osmosis membrane

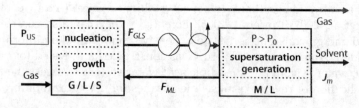

FIGURE 1
Structure of task based crystallizer with 3 tasks.

is used here to selectively remove solvent. The system is operated in semi-batch mode.

For the development of the model special attention has been paid to proper scaling of the state variables. Furthermore, instead of solving the full population balance describing the dynamic development of the crystal size distribution, which results in a system of PDAE equations, the moment transformation has been used to reduce the system to a set of DAE equations. The crystal growth rate depends linearly on supersaturation:

$$G = k_G \cdot \left(w_{GLS} - w_{GLS}^* \right) \tag{1}$$

The moment equations are given as follows:

$$\frac{dx_0}{dt} = \eta_U \alpha \tag{2}$$

$$\frac{dx_i}{dt} = i \cdot \left(w_{GLS} - w_{GLS}^* \right) \cdot x_{i-1} \qquad \text{for } i = 1,2,3,4 \tag{3}$$

Where the variable η_U represents the utilization of ultrasound. It can be imagined as the fraction of the total volume that is insonated. Furthermore, $x_0, \ldots, x_4$ represent the scaled moments ($m_0, \ldots, m_4$) and α represents the scaled maximum nucleation rate defined as follows:

$$x_0 = m_0 k_G^3, \; x_1 = m_1 k_G^2, \; x_2 = m_2 k_G, \; x_3 = m_3, x_4 = m_4 k_G^{-1}$$

$$\alpha = B_{max} \cdot k_G^3$$

The component (Eq. (4)) and total mass balance for the GLS compartment are given as follows respectively:

$$V_{GLS} \varepsilon \frac{dw_{GLS}}{dt} = F_{ML} w_{ML} - F_{GLS} w_{GLS} + V_{GLS} k_V \left(w_{GLS} - \frac{\rho_S}{\rho_L} \right) \frac{dx_3}{dt}$$

$$-k_V \frac{dx_3}{dt} = \frac{\rho_L (F_{ML} - F_{GLS})}{V_{GLS}(\rho_L - \rho_S)} \tag{5}$$

The liquid volumetric flow F_{ML} is flow controlled. The liquid fraction ε is connected to the third moment

$$\varepsilon = 1 - k_V x_3 \tag{6}$$

The total mass and component balance for the ML compartment complete the model:

$$\frac{dw_{ML} V_{ML}}{dt} = F_{GLS} w_{GLS} - w_{ML} F_{ML} \tag{7}$$

$$\rho_L \frac{dV_{ML}}{dt} = \rho_L F_{GLS} - \rho_L F_{ML} - J_m A_{mem} \tag{8}$$

The energy consumption of the system is the sum of the heating capacity, pump duty, and ultrasound duty

$$E_{tot} = \int_{t=0}^{t=t_f} \left(P_{US} + F_{GLS}(t) \left[\frac{\Delta P(t)}{\eta} + Cp_L \rho_L (T_{ML} - T_{GLS}) \right] \right) \cdot dt \tag{9}$$

The mass flux over the membrane J_m can be manipulated by changing the pressure in the membrane module, which depends linearly on the mass flux. The model was implemented in general PROcess Modeling System (gPROMS Modelbuilder 3.1.4, PSE Ltd., London, UK). The parameter settings corresponded to an NH_4SO_4 water system with initial values in both compartments according to a clear liquid saturated at 25°C. The manipulated variables were the mass flux over the membrane (J_m) and the ultrasound utilization (η_U), which both change the rate at which a specific task is executed. Furthermore, the flow rate F_{ML} between the two compartments was varied, which manipulated the connection between tasks. The fixed time space was discretized into 100 intervals. The operational policy could be optimized to maximize mean size with minimum energy consumption by introducing two objective functions in a Pareto-like optimization and the following constraints

$$\max_{J_M(t),\eta_U,F_{ML}(t),T_{ML}} \frac{m_4(t=t_f)}{m_3(t=t_f)} \tag{10}$$

Subject to:

$$\min_{J_M(t),\eta_U,F_{ML}(t),T_{ML}} E_{tot} \tag{11}$$

$$0 < J_M(t) < 2.5 \quad t \in [0, t_f] \tag{12}$$

$$1 \cdot 10^{-4} < F_{ML}(t) < 0.1 \quad t \in [0, t_f] \tag{13}$$

$$T_{GLS} < T_{ML} < 373 \quad t \in [0, t_f] \tag{14}$$

$$w_{GLS}^{SAT}(T_{GLS}) < w_{ML}(t) < w_{ML}^{SAT}(T_{ML}), \quad t \in [0, t_f] \tag{15}$$

$$0.05\% \le \frac{w_{GLS}(t) - w_{GLS}^{SAT}}{w_{GLS}^{SAT}} \le 1.0\%, \quad t \in [0, t_f] \tag{16}$$

$$\varepsilon(t = t_f) < 0.85 \tag{17}$$

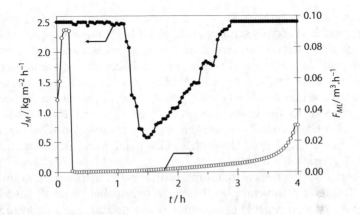

FIGURE 2
Optimized profile J_M (•) and F_{ML} (o).

The optimal profile for the mass flux over the membrane, which is determined by the pressure difference over the membrane ΔP, is depicted in Figure 2 together with the optimal trajectory of the flow rate between both compartments F_{ML}. The trajectory of supersaturation and concentration in ML compartment, which are both constrained state variables, is shown in Figure 3.

From Figures 2–3 it can be seen how both the solvent mass flux over the membrane, flow rate between both compartments and utilization of ultrasound work together to maximize the final volume based mean size. Ultrasound is active only for 7% during the first 2 minutes of the batch (not shown) to generate a certain amount of initial nuclei. The mass flux and flow rate between compartments are high in the beginning (Figure 2) to quickly raise the supersaturation to the constrained value (Figure 3) representing maximum growth (Eq. (1)). The flow rate is reduced as soon as the supersaturation hits the constraint. From this point

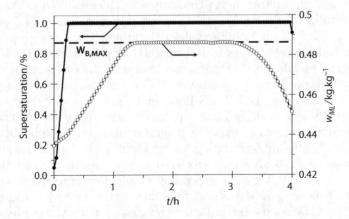

FIGURE 3
Relative supersaturation (•) and solute fraction ML compartment (o).

the concentration in the ML compartment increases more rapidly (Figure 3) until this concentration becomes saturated (T_{ML} = 81.8°C). To prevent super-saturation, and therefore scaling on the membrane, the mass flux is reduced quickly at this point to reduce the increase in concentration. As the crystal surface area increases over time more supersaturation can be delivered to the GLS compartment and the mass flux can gradually increase. Note that for minimization of energy the optimizer seeks to minimize the product of T_{GLS} and F_{ML} as heating is the most significant contribution to the energy consumption. This is reflected by the profile of W_{ML} (Figure 3), which is at the maximum value, determined by T_{ML}, for a large part of the batch.

It is important to notice that supersaturation is at the maximum value almost throughout the complete batch, showing that the task based crystallizer is very well capable of manipulating supersaturation and therefore product quality (size distribution). Using the same crystallizer design, also a small product with narrow size distribution can be produced (Lakerveld et al., 2008d). It illustrates increased flexibility in design and operation.

Conclusion

A TBD approach is applied for the design of a crystallization process unit. A two track approach has been followed. First of all, small scale experiments are designed and tested to evaluate technology that is capable of isolating single crystallization tasks. It delivers a proof of principle that dedicated equipment is capable to execute the fundamental crystallization tasks Growth, Nucleation and Supersaturation Generation. Modeling delivers the building blocks for design, which are used in the second part of this paper dealing with dynamic optimization of a task based batch crystallizer for optimization of the final mean size with minimum energy consumption. The results show that a task based crystallizer is flexible and therefore capable of maintaining supersaturation very closely to an optimum value determined by the designer.

Although the present work is an important step forward in proving the technological viability of TBD for solution crystallization processes, many follow-up questions must be answered. Out of the general engineering abilities reliability and availability have not been discussed. This will be a challenge to the approach since unconventional technology is used. The modeling of individual tasks will also contribute to an improved reliability of the approach and better understanding how to isolate tasks for various systems. Furthermore, the task structure was fixed for the optimization studies performed in this work as it represented the combination of tested experimental setups. However, optimization of the task structure itself could further improve the design of the crystallization process unit. It involves postulation of a task superstructure that is rich enough to incorporate interesting process alternatives on the one hand and removing abundant alternatives on the other hand. Furthermore, optimization of the task superstructure would transform the NLP optimization problem of the current work into an MINLP optimization problem, which is computationally more demanding. Those are challenging questions for future research.

References

Azouy, R., Garside, J., Robertson, W.G. (1986). Crystallization processes using reverse osmosis, *J. Cryst. Growth*, 654

Curcio, E., Criscuoli, A., Drioli, E. (2001). Membrane crystallizers, *Ind.Eng.Chem.Res.*, 40, 2679

Tun, C.M., Fane, A.G., Matheickal, J.T., Sheikholeslami, R. (2005). Membrane distillation crystallization of concentrated salts *J. Membr. Sci.*, 257, 144

Hojjati, H., Rohani, S., (2005). Cooling and seeding effect on supersaturation and final crystal size distribution (CSD) of ammonium sulphate in a batch crystallizer, *Chem. Eng. Process.*, 44, 949

Kondili, E., Pantelides, C.C., Sargent R.W.H. (1993). A general algorithm for short-term scheduling of batch operations –I. MILP formulation. *Comput.chem. Eng.*, 17, 211

Lakerveld, R., Kalbasenka, A.N., Kramer, H.J.M. Jansens, P.J., Grievink, J, The application of different seeding techniques for solution crystallization of ammonium sulphate, *in Proceedings of 14th Internationcal Workshop on Industrial Crystallization* 2007, 221–228

Lakerveld, R., Kramer,H.J.M., Jansens, P.J., Grievink, J. (2008a). Solution Crystallization in a Bubble Column Optimization of the task: Crystal Growth, *in ISIC17 Conference Proceedings*, Maastricht, Netherlands, 819.

Lakerveld, R., Kuhn, J., Bosch, M.A., Kramer, H.J.M., Jansens, P.J., Grievink, J. (2008b). Membrane Assisted Evaporative Crystallization: Optimization of Task Supersaturation Generation, *in ISIC17 Conference Proceedings*, Maastricht, Netherlands, 827

Lakerveld, R., Verzijden, P.G., Kramer, H.J.M., Jansens, P.J., Grievink, J. (2008c) The Application of Ultrasound for Seeding Purposes Optimization of the task: Nucleation, *in ISIC17 Conference Proceedings*, Maastricht, Netherlands, 835

Lakerveld, R., Kramer, H.J.M., Jansens, P.J., Grievink, J. (2008d) A task based design approach for solution crystallization, *in ISIC17 Conference Proceedings*, Maastricht, Netherlands, 27

Menon, A.R., Pande, A.A., Kramer, H.J.M., Grievink, J., Jansens, P.J. (2007). A task-based synthesis approach toward the design of industrial crystallization process units *Ind.Eng.Chem.Res.* 46, 3979.

Li, H., Li, H., Guo, Z., Liu, Y. (2006). The application of power ultrasound to reaction crystallization. *Ultrason. Sonochem.*, 13, 359

Virone, C., Kramer, H.J.M., Van Rosmalen, G.M., Stoop, A.H., Bakker, T.W. (2006). Primary nucleation induced by ultrasonic cavitation. *J. Cryst. Growth*, 294, 9

Yeomans, H., Grossmann, I.E. (1999). A systematic modeling framework of superstructure optimization in process synthesis. *Comput.chem. Eng.*, 23, 709

29

Solar Salt Harvest Planning

K. Ossandón[1], P. Pinto[2] and L. Cisternas[1,3]

[1]*Depto. de Ing. Química, Universidad de Antofagasta, Casilla 170, Antofagasta, Chile*
[2]*SKM MinMetal, Santiago, Chile*
[3]*Centro de Investigación Científico y Tecnológico de la Minería, CICITEM, Chile*

CONTENTS

ABSTRACT Several chemicals are produced from natural and artificial brines by solar crystallization using solar ponds. The process of salt harvest consists on mechanically retiring the salts precipited in the solar evaporation ponds and to leave them in its respective stockpile. In practice this operation is carried out by contractors that for optimal condition should carry out during the whole year. In an industrial operation several ponds are used for the fractional crystallization of several salts, and therefore the harvest planning can be a nontrivial task. Consequently, the objective of this work is to plan the feeding flow to each of the solar ponds, the manipulation of solution, and the solutions and solids inventories in each pond that maximizes the production and the harvest periods. The model developed corresponds to a mixed inter nonlinear programming problem, that includes the mass balances in each ponds, equilibrium conditions base on hyperplanes, and planning & operational restrictions. The problem was solved in two steps, first the maximization of the salt harvest was determined and then, using this maximum harvest, the maximum availability of the constractor was determined. Several cases have been studied, including: ternary ($NaNO_3$-KNO_3-H_2O) and quaternary systems (KCl-KNO_3-K_2SO_4-H_2O), ponds systems with 3 and 4 ponds, ponds systems with differents areas, and considering 12 and 26 operation periods per year.

KEYWORDS *Solar ponds, fractional crystallization, solar salt harvest*

Introduction

The solar evaporation ponds are used for the production of salts or for the generation of heat energy (Lior et al., 2001). The solutions processed in ponds for the production of salts are obtained from different origins like seawater, natural lakes, underground brines and mining solutions. Obtained products from brines include KCl, K_2SO_4, Na_2SO_4, Li_2SO_4, H_3BO_4 (Thieme et al., 2001; Flotz et al., 1993; Butts et al., 2002). The operation of solar pond is favorable if big area and arid climate are available, reason why the production depends on the location of the ponds.

At the present time the solar evaporation pond are formulated and operated combining art and science. However, due to increases in the demand of some salts, space limitations and the increase in operation and capital costs, it has been forced to give a more scientific focus for the design and operation of solar ponds. For example a thermodynamic model has been developed to study the behavior of solid liquid equilibrium in aqueous electrolyte systems (Kwok et al., 2008; Song et al., 2003). A model to optimize the process has been created together with experimental simulations that determine the best pond height depending on the climatic conditions (Murthy et al., 2003). Also a methodology has been developed based on experimental data to increase the salt production (Hamzaoui et al., 2003). In spite of the those advances, few efforts have been carried out to improve the planning (what has to be done) and scheduling (when this has to be done) of salt harvest in solar pond systems to maximize the salt production and minimize the cost.

The objective of this work consists on determining by means of planning and scheduling, the maximum precipitation of salts in solar pond systems and to distribute its harvest in such way to take advantage of the maximum use of the contractor in a year period horizon. With this objective a mathematical model is developed whose results indicate the planning and scheduling of brine flows between ponds, the brine and solid inventory in each particular pond. All of this having as data the evaporation rate, concentration operation range, concentration of the feed and the initial pond conditions.

Mathematical Model

In this section is presented the fundamental aspects of the MINLP model to optimize the planning and scheduling of the salt production and harvest in solar evaporation pond systems. The outlined problem consists on to determine a maximum production of salts precipitated in the solar evaporation ponds and to distribute its harvest in a such way of taking advantage of the maximum contractor's use that carries out the task. Also, it is wanted to determine the flow distribution among the evaporation ponds in each operation period. The solution strategy consists on using mathematical programming based on a superstructure of the solar evaporation process.

The formulation of the mathematical model is constituted by two objective functions subject to restrictions. The first function objective considers the maximization of salt harvest and the second objective function maximize the periods of salt harvest with the objective of minimizing the costs associated to this operation.

The objective function that maximizes the salt harvest is

$$Max \sum_{i \in I} \sum_{t \in T} \sum_{k \in K} CS_{i,t,k} \qquad \forall i \in I, t \in T, k \in K \qquad (1)$$

Where $CS_{i,t,k}$ represents the salt harvest in the pond i in the time period t for the component k. Once the maximum harvest, CM, is determined, this is introduced as a restriction in the second model that maximizes the harvest periods, that is to say

$$Max \sum_{i \in I} \sum_{t \in T} \left(1 - y_{i,t}^P\right) \qquad \forall i \in I, t \in T \qquad (2)$$

$$\sum_{i \in I} \sum_{t \in T} \sum_{k \in K} CS_{i,t,k} \geq \delta \cdot CM \qquad \forall i \in I, t \in T, k \in K \qquad (3)$$

where $y_{i,t}^P$ is a binary variable that takes the value 1 if the pond i is in operation in the period t, and takes the value 0 if the pond i is in harvest in the period t. δ is the minimum fraction of CM that is allowed (in this work a value of $\delta = 0.9$ was used).

The superstructure of the solar evaporation pond system is built including a mixer in the input of each pond and a divider in the output, as it is shown in figure 1, and allowing the transfer of brines among all the pond or among those that the designer want to consider. The pond model should include the mass balances and the solid-liquid equilibrium conditions between the precipitate solids and the brine. To develop this model, each pond is represented as it is shown in figure 2.

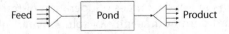

FIGURE 1
Solar evaporation pond representation in the superstructure.

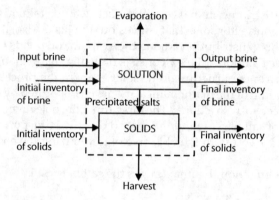

FIGURE 2
Solar evaporation pond representation for modeling.

Each pond has two possible states in every period: operation or harvest, then its modeling is carried out using a disjunction in the following way

$$\begin{bmatrix} Operation \\ A_O x = c_O \\ f_O(x) = 0 \end{bmatrix} \vee \begin{bmatrix} Harvest \\ A_C x = c_C \\ f_C(x) = 0 \end{bmatrix} \quad (4)$$

where the equations represent the mass balances, equilibrium conditions, and restrictions of operation conditions, scheduling and planning. The mass balance is developed using figure 2; therefore input and output brine streams, evaporation, initial and final inventories, and precipitated salts are considered. The equilibrium conditions between the solution and the precipitate solids are represented by hyperplanes according to the model developed by Pressly and Ng (1999). That is to say,

$$\sum_{k \in K} a_{n,i,k} x_{i,t,k} = b_{n,i,k} \quad \forall\, n \in N, i \in I, t \in T, k \in K \quad (5)$$

In practice the operation conditions of evaporation ponds, in a given period, as temperature and composition, change, and therefore the relationships of the equation 5 represent medium operation conditions. Its application to pond modeling was previously studied, concluding that the hyperplanes is an appropriate tool (Cisternas et al., 2006; Cisternas and Montenegro, 2007).

The operational restrictions include among others: ranges of concentrations of the solutions in ponds, ranges of operation of transfer flows among pond, maximum and minimum heights of solution and solid inventories.

When the harvest is carried out in more than a period of operation, logical expressions were included to condition that the harvest is carried out in continuous periods, and to unite the last period with the first one in case that harvest is carried out in the last period.

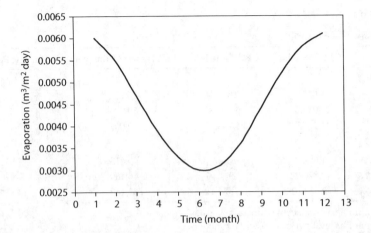

FIGURE 3
Solar evaporation rate in north of Chile.

Examples

To validate the proposed methodology, two systems of solar evaporation ponds were analyzed, studying diverse scenarios. The problems outlined for each case, contain common operational data in the industry, including the solid and brines densities, superior and inferior solid and brine height limits, and solar evaporation rate. For example, figure 3 shows the typical annual solar evaporation rate in the north of Chile.

The first example considered corresponds to the production of sodium nitrate in solar evaporation pond that contains $NaNO_3$, KNO_3 and H_2O operating at temperature of 25°C. Several cases were included: the first one with a system of 3 ponds of 40,000 m² and 26 periods of annual operation; the cases 2 and 3 operate with 4 ponds, the second with ponds of 30,000 m² and the third with 40,000 m², both with 26 annual periods of operation. The second example corresponds to the precipitation of potassium sulfate in solar evaporation ponds that contains KCl, KNO_3, K_2SO_4 and H_2O, operating at 25°C. In this example three cases were studied with the same characteristic of the first example, but with 12 annual operation periods.

The models of the cases were implemented in GAMS using BARON as solver. A total of 24 hours maximum was allowed using a processor of 2.39 GHz. The table 1 shows some of the obtained results.

As it is logical, the increase in the number of ponds and evaporation area increases the harvest salts and the harvest periods. Besides these results, the planning of the solution flow handling among ponds was obtained, the inventories of brine and solids in every period and pond was determined. The figure 4 shows the planning of solar salt harvest for the case 3 of the

TABLE 1

Result for the Examples.

Cases	CPU Time [s][†]	CPU Time [s][‡]	Salt Harvest [ton/year]	Harvest Periods
Example 1				
Case 1	8,613.6	5,876.4	36,232	3
Case 2	23,997.3	15,041.8	50,203	5
Case 3	11,167.7	24,186.7	82,007	9
Example 2				
Case 1	1,672.2	8,64.9	45,413	5
Case 2	2,870.02	944.61	46,489	9
Case 3	2,584.61	876.11	62,145	8

[†] Maximum harvest objective function.
[‡] Maximum harvest periods objective function.

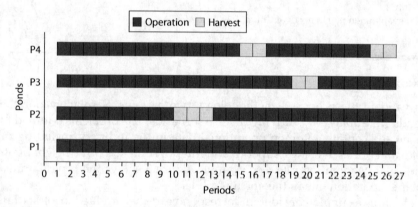

FIGURE 4
Planning of solar salt harvest for one year (case 3, example 1).

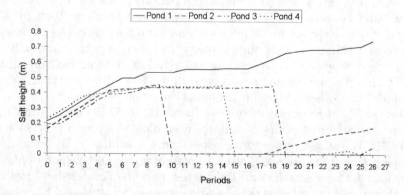

FIGURE 5
Salt height variation for one year (case 3, example 1).

example 1 and the figure 5 shows the salt height variation for the same case, both figures for one year of operation.

Several other examples and cases have been studied, but because of space are not shown in this work. For example, different values of δ in equation 3 have been used, and for case 1, for values of δ equal to 0.95, 0.9, 0.85, 0.8, 0.75 and 0.7, the same harvest production and harvest period have been obtained.

Conclusion and Future Work

A procedure for the planning and scheduling of salt production and harvest in solar evaporation pond systems was developed. The use of two objective functions, one to maximize the harvest and another to maximize the harvest periods, solved in sequential form, showed to be a good strategy for the solution of this type of problems. Industrial applications and simplifications to the model to reduce the CPU times correspond, at the moment, the future work.

Acknowledgments

We thank the CONICYT for financing and support of the work reported in this manuscript (FONDECYT 1060342).

References

Butts E., Chemicals from Brine, in "Kirk-Othmer Encyclopedia of Chemical Technology", (2001).

Cisternas LA., MM Montenegro (2006). Simulation of solar evaporation pond using hyperplanes. Part I Hyperplanes, Ingeniería Química (in spain), 30, 29–38.

Cisternas LA., JJ Cangana, R. Aravena, PL. Vargas (2007). Simulation of solar evaporation pond using hyperplanes. Part II Solar Evaporation Ponds, Ingeniería Química (in spain), 31, 61–69.

Flotz G.E., Lithium and Lithium Compounds, in "Inorganic Chemicals Handbook", J.J. McKetta, (1993), Marcel Dekker, New York.

Hamzaoui A.H., A. M'nif, H. Hammi and R. Rokbani, (2003), Contribution to the Lithium Recovery from Brine, Desalination, 158, 221–224.

Kwok K.S., Ka M. Ng, M.E. Taboada and L.A. Cisternas, (2008), Thermodynamics of Salt Lake System: Representation, Experiments, and Visualization, AIChE J., 54, 707–727.

Lior N., and R. Bakish, Supply and Desalination, in Kirk-Othmer Encyclopedia of Chemical Technology, (2001).

Murthy G.R. Ramakrishna and K.P. Pandey, (2003), Comparative Performance Evaluation of Fertiliser Solar Pond Under Simulated Conditions, Renewable Energy, 28, 455–466.

Pressly T.G., Ka M. Ng, (1999), Process Boundary Approach to Separations Synthesis, AIChE J., 45, 1939–1952.

Song P., and Y. Yan. (2003), Thermodynamics and Phase Diagram of the Salt Lake Brine System at 298.15 K, Computer Coupling of Phase Diagrams and Thermochemistry, 27, 343–352.

Thieme C., and R.J. Bauer, Sodium Carbonates, in "Ullmann's Encyclopedia of Industrial Chemistry", (2002).

30

Development of a New Group Contribution Method for the Product Design of Biofuels

Sven Kossack, Manuel Hechinger, Kathrin Stephan and Wolfgang Marquardt*

AVT-Process Systems Engineering, RWTH Aachen University
52064 Aachen, Germany

CONTENTS

ABSTRACT   A number of molecules are already identified as biofuel candidates. Additional candidates have been proposed in the literature and still more may be found through Computer Aided Molecular Design (CAMD). These designed molecules can be tailored to the numerous requirements posed by combustion engines. To this end, group contribution methods are required to describe the combustion properties so that the inverse problem of designing a molecule that meets specified targets can be solved. A new group contribution method for the laminar burning velocity is presented here, but due to the small amount of measured data, only few group contributions can be determined. The accuracy and the usefulness of the method could nevertheless be demonstrated by the correct prediction of the laminar burning velocity of ethanol. Here the experimental value was estimated with an error of 5.9%.

Besides more measurement data, for which CAMD can make a big contribution in optimal experimental design (OED), an even better physical

---

* To whom all correspondence should be addressed: wolfgang.marquardt@avt.rwth-aachen.de

understanding of the kinetic phenomena inside a combustion chamber is needed to develop a predictive method. Some steps towards this goal were taken, but measurements are needed to achieve this end.

KEYWORDS   *Group contribution methods, laminar burning velocity, biofuels, CAMD, product design*

Introduction

Ethanol, butanol, gamma-valerolactone (Horvath et al., 2008) and dimethylfuran (Roman-Leshkov et al., 2007) are among those molecules that have been proposed as biofuel candidates. Some of these candidate molecules have already been tested successfully in real engines and cars. An even larger number of molecules that can be derived from lignocellulosic biomass have been suggested in a recent study by Werpy and Peterson (2004). They identify 12 building blocks that can be used to derive 300 chemicals with a number of interesting properties. Some of these might even be useful as a fuel component. An even larger number of candidates may be found through Computer Aided Molecular Design (CAMD).

Here, a molecule is constructed from a set of building blocks to match or come close to a specified set of target properties (Gani et al., 2003). It is thus the inverse of a property prediction problem, where the properties are computed from the molecular structure (Gani et al., 2003). These computer generated molecules offer the possibility that they can be optimized for a high yield in the chemical conversion from the raw material to the final biofuel molecule. Furthermore, these molecules can be tailored to the numerous requirements posed by today's combustion engines or new combustion concepts.

Group contribution methods for combustion specific parameters are thus needed for three purposes. First of all, the physical properties of the large number of fuel candidates cannot be measured in a reasonable time period. To this end, group contribution methods are required to predict the combustion properties of likely fuel candidates. They can also be used to guide the experimental effort towards the most relevant experiments by identifying experimental conditions where the knowledge gain is the greatest (Bardow et al., 2008). Thirdly, they are needed for a computer aided molecular design approach, so that the inverse problem—to find a molecular structure that reaches targets for certain physical parameters—can be solved.

For some of the properties relevant for combustion, group contribution methods are already available. These are mostly simple thermodynamic properties such as boiling points, vapour pressures or viscosities. For the more combustion specific properties, however, no group contribution methods have yet been defined.

In this work, a group contribution method for the laminar burning velocity is presented. In general higher laminar burning velocities in the engine lead to higher efficiencies of the engine itself (Farrell and Johnston, 2003). The laminar burning velocity further allows an insight into the kinetic mechanism of combustion itself.

To predict the laminar burning velocity, a combination of a physically motivated model and a data driven approach is used to develop a group contribution method. A lack of sufficient measurement data, however, limits this model to more conventional components such as alkanes and alcohols. Therefore additional physical understanding of the combustion process can be used to improve the extrapolation capabilities of the group contribution method. It is shown that the predicted laminar burning velocities are in good agreement with the most recent measurement data.

Laminar Burning Velocity

The laminar burning velocity is an important characteristic value which is used in CFD simulations of spark ignition engines (Gülder, 1984). The laminar burning velocity is the velocity at which unburned fuel-air gas mixture moves through the combustion wave in the direction normal to the wave surface (Gülder, 1984). From its value, the kinetic mechanisms of the reactions taking place inside the engine can be deduced (e.g. Jerzembeck et al., 2008). This understanding of the reactions themselves help in designing new fuels for new combustion processes. While a number of measurement techniques have been designed for the laminar burning velocity (Gülder, 1984), closed vessel explosion techniques combined with an imaging technology for the evaluation of the burning velocity are accepted as a good measurement technique (Jerzembeck et al., 2008). Although the measurement itself is very quick, a change between different fuels and the large region of different operating regimes at which measurements are done, only a limited amount of candidate molecules can be screened in a reasonable time frame.

Because of the fundamental importance of the laminar burning velocity for the underlying kinetic mechanisms, a large number of publications describing experimental investigations for conventional fuel components as well as theoretical work on the kinetics of combustion have been published. From these publications the laminar burning velocity of a stoichiometric methane-air flame is given by

$$s_L = A(T^0)Y_{F,u}^m \frac{T_u}{T^0}\left(\frac{T_b - T^0}{T_b - T_u}\right)^n \tag{1}$$

(Müller et al., 1997). Here, T^0 is the inner layer temperature which can be calculated from the pressure p using

$$T^0 = -\frac{E}{\ln(p/B)}. \tag{2}$$

FIGURE 1
Adiabatic combustion of fuel F and air A entering at T_0 to exhaust gas E exiting at T_b.

Furthermore, $Y_{F,u}$ in Eq. (1) denotes the mass fraction of fuel in the unburned gas. The temperatures T_u and T_b are the temperatures in the unburned and burned gas. The function $A(T^0)$ accounts for the kinetic effects and is given by

$$A(T^0) = F \exp\left(-\frac{G}{T^0}\right). \tag{3}$$

The temperature in the unburned fuel, T_u, is set by the experiment while the temperature in the burned fuel, T_b, is defined to be the adiabatic flame temperature. The adiabatic flame temperature is determined from the energy balance of an adiabatic and stoichiometric combustion of fuel (F) with air (A) to exhaust gas (E) according to Fig. 1 given by

$$0 = v \cdot [h_E(T_b) - h_E(T_0)] - [h_F(T_F) - h_F(T_0)]$$
$$- l \cdot [h_A(T_A) - h_A(T_0)] - H_u(T_0). \tag{4}$$

Here, v and l are the molar amounts of exhaust gas and air per mole of fuel entering the combustion chamber, h_i is the specific enthalpy of stream i, T_F and T_A are inlet temperatures of fuel and air, T_0 is a reference temperature set to 298.15 K and H_u is the lower heating value of the fuel which is given by

$$H_u(T_0) = -\sum_i v_i \cdot \Delta h_i^{f,0}. \tag{5}$$

Here, v_i and $\Delta h_i^{f,0}$ are the stoichiometric coefficient and the enthalpy of formation of component i in a complete combustion reaction of the fuel and pure oxygen, respectively.

In CAMD, the molecular composition of the fuel is always known and, for the components considered here, it consists of carbon, oxygen and hydrogen only. Due to the combustion with air, the exhaust gas contains water vapor, carbon dioxide and nitrogen throughout. The enthalpy departure function for the exhaust gas in Eq. (4) can thus be described by

$$h(T_b) - h(T_0) = \int_{T_0}^{T_b} c_p^{ig}(T)dT, \tag{6}$$

where the ideal gas heat capacities for the constituents are determined via the Shomate equation parameters (NIST, 2008). Assuming that fuel and air enter the combustion at reference temperature T_0, the remaining enthalpy departures drop out.

This way, the adiabatic flame temperature can be iteratively obtained from Eq. (4) from a physical point of view, hence avoiding additional parameters and increasing the model predictability.

The model of Müller et al. (1997) also contains the parameters B through n, where a methane-like combustion requires theoretical values of $m = 0.5$ and $n = 2$. While fixing parameter n to its theoretical value, Eqs. (2) through (4) are combined and several algebraic modifications are performed. This results in the expression

$$s_L = Y_{F,u}^m \cdot L \cdot p^K \frac{T_u \cdot T_b}{(T_b - T_u)^2} \cdot \frac{(2 - \ln p)}{E} ,$$ (7)

where L and K are newly introduced parameters. It has to be pointed out that Eq. (7) together with the previously presented calculation of the adiabatic flame temperature is based only on 4 parameters, whereas the model of Müller et al. requires 11 parameters.

Development of a Group Contribution Method for the Laminar Burning Velocity

The modified model presented in the previous section now needs to be related to group contributions. This affects the enthalpy of formation $\Delta h_i^{f,0}$ of the fuel, whereas all other species involved in the combustion are the same for any organic fuel consisting of carbon, oxygen and hydrogen. Also, the four parameters m, L, K and E need to be related to group contributions.

A group contribution method for the enthalpy of formation is given by Joback and Reid (1987). The remaining parameters are fitted to the measured laminar burning velocities as they are provided by Müller et al. (1997). However, we do not fit the unsaturated hydrocarbons ethene, ethyne and methane, since their characteristic groups occur only once in the sample and would thus allow a perfect estimation of the laminar burning velocity for these components. The relation

$$P = \sum_i n_i X_i$$ (8)

is then assumed for each parameter $P \in \{m, L, K, E\}$, where n_i and X_i denote the number of occurrence of group i in the molecule and its contribution

TABLE 1

Regressed Contributions X_i for the Differrent Molecular Groups; $p = 10$ bar, $T_u = 373\ K$ and $Y_{F,u}$ ($\phi = 1$)

Group	CH$_3$	CH$_2$	CH	C	OH
L	−2,632	−0,551	−0,315	−0,314	−1,954
K	1,598	−0,207	−2,470	−2,470	1,012
E	1,970	−0,687	−1,615	−1,615	2,167
m	0,244	0,002	−0,358	−0,358	−0,244

to parameter P, respectively. The groups considered were CH$_3$, CH$_2$, CH, C and OH.

The fit of the parameters to the groups is done by least-squares minimization in Excel. Here, the laminar burning velocity is fitted to the values of Müller et al. (1997) according to

$$\min \sum_i (s_{L,\ Müller,i} - s_{L,\ fit,i})^2,\qquad (9)$$

where the summation is done over all measurements i. The results are given in Table 1. The data regression was performed including the branched iso-octane, though the simple parameter model presented in Eq. (8) is not able to account for isomeric effects.

As can be seen in Table 1, due to the lack of experimental data, only a few group contributions for the laminar burning velocity could be regressed. In addition, since the groups CH and C are only contained in iso-octane, their contributions are identical. This clearly shows the need for more measurements, in particular of uncommon fuel components such as gamma-valerolactone and dimethylfuran, which are made up of different oxygen containing groups.

Despite these drawbacks, the developed model in Eq. (7) matches the measured values of Müller et al (1997) perfectly for $p = 10$ bar and $T_u = 373$ K. Although the influence of pressure is not pictured correctly by Eq. (7), the temperature effects are reflected correctly with a maximum error of 0.8% in the range from 350K through 400K.

The quality of the group contribution method can be tested with the latest measurements of the laminar burning velocity of ethanol carried out by Jerzembeck et al. (2008), because all groups included in ethanol (CH3, CH2 and OH) have been included in the original molecules that were the basis for the group contributions in Table 1. The predicted burning velocity is 28.6 cm/s which compares very well to the measured value of 27 cm/s (Jerzembeck et al. 2008). This is an error of the group contribution method of 5.9% for this measured value.

Application of the Laminar Burning Velocity in Premixed Turbulent Flames

Apart from the significance for the assessment of spark ignited fuels, the laminar burning velocity is also indispensable in non-equilibrium flame chemistry such as turbulent jet flames. The so-called flamelet concept encounters the problem of non-equilibrium by considering the turbulent flame as an ensemble of thin, laminar, locally one-dimensional flamelet structures embedded within the turbulent flow field. The advantage of the laminar flamelet approach is that realistic chemical kinetic effects can be incorporated into turbulent flames (Fluent, 2006).

Within this approach, one describes the conditions at each point of the reacting flow field and hence the flame structure by a scalar parameter G which is related to the reaction progress (Peters, 2000). The governing equations to obtain G require the assumption that the laminar flame thickness is smaller than the smallest turbulent length scale (the so called Kolmogorov scale). This allows one to keep a locally well defined laminar burning velocity for flamelet propagation, which is identical to that predicted by the methods mentioned in the previous sections.

Hence, the laminar burning velocity is shown to have a broad field of applications, not only for our main purpose of designing gasoline fuel components.

Summary and Conclusion

A new group contribution method for the laminar burning velocity has been presented. Due to the small amount of measured data, only very few group contributions have been determined. The developed method can therefore not yet be used to predict the burning behavior of all new bio-based fuels. The accuracy and the usefulness of the method could nevertheless be demonstrated by the correct prediction of the laminar burning velocity of ethanol. Here the experimental value was estimated with an error of 5.9%.

Besides the provision of more measurement data, which should be obtained using optimal design of experiments (OED) assisted by CAMD (Bardow et al., 2008), an even better physical understanding of the kinetic phenomena inside the combustion chamber is needed to develop a predictive method. A first step towards this goal has been taken in this work with the elimination of 7 parameters in a semi-empirical equation for the laminar burning velocity reported in lituerature, such that only 4 parameters need to be determined experimentally in the future. Current work extends the modeling approach to cover the physcio-chemical phenomena in more detail.

In the future, a different strategy accounting for isomers needs to be included in the group contribution method but again more measurements are needed to achieve this end.

Acknowledgments

This work was performed as part of the Cluster of Excellence "Tailor-Made Fuels from Biomass", which is funded by the Excellence Initiative by the German federal and state governments to promote science and research at German universities. The authors would also like to thank N. Peters, J. Beeckmann and O. Röhl of ITV, Institute for Technical Combustion, at RWTH Aachen University for fruitful discussions and for providing us with their latest experimental data.

References

Bardow, A., Kossack, S., Kriesten, E., Marquardt, W. (2008). MEXA goes CAMD—computer-aided molecular design for physical property model building. *In Proceedings of ESCAPE 18*. Elsevier, Amsterdam, Netherlands, 733.

Gani, R., Achenie, L.E.K., Venkatsubramanian, V. (2003). Introduction to CAMD. *In Computer Aided Molecular Design: Theory and Practice*. Elsevier, Amsterdam, Netherlands, 3.

Gülder, Ö.L. (1984). Correlations of laminar combustion data for alternative S.I. engine fuels. *SAE Technical Paper Series*, 841000.

Farrell, J.T., Johnston, R.J. (2003). Hydrocarbon fuel with improved laminar burning velocity and method of making. *US Patent*, 60/485,001.

Fluent (2006), Fluent 6.3 Product Documentation

Horvath, I.T., Mehdi, H., Fabos, V., Boda, L., Mika, L.T. (2008). Gamma-valerolactone—a sustainable liquid for energy and carbon-based chemicals. *Green Chem.*, 10, 238.

Jerzembeck, S., Röhl, O., Glawe, C., Peters, N. (2008). Development of a high temperature ethanol-gasoline mechanism and experimental validation. submitted.

Joback, K.G. and Reid, R.C. (1987). Estimation of pure component properties from group contributions, *Chem. Eng. Comm.*, 57, 233.

Müller, U.C., Bollig, M., Peters, N. (1997). Approximations for burning velocities and Markstein numbers for lean hydrocarbon and methanol flames. *Combustion and Flame.*, 108, 349.

NIST Chemistry Webbook, NIST Standard Chemistry Database Number 69, http://webbook.nist.gov/chemistry, 2008

Peters, N. (2000). *Turbulent Combustion*, Cambridge University Press

Roman-Leshkov, Y., Barrett, C.J., Liu, Z.Y., Dumesic, J.A. (2007). Production of dimethylfuran for liquid fuels from biomass-derived carbohydrates, *Nature*, 447, 982.

Werpy T., Petersen, G. (2004). *Top Value Added Chemicals from Biomass*. US DOE. Oak Ridge, TN. Available electronically at http://www.osti.gov/bridge

31

Developing Sustainable Chemical Processes to Utilize Waste Crude Glycerol from Biodiesel Production

Robert Jernigan[1], Sean Hansrote[2], Kandace Ramey[2], Luke Richardson[2] and Jeffrey Seay[2*]

[1]Department of Chemical Engineering, Auburn University, Auburn, Alabama 36849
[2]Department of Chemical and Materials Engineering, University of Kentucky, Paducah, Kentucky 42002

CONTENTS

ABSTRACT Biodiesel is becoming increasingly important as a renewable motor fuel. However, the production of biodiesel also generates a substantial quantity of glycerol as a side product. Recent trends in biodiesel production have led to increased interest in processes utilizing this glycerol side product. Finding uses for this glycerol will not only improve the profitability of biodiesel, it also will increase the overall carbon utilization of the process. Therefore, the focus of this research is to layout the groundwork for a research program based on identifying the most cost effective routes for generating industrially important C3 compounds utilizing bio-based glycerol as a feedstock.

KEYWORDS *Sustainability, glycerol dehydration, process design, biodiesel*

---

* To whom all correspondence should be addressed

Introduction

Biodiesel is an important renewable motor fuel. The use of biofuels such as biodiesel is important from the perspective of sustainability since it is produced from renewable feedstocks such as vegetable oils or animal fats. Due to recent tax incentives, production of biodiesel in the United States and Europe has increased substantially in the last few years. In fact, recently published estimates predict that the demand for biodiesel will grow from 6 to 9 million metric tons per year in the United States and from 5 to 14 million metric tons per year in the European Union in the next few years.

However, for every 9 kg of biodiesel produced, 1 kg of crude glycerol is produced as a byproduct. Developing economically viable processes to generate commercially important chemical products based on using crude glycerol as a feedstock can potentially improve not only the economic performance, but also the overall carbon utilization of biodiesel production.

Numerous chemical products currently produced from crude oil derived propylene can also be produced from bio-based crude glycerol. Products based on C3 chemistry such as acrylic acid, acrolein, 1,2-propanediol, 1,3-propanediol, propionaldehyde and hydroxyacetone among others can be manufactured from crude glycerol. Many novel chemical processes based on acid catalyzed glycerol dehydration have been described to synthesize these products; however, in order to justify pursuing the development of manufacturing processes for these chemical species, cost targets must be identified for which these processes are economically viable.

This contribution will describe the development of models for identifying such cost targets. This work will serve as a basis for developing a research program to design and optimize conceptual processes based on glycerol dehydration chemistry. This program will integrate laboratory experimentation with process simulation and environmental impact assessment to develop optimized conceptual processes.

Reaction Pathways

Previous research has suggested that the catalytic dehydration of glycerol can be successfully carried out in either the liquid or vapor phase (Neher *et al.*, 1995). A wide variety of industrially important chemical products are possible from this dehydration reaction. Hydroxyacetone, 3-hydroxypropionaldehyde, acrolein, 1,2-propanediol, ethylene glycol, acetaldehyde and formaldehyde are all potential chemical products (Neher *et al.*, 1995) (Chiu *et al.*, 2006)(Antal *et al.*, 1985)(Feng *et al.*, 2008)(Dasari, *et al.*, 2005).

Further reactions with these products can lead to other industrially important chemical species. Hydrogenation of hydroxyacetone or

3-hydroxypropionaldehyde can produce 1,2-propanediol (also called propylene glycol) and 1,3-propanediol respectively (Zhu and Hofmann, 2004)(Miyazawa *et al.*, 2006)(Casale and Gomez, 1993)(Casale and Gomez, 1994)(Chiu *et al.*, 2006). Hydrogenation of glycerol itself can generate ethylene glycol and methanol (Miyazawa *et al.*, 2006). This is a potentially interesting reaction because the methanol can be directly recycled into the biodiesel process.

Hydrogenation of acrolein can produce propanol, propionaldehyde or allyl alcohol (Smith, 1963). Oxidation of acrolein can produce acrylic acid (Neher *et al.*, 1995)(Weigert and Haschke, 1976)(Etzkorn *et al.*, 2001)(Ramayya *et al.*, 1987)(Bub *et al.*, 2006). The identity and yield of the dehydration products are a function of the catalyst used and of the operating conditions of the reactor. A possible mechanism for these reaction pathways is illustrated in Figure 1.

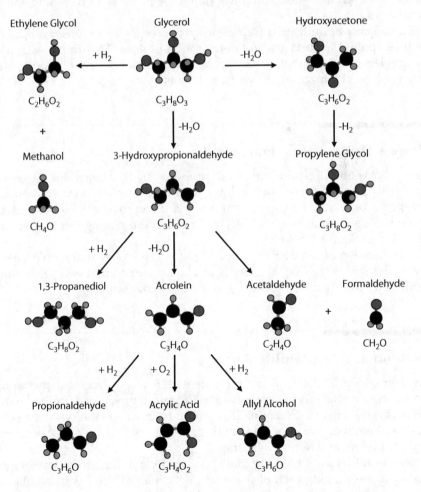

FIGURE 1
Potential reaction pathways for glycerol dehydration products (Seay, 2008-1).

Since glycerol is a side product of biodiesel manufacture, turning this glycerol into a higher value chemical product can improve the profitability of the manufacturing process. However, each potential product requires different operating conditions and equipment configurations for production. Therefore, the goal is to apply a systematic optimization framework to determine which potential products of glycerol dehydration are the most viable for a given set of market conditions.

The ultimate goal is to design a manufacturing process than can potentially produce numerous glycerol dehydration products. By integrating a process simulation with economic optimization, it is possible to first identify a slate of products that has the most potential for profitability, and second, once the process has been designed and implemented, continue to use the model to determine which products should be made at a given time and at what capacity.

The process optimization framework proposed by Sammons *et al.* (2007), will be applied for this glycerol dehydration process. This framework will be applied to ensure that as process designs are developed, economic and environmental impact targets will also be met.

Process Optimization Framework

The process optimization framework proposed by Sammons *et al.*, is based on decoupling the process modeling from the decision making framework for the purpose of reducing the complexity and improving the robustness of the solutions (Sammons *et al.*, 2007). A flowchart illustrating this framework is shown in Figure 2, below.

This framework can easily be applied to the glycerol dehydration process. By applying this technique, a model can be developed that will predict the optimum product allocation for a given set of market prices.

Preliminary Profitability Analysis

As a first step in determining the potential slate of glycerol dehydration products to include in the optimization process, a gross profitability analysis was conducted to determine if potential economic viability exists. Four chemical products were included in this analysis: propylene glycol, ethylene glycol, acrolein and hydroxyacetone.

Based on literature data for conversion and yield, the gross profitability for a process making each of these products from glycerol was calculated. For comparison purposes, the profitability was reported using both crude glycerol from a biorefinery and refined tallow based glycerol. The results of

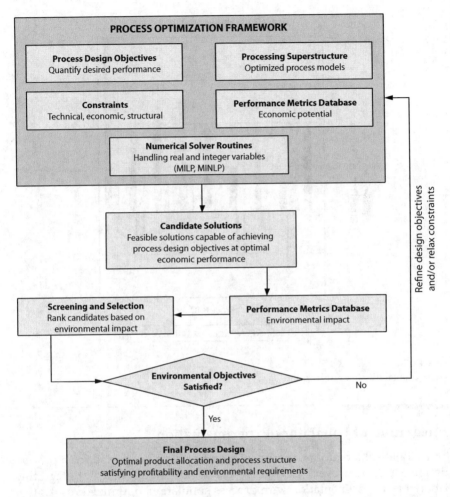

FIGURE 2
Process optimization framework (Sammons *et al.*, 2007).

this analysis are illustrated in Figure 3. Data from multiple sources for each product are included in the gross profitability analysis.

The reference for each case considered in the profitability analysis is listed in Table 1. From these results, it is clear that not only does the use of crude glycerol represent a significant improvement over refined tallow based glycerol, but there is also enough gross profit available to warrant further study.

However, despite these encouraging results, there is insufficient data in the literature to begin the conceptual design of a process based on glycerol dehydration chemistry. Additional laboratory experimentation is required to determine the optimum operating conditions for a process based on this chemistry.

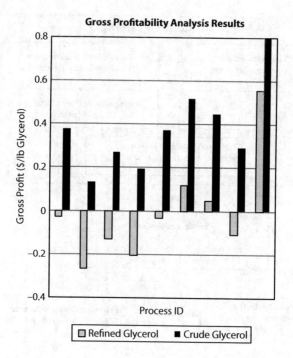

FIGURE 3
Gross profitability analysis results.

Integration of Laboratory Experimentation

Traditionally, laboratory experimentation has been decoupled from process design work. However, previous research has shown that by integrating laboratory experimentation with process simulation, optimized results that

TABLE 1

Profitability Analysis Case References

Process ID	Product	Reference
I	Propylene Glycol	Dasari *et al.*, 2005
II	Propylene Glycol	Kusunoki *et al.*, 2005
III	Ethylene Glycol	Feng *et al.*, 2008
IV	Propylene Glycol	Wang *et al.*, 2007
V	Acrolein	Atia *et al.*, 2008
VI	Acrolein	Tsukuda *et al.*, 2007
VII	Acrolein	Watanabe *et al.*, 2007
VIII	Acrolein	Corma *et al.*, 2008
IX	Hydroxyacetone	Chui *et al.*, 2006

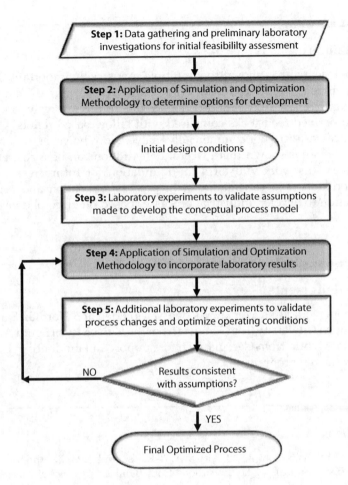

FIGURE 4
Process integration flowchart (Seay and Eden, 2009).

might have otherwise been overlooked can be identified (Seay *et al.*, 2008-2). This integration can be illustrated with another simple flowchart.

For this research project, steps 2 and 4, will be based on the process optimization framework of Sammons *et al.*, illustrated in Figure 2.

Previous research has indicated that the catalyst life for the glycerol dehydration process is too short for a viable industrial process (Seay, 2008-1). Therefore, one of the principle goals of the laboratory experimentation will be to test less active catalysts to determine if catalyst life can be extended, while not sacrificing economic and environmental viability.

Conclusions

Previous research has shown that multiple industrially important C3 compounds, currently produced from crude oil derived resources, can also be made from bio-based glycerol. Previously, due to the relatively low cost of crude oil derived feedstocks compared with tallow based refined glycerol, these processes were not commercially viable. Now, however, this preliminary research has shown that glycerol dehydration can be a cost effective alternative. This work will form the foundation for future research coupling laboratory experimentation with process simulation and economic optimization to develop industrial processes based on glycerol dehydration chemistry.

Acknowledgments

The authors would like to acknowledge Mario Eden and Norman Sammons from Auburn University, Tom Thomas from the University of South Alabama and Charles Liotta from Georgia Tech for support and guidance in developing this research program.

References

Antal, Jr., M.J., W.S.L. Mok, J.C. Roy, A. T-Raissi and D.G.M. Anderson, 1985, "Pyrolytic Sources of Hydrocarbons from Biomass", *Journal of Analytical and Applied Pyrolysis*.

Atia H, Armbruster U, Martin A., 2008. Dehydration of glycerol in gas phase using heteropolyacid catalysts as active compounds. *Journal of Catalysis* [serial online]. Vol. 258(1): pp. 71–82.

Bub, G., J. Mosler, A. Sabbagh, F. Kuppinger, S. Nordhoff, G. Stochniol, J. Sauer and U. Knippenberg, 2006, "Acrylic Acid, water-absorbent polymer structures based on renewable resources and method for producing said structures", International Patent WO 2006/092272 A2.

Casale, B. and A. Gomez, 1993, "Method of hydrogenating glycerol", U.S. Patent 5,214,219.

Casale, B. and A. Gomez, 1994, "Catalytic method of hydrogenating glycerol", U.S. Patent 5,276,181.

Chiu, C., M.A. Dasari and G.J. Suppes, 2006, "Dehydration of Glycerol to Acetol via Catalytic Reactive Distillation", *AIChE Journal*, Vol. 52.

Corma A, Huber G, Sauvanaud L, O'Connor P., 2008, Biomass to chemicals: Catalytic conversion of glycerol/water mixtures into acrolein, reaction network. *Journal of Catalysis* [serial online]. Vol. 257(1): pp. 163–171.

Dasari M, P. Kiatsimkul, W. Sutterlin, G. Suppes, 2005, Low-pressure hydrogenolysis of glycerol to propylene glycol. *Applied Catalysis A: General* [serial online]. March 18, Vol. 281(1/2), pp.225–231.

Etzkorn, W.C., S.E. Pedersen and T.E. Snead, 2001, "Acrolein and Derivatives", Kirk-Othmer Encyclopedia of Chemical Technology, John Wiley and Sons, Inc.

Feng J, Fu H, Wang J, Li R, Chen H, Li X., 2008. Hydrogenolysis of glycerol to glycols over ruthenium catalysts: Effect of support and catalyst reduction temperature. *Catalysis Communications* [serial online]. Vol. 9(6):1458–1464.

Kusunoki Y, Miyazawa T, Kunimori K, Tomishige K., 2005. Highly active metal–acid bifunctional catalyst system for hydrogenolysis of glycerol under mild reaction conditions. *Catalysis Communications* [serial online]. Vol. 6(10): pp. 645–649.

Miyazawa, T., Y. Kusnoki, K. Kunimori and K. Tomishigq, 2006, "Glycerol conversion in the aqueous solution under hydrogen over Ru/C + an ion-exchange resin and its reaction mechanism", *Journal of Catalysis*, Vol 240.

Neher, A., T. Haas, D. Arntz, H. Klenk and W. Girke, 1995, "Process for the production of acrolein", United States Patent 5,387,720.

Sammons, N., M. Eden, H. Cullinan, L. Perine and E. Connor, 2008, "A flexible framework for optimal biorefinery product allocation", *Computer Aided Chemical Engineering* Volume 21, Part 2, Pages 2057–2062. (Reprinted w/permission from Elsevier)

Seay, J., 2008-1, *A Methodology for Integrating Process Design Elements with Laboratory Experiments*, PhD Thesis, Auburn University.

Seay, J., H. Werhan, M. Eden, R. D'Alessandro, T. Thomas, H. Redlingshoefer, C. Weckbecker, and K. Huthmacher, 2008-2. Integrating laboratory experiments with process simulation for reactor optimization. In B. Braunschweig and X. Joulia (Eds.), *Computer-Aided Chemical Engineering*. (Paper 168, CD Volume).

Seay, J.R. and M.R. Eden. (2009). "Incorporating Environmantal Impact Assessment into Conceptual Process Design: A Case Study Example" *Journal of Environmental Progress and Sustainable Energy*, Published Online: Dec 10 2008, DOI: 10.1002/ep.10328.

Smith, C.W., 1963, *Acrolein*, John Wiley and Sons, Inc. New York.

Tsukuda E, Sato S, Takahashi R, Sodesawa T., 2007. Production of acrolein from glycerol over silica-supported heteropoly acids. Catalysis Communications [serial online]. Vol.8(9): pp.1349–1353.

Wang S, Liu H., 2007. Selective hydrogenolysis of glycerol to propylene glycol on Cu–ZnO catalysts. Catalysis Letters [serial online]. Vol.117(1/2): pp. 62–67.

Watanabe M, Iida T, Aizawa Y, Aida T, Inomata H., 2007. Acrolein synthesis from glycerol in hot-compressed water. *Bioresource Technology* [serial online]. Vol. 98(6): pp. 1285–1290.

Weigert, W. M. and H. Haschke, 1976, "Acrolein and Derivatives", *Encyclopedia of Chemical Processing and Design*, J, McKetta, Ed., Marcel Dekker, Inc.

Zhu, X. and H. Hofmann, 2004, "Intraparticle diffusion in hydrogenation of 3-hydroxypropanal", *AIChE Journal*, Vol. 43, Issue 2.

32

Extraction of Biofuels and Biofeedstocks Using Ionic Liquids

Alexandre Chapeaux, Luke D. Simoni, Mark A. Stadtherr and Joan F. Brennecke*

Department of Chemical and Biomolecular Engineering,
University of Notre Dame
Notre Dame, IN 46556

CONTENTS

ABSTRACT Biomass production of chemicals and fuels by fermentation, biocatalysis, and related techniques implies energy-intensive separations of organics from dilute aqueous solutions, and may require use of hazardous materials as entrainers to break azeotropes. We consider the design feasibility of using ionic liquids as solvents in liquid-liquid extractions for separating organic compounds from dilute aqueous solutions. As an example, we focus on the extraction of 1-butanol from a dilute aqueous solution. We have recently shown [Chapeaux et al. (2008), *Green Chemistry, 10*, 1301] that 1-hexyl-3-methylimidazolium bis(trifluoromethylsulfonyl)imide shows significant promise as a solvent for extracting 1-butanol from water. We will consider here two additional ionic liquids, 1-(6-hydroxyhexyl)-3-methylimidazolium bis(trifluoromethylsulfonyl)imide and 1-hexyl-3-methylimidazolium tris (pentafluoroethyl)trifluoro-phosphate, as extraction solvents for 1-butanol. Preliminary design feasibility calculations will be used to compare the

---

* To whom all correspondence should be addressed

three ionic liquid extraction solvents considered. The ability to predict the observed ternary liquid-liquid equilibrium behavior using selected excess Gibbs energy models, with parameters estimated solely using binary data and pure component properties, will also be explored

KEYWORDS *Ionic liquids, 1-butanol, extraction, liquid-liquid equilibrium, excess gibbs energy models*

Introduction

1-Butanol is used as a feedstock for making many common chemicals. It is also used as solvent in many applications, such as re-crystallization processes in the pharmaceutical industry. Furthermore, there is growing interest in 1-butanol as a fuel. For any current and potential use of 1-butanol, it is important to look to renewable sources for its synthesis. Thus, there is much current research focused on the fermentation of biomass, resulting in a broth composed of mainly water and alcohols, from which 1-butanol can be separated. Conventionally, separating alcohols and water requires a series of distillation columns. This method is energetically costly, and much room for improvement exists. It has been shown that ionic liquids (ILs) have the potential for separating alcohol/water mixtures with simple liquid-liquid extraction (Fadeev and Meagher, 2001; Chapeaux et al., 2008), which could be less energetically costly than distillation.

ILs are salts with a melting point below 100°C, and which are usually composed of a poorly coordinating, bulky organic cation, and an organic or inorganic anion. Some of the properties that make them advantageous for this application are a negligible vapor pressure, which allows for recovery and reuse of the IL; a large liquid range, which allows for ease of separation; and, finally, a tunability that allows the creation of ILs that preferentially select alcohols from water.

In this study we will determine, based on experimental observations, the distribution coefficients, selectivities, and number of equilibrium stages for multicomponent liquid-liquid extraction of 1-butanol from water using 1-(6-hydroxyhexyl)-3-methylimidazolium bis(trifluoromethylsulfonyl)imide ([HOhmim][Tf$_2$N]) and 1-hexyl-3-methylimidazolium tris(pentafluoroethyl)trifluoro-phosphate ([hmim][eFAP]), and compare these results to previously published results for 1-hexyl-3-methylimidazolium bis(trifluoromethylsulfonyl) imide ([hmim][Tf$_2$N]) (Chapeaux et al, 2008).

Predicting an IL's capability as a separation solvent is important as experimental observation of every system of interest is time consuming. Clearly, a model that predicts ternary liquid-liquid equilibrium (LLE) behavior *a priori* is desired; however, at the present time models based on first principles are both computationally expensive and inaccurate for multicomponent LLE. For example, COSMO-RS has been applied to predictions of binary LLE upper critical solution temperature behavior and of a ternary LLE system, but without satisfactory results (Freire et al., 2007; Jork et al., 2005). More recently, COSMO-RS has been modified for LLE (COSMO_LL), yielding better ternary predictions involving ILs, but still with much room for improvement (Banerjee et al., 2008). Molecular descriptor and group contribution methods, e.g., NRTL-SAC (NRTL Segment Activity Coefficient) and UNIFAC (Dortmund), respectively, also provide qualitatively inaccurate predictions in many cases (Chen et al., 2008; Chapeaux et al., 2008).

In this work, we use a semi-predictive method in which activity coefficient models are used to make ternary LLE predictions based on only binary and pure component data. It was shown previously (Chapeaux et al., 2008; Simoni et al., 2008) that ternary systems containing ILs and water were more difficult to qualitatively predict using this approach. In such systems, it is likely that there are different degrees of ionic dissociation in different phases. To account for this, we have developed a novel asymmetric framework in which different activity coefficient models are used in different liquid phases. Although ILs most likely partially ionize, as a first approximation, we assume the IL is *completely dissociated* in a dilute aqueous phase and completely paired (*molecular*) in an alcohol/IL-rich phase (Simoni et al., 2009a,b). We will apply this modeling approach here to one of the ternary systems investigated experimentally, namely [hmim][Tf$_2$N]/1-butanol/water.

Methodology

Experimental

1-Butanol (71-36-3, 99.8+% purity, anhydrous, 600 ppm water) was purchased from Sigma-Aldrich and used as received. The water was deionized by a Millipore purification system (>18 MΩ·cm resistivity). [HOhmim][Tf$_2$N] was made and purified according to previously described methods (Bonhote et al., 1996; Cammarata et al., 2001; Crosthwaite et al., 2004; Crosthwaite et al., 2005; Fredlake et al., 2004). [hmim][eFAP] was received from Merck KGaA and used as received. All the ILs used were dried under vacuum (~1.3 Pa) for 24 hr at 70°C.

All experimental procedures have been described in other publications (Chapeaux et al., 2008; Chapeaux et al., 2009b). In short, we mixed IL, water and 1-butanol in a vial, and then allowed the phases to separate. We analyzed all three components in each phase using high-performance liquid chromatography, gas chromatography, Karl-Fischer titration, and UV-Vis spectroscopy.

Modeling

In the asymmetric framework, we assume that the IL is *completely dissociated* in dilute aqueous phases (high average dielectric constant), and that the IL is *completely associated,* or molecular, as ion pairs in IL or solvent-rich phases (low average dielectric constant). Accordingly, we use electrolytic and conventional activity coefficient models to represent the dissociated and molecular phases, respectively, in particular the electrolyte-NRTL (eNRTL) (Chen et al., 2004) and NRTL models. Note that a complete, general formulation of this asymmetric framework, together with discussion of standard state definitions and phase stability analysis, for general mixed-salt/mixed-solvent systems will be presented elsewhere (Simoni et al., 2009a).

The degree of dissociation depends on the ability of the phase's components (mixed solvent) to screen the electrostatic forces of the ions. This implies that the molecular state of the electrolyte depends on its concentration and on the dielectric constant of the mixed solvent. The asymmetric framework uses a composite Gibbs free energy surface, in which model domains are defined by IL concentration and the dielectric constant of the mixed solvent (1-butanol/water in this case). In order for a phase to be considered as dissociated, the observable mole fraction of electrolyte (IL) must be less than some critical value (0.10 is used here) and the average mixed-solvent dielectric constant must be greater than some critical value (50 is used here). Otherwise, a phase will be treated as molecular.

The models used contain two energetic binary interaction parameters for each pair of components. These are determined based on *binary* data and are the only adjustable model parameters. Furthermore, we make the key assumption that these parameters are the same in both the dissociated-phase and molecular-phase models that are combined in the asymmetric framework. This follows from the assumption of Chen et al. (2004), based on local electroneutrality and the symmetry of interaction energies, that short-range cation-solvent and anion-solvent interaction energies are the same. For immiscible binaries, the binary parameters are determined from mutual solubility data by solving the equal chemical potential conditions for binary LLE. For miscible binaries, vapor-liquid equilibrium (VLE) data is used for parameter estimation. Details of the procedures used to determine binary parameters are given by Simoni et al. (2007; 2009a).

In using the asymmetric framework to compute multicomponent LLE, one must ensure the resultant equilibrium phases are thermodynamically stable. The conditions for phase stability in the context of the mixed-salt, mixed-solvent asymmetric model have been developed by Simoni et al. (2009a), based on an extension of tangent plane analysis (Baker et al., 1982; Michelsen, 1982). To implement this, we use an approach, similar to that described by Tessier et al. (2000), based on rigorous global optimization, accomplished using an interval-Newton approach.

Results

Experimental

Based on our experimental observations of the phase behavior of the ternary IL/1-butanol/water systems, we calculated the selectivity (S) and distribution coefficient (D) for 1-butanol in each of the three systems considered. These quantities are defined by

$$D = \frac{n^{\beta}_{\text{Alcohol}}}{n^{\alpha}_{\text{Alcohol}}} \tag{1}$$

$$S = \frac{n^{\beta}_{\text{Alcohol}} \Big/ n^{\beta}_{\text{H}_2\text{O}}}{n^{\alpha}_{\text{Alcohol}} \Big/ n^{\alpha}_{\text{H}_2\text{O}}} \tag{2}$$

where n^{β}_i indicates moles of i in the IL-rich phase and n^{α}_i in the aqueous phase. A multistage liquid-liquid extraction calculation was performed using the Hunter-Nash method (Seader and Henly, 1998) to determine the number of equilibrium stages required, based on the assumptions: 1) feed composition of 5 wt% of 1-butanol in water, 2) pure IL as the solvent, 3) equal feed and solvent mass flow rates, and 4) requiring 99 wt% water in the final raffinate.

Table 1 shows the results for all three IL systems. We can see that [hmim][eFAP] provides the highest selectivity with almost the highest distribution coefficient. [hmim][eFAP] is extremely hydrophobic (Chapeaux et al 2009a), and therefore repels water more than it attracts alcohol. The results for [HOhmim][Tf$_2$N] indicate that adding a hydroxyl on the cation chain attracts more water than alcohol and therefore reduces the selectivity and the distribution coefficient.

Modeling

Experimentally (Chapeaux et al, 2008), the ternary system [hmim][Tf$_2$N]/1-butanol/water at 295 K exhibits Type 2 ternary LLE behavior (Figure 1).

TABLE 1

Distribution Coefficient (D), Selectivity (S) and Number of Equilibrium stages for IL/water/1-butanol

IL	D	S	Number of Stages
[hmim]Tf$_2$N]*	7.5	90	4
[HOhmim][Tf$_2$N]	1.5	35	5
[hmim][eFAP]	6	300	3

* Chapeaux et al., 2008

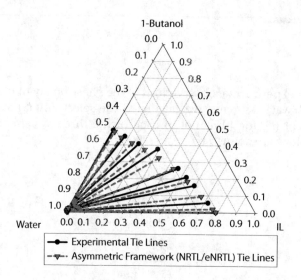

FIGURE 1

Predicted ternary LLE for [hmim][Tf$_2$N]/1-Butanol/Water at 295 K (in mol fraction), compared to experimental measurements (Chapeaux et al., 2008).

That is, there are two binary miscibility gaps and a single two-phase envelope that spans from one binary miscibility gap to the other. Mutual solubility data for [hmim][Tf$_2$N]/water (Chapeaux et al., 2007) and 1-butanol/water (Sørensen and Arlt, 1979-1980) at 295 K were used to calculate model parameters for these two immiscible binaries. For the completely miscible [hmim][Tf$_2$N]/1-butanol binary, there is no VLE data available at the system temperature. Therefore, to estimate the model parameters for this binary, LLE data for [hmim][Tf$_2$N]/1-butanol at lower temperatures (Łachwa et al., 2006) were linearly regressed and extrapolated to the system temperature.

Figure 1 shows the ternary LLE predicted by the asymmetric NRTL/eNRTL model, with the model parameters obtained from binary data only, as described above. Comparison to the experimental data indicates that this provides a prediction that is quite good, in terms of both the phase envelope and the slope of the tie lines. However, Type 2 diagrams are generally considered the easiest type to predict. In fact, for this particular system, the standard (symmetric) NRTL, UNIQUAC and eNRTL models also provide reasonably accurate predictions (Chapeaux et al., 2008). The asymmetric approach has also been tested on more difficult problems and, in such cases, found to be superior to conventional symmetric models (Simoni et al., 2009b).

Conclusions

Separating alcohols from fermentation broths is a critical step in producing the building blocks for renewable fuels and feedstocks. In this work, we have shown that ILs are apt solvents for liquid-liquid extraction of 1-butanol from water, with selectivities ranging from 40 to 300, and high distribution coefficients. From this standpoint, [hmim][eFAP] appears to be an especially a good solvent for this separation. Detailed design studies, including economic analysis and life cycle analysis, are now needed.

We also have shown that we can use a semi-predictive method, based on a novel asymmetric framework, to model IL/water/1-butanol systems. The asymmetric framework allows for the use of different excess Gibbs models in different phases. This framework, with the assumption that the IL is completely dissociated in a dilute aqueous phase and completely paired in an alcohol/IL-rich phase, allows for good predictions of ternary systems based solely on binary experimental data.

Acknowledgments

We thank Merck KGaA for the sample of [hmim][eFAP] We also acknowledge Thomas R. Ronan for his contributions to the measurements of experimental data.

References

Baker, L. E., Pierce, A. C., Luks, K. D. (1982). Gibbs energy analysis of phase equilibria. *Soc. Petrol. Engrs. J.*, 22, 731.

Banerjee, T., Verma, K. K., Khanna, A. (2008). Liquid-liquid equilibrium for ionic liquid systems using COSMO-RS: Effect of cation and anion dissociation. *AIChE J.*, 54, 1874.

Bonhote, P., Dias, A. P., Papageorgiou, N., Kalyanasundaram, K., Gratzel, M. (1996). Hydrophobic, highly conductive ambient-temperature molten salts. *Inorg. Chem.*, 35, 1168.

Cammarata, L., Kazarian, S. G., Salter, P. A., Welton, T. (2001). Molecular states of water in room temperature ionic liquids. *Phys. Chem. Chem. Phys.*, 3, 5192.

Chapeaux, A., Simoni, L. D., Stadtherr, M. A., Brennecke, J. F. (2007). Liquid phase behavior of ionic liquids with water and 1-octanol and modeling of 1-octanol/water partition coefficients. *J. Chem. Eng. Data*, 52, 2462.

Chapeaux, A., Simoni, L. D., Ronan, T.; Stadtherr, M. A., Brennecke, J. F. (2008). Extraction of alcohols from water with 1-hexyl-3-methylimidazolium bis(trifluoromethylsulfonyl)imide. *Green Chem., 10,* 1301.

Chapeaux, A., Andrews N., Brennecke, J. F. (2009a). Liquid-liquid behavior of ionic liquids with water and 1-hexanol. *In preparation.*

Chapeaux, A., Simoni, L. D., Ronan, T., Stadtherr, M. A., Brennecke, J. F. (2009b). Separation of water and alcohols using 1-(6-hydroxyhexyl)-3-methylimidazolium bis-(trifluoromethylsulfonyl)imide and 1-hexyl-3-methylimidazolium tris(pentafluoroethyl)trifluoro-phosphate. *In preparation.*

Chen, C.-C., Song, Y. (2004). Generalized electrolyte-NRTL model for mixed-solvent electrolyte systems. *AIChE J., 50,* 1928.

Chen, C.-C., Simoni, L. D., Brennecke, J. F., Stadtherr, M. A. (2008). Correlation and prediction of phase behavior of organic compounds in ionic liquids using the nonrandom two-liquid segment activity coefficient model. *Ind. Eng. Chem. Res., 47,* 7081.

Crosthwaite, J. M., Aki, S. N. V. K., Maginn, E. J., Brennecke, J. F. (2004). Liquid phase behavior of imidazolium-based ionic liquids with alcohols. *J. Phys. Chem. B, 108,* 5113.

Crosthwaite, J. M., Aki, S., Maginn, E. J., Brennecke, J. F. (2005). Liquid phase behavior of imidazolium-based ionic liquids with alcohols: effect of hydrogen bonding and non-polar interactions. *Fluid Phase Equilib., 228–229,* 303.

Fadeev, A.G., Meagher M.M. (2001) Opportunities for ionic liquids in recovery of bio-fuels. *Chem. Comm., 3,* 295

Fredlake, C. P., Crosthwaite, J. M., Hert, D. G., Aki , S. N. V. K., Brennecke, J. F. (2004). Thermophysical properties of imidazolium-based ionic liquids. *J. Chem. Eng. Data, 49,* 954.

Freire, M. G., Santos, L. M. N. B. F., Marrucho, I. M., Coutinho, J. A. P. (2007). Evaluation of COSMO-RS for the prediction of LLE and VLE of alcohols + ionic liquids. *Fluid Phase Equilib., 255,* 167.

Hu, X. S., Yu, J., Liu, H. Z. (2006). Liquid-liquid equilibria of the system 1-(2-hydroxyethyl)-3-methylimidozolium tetrafluoroborate or 1-(2-hydroxyethyl)-2,3-dimethylimidozolium tetrafluoroborate plus water plus 1-butanol at 293.15 K. *J. Chem. Eng. Data, 51,* 691.

Jork, C., Kristen, C., Pieraccini, D., Stark, A., Chiappe, C., Beste, Y. A.; Arlt, W. (2005). Tailor-made ionic liquids. *J. Chem. Thermodyn., 37,* 537.

Łachwa, J., Morgando, P., Esperança, J. M. S. S., Guedes, H. J. R., Lopes, J. N. C.; Rebelo, L. P. N. (2006). Fluid phase behavior of {1-hexyl-3-methylimidazolium bis(trifluoromethylsulfonyl)imide, [C$_6$mim][NTf$_2$], + C$_2$-C$_8$ *n*-alcohol} Mixtures: Liquid-liquid equilibrium and excess volumes. *J. Chem. Eng. Data, 51,* 2215.

Michelsen, M. L. (1982). The isothermal flash problem. Part I: Stability. *Fluid Phase Equilib., 9,* 1.

Seader, J. D., Henley, E. J. (1998). *Separation Process Principles,* Wiley, New York, NY.

Simoni, L. D., Lin, Y., Brennecke, J. F., Stadtherr, M. A. (2007). Reliable computation of binary parameters in activity coefficient models for liquid-liquid equilibrium. *Fluid Phase Equilib., 255,* 138.

Simoni, L. D., Lin, Y., Brennecke, J. F., Stadtherr, M. A. (2008). Modeling liquid-liquid equilibrium of ionic liquid systems with NRTL, electrolyte-NRTL, and UNIQUAC. *Ind. Eng. Chem. Res., 47,* 256.

Simoni, L. D., Brennecke, J. F., Stadtherr, M. A. (2009a). Asymmetric framework for predicting liquid-liquid equilibria of ionic liquid-mixed solvent systems: I. Theory and parameter estimation. Submitted for publication.

Simoni, L. D., Brennecke, J. F., Stadtherr, M. A. (2009b). Asymmetric framework for predicting liquid-liquid equilibria of ionic liquid-mixed solvent systems: II. Ternary system predictions. Submitted for publication.

Sørensen, J. M., Arlt, W. (1979-1980) *Liquid-Liquid Equilibrium Data Collection*, DECHEMA, Frankfurt/Main, Germany.

Tessier, S. R., Brennecke, J. F., Stadtherr, M. A. (2000). Reliable phase stability analysis for excess Gibbs energy models. *Chem. Eng. Sci., 55*, 1785.

33

Influence of the Thermodynamic Model on the Energy Consumption Calculation in Bio-Ethanol Purification

N. Ferrari[1], S. Gamba[1], L. A. Pellegrini[1,*] and G. S. Soave[2]

[1] *Dipartimento di Chimica, Materiali e Ingegneria Chimica "G. Natta", Politecnico di Milano Piazza Leonardo da Vinci 32, I-20133 Milano, Italy*
[2] *Via Europa 7, I-20097 San Donato Milanese, Milano, Italy*

CONTENTS

ABSTRACT   A thermodynamic model, suitable for an accurate description of phase equilibrium of strongly non-ideal mixtures, is fundamental for the simulation and, consequently, the process optimization of the extractive distillation of the azeotropic ethanol–water mixture, using ethylene glycol as entrainer.

Aspen PLUS® has been used to simulate the process for the concentration and dehydration of the hydro-alcoholic solution, coming from agricultural biomass fermentation, to obtain high purity ethanol. The application of the proposed thermodynamic model (that belongs to the class of the direct methods) leads to more reliable results and thus allows the correct choice of the best alternative for energy saving in order to make the production of bio-ethanol industrially competitive.

KEYWORDS   *Bio-ethanol purification, non-ideal mixture, thermodynamic model*

---

* To whom all correspondence should be addressed. Tel.: +39 02 2399 3237. Fax: +39 02 7063 8173. E-mail: laura.pellegrini@polimi.it.

Introduction

The current effort to reduce fossil fuel consumption using naturally renewable energy sources has increased the interest in fermentation of agricultural biomasses in order to produce bio-ethanol, which represents one of the main biorefinery products and can be used as both a bio-fuel and a reactant in chemical industry.

Motor fuel ethanol, owing to the high octane number (RON = 108), can be directly used as fuel in internal combustion engines or as an additive to gasoline, producing gasohol with an increased octane number. In Europe, ethanol is primarily used as reactant in the synthesis of ETBE, additive in gasoline blending. Ethanol is also one of the principal reactants in chemical industry for the synthesis of ethylene and acetaldehyde, base products for many commodities.

Mainly produced for industrial use from hydration of ethylene, petroleum or coal-based feedstocks, high purity ethanol can be obtained from fermentation of agricultural biomasses. The main limit to the production of bio-ethanol is the large consumption of energy required for concentration and dehydration of the hydro-alcoholic solution coming from the fermentation bio-reactor. The purification of bio-ethanol consumes 50 to 80% of the energy used in a typical fermentation ethanol manufacturing process owing to the azeotrope in the ethanol–water solution. The aim of this study is the energy optimization of the overall process including extractive distillation of the azeotropic ethanol–water mixture, using ethylene glycol as entrainer, in order to obtain a profitable industrial process.

The system involves a strongly non-ideal mixture of polar and associated compounds, badly described by most of the thermodynamic models implemented in process simulators. A thermodynamic model for calculating phase equilibrium has been developed in this work in order to assure an accurate description of VLE for the mixture under study and, consequently, more reliable results of the process simulation and optimization.

Thermodynamic Model

The thermodynamic model, developed in this work, calculates phase equilibrium through the direct method (ϕ/ϕ): liquid and vapor fugacities are calculated by means of a modified RK equation of state using the modified Huron-Vidal mixing rules with activity coefficients derived from the NRTL model.

The equation of state used for fugacity calculation is a RK-type equation (RK, SRK) in which the original alpha-function has been replaced by the Mathias-Copeman alpha function (Mathias and Copeman, 1983):

$$\sqrt{\alpha} = 1 + m_1 \cdot (1 - \sqrt{T_R}) + m_2 \cdot (1 - \sqrt{T_R})^2$$
$$+ m_3 \cdot (1 - \sqrt{T_R})^3 \qquad T_R < 1 \tag{1}$$

$$\sqrt{\alpha} = 1 + m_1 \cdot (1 - \sqrt{T_R}) \qquad T_R > 1 \tag{2}$$

This three-parameter function allows an increased accuracy in the prediction of pure compound vapor pressure especially at low temperatures for both non-polar (e.g., hydrocarbons, Gamba et al., 2009) and polar substances.

The modified Huron-Vidal mixing rules, reported in Eqs. (3) and (4), apply the common linear mixing rule for volume parameter B, while introduce a non-linear expression for energy parameter A/B, which shows a linear dependence on the excess Gibbs energy.

$$B = \sum_i x_i B_i \tag{3}$$

$$\frac{A}{B} = \sum_i x_i \frac{A_i}{B_i} - \frac{g^E}{Q_1 RT} + \frac{1}{Q_1} \sum_i x_i \ln \frac{B_i}{B} \tag{4}$$

The NRTL model for activity coefficients presents two parameters for every binary mixture. Renon and Prausnitz (1968) introduced the parameter α_{ij} characteristic of the non-randomness of the binary mixture in addition to the asymmetric parameter τ_{ij} expressing the energy of interaction between two different molecules.

$$\ln \gamma_i = \frac{\sum_j x_j \tau_{ji} G_{ji}}{\sum_k x_k G_{ki}} + \sum_j \frac{x_j G_{ij}}{\sum_k x_k G_{kj}} \left(\tau_{ij} - \frac{\sum_k x_k \tau_{kj} G_{kj}}{\sum_k x_k G_{kj}} \right) \tag{5}$$

$$G_{ij} = \exp(-\alpha_{ij} \tau_{ij})$$

The non-randomness parameter determines the higher flexibility of the NRTL model compared to the other activity coefficient models. In the proposed model τ_{ij} parameters express a linear dependence on the inverse of the temperature.

$$\tau_{ij} = a_{ij} + \frac{b_{ij}}{T} \tag{6}$$

The accurate and reliable description of VLE depends on the adaptable parameters of the model, estimated by fitting experimental vapor–liquid equilibrium data.

The model can be easily implemented in Aspen PLUS® process simulation software, through the selection of proper equations and the introduction of the fitted parameters. The compatibility between the proposed model and a widely diffused simulation software is fundamental for its use in process simulation and optimization.

Data Fitting

Both the pure compound coefficients (i.e., parameters for the Mathias-Copeman alpha function) and the binary interaction parameters (i.e., NRTL model parameters) have been evaluated by fitting vapor–liquid equilibrium data (Ambrose and Sprake, 1970; Haar et al., 1984; Hughes and Maloney, 1952; Johnson and Furter, 1957; Kretschmer and Wiebe, 1949; Rieder and Thompson, 1949) by minimizing the sum of the least squares.

The obtained parameters allow the cubic equation of state to describe phase equilibrium of pure compounds involved in the distillation process with relative errors in the vapor pressure lower than 1% as shown in Figure 1.

The proposed model, with binary parameters estimated by fitting experimental equilibrium data, significantly improves the description of the system when compared with a predictive group contribution method based on UNIFAC.

The results of the comparison between the predictive PSRK model by Holderbaum and Gmehling (1991) that uses UNIFAC for computing the excess Gibbs energy (such a method is directly implemented in Aspen PLUS®) and the proposed method based on adaptive parameters that applies the NRTL model for the calculation of the activity coefficients are reported in the following. The PSRK approach has been assumed as reference for the

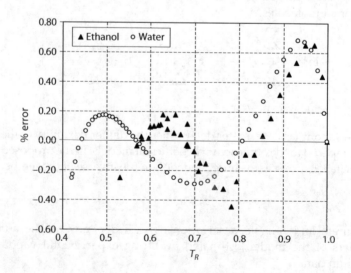

FIGURE 1
Percent errors in the vapor pressure for water and ethanol.

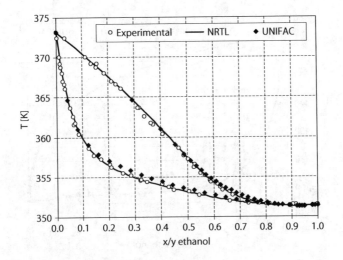

FIGURE 2
Equilibrium diagram for the ethanol–water system at 1 atm.

comparison because has proved to be the most reliable of the ready to use thermodynamic methods implemented in Aspen PLUS®.

Figure 2 represents the equilibrium diagram for the ethanol-water system at 1 atm. The curve computed by using the adaptive parameters from NRTL method shows a better fit of experimental data. In particular the azeotrope corresponds to an experimental value of the ethanol molar fraction ranging from 0.904 to 0.915; the proposed model evaluates a molar fraction between 0.91 and 0.92 while the PSRK with parameters from UNIFAC overestimates the azeotrope between 0.93 and 0.94 (Figure 3).

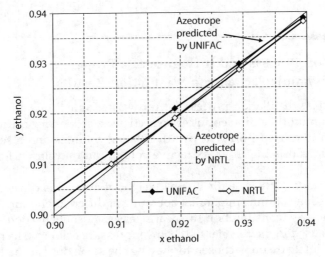

FIGURE 3
x/y diagram for ethanol-water system at 1 atm.

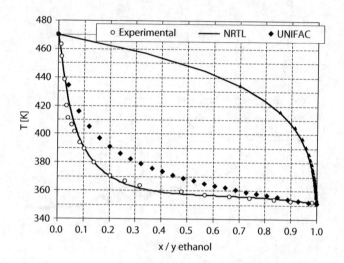

FIGURE 4
Equilibrium diagram for the ethanol–glycol system at 1 atm.

For the ethanol–glycol system, Figure 4 shows that the predictive method is not able to represent the correct boiling point curve also at usual operating conditions (1 atm).

This is due to the fact that PSRK/UNIFAC calculates activity coefficients close to unity while the proposed method with adaptive parameters is able to better reproduce the system non ideality as from experimental evidences.

Distillation Process and Influence of a Proper Thermodynamic Package on Simulation Results

The extractive distillation process for recovering anhydrous ethanol from fermentation broth is divided into two sections: a section for concentrating ethanol up to a fixed purity in two distillation columns and a section for dehydration consisting in an extractive column and a column for solvent recovery.

In the former a stream with 10% by weight of ethanol is concentrated in two distillation columns in parallel operating at different pressure. In order to reduce energy consumption the heat released in condensation at higher pressure is used in the reboiler of the lower pressure column. The split fraction of the inlet stream is chosen for having the same duty at the low pressure reboiler and the high pressure condenser.

The distillate product of the two concentrating columns represents the feed of the dehydration section that consists of the extraction column followed by the regeneration one. The ethylene glycol coming from regeneration enters at the top of the first column while the feed tray location for ethanol solution (at the bottom) derives from optimization for a fixed concentration of ethanol in the distillate.

The concentration and dehydration sections have been simulated both by using Aspen PLUS® with predictive PSRK and Aspen PLUS® with PSRK adjusted with adaptive parameters from NRTL.

The heat duty required by the concentration section increases with the molar fraction of ethanol in the hydro-alcoholic solution sent to the dehydration section. On the contrary the heat duty required by the dehydration section decreases for increasing molar fractions of ethanol in the concentrated solution. Thus there is a value of the molar fraction of ethanol in the concentrated solution that leads to the minimum global energy requirement. The qualitative trend of the heat duty necessary for the separation is reported in Figure 5.

It is evident how the energy consumption rapidly increases after the minimum value, that is when approaching the azeotrope in the concentration section (i.e., the controlling section). Thus the correct prediction of the azeotropic concentration of the ethanol-water system is necessary in order to obtain reliable results from an optimization procedure.

In this case, UNIFAC, that overestimates the molar fraction at the azeotrope, predicts significantly lower reflux ratios and energy consumption, mainly the duty at the reboiler.

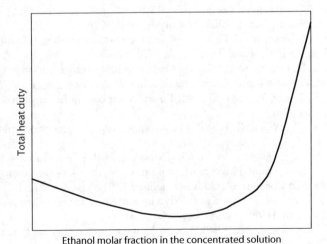

Ethanol molar fraction in the concentrated solution

FIGURE 5
Qualitative trend of the energy requirement for the bio-ethanol purification.

Conclusions

A thermodynamic model for predicting VL equilibrium for non-ideal mixtures has been proposed. By means of experimental data fitting a reliable model easily usable in Aspen PLUS® can be obtained. When simulating the separation of an azeotropic mixture a thermodynamic model that allows a correct prediction of the azeotropic concentration is needed since an error as low as 2% on this concentration heavily influences the energy requirement calculation (Pellegrini et al., 2009).

References

Ambrose, D., Sprake, C. H. S. (1970). Thermodynamic Properties of Organic Oxygen Compounds XXV. Vapour Pressures and Normal Boiling Temperatures of Aliphatic Alcohols. *J. Chem. Thermodyn.*, *2*, 631.

Aspen PLUS®, Copyright© 1981–2006, Aspen Technology, 2006.

AspenTech, *Aspen Physical Property System. Physical Property Methods and Model*, AspenONE™ Documentation, 2006.

Gamba, S., Soave, G. S., Pellegrini, L. A. (2009). Use of Normal Boiling Point Correlations for Predicting Critical Parameters of Paraffins for Vapour–Liquid Equilibrium Calculations with the SRK Equation of State. *Fluid Phase Equilib.*, *276*, 133.

Haar, L., Gallagher, J. S., Kell, G. S. (1984). *NBS/NRC Steam Tables: Thermodynamic and Transport Properties and Computer Programs for Vapor and Liquid States of Water in SI Units*. Hemisphere Publishing Corporation: Washington, DC.

Holderbaum, T., Gmehling, J. (1991). PSRK: A Group Contribution Equation of State Based on UNIFAC. *Fluid Phase Equilib.*, *70*, 251.

Hughes, H. E., Maloney, J. O. (1952). Application of Radioactive Tracers to Diffusional Operations. Binary and Ternary Equilibrium Data. *Chem. Eng. Prog.*, *48*, 192.

Johnson, A. I., Furter, W. F. (1957). Salt Effect in Vapor-Liquid Equilibrium. *Can. J. Technol.*, *34*, 413.

Kretschmer, C. B., Wiebe, R. (1949). Liquid-Vapor Equilibrium of Ethanol-Toluene Solutions. *J. Am. Chem. Soc.*, *71*, 1793.

Mathias, P. M., Copeman, T. W. (1983). Extension of the Peng-Robinson Equation of State to Complex Mixtures: Evaluation of the Various Forms of Local Composition Concept. *Fluid Phase Equilib.*, *13*, 91.

Pellegrini, L. A. et al. A Proper Thermodynamic Method for Simulation of Systems with Associated Compounds. In preparation.

Renon, H., Prausnitz, J. M. (1968). Local Composition in Thermodynamic Excess Functions for Liquid Mixtures. *AIChE J.*, *14*, 135.

Rieder, R. M., Thompson, A. R. (1949). Vapor-Liquid Equilibria Measured by a Gillespie Still - Ethyl Alcohol - Water System. *Ind. Eng. Chem.*, *41*, 2905.

34

Choice of a Sustainable Forest Biorefinery Product Platform Using an MCDM Method

Matty Janssen, Virginie Chambost and Paul Stuart
NSERC Environmental Design Engineering Chair in Process Integration
Department of Chemical Engineering, École Polytechnique Montréal
Montréal (Québec), Canada, H3T 1J7

CONTENTS

ABSTRACT   In practice, design decision making overwhelmingly considers economic objectives for selection of a design alternative, and the environmental and social aspects of alternatives are considered as constraints. Multi-Criteria Decision Making (MCDM) methods enable decision makers to make complex decisions that take into account the different dimensions of sustainability. In this paper, two design case studies are presented where MCDM is applied to evaluate retrofit design alternatives at a pulp and paper mill. Furthermore, an MCDM method is proposed for the choice of a forest biorefinery (FBR) product platform that can be incorporated sustainably into an existing company product portfolio. The decision criteria are the outcomes of various systems analyses, and are provided to a decision panel including stakeholders with expert knowledge relevant to the different sustainability dimensions. Using MCDM techniques in an expert panel setting raises the awareness of all decision makers concerning the complexity of a breadth of critical outcomes. It also helps them to systematically consider their preferences and the relative importance of the criteria in order to make sustainable decisions.

KEYWORDS *Sustainability, multi-criteria decision making (MCDM), retrofit process design, forest biorefinery*

Introduction

The Canadian pulp and paper industry currently faces many challenges. Due to its capital intensive, commodity-focused nature, this industry's competitive position in the global market place is challenged. On the other hand, it employs renewable materials which offers the industry an opportunity to re-think its strategic goals and consider sustainability as a new basis for competitiveness.

Sustainable development was defined by the Brundtland commission as "development that meets the needs of the present without compromising the ability of future generations to meet their own needs." (World Commission on Environment and Development, 1987). Sustainable development has three dimensions: the economic, the environmental and the social dimension. In process design, traditionally only the economic dimension has been targeted. More recently, however, establishing a trade-off between the environmental and economic dimensions has received more attention.

Considering sustainability in process design offers many opportunities for applying process systems engineering (PSE) tools (Bakshi & Fiksel, 2003). For instance, more advanced cost analysis can be carried out for a better economic assessment (Janssen, 2007). Furthermore, Life Cycle Assessment (LCA) and supply chain modeling can be used to expand the boundaries of the system under study. Instead of only considering an individual subsystem such as one production plant, larger systems such as complete supply chains or enterprises should be considered for sustainability assessment (Bakshi & Fiksel, 2003). PSE tools can also be used to measure the level of sustainability of a process design (Sikdar, 2003). Multi-Criteria Decision Making (MCDM) may be used to consolidate economic, environmental and social metrics in order to provide an overall sustainability metric. MCDM can thus be used to take into account all aspects for more sustainable design decision making without losing the detailed information that results from preceding analyses. Using MCDM methods, the trade-offs between the sustainability aspects can be found based on the decision maker's preferences. Furthermore, these methods raise the awareness of the decision maker(s) about the problem complexity and they can be used to address decision uncertainty.

This paper presents two previous case studies that propose frameworks for environmental and sustainable retrofit design decision making using MCDM methods. Furthermore, a new case study is described that considers the sustainable choice of a forest biorefinery (FBR) product platform to be incorporated into a forestry company's existing product portfolio.

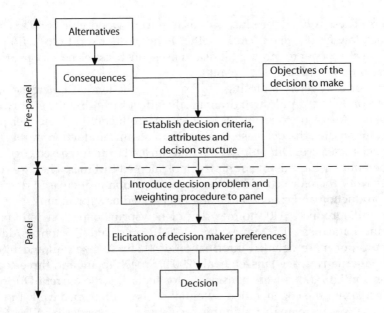

FIGURE 1
General procedure for carrying out a Multi-Criteria Decision Making (MCDM) panel.

Retrofit Design Case Studies

General MCDM Methodology

Expert decision making panels were formed in the two previous case studies to reflect the decision making process in "the real world". The general MCDM methodology that was followed in these studies consisted of the following steps (Figure 1):

1. The objectives, the decision criteria and their attributes, and decision structure were defined before the assembly of the decision panel
2. The panel was introduced to the background of the decision problem and the procedure for eliciting preferences was clarified and discussed
3. The overall preference for each alternative was obtained and a final decision was made.

Retrofit Design Problem at a Newsprint Mill

Newsprint production is an energy-intensive process because it requires large amounts of steam and electricity. Steam is produced in a boiler plant by burning fossil fuel and/or biomass. Thermo-mechanical pulp (TMP) production is power-intensive due to its high electricity consumption by the TMP refiners. These refiners produce wood fibre from wood chips and, as a by-product, low grade steam. On the other hand, a de-inked pulp (DIP) plant uses recycled paper to produce fibre and uses much less electricity per tonne of fibre produced than a TMP plant.

Partly or completely replacing TMP pulp production with increased production of de-inked pulp will dramatically affect the mill-wide energy consumption. Also, due to an increase of steam production in the boiler plant that compensates the decrease or absence of steam production in the TMP plant, the increase of DIP pulp production can lead to an increased cogeneration potential using existing or newly implemented turbines. The case study in this work therefore considers the implementation of increased de-inked pulp production and cogeneration at an integrated newsprint mill.

The mill produces 1100 tonnes/day of newsprint paper on four paper machines, and uses 925 tonnes/day of TMP pulp and 175 tonnes/day of DIP pulp (for more detailed information on the base case mill and retrofit design alternatives, see Janssen et al. (2006). Six DIP plant and three cogeneration configurations were considered for replacing the current DIP plant and increasing the cogeneration potential of the mill, respectively. In total, 18 alternatives were analyzed in this case study by considering all possible combinations of the DIP and cogeneration configurations.

Environmental Design Decision Making

A methodology was proposed for improving Environmental Impact Assessment (EIA) by integrating LCA considerations (Cornejo-Rojas et al., 2005). Using the classical EIA, facility-based environmental impacts are addressed (e.g. NO_x, SO_2 emissions), whereas LCA takes a product-based viewpoint to address potential environmental impacts (e.g. global warming, ozone depletion). Retrofit design alternatives may change the environmental impact at the product level due to changes in upstream and downstream activities. Therefore, integrating LCA into EIA provides opportunities for better evaluating and comparing such alternatives.

Instead of only evaluating the most profitable design alternative for its environmental regulatory compliance, the environmental impact of a set of profitable alternatives was characterized using EIA and LCA (Figure 2). In this case study, the TRACI Life Cycle Impact Assessment (LCIA) method was used to characterize the emissions from the system. The relative importance of the resulting EIA and LCA metrics, that are used as decision criteria, were established using the Analytic Hierarchy Process (AHP) (Saaty, 1980). AHP is a MCDM method that uses pair-wise comparisons of all decision criteria to establish this importance. An expert panel carried out the decision weighting, while taking into account the performance of the alternatives relative to environmental targets. These targets were based on regulatory levels for the EIA criteria; the LCA target levels were based on Canadian reduction targets.

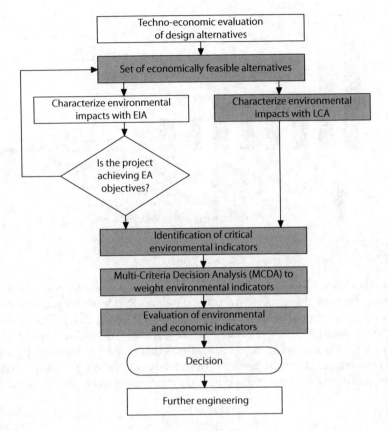

FIGURE 2
Improved methodology for environmental design decision making (highlighted boxes are additions to the standard methodology).

All LCA criteria were more important according to the preferences of the expert panel (Figure 3), because all the EIA decision criteria met regulations and were therefore less important. On the other hand, the panelists had problems with the interpretation of the LCA criteria: the LCA target levels were not regulated and the units to express the LCA criteria were not clear. Next, the decision weights were used to select the environmentally preferred alternative. However, a trade-off analysis between the environmental and economic performance of the alternatives still needed to be done.

Sustainable Design Decision Making

In recent years, the collection and use of process and cost data has increased tremendously in the pulp and paper industry due to the implementation of mill-wide information management systems. This has led to the need for new and practical methodologies in order to better exploit these data for decision making (Janssen et al., 2004).

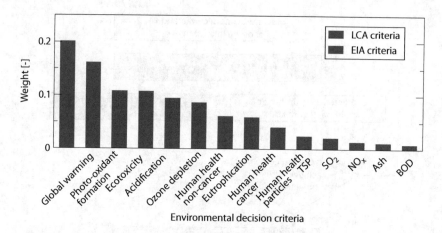

FIGURE 3
Decision weights for the EIA and LCA decision criteria.

A methodology based on process and product modeling was proposed that uses the available process and cost data and PSE tools for more sustainable retrofit design decision making (Figure 4) (Janssen, 2007). First, the retrofit design alternatives were characterized at the process level using mass and energy balances and a novel operations-driven cost modeling approach that is based on the principles of Activity-Based Costing. The profitable design alternatives were retained and their consequences were modeled at

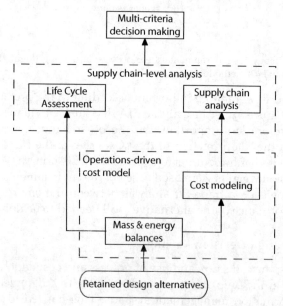

FIGURE 4
Methodology for sustainable retrofit design decision making.

the supply chain level using supply chain analysis and LCA. Impact2002+ was used as the LCIA method. The last step in this methodology determined the preferred design alternative based on economic and environmental criteria by conducting an expert panel. This was done using Multi-Attribute Utility Theory (MAUT) in which the decision criteria are quantified or characterized using attributes and utility measures the preference of the decision maker for a certain attribute level by means of an utility function (Keeney & Raiffa, 1976). The decision weights were determined by trading off all decision criteria to the most important criterion, profitability, which was established by the panel members prior to the weighting.

The selected economic (profitability, investment, energy economics and supply chain profit) and environmental (human health, ecosystem quality, climate change, resource use) decision criteria were placed at the same hierarchical level. This was done in order to increase the awareness of the expert panel about the different trade-offs between the criteria, and to avoid a trade-off between overly aggregated economic and environmental decision criteria. The panel members had trouble trading off the environmental criteria against profitability due to the difficulty of interpreting these criteria and consequently, monetizing them. A surprising result was that the investment criterion (with capital cost as the attribute) was not given any weight (Figure 5). It was argued that smaller investment projects are often more risky and therefore require a higher IRR.

Taking into account more decision criteria besides profitability led to a change in the final decision. Two alternatives had a similar overall utility value, whereas one of these two clearly had a higher utility value for profitability. The other alternative compensated this by outperforming the first alternative for all decision criteria. Due to including supply chain-level

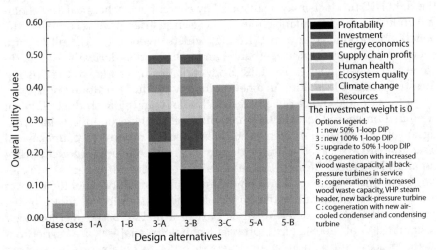

FIGURE 5
Overall utility values for the design alternatives.

criteria (supply chain profit and the LCA impact categories) which expanded the system boundaries, a more sustainable final design decision was made.

Sustainable Choice of a Forest Biorefinery Product Platform

In order to cope with the current challenges, forestry companies have typically focused their corporate strategy on consolidation through merger and acquisition activities, and on minimizing manufacturing costs through cost cutting programs and selling non-core assets (Stuart, 2006). On the other hand, the forest biorefinery (FBR) offers an alternative strategy to forestry companies for utilization of ligno-cellulosic biomass and diversification of their core business.

The FBR has been defined as the "full integration of the incoming biomass and other raw materials, including energy, for simultaneous production of fibres for paper products, chemicals and energy" (Axegård, 2005). It may give forestry companies an opportunity to diversify their revenues by producing a set of new products and penetrating new markets. Process technologies, based on chemical, thermo-chemical or biochemical pathways, that convert the biomass into these products are currently being developed by many different technology providers. These process technologies can be integrated at existing pulp and paper mills in order to enlarge the mills' product portfolio that then will consist of traditional pulp and paper and FBR products. However, many forest biorefinery configurations are possible and it is not obvious which combination of process technology and product portfolio will best serve a company's biorefinery strategy over the longer term. Selecting the right FBR products may be inspired by current petrochemical and chemical industry product management. These industries have recognized the benefits of product platform thinking, which involves the definition of a chemical building block (e.g. ethanol) and value-added derivatives (e.g. ethylene and poly-ethylene). The FBR product platform is added to the existing forestry products, resulting in a new company product portfolio (Figure 6).

The sustainable choice of a process technology and FBR product platform combination should lead to the maximization of the economic value from ligno-cellulosic material. Such a choice will need to take into account several risks that are associated with the design of a new company product portfolio and FBR process design. These risks can be grouped under technological, economic, commercial and environmental risks. Subsequently, metrics are established to quantify these risks. Metrics related to the choice of the product platform are determined as part of a market-based methodology (Chambost et al., 2008).. Next, process technology analyses are carried out at the process and supply chain levels in order to obtain process metrics (Figure 4). Metrics that may be considered are profitability, product demand, GHG emissions and maturity of the technology.

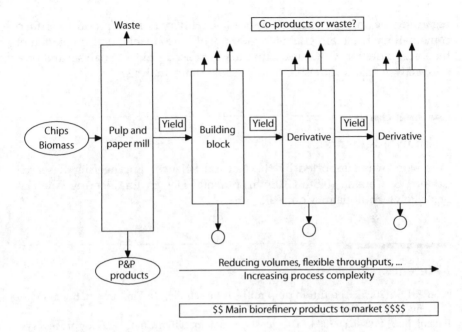

FIGURE 6
Forestry company product portfolio including a FBR product platform.

These performance metrics are used as decision criteria in order to identify the preferred process/portfolio combination using a MCDM method. Both the process and product platform metrics will be used at the same hierarchical level. Thus, the choice of the most promising company product portfolio is based on an understanding of market potential of FBR products, coupled with the application of techno-economic and process selection criteria.

Conclusions

This paper presented two retrofit design case studies that employed MCDM for design decision making. There are several advantages related to the use of MCDM. It rationalizes the design decision process, provides a systematic approach to design decision making, and guides the decision maker(s) towards a rigorous and more balanced decision. Furthermore, the final decision is more sustainable if the system boundaries are expanded by using LCA and supply chain analysis. Finally, using MCDM makes the decision maker more aware of the trade-offs between economic and environmental criteria that need to be made. The final decision changed due to the inclusion of supply chain-level criteria besides profitability. MCDM can also be used to make decisions regarding the implementation of a forest biorefinery process/product portfolio combination. The main goal is to guarantee a sustainable implementation of the forest biorefinery by conducting market-based product analysis and process analysis that includes LCA and supply chain analysis.

Acknowledgments

This work was supported by the Natural Sciences Engineering Research Council of Canada (NSERC) Environmental Design Engineering Chair at École Polytechnique in Montréal.

References

Axegård, P. (2005). The future pulp mill—A biorefinery. In First international biorefinery workshop.

Bakshi, B., & Fiksel, J. (2003). The quest for sustainability: Challenges for process systems engineering. AIChE Journal, 49(6), 1350–1358.

Chambost, V., McNutt, J., & Stuart, P. (2008). Guided tour: Implementing the forest biorefinery (FBR) at existing pulp and paper mills. Pulp and Paper Canada, 109(7–8), 19–27.

Cornejo-Rojas, F., Janssen, M., Gaudreault, C., Samson, R., & Stuart, P. (2005). Using Life Cycle Assessment (LCA) as a tool to enhance Environmental Impact Assessments (EIA). Chemical Engineering Transactions, 7, 521–528.

Janssen, M. (2007). Retrofit design methodology based on process and product modeling. Unpublished doctoral dissertation, École Polytechnique de Montréal.

Janssen, M., Cornejo, F., Riemer, K., Lavallee, H., & Stuart, P. (2006). Techno-economic considerations for DIP production increase and implementation of cogeneration at an integrated newsprint mill. Pulp & Paper Canada, 107(9), 33–37.

Janssen, M., Laflamme-Mayer, M., Zeinou, M.-H., & Stuart, P. (2004). Survey indicates mills' need to exploit it systems with new business model. Pulp & Paper, 78(6), 46–51.

Keeney, R. L., & Raiffa, H. (1976). Decisions with multiple objectives: Preferences and value trade-offs. Cambridge, MA, USA: Cambridge University Press.

Saaty, T. L. (1980). The analytic hierarchy process. New York, NY, USA: McGraw-Hill.

Sikdar, S. (2003). Sustainable development and sustainability metrics. AIChE Journal, 49(8), 1928–1932.

Stuart, P. (2006). The forest biorefinery: Survival strategy for Canada's pulp and paper sector? Pulp and Paper Canada, 107(6), 13–16.

World Commission on Environment and Development. (1987). Our common future (G. Brundtland, Ed.). Oxford, UK: Oxford University Press.

35

Sustainability Metrics for Highly-Integrated Biofuel Production Facilities

H. Dennis Spriggs, Benjamin Brant, Daniel Rudnick
Byogy Renewables, Inc.
150 Almaden Blvd., Suite 700, San Jose, CA 95113

Kenneth R. Hall and Mahmoud M. El-Halwagi
Artie McFerrin Department of Chemical Engineering
Texas A&M University
College Station, Texas 77843-3122

CONTENTS

ABSTRACT Traditionally, energy processing facilities have used a number of metrics and indices to assess process sustainability. Among these indicators is the ratio of output energy of the facility to the input energy. With the growth of the biofuels industry and the increasing emphasis upon integrating mass and energy resources associated with the life cycle analyses of these facilities, it is desirable to "close the loop" for energy usage (and other resources such as water, nutrients and air emissions) of a biofuel product. As the infrastructure and process economics associated with biofuels evolve, it is possible to reduce (or eliminate) substantially the use of fossil fuels in the life cycle of a biofuel product. Sufficient amounts of a biofuel from a given facility can provide the energy needed for the process and for the other life-cycle activities (e.g., transportation) and sell the excess quantity as the net product. This poses a dilemma for the definition of energy efficiency (output energy of the process/input energy to the process) as it becomes infinite when the material balance loop is closed. The paper addresses this issue and introduces two new concepts. The first one is to involve the mass flows of the feedstocks as energy carriers and identify energy equivalents.

The second new concept is the "incremental return on sustainability" which is analogous to the incremental return on investment.

KEYWORDS   *Sustainability, metrics, biofuels, biorefineries, process integration*

Introduction

First generation biofuels such as ethanol or biodiesel produced from food-based resources such corn, sugar cane or soybean are believed to be responsible in part for increases in food costs and to cause a net increase in greenhouse gas (GHG) emissions. Moreover, expanding U.S. corn ethanol production to meet the renewable fuels standards mandated under the Energy Independence and Security Act of 2007 by 2022 will cause an indirect change in land use, with the new farmlands opened to replace corn diverted to fuel uses. As such, there is the prospect of deforestation and global climatic changes. Clearly there is need to direct the development of biofuels in more sustainable directions, using feedstocks such as farm, forest and municipal waste streams; energy crops grown on marginal lands, and algae that minimize competition for prime croplands. These second generation biofuels feedstocks are expected to dramatically reduce GHGs and also increase storage of carbon in soils. One of the key challenges in assessing the true impact of biofeuls on the environment is how to define what constitutes a sustainable biofuel and what are the associated environmental impacts. What is needed is a new fresh approach to defining and quantifying and guiding the development of advanced biofuels in a more sustainable manner than done in the past.

Metrics for Measuring Sustainability of Biofuels

Today, there is no standard method of measuring the sustainability of biofuels, such as the principle of "triple bottom line" (or People, Planet, Profit). Therefore, it is difficult to compare one biofuels process with another. How do you measure performance, e.g. in terms of energy efficiency, carbon footprint, water use? How large of a circle do we draw to assess performance or impact? Do we just focus on the biofuels conversion process itself or do we enlarge the circle of impact to include the production and supply of feedstocks to the end use of the biofuel? These are the challenges faced in coming up with a universally acceptable set of metrics for measuring the sustainability of biofuels. To this end, we will examine the issue from a historical perspective and look at examples where various metrics have been used to measure sustainability.

In the review of various measurement techniques and assessment tools described below, many of the approaches considered the circle of impact to include the complete "value chain" enterprise of developing biofuels from "Crop to Wheel". The following flow diagram below illustrates the scope of measurement for determining the sustainability and/or environmental impact of biofuels development:

There has been an expansion of interest in developing Sustainable Development Indicator (SDIs) systems to analyze the affect or impact of industrialization. In the effort to develop standardized metrics for measuring biofuel sustainability a number of assessment tools have been utilized, both quantitative and qualitative. These include: 1) Life Cycle Assessment (LCA); 2) Exergy; and 3) Material Flow Analysis (MFA). A new framework for sustainability metrics to industrial chemical processes using a set of 3D indicators that represent all three dimensions of sustainability: economic, environmental, and societal (Martins et al., 2007). Four 3D metrics are used: 1) material intensity, 2) energy intensity, 3) potential chemical risk, and 4) potential environmental impacts that are applied to a wide range of process systems. The first two metrics are associated with the process operation and the remaining two metrics represent chemical risk to human health and the potential impact on the surrounding environment. To illustrate this framework two case studies are presented for a chlorine production process and the separation of acetone/chloroform. The ecological input/output analysis (EIOA) is a method used in combination with established environmental (mass intensity) and economic (gross profit) sustainability metrics to (i) create a systematic analysis methodology capable of evaluating alternatives, and (ii) determine which option results in the best route for improving sustainable development (Piluso et al., 2008). Energy Efficiency Ratio (EER) and Environmental Impact (EEER) are metrics developed to measure the total energy inputs and outputs of alternative biofuels processes and is a ratio of the energy output to the fossil fuel input as expressed as the ratio between total output energy to total input energy of the process (Granda et al, 2007). Factored into the calculation is the source and amount of energy used for crop/feedstock production; feedstock transportation; process conversion; and, fuel distribution. To illustrate this concept, EER's were calculated for both 1st and 2nd generation biofuels conversion processes. First generation processes included corn and sugarcane based ethanol, cellulosic ethanol and biodiesel from soybeans. These were compared with 2nd generation processes that included cellulosic ethanol technologies including advanced gasification, enzymatic fermentation and mixed acid fermentation. As a secondary measurement, an LCA calculation was made for associated air pollution of "tailpipe" emissions for alternative biofuels which included an analysis of CO_2 greenhouse gases, CO, VOCs, PM_{10}, NO_x and SO_2 as comparison with reformulated gasoline (RFG).

Approach

One of the common sustainability metrics is energy efficiency:

$$Energy\ Efficiency = \frac{Useful\ energy\ out}{Total\ energy\ in} \tag{1}$$

To improve the energy efficiency metric of the process, energy and mass integration techniques can:

- Induce heat integration by the synthesis of heat exchange networks to minimize heating and cooling utilities
- Employ process cogeneration to optimize the combined heat and power aspects of the process
- Recycle byproducts and waste streams to boilers and industrial furnaces; thereby producing energy
- Use process modification and mass integration to reduce the energy requirements of various units
- Use stream re-allocation to enhance the energy efficiency of the process
- Use process products (e.g., energy products) to substitute for external energy inputs

The result of the higher levels of integration is the possibility to eliminate completely the total external energy input to the process as shown in Figure 1. This drives the value of the metric to infinity. However, the question is whether or not the impact upon sustainability has maximized or even if the impact upon sustainability has improved? The answer is: not necessarily! In closing the energy loop, it is possible to substitute energy forms while worsening the impact upon the environment. As such, a need exists to define a new metric that accounts for a meaningful impact upon the environment at higher levels of process integration. We refer to this metric as the *incremental return on sustainability "IROS"*.

It is instructive to recall a useful concept in the area of process economics commonly used in techno-economic feasibility studies for the chemical process industries. This concept is referred to as *incremental return on*

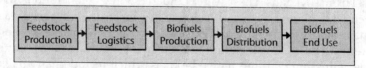

FIGURE 1
Biofuels development process used in measuring sustainability.

investment "IROI". Suppose it is conceivable to add investment options to a base design. Each additional dollar spent must provide sufficiently-large savings. Therefore, the IROI is defined as the ratio of the additional savings to the additional cost. This ratio must be greater than a certain attractive range to justify the proposed investment. The IROI concept is particularly useful when comparing alternatives. It also can be the basis for defining the IROS:

$$IROS = \frac{Change\ in\ environmental\ impact}{Change\ in\ net\ energy\ usage} \tag{1}$$

Therefore, the objective is to assess the true impact of an integration activity within the process by comparing the change in environmental impact (e.g., emission of greenhouse gases "GHGs") from the nominal case to the integrated case with the change in the net energy usage in the process. As such, for an energy-reduction project to be acceptable, it must meet a minimum value of the IROS which guarantees a basic level of environmental performance. For instance, a minimum limit may be the best in class (e.g., gm eq. CO_2 emission per MM Btu).

In addition to the energy integration/closure scenario described by Figure 2, the IROS also finds use in two other applications: (1) alternate pathways and (2) extent of processing. When a biomass can have alternate process pathways (e.g., chemical, biochemical thermal, etc.) or different variations of the same technology, the IROS can compare these alternatives by taking the most energy demanding alternative as the basis and comparing the other alternatives based upon the difference in energy and sustainability metrics (e.g., GHG emissions). The other category deals with extent of processing,

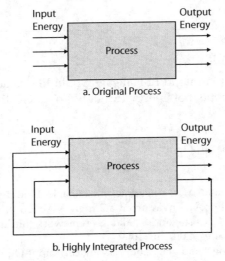

a. Original Process

b. Highly Integrated Process

FIGURE 2
Closing the energy loop via process integration.

which attempts to answer the question of whether or not additional process steps leading to better energy output are justifiable from a sustainability perspective. To illustrate this scenario, consider the production of algae in photo-reactors or ponds by sequestering CO_2 from power plants or industrial processes. The generated algae may be burnt to obtain thermal energy. Alternatively, additional processing of the algae may extract the oil for combustion at higher efficiency and yield more energy. Additionally, the extracted oil may be fed to a transestrification process to produce biodiesel which provides a higher energy output that the previous two alternatives. In fact, it is possible to process the algae *in toto* into biofuels without the intermediate steps. Notwithstanding the improvement in the energy output with further processing, there are implications pertaining to GHG emissions. In this case, the thermal alternative is the basis and the IROS rates for each additional step to assess whether or not further processing is environmentally justifiable.

Conclusions

This paper has introduced the new sustainability metrics IROS. It is based on measuring incremental benefits compared to the energy expenditure. The new metric overcomes the limitations encountered when closure of the energy cycle is achieved and reconciles the objectives of mass (biofuels) and energy. It is also applicable to grassroot designs as well as retrofitting projects. Examples of the application of this approach can be developed which provide clarity and consistency of efficiency assessments.

Acknowledgments

Kenneth Hall and Mahmoud El-Halwagi would like to acknowledge funding from the Department of Energy through Project GO88075.

References

Granda, C., L. Zhu and M. Holtzapple, Sustainable Liquid Biofuels and Their Environmental Impact, Environmental Progress (Vol.26, No.3) October 2007

Martins, A., T. Mata, C. Costa, and S. Sikdar, Framework for Sustainability Metrics, Ind. Eng. Chem. Res., 2007, 46 (10)

Piluso, C., Y. Huang, and H. Lou, Ecological Input-Output Analysis-Based Sustainability Analysis of Industrial Systems,Ind. Eng. Chem. Res., 2008, 47 (6)

36

Capacity Optimization of Dry Grind Bio-Ethanol Separation Process Using Aspen Plus Based Computer Simulations

Y. Tom Xu*, **Bruce M. Vrana and David A. Culver**

DuPont
Wilmington, DE 19898

CONTENTS

ABSTRACT   Flowsheet modeling has been used extensively for the design of new plants, but modeling of existing processes has also proved to be valuable, and with fewer new plants being built, the modeling of existing plants is becoming even more important. Typical objectives for modeling existing plants include capacity debottlenecking, planning plant expansions and troubleshooting process problems. In this application, modeling was used to quantify the potential benefits from a proposed modification to the ethanol recovery area in a dry grind fuel ethanol process. The ethanol recovery area removes solids and concentrates ethanol from 10–16% to 99.5%. This is generally accomplished in a three column distillation train followed by a final drying with molecular sieves. The sieves are normally regenerated using 99.5% ethanol. The process modification proposed was to regenerate the sieves using a non-condensing gas, like CO_2, or alternatively, using a vacuum for regeneration and eliminating the need for a carrier gas stream altogether. This change would greatly reduce the amount of ethanol in the regeneration effluent which is returned to the distillation train for reprocessing. The

---

* To whom all correspondence should be addressed

capacity of the distillation train should increase, but by how much? How can we best take advantage of the increased capacity, and would the benefit justify the cost of development and implementation? Modeling was used to answer these key questions. Results indicated that the capacity of the entire ethanol recovery area could be increased by 10% which would translate to a 10% productivity boast if no other bottlenecks were encountered in the rest of the process. The modeling study also indicated that the benefits could be targeted to specific parts of the ethanol recovery area, and identified process variables that can be used to optimize the impact for a specific process.

KEYWORDS *Bio-ethanol process modeling and design, bio-fuel product recovery, distillation capacity optimization*

Introduction

Ethanol is an important source of energy and is useful both as a fuel and as an additive to gasoline. Ethanol can be produced by fermentation of a wide variety of organic feedstocks. The dilute ethanol produced is recovered by distillation and dehydration to produce a high purity product (Aden et al., 2002, Madson and Monceaux, 1999). The ethanol recovery area of a dry grind ethanol plant has been modeled in Aspen Plus simulation software (McAloon et al., 2004). The function of this area is to remove solids and concentrate ethanol from 10-16% to 99.5%. This function is normally accomplished in a train of three distillation columns followed by final drying with molecular sieves. The first column, commonly called the Beer Column, removes solids and does preliminary concentration of ethanol. The second column, generally referred to as the Rectification Column, further distills the ethanol to a concentration just below the H_2O/ethanol azeotrope. A third column processes a side-stream from the Rectification Column to concentrate and remove minor impurities. The overhead product from the Rectification Column is dried to 99.5% ethanol product using beds of molecular sieves. Regeneration of the molecular sieve beds is generally carried out with 99.5% product ethanol. The resulting spent regeneration stream containing a mixture of ethanol and water must be returned to the Rectification Column to recover the ethanol. The addition of this spent regeneration stream consumes a significant amount of Rectification Column capacity.

It has been proposed that the molecular sieves could be regenerated with a non-condensing gas, like CO_2 or at a vacuum sufficiently deep to eliminate the need for dry ethanol or a carrier gas like CO_2 (Sylvester et al. 2007). This modification would minimize or nearly eliminate the ethanol in the spent regeneration stream. This resulting stream could be sent the scrubbers, returned to the rectification column, or added back into the process at some other point. By eliminating the ethanol rich spent regeneration stream feed to the Rectification Column, column capacity should increase. This would be very beneficial for any existing "dry grind" process with its capacity bottleneck located in the Rectification Column.

If the process bottleneck were located in the Beer Column instead, improvements to the Rectification Column would not directly translate to an increase in overall capacity in the product recovery area. However, some or all of the capacity benefits could potentially be shifted to the Beer Column if a way could be found to transfer some of the Beer Column workload to the Rectification Column. If this could be done, the new proposed molecular sieve regeneration could be applied to increase overall distillation area capacity in plants where the bottleneck is located in the Beer Column. This capacity optimization can be accomplished by using column targeting techniques frequently used in revamping distillation columns, for instance, partially flashing the Beer Column feed and feeding the vapor portion to the Rectification Column. The vapor feed can be fed directly or partially condensed first to reduce vapor traffic in the Rectification Column.

When there are bottlenecks in both the Beer and the Rectification Columns, the amount of Beer Column vapor flashed and the amount subsequently condensed can be adjusted to target specific bottleneck sites. Beer Column capacity can be selectively increased by flashing and bypassing more Beer Column feed. If the increased vapor feed to the Rectification Column threatens to reduce capacity in the upper section of that Column, it can be partially condensed to reduce vapor traffic and help achieve throughput balance.

Proposed Design Changes in Product Recovery Area

The base case process model diagram of the ethanol recovery area is shown in Figure 1. The Rectification Column, a conventional sieve tray column with 16 theoretical stages plus the condenser and re-boiler was described in the previous study (McAloon et al., 2004). The Beer Column is a stripper column with sieve trays equivalent to 8 theoretical stages plus a re-boiler, but no condenser. The primary feed (28BEER), containing ethanol, water, residual sugars and solids from the fermenter, is fed to the top tray of the Beer Column, along with a smaller ethanol/water feed (PCOND) which has been degassed to remove CO_2. Concentrated ethanol and water are removed overhead as a vapor (30BOV) which is fed directly to the Rectification Column. Ethanol concentration in the bottoms is controlled at 500 parts per million by weight by manipulating the liquid boil-up ratio. Most of the bottom product is water but it also contains residual sugars and solids from the fermenter.

The process flow diagram with proposed changes is shown in Figure 2. The changes include the addition of a Beer Feed flash tank and partial condenser, and modifications to the molecular sieve spent regeneration return.

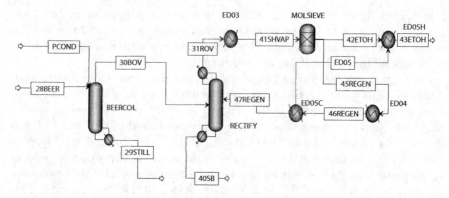

FIGURE 1
Base case aspen process diagram.

De-bottlenecking Rectification Column (Case 1)

When the bottleneck is located in the Rectification Column, the proposed change in the molecular sieve regeneration will have a direct impact on capacity. The new spent regeneration stream, containing water and very little ethanol, will be fed to the re-boiler of the Rectification Column. It could potentially be added at some other point in the process, but feeding it to the re-boiler would ensure recovery of any ethanol present, and it should have little impact on the operation of the Column. Comparing internal column flowrates for Case 1 with the Base Case in Table 1, we see that the proposed modifications caused decreases in liquid and vapor traffic, as well as condenser and re-boiler duties.

From these results, it is clear that the capacity of the Rectification Column upper section limited by flooding would be greater using the proposed molecular sieve regeneration. For a new plant, a smaller column could be specified

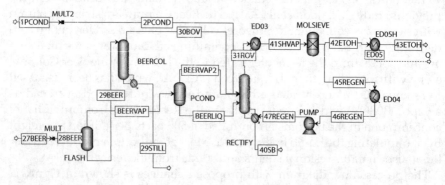

FIGURE 2
Improved aspen process diagram.

TABLE 1

Column Capacity Comparisons Between Base Case
and Case 1

	Net Differences Between Base Case and Case 1			
Stage	Liquid Flow	Vapor Flow	Required Tray Area	Duty
condenser	−9.9%	−16.2%		−9.9%
2	−10.0%	−11.9%	−12.2	
3	−10.1%	−12.0%	−12.3	
4	−10.3%	−12.1%	−12.4	
5	−10.7%	−12.3%	−12.6	
6	−11.7%	−12.6%	−13.0	
7	−25.8%	−13.2%	−15.3	
8	−33.0%	−16.0%	−18.1	
9	−29.6%	−20.0%	−15.5	
10	−29.9%	−53.6%	−48.3	
11	−30.4%	−54.3%	−48.9	
12	−30.6%	−55.6%	−49.4	
13	−29.0%	−57.3%	−48.2	
14	−25.3%	−56.6%	−44.8	
15	−23.0%	−52.0%	−42.8	
16	−22.3%	−48.4%	−42.3	
17	−22.2%	−47.3%	−42.2	
Reboiler	0.1%	−47.1%		−44.0%

and less heat duty would be required. For an existing plant, more material could be fed to the column and internal vapor and liquid traffic in the column could be increased to maintain required separation. This could be done without increasing the column diameter or altering the tray design. As long as the increases were calibrated to match the extra capacity gained by the new regeneration technique, the higher feedrate could be handled without flooding. The extra capacity gained could be very valuable to a sold out plant.

To determine how much extra capacity could be gained in the existing plant, the simulated Rectification Column feed rate was increased until the required column diameter at any stage based on Aspen flooding calculations matched the maximum column diameter for the Base Case, 2.46 meters. If the capacity bottleneck is in the upper section, the feed rate can be increased from 17682 kg/h to 20,000 kg/h, a 13.1% improvement, before the limit was reached on the top stage. In addition, a 36.6% reduction in the re-boiler duty can be realized. Also, the regeneration stream in Case 1 no longer needs process to process heat exchange from the first ethanol product cooler which amounts to approximately 430 kcal/sec. This energy, available between 115 and 85°C, could be used elsewhere in the plant. If the capacity bottleneck is located in the lower section, the model indicates that the Column and its re-boiler could potentially handle feedrates up to 30,000 kg/h, a 70% increase. Of course, other process bottlenecks would certainly be encountered to limit capacity well below this point. Furthermore, there will be a net reduction in re-boiler duty of 36.6% and 5% respectively if capacity were increased 13.1% and 70% as described above. Overall energy demand per kg of ethanol produced is roughly 10.3% lower compared to the based case.

De-bottlenecking Beer Column (Case 2)

Case 2 represents a scenario where the capacity bottleneck is located in the Beer Column rather than in the Rectification Column. This would occur when extra capacity is available in the Rectification Column due to improved molecular sieve regeneration. One way to utilize extra capacity in the Rectification Column would be to send a portion of the Beer feed to it, for example by partially flashing the Beer feed and feeding the vapor to the Rectification Column. This approach keeps all the solids in the Beer Column and adds an extra stage to concentrate the ethanol feed. Although Beer Column bottleneck can be relieved this way, the vapor feed can increase vapor traffic in the upper section of the Rectification Column and reduce capacity there. Partial condensation of the vapor feed counteracts this effect by reducing vapor traffic in the Rectification Column.

The Aspen model was used to explore these issues. Results are summarized in Table 2. Case 2a and Case 2b represent a 4% and 8% vaporization of Beer feed respectively with no condensation. Vapor is fed to stage 9 of the Rectification Column. Case 2c and Case 2d also represent 4% and 8% vaporization, but with 50% condensation. The condensed liquid is sent to stage 14 of the Rectification Column for both cases.

It is evident from Table 2 that Beer Column capacity can be substantially enhanced if 4% of the Beer feed is flashed and redirected to the Rectification Column. It can be further increased if a higher portion of Beer feed is flashed

TABLE 2

Projected Capacity Changes at Several Potential Bottleneck Sites when a Portion of Beer Feed is Vaporized and Bypassed

Capacity Limiting Locations	Case 2a	Case 2b	Case 2c	Case 2d
Beer Column	14.7%	25.7%	14.7%	25.7%
Rectification Column upper section	−3.0%	−14.8%	3.8%	−2.8%
Rectification Column lower section	108.0%	104.0%	71.7%	52.7%
Rectification Column re-boiler	89.3%	86.9%	49.2%	28.2%

and bypassed. Heat is required for the partial flash step but the reduction in re-boiler duties in the columns offsets most of the extra duty for the flash. This is especially true for Cases 2a and 2b where no partial vapor condensation is performed. With a 4% Beer feed flash, only 3% of the upper Rectification Column capacity is lost. However, as the flash is increasesd to 8%, nearly 15% additional capacity in the upper section of the Rectification Column is needed to handle the extra vapor load. This creates a new bottleneck at that point. On the other hand, there are large gains in the lower section and re-boiler because, with the added vapor feed, less boil-up is needed to achieve product specifications. The loss of capacity in the upper section can be countered by condensing a portion of the flashed vapor as was done in Case 2c and Case 2d. With 8% vaporization and 50% condensation, Beer Column capacity could be increased by 25.7% without sacrificing Rectification Column capacity significantly, as shown in Case 2d.

The amount of Beer Column vapor flashed and the amount subsequently condensed are important parameters which can be used to target specific bottlenecks for improvement. Beer Column capacity can be selectively increased by flashing and bypassing more Beer Column feed. If the increased vapor feed to the Rectification Column threatens to reduce capacity in the upper section of that column, the vapor can be partially condensed to reduce vapor traffic and achieve throughput balance.

De-bottlenecking Both Distillation Columns (Case 3)

Balanced capacity improvements are usually sought because the Beer and Rectifications Columns in a given process are normally designed to have the same capacity. From the previous discussion, it is reasonable to conclude that flooding in the Beer Column (or limitations in the Beer Column re-boiler) and flooding in the upper section of the Rectification Column (or limitations in the Rectification Column condenser) will most likely limit process capacity of the ethanol recovery area. Both bottlenecks can simultaneously be improved by adopting the improved molecular sieve regeneration method, and generally, the benefits should be distributed as equally as possible between the two columns.

The Aspen model was used to explore conditions needed to achieve a maximum, balanced capacity increase. The amount of Beer feed flashed and the fraction subsequently condensed were varied by trial and error. Optimum results were found with the Beer Column feed flash set to 2.3% and with all resulting vapor condensed before being fed to the Rectification Column. The resulting bypassed feed ethanol concentration was substantially higher than in previous cases where a partial condenser was used, so the feed location is raised from stage 14 to the new optimum location on stage 8. The results of this scheme are shown in Table 3.

TABLE 3

Projected Balanced Capacity Increase at Potential Bottleneck
Locations

Capacity Limiting Locations	Capacity Gain
Beer Column	10.0%
Rectification Column upper section	10.4%
Rectification Column lower section	44.2%
Rectification Column re-boiler	35.4%

Results summarized in Table 3 indicate that an across-the-board capacity increase of 10% can be realized by properly balancing the benefits gained through the improved molecular sieve regeneration method. This was accomplished while maintaining a large margin of extra capacity in the Rectification Column lower section and re-boiler. At this increased capacity, overall energy demand is 1% lower because the extra duty in the flasher is offset by lower duties in Beer and Rectification Column re-boilers.

Situations could arise when impurity level, such as fusel oil composition increases in the Beer Column feed, or ethanol composition increases due to higher titer in the fermenter. Expected capacity gains should be adjusted accordingly based on common design practices.

Conclusions

Using computer aided process design tools, operation of an existing plant can be explored, understood, and improved. In this application, process modeling was used to quantify the benefits of proposed process changes to a dry grind corn ethanol process. The impact of an improved method for regenerating molecular sieves on ethanol recovery area capacity was studied. Column targeting techniques were used to evaluate the potential capacity improvement for each column and re-boiler in the target area. Capacity optimization was realized by manipulating newly created process parameters. Overall ethanol recovery area capacity was increased by 10% without energy penalty using the improved regeneration method which minimized ethanol recycled to Rectification Column in the spent regeneration stream from the molecular sieves. This was accomplished by flashing 2.3% of the Beer Column feed, condensing the vapor, and feeding it to the Rectification Column. While revamp for more capacity is the objective here, the tools and workflow also apply to new column design or to reducing energy and utility consumption of columns with or without a capacity increase.

Acknowledgments

The authors would like to thank our colleagues R.W. Sylvester and S.T. Breske for ideas and discussions toward this work. The authors would also like to thank DuPont for support in publishing this article.

References

Aden, A., Ruth, M., Ibsen, K., Jechura, J., Neeves, K., Sheehan, J, Wallace, B., Montague, L., Slayton, A. (2002). Lignocellulosic Biomass to Ethanol Process Design and Economics Utilizing Co-Current Dilute Acid Prehydrolysis and Enzymatic Hydrolysis for Corn Stover. *NREL Report No. TP-510-32438*.

Sylvester, R. G., Breske, S. T., Culver, D. A., Vrana, B. M., (2007). Process For Providing Ethanol. U.S. Patent Application, Publication Number 2007-0144886 A1.

Madson, P. W., Monceaux, D. A. (1999). Fuel Ethanol Production. *Fuel Ethanol Textbook*, Alltech Inc.

McAloon, A. J., Taylor, F., Yee, W.C. (2004). A Model of the Production of Ethanol by the Dry Grind Process. Proceedings of the Corn Utilization & Technology Conference, Indianapolis, IN., June 7–9, 2004.

37

Design and Control of Fermentation Processes with Gas Stripping of Product

Keith A. Riegel and Michael L. Luyben[*]
DuPont Engineering Research & Technology
1007 Market St
Wilmington, DE 19898

CONTENTS

ABSTRACT Industrial fermentation products often have an undesirable effect on the biocatalyst that can lead to a loss of productivity at high concentrations. These physiological constraints can greatly affect the economics of the process design. Several separation technologies have been proposed and studied to reduce the negative effects of product inhibition on fermentation performance. Some of these technologies add while others remove control degrees of freedom. This work studies the design and control of a fermenter with product removal via gas stripping. A dynamic process model is developed to simulate a batch fermentation as well as mass transfer of the volatile product from the broth to the gas phase, either with or without external gas feed. The superficial gas velocity plays a critical role in the design and control of the system since it determines the mass transfer coefficient and affects the product concentration gradient between the bulk liquid and liquid-gas interface. It also is the single manipulator available to control product composition in the broth if the fermenter has no mechanical agitation.

---

[*] To whom all correspondence should be addressed

KEYWORDS *In situ product removal, design and control, dynamic modeling*

Introduction

The use of fermentation has been widespread in the pharmaceutical, food processing, and wastewater treatment industries. The chemical industry has grown more focused on the development and commercialization of biologically-based processes in which biocatalysts convert renewable feedstocks into one or several desired chemicals of commercial value. Hence fermenter design, control, and operation will be growing increasingly important since they play a critical role in achieving economic objectives.

Although fermenters typically operate under dilute aqueous conditions, fermentation products often have an undesirable effect on the biocatalyst. This can lead to a loss of productivity at high concentrations with these physiological constraints greatly affecting the economics of the process design. Several separation technologies have been proposed and studied to reduce the negative effects of product inhibition on fermentation performance. Stark and von Stocker (2003) have compiled an extensive review of these various technologies that are generically referred to as *in-situ product removal* (ISPR).

ISPR techniques remove the inhibitory product from the biocatalyst as soon as it is formed, either within the fermenter or external to the fermenter. Methods for product removal cover a wide spectrum including gas stripping, liquid extraction, vacuum distillation, permeation, pervaporation, adsorption, etc.

In this work, we examine the design and control of a fermenter that has product removal from within the fermenter via gas stripping. For gas stripping to be an effective technique the product component must be more volatile than water under fermentation conditions. A dynamic process model is developed to simulate the fermentation as well as mass transfer of the product from the broth to the gas phase.

Dynamic Fermenter Model

Some previous ISPR modeling work has been published (e.g. Sun et al., 1999), although seemingly few connections have been made between the modeling results and an analysis of actual fermenter control. Kollerup and Daugulis (1985) studied ethanol production using extractive fermentation in a continuous stirred tank fermenter with a steady-state model.

Here a dynamic model is linked to fermenter control and operation. A simple batch fermenter with no mechanical agitation is used where all process conditions change dynamically over time. Mixing is assumed to be provided either by external vapor that enters the fermenter base via a gas sparger or

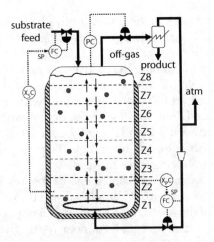

FIGURE 1
Batch or fed-batch bubble column.

by gas generated internally by the biocatalyst (Figure 1). The fermenter head space pressure is controlled via a valve in the off-gas line. In the following, fermenter volume is 10,000 L with an aspect ratio of 2:1 height to diameter.

For purposes of example, the following generic stoichiometry is assumed

$$S \rightarrow 0.18X + 1.656P + 1.8CO_2 \tag{1}$$

where S is substrate, X is biomass, and P is product. The rate of biomass production is governed by

$$r_X = \mu C_X \frac{C_S}{C_S + K_S}\left(1 - \frac{C_P}{K_P}\right) \tag{2}$$

where substrate is assumed to have a Monod-type kinetic effect and product is assumed to have a simple linear inhibition effect. The rate of product formation is given by

$$r_P = k_P C_X \frac{C_S}{C_S + K_S}\left(1 - \frac{C_P}{K_P}\right) \tag{3}$$

A dynamic fermenter model involves a set of standard differential equations for material and energy balances. A first approximation is to assume one theoretical stage for vapor-liquid equilibrium, but in reality the transfer of product from the broth into the vapor phase is governed by a mass transfer coefficient and product concentration gradient between the bulk and interface. The interface concentration is related by vapor-liquid equilibrium to the partial pressure of product in the vapor.

$$MT_P = k_L a\left(C_P - C_P^*\right) \tag{4}$$

In a bubble column fermenter, the liquid broth phase is assumed to be well mixed as the vapor phase travels upward in plug flow. A rigorous description involves partial differential equations for both time and axial position. One standard approach to handle the distributed nature of the problem is to divide the fermenter into discrete zones. Here we use 8 liquid zones. Each zone is perfectly mixed and interchanges liquid with the zones above and below at rates calculated from mixing correlations (Van't Riet and Tramper, 1991). The gas flows upward with no backmixing. In this way, we capture the change in vapor product composition and partial pressure, which affects the concentration gradient driving force in eq. (4).

The mass transfer coefficient becomes one of the key design parameters. It typically needs to be determined empirically and experimentally since it depends upon many of the specific system fluid and physical properties, along with the flow regime. Van't Riet and Tramper (1991) give a correlation that relates the mass transfer coefficient to the superficial gas velocity, u_G, in cm/s:

$$k_L a = \alpha \, u_G^{0.7} \tag{5}$$

This empirical relationship applies to bubble columns with gas sparging where the bubble diameter is independent of process conditions and the sparger design. Here such a simplified correlation is adopted for $k_L a$, even without gas sparging, and the effect on product removal over a range of α is shown. Superficial gas velocity is the sole manipulator for control of product composition in the broth. If the internal CO_2 generation rate is inadequate, the design must be modified to include an external gas feed (using the off-gas separated from the product) to achieve the desired mass transfer rate. Additionally, temperature has an effect on u_G, but is often constrained by biocatalyst requirements. The fermenter aspect ratio (diameter) can also affect u_G and may be considered as a design variable.

Fermentation Design and Control

Anaerobic Batch

With no mass transfer ($\alpha = 0$) and no external gas feed, the effect of product inhibition prevents complete consumption of substrate in the required batch time of 35 hr (Table 1). If the value of α is higher, then the batch can produce a higher effective titer (amount of P in the liquid plus the amount removed in the vapor divided by the total liquid volume) and the batch completion time is reduced. Figure 2 shows how effective product concentration can be increased with an increase in mass transfer rate.

With less product inhibition, higher specific substrate consumption rates (Figure 3) can be sustained (this is the instantaneous rate of substrate

TABLE 1

Batch Results

α	T (hr)	C_{Peff} (g/l)	C_S (g/l)	C_X (g/l)	P_{yield}	P (kg)	P_{vap} (kg)
0	35	89	67	3.3	0.23	668	0
0.1	35	99	54	3.6	0.24	612	120
1	30	146	0	4.9	0.28	196	766
10	25	150	0	5	0.28	32	931

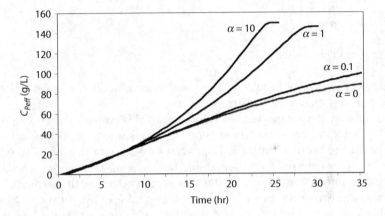

FIGURE 2
Effect of α on effective product titer.

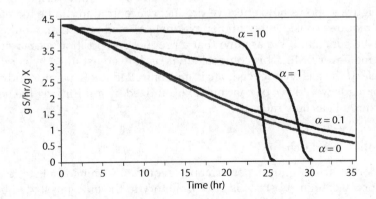

FIGURE 3
Effect on specific S consumption rate.

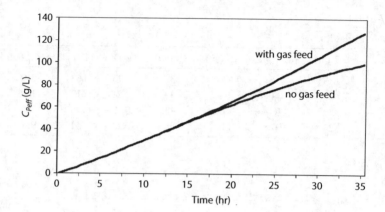

FIGURE 4
Effective product titer $\alpha = 0.1$.

consumed per unit of biomass). The yield of product is the mass of P produced per unit mass of S consumed.

Without an external gas feed, product removal is entirely dependent upon the CO_2 generation rate and there is no mechanism to control product composition in the broth. Control is improved by adding a degree of freedom, which involves external gas sparged into the fermenter. This gas comes from the off-gas once product and water are separated from this stream. This separation and recycle, however, increase capital and operating costs. Gas recycle requires a compressor. Additionally a measurement (either on or off-line) of product composition in the broth or an indirect calculation of the production rate is required to set the gas feed flow as shown in Figure 1. Hence, the trade-off between improved productivity and increased costs must be individually assessed.

Assuming a value of $\alpha = 0.1$, we can simulate the use of external gas feed flow. Figure 4 shows how effective product concentration can be increased, which requires considerably higher external gas feed rates than would otherwise be generated. We assume that a product composition measurement is obtained every 2 hr. This composition is used to adjust the off-gas recycle feed rate. If the gas recycle feed rate is not controlled, the separator and compressor loads would not be continually optimized by minimizing the necessary external gas flow rate.

Anaerobic Fed-Batch

When optimal biocatalyst performance requires a limited substrate concentration (i.e. high substrate is also inhibitory to the biocatalyst), fed-batch operation can be beneficial and is commonly employed. During the batch, substrate is fed into the fermenter to control the substrate concentration in the broth (Figure 1).

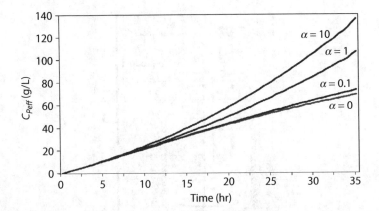

FIGURE 5
Effect of α on effective product titer.

We can dynamically simulate the fed-batch system, comparing the effects of $k_L a$ from an initial volume of 7400 L going out to a batch time of 35 hr with substrate feed stopped at 34 hr. The amount of product made is generally lower than the batch results because of the constraints in fermenter volume and batch time. These batch results have no substrate inhibition parameters in the kinetics.

Increasing the mass transfer rate increases the effective product titer (Figure 5). With higher mass transfer rates the substrate feed flow, manipulated to control substrate composition, is much higher (Figure 6). In the fed-batch case, we can make more product than in the batch case at high product removal rates because we can run the batch longer and are not constrained by the amount of substrate charged initially into the fermenter.

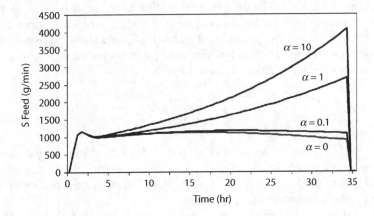

FIGURE 6
Effect of α on S feed rate.

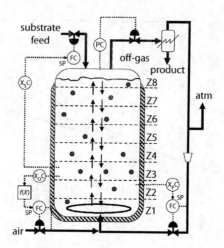

FIGURE 7
Aerobic bubble column control.

Aerobic Fed-Batch

Aerobic systems have an additional control requirement as the dissolved oxygen (DO) concentration is important to cell vitality. The same process design issues apply here related to the mass transfer coefficient as in the anaerobic case, however the value of $k_L a$ affects simultaneously the transfer of product into the vapor phase and the transfer of oxygen into the liquid phase for biocatalyst respiration. A proposed control strategy for DO and product concentration (Figure 7) is more complex than the anaerobic system.

Here we have two manipulators, an off-gas recycle stream and a supplemental air feed stream. The air fed ensures that adequate oxygen is supplied to the biocatalyst. The off-gas recycle stream can be utilized to increase the product mass transfer to the gas phase. A simple oxygen balance can be derived to calculate the supplemental air feed rate required (F_{air}) from the DO controller output, the recycle off-gas flow (F_{rec}) and oxygen mole fraction ($y_{O_2}^{rec}$), and the oxygen mole fraction in the air ($y_{O_2}^{air}$).

$$F_{air} = \frac{DO_{out} - F_{rec} y_{O_2}^{rec}}{y_{O_2}^{air}} \tag{6}$$

The off-gas composition must be analyzed to determine the oxygen content. If an overall oxygen balance is not implemented into the control strategy, the DO concentration may not be well controlled. This design requires that the broth product concentration is measured.

Overall, any control strategy must be specifically tailored to the fermenter design and to the type of ISPR considering also biocatalyst requirements.

Conclusions

Simultaneous fermentation and product separation can increase fermenter productivity. However, trade-offs between economics and controllability must also be considered when such process integration occurs. A dynamic controllability analysis can show whether additional control degrees of freedom are required to achieve the process objectives. In the specific case where we have a volatile product and can use gas stripping within the fermenter, we may not achieve the design objectives if we rely only on the CO_2 generation rate. By adding a manipulator of the external gas feed rate (which also adds cost), we can more effectively control the product concentration in the broth. For aerobic fermentations in bubble columns, where we use the air feed flow to control dissolved oxygen concentration and mixing, the product gas stripping rate does not have any independent control unless we add another degree of freedom involving some separate gas feed flow rate.

List of Symbols

C_P	g/L	product concentration in bulk
C_P^*	g/L	product interface concentration
C_S	g/L	substrate concentration
C_X	g/L	biomass concentration
k_P	g P/hr·g X	product rate constant ($= 1.5$)
K_S	g/L	substrate saturation constant ($= 5$)
K_P	g/L	product saturation constant ($= 100$)
α	hr^{-1}	gas velocity coefficient
μ	hr^{-1}	specific biomass growth rate ($= 0.35$)

References

Kollerup, F., Daugulis, A.J. (1985). A Mathematical Model for Ethanol Production by Extractive Fermentation in a Continuous Stirred Tank Fermentor. *Biotechnol. Bioeng.*, *27*, 1335.

Stark, D., von Stocker, U. (2003). In Situ Product Removal (ISPR) in Whole Cell Biotechnology During the Last Twenty Years. *In Advances in Biochemical Engineering/Biotechnology, Vol. 80*, 149.

Sun, Y., Li, Y.-L., Bai, S., Hu, Z.-D. (1999). Modeling and Simulation of an In Situ Product Removal Process for Lactic Acid Production in an Airlift Bioreactor. *Ind. Eng. Chem. Res.*, *38*, 3290.

Van't Riet, K., Tramper, J. (1991). Basic Bioreactor Design. *Marcel Dekker, Inc.* New York, NY.

38

A Hierarchical Approach to the Synthesis and Analysis of Integrated Biorefineries

Denny K.S. Ng[*1], Viet Pham[2], Mahmoud M. El-Halwagi[2],
Arturo Jiménez-Gutiérrez[3] and H. Dennis Spriggs[4]

[1]Department of Chemical and Environmental Engineering,
University of Nottingham Malaysia,
Broga Road, 43500 Semenyih, Selangor, Malaysia
[2]Department of Chemical Engineering, Texas A&M University
College Station, TX 77843-3122, USA
[3]Instituto Tecnológico de Celaya, Departamento de
Ingeniería Química, Celaya, Gto. México
[4]Matrix Process Integration, Virginia, USA

CONTENTS

ABSTRACT Biorefineries are industrial processes that have been used to produce biofuels. Nonetheless, they could be defined in a broader sense as processing facilities that transform biomass into any value-added products. Because of the large number of possible feedstocks and products, there is a need to develop systematic procedures for the quick synthesis and analysis of biorefineries. Standard design techniques for the process industries are not directly applicable to biorefineries because of the specific nature of feedstocks, technologies, and products, and because of the numerous components included in most biomass. The objective of this paper is aimed at developing a hierarchical procedure for the efficient generation of process alternatives of biorefineries and for their

---

[*] To whom all correspondence should be addressed

analysis. Attention is given to distinguishing the promising alternatives early enough in the process. The procedure involves steps pertaining to reaction pathway synthesis, estimation of theoretical targets for conversions and yields, identification of key energy consumption and targets, overall preliminary cost analysis (including possible subsidies, and carbon credit), process integration, and process simulation. Different levels of details are provided for the various tasks.

KEYWORDS   *Biorefineries, biofuels, design, hierarchical, process integration*

Introduction

Due to the depletion of fossil-fuel resources, there is an increased attention to the issues of energy security, environmental protections, and sustainable development. In addition, the increase in global energy demands and the desire to reduce damages to the environment motivate a shift to renewable energy sources. Biofuels are among the promising forms because of the renewable nature of biomass and the sequestration of carbon dioxide during photosynthesis and growth of energy crops. Furthermore, biomass has versatility in providing a wide variety of feedstocks (e.g. traditional agriculture crops, energy crops, forestry products, municipal solid waste, etc.). Biorefineries are processing facilities that convert biomass into value-added products such as biofuels, specialty chemicals, and pharmaceuticals. One of the key challenges in designing biorefineries is the synthesis of diverse, innovative, and cost-effective process flowsheets.

Much progress has been made in the area of chemical process synthesis including research in the areas of synthesizing heat-exchange networks, mass-exchange networks, separation networks, material recycle/reuse networks, reaction pathways, reactor networks, and property-based networks. Coverage of the subject may be found in literature (e.g., Douglas, 1988; Biegler et al., 1997; El-Halwagi, 1997; Seider et al., 2003; Westerberg, 2004; Smith, 2005; El-Halwagi, 2006; and Kemp, 2007). While significant work has been carried out in the area of synthesizing and designing chemical processes, much less research has been carried out in the area of synthesizing biorefineries.

It is important to note that biorefineries have unique features which make them more difficult than conventional chemical processes. For example, the thermodynamic properties (e.g. Gibbs free energy, enthalpy, entropy, etc.) of biomass are not well established compared to conventional chemicals. In addition, there is lack of information related to the rate of reaction for biomass conversion especially biological conversion (i.e. fermentation, hydrolysis etc.). This is mainly due to the complexity of the structure and variety composition of biomass. Therefore, the previous contributions in the field process synthesis and design for chemical processes are not directly applicable to synthesizing biorefineries. As such, there is a need to develop a systematic procedure for the design and synthesis of biorefineries. This is the goal of this paper.

Approach

Hierarchical design procedures developed by Douglas and co-workers (e.g., Douglas, 1988) have proven to be effective in generating conceptual designs for chemical processes. In this work, a hierarchical approach that is geared towards the synthesis and design of biorefineries is presented. The following are the key steps in the approach:

- Identify inputs and outputs of the process (primary feedstocks and products)
- Define process objectives and overall constraints
- Determine mode of operation (e.g., batch, continuous)
- Synthesize reaction pathways:
  - o **Experimentally:** Using verified biomass conversion pathway
  - o **Theoretically:** Synthesizing reaction pathways
- Estimate theoretical targets for conversions and yields (e.g., equilibrium calculations, stoichiometric limits.)
- Identify key energy consumption steps (e.g., pretreatment, drying, heat of reaction, etc.)
- Conduct overall preliminary cost analysis (cost of feedstocks, value of products, subsidies, carbon credit, etc.)
- Synthesize a separation system
- Perform top-level simulation
- Perform preliminary process integration
- Carry out process optimization and cost estimation.

Generation Pathways

One of the key challenges in the synthesis of biorefineries is the need to synthesize, track pathways, and separate the numerous compounds included in biomass. Synthesis of reaction pathways leading to desired chemicals may be carried out using a variety of methods. In the early 1980's, two main categories (logic-centered and direct associative methods) were developed (Agnihotri and Motard, 1980; Nishida et al., 1981; Yang and Shi, 2000). The logic centered method use abstract representations and synthesize reaction pathways employing mathematical techniques, for example, the matrix synthesis approach (Ugi and Gillespie, 1971) and symbol triangle approach (Hendrickson, 1971). These methods generate sets of intermediates which can be converted to the target molecules. On the other hand, the direct associative methods such as the geometry synthesis approach (May and Rudd, 1976; Rudd, 1976) and free energy chart approach (Rotstein et al., 1982; Fornari et al., 1989; Fornari and Stephanopoulos, 1994a, 1994b) require vast information for all possible chemical modifications. These methods recognize the fact that a number of structural subunits are available and, if they are brought together using some standard reactions, the target molecule can be synthesized.

Later, economic and environmental objectives were introduced by Crabtree and El-Halwagi (1994) for the synthesis of environmentally-acceptable reactions. Subsequently, Buxton et al. (1997) incorporated environmental considerations and life cycle analysis into the methodology for environmental impact minimization and Pistikopoulos et al. (1994) developed a procedure for the environmental evaluation of alternatives.

Recently, optimization-based approaches for reaction path synthesis were introduced (Li et al., 2000; Hu et al., 2004). Mathematical-programming methods include the construction of reaction networks that embed pathways of interest and select from among those pathways.

Because of the special nature of biomass-related species and reaction pathways, a new approach has to be developed. Two methods are introduced in this work.

1. **Evolutionary technique:** Different alternative pathways are generated based on known pathways, and then the pathways are evolved to the desired pathways. For example, Figure 1a illustrates two known pathways (hydrolysis and gasification followed by fermentation and Fischer Tropsch) for converting biomass to alcohols and gasoline respectively. Next, the evolution of the initial pathways

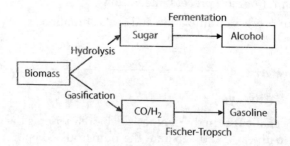

FIGURE 1A
Initial pathways.

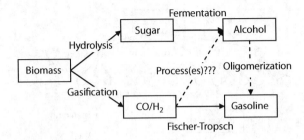

FIGURE 1B
Evolved pathways.

using known pathways or by defining tasks for new processes are developed (e.g., alcohols to gasoline via oligermization, syngas to alcohols) as shown in Figure 1b. It is worth noting that the proposed technique requires enormous information in generating alternative pathways; therefore, this method is categorized as direct associative method.

2. **Forward-Reverse Synthesis Tree:** This method extends the concept of logic-centered methods for reaction pathway synthesis of biorefineries. This method involves forward synthesis of biomass to possible intermediates and reverse synthesis starting with desired products and identifying necessary species and pathways leading to them (Figure 2a). Once the feedstock-forward and the product-reverse pathways are synthesized, two activities are carried out: *matching* (if one of the species synthesized in the forward step is also generated by the reverse step) or *interception* (a task is determined to take a forward-generated species with a reverse-generated species). This

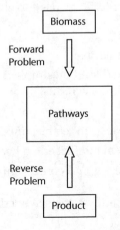

FIGURE 2A
Forward-Reverse synthesis approach.

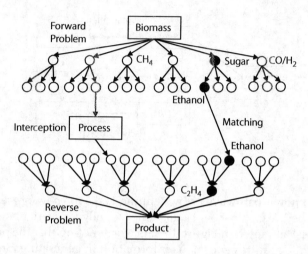

FIGURE 2B
Matching and interception of the species generated by the forward and the reverse problems.

task may be detailed by identifying known process to achieve such interception or by using reaction pathways synthesis to link to two species. Figure 2b is a schematic illustration of the approach.

Screening

Since the number of synthesized pathways may be potentially substantial, it is important to use quick screening techniques to reduce the number of alternatives. The following criteria are used to eliminate a pathway if it happens to be:

- Thermodynamically infeasible
- Economically infeasible (based on simple data: e.g., product cheaper than feedstock even with tax credits)
- Too complex
- Too difficult to operate (e.g., inflexible, unreliable, etc.)
- Require high energy demand (based on heat of reaction)
- Involve too many processes and equipments

Next, the remaining set of pathways is screened using more detailed tools that include:

o Techno-economic analysis (data, modeling, simulation, targeting, track record, etc.)

o Determination of attractive pathways from the superstructure (using optimization techniques such as mixed-integer nonlinear programming).

Work Process

In order to transform the synthesis and design approach to an industrial practice, it is necessary to develop a systematic work process. The following are the key components on the proposed work process:

- **Objectives**—identify business, financial, technical and operational objectives
- **Data**—gather data including (1) plant/utility/technical and (2) business/financial
- **Models/pictures**—develop models/graphical representations of systems
- **Insights**—develop understanding of systems/process behavior and identify inefficiencies
- **Targets**—use benchmarking techniques and experienced based rules to target scope for savings/efficiency improvements
- **Opportunities**—use data, models, insights and targets together with integration principles to extract optimization opportunities
- **Strategy**—develop optimum energy strategy for implementation.

Conclusion

This paper has presented a hierarchical procedure for the synthesis and design of biorefineries. The procedure involves a sequence of interconnected activities that use the appropriate level of details for initial design.

References

Agnihotri, R. B., Motard, R. L. (1980). Reaction Path Synthesis in Industrial Chemistry. *Computer Applications to Chemical Process Design and Simulation, ACS Symposium Series, 124*, 193.

Biegler, L. T., Grossmann, I. E., Westerberg, A. W. (1997). Systematic Methods of Chemical Process Design. *Prentice Hall*, New Jersey.

Buxton, A., Livingston, A. G., Pistikopoulos, E. N. (1997). Reaction path synthesis for environmental impact minimization, *Computers Chem. Engn., 21*, 959–964.

Crabtree, E. W., El-Halwagi, M. M. (1994). Synthesis of Environmentally Acceptable Reactions. *Pollut. Prevent. Process Product Modifications, AIChE Symp.*, 117–127.

Douglas, J. M. (1988). Conceptual Design of Chemical Processes. *McGraw Hill*, New York.

El-Halwagi, M. M. (1997). Pollution Prevention through Process Integration: Systematic Design Tools. *Academic Press*, San Diego.

El-Halwagi, M. M. (2006). Process Integration. *Elsevier*, Amsterdam.

Fornari, T., Rotstein, E., Stephanopoulos, G. (1989). Studies on The Synthesis of Chemical Reaction Paths — II. Reaction Schemes with Two Degrees of Freedom. *Chem. Eng. Sci.*, 44(7), 1569–1579.

Fornari, T., Stephanopoulos, G. (1994a). Synthesis of Chemical Reaction Path Synthesis: The Scope of Group Contribution Methods, *Chem. Eng. Comm.*, 129, 135–137.

Fornari, T., Stephanopoulos, G. (1994b). Synthesis of Chemical Reaction Path Synthesis: Economic and Specification Constraint, *Chem. Eng. Comm.*, 129, 159–182.

Hendrickson, J. B. (1971). Systematic Characterization of Structures and Reactions for Use in Organic Synthesis. *J. Am. Chem. Soc.*, 93 (25), 6847–6854.

Hu, S., Li M., Li, Y., Shen, J., Liu Z. (2004). Reaction Path Synthesis Methodology for Waste Minimization. *Science in China, Series B: Chemistry*, 47(3), 206–213.

Kemp, I. C. (2007). Pinch Analysis and Process Integration: A User Guide on Process Integration for the Efficient Use of Energy. *Elsevier*, Burlington.

Li, M., Hu, S., Li, Y., Shen, J. (2000). A Hierarchical Optimization Method for Reaction Path Synthesis. *Ind. Eng. Chem. Res.*, 39, 4315–4319.

May, D., Rudd, D. F. (1976). Development of Solvay Clusters of Chemical Reactions. *Chem. Eng. Sci.*, 31, 59–69.

Nishida, N., Stephanopoulos, G., Westerberg, A. W. (1981). A Review of Process Synthesis. *AIChE J.*, 27(3), 321–351.

Pistikopoulos, E. N., Stefanis, S. K., Livingston, A. G. (1994). A Methodology for Minimum Environmental Impact Analysis, *AIChE Symposium, Volume on Pollution Prevention via Process and Product Modifications*, 90(303): 139–150.

Rotstein, E., Resasco, D., Stephanopoulos, G. (1982). Studies on the Synthesis of Chemical Reaction Paths — I. Reaction Characteristics in the (ΔG, T) Space and a Primitive Synthesis Procedure. *Chem. Eng. Sci.*, 37(9), 1337–1352.

Rudd, D. F. (1976). Accessible Designs in Solvay Cluster Synthesis. *Chem. Eng. Sci.*, 31, 701–703.

Seider, W. D., Seader, J. D., Lewin, D. R. (2003). Product and Process Design Principles. *Wiley*, New York.

Smith, R., (2005). Chemical Process Design and Integration. *Wiley*, New York.

Ugi, I., Gillespie, P. (1971). Chemistry and Logical Structure. 3. Representation of Chemical Systems and Interconversions by BE Matrices and Their Transformational Properties. *Angew. Chem. Int. Ed. Engl.*, 10, 914–915.

Westerberg, A. W. (2004). A Retrospective on Design and Process Synthesis. *Comp. Chem. Eng.*, 28, 447–458.

Yang, Y., Shi, L. (2000). Integrating Environmental Impact Minimization into Conceptual Chemical Process Design A Process Systems Engineering Review. *Comp. Chem. Eng.*, 24, 1409–1419.

39

Conceptual Design of a Commercial Production Facility with New Biofuel Technology and Fast-Track Approach: Challenges and Strategies

Edi D. Eliezer

BioPrizM
Cherry Hill, NJ 08034

CONTENTS

ABSTRACT   The key project drivers for introducing new technologies in the market place and especially for today's alternative biofuels are: minimization of capital investment, manufacturing costs, environmental impact and energy utilization. The present paper discusses an industrial case facing all these challenges in a very competitive climate where technology evolution pace is not as fast as the demands by world energy markets and especially ever changing availability of capital investment funds. A new generation microbial biofuel technology with promising commercial potential warranted a conceptual design of the first commercial manufacturing facility. This fast-track project had some challenges: the urgent market and timeline demands on one side, and on the other side, only very preliminary lab-scale process definition. The conceptual design study used simple process modeling and facility engineering tools to provide a high level assessment of process scale-up, reliability, risk factors and capital costs for various process design scenarios. In addition to the optimal, cost-effective selection of commercial technologies and equipment, the design also focused on minimization of energy, water use and environmental emissions through various scenarios simulation. This paper demonstrates the methodology and outcomes of a fast-track conceptual design for a new biofuel technology and facility.

KEYWORDS *Industrial biotechnology, biofuels, facility conceptual design, process modeling, scale-up, risks, costs*

Introduction

Today's industrial and societal focus on energy and environment is presenting some new opportunities for process industries and challenges in achieving goals with some aggressive timelines or other conflicting factors. The key project drivers for introducing new technologies in the market place are: minimization of capital investment, manufacturing costs, environmental impact and energy utilization. Also, today's alternative fuels and especially biofuels –focus on this presentation- are introducing new additional drivers such as choice of the most economical feedstock without major global societal impacts and regulatory considerations with genetically engineered microorganisms ('gems' for some entrepreneurs). The present paper discusses an industrial case facing all these challenges in a very competitive climate where technology evolution is not at the same pace as the demands by world energy markets and especially ever changing availability of capital investment funds. A new generation microbial biofuel technology, developed only by few global companies, has promising commercial potential and warranted a conceptual design of the first commercial manufacturing facility. The project had some challenges: the market and timeline pressures on one side, i.e. build now to be first in the market, and on the other side, only preliminary lab-scale process definition, i.e. need more time for process optimization and scale-up. The conceptual design study used simple process modeling and facility engineering tools to provide a high level assessment of process scale-up, reliability, risk factors and capital costs for various process design scenarios. In addition to the optimal or cost-effective selection of commercial technologies and equipment, the design also focused on minimization of energy, water use and environmental emissions through various scenarios simulation. The overall project assessment demonstrates that new biofuels costs are not only sensitive to a high technology process design, but also to facility design scenarios and business risk factors.

Project Challenges & Drivers

The implementation of 'new biofuel' (here, non-Ethanol) production processes at large scale share the same project challenges as other 'industrial biotechnology' technologies and projects (Eliezer, 2005). The first challenge is the design of these processes and facilities, in terms of quality and costs. Indeed they combine biotechnology processes involving genetically engineered microorganisms ('gems') with cGMP (current Good Manufacturing Practices) design or operation guidelines with, on the other hand, bulk commodity chemical processes involving different and sometimes less stringent design or cost implications. Both industry practices may have opposite requirements but an experienced professional will have to make rational compromises in cost and quality of design concepts and specifications (e.g. materials of construction, SIP vs. CIP in next sections).

The second challenge revolves around the project execution strategy considering the status of technology development and business risks factors. One can say that the facility design is mostly capital-driven, i.e. we force the lowest cost solution to the current non-optimized process and biocatalyst ('gem'). Another approach is to accept that it is a science or technology driven project, i.e. we have to fit the bioprocess or bioreactor design to what the biocatalyst requires within project life-cycle. The third challenge is how to address the regulatory aspects, in terms of design and operation, of the industrial biotechnology projects with gems. This is a relatively new modus operandi in unchartered territories between the NIH, FDA, cGMP and EPA guidelines[1].

The present project involved a very new technology where biocatalysts or 'gems' are able to produce biodiesel-type biofuels in a fed-batch fermentation process. The process has been proven at a lab scale (<10L) and needed a conceptual design and feasibility study with cost estimates for a first large-scale production facility. This project's challenges were multiple: limited scale-up data for the fermentation, very limited definition of the downstream process for product recovery and purification, not well-defined capacity for the first large scale and as discussed earlier, limited time and capital funds.

The concept design started by integrating all pertinent information obtained through focused sessions involving scientists, engineers and business managers. Then, the preliminary lab process has been conceptually developed and modeled into a large commercial scale production facility as shown in Figure 1. The main project challenges lead to the following process design and operational project drivers: first, what are the optimal sizes of the plant and bioreactors and second, what is the best bioreactor design fitting size and performance criteria. Process modeling has been developed with first assumed scenarios to establish a design basis that later on helped to simulate other scenarios. The conceptual design basis and project approach have been initially to scale-up from lab to an intermediary 'demo' production facility while waiting for pilot data and before a full-scale commercial facility. The next sections discuss major areas with impact on facility design scope and costs.

---

[1] NIH: National Institutes of Health; FDA: Food and Drug Administration; EPA: Environmental Protection Agency

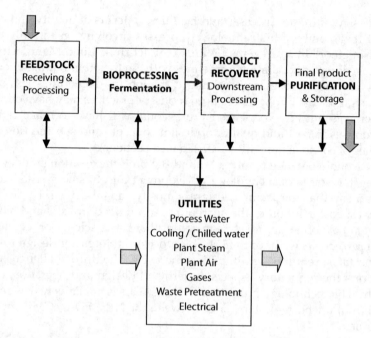

FIGURE 1
Biofuel fermentation process and facility block flow diagram.

Feedstock

This area has been designed and modeled assuming relatively flexible design to process a variety of carbohydrate-type feedstocks and a fermentation facility next to or integrated with the feedstock processing or supply plant. Some of the feedstocks include raw sugar, sugar cane, hydrolyzed corn starch or syrups. The major operations here consist of mechanical conveying, processing and media sterilization in batch or continuous mode, the latter providing energy savings.

Fermentation Scale-up & Plant Design Basis

The main technology transfer project drivers, as mentioned earlier, are in the fermentation process design, and operations scale-up: what are the optimal size of the plant and the bioreactors and, what is the best bioreactor design option fitting size and performance criteria? The project team decided first on a demonstration, 'demo', plant capacity (in MM gals) and had to evaluate

TABLE 1

Bioreactor Design Types and Comparison

Bioreactor or Fermentor Design	Mechanically Agitated, STR	Pneumatic, Air-Lift/ Bubble Column
Height-to-Diameter ratio	< 3.5	> 4
Maximum OTR, mM/lhr	> 200	< 100
Mixing Power, kW/m3	> 0.6	< 0.5
Aeration, VVM	Lower	Higher
Cooling area/volume, m2/m3	Moderate	Higher
Flexibility	Higher	Lower
Scale-up risk	Lower	Higher
Relative Costs	Higher	Lower

various options in terms of bioreactor sizes, quantities and design types. This study selected two fermentor design types, as applicable to current bioprocesses, and performed a preliminary relative comparison, summarized in Table 1. Combining the latter with the specific biofuel process performance and capacity model, the concept design evaluated various facility design scenarios (Figure 2).

The mechanically agitated or stirred tank reactor (STR) has been the workhorse of both the chemical and fermentation industries for many decades. It can provide very high mass transfer and oxygen transfer rates (OTR) well above 100 millimoles/liter.hour (mM/l.hr) critical for highly aerobic fermentation processes. On the other hand the non-mechanical, pneumatic (i.e. only air mixed) reactors such as airlift and bubble column (BC) reactors are well known to be more energy efficient than the STR. The latter, in addition to the mechanical agitation power, requires aeration (i.e. air compressor power) that is slightly lower than for BC in terms of volume of gas per unit of liquid volume per minute (VVM). For the present and related fermentation processes, the BC reactor's total mixing energy requirements per unit liquid volume (kW/m3) has been estimated to be 5 to 20% lower than the STR design. This difference

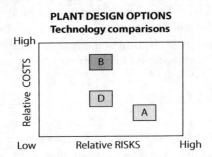

FIGURE 2
Plant design options: relative technology risks vs. costs.

in energy consumption can be significant in large-scale fermentation facilities both in terms of capital and operating costs. In spite of these advantages, the BC bioreactors have some limitations in terms of large-scale applications. First, the maximum OTR capability of BC reactors, in typical fermentations is less than 100mM/l.hr. Second, although there is sufficient literature for chemical reactions, there is less published data on scale-up of aerobic fermentation processes with BC designs at very large scales. Third, the BC design has only one variable parameter - air - to control the whole process, whereas the STR is a more flexible design in terms of decoupling mixing from oxygen/air supply. There are other special bioreactor designs either in published literature or in industrial fermentations (Castro et al., 1987; Eliezer and Jones, 1984; Eliezer, 1987) that can provide advantages of both STR and BC designs; these were not evaluated due to the fast-track project timelines.

The design of the overall production facility involved process modeling, operations schedule analysis and evaluation of energy and utilities requirements for various options with STR and BC designs. The overall plant design options A, B, and D have been rated in terms of relative risks and costs as shown in Figure 2. The analysis concluded to proceed with the concept design option D.

Downstream Processing & Purification

This area's operations have first to recover the biofuel from the fermentation broth mixture and then purify it to a level acceptable to industry biodiesel standards. The first task, involving a three-phase separation (solids, aqueous and organic liquid streams) is more challenging than at first sight in the lab. Indeed based on the current process and economic modeling, these operations needed to be performed in a continuous fashion in the large scale with different technologies than the lab scale. The design and final selection of technology or equipment for this area relied mostly on pilot testing of equipment that can be economically scaled-up to the large production scale. Various process scenarios have been modeled in order to develop a cost-efficient conceptual design. This was achieved by water recovery (>30%; relatively high), reductions in environmental emissions and overall principles of integration between upstream and downstream processes (Eliezer, 1993).

Utilities

Among the typical utilities used in fermentation, plant steam can be a significant one depending on process sterility requirements or sensitivity to contamination. Some control of the latter can be achieved in the large plant by using various combinations of steam-in-place (SIP), clean-in-place (CIP) and proprietary chemicals. This design is more cost-effective than typical fermentation plants for biopharmaceuticals. In addition to compressed air and electrical power, utilities with significant cost impact at large scale are the cooling and waste treatment. Process modeling evaluating various heat exchanger types, coolants and operation timelines has optimized the design of cooling or chilled water. The waste treatment in this biofuel project had also to deal with special considerations related to the presence of 'gems' in all solid, liquid and gaseous streams.

Facility Costs, Risk Analysis & Project Strategies

The first biofuel manufacturing facility has been designed with the selection of most appropriate types of technologies in all sections and optimal sizes of specific process and utility equipment. Then a plant layout has been modeled, keeping in mind cost-efficient and functional adjacencies and infrastructures. At the end of this conceptual design, a complete equipment list with major specifications, actual vendor quotes or cost estimates, complemented the facility engineering and construction costs to come up with final project total installed cost, TIC. The first biofuel 'demo' production plant concept design and TIC provided a basis for further assessment of the next full scale commercial facility. As for many new technology and business ventures, a preliminary risk and cost analysis have been performed for various scale-up strategies as shown in Figure 3 and Figure 4. The TIC difference between the two strategies has been estimated to be between 10 and 20%.

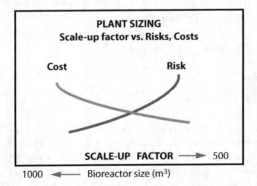

FIGURE 3

Plant and bioreactor sizing: scale-up factor vs. relative risks and costs.

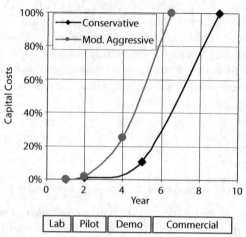

FIGURE 4
Capital expenditures for various scale-up strategies.

Conclusions

The concept design offered new options for cost-efficient biofuel facilities. The overall project assessment demonstrated that new biofuels costs are sensitive to both process technology and to facility design scenarios and business risk factors. In view of increasing availability of new but unused Ethanol plants, there are opportunities to retrofit these to new generation biofuel plants by careful evaluation and re-design of plant, equipment and operations. This approach (subject of a 'white paper' by the author) can save project time and capital expenditures.

References

Castro M.D., Goma. G., Durand G. (1987) Transfer oxygen potential of an air pulsed continuous fermentor. In *Biotechnology Processes: Scale-up and Mixing*, Ed. by C.S. Ho & J. Y. Oldshue, AIChE Symposium Series, New York (NY), p.135.
Eliezer, E. D. and Jones, R.T. (1984). Application of a new bioreactor design to mass transfer limited fermentation. In *The World Biotech Report 1984, Volume 2: USA*, Online Publications, Pinner (U.K.), p.303.

Eliezer, E. D. (1987) Power absorption by new and hybrid mixing systems under gassed and ungassed conditions. In *Biotechnology Processes: Scale-up and Mixing,* Ed. by C.S. Ho & J. Y. Oldshue, AIChE Symposium Series, New York (NY), p.22.

Eliezer, E. D. (1993). Bioprocess development for cost-effective facility design. BioPharm, 6(4), 24

Eliezer, E. D. (2005). The upcoming Bio-PDO™ production by Dupont Tate & Lyle Bioproducts in Loudon,TN. *Biotechnology Association (BIO), Tennessee Annual Meeting,* Oak Ridge, TN, September 23.

40

An Optimization Approach to Heat
Integration with Incorporation of
Solar Energy and Biofuels

**Eman A. Tora[1], Amro El-Baz[2], Diaa El-Monayeri[2]
and Mahmoud M. El-Halwagi[1]**
[1]*Chemical Engineering Department, Texas A&M
University, College Station, TX 77843, USA*
[2]*Department of Environmental Engineering, Zagazig University, Zagazig, Egypt*

CONTENTS

ABSTRACT   The escalating energy prices and the increasing environmental impact posed by the industrial usage of energy have spurred industry to adopt various approaches to conserving energy and mitigating negative environmental impact. This work is aimed at the development of a systematic procedure for energy conservation and incorporation of two renewable sources of energy into industrial usage: biofuels and solar energy. First, heat integration is carried out to minimize industrial heating and cooling utilities. Next, different types of biomass are processed to produce thermal energy and biofuels (such as biodiesel) to be used in supplying an appropriate portion of the needed utilities. Additionally, the solar system is included as a candidate source of energy. To optimize the cost and to overcome the dynamic fluctuation of the solar energy and biofuel production systems, fossil fuel is used to supplement the renewable forms of energy. An optimization approach is adopted to determine the optimal mix of energy forms (fossil, biofuels, and solar) to be supplied to the process, the system specifications, and the scheduling of the system operation. A case study is solved to demonstrate the effectiveness and applicability of the devised procedure.

KEYWORDS   *Solar, biofuels, hybrid energy, energy integration, process integration*

Introduction

The dwindling fossil energy sources and the increasing concern about the consequences of greenhouse gas emissions on global climatic changes provide much motivation for the use of renewable energy sources. As such, there is a significant need to incorporate renewable energy sources in industrial processes. In this work, we consider solar energy and biofuels in addition to fossil fuels. The objective is to develop a systematic procedure for integrating the three forms of energy to meet the thermal needs of an industrial process. The problem is posed as an optimization program that seeks to minimize the overall cost subject to technical and environmental constraints.

Problem Statement

The problem to be addressed by this work is stated as follows: Given a continuous process with:

- A number N_H of process hot streams (to be cooled) and a number N_C of process cold streams (to be heated). Given also are the heat capacity (flowrate x specific heat) of each process hot stream, $FC_{p,u}$; its supply (inlet) temperature, T_u^s; and its target (outlet) temperature, T_u^t, where $u = 1,2,...,N_H$. In addition, the heat capacity, $fc_{P,v}$ supply and target temperatures, t_v^s and t_v^t, are given for each process cold stream, where $v = 1,2,.,N_C$.

Available for service are the following:

- A set $C\_UTILIITY = \{c|c = 1,2,..., N_{CU}\}$ of cooling utilities whose supply and target temperatures (but not flowrates) are known.

- A set $FOSSIL\_UTILITY = \{f|f = 1,2,..., N_{FHU}\}$ of fossil-based heating utilities. For each fossil-based heating utility, the temperature (T_f^{Fossil}) and the cost (C_f^{Fossil}) are known.

- A number of candidate biofuels that can be fed to specific boilers and furnaces to produce heating utilities given by a set $BIO\_UTILITY = \{b|b = 1,2, ..., N_{BHU}\}$ of biofuel-based heating utilities. For each Biofuel-based heating utility, the temperature ($T_b^{Biofuel}$) is known and the cost ($/MM Btu) ($R_b^{Biofuel}$) is also known as a function of time and demand. There is also $R_b^{Biofuel}$ carbon credit ($/MM Btu) for using the bth biofuel utility which is based on the net reduction in greenhouse gas (GHG) emissions when the biofuel is used in lieu of the fossil fuel.

- A candidate solar energy plant. The size and cost of the solar plant are unknown and its delivery of thermal energy and temperature of the heating utility provided by the solar plant will vary over time depending on the variations of solar intensity. There is also R^{Solar} carbon credit (\$/MM Btu) for using the solar energy to replace the fossil fuel.

- A given decision-making time horizon (t_h). Within this horizon, the variations in the energy availability, cost, and conditions are anticipated and expressed in terms of time-dependent changes in quantities of the energy supply.

It is desired to develop a systematic procedure that can determine the usage of the various forms of energy over time and the size and characteristics of the solar plant.

Approach

To simplify the problem, the following assumptions are introduced:

- The decision-making time horizon (e.g., a year) is discretized into N_t periods leading to a set of operating periods: PERIODS = {t|t = 1,2,...,N_t}. Within each time period, the values of energy usage are averaged and the analysis within that period is carried out for the averaged energy values of each energy type.

- It is only allowed to have intra-period integration (i.e., no energy is stored, integrated, or exchanged over more than one period). In selecting the number and duration of the periods, it is important to reconcile the accuracy of fluctuations (e.g., in solar energy) versus the computation efforts.

The approach decomposes the solution into two stages: heat integration followed by energy assignment, design, and scheduling. In the first stage, heat integration is carried out (graphically using thermal pinch analysis or mathematically using linear programming) to determine minimum heating and cooling utilities and optimal level of utility usage (e.g., Kemp, 2006; El-Halwagi, 2006). This includes the identification of heat to be provided to each steam header characterized by heat and temperature (Q_d^{Header} and T_d^{Header}). In the second stage, the optimal quantity of each type of energy is determined for each operating period and the overall design of the system

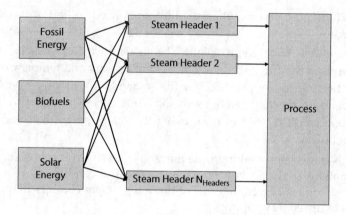

FIGURE 1
Structural representation of the problem.

is synthesized. For each period t, a utility-header-process structural representation (Fig. 1) is developed to embed potential configurations of interest. Outputs from each energy type is split into fractions and assigned to different steam headers which subsequently feed the process to satisfy the optimal requirements from the heat integration analysis.

The objective function is given by:

Minimize total annualized cost =

$$\sum_f C_f^{Fossil} \sum_t Q_{f,t}^{Fossil} * D_t + \sum_b \sum_t C_{b,t}^{Biofuel} * Q_{b,t}^{Biofuel} * D_t +$$

$$+ \sum_s AFC_s^{Solar} + \sum_s \sum_t C_{s,t}^{Solar} * Q_{s,t}^{Solar} * D_t -$$

$$- \sum_b \sum_t R_b^{Biofuel} * Q_{b,t}^{Biofuel} * D_t - \sum_s \sum_t R_s^{Solar} * Q_{s,t}^{Solar} * D_t \qquad (1)$$

where C_f^{Fossil} is the cost of the fth fossil fuel, $Q_{f,t}^{Fossil}$ is the rate of heating of the fth fossil-based utility during period t, D_t is the duration of period t, $C_{b,t}^{Biofuel}$ is the cost of the bth biofuel, $Q_{b,t}^{Biofuel}$ is the rate of heating of the bth biofuel-based utility during period t, AFC_s^{Solar} is the annualized fixed cost of the sth solar plant, $C_{s,t}^{Solar}$ is the operating cost of the sth solar heating utility, $Q_{s,t}^{Solar}$ is the rate of heating of the sth solar-based utility during period t, $R_b^{Biofuel}$ is the carbon credit of the bth biofuel, and R_s^{Solar} is the carbon credit of the sth solar utility.

The heat balance around the dth steam header in period t is given by:

$$Q_{f,d,t}^{Fossil} + Q_{b,d,t}^{Biofuel} + Q_{s,d,t}^{Solar} = Q_d^{Header} \qquad \forall d, \forall t \qquad (2)$$

$Q_{f,d,t}^{Fossil}, Q_{b,d,t}^{Biofuel}$,and $Q_{s,d,t}^{Solar}$ are the heat rates provided from the fossil, the biofeul, and the solar sources, respectively, to the dth header. In assigning the heat flows from sources to headers, attention is given to allowing only allocations with the temperature of the source greater than or equal to the temperature of the header.

The total rate of a given fuel is calculated by summing the splits over the headers:

$$Q_{f,t}^{Fossil} = \sum_d Q_{f,d,t}^{Fossil} \quad \forall f, \forall t \tag{3}$$

$$Q_{b,t}^{Biofuel} = \sum_d Q_{b,d,t}^{Biofuel} \quad \forall b, \forall t \tag{4}$$

$$Q_{s,t}^{Solar} = \sum_d Q_{s,d,t}^{Solar} \quad \forall s, \forall t \tag{5}$$

Although the heat collected and delivered by the solar plant will vary from one period to another, for sizing the solar power plant, it should be large enough to accommodate the largest solar heat provided to the headers (which is still unknown to be optimized), i.e.:

$$Q_s^{Design\_Solar} = \arg\max\{Q_{s,t}^{Solar} \mid t = 1,2,...,N_t\} \quad \forall s \tag{6a}$$

Constraint (6b) may be equivalently expressed as:

$$Q_s^{Design\_Solar} \geq Q_{s,t}^{Solar} \quad \forall s, \forall t \tag{6b}$$

The annualized fixed cost of the sth solar power plant is given as a function (ψ_s) of the size:

$$AFC_s^{Solar} = \psi_s(Q_s^{Design\_Solar}) \quad \forall s \tag{7}$$

The foregoing expressions constitute the mathematical program for the problem. In general, it is a nonlinear program (NLP) and becomes a linear program when the function ψ_s is linear (which may be a very good approximation in many solar applications). The solution of the optimization program provides the optimal values for each form of energy over each period, the distribution to the steam headers, and the design of the solar power plant.

Case Study

An industrial facility is located at the city of Daggett which is located in San Bernardino County in California with the following coordinates {N 34° 52'} {W 116° 46'}. Upon heat integration, the minimum heating utility of the

TABLE 1

Cost and Availability of Biofuel

Month	Cost of Biofuel $/10⁹ J	Maximum Available Heat Rate kW
January	3.5	2,970
February	3.8	2,850
March	3.8	2,850
April	3.4	2,490
May	4.8	2,160
June	5.8	1,920
July	5.8	1,920
August	5.8	1,920
September	5.3	2,120
October	4.9	1,990
November	4.3	2,280
December	3.9	2,610

process is determined to be 5,600 kW. On an annual basis, it is desired to fulfill at least 75% of the process heating utility requirement using renewable sources (solar and biofuel). The biofuel is in the form of biodiesel which is produced from a feedstock of waste cooking oil. The biodiesel process is owned by the company and is based on the oil transestrification technology (e.g., Myint and El-Halwagi, 2009). The cost and availability of the biofuel are listed in Table 1.

Solar energy is collected using the parabolic trough collectors that produce high pressure superheated steam. The parabolic trough collectors are among the most commonly used collectors and are recognized with a proven track-record of providing high efficiency and ability to operate at high temperature (Eck and Zarza, 2006; Cohen et al, 1999). The data for the collected solar energy are reported as monthly averages in Table 2. The total annualized cost ($/yr) of the solar system is given by:

$$TAC_{Solar} = 15.3A_C + 1,085A_C^{0.6} + 0.012Q_{Solar}^{Annual} \qquad (8)$$

Where A_C is the area of the solar collector (m²) and Q_{Solar}^{Annual} is the annual energy collected by the solar system (kWhr/yr). The cost of the fossil fuel is $5.0/10⁹ J.

Using the aforementioned approach and coding the optimization formulation using the software LINGO, the minimum cost of the heating utility was found to be $1,211,875/yr. The optimal area of the solar collectors was found to be 11,300 m². Table 3 summarizes the optimization results of the monthly distribution of the three forms of energy.

TABLE 2

Monthly Average of Collected Solar Energy

Month	Monthly Average of Collected Solar Energy kWhr/(m² · Month)
January	60.6
February	69.5
March	115.8
April	162.3
May	186.2
June	169.5
July	159.1
August	159.0
September	138.9
October	104.6
November	67.5
December	51.1

TABLE 3

Fractional Contribution of Each Form of Energy to the Monthly Heating Utility Requirement of the Process

Month	Solar	Biofuel	Fossil
January	0.16	0.53	0.31
February	0.21	0.51	0.28
March	0.31	0.51	0.18
April	0.45	0.45	0.10
May	0.50	0.39	0.11
June	0.48	0.34	0.18
July	0.43	0.34	0.23
August	0.43	0.34	0.23
September	0.39	0.38	0.23
October	0.28	0.36	0.36
November	0.19	0.41	0.40
December	0.14	0.47	0.39

Acknowledgments

The authors would like to acknowledge support from the NSF (Project Number OISE 0710936), the US-Egypt Board of Science and Technology, and the Egyptian Government.

References

Cohen G.E, Kearney D.W. and Price H. W., Performance history and future costs of parabolic trough solar electric systems, J. Phys. IV France, Vol (9), 1999.

Eck M., Zarza E., Direct steam process with direct steam generating parabolic troughs, Solar Energy, Vol(80), 2006

El-Halwagi, M. M. Process Integration, Vol. 7, Process Systems Engineering, Elsevier, 2006

Kemp, I., Pinch Analysis and Energy Integration: A user guide on process integration for the efficient use of energy, Elsevier, 2006

Myint, L. and M. M. El-Halwagi, Process analysis and optimization of biodiesel production from soybean oil, Clean Tech. & Env. Policy, in press (2009), DOI: 10.1007/s10098-008-0156-5

41

A Systems Approach Towards the Identification and Evaluation of Hydrogen Producing Thermochemical Reaction Clusters

R. J. Andress, L. L. Martin* and B. W. Bequette

Rensselaer Polytechnic Institute
Troy, NY 12180

CONTENTS

ABSTRACT Here we present and demonstrate a process systems engineering-based methodology for the initial evaluation of alternative thermochemical cycles for hydrogen production. The five major components of the proposed strategy are (i) conceptualization, (ii) reaction cluster synthesis, (iii) flowsheet design, simulation, and analysis, (iv) process integration and (v) performance evaluation. Specifically, the resulting approach involves selecting atomic and molecular species with desired properties, identifying suitable thermodynamically feasible clusters of reactions based on these species, screening and sequencing the most promising of the identified reaction clusters, designing a process of unit operations to carry out the reaction cluster, and integrating those processes to maximize thermodynamic efficiency and resource utilization. An integrated suite of resources is employed to aid

---

* To whom all correspondence should be addressed

in implementation of the algorithm. The strength of the proposed methodology is demonstrated through an illustrative example involving the commonly studied Fe-Cl system.

KEYWORDS *Energy, optimization, sustainability*

Introduction

Hydrogen is considered to be an environmentally attractive potential alternative to fossil based fuels. It is clean burning and has a large capacity as a carrier of energy. These features characterize an ideal sustainable energy source that can be used to meet increasing residential, commercial, and industrial demands (see Marshall and Blencoe (2005), Balat (2008), and Stiller et al. (2008)).

Electrolysis of water, a widely used method for hydrogen production, is very inefficient and involves the conversion of primary energy to electricity and electricity to hydrogen (Berry and Aceves, 2005). This process, consisting of a bottoming cycle and electrolysis, has an overall efficiency between 27% and 32%, as reported by Hake et al. (2006). Other methods also pose distinct challenges and limitations, as noted by Ersoz et al. (2006), Hake et al. (2006), Pinto et al. (2005), Saxena et al. (2008), Toonssen et al. (2008), and Tugnoli et al. (2008). The primary challenge with hydrogen production from water is the following reaction:

$$2H_2O_l \rightarrow 2H_{2(g)} + O_{2(g)} + \Delta G^o(298K) = 237\,kJ/mol\,O_2 \tag{1}$$

For a reaction to occur the change in Gibbs free energy (ΔG^o) must be negative, therefore reaction (1) does not occur at normal temperatures, it will only occur at temperatures greatly exceeding 2000 K (Gaykara, 2005), resulting in very energy intensive processes to generate hydrogen from water. To overcome this energy barrier a sequence of thermodynamically feasible reactions, or thermochemical cycle (see Marshall and Blenco (2005)), whose sum is the overall reaction of splitting water (1), is developed based on non-toxic, abundant compounds.

The appeals of the thermochemical cycle are that the primary heat from a nuclear reactor can be used to supply the necessary hot utility (Yildiz and Kazimi, 2006), and purified oxygen, a much sought after commodity, is produced as a by-product. This paper specifically deals with alternative thermochemical cycles (ATCs) (Lewis, 2006), or cycles that do not contain sulfur. The difficulty with ATCs is with the selection and evaluation of reaction clusters.

Traditional assessment tools have primarily included literature searches (Brown et al., 2002), inspection, and evaluation by economic comparison (see Shinnar et al. (1981) and Graf et al. (2008)) to screen possible clusters.

Few attempts have been made to systematically identify, screen, and evaluate hydrogen producing ATCs. A proper study should be thorough, exhausting all possible thermodynamically feasible options.

In this work a systems engineering approach is taken to develop a systematic methodology, based on thermodynamic feasibility, for the initial evaluation of ATCs for hydrogen production. This proposal consists of five major components, (i) conceptualization, (ii) reaction cluster synthesis, (iii) flowsheet design, simulation, and analysis, (iv) process integration and (v) performance evaluation. After the strategy is outlined, its strengths and flexibility are demonstrated through a case study involving the Fe-Cl thermochemical cycle.

The Methodology

Presented here is our five part systematic methodology for identifying, screening, and evaluating ATCs.

I. Conceptualization

The foundation of our ATC identification method is thermodynamic feasibility, which we define below (for a detailed derivation see Andress et al. (2009)).

$$\Delta G^{o}(T) + zRT \leq 0 \quad z > 0 \tag{2}$$

In Eq. (2) $\Delta G^{o}(T)$ is the standard change in Gibbs energy of reaction at temperature T, z is a positive driving force, and R is the universal gas constant. With the criterion for thermodynamic feasibility established, we can now create numerical rules for identifying ATCs.

The space containing the reactions involved in a desired cycle is defined as follows, which is adapted from the work of Holiastos and Manousiouthakis (1998), Mavrovouniotis and Stephanopoulos (1992), and Fornari and Stephanopoulos (1994).

A cycle consists of N^{R} reactions, N^{S} molecular species, and N^{A} atomic species. γ_{ij} is the reactant coefficient for species Λ_{i} in reaction j and δ_{ij} is the product coefficient for species Λ_{i} in reaction j. The overall desired reaction has coefficients of v ($v_{i} > 0$ is a product) which the cluster must be constrained to.

$$\sum_{i=1}^{N^{S}} \gamma_{ij} \Lambda_{i} \rightarrow \sum_{i=1}^{N^{S}} \delta_{ij} \Lambda_{i} \qquad j \in \{1, 2, \ldots, N^{R}\} \tag{3}$$

$$\sum_{j=1}^{N^{R}} (\delta_{ij} - \gamma_{ij}) = v_{i} \qquad i \in \{1, 2, \ldots, N^{S}\} \tag{4}$$

Each reaction must abide by an atomic species balance, where B_{ki} is the number of atoms of k in species i.

$$\sum_{i=1}^{N^S} B_{ki}(\delta_{ij} - \gamma_{ij}) = 0 \tag{5}$$

$$k \in \{1, 2, \ldots, N^A\} j \in \{1, 2, \ldots, N^R\}$$

Limits on allowable coefficients must also be placed, constraining the coefficients to assume only integer values subject to several upper and lower bounds.

$$\delta_j^l \leq \sum_{i=1}^{N^S} \delta_{ij} \leq \delta_j^u \quad j \in \{1, 2, \ldots, N^R\} \tag{6}$$

$$\gamma_j^l \leq \sum_{i=1}^{N^S} \gamma_{ij} \leq \gamma_j^u \quad j \in \{1, 2, \ldots, N^R\} \tag{7}$$

$$\delta_{ij} \in \{\phi_1, \ldots, \phi_n\} \quad i \in \{1, 2, \ldots, N^S\} \ j \in \{1, 2, \ldots, N^R\} \tag{8}$$

$$\gamma_{ij} \in \{\phi_1, \ldots, \phi_m\} \quad i \in \{1, 2, \ldots, N^S\} \ j \in \{1, 2, \ldots, N^R\} \tag{9}$$

$$0 \leq \delta_{ij} \leq \delta_{ij}^u \quad i \in \{1, 2, \ldots, N^S\} \ j \in \{1, 2, \ldots, N^R\} \tag{10}$$

$$0 \leq \gamma_{ij} \leq \gamma_{ij}^u \quad i \in \{1, 2, \ldots, N^S\} \ j \in \{1, 2, \ldots, N^R\} \tag{11}$$

Flag variables are then defined to control the number of products and reactants in a reaction and to allow a species to only participate as product or a reactant.

$$\beta_{ij} \in \{0, 1\} \quad i \in \{1, 2, \ldots N^S\} \ j \in \{1, 2, \ldots, N^R\} \tag{12}$$

$$\theta_{ij} \in \{0, 1\} \quad i \in \{1, 2, \ldots N^S\} \ j \in \{1, 2, \ldots, N^R\} \tag{13}$$

$$0 \leq \delta_{ij} \leq \delta_{ij}^u \beta_{ij} \quad i \in \{1, 2, \ldots N^S\} \ j \in \{1, 2, \ldots, N^R\} \tag{14}$$

$$0 \leq \gamma_{ij} \leq \gamma_{ij}^u \theta_{ij} \quad i \in \{1, 2, \ldots N^S\} \ j \in \{1, 2, \ldots, N^R\} \tag{15}$$

$$\sum_{i=1}^{N^S} \beta_{ij} \leq s_P \quad j \in \{1, 2, \ldots, N^R\} \tag{16}$$

$$\sum_{i=1}^{N^S} \theta_{ij} \leq s_R \quad j \in \{1, 2, \ldots, N^R\} \tag{17}$$

$$\beta_{ij} + \theta_{ij} \leq 1 \quad i \in \{1, 2, \ldots, N^S\} \quad j \in \{1, 2, \ldots, N^R\} \tag{18}$$

The thermodynamic feasibility criterion, Eq. (2), is now structured into our formulation. To capture non-linear thermodynamics we define a discrete temperature set, which a reaction temperature T_j that falls within that set, and assume that we know corresponding Gibbs energies ΔG_{iT}^o, for each chemical species at those temperatures, from a thermodynamic database.

$$T^L \leq T_j \leq T^U \quad j \in \{1, 2, \ldots, N^R\} \tag{19}$$

$$T_j \in \{\psi_1, \ldots, \psi_n\} \quad j \in \{1, 2, \ldots, N^R\} \tag{20}$$

$$\sum_{i=1}^{N^S} \Delta G_{iT_j}^o (\delta_{ij} - \gamma_{ij}) + zRT_j \leq 0 \quad j \in \{1, 2, \ldots, N^R\} \tag{21}$$

II. Reaction Cluster Synthesis

The following mixed integer program (P1), with user defined weights d_i and c_i, can be used to generate exhaustive sets of feasible ATCs for a given system of atomic species.

$$\min \sum_{i=1}^{N^S} \sum_{j=1}^{N^R} (d_i \delta_{ij} + c_i \gamma_{ij})$$

$$\text{s.t. } (4) - (21) \tag{P1}$$

III. Flowsheet Design, Simulation, and Analysis

The set of identified cycles is screened by an initial evaluation, which is executed by constructing the cycle with simulation software. The reactions are assumed to go to completion, separations are assumed to be perfect, and heat exchanges are assumed to have zero heat loss. These assumptions allow for a quick and efficient screening of ATCs, while maintaining conservative evaluation techniques.

IV. Process Integration

Using the latent, sensible, and reaction heats, from the simulation, heat integration is performed with the aid of a pinch diagram. Hot and cold curves are plotted as to approach a minimum pinch temperature, which determines the maximum attainable heat integration, and the corresponding required hot and cold utility. Their sum is the heat duty (Q) of the cycle.

V. Performance Evaluation

Cycle evaluation is done following Argonne National Laboratories prescribed method (see Lewis, 2006, Lewis and Masin, 2008, and Lewis et al. 2008 A and B). The performance measure is cycle efficiency (η), defined below in terms of the energy contained in hydrogen, the heat duty, and the required work with an assumed efficiency of 50%. For an initial evaluation the work is equal to the Gibbs energy of separation for separating the hydrogen and oxygen, where n_i is the molar flow rate of i, and y_i is mole fraction of species i. A promising cycle has an efficiency greater than 35%, as to be comparable to electrolysis.

$$\eta = \frac{\Delta H^\circ_{298K}(H_2O)}{Q + \frac{W}{0.5}} \tag{22}$$

$$W = \Delta G_{SEP} = -RT \sum_i n_i \ln y_i \tag{23}$$

Results and Discussion : A Case Study

A case study involving the commonly studied iron-chlorine system (see Gahimer et al. (1976) and van Velzen and Langenkamp (1978)) is conducted using the presented methodology; details of this can be found in Andress et al. (2009). Six cycles are identified that exceed the minimum threshold efficiency of 35%, shown in Table 1, the first of which is a well noted cycle in the literature.

It should be stressed that we only present an initial evaluation and that further, more rigorous analysis should be conducted on promising cycles. Such considerations should include thermodynamic yields, competing products, kinetics, separations, heat and pressure losses, and operability.

Conclusions

In this paper we outline a methodology to identify, screen, and evaluate hydrogen producing reaction clusters. The potential of the methodology is demonstrated through a case study involving the Fe-Cl thermochemical cycle. Six cycles are identified that exceed a minimum threshold base efficiency of 35%. This approach allows for a standard and efficient procedure for the future evaluation of alternative thermochemical cycles and can be extended for a more comprehensive analysis of promising cycles.

TABLE 1

Identified Fe-Cl ATCs

Cycle	η (%)
$6FeCl_2 + 8H_2O \rightarrow 2Fe_3O_4 + 12HCl + 2H_2$	35.6
$2Fe_3O_4 + 16HCl \rightarrow 4FeCl_3 + 2FeCl_2 + 8H_2O$	
$4FeCl_3 \rightarrow 4FeCl_2 + 2Cl_2$	
$2Cl_2 + 2H_2O \rightarrow 4HCl + O_2$	
$6H_2O + 10FeO \rightarrow 2Fe_3O_4 + 2H_2 + 4Fe(OH)_2$	42.5
$2Fe_3O_4 + 3Fe(OH)_2 \rightarrow 3H_2O + O_2 + 9FeO$	
$Fe(OH)_2 \rightarrow H_2O + FeO$	
$8H_2O + 4Fe \rightarrow 4H_2 + 4Fe(OH)_2$	44.1
$2H_2 + 4Fe(OH)_2 \rightarrow 6H_2O + O_2 + 4Fe$	
$8H_2O + 4Fe \rightarrow 5H_2 + 2FeO + 2Fe(OH)_3$	48.8
$H_2 + 2Fe(OH)_3 \rightarrow 4H_2O + O_2 + 2Fe$	
$H_2O + 2FeO \rightarrow H_2 + Fe_2O_3$	
$3H_2 + Fe_2O_3 \rightarrow 3H_2O + 2Fe$	
$6H_2O + 10FeO \rightarrow 2Fe_3O_4 + 2H_2 + 4Fe(OH)_2$	43.0
$2Fe_2O_3 + 3Fe(OH)_2 \rightarrow 3H_2O + O_2 + 7FeO$	
$Fe_3O_4 + Fe(OH)_2 \rightarrow H_2 + 2Fe_2O_3$	
$Fe_3O_4 + H_2 \rightarrow H_2O + 3FeO$	
$6H_2O + 10FeO \rightarrow 2Fe_3O_4 + 2H_2 + 4Fe(OH)_2$	41.7
$2Fe_3O_4 + 3Fe(OH)_2 \rightarrow 3H_2O + O_2 + 9FeO$	
$FeO + Fe(OH)_2 \rightarrow H_2 + Fe_2O_3$	
$H_2 + Fe_2O_3 \rightarrow H_2O + 2FeO$	

Acknowledgments

Financial support from the IGERT fellowship, through NSF Grant No. DGE-0504361, from Argonne National Laboratories, through Grant No. W-31-109-ENG38, and from Rensselaer Polytechnic Institute are gratefully acknowledged.

References

Andress, R. J., Huang, X., Bequette, B. W., Martin, L. L. (2009). A systematic methodology for the evaluation of alternative thermochemical cycles for hydrogen production. *Int. J. Hydrogen Energy, to appear.*

Balat, M. (2008). Potential importance of hydrogen as a future solution to environmental and transportation problem. *Int. J. Hydrogen Energy, 33, 4013.*

Berry, G. D., Aceves, S. M. (2005). The Case for Hydrogen in a Carbon Constrained World. *J. Energy Resour. Technol., 127(2), 89.*

Brown, L. C., Besenbruch, G. E., Schultz, K. R., Showalter, S. K., Marshall, A. C., Pickard, P. S., Funk, J. F. (2002). High Efficiency Generation of Hydrogen Fuels using Thermochemical Cycles and Nuclear Power. *Prepared for Presentation at the AIChE 2002 Spring National Meeting,* New Orleans, LA.

Ersoz, A., Olgun, H., Ozdogan, S. (2006). Reforming options for hydrogen production from fossil fuels for PEM fuel cells. *J. Power Sources, 154(1), 67.*

Fornari, T., Stephanopoulos, G (1994). Synthesis of Chemical Reaction Paths: Economic and Specification Constraints. *Chem. Eng. Commun., 129, 159.*

Gahimer, J., Mazumder, M., Pangborn, J. (1976). Experimental Demonstration of an Iron Chloride Thermochemical Cycle for Hydrogen Production. *Eleventh Intersociety Energy Conversion Engineering Conference, 1, 933.*

Gaykara, S. Z. (2005). Experimental solar water thermolysis. *Int. J. Hydrogen Energy, 29(14), 1459.*

Graf, D., Monnerie, N., Roeb, M., Schmitz, M., Sattler, C. (2008). Economic comparison of solar hydrogen generation by means of thermochemical cycles and electrolysis. *Int. J. Hydrogen Energy, 33, 4511.*

Greenberg, M. D. (1998). Advanced Engineering Mathematics, Second Edition. *Prentice Hall.* Upper Saddle River, NJ.

Hake, J.-F, Linssen, J., Walbeck, M. (2006). Prospects for hydrogen in the German energy system. *Energy policy, 34, 1271.*

Holiastos, K., Manousiouthakis, V (1998). Automatic Synthesis of Thermodynamically Feasible Reaction Clusters. *AIChE Journal, 44(1), 164.*

Lewis, M. A. (2006). An Assessment of the Potential of Alternative Cycles. *Prepared for U.S. Department of Energy Office of Nuclear Energy, Science and Technology.*

Lewis, M. A., Masin, J. G., O'Hare, P. A. (2008 A). Evaluation of alternative thermochemical cycles—Part I: The methodology. *Int. J. Hydrogen Energy, Available online.*

Lewis, M. A., Masin, J. G. (2008). Evaluation of alternative thermochemical cycles—Part II: The down-selection process. *Int. J. Hydrogen Energy, Available online.*

Lewis, M. A., Ferrandon, M. S., Tatterson, D. F., Mathias, P. (2008 B). Evaluation of alternative thermochemical cycles—Part III: Further development of the Cu-Cl cycle. *Int. J. Hydrogen Energy, to appear.*

Marshall, S. L., Blencoe, J. G. (2005). Equilibrium analysis of thermochemical cycles for hydrogen production. *Separation Science and Technology, 40(1–3), 483.*

Mavrovouniotis, M., Stephanopoulos, G. (1992). Synthesis of Reaction Mechanisms Consisting of Reversible and Irreversible Steps. 1. A Synthesis Algorithm. *Ind. Eng. Chem. Res., 31(7), 1625.*

Pinto, F., Franco, C., Lopes, H., Andre, R., Gulyurtlu, I., Cabrita, I. (2005). Effect of used edible oils in coal fluidised bed gasification. *The 5th European Conference on Coal Research and its Applications. Fuel, 84(17), 2236.*

Saxena, S.K., Drozd, V., Durygin, A. (2008). A fossil-fuel based recipe for clean energy. *Int. J. Hydrogen Energy, 33, 3625.*

Shinnar, R., Shapira, D., Zakal, S. (1981). Thermochemical and Hybrid Cycles for Hydrogen Production: A Differential Economic Comparison with Electrolysis. *Ind. Eng. Chem. Proc. Des. Dev., 20, 581.*

Stiller, C., Seydel, P., Bunger, U., Wietschel, M. (2008). Early hydrogen user and corridors as part of the European hydrogen energy roadmap (HyWays). *Int. J. Hydrogen Energy, 33, 4193.*

Toonssen, R., Woudstra, N., Verkooijen, A. H. M. (2008). Exergy analysis of hydrogen production plants based on biomass gasification. *Int. J. Hydrogen Energy, 33, 4074.*

Tugnoli, A., Landucci, G., Cozzani, V. (2008). Sustainability assessment of hydrogen production by steam reforming. *Int. J. Hydrogen Energy, 33, 4345.*

Van Velzen, D., Langenkamp, H. (1978). Problems around Fe-Cl Cycles. *Int. J. Hydrogen Energy, 3, 419.*

Yildiz, B., Kazimi, M. S. (2006). Efficiency of hydrogen production systems using alternative nuclear energy technologies. Int. J. Hydrogen Energy, 31, 77.

42

Energy Saving Potential Identification in the Batch Chemical Industry

Andrej Szíjjarto*, Stavros Papadokonstantakis and Konrad Hungerbühler
Safety & Environmental Technology Group
Swiss Federal Institute of Technology (ETH), CH-8093 Zürich, Switzerland

CONTENTS

ABSTRACT   Presently, energy saving motivation of the chemical manufacturers is growing due to high prices of the primary energy sources. Therefore, efficient and reliable methodologies and tools allowing analysis and optimization of the energy demand are of high importance for the industry.

Developed methodology for the identification of the optimization potential significantly simplifies the energy-saving efforts in multipurpose chemical batch plants. The investigation of the energy-saving potential was carried out in two stages comprising efficiency screening on the plant level followed by a detailed analysis of the unit operations (UOs) in the production lines.

The proposed efficiency screening of the steam usage in the UOs on the plant level facilitates the selection of the targets for the further detailed analysis. The efficiency of a particular UO was calculated batch-wise and averaged over a period of 60 days for 19 different UOs. The UOs with low batch-averaged thermal efficiency in combination with high standard deviation of the efficiency values were identified as most important optimization spots. Two production lines were investigated in a detailed analysis. Significant energy consumption was identified within the synchronization steps which are representing a non-productive part of the batch production. It was identified that the high energy consumption in the synchronization steps was caused by overheating/

---

* To whom all correspondence should be addressed, E-mail: andrej.szijjarto@chem.ethz.ch overcooling problems of the temperature control loop together with a long duration of the synchronization steps in the low-performing batches.

KEYWORDS *Energy saving, energy efficiency, process monitoring, batch chemical production*

Introduction

The process-side model of the chemical production processes was used in order to quantify energy demand of particular unit operation (UO) (Szíjjarto *et al.*, 2008, Bieler *et al.*, 2004). The basic principle of the PSM is the energy balance of the process carried out inside the UO using standard or measured process data (e.g. temperature and reaction mass of the material being processed in the UO), physicochemical data of the substances that are processed in the UO (heat capacity, heat of evaporation, and heat of reaction) and the process description (control recipe) as input data.

Generally, the energy utility required for a dynamic H/C operation defined by a recipe in a differential time dt can be expressed by the heat balance of the UO as shown in Eq. (1).

$$\frac{dE_{HCU,UO}^{PSM,TOT}}{dt} = \frac{dE_{HCU,UO}^{PSM,THEO}}{dt} + \frac{dE_{HCU,UO}^{PSM,LOSS}}{dt} - \frac{dE_{HCU,UO}^{PSM,DISS}}{dt} \qquad (1)$$

where $E_{HCU,UO}^{PSM,TOT}$ is total required energy supplied by the H/C utility, $E_{HCU,UO}^{PSM,THEO}$ is the theoretical use of the energy for thermal operations as defined in the recipe, $E_{HCU,UO}^{PSM,LOSS}$ is the term representing the thermal losses and $E_{HCU,UO}^{PSM,DISS}$ represents the dissipation of the mechanical energy of the stirrer.

The thermal efficiency of the steam usage in the UO for a product specific batch, used as a main indicator for the efficiency screening is defined as the ratio between the theoretically required steam defined by thermodynamics and the overall used steam including the losses

Analysis of Potential Energy Savings

Ranking the UOs according to their steam usage efficiency can help to detect optimization potential in the plant. The steam usage efficiency analysis allows a fast screening of the UOs in the case study plant leading to the selection of the units with low efficiency for a more detailed investigation of the energy consumption and finally an optimization. The thermal efficiencies of observed UOs with an average steam consumption of >200 kWh per batch are depicted in Figure 1.

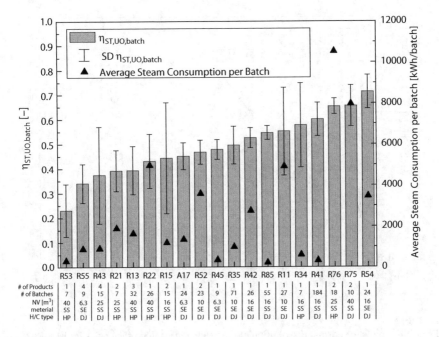

FIGURE 1

Batch aggregated average efficiency of the steam usage including standard deviation (depicted by error bars) and average steam consumption in particular UOs, considering different specifications of the UOs (Nominal Volume (NV), Material—Stainless Steel (SS), Glass-lined Steel (SE) and type of H/C system—Half-pipe Coil(HP) and Double Jacket (DJ)).

The most promising UOs from the steam consumption optimization point of view are those with low mean value and high standard deviation of the efficiency as well as high average steam consumption per batch. It can be seen that the average efficiency ranges, according to Figure 1, from 25 to 70% and it is mainly influenced by the number of products produced in the particular UO. The UOs with higher product variability tend to have lower mean value and higher standard deviation of the efficiency. One reason for the lower efficiencies compared to the conventional heat exchangers are also caused by the batch operation mode of the UOs. In this mode the operations of heating and cooling are ordered accordingly, spending a lot of supplied energy to reheat the media in the H/C system. This part of energy is modeled by the loss model and decreases the efficiency of the UO. Another reason can be the lower heat exchange area to volume ratio of the investigated UOs compared to the conventional heat exchangers.

As an example, two monoproduct UOs with comparable number of produced batches, R22 and R54, were chosen for further analysis.

In Figure 2, the probability plots for normal distributions of observed UOs are depicted.

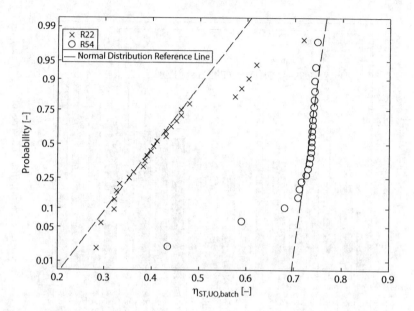

FIGURE 2

Probability plots for normal distribution of the batch aggregated efficiencies for UOs R22 and R54; dashed lines refer to the reference lines of the normal distribution.

Detailed Investigation of the Energy Use

In Figure 2, the cumulative steam consumption of the UO R22 within one batch is depicted. It can be seen that almost 40% of the overall steam required for the production of this batch is consumed during the synchronization step, when the temperature is maintained constant at 50°C. During this step the UO R22 is waiting for the end of the reaction step in the previous UO. Long duration of the synchronization step indicates the problems with the synchronization between 2 UOs.

Furthermore, analysis of this UO in Figure 3 shows the CO and temperature course during the same batch as depicted in Figure 3. It is obvious, that during the synchronization step heating of the reaction mass occurs followed by cooling. This indicates a problem with the control valve, which represents the main optimization potential concerning this particular UO.

The effect of overheating and overcooling was considered as a main source of the energy losses. The oscillation of the temperature around the setpoint can be significantly high during the synchronization step as shown in Figure 4. Therefore the batch-wise investigated relation between the steam power consumption and the controller performance (assessed by *Averaged Absolute Error (AAE)* of controller) is crucial for the analysis of the steam power consumption in the synchronization steps.

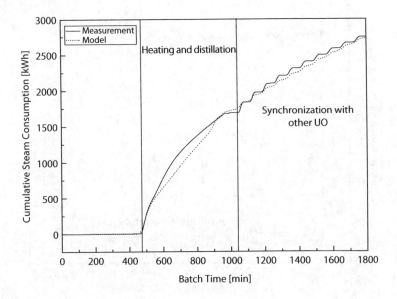

FIGURE 3
Cumulative steam consumption of UO R22 within one batch.

The analysis of the relation between the steam power consumption and the *AAE* for particular UO R22 is shown in Figure 5.

It can be clearly seen, that the power consumption in the synchronization step is strongly correlated with the *AAE*. The power steam consumption

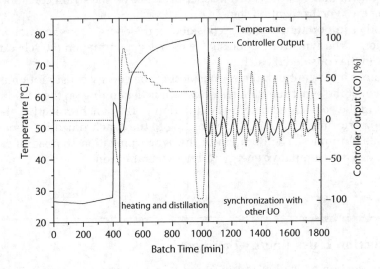

FIGURE 4
Detailed analysis of the steam consumption; oscillation of the temperature around the setpoint 50°C indicates a problem with the controller; positive and negative values of the controller output indicate heating and cooling respectively.

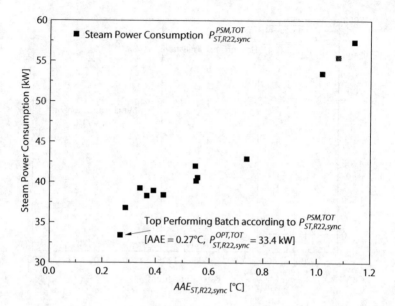

FIGURE 5
Influence of the average absolute error on the steam consumption during the synchronization step in UO R22.

varies between the different batches by a factor of 1.75. This behavior confirms the influence of the controller performance on the steam consumption during the synchronization step.

Another important parameter considering the energy consumption is the duration of the synchronization step, which is depending on the scheduling issues in the considered batch.

Another important parameter considering the energy consumption is the duration of the synchronization step, which is depending on the scheduling issues in the considered batch. The duration of the synchronization step of the top performing batch in R22 according to the synchronization time was below 300 minutes. The duration of the synchronization step significantly influences the cumulative energy utilities consumption

Production Lines Energy Savings

The production lines of products A and B include in their equipment pools UOs which were considered in the detailed investigation of the energy consumption. Products A and B were produced in two different equipment pools each one consisting of 6 main UOs.

Modeled batches are not considered to be starting or ending batches of the campaign, which are produced according to a special control recipe. Therefore, the control recipe within the modeled period is considered to be constant. However, variations between the batches arise. The variations in step durations between different UOs are compensated by synchronization steps, remedying time lag according to a planned schedule.

In order to evaluate the effect of the particular UO energy savings, the energy consumption of the synchronization steps was recalculated according to Eq. 2.

$$E^{OPT,TOT}_{ST,UO,sync} = P^{OPT,TOT}_{ST,UO,sync} \cdot t^{OPT}_{EU,UO,sync} \tag{2}$$

where $P^{OPT,TOT}_{ST,UO,sync}$ and $t^{OPT}_{EU,UO,sync}$ represent the optimal steam power consumption and optimal duration of the synchronization step, respectively.

Moreover, the electricity consumption can be achieved. In this case, only the duration of the step is considered as variable, since the power of the motor is considered to be independent from the process variables

Potential energy savings were estimated based on the optimal heating power consumption during the synchronization step and the minimal duration of this step considering several batches produced in these two investigated production lines. These energy savings were between 1 and 20% for the steam and 10 and 14% for the electricity of the average batch consumption respectively within the investigated time period. These energy savings demonstrate the suitability of the proposed approach for monitoring and optimization targets setting in the production line level.

Conclusions

The case study focused on the analysis of the energy saving potential was carried out based on the Process-side Model results and was performed in two stages. First, the plant-wise screening of the steam usage efficiency was carried out. This identified the most interesting UOs and issues from energy saving potential point of view, which have to be addressed in the second part, namely in the detailed investigation of steam and electricity use. This was carried out in two different production lines, where different equipment dependencies and scheduling issues were considered. Especially the batch-to-batch variability of the process parameters and the controller performance was identified as a main source of the fluctuating energy consumption between different batches. Namely, the synchronization and waiting steps were further analyzed and energy saving potential was revealed for the two modeled products. The batch variation of the energy consumption tends to be correlated with the batch cycle-time, especially for the products, where

UOs with including synchronization steps at temperatures well above the ambient temperature are involved.

Additional energy savings could be achieved by the pinch analysis and heat integration. Although these methods were also applied for the batch chemical processes, they have not been proved as appropriate for the batch as for the continuous processes. This was mainly due to the necessary installation of heat storage tanks that increase the costs of the batch process integration. Therefore, most batch processes cannot be energy integrated to the same extent that is possible for the continuous case without a very complex set-up.

Acknowledgments

The funding of this research project was covered by the Swiss Federal Office of Energy (Project No. 100536) and is gratefully acknowledged.

References

Bieler, P., Fischer, U., Hungerbuhler, K. (2004). Modeling the Energy Consumption of Chemical Batch Plants: Bottom-Up Approach. *Ind. Eng. Chem. Res.*, 43, pp 7785–7795.

Szíjjarto, A., Papadokonstantakis. S., Fischer, U., Hungerbuhler, K. (2008). Bottom-up Modeling of the Steam Consumption in Multipurpose Chemical Batch Plants Focusing on Identification of the Optimization Potential. *Ind. Eng. Chem. Res.*, 47 (19), pp 7323–7334.

43

Synthetic Production of Methanol Using Solar Power

Carlos A. Henao[1], Christos T. Maravelias[1*],
James E. Miller[2] and Richard A. Kemp[2]

[1]*Department of Chemical and Biological Engineering, University of Wisconsin Madison–WI 53706*
[2]*Advanced Materials Laboratory, Sandia National Laboratories, Albuquerque–NM 87106*

CONTENTS

ABSTRACT Energy security and global climate change are two intertwined problems that demand attention. The vision for the "hydrogen economy" is a proposed solution that is based on the application of sustainable energy sources to split water. However, many technical and infrastructure challenges remain for hydrogen that do not exist for hydrocarbon fuels. Integrating CO_2 capture and conversion into liquid fuels produces a new vision that promises the benefits of hydrogen while preserving many of the advantages of the hydrocarbon economy. In this paper, we study the production of methanol from H_2/CO_2 and H_2O/CO mixtures. We present two alternative processes which are based on the combined action of two reversible reactions: water gas shift (WGS) and methanol synthesis (MS) on a $Cu/ZnO/AlO_3$ catalyst. Detailed flowsheet simulations and economic evaluations under multiple scenarios indicate that both processes can be economically feasible in the near future, while having energy efficiencies which are significantly better

---

* To whom all correspondence should be addressed.

than their biological counterpart. Finally, the conversion and energy efficiency of both processes are better than previously proposed designs such as the so called CAMERE process.

KEYWORDS *Renewable energy, methanol production, process synthesis and evaluation*

Introduction

Energy resources are the foundation for developed economies and are inextricably linked to national security, social stability, and quality of life. Hence, global demand and competition for petroleum as a transportation fuel is projected to continue to climb even as supplies of conventional oil decline. Less-conventional resources such as coal, oil-shale and tar-sands can be converted to liquid fuels and help fill the gap. However, tapping into and converting these resources into liquid fuels exacerbates green house gas emissions as they are carbon rich, but hydrogen deficient. Revolutionary thinking is required if the coupled problems of energy (transportation) security and climate change are to be addressed. Hydrocarbon fuels are ideal energy carriers, but they can no longer be thought of as primary energy sources. Rather, it is necessary that we take the realistic view that our conventional hydrocarbon fuels are in fact "stored sunlight" and "sequestered carbon." That is, petroleum, coal and other fossil fuels are the end result of a long process that began with a biological organism capturing sunlight and using it to drive chemical conversions of CO_2 and H_2O to carbohydrates and oxygen (photosynthesis).

Biofuels, e.g. bio-ethanol, can be thought of as a modern approach to improving upon the overall (sunlight to fuel) efficiency of this process and shortening the time scale. As before, the starting point is the photosynthetic conversion of CO_2 and H_2O to organic matter. Additional chemical or biological steps are then undertaken to produce a hydrocarbon fuel. The overall sunlight to fuel efficiency is dependent on location and the process specifics and is thus difficult to define precisely or to generalize. However, it is still generally quite low, although significantly better than that for oil. As an example, it is commonly accepted that the solar to ethanol efficiency from corn kernels is less than 1% (Smith et al.). One can put an upper limit on the biomass approach by considering the efficiency of the photosynthetic step alone. Photosynthesis is generally measured to be 2.5% efficient at best (Dukes, 2003). The maximum possible efficiency is estimated to be 4.6% for C3 photosynthesis and 6% for C4 photosynthesis, under current atmospheric conditions (Zhu et al., 2008).

Given the limits on overall sunlight to hydrocarbon efficiency imposed by photosynthesis, it is reasonable to consider other, more direct, chemical approaches for "re-energizing" CO_2 and H_2O and ultimately converting them to transportation fuels. Solar-driven thermochemical processes have the potential to split CO_2 and H_2O to yield CO and H_2 at high solar efficiencies (Diver et al., 2008; Miller et al., 2008). In this paper we consider two process alternatives for converting the products of this and similar processes to methanol, as starting point for determining the viability of this approach. We consider methanol because it can converted to liquid fuels and chemicals, and used in direct methanol fuels cells.

Process Alternatives

The first process converts H_2 and CO_2 into methanol (70,000 MT_MeOH/yr), while the second one converts CO and H_2O into methanol (85,000 MT_MeOH/yr). All reactors in this study are multi-tube packed with a commercial Cu/ZnO/Al_2O_3 catalyst.

Process Alternative # 1: Methanol from H_2 and CO_2

The proposed process includes two reaction systems and one separation system. In the first reaction system H_2 and CO_2 are partially converted according to the reaction:

$$H_2 + CO_2 \leftrightarrow H_2O + CO \quad \text{Reverse WGS (RWGS)}$$

After removing most of the produced water, the resulting H_2/CO_2/CO mixture is then fed to a second reaction system where two reactions take place:

$$CO_2 + 3H_2 \leftrightarrow CH_3OH + H_2O \quad \text{Methanol Synthesis (MS)}$$
$$CO + H_2O \leftrightarrow CO_2 + H_2 \quad \text{Water Gas Shift (WGS)}$$

The purpose of the first reaction system is to produce enough CO to eliminate, via WGS, the water produced by MS. This is beneficial because water has been proven to block active sites in the MS catalyst. The separation system consists of a distillation column to separate the heavy products methanol and water.

The flowsheet of this process is shown in Figure 1. It is an improvement over the CAMERE processes (Joo et al. 2004). The main difference between this alternative and the original CAMERE process is the presence of the recycle loop from the exit of the second reactor to the entrance of the first. The optimization of the reactor conditions and major recycle streams leads to a significant improvement of the overall H_2-to-methanol yield, from 53% in the original CAMERE process to 87%.

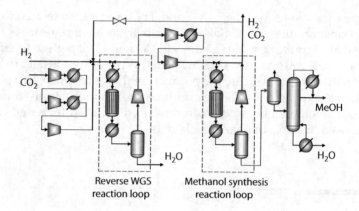

FIGURE 1
MeOH production from H_2 and CO_2.

Process Alternative # 2: Methanol from H_2O and CO

In this case, H_2O and CO coming from the thermo-chemical splitting of CO_2 are converted to methanol in the process shown in Figure 2. As in the previous case, this alternative includes two reaction systems. In the first one, H_2O and CO are partially converted according to the reaction:

$$H_2O + CO \leftrightarrow H_2 + CO_2 \quad \text{Water Gas Shift (WGS)}$$

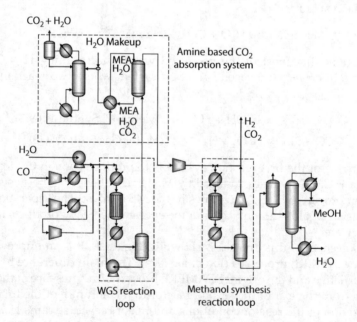

FIGURE 2
MeOH production from H_2O and CO_2.

The resulting $H_2/CO/CO_2$ mixture is then passed to an absorption system which selectively removes part of the CO_2, adjusting the CO/CO_2 ratio before feeding it to the second system where methanol is produced according to:

$$CO_2 + 3H_2 \leftrightarrow CH_3OH + H_2O \quad \text{Methanol Synthesis (MS)}$$
$$CO + H_2O \leftrightarrow CO_2 + H_2 \quad \text{Water Gas Shift (WGS)}$$

The purpose of the first reactor is to produce enough H_2 and CO_2 to drive the production of methanol in the second reactor, while leaving enough CO to eliminate the poisoning from the produced water.

There are some fundamental differences between the two alternatives. First, the CO_2 used in the methanol synthesis is not fed to the process, but produced by partial transformation of the CO feed. Second, the control of the CO/CO_2 ratio is achieved by an additional separation system. Note that in the integrated process this separation system will be used for the recycling of CO_2 back to the thermochemical splitting reactor.

Conversion and Energy Efficiency

The two processes were modeled in ASPEN PLUS®. Some of the most important characteristics of the two alternatives are presented in Table 1. The H_2, CO to MeOH yields are based on the stoichiometry of the combined reactions discussed earlier (i.e. $CO_2 + 3H_2 \rightarrow MeOH + H_2O$ for the first process, and $3CO + 2H_2O \rightarrow MeOH + 2CO_2$ for the second process). In this study, the energy efficiency is defined as the ratio of the energy released by burning the produced MeOH, and the total energy that enters the process as chemical energy in the main raw material (reenergized H_2O in the form of H_2 or reenergized CO_2 in the form of CO) plus the energy supplied by the utility system.

TABLE 1
Major Process Characteristics

Process Alternative	1	2
Material flows [kmol/h]:		
H_2	900	—
CO	—	1050
MeOH	262	323
Capacity (MT_MeOH/yr)	70,000	85,000
H_2 to MeOH yield	87%	—
CO to MeOH yield	—	92%
Reactor #1 T-P [°C]-[bar]	310–15	270–22
Reactor #2 T-P [°C]-[bar]	240–46	205–46
Reactor catalysts	Cu/ZnO/Al$_2$O$_3$	Cu/ZnO/Al$_2$O$_3$
Energy efficiency	49%	62%

TABLE 2

Economic Evaluation Parameters.

Project's Economic Life [yr]	30
Working Capital/Capital Expense	5%
Operating Charged /Operating Labor	15%
Plant Overhead/Operating Labor	40%
Desired Rate of Return[%/yr]	8%
Tax Rate[%/yr]	40%
Salvage Value/Capital Cost	20%
Depreciation	Straight line
Capital Escalation [%/yr]	5%
Raw Material Escalation [%/yr]	1.5%
Product Escalation [%/yr]	5%
Utility Escalation [%/yr]	3%

Economic Evaluation

Detailed economic evaluation of both process alternatives was conducted using ICARUS PROCESS EVALUATOR ® based on the detailed process simulation models, and the evaluation parameters given in Table 2.

The main objective of this analysis was to determine the break-even prices of H_2 and CO that would allow the processes to be economically feasible. In order to accomplish this, prices of CO_2, H_2O and MeOH according to recent technology analysis and market trends were used. Detailed Net Present Value (NPV) sensitivity analysis studies were also performed for both projects. Material prices were selected as follows:

- CO_2 price: Several studies identify amine absorption as one of the most economic systems for CO_2 sequestration (Mignard and Pritchard 2003). Using this technology the price of CO_2 coming from a sequestration unit in a power station is around 35 USD/MT.

- H_2O price: The price considered here was the standard value of de-ionized water: 1 USD/MT.

- Methanol price: Methanol price has a highly fluctuating behavior. For this particular study, the most recent value 330 USD/MT was considered (www.methanex.com).

The capital expenditures for both projects were around 17.5 million USD. The NPV sensitivity analysis studies indicate that the maximum raw material prices are 1.12 USD/kgH$_2$ for process #1, and 0.17 USD/kgCO for process #2 (see Figures 3 and 4). However, even if the prices are as high as 2.2 USD/kgH$_2$ and 0.275 USD/kgCO, respectively, both processes can still be economically

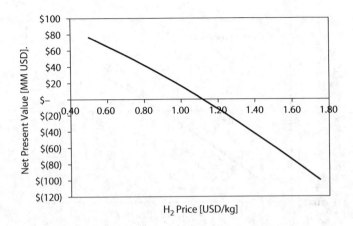

FIGURE 3
Effect of H_2 price on process #1 NPV.

viable if the price of methanol is around 550 USD/MT (see Figure 5), which was the price of methanol as per Sept 2008. Note that the price of methanol is expected to increase in the future reaching at least 500 USD/MT.

Furthermore, in the current study we assumed that we pay 35 USD/MT CO_2 for sequestration. However, if emission regulations or emission trading schemes are introduced, CO_2 consumers will get credits, which means that the NPV of process #1 will increase.

Finally, we carried sensitivity analysis studies that show that an improvement of less than 5% in the overall yield of H_2 and CO in processes #1 and #2, respectively, have a substantial impact on the profitability of the two processes. Therefore, we believe that research efforts in the area of new catalyst

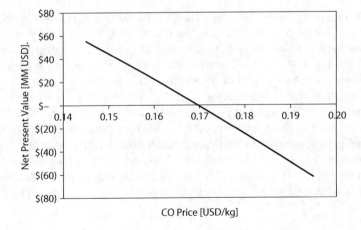

FIGURE 4
Effect of CO price on process#2 NPV.

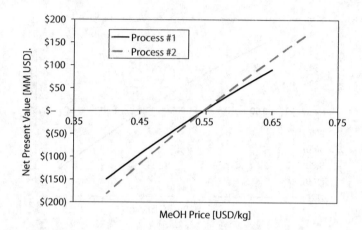

FIGURE 5
Effect of MeOH price on NPV, assuming H_2 price = 2.5 USD/kg and CO price = 0.275 USD/kg.

development and process optimization will result in significant improvements. We are currently working towards these two goals, as well as the development of an integrated process that includes the thermochemical splitting of CO_2 and the conversion of CO/H_2O to methanol.

Conclusions

We presented two process designs for the production of methanol from H_2/CO_2 and H_2O/CO. The two alternatives can be integrated with thermochemical processes for the splitting of H_2O and/or CO_2, leading to technologies that can change the way we view renewable energy. The integrated processes satisfy the twofold objective of fomenting the use of renewable energy (in this case concentrated solar power) while reducing CO_2 emissions trough recycling. Simulations of the proposed alternatives feature methanol yields significantly better than those reported in the literature. Based upon current methanol prices, sensitivity analysis indicates economic feasibility, if prices do not exceed 1.12 USD/kgH_2 (7.88 USD/GJ) and 0.17 USD/kgCO. However, even if the cost is twice as high, the processes can be economically attractive if the price of methanol increases moderately.

Acknowledgments

The authors would like to acknowledge financial support provided by the Laboratory Directed Research and Development program (LDRD grant 113486) at Sandia National Laboratories. Sandia is a multiprogram laboratory operated by Sandia Corporation, a Lockheed Martin Company, for the United States Department of Energy under Contract No. DE-AC04-94AL85000.

References

Diver, R.B., Miller, J.E., Allendorf, M.D., Siegel, N.P. Hogan, R.E. (2008). Solar Thermochemical Water-splitting Ferrite-cycle Heat Engines. *Journal of Solar Energy Engineering*, 130 (4), 041001.

Dukes, J. S. (2003). Burning Buried Sunshine: Human Consumption of Ancient Solar Energy. *Climatic Change*, 61, 31.

Galindo Cifre, P., Badr, O. (2007). Renewable Hydrogen Utilization for the Production of Methanol. *Energy Conversion and Management*, 48, 519.

Joo, O. S., Jung, K. D., Jung, Y. S. (2004) CAMERE Process for Methanol Synthesis from CO_2 Hydrogenation. *Carbon Dioxide Utilization for Global Sustainability*, 153, 67.

Mignard, D., Sahibzada, M., Duthie, J. M., Whittington, H. W. (2003). Methanol Synthesis From Flue-Gas CO_2 and Renewable Electricity: a Feasibility Study. *Int J Hydrogen Energy*, 28, 455.

Miller, J.E., Allendorf, M.D., Diver, R.B., Evans, L.R., Siegel, N.P., Stuecker, J.N. (2008). Metal Oxide Composites and Structures for Ultra-High Temperature Solar Thermochemical Cycles. *Journal of Materials Science*, 43, 4714 **DOI:** 10.1007/s10853-007-2354-7.

Smith, O. H., Friedman, R., Venter, J. C. Biological solutions to renewable energy. National Academy of Engineering website:http://www.nae.edu/nae/bridge-com.nsf/BridgePrintView/MKUF-5NTMX9?OpenDocument

Zhu, X-G., Long, S.P., Ort, D. R. (2008). What is the Maximum Efficiency with Which Photosynthesis Can Convert Solar Energy into Biomass? *Biotechnology*, 19, 153.

44

Multi-Objective Optimization Analysis of the IGCC System with Chemical Looping Combustion

Yogendra Shastri, Juan Salazar and Urmila Diwekar*
Vishwamitra Research Institute
Center for Uncertain Systems: Tools for Optimization and Management
Clarendon Hills, IL 60514

CONTENTS

ABSTRACT Integrated Gasification Combined Cycle (IGCC) system using coal gasification is an important approach for future energy options. This work focuses on understading the system operation and optimizing it in the presence of uncertain operating conditions using ASPEN Plus and CAPE-OPEN compliant stochastic simulation and multiobjective optimization capabilities developed by Vishwamitra Research Institute. The feasible operating surface for the IGCC system is generated and deterministic multiobjective optimization is performed. Since the feasible operating space is highly non-convex, heuristics based techniques that do not require gradient information are used to generate the Pareto surface. Accurate CFD models are simultaneously developed for the gasifier and chemical looping combustion system to characterize and quantify the process uncertainty in the ASPEN model.

KEYWORDS IGCC, gasifier, chemical looping combustion, uncertainty, multi-objective optimization

---

* To whom all correspondence should be addressed

479

Introduction

The U.S. Department of Energy (DOE) is investing heavily in Fossil Energy R&D programs to promote the development of advanced power generation systems that meet the Nation's energy needs while achieving a sustainable balance between economic, environmental, and social performance. Integrated Gasification Combined Cycle (IGCC) technology is becoming increasingly important in this effort, where low-cost opportunity feedstock such as coal, heavy oils and pet coke are the fuels of choice. IGCC technology produces low-cost electricity while meeting strict environmental regulations. Efficient gasification process and combined with excellent post-gasification processing such as chemical looping combustion makes this an exciting option.

To achieve performance targets and at the same time reduce the number of costly pilot-scale and demonstration facilities, the designers of these systems increasingly rely on high-fidelity process simulations to design and evaluate virtual plants. Developed by the DOE's National Energy Technology Laboratory (NETL), the Advanced Process Engineering Co-Simulator (APECS) is a virtual plant simulator that combines process simulation, equipment simulations, immersive and interactive plant walk-through virtual engineering, and advanced analysis capabilities (Zitney et al. 2006, Zintey 2006a). The APECS system uses commercial process simulation software (e.g., Aspen Plus®) and equipment modeling software (e.g., FLUENT® computational fluid dynamics) integrated with the process-industry CAPE-OPEN (CO) software standard (Braunschweig 2002, Zitney 2006b). Plug-and-play interoperability of analysis tools in APECS is also facilitated by the use of the CO standard.

This work presents the application of the CO-compliant stochastic modeling and multi-objective optimization framework for APECS for the analysis of an IGCC system with single-stage gasification and chemical looping combustion. This framework enables optimizing model complexities in the face of uncertainty and multiple and sometimes conflicting objectives of design. It also provides a decision support tool to address some of the key questions facing designers and planners of advanced process engineering systems

The paper is arranged as follows. The next section gives a brief review of the IGCC system analyzed in this work. The theory behind the multi-objective analysis and the analysis results are presented in the subsequent sections. The last section draws the important conclusions.

IGCC System

The system is based on the General Electric (GE) Energy gasifier (single stage entrained-flow), two Advanced F-Class gas turbines partially integrated with an elevated pressure Air Separation Unit (ASU). Syngas desulfurization is provided by a Selexol Acid Gas Removal (AGR) system and a two-bed Claus Unit with Tail Gas Recycle to Selexol, and a chemical looping combustion system. In chemical looping combustion, metal oxide particles (oxygen carrier) are used for the transfer of oxygen from the combustion air to the fuel, thus the combustion products CO_2 and H_2O (or pure H_2) are obtained in a separate stream (Lyngfelt et al. 2001). The flowsheet for chemical looping combustion combined cycle system is separated into sections like coal gasification, water gas shift reaction, sorbent energy transfer system and power generation. The coal gasification the process is modeled with two reactors, the first reactor decomposes the stream of coal into its elemental constituents (C, H_2, N_2, O_2, S, H_2O and Ash as a yield reactor) and the second reactor generates the fuel comprising CH_4, H_2, CO, NH_4, H_2S and CO_2 (plus unreacted N_2, O_2, H_2 and H_2O) by minimizing the total Gibbs free energy of the system (Gibbs reactor). The products of the gasification process are fed to the water gas shift (WGS) reactor section to complete the oxidation of CO with the simultaneous production of hydrogen. The reaction takes place in two adiabatic reactors in series simulated as Gibbs reactors with intermediate heat exchangers. Sulfur and Ammonia removal unit is located downstream the WGS section. H_2S is removed by an absorber using methanol or glycol. The removal of Ammonia is carried out with a reactive absorption process on sulfuric acid. Both separations are modeled in one single unit as a simple component separator. This section of the process is finished with a pressure swing adsorption (PSA) unit to purify the hydrogen that is fed to the gas turbine. The pure hydrogen stream and the fuel gas stream from the PSA are fed to the sorbent energy transfer system (SETS) which uses a chemical looping principle to purify CO_2 and to recover the energy from the oxidation processes. The reduction reactor, the oxidation reactor and the combustion chamber are modeled as adiabatic Gibbs reactors. The metal carrier used in this case is 25% NiO in Al_2O_3. Please refer to DOE/NETL (2006) and Maurstad (2005) for further details related to the process. Aspen Plus model for the IGCC system has been developed by the Department of Energy in order to conduct system level analysis of the process. The Aspen Plus modeling details (including the modeling approximations and configuration) is explained in DOE/NETL (2006) and not discussed here for the sake of brevity.

Multi-objective Analysis: Theory

Multi-objective problems appear in virtually every field and in a wide variety of contexts (Diwekar 2003). Conventional process models (such as the IGCC Aspen Plus model) now in use are largely based on a deterministic computational framework used for simulation of a specified flowsheet. An important shortcoming of these models is their inability to analyze uncertainties rigorously. Uncertainty analysis is especially important in the context of advanced energy systems, since available performance data typically are scant, accurate predictive models do not exist, and many technical as well as economic parameters are not well established. The work, therefore, goes beyond a deterministic analysis and studies multi-objective optimization in the presence of uncertainty. A generalized multi-objective optimization problem is of the form:

Minimize :

$$f_i(\vec{x}), i = 1, \ldots, k, k \geq 2$$

Subject to :

$$h_I(\vec{x}) = 0, I \geq 0$$

$$g_J(\vec{x}) \leq 0, J \geq 0 \tag{1}$$

$$l_j \leq x_j \leq u_j, \quad j = 1, \ldots, n,$$

$$\vec{x} = (x_1, \ldots, x_n)$$

The problem under consideration involves a set of n decision variables represented by the vector $\vec{x} = (x_1, x_2, \ldots, x_n)$. The equality constraints $h_I(\vec{x})$, $I \geq 0$: $\mathbf{R}^n \to \mathbf{R}$, and inequality constraints $g_J(\vec{x}) \leq 0$, $J \geq 0$: $\mathbf{R}^n \to \mathbf{R}$ are real-valued (possibly nonlinear) constraint functions, and l_j and u_j are given lower and upper bounds of decision variable x_j (allowed to be $-\infty$ and/or $+\infty$). If $I = 0$ and $J = 0$, the problem becomes unconstrained. The problem involves k (≥ 2) continuously differentiable nonlinear objective functions $f_k : \mathbf{R}^n \to \mathbf{R}$. Without loss of generality, we assume that all the objective functions are to be minimized simultaneously.

The solution of a multi-objective optimization problem is a set of solution alternatives called the Pareto set. For each of these solution alternatives, it is impossible to improve one objective without sacrificing the value of another relative to some other solution alternatives in the set. There are usually many (infinite in number) Pareto optimal solutions. The collection of these is called the Pareto set. The result of the application of a nonlinear multi-objective technique to a decision problem is the Pareto set for the problem, and it is from this subset of potential solutions that the final, preferred decision is chosen by the decision-makers.

The most commonly used analytical techniques for multi-objective optimization problems (and to generate the Pareto surface) are: preference-based methods and generating methods (Diwekar 2003). Preference based methods require a-priori knowledge of the weights on different objectives and solves a single objective problem. Generating methods provide a great deal of information, emphasizing the Pareto optimal set or the range of choice available to decision-makers, and providing the trade-off information of one objective versus another. Generating techniques can be further divided into two sub-classes: no-preference methods and a posteriori methods. The selection of the appropriate method often depends on the optimization problem formulation.

It has been mentioned that this work performs multi-objective optimization in the presence of uncertainty. Uncertainty analysis consists of four main steps: (1) characterization and quantification of uncertainty in terms of probability distributions, (2) sampling from these distributions, (3) propagation through the modeling framework, (4) analysis of results (Diwekar et al. 1997). Once the uncertain parameters of a given model are identified in terms of their probability distributions, an efficient sampling techniques, such as the Hammersley Sequence Sampling (Kalagnanam and Diwekar 1997), is used to sample the uncertain space. The next important step is the propogation of uncertainty through the model. This work uses the Cape-Open compliance stochastic simulation capability to perform uncertainty analysis of the IGCC system and compute the feasible solution space.

IGCC System: Multi-objective Optimization Results

In this work, the non-dominated (Pareto) surface for the IGCC system is computed for following three important performance measures of the system, constituting the objectives for the system: Total plant efficiency (based on HHV of coal); Total CO_2 emissions measures in Ib/hr; and Total SOx emission measured as volumetric fraction of the total flue gas volumetric flow rate. For multi-objective optimization, model parameters that have a significant impact on the system performance are first identified using the stochastic modeling PRCC (Partial Rank Correlation Coefficient) analysis for 11 different model parameters (Diwekar and Rubin, 1991). PRCC provides a major or unique or unshared contribution of each variable, and explains the unique relationship between two variables that cannot be explained in terms of the relations of these variables with any other variable. PRCC analysis identifies following critical parameters for the IGCC system: Gasifier operating temperature; Gasifier operating pressure; and Claus burner temperature. These parameters are used to analyze results of multi-objective optimization (used as decision variables).

The first step in the multi-objective analysis is to identify the feasible solution surface for the IGCC system for variations of the given decision variables. The feasible surface is generated through stochastic simulation by using 1000 samples of eleven different model parameters and propagating the uncertainty through the Cape-Open compliance stochastic simulation capability. The feasible surface is plotted as the values of the three objectives identified before and is shown in Figure 1. Here, the contours of one of the objectives

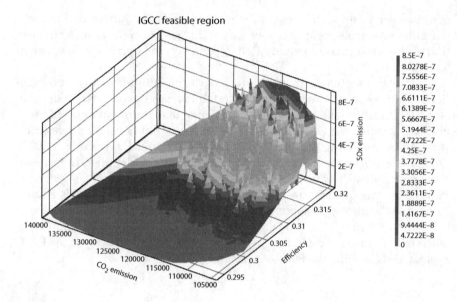

FIGURE 1
Feasible surface for IGCC system.

(SOx) are plotted with respect to the values of the other two objectives for the 1000 samples. Please note that CO_2 emissions are measures in Ib/hr while the SOx emissions are measures in volumetric fraction of the total exit gas volume as mentioned before. From the figure it is clearly evident that the trade-off surface is highly non-convex. One can observe multiple peaks and valleys in terms of one of the objectives when the other two objectives are varied.

Although constraint based methods or preference based methods give a good estimate of the Pareto surface, they require the use of gradient based techniques to determine the surface. However, since the feasible surface for the IGCC system is non-convex, it makes the use of the previous mentioned techniques very difficult. The 'No preference based methods' can successfully tackle such problems and hence are used in this case. The 'No preference based methods' include compromise programming (Yu 1985), Multi-objective Proximal Bundle (MPB) (Miettinen 1999), and feasibility-based methods, such as the parameter space investigation (PSI) methods (Osyczka 1984). They focus on generating a feasible solution or all the feasible solutions instead of the Pareto set (the best feasible solutions). In PSI methods the continuous decision space is first uniformly discretized using the Monte Carlo sampling technique; next a solution is checked with the constraints. If one of the constraints is not satisfied, the solution is eliminated and the objective values are finally calculated, but only for those feasible solutions. Therefore, a *discretized approximation* of the feasible objective region, instead

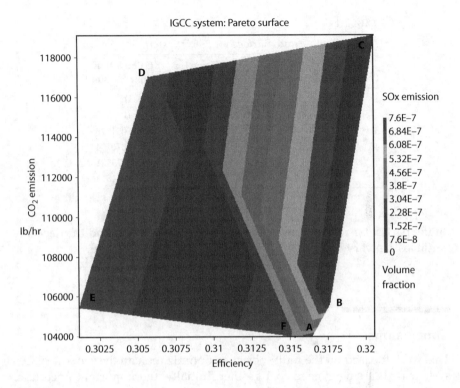

FIGURE 2
IGCC Pareto surface.

of the Pareto set, is retained by the PSI method. The solutions of this feasibility-based method cover the whole feasible objective region rather than covering only the optimal solutions in the Pareto set. Because most of the feasible solutions are not Pareto optimal, a relatively small number of the non-dominated (relatively better, but not necessarily Pareto optimal) solutions must be extracted from the whole feasible solution set to formulate an approximate representation of the Pareto set for feasibility-based methods. This justifies the use of PSI method to generate the approximate Pareto surface for the IGCC system. The feasible space shown in Figure 1 constitutes the output of the stochastic simulation which must be used for approximate identification of the Pareto surface.

Figure 2 shows the approximate Pareto surface in a 2-dimensional space where the Pareto surface is constituted by points A-B-C-D-E. The values of different objectives for the Pareto surface are reported in Table 1. The values of decision variables corresponding to this Pareto surface are shown in Table 2. The results illustrate that the optimal operating point changes based

TABLE 1

Approximate Pareto Surface for IGCC System

Point	Efficiency	CO_2 Emission (Ib/hr)	SOx Emission (Volume Fraction)
1.	0.30111	105430	0
2.	0.30578	117000	0
3.	0.31533	103950	0.00000031028
4.	0.32057	119110	0.00000076794
5.	0.31757	105580	0.00000073921
6.	0.31641	103960	0.00000024813

on the particular realizations of the uncertain parameters, and the decision variables can vary significantly for the optimal operating point.

Conclusions

This work focused on the multi-objective optimization analysis of the IGCC system with a single stage coal gasifier. Initially, the important uncertain model parameters and the critical system objectives (performance indicators) are identified. The stochastic simulation of this system shows that the feasible solution is highly non-convex. Therefore, the PSI method is used to determine the approximate Pareto surface. The analysis of the results shows that uncertainty has a significant effect on the Pareto surface. It is also observed that there is a trade-off between the different objective functions of the system. Thus, Low SO_x and CO_2 emission often lead to low efficiency of the IGCC system.

TABLE 2

Decision Variable Values for the Pareto Surface

Point	Gasifier Operating Temperature (F)	Gasifier Operating Pressure (psia)	Claus Burner Temperature (F)
1.	2295.7043	741.1281	2245.6123
2.	2376.7432	832.2906	2166.2783
3.	2589.1709	759.775	2307.2532
4.	2639.041	773.8	2590.3447
5.	2636.6433	886	2507.5862
6.	2615.0649	786.55	2242.1877

Acknowledgments

This work has been funded by the National Energy Technology Laboratory (NETL) under contract RDS SUBTASK 41817.312.01.06.

References

Braunschweig, B. L., & Gani, R. (2002). *Software Architectures and Tools for Computer Aided Process Engineering*. 1st ed., Elsevier Science, Amsterdam (2002).

Diwekar, U.M. & Rubin, E.S. (1991). Stochastic modeling of chemical processes. *Computers & Chemical Engineering*, 15(2), 105–114.

Diwekar, U. M., Rubin, E. S., & Frey, H. C. (1997). Optimal Design of Advanced Power Systems Under Uncertainty. *Energy Conversion and Management Journal*, 38, 1725–1735.

DOE/NETL (2007). Model documentation: IGCC with GE Energy gasifier and CO_2 capture. DOE/NETL-401/042606, February.

Kalagnanam J. R. & Diwekar, U. M. (1997). An efficient sampling technique for off-line quality control. *Technometrics*, 39(3), 308.

Lyngfelt A. , B. Leckner, and T. Mattison (2001). A Fluidized-bed combustion process with inherent CO_2 separation; application of chemical-looping combustion, *Chemical Engineering Science*, 56, 3101–3113.

Maurstad, O. (2005). An overview of coal based Integrated Gasification Combined Cycle (IGCC) technology. Laboratory for Energy and Environment, Massachusetts Institute of Technology, MIT LFEE 2005-002 WP.

Miettinen, K.M. (1999). Nonlinear multiobjective optimization. Kluwer Academic Publishers, Norwell, Massachusetts, MA.

Osyczka, A. (1984). Multicriteria optimization in engineering with FORTRAN programs. Ellis Horwood Limited.

Yu, P.L. (1985). Multiple-criteria decision making concepts, techniques, and extensions. Plenum Press, New York, NY

Zitney, S. E. (2006a). High-Fidelity Process Co-Simulation of Advanced Power Generation Systems. *Clean Coal Today*, Office of Fossil Energy, U.S. Department of Energy, 68, 6–7, Fall.

Zitney, S. E. (2006b). CAPE-OPEN Integration for CFD and Process Co-Simulation. *Proc. of the AIChE 2006 Annual Meeting*, 3rd Annual U.S. CAPE-OPEN Meeting, November 12–17, San Francisco, CA.

Zitney, S. E., Osawe, M. O., Collins, L., Ferguson, E., Sloan, D. G., Fiveland, W. A. & Madsen, J. M. (2006). Advanced Process Co-Simulation of the FutureGen Power Plant. *Proc. of the 31st International Technical Conference on Coal Utilization & Fuel Systems*, May 21–25, Clearwater, FL.

45

Design of Compact Heat Exchangers Using Parameter Plots

M. Picón-Núñez, G. T. Polley and D. Jantes-Jaramillo

Department of Chemical Engineering, University of Guanajuato
Guanajuato, Mexico

CONTENTS

ABSTRACT This paper presents a graphical approach for the selection of compact heat exchangers of the spiral type. The graphical representation used in this work is derived from the concept of parameter plot that describes the design space where a set of exchanger geometries meet the main process specifications, namely heat duty, cold side allowable pressure drop and hot side allowable pressure drop. A set of design options are provided to the designer and they open the way for reconciling exchanger specification stages with actual exchanger design.

KEYWORDS *Compact heat exchangers, spiral heat exchangers, parameter plot*

Introduction

The design of a spiral heat exchanger refers to the definition of the geometry that will transfer the required heat duty within the limitations of the allowable pressure drop. This geometry includes: plate spacing for the two streams, plate width, inner diameter, outer diameter and passage length. In this work, a simple methodology (Picon et. al., 2007) for the design of spiral plate exchangers is used for the production of a pictorial representation of design options available to the designer.

The most important geometrical features of a spiral heat exchanger are: plate spacing for the hot and cold streams (b_h and b_c), plate width (H), plate

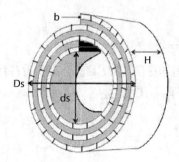

FIGURE 1
Geometrical features of a spiral heat exchanger.

length (L), spiral internal diameter (d_s) and spiral external diameter (D_s) as shown in Figure 1.

Standard geometry for the construction of spiral exchangers is characterized by discrete values. Therefore, a graphical design tool must take into consideration such features. A parameter plot is a pictorial representation of the design space where a set of exchanger geometries meet the main process specifications, such as heat duty, cold side allowable pressure drop and hot side allowable pressure drop (Poddar and Polley, 2000). For this concept to be applied to the case of spiral heat exchangers, a bar chart is used to show all available design options at a glance. The assumptions made in the development of the methodology are: the effects of the entrance regions are not considered; the heat transfer coefficient is constant along the length of the exchanger; losses to ambient are negligible; the bolts that separate the plates are considered not to affect the fluid movement

Design Methodology

The design of a spiral heat exchanger starts with the specification of three variables: plate spacing for the hot and cold (b_h and b_c) and the plate width (H). In the spiral geometry, the number of spiral turns Np is 0.5 more than that of the stream N, i.e Np is equal to $N + 0.5$ (Dongwu, 2005). With these parameters a first approximation to the free flow area (Ac) and the Reynolds number is possible.

$$A_c = bH \tag{1}$$

$$Re = \frac{d_h m}{\mu A_c} \tag{2}$$

where d_h is the hydraulic diameter and is given by:

$$d_h = \frac{2bH}{b+H} \tag{3}$$

For a Reynolds number in the range of 400 to 30,000, Holger (1992) presents the following equation for heat transfer:

$$Nu = 0.04\, Re^{0.74}\, Pr^{0.4} \tag{4}$$

where the Prandtl number is given by:

$$Pr = \frac{c_p \mu}{k} \tag{5}$$

The pressure drop across the core of the exchanger can be calculated from:

$$\Delta P = \frac{2\, f L m^2}{\rho\, d_h A_c^2} \tag{6}$$

where f is the friction factor, L is the length of the stream, m is the mass flow rate and ρ is the density. The friction factor for laminar, transitional and turbulent flow can respectively be estimated using the following expressions (Hesselgreaves, 2001):

$$f\, Re = 24\left(1 - 1.3553\frac{b}{H} + 1.9467\left(\frac{b}{H}\right)^2 - 1.7012\left(\frac{b}{H}\right)^3 \right.$$

$$\left. + 0.9564\left(\frac{b}{H}\right)^4 - 0.2537\left(\frac{b}{H}\right)^5\right) \tag{7}$$

$$f = 0.0054 + \frac{2.3 \times 10^8}{Re^{-\frac{3}{2}}} \tag{8}$$

$$\frac{1}{\sqrt{f}} = 1.56\ \ln(Re) - 3.00 \tag{9}$$

The exchanger outer diameter (D_o) and the number of spiral plate turns N_p is computed according to Dongwu (2005) as follows:

$$D_o = r_1 + r_2 \tag{10}$$

$$N_p = N + 0.5 \tag{11}$$

$$N = n + n_o \tag{12}$$

$$n + n_{odi} = \frac{-(d_1 - t/2) + \sqrt{(d_1 - t/2)^2 + 4tL/\pi}}{2t} \tag{13}$$

When $0 \le n_{odi}$ (or n_o) ≤ 0.5

$$L = \pi t n^2 + \pi(d_1 - t/2)n + \pi(d_1 + 2nt)n_o \tag{14}$$

$$R = \frac{d_1}{2} + nt \tag{15}$$

$$r = \sqrt{R^2 + \left(\frac{t}{4}\right)^2 - \frac{Rt}{2}\cos(2n_o\pi)} \tag{16}$$

When $0.5 \le n_{odi}$ (or n_o) ≤ 1

$$L = \pi t n^2 + \pi(d_1 - t/2)n + \pi[d_1 + (2n+1)t]n_o\pi t/2 \tag{17}$$

$$R = \frac{d_1}{2} + \left(n + \frac{1}{2}\right)t \tag{18}$$

Where the terms: t, d_{11} and df_1 are computed according to:

$$t = b_1 + b_2 + 2\delta \tag{19}$$

$$d_{11} = d_{21} + b_1 - b_2 \tag{20}$$

$$d_{f1} = d_{21} + b_1 + 2\delta \tag{22}$$

Where δ is the plate thickness. When L and R are known, d_1 is computed according to the following expressions:

$$L: \quad d_1 = d_{f1} \tag{23}$$

$$L_1: d_1 = d_{11} + \delta \tag{24}$$

$$L_2: \quad d_1 = d_{21} + \delta \tag{25}$$

$$R_1: \quad d_1 = d_{11} + 2\delta \tag{26}$$

$$R_2: \quad d_1 = d_{21} + 2\delta \tag{27}$$

Parameter Plots in the form of bar charts are used to determine the design space for the case study given in Table 1. Figure 2 shows a flow diagram of the design methodology. The design space shows the plate length that meets

TABLE 1

Stream Data for Case Study

	Hot Stream	Cold Stream
Flow rate (kg/s)	0.7833	0.7444
Inlet temperature (°C)	200	60
Outlet temperature (°C)	120	150.4
Heat capacity (J/kg°C)	2,973	2,763
Thermal conductivity (W/m°C)	0.348	0.322
Density (kg/m³)	843	843
Pressure drop (Pa)	6.89×10^{-2}	6.89×10^{-2}
Viscosity (kg/m s)	3.35×10^{-3}	8.0×10^{-3}
Plate thickness (m)	3.175×10^{-3}	
Internal diameter (m)	0.203	
Thermal conductivity of material of construction (W/m°C)	17.3	

the heat duty and allowable pressure drops as a function of the plate width for a given plate spacing. In Figure 3 it is seen that for the chosen plate spacing of 0.004762 m the design is pressure drop constrained. For instance, take the plate width of 0.102; a length of 18 m is required to meet the heat duty, whereas the pressure drop on the hot and cold side are fully absorbed in 2m and 1m respectively. In other words, for the heat duty to be achieved, the pressure drop on each side reaches values much higher than the ones permitted. In order to shift to a heat load controlled design, plate spacing is increased. In Figure 4, its value is increased to 0.00635 m; in this case, acceptable designs are obtained for plate widths starting from 0.762 m where the heat duty is achieved with a length of 12 m and the corresponding pressure drops within this length are less than permitted.

Conclusions

This work has shown the use of the concept of parameter plot for the pictorial representation of a design space that aids in the selection of spiral heat exchangers. A parameter plot is a graphical representation of the design space that shows at a glance, all possible geometrical combinations that meet heat duty within the restrictions of pressure drop. The information provided by these plots guides the user in choosing the geometry that best suits the required specifications.

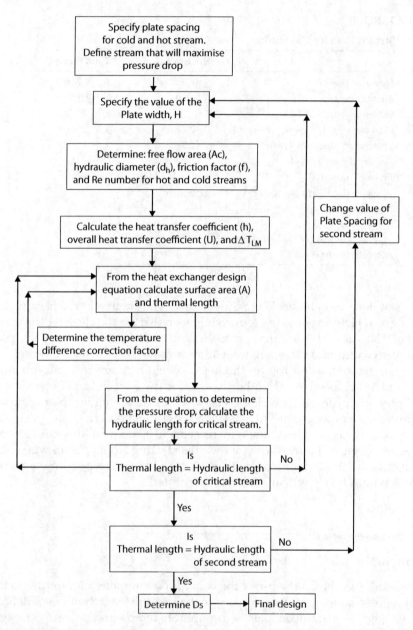

FIGURE 2
Flow diagram of design methodology.

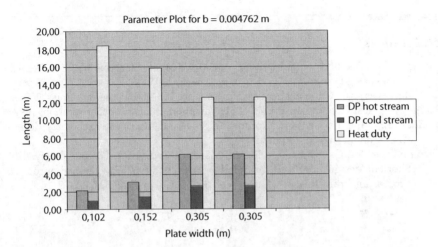

FIGURE 3
Design space for a plate spacing of 0.004762 m.

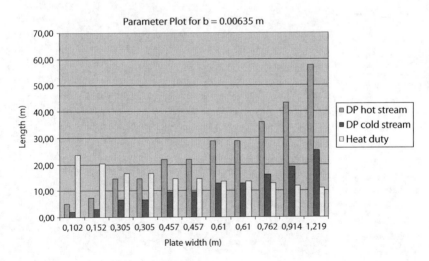

FIGURE 4
Design space for a plate spacing of 0.00635 m.

References

Dongwu, W, Method of computing outside diameter of the SHEs, Chemical Engineering Design, vol. 15, No. 4, pp. 36–42, 2005.

Hesselgreaves, J. E., Compact heat exchangers, pp. 155–200, Pergamon Press, 2001.

Holger, M., Heat Exchangers, pp. 73–82, Hemisphere Publishing Corporation, 1992.

Poddar, T.K. and Polley, G.T., Optimising shell-and-tube heat exchanger design, Chem. Eng. Prog. September 2000.

Picón-Núñez, M., Canizalez-Dávalos, L., Martínez-Rodríguez, G., and Polley, G.T., Shortcut design approach for spiral heat exchangers, Food and Bioproducts Processing, Trans IChemE, Part C, vol. 85, no. C4, pp. 322–327, 2007.

46

Design Challenge—Towards Zero Emissions Fossil Fuel Power Plants

V. Mahalec[1], M. El-Halwagi[2], P. Medellin[3], G. Pokoo-Akins[2], Y.L. Maldonado[3] and A. Nadim[1]

[1]*McMaster University, Hamilton, Ontario, Canada*
[2]*TAMU, College Station, TX, USA*
[3]*Universidad Autónoma de San Luis Potosí, Mexico*

CONTENTS

ABSTRACT In this paper we examine a promising pathway involving the elimination of CO_2 emissions by growing oil-containing algae which naturally absorb solar energy during the growth phase. The dry algae or algal oil can be used as a fuel in the power plant boilers or to make bio-diesel. The remaining biomass can also be used for producing thermal energy.

If we use dry algae or algae oil as a fuel in the power plant boiler, then in the extreme one can envision the following process:

1. Use enough fossil fuel to produce sufficient amount of CO_2 to produce as much algae oil as needed for electricity production and for algae processing.
2. Use algae oil as fuel in the boilers and grow again algae from CO_2 in the flue gas.
3. Etc. (repeat from 2).

Theoretically, if there is 100% capture of CO_2 via algae and if there is efficient capture of solar energy in the growth of algae, we could continue production of electricity without consuming any fossil fuel (other than the initial "charge"). We present a preliminary process design for 1,000 MW plant using algae oil as fuel.

KEYWORDS   *Zero CO_2 emissions, power plant, algae*

Introduction

We present a design for a 1,000 MW power plant that uses algae as fuel. Our assumption is that the climate is temperate throughout a year, thereby enabling algae to grow without being sheltered within buildings. Advantage of using algae over other forms of sources of biofuels are: most algae can grow in saline water and do not have to compete for fresh water, productivity per unit area of land is much higher, and the overall uses of water is much less than growing oilseeds.

Our goal is to capture all CO_2 emissions from a power plant. Hence, we cannot use open ponds to grow algae.

Flue gases from the power plant boilers contain 13–14% CO_2 which is absorbed in H_2O in a pressurized absorption column. CO_2 absorbed in water is a feed to photo-bioreactors to grow algae. Algae in the reactor effluent is separated by centrifuging, dried and then either used as a fuel in the power plant boiler or to produce algal oil. Algal oil can be used again as a fuel or to produce bio-diesel.

Choice of alternative uses for algae depends on the price of different forms of energy (e.g. power plant fuel vs. biodiesel) and on the efficiency of boilers that burn different forms of fuel. If we can use a biomass boiler that has the same efficiency as boilers for conventional fuels (e.g. natural gas or powdered coal), then the most energy efficient alternative is to burn dried algae in the boilers, thereby displacing fossil fuels.

Once-through process captures all CO_2 from the flue produced by burning fossil fuel, converts it to algae and burns the algae in the boilers. Assuming 50% algae oil content and 80% biomass boiler efficiency, such process produces net 250 MW from the algae. Therefore, to produce 1,000 MW entirely from algae, i.e. without using any fossil fuel, one would need to process 4 times larger amount of CO_2. This would require a modification of the combustion subsystem in the power plant in order to handle 4 times larger amounts of flue gases.

Process Structure

Simplified process flowsheet is shown in Fig. 1. Flue gas from the power plant has to be compressed in order to absorb all of CO_2. Prior to entering the absorption section, the compressed flue gas is used to dry algae that was

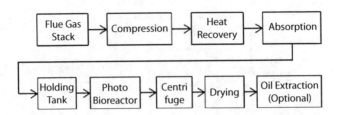

FIGURE 1
Algae growth from flue gas.

separated from the reactor effluent. The absorbers operate at a pressure sufficiently high to absorb all CO_2 from the flue gas. Following the absorption section is a holding tank which operates at a pressure sufficiently high to keep all of the absorbed CO_2 in the liquid phase. Photobioreactors are horizontal tubes made of transparent plastic (plexiglass). Effluent of the photobioreactors is divided into a recycle (which brings the required seed concentration of algae to the reactor inlet) and the product stream which is sent to a bank of centrifuges to separate algae from water. Dryers pass hot flue gas above the centrifuged algae until the water content is reduced sufficiently to send the algae to spray dryers.

Dry algae can be used either as a fuel in the boilers or it can be processed further to extract the oil from it.

CO_2 Capture by Absorption in Water

In order to absorb the entire CO_2 present in the flue gas, the design uses 10 absorption columns operating under an elevated pressure. From the absorption column, the solution is pumped to a holding tank which is kept at 3 atm in order to retain all CO_2 in the liquid phase.

Selection of Algae and Photobioreactor Design

Production of algae occurs in the presence of sunlight as described by Eq. (1):

$$\text{Sunlight} + 6\,CO_2 + 4.75\,H_2O + \text{Nutrients} \rightarrow C_{6.5}H_{11.5}O_{2.9}N + 7.18\,O_2 \quad (1)$$

Various strains of algae have been studied (Chisti, 2007) with respect to their oil content and the conditions optimal for their growth. For instance, Chlorella have 28–32% of algae oil, while Schizochytrium and

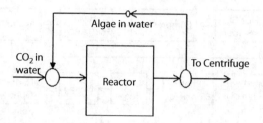

FIGURE 2
Recycle enables control of algae concentration at the reactor inlet.

Nannochloropsis can have more than 50% of oil. The oil content of Chlorella sp. can reach 46% dry weight under stress conditions (Hu et al, 2008). It was found that chlorella is tolerant to CO_2 concentrations of up to 40% by volume (Hanagata et al, 1992). Similar results reported that maximum growth occurred at a concentration of 10% and found that the strain could grow under various combinations of the trace elements NO_x and SO_2 in flue gas (Maeda et al, 1995). Finally, chlorella was found to grow in conditions of up to 40 degrees Celsius (Sung et al, 1998). Chlorella doubles every 8 hours on average. All of these characteristics may Chlorella a suitable algae for the proposed design.

In order to initiate the growth of algae, there must be some amount of algae at the reactor inlet. The initial concentration, rate of algae growth, fraction of the reactor effluent that is recycled, and the reactor residence time are design parameters that determine the length of the reactor.

Reactor diameter is limited by the requirement that the sun light must penetrate through the liquid in order to enable the growth of algae. It has been found (Edwards, 2008) that the reactor diameter can be up to 0.6m, with maximum productivity per ground surface area occurring at diameter of approx. 0.3m.

Reactor residence times from 2hr to 24hrs and the recycle fractions from 0.8 to 0.12 respectively, have been evaluated. Concentration of algae in the reactor effluent has been kept at $1kg/m^3$ in all cases. For 0.3m reactor diameter, the length of the reactors is 2.2 to 26km. Residence time of 8 hrs requires a reactor length of 8,600m with total number of reactors being 1,722 for 1,000 MW plant.

Oxygen produced during the growth of algae impedes algae growth. Hence, the reactor design needs to accommodate removal of oxygen. Molina and Chisti (2007) have determined that the liquid velocities of 0.3–0.5 m/s are required to have a steady growth and for the culture not to collapse in the tubular reactor. Moreover, they determined that the oxygen concentration should not exceed 300% of the saturation in order not to impede the growth of algae. This study uses 0.3 m/s velocity which results in the requirement that the oxygen be removed approximately every 60m of reactor length.

Algae Harvesting and Drying

The least costly but also the least efficient method to harvest algae is through filtration. This method is not very efficient since the large flow rates associated with harvesting can create clogging and backwash (Boersma, 1978). Centrifugation and chemical flocculation are more expensive than filtration but they are also more effective ways to harvest algae. Chemical flocculation has the main disadvantage of introducing chemicals to our process, which can be costly, but more importantly it can affect the quality of the biomass produced. Flocculation can add toxins to the biomass which would be a disadvantage in the case of using the biomass as feed or fertilizer. Benemann (1996) tested various ways of harvesting algae that included centrifugation, chemical flocculation, filtering, and sedimentation. He has found that harvesting using centrifugation is the most efficient as it concentrates the paste to about 20% solids compared to less than 10% for all the other methods that he has evaluated. For our proposed design, a solid ejecting nozzle separator centrifuge will be used that has a maximum capacity throughout of 200m³/hr. Therefore we will require aprox. 220 centrifuges to be used to centrifuge the entire production per hour for the absorption at 15 atm.

Drying of the algae is to be accomplished in two stages: countercurrent flow of flue gas against the slurry containing algae. Since the compression of the flue gas is a employs multistage centrifugal compressors, flue gas after each stage is passed through a dryer with the target exit temperature of the flue gas being sufficiently low that it can enter the absorption columns. The amount of energy available from the flue gas provides only about 2/3 of the energy required to dry the algae. Following energy recovery section is a set of spray dryers which reduce the moisture content in the algae to 5%.

Process Flowsheet

Simplified process flowsheet is shown in Fig. 3. Due to a large volume of flue gas, it is required to have parallel multi-stage stage compression trains. After each compression stage there is an inter-stage heat recovery through algae dryers. The absorption section consists of 10 parallel absorption columns due to the large volume of flue gas.

Economics

Absorption at increased pressures enables the entire CO_2 to be removed from the flue gas. Moreover, the higher the pressure, the less water is needed to absorb CO_2. However, as the CO_2 concentration increases so does the required

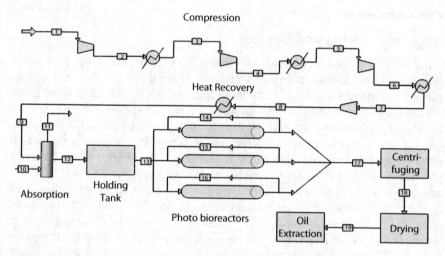

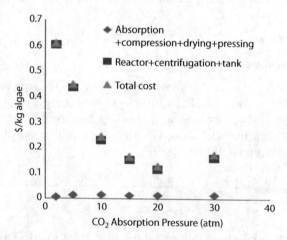

FIGURE 3
CO_2 capture via absorption and growing algae for fuel or algae oil.

pressure in the reactors in order to keep the CO_2 in the liquid phase. Higher reactor pressures require thicker reactor walls, thereby increasing the capital cost of the reactors.

The proposed design has been evaluated at a range of pressures (from 2 to 30 atm). Shown in Fig 4. is the total capital cost per kg of algae produced. The optimum absorption pressure is approx. 20 atm.

Algae biomass can be burnt as fuel in the boilers (current biomass boiler efficiency is 35%), or one can extract oil and then burn the oil in 80% efficient boiler or one can produce biodiesel. Table 1. summarizes net energy

FIGURE 4
Capital cost as a function of absorption pressure.

TABLE 1

Comparison of Algae Usage Options

Power [M W]	Combust Biomass, No Oil Extraction	Combust Algal Oil & Biomass Byproduct	Biodiesel & Combust Biomass Byproduct
30% oil in algae, 35% boiler eff.			
Consumed	152	168	168
Produced	229	195	157
Net	77	27	−11
50% oil in algae, 35% biomass boiler eff.			
Consumed	152	168	168
Produced	305	276	214
Net	153	108	46
30% oil in algae, 80% biomass boiler eff.			
Consumed	152	168	168
Produced	303	232	194
Net	151	64	26
50% oil in algae, 80% biomass boiler eff.			
Consumed	152	168	168
Produced	404	318	256
Net	252	150	88

produced by burning algal oil, including a case where algae contains 50% oil and it is burnt in an 80% efficient boiler for algae produced from natural gas flue. Such boiler design should not be out of reach, since pulverized coal boilers have roughly the same efficiency.

Zero CO_2 Emissions Plant

Our best case scenario is to use algae with 50% oil content and 80% biomass boiler efficiency. Both of these assumptions are within reach and should be attainable with more research effort. In order to produce 1,000 MW we need to process 4 times as much CO_2 as the amount of CO_2 from the 1,000 MW plant that burns natural gas (see Fig. 5). This enables us to operate a closed system that recirculates all of CO_2. Incremental cost of producing electricity incurred by using the algae is approx. \$0.06/kWh. To this price one needs to add the amortization of the traditional power plant equipment. There is no cost for purchasing fossil fuel (e.g. natural gas) since all required energy is produced from algae.

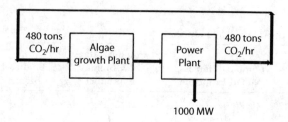

FIGURE 5
1,000 MW plant with CO_2 recirculation.

Conclusions

Electricity production in a closed system, with zero CO_2 emissions is possible in fossil fuel power plants if algae produced by capturing flue gas CO_2 are used as fuel. Incremental costs for retrofitting existing power plants are estimated to be lower than building equivalent capacity, solar powered electricity generating plants. Our challenge is to verify this by more detailed design and prototyping.

References

Benemann, J.; Oswald, P.I.: Systems and Economic Analysis of Microalgae Ponds for Conversion of CO2 to Biomass. Department of Energy Pittsburgh Energy Technology Center, 1996

Boersma, L, "Management of swine manure for the recovery of protein and biogas: final report", Corvallis, Or.: Agricultural Experiment Station, Oregon State University 1978

Chisti, Y. (2007) "Biodiesel from microalgae". Biotechnol. Adv. 25, 294–306

Hanagata, N., Takeuchi, T., Fukuju, Y., Barnes, D. J., Karube, I. (1992) "Tolerance of microalgae to high CO2 and high temperature". Phytochemistry 31(10), 3345–3348.

Hu, Q. et al., 2008. "Microalgal triacylglycerols as feedstocks for biofuel production: perspectives and advances". The Plant Journal, 54, 621–639.

Maeda, K., Owada, M., Kimura, N., Omata, K., Karube, I. 1995. "CO2 fixation from the flue gas on coalfired thermal power plant by microalgae". Energy convers. Mgmt 36, 717–720

Molina, E.M and Chisti, Y, Scale-up of tubular photo bioreactors, *J. Appl. Phycol.* 12 (2000), pp. 355–368

Sung KD, Lee JS, Shin CS, Park SC, Choi MJ "CO2 fixation by Chlorella sp. KR-1 and its cultural characteristics". *Biores Tech.* 68 (1999) pp. 269–273

47

Web-Based Modules for Product and Process Design

Paul Stuart[1], Mario Eden[4], Mahmoud El-Halwagi[3], Jeff Froyd[3], Vladimir Mahalec[2], Mario Moscosa[6], Pedro Medellín Milán[6] and Martín Picón-Núñez[5]

[1]*Ecole Polytechnique—Montréal QC, Canada,*
[2]*McMaster University—Hamilton ON, Canada*
[3]*Texas A&M University—College Station TX, USA,*
[4]*Auburn University—Auburn AL, USA*
[5]*University of Guanajuato—Guanajuato, Mexico,*
[6]*University of San Luis Potosí—San Luis Potosí, Mexico*

CONTENTS

ABSTRACT Integrative, complex, design-oriented learning outcomes involve multidisciplinary knowledge and procedures, engineering know-how related to practice and experience, and professional or "soft" skills. Achieving these outcomes in a web-based environment is a significant challenge (Brault *et al*, 2007). The paper describes an ongoing project geared toward the development of web-based modules on Product and Process

Design as part of a North American Mobility Program (NAMP) involving 6 universities in Canada, the US, and Mexico. The mechanism for developing the modules involves exchanging students and faculty between the 6 universities, and creating cooperative and collaborative environments. A total of 42 student exchanges will be made to develop innovative web-based educational materials ("modules") using a unique vision for Product and Process Design, and using the biorefinery for the design case study context. The objective for some of the initial exchanges was to identify and apply state-of-the-art pedagogical and web-based techniques, in order to produce user-friendly and effective distance-learning modules which incorporate open-ended design problems. Another more ambitious objective of this project will be to set the foundation for establishing a distance-learning graduate program in Product and Process Design which is international, multilingual, and multi-institutional in character.

KEYWORDS *Product and process design, web-based design education, open-ended problems*

Introduction

Increasing global demand for improved product and process functionality requires higher quality, lower costs, product customization, conservation of natural resources, reuse and recycling of materials, and meeting new energy and environmental challenges. To respond and stay competitive internationally, from a manufacturing perspective North America must be more innovative and responsive—and currently the biorefinery represents an important potential opportunity to pursue these design objectives. The design competency developed in engineers at the undergraduate and graduate levels is critical to achieving these goals.

Contrasting approaches are used to develop the design competency, which depend on many factors including in particular the previous industrial design experience of the instructor highlighting the empirical nature of design. Furthermore, teaching the engineering design competency can involve various methods such as project-based learning and interdisciplinary approaches.

The definition and description of design and design processes have been given considerable attention in the literature. Crain *et al* (1995) describe the TIDEE initiative (Transferable Integrated Design Engineering Education), which was aimed at developing and integrating design education in the first two years of the engineering curriculum. More recently, the National Science and Engineering Research Council of Canada (NSERC) Chairs in Design Engineering (2004) developed a definition of the engineering design competency. Engineering design has been characterized in many ways, including the following:

(1) **Creative process.** Design is supported by knowledge and methods as well as by analysis and synthesis capabilities.

(2) **Iterative.** Designers need to access different sources of information during the design process and this information must be refined by negotiations, clarifications, discussions and evaluations until the information and context are appropriate for the design problem.

(3) **Interactive and social.** Interaction between experienced designers and teamwork are critical for design, regarding in particular the information acquired through case-based and problem-based learning. This reflects to some extent the empirical nature of design.

(4) **Open-ended.** Questions that are posed in design situations typically have no unique answer. Design questions solicit divergent thinking, based on deduction incorporating facts and extending these to the possibilities that can be created from them (Dym *et al*, 2005).

(5) **Multidisciplinary.** The design competency is considered by some to be similar for different engineering disciplines. One key to improved technology development and innovation lies in cooperation amongst disciplines in order to identify unique designs.

These important characterizations are increasingly recognized, and at many universities they are increasingly incorporated into engineering undergraduate curricula. However this emphasis on multidisciplinary synthesis and design coupled with a greater emphasis to deeper inquiry and open-ended problem solving, development of management and communication skills, international exposure and preparation for continuing professional development and career-long learning are all mounting pressures on already overburdened engineering curricula (Fromm, 2003). Institutions are being confronted with having to cope with the need for improved know-how and interdisciplinary design, at the same time as the need for curriculum renewal associated with new content resulting from technology advances and knowledge growth.

Objective of This Paper

The paper describes an ongoing project whose goal is to develop web-based modules on Product and Process Design, as part of a North American Mobility Program (NAMP) involving 6 universities in Canada, the USA, and Mexico.

The web-based modules are being developed through student exchanges between the participating organizations, which can be incorporated into existing and future design and design-oriented courses to improve the design competency and ability for innovation of the engineering student. Once completed, the set of web-based modules may form the basis for a distance-learning Masters program in Product and Process Design.

Design and the Biorefinery

Engineering design strives to formulate general principles on which the design of an engineering product is based, regardless of the engineering discipline. The Institution of Chemical Engineers (IChemE) in Great Britain state the following on their website: "If you do not wash, use deodorant, shave or wear makeup, eat, feed your pets, wear shoes, drive a car, play CD's, go on holidays, stay at home, sleep on a mattress, take pills, brush your teeth or wear false ones, go to the movies, watch television, listen to the radio, buy books or read newspapers—then Chemical Engineering design does not affect your life". In the field of Chemical Engineering, the range of commodity and specialty products and processes that must be considered is extremely broad.

In many cases, Chemical Engineering design instructors emphasize engineering design content and analysis. Instructors adopt different approaches to teaching design, depending on their background and experience, which highlights the empirical nature of design. Teaching the synthetic, evaluative, and integrative capabilities required for engineering design is less straightforward. Additionally, there is a significant need to foster the students' creativity and innovation in design. Learning the engineering design competency requires innovative, interdisciplinary teaching strategies such as project-based and inquiry-guided learning. How these can be incorporated into web-based modules used for distance learning remains a significant challenge.

The project seeks to develop distance learning or web-based teaching materials that enhance the teaching of the Product and Process Design competency from a Chemical Engineering perspective, including engineering design know-how, and using a multidisciplinary case study context of the biorefinery. There are certain essential characteristics of the web-based modules:

- Design of a product or process begins with the identification of a specific societal need and the generation of potential solutions. In order to select an appropriate solution, an analytical process is employed where the set of solutions are evaluated against criteria related to the societal need and other factors. Finally, the engineer must verify that the selected solution satisfies the original requirements. This basic structure is essential to product and process design activities, and should be reflected in case study based learning.

- Whereas many design programs and much of the design literature was initiated within Mechanical Engineering, Chemical Engineering has its own particular perspective on the Product and Process Design that emphasizes sustainability and environmental impact for the production of chemicals.

- The web-based modules will emphasize a multidisciplinary case study approach that uses the context of the biorefinery.

There are similarities between the biorefinery and the petroleum refinery, however there are also important differences. For example there is more oxygen present in lignocellulosic feedstocks, which leads to opportunities for production of certain carbohydrate-based bioproducts as substitutions having somewhat different functionality than the competing petroleum-based product. Separations will be as critical for the biorefinery as they are for the petroleum refinery however since bio-based chemicals are generally less volatile, it is possible that solvent-based extraction may be more attractive than distillation and membrane technology may play an ever-more important role. Developing innovative biorefinery solutions that are unique to the North American competitive landscape in a global economy will require bold fast-track research into new conversion processes, coupled with innovative design techniques.

Distance Learning Modules

Web-based educational content has significant advantages including global access through the Internet infrastructure, rich media, easy updating and content maintenance, and an interactive communications environment through the use of external links and discussion groups. On the other hand, web-based education also poses several challenges including the dislike for on-screen learning, the lack of verbal cues and an appropriate navigation structure, the size of the educational files and the lack of instructor familiarity with web-based technology. Design instructors might be competent at teaching engineering design, but might be less capable of incorporating web-based technologies to their traditional teaching methods to answer the need for innovative strategies for teaching.

Challenges associated with web-based teaching are further complicated by the complex learning outcomes that support engineering design. For example in a web-based format, how can students best be guided to establish and evaluate design options for an open-ended problem? How can a student's teamwork capabilities and other soft skills be enhanced and evaluated using a web-based format? These issues must be addressed in modules developed as part of this project, recognizing that it is essential to use a hybrid model employing web-based technologies to enhance rather than replace face-to-face education.

One approach has been proposed by the Canadian Design Engineering Network (CDEN) (Yellowley *et al*, 2001), using a unique three-tiered structure that covers the scope from fundamental engineering science principles to open-ended engineering design problems. The first web-based module tier (Tier I) addresses fundamental principles such as basic qualitative and quantitative laws and empirical design relations. The second tier (Tier II) addresses different technologies, methods and engineering practices through the directed (close-ended) analysis of case studies. In these first two tiers, the design experience is presented to emphasize interrelationships of fundamental principles, and to provide context. Finally, the third tier (Tier III) integrates information taught in both Tiers I and II via the resolution of open-ended problems in which the student's ability to integrate methods and technologies is challenged. The objective in using this tiered structure is to allow the student to evolve from knowledge acquisition to an application of engineering design know-how.

Basic Approach for Developing Web-Based Modules in Product and Process Design of the Biorefinery

The definition of engineering design encompasses many types of interactions with technology, including the following:

- *Product Design*—Creating new product concepts; innovative products in new market segments; integration of multiple technologies; decision processes in product development; new materials and manufacturing processes; life-cycle analysis; quality management through inherent design; business and cultural issues in producing and using products. In the context of product design for the biorefinery, a particular emphasis is needed related to the supply chain, to ensure that the proposed biorefinery product integrated into a product portfolio will lead to a sustainable business model for the longer term.

- *Process Design*—Synthesizing new processes; structurally and parametrically optimizing processes; environmentally sustainable processes; processes with minimal environmental impact; design paradigms to support process innovation; high reliability processes. Process integration techniques (e.g., El-Halwagi, 2006) provide an attractive framework for systematizing many of the decision involved in process design, benchmarking process performance, and reconciling various objectives. In the context of process design of the biorefinery, critical issues concern the performance of emerging biorefinery processes, today but very importantly as the processes mature, and as well in process intensification and innovative technologies for separation and purification of complex organic mixtures to optimize the production of by- and waste products.

- *Simultaneous Product and Process Design*—Integrating design cycles for products and processes, interacting objectives and constraints of products and process design, assessing combined impact of products and processes, optimizing overall enterprises and supply chain management. In the context of the biorefinery, this aspect is especially important, requiring a multi-criteria decision-making (MCDM) activity in order to simultaneously and appropriately consider the range of design assessments related to such factors as competitive position, process performance uncertainty, environmental performance using a product-chain perspective, unique supply chain for product portfolios, etc. Systems analysis and synthesis tools are properly suited to extract the knowledge from the molecular level and introduce it in the various scales of the process (Armstrong, 2006). Recent advances in the area of property integration (Eljack et al., 2007, Kazantzi et al., 2007) provide effective framework for integrating product and process design.

The project recognizes that engineering curricula need to embrace the multidisciplinary nature of the modern design context, including sophisticated decision-making related to a range of factors including process performance, process integration, profitability and environmental performance. Process systems engineering (PSE) involving the set of process integration (PI) design tools are thus critical for the web-based modules.

The module content focuses on development of student capabilities related to comprehension, application, evaluation, and synthesis of the design concepts in the context of the biorefinery. This approach is a departure from content-oriented development of materials, and is consistent with the growing recognition of learning as provided by cognitive learning theory. Implementation of this strategy requires that the project team engage in extended conversations to synthesize the core concepts, and empirical knowledge of key biorefinery issues.

Ideally, no research should be required to establish the content of the web-based modules, however due to the nature of the emerging biorefinery this is not the case. Modules are requiring that original calculations be performed in order to establish the particular and novel design approaches appropriate for the biorefinery, and then module development activities focus on developing the ability to construct, analyze, evaluate, and synthesize solution alternatives through approaches such as case studies and problem-based learning in which students apply content to address open-ended biorefinery scenarios. Effective teaching tools for open-ended design problems.

Modules Under Development

At the outset of the grant, the project leads from each institution and other stakeholders assembled a list of "critical" and "other" characteristics sought in the modules, as follow.

Critical Characteristics of Web-Based Modules

1. The set of modules should be structured so that they are coherent with each of the following (1) the design process, (2) the design tools, and (3) the biorefinery case studies,
2. The context should be as real as possible, considering the industry context for the biorefinery,
3. They must be structured to integrate different perspectives and tools, e.g. simulation, economics, optimization, life cycle analysis.

Other Important Characteristics of Web-Based Modules

- Recommendations should be made in the case study for ways to use in teaching,
- There should be an elaboration of alternatives, and why specific alternatives are chosen,
- Each module should contain a knowledge component, that links between coursework and the module case study,
- The modules should clearly state requirements or needs and how these were developed,
- The modules should have an emphasis on definitions and terminology,
- Closed-ended problems that serve as illustrative example should be included between knowledge component and open-ended design component, where appropriate,
- There should be a module developed that presents the context of Product Design and Entrepreneurial aspects.

It was felt that each module will reflect the particular biorefinery and other design experience of the faculty involved from the host institutions of those receiving exchange students, and that additional characteristics would be constraining. On the other hand, the team has had the important benefit of a team member specialized in engineering pedagogy.

Near the outset of the grant, the project team also developed the list of modules to be developed. In the context of the process and product design approach, the module themes typically make reference to the industry context and not the process integration tools to be employed in resolving the subject. In this context, how the modules would assemble into a cohesive syllabus has not been made explicit.

The list of modules developed and under development is as follows.

Overview of Biorefineries

   I. *Viable alternatives for biorefineries in North America*: Analysis of supply and demand, survey, feedstocks, products, policies, decision making, initiatives, and models

   II. *Technology alternatives for forestry feedstocks: different pathways and platforms*: chemical, thermal, biochemical, etc. Typical processes, basic mass and energy balances, and performance criteria.

   III. *Technology alternatives for biorefinery feedstocks*: different pathways, platforms: chemical, thermal, biochemical, etc. Typical processes, basic mass and energy balances, and performance criteria.

   IV. *Economic and environmental issues and constraints*: availability and displacement of land, water, competition with food supply chain, LCA, cost/subsidies/taxes, plant location.

The Forest Biorefinery

   I. *Designing the Forest Biorefinery*

   II. *Innovation and entrepreneurship*

Biodiesel

   I. *Overall assessment of biodiesel production in North America*: alternatives, innovation in pathways, strategic maps, LCA, global impact.

   II. *Large-Scale Biodiesel Plant*: Simulation, design, integration, and economic analysis. Sensitivity analysis. Process safety and risk management.

   III. *Small-Scale Biodiesel Plant*: Design, use of alternative energy sources (e.g., solar), economics.

Bioethanol

I. *Overall assessment of ethanol production in North America*: alternatives, innovation in pathways, strategic maps, LCA, global impact.

II. *Large-Scale Cellulosic Ethanol Plant*: Simulation, design, integration, and economic analysis. Sensitivity analysis. Process safety and risk management.

III. *Large-Scale Non-Cellulosic Ethanol Plant*: Simulation, design, integration, and economic analysis. Sensitivity analysis. Process safety and risk management.

Conclusion

Innovative teaching methods offer good prospects to enhance engineering design competency teaching and learning in an environment of continuously changing knowledge and skills.

Besides the technical challenges, students involved In the project undertake cultural and language exchanges with partner institutions, during which they have a rich professional experience addressing issues related to Product and Process Design with an emphasis on innovation and sustainability.

Acknowledgements

This work has been supported by the Natural North American Mobility Program (NAMP) program.

References

Armstrong, R.C. (2006) 'A vision of the curriculum of the future', Chemical Engineering. Education, Vol. 40, No. 3, Spring, pp.104–109.

Brault, J. M., Medellín-Milán, P. M., Picón-Núñez, M. C., El-Halwagi, M. M., Heitmann, J., Thibault, J., and Stuart, P., "Web-Based Teaching of Open-Ended Multidisciplinary Engineering Design Problems," Trans. IChemE, Part D Education for Chemical Engineers, 2(1), 1–13 (2007)

Crain, R.W., Davis, D.C., Calkins, D.E., Gentili, K., 1995, Establishing engineering design competencies for freshman/sophomore students, *Proceedings of the 1995 25th Annual Conference on Frontiers in Education. Part 2 (of 2)*, Atlanta, GA, USA, November 1–4, 1995, pp. 937–939.

Dym, C.L., Agogino, A.M., Eris, O., Frey, D.D., Leifer, L.J., 2005, Engineering design thinking, teaching, and learning, *Journal of Engineering Education*, Vol. 94, No. 1, pp. 103–119.

El-Halwagi, M. M., "Process Integration", Elsevier (2006)

Eljack, F. M. Eden, V. Kazantzi, X. Qin, and M. M. El-Halwagi, "Simultaneous Process and Molecular Design—A Property Based Approach"; AIChE J., 35(5), 1232–1239 (2007)

Fromm, E., The changing engineering educational paradigm, *Journal of Engineering Education*, Vol. 92, No. 2, 2003, pp. 113–121.

Kazantzi, V., X. Qin, M. El-Halwagi, F. Eljack, and M. Eden, "Simultaneous Process and Molecular Design through Property Clustering—A Visualization Tool", Ind. Eng. Chem. Res., 46, 3400–3409 (2007)

NSERC Chairs In Design Engineering, The Engineering Design Competency, *The Inaugural CDEN Design Conference*, Montreal, Quebec, Canada, July 29–30, 2004.

Ragauskas *et al*, 2006, The Path Forward for Biofuels and Biomaterials, Science Magazine, Vol. 311 (27 January), pp. 484–489

Yellowley, I., Venter, R.D., Salustri, F., 2001, The Canadian Design Engineering Network, *NSF Conference on Design and Manufacturing*. Clearwater, Florida, United States.

48

Teaching "Operability" in Undergraduate Chemical Engineering Design Education

Thomas E. Marlin

Department of Chemical Engineering, McMaster University,
Hamilton, Ontario, Canada L8P2E3

CONTENTS

ABSTRACT This paper presents a proposal for increased emphasis on operability in the Chemical Engineering capstone design courses. Operability becomes a natural aspect of the process design courses for a project that is properly defined with variation in operations and model uncertainty. Key topics in operability are operating window, flexibility, reliability, safety, efficiency, operation during transitions, dynamic performance, and monitoring and diagnosis. The key barrier to improved teaching and learning of operability is identified as easily accessed and low cost educational materials, and a proposal is offered to establish a portal open to all educators.

KEYWORDS   *Engineering education, process design, operability, problem solving*

Introduction

Engineering instructors and practitioners agree that a design must be "operable"; however, the term operability is not used consistently. Since a concise definition of operability is difficult, a taxonomy of eight operability topics is proposed to define the topic: operating window, flexibility (and controllability), reliability, safety (and equipment protection), efficiency (and profitability), operation during transitions, dynamic performance, and monitoring and diagnosis. Theses topics have been selected to cover the most common issues in process plants and to reinforce prior learning. These topics are not addressed in standard engineering science courses and are not typically addressed thoroughly in the design course.

The intension of this paper is twofold; the first is to encourage greater coverage of operability topics, and second to begin collaboration among educators that will result in a consensus on the key operability topics and the development of essential resources to assist instructors in tailoring the topics to their courses.

This paper begins with a design project definition that explicitly includes variation and directs attention from a design point to a design range. Then, the paper presents each of the operability topics briefly, giving examples of their impact on important design decisions. The paper concludes with a proposal to promote the development and sharing of educational materials to facilitate teaching process operability.

Designing for Realistic Scenarios

The traditional process design course is centered on a major project, in which students perform specific tasks, including (but not limited to) process synthesis, process flowsheeting, selection of materials of construction, rough equipment sizing, and economic analysis. Typically, the final report gives the process design for a *single operating point* that is not adequate for engineering practice and limits the educational experience of the student. The expansion of the design goals introduces many related topics, which are here combined under the term "operability".

The reason for considering a range of operations is often given as "uncertainty"; however, many factors are certain to occur, such as changes in feed properties, productions rates, and product specification, as well as larger changes for startup and shutdown and removal of equipment for maintenance. The equipment should be designed to operate as specified during these transitions, using the *known* variation in operating conditions and performance requirements.

In spite of our best efforts, substantial uncertainty also exists in, for example, correlations for rate processes, physical properties, and efficiencies of equipment performance. Students should be encouraged to understand and quantify the likely range of uncertainty, which they can do by accessing the original references. They will appreciate the importance of uncertainty on their designs, and they should be required to report errors bars and uncertainty estimates with their results, especially their economic analysis. By raising the issue of uncertainty explicitly, students will be aware of the importance of knowing the basis for the models and data being used and for limiting designs to regions supported by the information.

A design must satisfy all of these variability and uncertainty issues. The students' first design task should be preparing the design specification. For example, a design task could be to "design a waste water treating facility for a town of 50,000 people, which will grow in 10 years to 100,000 people, in southern Ontario, Canada". While performing this task, the students recognize the fallacy of the "single-point design" approach.

The Eight Topics of Operability

The following topics were selected to concentrate on the most important issues and to provide the students with a structure or checklist of major categories. Naturally, additional topics can be included, and some issues can be located in more than one topic. Also, the topics can accommodate issues not covered in this review; for example, safety can include clean-in-place operations or consumer safety; students are encouraged to expand their investigations and design decisions beyond this introductory coverage.

1. The Operating Window

An important objective of process design is ensuring that the range of operating conditions defined in the specification can be achieved. The typical goal of this analysis is to ensure that process equipment has a large enough capacity to achieve all expected operations. However, students must be aware that process equipment has *minimum as well as maximum* limits on operating variables, for example, a minimum fuel rate to a boiler and a minimum flow to a fluidized bed.

Naturally, we must consider key variabilities and uncertainties, which must be based on a comprehensive design specification and a thorough knowledge of the process models. Students should understand the accuracy of constitutive models and the assumptions that limit the regions of application, for

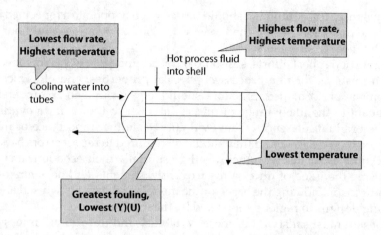

FIGURE 1
Take the worst-case conditions when determining the area for heat transfer.

example, laminar or turbulent flow, horizontal or vertical tubes, etc. Also, they should acknowledge the uncertainty in the model structure and the danger in extrapolation beyond the data used in model building.

Students can use first principles to determine the limiting, or "worst case", conditions over a range parameter values. For example, the area required for the heat exchanger in Figure 1 can be determined for the base case data. However, we see that the cooling water temperature and the process flow rate vary over a range. In addition, the fouling factor and the film heat transfer coefficients have uncertainty associated with their values. In many cases, the heat capacities of the fluids and the metal thermal conductivities are known with little error. Therefore, the design is based on the "worst case" values of the uncertain variables, i.e., the values that result in the largest area: these are the highest process inlet flow rate, highest cooling water temperature, highest fouling factor.

Students should come away from a design course with disdain for gross overdesign of plants that simply add an arbitrary amount of excess capacity. Safety factors should be small and "for well tested processes, safety factors can approach zero percent" (Valle-Riestra, 1983).

2. Flexibility and Controllability

Once students recognize that the process will be required to achieve a range of operating conditions, they will accept the need to adjust selected (manipulated) variables to achieve key objectives such as safety, product quality, production rate and profit. Therefore, the process must have *flexibility*, i.e., it must have a sufficient number of manipulated variables located so that the operating objectives can be achieved. The selection of the proper manipulated

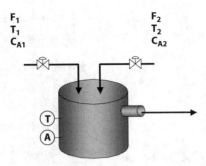

FIGURE 2
Overflow mixing tank in which the effluent composition and temperature cannot be controlled by manipulating the two inlet flow rates.

variables is not obvious, so that students should be taught to rely on fundamentals and innovation when providing flexibility.

Students also need to recognize that simply having many adjustable flows and power inputs does not ensure that a specific set of variables can be influenced independently. We will term this controllability, which is generally defined as "the ability of a system to achieve a specified dynamic behavior for specified controlled variables by the adjustment of specified manipulated variables".

In Figure 2, two streams are mixed in a tank. Each stream has a different temperature and percentage of component A, and the composition and temperature of the tank effluent is to be controlled by adjusting the two inlet flow rates. (Note that the tank effluent exits by overflow; thus, the volume is constant.) Again, either individual effluent variable could be controlled, but both depend on the same ratio of inlet flow rates; therefore, the two effluent variables cannot be controlled simultaneously.

3. Process Reliability

Designs must provide reliability to prevent equipment malfunctions from causing costly product stoppage. Standard approaches to increase reliability include material selection, rugged equipment design, break-in periods, and maintenance. In addition, the structure of the design can have a strong impact on process reliability. A few typical process structures are given in Figure 3 that are guided by the principle that parallel structures typically have much higher reliability than series structures (Wells, 1980). We note that process structure can increase reliability at the cost of additional capital investment.

Another important design feature that increases reliability is the ability to by-pass equipment so that the process can continue operation with the equipment removed (at least for a short time at a lower production rate or

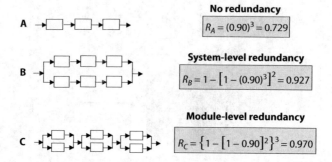

FIGURE 3

Reliability for several structures of elements. Each element has a reliability of 0.90, and all faults are independent.

efficiency). To accommodate this situation, by-pass lines are provided for many pieces of process equipment, such as control valves (leaking), heat exchangers (fouling), and filters (backwashing).

Some equipment is especially critical for plant operation because it affects the entire plant. For example, fuel, steam, compressed air and cooling water must be supplied reliably to the plant and if even one of these failed, a total plant shutdown could occur. Therefore, these utility steams are provided by "distribution systems" in which multiple sources can supply any of multiple consumers.

The stoppage of major equipment for repair or replacement is inevitable because duplicate equipment can be too costly. Acceptable plant reliability can be achieved by maintaining inventory before and after units, so that both upstream and downstream units can continue operation. However, increased inventory brings negative aspects, such as capital and operating costs, possible material degradation, and potential safety hazards. Nevertheless, inventories are here to stay for many processes, and the negative aspects of inventory have to be balanced against the reliability improvement.

4. Process Safety

The students need to recognize that equipment can fail and that people make mistakes. They would benefit from the review of at least one industrial accident case study from the many available (King, 1990; Lees, 1996), and good case study documentation is available from the AIChE (2009) for member universities and companies.

Safety involves a vast range of topics, and the topic selection here is guided by the desire for general applicability. Safety is explained using the six layers (AIChE, 1993) shown in Figure 4. The first four layers are addressed in some detail. The coverage of the lower level basic process control system, BPCS, (alarms, valve failure positions, automatic control loops) can be covered

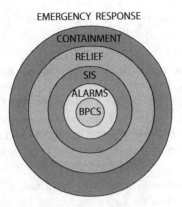

FIGURE 4
Layers of control of safety.

quickly relying on exercises to refresh the students' memories. Typically, the safety instrumented systems (SIS) and pressure relief are new to students, and in this author's opinion, the topics are essential for every practicing engineer. Pressure relief should introduce typical devices, guidelines for their selection, and most importantly, identify locations where the devices should be located (Crowl and Louver, 1990). Automated "Safety Interlock Systems" or "Safety Instrumented Systems" (SIS) perform extreme actions to prevent unsafe conditions from occurring (AIChE, 1993).

All of the prior topics, especially flexibility, reliability and safety, are integrated through lessons and exercises using the hazards and operability "HAZOP" method (Kletz, 1986; Wells, 1996). This method provides a structure for small teams of students to apply their knowledge and creativity to realistic process problems. The structure enables everyone in the group to concentrate on the same unit/node/parameter/guideword at the same time. This focus enables everyone to benefit from the insights of their colleagues as they work in HAZOP groups.

5. Operation During Transitions

Many processes are designed to operate at steady-state conditions during normal operation, but they experience transitions during important situations, such as start-up shutdown, and regeneration. Therefore, process equipment and operating procedures are required to be suitable for operating during important transitions, such as the following.

- **Startup and shutdown**—Integrated process designs face a conundrum, because many sources of material and heat transfer are only available when the process is in operation. Thus, how can the process be started-up?

- **Regeneration**—Transitions for "regeneration" provide special challenges. Here, we use regeneration to denote a wide range, such as regenerating a catalyst or cleaning equipment to maintain hygiene.

- **Short steady-state runs**—In some processes, many different products (levels of purity or material properties) are produced, and each product is produced for a relatively short time by operating the equipment at steady state. This frequent switching is required because of the need to supply the market for all products combined with limited product storage due to cost, safety and product quality requirements.

- **Load following**—Some units act as "utilities', in that they must provide material when other units require it. Examples include steam boilers, fuel vaporizers and hot oil flow for heat exchange (hot oil belt). These units have to respond quickly and without prior warning to large increases or decreases in demand.

Batch processes are essential for many industries. We have seen the importance of the design definition, which is more complex for batch systems than for steady-state systems. Only two key batch definition issues will be noted here.

- **Batch policy**—Since the operation is not at steady state, the design requires knowledge of the best values for key variables during the batch. Typically, flow rates, temperatures, and pressures are adjusted to follow a *desired trajectory or path* from the initial to final states of the batch. Generally, the best operation is determined by solving an optimization problem, and several nice examples are given in Seider et. al. (2004).

- **Production policy**—The integrated operation of the entire plant must be determined to be sure that the desired process operation is feasible. The integrated operation provides information of how long and when each equipment is in operation, the required inventories, and the required flows of materials. Methods for determining the policies are introduced in Biegler et. al. (1997).

6. Dynamic Behavior

Rapid responses to disturbances and to set point changes are required for critical process variables. This requirement can be achieved only if the process equipment and control system are capable of providing fast compensation. As is well recognized, no control algorithm can provide acceptable control performance for a poorly designed plant. While this topic builds on the prior process control course, it provides more emphasis on an issue usually not given sufficient attention in the first course; the effects of process design on control. For example, Figure 5 presents some general guidelines

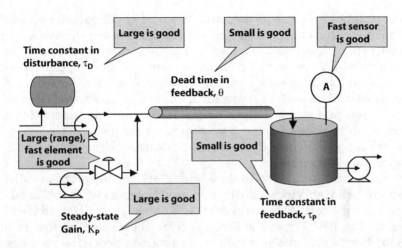

FIGURE 5
Guidelines for good and poor process features for single-loop feedback control.

on the effects of process design on the performance of a single-loop control loop (Marlin, 2000).

7. Efficiency

The goal of the plant is to be profitable, and economic analysis using standard methods for time-value-of money, profitability, and sensitivity analysis is part of every design project (Blank and Tarquin, 2002). After the plant has been constructed and is in operation, economics remains a high priority issue. The plant equipment cannot be changed, at least in the short term. However, because of extra equipment provided for operability many degrees of freedom exist that enable plant personnel to influence profit while achieving other, higher objectives like safety, product quality, production rates, and so forth.

When several parallel equipments are in operation for reliability reasons, process efficiency is affected by the selection of equipment placed in service and the relative "load" on each. For example, the steam demand in a plant can be satisfied by a multitude of loadings of parallel boilers; only one loading has the lowest operating cost because of differences in the efficiencies of the boilers. Importantly, we should recognize all benefits for increased efficiency, including reduced capital investment and reduced effluents.

8. Monitoring and Diagnosis

Undergraduate education properly places emphasis on control via closed-loop systems. However, we often fail to address many important aspects of plant operation involving open-loop decisions made by plant personnel.

Students should recognize the importance of continual monitoring of equipment performance, operating variables, and even the sensors and control system itself. We start with decisions requiring *rapid analysis and action*, which are typically made by plant operators, because people have special advantages in problem solving for complex situations and must monitor the control system for failures in sensors, final elements, or algorithm performance. Many more monitoring and diagnosis challenges occur because of *longer-term changes* in the process. For example, heat exchangers experience fouling, reactor catalyst deactivates, and flow systems can slowly plug.

Students need guidance in building a systematic problem solving approach for identifying and diagnosing root causes of faults. Excellent support using examples from chemical engineering is available in the literature (Woods, 1994; Fogler and LeBlanc, 1995). The six-step approach advocated by Woods provides an excellent learning experience and provides an understanding of the need for substantial design investment for monitoring and diagnosis.

After performing trouble-shooting exercises, the students become aware of the importance of the multitude of sensors and sample collection ports that are required for adequate process monitoring. They can then develop a number of likely faults and the sensors required to identify each of the faults.

Course Design and Delivery

Typically, the design course involves a problem-based approach with the instructor providing only general design goals and the students developing solutions under the guidance of the instructor. Nearly every student group arranged at least one plant visit; some had frequent access to a plant and operating personnel. They reported on all aspects of operability in their written reports and presented highlights of their studies during a class presentation.

The students work in groups of 4–5 people per group. The instructor should provide guidance on group work, specifically developing group behavior practices, meeting agendas, project schedules, and standards for work quality.

Process Case Study

Operability is a generic topic of importance to essentially all processes. Perhaps, the only requirement for a case study is a minimum level of complexity in the equipment and operation; for example, a case study would be

TABLE 1

Example Process Operability Projects

Ammonia reactor and separation loop
Milk powder evaporators and fluid bed drier
Municipal water purification plant
Desalination plant by reverse osmosis
Ethanol Production from corn
Penicillin production (reactor and separation)
Vapor-recovery refrigeration and cooling tower
Boiler feed water treatment and storage
Boiler and condensate return
Wine production
Candy Manufacture
Bio-diesel

too limited if only a single reactor or distillation tower were considered. In recent years, the author's class selected their own case studies, ranging from basic chemicals to pharmaceuticals to food and even medical devices; see Table 1 for some examples.

Experiences and Observations

Students especially enjoyed the safety analysis (HAZOP) and the trouble shooting workshops. They learned to apply engineering principles to solve qualitative problems using insights about causality, dynamic responses, and "order of magnitude" of effects. However, they would not have performed well without instruction on structured problem solving and the opportunity to participate in guided workshops with instructor feedback.

Importantly, their discussions with plant personnel (operators and engineers) reinforced the importance of operability issues. Group work improved their organizational, time management and meeting skills. The usual issues rose in apportioning grades fairly to individuals; this issue was addressed through frequent progress meetings with instructors, documentation of contributions in the final report, and a peer evaluation scheme based on Kaufman et. al. (2000).

The students struggled with the concepts of variability and uncertainty that are inherent in establishing the operating window; they must "unlearn" the tacit assumptions in many previous courses that models and data are exact and conditions (product rate, feed composition, etc.) are static.

A Proposal for Sharing Resources

The following is proposed, "The teaching community would benefit from a repository of teaching and learning materials on process operability supported by technical references to be used by students in problem-based learning". To begin this task, we are assembling a team of instructors to develop teaching and learning materials to aid students and instructors. These materials will be available without charge for university use via the Internet, and they will be supported by CACHE. Interested instructors are encouraged to contact the author.

What might the resources look like? The author has developed operability lessons with power point visual aids that are available on the WWW for review by other instructors. This material is available at Marlin (2008) along with an extended version of this paper containing expanded discussions on each topic and many more process examples.

Conclusions

In this paper, an argument has been presented for greater emphasis to be placed on topics related to process operability, which includes decisions on process structure, equipment capacity, and many issues in control and operation. These issues reinforce prior learning and complement the flowsheeting and equipment sizing traditionally emphasized in design courses. Experience demonstrates that operability applies to essentially all process industries and the topics provide an excellent learning experience for upper-level students. However, challenges remain. A principal challenge is lack of teaching resources, including accessible materials on every topic and solved case studies. For the operability topic to flourish, a central, easily accessed repository of teaching resources is required. The teaching community is invited to participate in discussing the value the operability topics, the resources required and the management of the resources.

Acknowledgments

The author would like to acknowledge Don Woods for his many contributions to engineering education that contributed to the development of the viewpoint expressed in this paper, especially his work on process trouble shooting.

References

AIChE, (1993). *Guidelines for safe automation of Chemical Processes*, AIChE, New York.

AIChE, (2009) *Safety and Chemical Engineering Education Program*, (www.sache.org), last accessed on January 26, 2009.

Biegler, L., I. Grossmann, and A. Westerberg (1997). *Systematic Methods of Chemical Process Design*, Prentice Hall, Upper Saddle River.

Blank, L. and A. Tarquin (2002) *Engineering Economy, 5th Edition*, McGraw-Hill, New York.

Crowl, D. and J. Louvar, (1990). Chemical Process Safety: Fundamentals with Applications, Prentice Hall, Englewood Cliffs.

Fogler, F.S. and S. LeBlanc (1995) *Strategies for Creative Problem Solving*, Prentice-Hall, Upper Saddle River, NJ.

Kaufman, D., R. Felder, and H. Fuller, (2000) *J. of Eng. Ed.*, April 2000, 133–140.

King, R., (1990) Safety in the Process Industries, Butterworth-Heineman, London.

Kletz, T. (1986). *HAZOP and HAZAN, Second Edition*, The Institute of Chemical Engineers, Warkwickschire, UK.

Lees, F. (1996). *Loss Prevention in the Process Industries, Volume 1*, Butterworth-Heinemann, Oxford, UK.

Marlin, Thomas (2000) *Process Control: Designing Processes and Control Systems for Dynamic Performance*, McGraw-Hill, New York.

Marlin, Thomas (2008) http://pc-education.mcmaster.ca/operability/operability_home.htm .

Seider, W., J. Seader, and D. Lewin (2004). *Product and Process Design Principles 2nd Edition*, Wiley, New York.

Valle-Riestra, J.F. (1983). *Project Evaluation in the Process Industries*, McGraw-Hill, New York, pg. 169.

Wells, G., (1980) *Safety in Process and Plant Design*, Godwin, London.

Wells, G., (1996). *Hazard Identification and Risk Assessment*, Institute of Chemical Engineers, Gulf Publishing, Houston.

Woods, D. (1994). *Problem-Based Learning: How to Gain the Most from PBL*, D.R. Woods, Waterdown, Ontario.

49

Assessment of Regional Sustainablility Associated with Biofuel Production

Xiang Li, Aditi Singh and Helen H. Lou*

Department of Chemical Engineering, Lamar University
Beaumont, TX 77710

CONTENTS

ABSTRACT Production of biofuels from cellulosic materials will supply renewable energy to the market and stimulate economic activities in rural areas. In the context of sustainability, the performance of the individual plant, as well as its impacts on the sustainability of the region needs to be evaluated simultaneously. In this work, an integrated study is conducted to assess the impacts of a potential biobutanol production facility in a region concentrated with refineries. The economic and societal impacts are calculated based on the supply-demand relationship depicted by a set of modified regional input-output multipliers. The environmental impact is assessed using the EIO-LCA method. This research provides valuable information for a comprehensive analysis of the potential impacts of the emerging biofuel industry to a region.

KEYWORDS *Sustainability, assessment, biofuel, input-output analysis, EIO-LCA*

---

* All correspondence should be addressed to Prof. Helen H. Lou (Phone: 409-880-8207; Fax: 409-880-2197; E-mail: Helen.lou@lamar.edu).

Introduction

Biobutanol provides a promising alternative to be used as an additive in gasoline. It can be produced by fermentation of biomass like corn, rice straw, sorghum etc. In US, rice is produced mainly in Arkansas Grand Prairie, Northeastern Arkansas and boot heel of Missouri, Mississippi River Delta in Arkansas, Mississippi and Northeast Louisiana, Southwest Louisiana, Coastal Prairie of Texas and Sacramento valley of California. According to the statistics from the US Department of Agriculture, 145,000 acres of rice were planted and harvested in Texas in 2007, producing a yield of 6600 lbs/acre, total production of 9,565,000 cwt sold for 11.3 $/cwt to generate a revenue of $108,085,000. Rice crop produces around 2–4.5 tonnes of rice straw per harvested acre. As a result, Arkansas, Mississippi, Louisiana, Texas and California has abundant rice straw which can be used as feedstock for biofuel production. Usually, rice straw is burnt, disposed off in the field itself to recycle the nutrients or used as cattle feed. It is required to develop alternate methods to use biomass with increasing regulations on the ways of disposing rice straw and limit on the amount of biomass that can be burnt and disposed. Due to the increasing demand of fuels and rapidly diminishing non-renewable resources, biofuel production using available biomass seems to be a promising choice.

With the broad scope of sustainability, not only the performance of the individual plant/process needs to be considered, but also its impacts on the sustainability of the region need to be evaluated as well. In this work, an integrated study is conducted to assess the feasibility of starting a biobutanol production facility in Southeast Texas, which is heavily concentrated with refineries. Furthermore, the economic and societal impact of this new industry on the existing regional economic infrastructure are calculated based on the supply-demand relationship depicted by a set of modified regional input-output multipliers. Then the EIO-LCA method is utilized to estimate the potentional environmental impacts to the region. Different scenarios are analyzed depending upon the change in overall demand of energy with the introduction of biofuels in the market. Figure 1 depicts the major steps in this methodology.

Process Development for Industrial Scale Production of Biobutanol

One of the oldest methods of producing biobutanol is by fermentation of biomass like corn, rice straw, sugar beet, sugar cane, etc. Typically, clostridium acetobuty licum converts the available carbon source, glucose, xylose etc. to

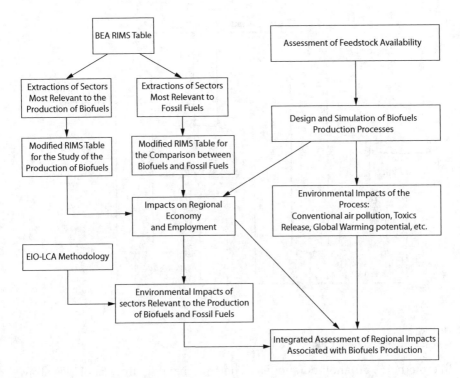

FIGURE 1
Schematic diagram of the methodology.

acetic acid and butyric acid. This procedure is known as acidogenesis. After acidogenesis, through a metabolic shift, it produces acetone, butanol and ethanol in 3:6:1 ration. This pathway is known as ABE fermentation pathway. An "Immobilized Cell Continuous Fermentation and Pervaporative Recovery" (ICCFPR) process is employed for the production of butanol in the current case study. Figure 2 shows the schematic of the ICCFPR process.

In this process, the feed from the feed tank, including corn steep liquor (CSL), glucose and other nutrients, is sent to the immobilized cell reactor. The reactor effluent stream is sent to the buffer tank from where the outlet stream is sent to the pervaporation membrane. This finally produces concentrated solution of acetone-butanol-ethanol. After fermentation, the solvents are recovered from the broth using silicone pervaporation membrane. These solvents are separated into butanol and by-products (acetone and ethanol) by distillation.

In order to develop a process that could use rice straw instead of a mixture of CSL and glucose, it is necessary to understand the chemical composition of rice straw and the procedure of breaking it down to monosachcharides. Rice straw and similar agricultural waste products contain cellulose and hemicelluloses which can be utilized for the production of useful products

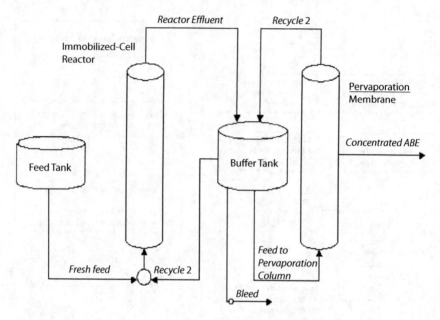

FIGURE 2
Schematic diagram for ICCFPR.

like methane, ethanol, butanol etc. The enzymatic hydrolysis of rice straw produces two types of mono-sachcharides, glucose and xylose, that can be converted to butanol through fermentation.

Regional Input and Output Multipliers

The Regional Input Output Modeling System (RIMS) quantitatively defines the inter-industry relationship within regions and is very useful for conducting regional economic impact analysis. This method was developed by Bureau of Economic Analysis (BEA). It utilizes an accounting framework named I-O table to quantify the relationship between the inputs and outputs. The I-O table was originated from Leontief (1986)'s seminal work of developing input-output models of the U.S. economy. From the Input-Output accounts, a matrix or table A is created to represent the intersectoral relationships. The rows of A indicate the amount of output from industry i required to produce one dollar of output from industry j. Next, consider a vector of final demand, y, of goods in the economy. The sector must produce I × y units of output to meet this demand. At the same time A × y units of output are produced in all other sectors. The resulting total output, X_{direct}, of the overall economy can be written as shown in the following equation:

$$X_{direct} = (I + A)y \qquad (1)$$

Typically, the data in RIMS is derived from BEA's National I-O table and its regional economic accounts. The national I-O table presents the input and output structure of almost five hundred industries in the nation while the regional accounts are used for adjusting the national level data to reflect the regional economy. The advantage of using RIMS is that it can be used for any region composed of one or more counties or for any industry listed in the national table. There are five types of multipliers available in RIMS: Final Demand Multipliers for Output, Final Demand Multipliers for Earnings, Direct Effect Multipliers for Earnings, Final Demand Multiplier for Employment, and Direct Effect Multiplier for Employment. The authors view that the output and earning date reflects the economic dimension of sustainability (the first three sets), the employment data (the last two sets) reflects a societal impacts.

EIO-LCA Method

Input-Output analysis is not only a powerful tool for economic study, but also a useful instrument for analyzing environmental impacts. The Economic Input-Output Life Cycle Assessment (EIO-LCA) method estimates the materials and energy resources required for, and the environmental emissions resulting from, activities in the economy. The EIO-LCA model is built upon the RIMS model. The environmental impact vectors include: Conventional Air Emission Pollutants, Toxics Release Inventory (TRI) Emissions and Global Warming Potential Emissions. Note that different regions may have different environmental issues. Therefore, the user may pick those impacts that are of most concern to the stakeholders.

Assessment of Regional Impacts Associated with Biobutanol Production

In order to conduct a regional sustainability assessment using the input-output model due to a specific industrial project, the detailed cost of the project needs to be provided. On the other hand, the original regional multipliers may need to be modified to fit into the specific need.

A comprehensive analysis was conducted for a production facility with a capacity of producing 150.32×10^6 liters of butanol. The detailed costs of building and running this biobutanol facility derived from a CSL-based production facility are listed in Table 1. These are used as the basic cost for calculating the regional impacts.

The complete list of RIMS multipliers for the target region consists of 19×473 multipliers. Nineteen column industries, which are most relevant

TABLE 1

Stage-wise Cost Details for Biobutanol Production using Rice Straw

Category	Cost (M$)
Rice Milling	319.26
Rice Straw Saccharification	5.97
Miscellaneous chemical product manufacturing	13.77
Tank manufacturing	3.45
Pump and pumping equipment Manufacturing	0.76
Grain farming	261.10
Professional and technical services	1.68
Electricity Consumption	5.74
Wastewater Treatment	3.51
Scientific research and development services	0.80
Office administrative services	0.16
Taxes/Depreciation	3.19
Insurance	2.00
Interest	3.95

to butanol production facility, are selected from these 473 columns. Then for each type of multipliers, a 19 x 14 matrix is subtracted from the original multipliers table and is used as the base for further calculations. Tables A.1~ A.3 in Appendix list the original RIMS multipliers for output, earnings and employment for all the industries/sectors closely associated with biobutanol production. Next, the multipliers corresponding to rice saccharification and biobutanol as a substitute of gasoline are derived based on a comprehensive analysis. The summary of economic and societal impact of biobutanol industry on the region under one scenario is listed in Table 2.

Conventional air pollution emissions may cause public health problems. In the Southeast Texas region, the major concern of pollutant in the category of "Conventional air pollutants" is NOx (nitrogen oxides). So in the environmental impact assessment, the emission of NOx is selected as the major concern in the category of "conventional air pollutants".

Conclusion and Discussions

This study was conducted to assess the potential regional impacts of launching an industrial biobutanol production facility in Southeast Texas. It is demonstrated that RIMS and EIO-LCA models can help predicting the potential economic, societal and environmental impacts caused by emissions.

TABLE A.1

Original RIMS Multipliers for Output Corresponding to Butanol Production

	Rice Milling	Misc. Chemical Product Manufacturing	Metal Tank/ Heavy gauge Manufacturing	Pump/ Pumping Equipment	Grain Farming	Corn Milling	Professional & Technical Services
	311212	325998	332420	333911	1111B0	311221	5419A0
Agriculture, forestry, fishing, and hunting	0.4771	0.0077	0.0012	0.0011	1.1128	0	0.0009
Mining	0.0092	0.0209	0.0057	0.0043	0.0128	0	0.0022
Utilities	0.0264	0.0277	0.0268	0.0216	0.0193	0	0.0109
Construction	0.0053	0.0045	0.0053	0.0048	0.0058	0	0.0037
Manufacturing	1.1078	1.2936	1.301	1.2127	0.1264	1	0.0146
Wholesale trade	0.1063	0.0568	0.0497	0.0483	0.0365	0	0.0075
Retail trade	0.0334	0.0354	0.0436	0.0398	0.0238	0	0.0285
Transportation and warehousing	0.088	0.0452	0.0315	0.0238	0.0273	0	0.013
Information	0.0175	0.016	0.017	0.0189	0.0094	0	0.0267
Finance and insurance	0.0267	0.0216	0.0256	0.0234	0.0219	0	0.0172
Real estate and rental and leasing	0.055	0.0361	0.047	0.0428	0.0649	0	0.0359
Professional, scientific, and technical services	0.0194	0.034	0.0231	0.0207	0.0123	0	1.0274
Management of companies and enterprises	0.0073	0.0182	0.0082	0.011	0.0021	0	0.0014

(Continued)

TABLE A.1

Original RIMS Multipliers for Output Corresponding to Butanol Production (*continued*)

	Rice Milling	Misc. Chemical Product Manufacturing	Metal Tank/ Heavy gauge Manufacturing	Pump/ Pumping Equipment	Grain Farming	Corn Milling	Professional & Technical Services
	311212	325998	332420	333911	1111B0	311221	5419A0
Administrative and waste management services	0.0119	0.0136	0.0122	0.011	0.0083	0	0.0201
Educational services	0.0032	0.0032	0.0042	0.0039	0.0022	0	0.0026
Health care and social assistance	0.0334	0.0328	0.0444	0.0417	0.0238	0	0.0277
Arts, entertainment, and recreation	0.0014	0.0014	0.0018	0.0017	0.0009	0	0.0011
Accommodation and food services	0.0161	0.0162	0.0208	0.0197	0.01	0	0.0121
Other services	0.0283	0.0219	0.0267	0.0205	0.0241	0	0.0147
Households	0.259	0.2546	0.3447	0.3243	0.1848	0	0.2153

TABLE A.1

Original RIMS Multipliers for Output Corresponding to Butanol Production (*continued*)

	Electricity/ Signal Testing	Waste Management & Remediation Services	Scientific Research & Development Services	Office Administrative Services	Funds, trusts, & Other Financial Vehicles	Insurance Carriers	Securities, Commodity Contracts, Investments
	334515	56200	541700	561100	525000	524100	523000
Agriculture, forestry, fishing, and hunting	0.001	0.0017	0.0024	0.0022	0.0008	0.0012	0.0019
Mining	0.0027	0.0189	0.0054	0.0044	0.002	0.0025	0.0048
Utilities	0.0138	0.0417	0.0255	0.018	0.0097	0.0103	0.0246
Construction	0.0048	0.0045	0.0099	0.0054	0.0039	0.0028	0.0068
Manufacturing	1.0509	0.0784	0.0496	0.0321	0.0118	0.0167	0.0267
Wholesale trade	0.0251	0.0433	0.0216	0.0175	0.0073	0.0101	0.0161
Retail trade	0.0347	0.0717	0.0646	0.0655	0.0271	0.042	0.0655
Transportation and warehousing	0.013	0.0374	0.0219	0.0216	0.0141	0.0105	0.0219
Information	0.0182	0.0293	0.0283	0.0285	0.013	0.0287	0.0301
Finance and insurance	0.0206	0.0311	0.0292	0.0339	1.1732	1.2364	1.0724
Real estate and rental and leasing	0.0389	0.0589	0.08	0.0762	0.0355	0.0557	0.0821
Professional, scientific, and technical services	0.026	0.025	1.0476	0.041	0.0323	0.019	0.0642

(*Continued*)

TABLE A.1

Original RIMS Multipliers for Output Corresponding to Butanol Production (*continued*)

	Electricity/ Signal Testing	Waste Management &Remediation Services	Scientific Research & Development Services	Office Administrative Services	Funds, trusts, & Other Financial Vehicles	Insurance Carriers	Securities, Commodity Contracts, Investments
	334515	56200	541700	561100	525000	524100	523000
Management of companies and enterprises	0.0101	0.0033	0.0031	0.0037	0.0211	0.0038	0.0033
Administrative and waste management services	0.0084	1.1608	0.0315	1.0441	0.013	0.0107	0.0203
Educational services	0.0035	0.0055	0.0063	0.0066	0.003	0.0043	0.0069
Health care and social assistance	0.0375	0.0585	0.0684	0.071	0.0297	0.0464	0.0721
Arts, entertainment, and recreation	0.0015	0.0025	0.0026	0.0027	0.0012	0.0017	0.0028
Accommodation and food services	0.0171	0.0292	0.0291	0.0312	0.0154	0.0187	0.033
Other services	0.0161	0.0887	0.0318	0.0293	0.0131	0.0181	0.0321
Households	0.2911	0.4544	0.5315	0.552	0.2308	0.3609	0.5605

TABLE A.2

Original RIMS Multipliers for Earnings Corresponding to Butanol Production

	Rice Milling	Misc. Chemical Product Manufacturing	Metal Tank/ Heavy gauge Manufacturing	Pump/ Pumping Equipment	Grain Farming	Corn Milling	Professional & Technical Services
	311212	325998	332420	333911	1111B0	311221	5419A0
Agriculture, forestry, fishing, and hunting	0.0434	0.0008	0.0002	0.0002	0.1011	0	0.0002
Mining	0.0009	0.0019	0.0005	0.0004	0.0012	0	0.0002
Utilities	0.0035	0.0036	0.0035	0.0029	0.0026	0	0.0014
Construction	0.0014	0.0012	0.0014	0.0013	0.0015	0	0.0010
Manufacturing	0.0941	0.1441	0.2384	0.2255	0.012	0	0.0021
Wholesale trade	0.0291	0.0156	0.0136	0.0132	0.01	0	0.0020
Retail trade	0.0096	0.0102	0.0126	0.0115	0.0069	0	0.0082
Transportation and warehousing	0.0182	0.0115	0.0080	0.0066	0.0071	0	0.0049
Information	0.0043	0.0042	0.0042	0.0047	0.0022	0	0.0062
Finance and insurance	0.0055	0.0044	0.0053	0.0048	0.0045	0	0.0036
Real estate and rental and leasing	0.0033	0.0014	0.0017	0.0014	0.0047	0	0.0017
Professional, scientific, and technical services	0.0074	0.0132	0.0091	0.0081	0.0051	0	0.1525

(Continued)

TABLE A.2

Original RIMS Multipliers for Earnings Corresponding to Butanol Production (*continued*)

	Rice Milling	Misc. Chemical Product Manufacturing	Metal Tank/ Heavy gauge Manufacturing	Pump/ Pumping Equipment	Grain Farming	Corn Milling	Professional & Technical Services
	311212	325998	332420	333911	1111B0	311221	5419A0
Management of companies and enterprises	0.0036	0.0090	0.0040	0.0054	0.0011	0	0.0007
Administrative and waste management services	0.0040	0.0044	0.0040	0.0036	0.0026	0	0.0078
Educational services	0.0014	0.0014	0.0018	0.0017	0.0001	0	0.0012
Health care and social assistance	0.0151	0.0148	0.0200	0.0188	0.0107	0	0.0125
Arts, entertainment, and recreation	0.0004	0.0004	0.0005	0.0005	0.0003	0	0.0003
Accommodation and food services	0.0058	0.0059	0.0076	0.0071	0.0036	0	0.0044
Other services	0.0076	0.0062	0.0076	0.0059	0.0064	0	0.0041
Households	0.0004	0.0004	0.0006	0.0005	0.0003	0	0.0004

TABLE A.2

Original RIMS Multipliers for Earnings Corresponding to Butanol Production (*continued*)

	Electricity/ Signal Testing	Waste Management & Remediation Services	Scientific Research & Development Services	Office Administrative Services	Funds, trusts, & Other Financial Vehicles	Insurance Carriers	Securities, Commodity Contracts, Investments
	334515	56200	541700	561100	525000	524100	523000
Agriculture, forestry, fishing, and hunting	0.0002	0.0003	0.0004	0.0004	0.0002	0.0002	0.0003
Mining	0.0003	0.0017	0.0005	0.0004	0.0002	0.0002	0.0004
Utilities	0.0018	0.0054	0.0034	0.0024	0.0013	0.0014	0.0033
Construction	0.0013	0.0012	0.0026	0.0014	0.0010	0.0008	0.0018
Manufacturing	0.2084	0.0085	0.0067	0.0041	0.0015	0.0022	0.0034
Wholesale trade	0.0069	0.0119	0.0059	0.0048	0.0020	0.0028	0.0044
Retail trade	0.0100	0.0207	0.0187	0.0189	0.0078	0.0121	0.0189
Transportation and warehousing	0.0042	0.0108	0.0074	0.0077	0.0054	0.0034	0.0072
Information	0.0047	0.0070	0.0069	0.0070	0.0035	0.0069	0.0080
Finance and insurance	0.0043	0.0064	0.0060	0.0070	0.1525	0.2788	0.4128
Real estate and rental and leasing	0.0012	0.0017	0.0032	0.0027	0.0015	0.0023	0.0031
Professional, scientific, and technical services	0.0097	0.0108	0.4014	0.0171	0.0143	0.0084	0.0298

(*Continued*)

TABLE A.2
Original RIMS Multipliers for Earnings Corresponding to Butanol Production (*continued*)

	Electricity/ Signal Testing	Waste Management & Remediation services	Scientific Research & Development Services	Office Administrative Services	Funds, trusts, & Other Financial Vehicles	Insurance Carriers	Securities, Commodity Contracts, Investments
	334515	56200	541700	561100	525000	524100	523000
Management of companies and enterprises	0.0050	0.0016	0.0015	0.0018	0.0104	0.0019	0.0016
Administrative and waste management services	0.0028	0.3006	0.0119	0.4198	0.0044	0.0036	0.0070
Educational services	0.0016	0.0025	0.0028	0.0029	0.0013	0.0019	0.0031
Health care and social assistance	0.0169	0.0264	0.0308	0.0320	0.0134	0.0209	0.0325
Arts, entertainment, and recreation	0.0004	0.0007	0.0008	0.0008	0.0004	0.0005	0.0008
Accommodation and food services	0.0062	0.0106	0.0105	0.0113	0.0056	0.0068	0.0120
Other services	0.0047	0.0248	0.0092	0.0086	0.0038	0.0052	0.0090
Households	0.0005	0.0008	0.0009	0.0009	0.0004	0.0006	0.0009

TABLE A.3

Original RIMS Multipliers for Employment Corresponding to Butanol Production

	Rice Milling	Misc. Chemical Product Manufacturing	Metal Tank/ Heavy Gauge Manufacturing	Pump/ Pumping Equipment	Grain Farming	Corn Milling	Professional & Technical Services
	311212	325998	332420	333911	1111B0	311221	5419A0
Agriculture, forestry, fishing, and hunting	3.1787	0.0752	0.0174	0.0160	7.4021	0.0000	0.0141
Mining	0.0082	0.0189	0.0051	0.0046	0.0111	0.0000	0.0019
Utilities	0.0418	0.0427	0.0417	0.0339	0.0308	0.0000	0.0171
Construction	0.0380	0.0322	0.0381	0.0350	0.0415	0.0000	0.0267
Manufacturing	2.1597	1.8126	5.5698	3.9680	0.1517	0.0000	0.0414
Wholesale trade	0.5165	0.2761	0.2415	0.2345	0.1774	0.0000	0.0362
Retail trade	0.4162	0.4409	0.5436	0.4958	0.2973	0.0000	0.3549
Transportation and warehousing	0.3936	0.2946	0.2014	0.1705	0.1755	0.0000	0.1121
Information	0.0791	0.0742	0.0749	0.0853	0.0402	0.0000	0.1131
Finance and insurance	0.1129	0.0902	0.1083	0.0990	0.0932	0.0000	0.0729
Real estate and rental and leasing	0.0991	0.0404	0.0491	0.0413	0.1539	0.0000	0.0523
Professional, scientific, and technical services	0.1301	0.2278	0.1537	0.1380	0.0885	0.0000	1.9944
Management of companies and enterprises	0.0786	0.1969	0.0884	0.1187	0.0232	0.0000	0.0156

(Continued)

TABLE A.3

Original RIMS Multipliers for Employment Corresponding to Butanol Production (*continued*)

	Rice Milling	Misc. Chemical Product Manufacturing	Metal Tank/ Heavy gauge Manufacturing	Pump/ Pumping Equipment	Grain Farming	Corn Miling	Professional & Technical Services
	311212	325998	332420	333911	1111B0	311221	5419A0
Administrative and waste management services	0.1722	0.1745	0.1674	0.1506	0.108	0.0000	0.3797
Educational services	0.0654	0.0643	0.0850	0.0798	0.0455	0.0000	0.0537
Health care and social assistance	0.4105	0.4030	0.5451	0.5126	0.2926	0.0000	0.3404
Arts, entertainment, and recreation	0.0245	0.0242	0.0312	0.0294	0.0161	0.0000	0.0194
Accommodation and food services	0.4127	0.4167	0.5379	0.5082	0.2571	0.0000	0.3112
Other services	0.2770	0.2426	0.3006	0.2393	0.2258	0.0000	0.1640
Households	0.0490	0.0481	0.0652	0.0613	0.0349	0.0000	0.0407

TABLE A.3

Original RIMS Multipliers for Employment Corresponding to Butanol Production (*continued*)

	Electricity/ Signal Testing	Waste Management & Remediation Services	Scientific Research & Development Services	Office Administrative Services	Funds, trusts, & Other Financial Vehicles	Insurance Carriers	Securities, Commodity Contracts, Investments
	334515	56200	541700	561100	525000	524100	523000
Agriculture, forestry, fishing, and hunting	0.0141	0.0231	0.0335	0.0332	0.0124	0.0171	0.0277
Mining	0.0025	0.0162	0.0048	0.0039	0.0017	0.0022	0.0042
Utilities	0.0217	0.0645	0.0402	0.0284	0.0153	0.0162	0.0390
Construction	0.0347	0.0321	0.0713	0.0390	0.0281	0.0203	0.0491
Manufacturing	3.3677	0.1135	0.1224	0.0769	0.0291	0.0431	0.0635
Wholesale trade	0.1221	0.2105	0.1050	0.0850	0.0356	0.0490	0.0784
Retail trade	0.4327	0.8941	0.8058	0.8171	0.3380	0.5241	0.8164
Transportation and warehousing	0.1073	0.2523	0.1688	0.1772	0.1247	0.0852	0.1972
Information	0.0850	0.1264	0.1260	0.1254	0.0616	0.1232	0.1422
Finance and insurance	0.0874	0.1307	0.1233	0.1418	2.2029	5.6821	7.6071
Real estate and rental and leasing	0.0387	0.0544	0.0998	0.0840	0.0451	0.0718	0.0968
Professional, scientific, and technical services	0.1752	0.1824	6.6330	0.3165	0.2523	0.1342	0.4842

(*Continued*)

TABLE A.3

Original RIMS Multipliers for Employment Corresponding to Butanol Production (*continued*)

	Electricity/ Signal Testing	Waste Management & Remediation Services	Scientific Research & Development Services	Office Administrative services	Funds, trusts, & Other Financial Vehicles	Insurance Carriers	Securities, Commodity Contracts, Investments
	334515	56200	541700	561100	525000	524100	523000
Management of companies and enterprises	0.1094	0.0358	0.0333	0.0403	0.2290	0.0409	0.0355
Administrative and waste management services	0.1256	7.3722	0.5751	7.8326	0.1832	0.1538	0.3104
Educational services	0.0720	0.1126	0.1290	0.1341	0.0599	0.0882	0.1401
Health care and social assistance	0.4604	0.7187	0.8400	0.8721	0.3653	0.5703	0.8858
Arts, entertainment, and recreation	0.0262	0.0439	0.0460	0.0479	0.0218	0.0300	0.0486
Accommodation and food services	0.4421	0.7510	0.7508	0.8063	0.3955	0.4829	0.8478
Other services	0.1947	0.9080	0.3744	0.3546	0.1562	0.2217	0.3624
Households	0.0550	0.0859	0.1005	0.1044	0.0436	0.0682	0.1060

TABLE 2

Summary of Economic and Societal Impact of a Biobutanol Industry on the region

	Output	Earnings	Employment
Agriculture, forestry, fishing, and hunting	25.19	2.29	168
Mining	0.39	0.04	0
Utilities	0.84	0.11	1
Construction	0.20	0.05	1
Manufacturing	25.19	2.23	48
Wholesale trade	2.65	0.72	13
Retail trade	1.08	0.31	14
Transportation and warehousing	2.15	0.47	11
Information	0.52	0.13	2
Finance and insurance	1.50	0.34	7
Real estate and rental and leasing	2.10	0.13	4
Scientific and technical services	0.78	0.28	5
Management of companies and enterprises	0.20	0.10	2
Administrative and waste management services	0.63	0.19	7
Educational services	0.10	0.05	2
Health care and social assistance	1.08	0.49	13
Arts, entertainment, and recreation	0.04	0.01	1
Accommodation and food services	0.50	0.18	13
Other services	0.95	0.26	9
Households	8.39	0.01	2
Total	74.46	8.39	322

This research will facilitate a comprehensive study of biofuel production on regional sustainability.

Note that this study has its own limitations. Due to the availability of data, transportation cost of cellulosic material to the plant is not included. Some impacts of biomass production to the ecosystem, such as nutrient cycling, erosion, watershed quality is not known yet. More experimental works are needed to accurately estimate the cost of rice straw saccharification. It is also noticed that the large area of land needed for rice production could be a challenging issue for supplying sufficient feedstock.

On the other hand, data uncertainties in both national and regional I-O tables will affect the accuracy of the assessment of regional economic and societal impacts. Regarding the environmental impact analysis, it was recognized the EIO-LCA method has its own limitations. As an LCA tool, the EIO-LCA models are incomplete in as much as a limited number of environmental effects are included. In most countries, regions and industries, the environment statistical data is too limited or unreliable to do robust empirical studies.

In EIO-LCA method, the environmental impacts are correlated to dollar values. This makes EIO-LCA method convenient to use, but also raises some arguments. A more accurate assessment of regional environmental impacts relies on the advancement of the LCA methodology. Effective uncertainty handling methodologies are needed for a robust assessment and decision-making. Statistical approaches based on probability theory could be helpful in reducing the uncertainties in the assessment.

Acknowledgments

This work is in part supported by the National Science Foundation under Grants CBET0731066 and DUE 0737104.

References

Carnegie Mellon University Green Design Institute. (2008) Economic Input-Output Life Cycle Assessment (EIO-LCA), US 1997 Industry Benchmark model (Internet), Available from: (http://www.eiolca.net) Accessed 1 January, 2008.

Daley, W. M., Ehrlich, E. M., Landefeld, J. S., Barker, B. L. (1997). Regional Multipliers: A user Handbook for the Regional Input-Output Modeling System, US Department of Commerce. Washington, DC.

Huang, W. C., Ramey, D. E., Yang, S. T. (2004). Continuous production of butanol by Clostridium acetobutylicum immobilized in a fibrous bed bioreactor. Appl. Biochem. Biotech., 113–116, 887–898.

National Agricultural Statistics Service. Texas state agriculture overview-2007, http://www.nass.usda.gov/Statistics_by_State/Ag_Overview/AgOverview_TX.pdf.

Qureshi, N., Hughes, S., Maddox, I. S., Cotta, M. A. (2005). Energy-efficient recovery of butanol from model solutions and fermentation broth by adsorption. Bioproc. Biosyst. Eng., 27, 215–222.

Schnepf, R. D., Just, B. (1995). Rice background for 1995 farm legislation, an economic research service report, http://www.ers.usda.gov/Publications/aer713.

Singh, Aditi. (2007). Multi-objective Decision Making In Design For Sustainability. Ph.D dissertation, Lamar University, Beaumont, TX.

50

Sustainable Supply Chain Planning for the Forest Biorefinery

Louis Patrick Dansereau[a], Mahmoud El-Halwagi[b] and Paul Stuart[a*]

[a]*NSERC Environmental Design Engineering Chair in Process Integration, Department of Chemical Engineering, École Polytechnique–Montréal, Montréal H3C 3A7, Canada*
[b]*Department of Chemical Engineering, Texas A&M University, College 3, Texas 77843-3122, USA*

CONTENTS

ABSTRACT The North American forestry industry is currently facing an economic stalemate situation due to mature and declining domestic markets, production overcapacity and global low cost competition. The transformation to the forest biorefinery represents an important and increasingly considered opportunity for this industry to address this situation. Biorefining implies a more complete utilization of renewable forest biomass and the diversification of the traditional core business into the production of green organic chemicals in addition to wood, as well as pulp and paper products. However, to ensure a sustainable transformation, forestry companies must examine changing their manufacturing culture and in that perspective, implementing innovative supply chain strategies will be a critical issue. To illustrate the impact of going from a manufacturing-centric operating policy to a margins-centric one, two supply chain models representing a pulp and paper mill under these policies have been compared under different market scenarios. The models incorporate procurement and manufacturing operations, and are formulated as mixed-integer linear programs. Results show that having a margins-based business model can increase profitability between 15%

---

* To whom all correspondence should be addressed

and 137%, depending of market scenarios. As biomass costs increase and product prices decrease, margins improvement increase. Even though transition and production costs are generally higher when operating under this policy, reduced inventory costs and fulfillment of better orders significantly counterbalance this increase. Operating under this policy can result in better profitability for paper manufacture, and for similar reasons will lead to even better results with the implementation of the forest biorefinery.

KEYWORDS   *Supply chain, planning, pulp and paper, forest biorefinery*

Introduction

The forestry industry is an important industry of North America. However, the latter has been experiencing significant financial problems for some time. Due to a mature and declining domestic markets, production overcapacity and global low cost competition, it is now facing an economic stalemate. To overcome this difficult period, pulp and paper (P&P) companies have implemented in the last fifteen years strategies such as drastic reduction in costs and mergers and acquisitions, resulting in plant closures, layoffs, as well as little investment in research and development. These strategies may have helped in the industry's survival in the shorter term, but is not enough to guarantee success over the long term (Orzechowska, 2005).

Recently, some major forestry companies have shown an interest in exploring opportunities for the implementation of the forest biorefinery. This concept relies in a more complete utilization of renewable forest biomass and in the diversification of the regular production of this industry, such as timber, pulp and paper products, to also produce energy, biofuels and green organic chemicals. This diversification provides an interesting opportunity for the forestry industry to overcome this difficult period by producing value added products. Moreover, the biorefinery could present a significant opportunity for reducing environmental impacts such as the production of green house gases. However, this transformation implies several strategic changes, particularly in their business model. According to Thorp (2005), the biggest challenge for this industry will be to move away from the commodity business mentality. Producing high volumes of undistinguished products with low margins is not a sustainable business model for the biorefinery.

Due to the capital intensiveness of the forestry industry and poor returns over the longer term, capital spending in recent years has been significantly reduced (to under 50% of depreciation in many cases) and has focused on reducing operating costs. Supply chain (SC) management concepts are of increasing importance relative to the cost effectiveness of overall operations and can greatly help the North American P&P industry compete globally (Eamer 2003, McLean 1999).

Compared to other industries, the P&P industry has been relatively slow to implement supply chain management practices, mainly because of the differences and special characteristics of this industry (Lail 2003). Carlsson et al. (2006) provide an overview of the supply chain literature and models developed so far for this industry.

Implementing innovative SC strategies will be a critical issue for the transformation to the biorefinery. To ensure the sustainable transformation, one of the first things to do for this industry is to change their manufacturing culture, to go away from the commodity business mentality. In this manufacturing-centric operating policy, P&P companies view the process as the main actor in profitability. It is believed that minimizing production costs will result in the highest profitability. However, SC costs are often neglected and lesser profit is generated. Hence, by looking more carefully at the SC, that is, by minimizing the costs over the whole SC and by selecting orders that provide the best returns, additional value could be created. This description will here be used as a definition for the margins-based operating policy. This strategy could provide the P&P industry an opportunity to overcome these difficult times and to have a more adapted business model for the transformation to the biorefinery.

The main objective of this paper is to show the benefits for a forestry company to change manufacturing culture, particularly in a constrained and difficult market environment. To achieve this, two SC models representing the manufacturing-centric and the margins-centric operating policies have been developed and tested under different order books, biomass and product prices scenarios.

Problem Description

The operations considered in the models consisted of the fibre supply chain of a hypothetical integrated newsprint mill of a capacity of 1000 tons of paper per day, including the fibre procurement, the manufacturing and the distribution/customer cycles. These operations are presented in figure 1. The manufacturing activities at the mill consist of four parallel thermo mechanical pulping (TMP) lines and one deinking pulping line (DIP). A total of six different grades and grammages of newsprint paper can be produced on two identical paper machines. Transitions of paper grades on paper machines are sequence dependant. The final product is sold to different customers in the form of paper rolls of various sizes. The manufacturer has an order book of different possible orders to fulfill. The fibre procurement for the TMP lines consists of chips and logs of various wood species. Old newsprint (ONP) and old magazines (OMG) are the main raw materials considered for the DIP line. Procurement of kraft pulp, which is an ingredient of certain paper grades, is also considered. The procurement of all these raw materials can be done every week, according to the suppliers' capacity and market costs.

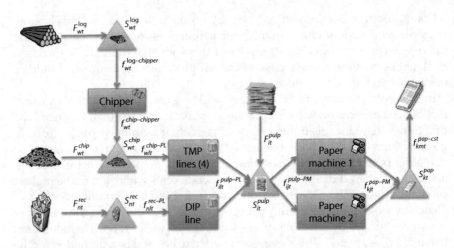

FIGURE 1
Supply chain modeled.

Models Formulation

Depending of the manufacturing strategy, the aim of the model differs. In the manufacturing-centric model, the production sequence on the paper machine is predetermined by a heuristic to minimize transition time and cost. However, the production campaign length can vary. The selection of paper grade to be produced at each time frame is therefore determined and all the capacity available is used. Thus, the aim of this model is to select customer orders that are possible to fulfill according to the production sequence.

In the margins-centric operating policy, any transition on the paper machine is allowed, even if it is longer and more expansive. Consequently, the aim of this model is to determine the production sequence that will allow the fulfillment of the highest margins customer orders.

Both models are formulated as mixed-integer linear programming problems with a discrete time horizon of 30 days. Decisions to be made, such as the production of a given recipe during a particular time period, are represented through several constraints. In addition, capacity constraints are used to represent limitations of available resources and manufacturing capability. Sequence-dependant transitions on paper machines are made possible through the introduction of additional logical conditions.

The objective function of both models is to maximize the total profit over the time horizon. Revenues are based only on satisfied customer orders. Paper prices differ according to the grades and orders. Fibre supply costs include cost for procurement of logs and chips of different wood species, recycled paper and pulp. Fixed and variable processing costs for the chipper, for each pulp line and for each paper machine are included, as well as transition costs. Inventory holding costs for logs, chips, recycled paper

TABLE 1

Variations of Test case Considered

Variations of Test Case	Description
Case # 1	Wood fibre cost 25% higher than test case
Case # 2	Paper price 10% lower than test case
Case # 3	Recycled paper cost 50% lower than test case
Case # 4	Recycled paper cost 50% higher than test case

and paper products are also considered. Pulp holding costs are not taken into account since it is assumed that the mill produces only what is needed for paper manufacturing. Finally, to represent the manufacturer's goodwill, penalty costs for unmet demand have been incorporated. Because of the fixed production sequence and finite number of possible orders to fulfill in the manufacturing-centric model, the order selection becomes fixed and thus, the objective function of the latter model is to minimize costs.

Results and Interpretation

To compare the two operating policies, both models were run under different market scenarios. These scenarios are described in table 1. Each of these cases were tested together for a total of 12 different scenarios.

Results presented have been generated using AMPL modeling system on an Intel 2.8 GHz computer and the MIP CPLEX 11.0 optimizer. The models developed contain 540 binary variables, 6590 continuous variables and a total of 14692 constraints. In the manufacturing-centric model, since the production sequence of paper machines is fixed, the model is simplified, and optimized profits with a relative optimality gap of less than 1% have been obtained in less than 2 seconds. However, because of the increased complexity of the margins-centric model, solutions with a relative optimality gap between 1% and 3% have been obtained in 2 hours.

As can be seen in figure 2, profits are always higher for the margins-based operating policy. Moreover, as market conditions become more difficult or constrained, the relative increase in profit of the latter strategy over manufacturing-centric increases. In the most favorable market environment (scenario 3), there is a 15% profit improvement when the margins-based operating policy is used over the manufacturing-centric one. This profit improvement goes up to 137% for the most difficult market conditions (scenario 1-2-4). For all cost scenarios, the margins (in dollar per ton of paper sold) are between 21$/ton and 27$/ton better for the margins-centric policy. This improvement can be of significant importance in the most difficult market conditions, as it can make the difference between selling at a loss, or not.

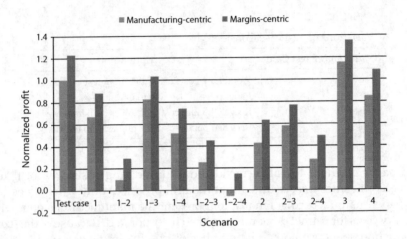

FIGURE 2
Normalized profit for different scenarios.

Processing, transition and procurement costs are presented in figure 3. These operating costs are typically higher (up to 3%) in the margins-based operating policy. More transitions and the production of more expansive grades are responsible for this increase. However, inventory costs are between 14% and 24% lower for the latter policy. Because the goal of this strategy is maximizing profit over the entire supply chain, the production sequence will be tightly aligned with the selected order due dates, thus minimizing costs. The savings in inventory costs and the opportunity of fulfilling more profitable orders counterbalance this increase in production costs.

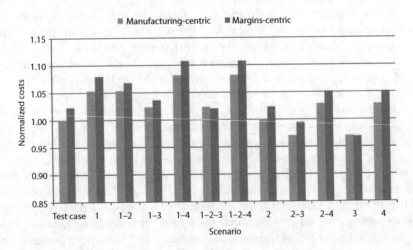

FIGURE 3
Normalized procurement, transition and processing costs for different scenarios.

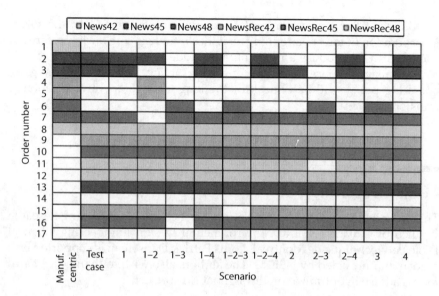

FIGURE 4
Selection of orders.

Selected customer orders for the different scenarios in the manufacturing-centric and margins-centric policies are presented in figure 4. If a cell is blank, the corresponding order is not selected. Colors represent the different kind of paper grade required in each order. Because of the sequence heuristic in the manufacturing- centric model, the selected orders are always the same, no matter what the market conditions are, and thus results are only presented for the test case (column 1). However, depending on market conditions, different orders and paper grades produced are selected for the margins-based operating policy. This is because margins differ according to the biomass cost and product prices. In other words, overall margins are a function of the costs over the supply chain and not only production costs.

Conclusion

Changing the manufacturing culture to operate with a margins-base SC policy can result in significant margins improvement because fibre procurement, manufacturing and demand cycle costs are simultaneously addressed. Moreover, Laflamme-Mayer et al. (2007 and 2008) have shown the benefits for a P&P company to implement a multi-scale planning approach and flexible manufacturing. Incorporating these SC strategies together with a margins-centric mentality could further increase the profitability of the activities of a P&P company.

This result has an even more far reaching implication. With the future incorporation of forest biorefining activities into their core business, forestry companies will produce a range of value-added products that they should manage using a margins-centric operating policy. This critical change in operating policy will be critical for the long term competitive position of transformed P&P mills.

Acknowledgments

This work was supported by the Natural Sciences and Engineering Research Council of Canada (NSERC) Environmental Design Engineering Chair at École Polytechnique de Montréal. Louis Patrick Dansereau is supported by a scholarship provided by NSERC. The authors also wish to thank Carl Laird for his advise in optimization throughout this project.

References

Carlsson, D., S. et al. (2006). Supply chain management in the pulp and paper industry, Working paper DT-2006-AM-3, Interuniversity Research Center on Enterprise Networks, Logistics and Transportation (CIRRELT), Université Laval, Québec, G1K7P4, Canada.

Eamer, R. J. (2003). The road to recovery facing the brutal facts. Paper presented at the International Finance Forum for the Pulp, Paper, and Allied Industries, Atlanta, USA.

Laflamme-Mayer, M. et al. (2007). Multi-scale on-line supply chain planning Part B: Model formulation and test case results. *submitted to AIChE Journal.*

Laflamme-Mayer, M. et al. (2008). Manufacturing Flexibility to Support Cost Effective Fibre Supply, *submitted to TAPPI Journal.*

Lail, P.W., Supply chain best practices for the pulp and paper industry, Atlanta, GA: Tappi Press, 2003

McLean, S. (1999). Finding strategic advantage through SCM. *PPI*, 41(10), pp.28

Orzechowska, A. (2005). Saving the Canadian Industry : Discussing Possible Solutions, *Pulp and Paper Canada*, 106(1), pp.25–27

Thorp, B., (2005) Biorefinery Offer industry Leaders Business Model for Major Change, *Pulp and Paper*, 79(110, pp.35–39

51

A Property-Integration Approach to the Design and Integration of Eco-Industrial Parks

Eva M. Lovelady and Mahmoud M. El-Halwagi
Department of Chemical Engineering, Texas A&M University
College Station, TX 77843-3122, USA

Irene M.L. Chew, Denny K.S. Ng and Dominic C.Y. Foo
Department of Chemical and Environmental Engineering
University of Nottingham Malaysia, Broga Road, 43500 Semenyih, Selangor, Malaysia

Raymond R. Tan
Center for Engineering and Sustainable Development Research,
De La Salle University-Manila
2401 Taft Avenue, 1004 Manila, Philippines

CONTENTS

ABSTRACT An eco-industrial park (EIP) represents an infrastructure that provides integrative services to multiple industrial processes. The primary objective is to induce integration of materials exchange, utility sharing, waste treatment, and discharge. Because of the high level of interaction provided by the EIP, a system integration approach must be adopted to design the common infrastructure and to make decisions on the optimal allocation of streams. The objective of this work is to develop a systematic approach to the design of EIPs using a property-integration framework. Interfaces among the various processing facilities are characterized in terms of key properties. Next, property integration techniques are used to synthesize the EIP. A structural representation is used to embed potential configurations of interest. The representation accounts for the possibilities of direct reuse/recycle, material (waste) exchange, mixing and segregation of different streams, separation and treatment in interception units, and allocation to process users (sinks). A case study is solved to illustrate the applicability of the devised approach.

KEYWORDS *Eco-industrial parks, property integration, process integration, sustainability*

Introduction

The concept of an eco-industrial park (EIP) is gaining much interest as an effective organizational framework and operating facilities involving a cluster of several processes that share a common infrastructure that is designed and operated primarily to induce integration of materials exchange, waste treatment, and discharge. A well-designed EIP provides numerous ecological and economic benefits (Gibbs and Deutz, 2007). They also lead to benefits for the local governments by generating more revenues and taxes and by lowering the burden on local treatment facilities (Lowe, 1997). Recently, there have been many successful cases of well-functioning EIPs. A commonly-referenced EIPs is Kalundborg Park, located in Denmark. Other successful examples span in various parts of the world.

Recent research contributions have been made in the area of locating and designing EIPs. Chew et al. (2008) as well as Lovelady and El-Halwagi (2009) developed an integrated framework and an optimization formulation for the design of a multi-facility EIP with composition-based constraints. Fernandez and Ruiz (2009) developed a mathematical model that can be used to locate the geographical location of sustainable industrial areas. Sendra et al. (2007) used a framework of materials flow analysis to keep track of the various inputs and outputs of the EIP. Zhao et al. (2007) developed a dynamic model to simulate and retrofit an EIP in Changchun Economic and Technological Development Zone in China. Spriggs et al. (2004) developed a framework to characterize the challenges associated with EIPs and grouped them into two classes: technical/economic challenges and organizational/commercial/political challenges.

Problem Statement

Given a set of multiple processes with the following:

- A set of process sinks (units): SINKS = $\{j|j = 1,2, ..., N_{sinks}\}$. Each sink requires a certain flow rate, G_j and a given constraint on inlet property for each property p:

$$p_{p,j}^{\min} \leq p_{p,j} \leq p_{p,j}^{\max} \qquad \forall j, \forall p \tag{1}$$

where $p_{p,j}^{\min}$ and $p_{p,j}^{\max}$ are given lower and upper bounds on permissible values of the pth property entering the jth sink.

- A set of process waste streams referred to as sources: SOURCES = $\{i \mid i = 1,2, ..., N_{sources}\}$. Each source has a given flow rate, F_i, and properties $p_{p,i}$.

The objective is to develop a systematic procedure for the optimal design of an EIP that treats various sources and assigns them to different sinks. The EIP requires the installation of a set of interception units: INTERCEPTORS = $\{k \mid k = 1,2, ..., N_{\text{Int}}\}$ to be used for treating the effluents by modifying the targeted properties to allow them to be assigned to various process sinks for further recovery or environmental discharge. The interceptors can be either a fixed outlet concentration or removal types. Available for service are f^{th} fresh (external) resources that can be purchased to supplement the use of process sources.

Optimization Model

A source-interception-sink superstructure (Figure 1) is used here, analogous to the composition-based EIP framework developed by Chew et al. (2008) as well as Lovelady and El-Halwagi (2009). Each source is split into several fractions that are fed to interceptors (Int k) which adjust properties. Intercepted streams are allowed to mix and be allocated to process sinks or to final discharge.

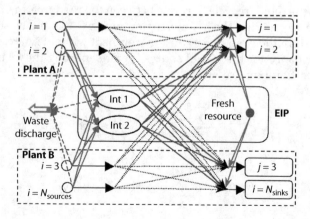

FIGURE 1
Superstructure for property-based EIP.

Before proceeding to the mathematical formulation, a mixing rule is needed to define all possible mixing patterns among individual properties. One such form for mixing is the following expression (Shelley and El-Halwagi, 2000):

$$\psi(\bar{p}_p) = \sum_i x_i \psi(p_{p,i}) \; \forall p \tag{2}$$

where $\psi(p_{p,i})$ and $\psi(\bar{p}_p)$ are linearized operators on the property p of stream i and the mixture property $\bar{p}_p$, respectively, and x_i is the fractional contribution of stream i of the total mixture flow rate.

The optimization objective is to minimize the total annualized cost of the interception devices in the EIP and the cost of fresh resource(s). Therefore, the objective function is given by:

Minimize total annualized cost =

$$\sum_{k=1}^{N_{int}} F_k^{Int} \, C_k^{Int} + \sum_{f=1}^{N_{fresh}} F_f^{Fresh} \, C_f^{Fresh} \tag{3}$$

where C_k^{Int} is the unit cost associated with the kth interceptor (including fixed capital cost) and C_f^{Fresh} is the unit cost of the fth fresh resource. F_k^{Int} is the total intercepted flow rate at interceptor k while F_f^{Fresh} is the total amount of the fth fresh resource.

The model is subject to the following constraints:

Distribution of sources for reuse/recycle, to interception devices and final waste discharge:

$$F_i = \sum_{j=1}^{N_{sinks}} H_{i,j} + \sum_{k=1}^{N_{int}} w_{i,k} + w_i^{waste} \qquad \forall i \in \{1 \ldots N_{Sources}\} \tag{4}$$

where $H_{i,j}$ is the reuse/recycle flow rate between the ith source and j^{th} sink, $w_{i,k}$ is the ith source entering the kth interceptor and w_i^{waste} is the waste flowrate discharged from the ith source to the environment without interception.

Material balance for the mixed sources before entering the kth interceptor and its property mixing rules:

$$W_k = \sum_{i=1}^{N_{sources}} w_{i,k} \qquad \forall \, k \in \{1 \ldots N_{int}\} \tag{5}$$

$$W_k \psi(\bar{p}_{p,k}) = \sum_{i=1}^{N_{sources}} w_{i,k} \psi(p_{p,i}) \qquad \forall k, \forall p \tag{6}$$

Distribution of intercepted streams from the EIP:

$$W_k = \sum_{j=1}^{N_{sinks}} g_{k,j} + w_k^{waste} \qquad \forall\, k \tag{7}$$

where W_k is the intercepted flowrate, and $g_{k,j}$ and w_k^{waste} are the flow rate of the kth source fed to the jth sink and wastewater discharge to the environment, respectively.

Mixing of the distributed streams before the jth sink and its property mixing rules:

$$G_j = \sum_{i=1}^{N_{sources}} H_{i,j} + \sum_{f}^{N_{fresh}} F_{f,j} + \sum_{k=1}^{N_{int}} g_{k,j} \quad \forall j \tag{8}$$

$$G_j \psi(\bar{p}_{p,j}) = \sum_{i=1}^{N_{sources}} H_{i,j} \psi(p_{p,i}) + \sum_{f}^{N_{fresh}} F_{f,j} \psi(p_{p,f}^{Fresh})$$

$$+ \sum_{k=1}^{N_{int}} g_{k,j} \psi(p_{p,k}^{int}) \quad \forall j \tag{9}$$

where $F_{f,j}$ is the flow rate of the fth fresh resource, while $p_{p,f}^{Fresh}$ and $p_{p,k}^{int}$ are the properties for the fth fresh resource and the kth intercepted source, respectively.

The total flow rate of the fth fresh resource is given as:

$$F_f^{Fresh} = \sum_{j} F_{f,j} \quad \forall j \tag{10}$$

Sink constraints:

$$p_{p,j}^{min} \leq p_{p,j} \leq p_{p,j}^{max} \quad \forall j, \forall p \tag{1}$$

This is a nonlinear program (NLP) that can be solved to determine the allocation of streams and design of the EIP. A property-based water minimization case study is next used to illustrate the proposed method.

Water Minimization Case Study

Two industrial wafer fabrication plants that possess similar process water characteristics, i.e. resistivity and heavy metal content are located within an EIP. Table 1 tabulates the process sinks and sources for both plants, adapted

TABLE 1

Limiting Data for Case Study

Plant	Process	Flow rate (t/h)	Resistivity, R (MΩ)		Operator, ψ (MΩ^{-1})		Heavy Metal Concentration (ppm)
			Lower Bound	Upper Bound	Lower Bound	Upper Bound	
	(Sink)						
	Wet (SK1)	500	7	18	0.1429	0.0556	—
	Litography (SK2)	450	8	15	0.1250	0.0667	—
	CMP (SK3)	700	10	18	0.1000	0.0556	—
	Etc (SK4)	350	5	12	0.2000	0.0833	—
Plant A	**(Source)**						
(Ng et al., 2009)	Wet I (SR1)	250	1		1		5
	Wet II (SR2)	200	2		0.5		4.5
	Litography (SR3)	350	3		0.3333		5
	CMP I (SR4)	300	0.1		10		10
	CMP II (SR5)	200	2		0.5		4.5
	Etc (SR6)	280	0.5		2		5
	(Sink)						
	Wafer Fab (SK5)	182	16	20	0.0625	0.0500	—
	CMP (SK6)	159	10	18	0.1	0.0556	—
Plant B	**(Source)**						
(Gabriel et al., 2003)	50% spent (SR7)	227	8		0.1250		5
	100% spent (SR8)	227	2		0.5000		11
	Ultra pure water (UPW)	?	18	—	0.0556		—

TABLE 2

Performance and Unit cost for Interceptor

Interceptor	Outlet Concentration of Heavy Metal (ppm)	Resistivity (MΩ)	Interception Cost ($/t)
I	2	5	0.9
II	2	8	1.5

from Ng et al. (2009) and Gabriel et al. (2003), respectively. Resistivity is taken as the main characteristic in evaluating water reuse/recycle opportunity between both plants and ultra pure water (UPW) is used to supplement the use of process sources (with a unit cost of $2/t). The mixing rule for resistivity is given as follows: (El-Halwagi, 2006).

$$\frac{1}{R} \sum_i \frac{x_i}{R_i} \qquad (11)$$

Two interceptors with fixed resistivity value and heavy metal concentration are given for use in the EIP, to treat process sources either for further reuse/recycle or environmental discharge. Each of them has different performance and unit treatment cost, as shown in Table 2. Meanwhile, heavy metal concentration is chosen as the main characteristic for final discharge and its limit is given as 2 ppm.

It is further assumed that the EIP is operated for 8760 hours per annum. The CPLEX linear solver of GAMS v2.5 was used to solve the optimization model. Minimum total annualized cost is determined at $ 28.69 million/year, associated with a total UPW flow rate of 473 t/h and discharge effluent of 166 t/h.

Fig. 2 shows the optimized EIP design for the case study. It is observed that, while both interceptors regenerate water for further reuse/recycle in Plants A and/or B, interceptor I also treat wastewater for final discharge. Note that there is one cross-plant pipeline that connects between SR7 of Plant B and SK1 of Plant A. Note also that this design is essentially a hybrid of direct and indirect integration schemes as proposed in the work of Chew et al. (2008).

Conclusion

This paper has presented a structural representation and optimization formulation for the design of EIPs with property-based constraints. Plants are allowed to exchange streams, intercept them, and discharge unused portions. It is observed that processes participating in an EIP are likely to gain

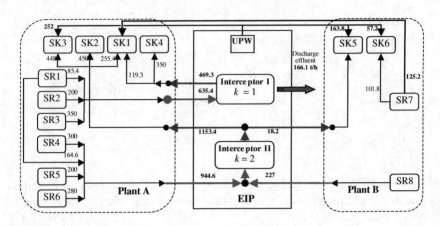

FIGURE 2
EIP design for case study.

savings in the forms of cost reduction for fresh resource(s) and overall reduction in waste discharge as well as treatment through material exchange and common infrastructure for interception and treatment. A case study was solved to illustrate these points.

References

Chew, I. M. L., Ng, D. K. S., Foo, D. C. Y., Tan, R. R., Majozi, T., Gouws, J. (2008). Synthesis of Inter-Plant Water Network. *Ind. Eng. Chem. Res.*, 47 (23), 9485–9496.

El-Halwagi, M.M. (2006). Process Integration, Amsterdam: Elsevier.

Fernandez, I., Ruiz, M. C. (2009). Descriptive Model and Evalutaion System to Locate Sustainable Industrial Areas. *J. Clean. Prod.*, 12, 87–100.

Gabriel, F. B., Harell, D. A., Dozal, E., El-Halwagi, M. M. (2003). Pollution Targeting via Functionality Tracking, *AICHE Spring Meeting*. New Orleans.

Gibbs, D., Deutz, P. (2007). Reflections on Implementing Industrial Ecology through Eco-Industrial Park Development. *J. Clean. Prod.*, 15, 1685–1695.

Lovelady, E. M., El-Halwagi, M. M. (2009). Design and Integration of Eco-Industrial Parks for Managing Water Resources. *Env. Prog.* (article in press)

Lowe, E. A. (1997). Creating By-Product Resource Exchanges: Strategies for Eco-Industrial Parks. *J. Clean. Prod.*, 5 (1–2), 57–65.

Ng, D. K. S., Foo, D. C. Y., Tan, R. R., Tan, Y. L., Pau, C. H. (2009). Automated Targeting for Conventional and Bilateral Property-Based Resource Conservation Network. *Chem. Eng. J.*, 149, 87–101.

Sendra, C., Gabarrell, X., Vincent, T. (2007). Material Flow Analysis Adapted to an Industrial Area. *J. Clean. Prod.*, 15, 1706–1715.

Shelley, M. D., El-Halwagi, M. M. (2000). Componentless Design of Recovery and Allocation Systems: A Functionality-Based Clustering Approach. *Comput. Chem. Eng.*, *24*, 2081–2091.

Spriggs, H. D., Lowe, E. A., Watz, J., Lovelady, E. M., El-Halwagi, M. M. (2004). Design and Development of Eco-Industrial Parks. *AIChE Spring Meeting*. New Orleans.

Zhao, Y., Shang, J. C., Chen C., Wu, H. (2007). Simulation and Evaluation of the Eco-Industrial System of Changchun Economic and Technological Development Zone, China. *Environ Monit Assess*. 139, 339–349.

52

Assessing the Risks to Complex Industrial Networks Due to Loss of Natural Capital and Its Implications to Process Design

Vikas Khanna, Jun-Ki Choi, Bhavik R. Bakshi* and Timothy Haab
The Ohio State University
Columbus, OH 43201

CONTENTS

ABSTRACT An Input-Output (IO) based framework is presented for studying the effect of sudden shocks and quantifying the associated risks on complex industrial networks. We are specifically using the IO model to understand the impact of changes in the availability of natural resources including natural capital on industrial systems. This includes understanding the potential impact of loss of services such as pollination, water scarcities, and soil fertility. The utility of the framework is highlighted using two case studies involving a reduction in the availability of crude oil and crop production. The approach is suitable for modeling the effect of sudden perturbations such as resource shortage on the complex industrial systems and identifying industrial sectors with greatest sensitivity to a given perturbation. This work is expected to complement the traditional biophysical models and methods by including the behavior of complex industrial networks under sudden shocks, quantifying the associated risks and support a decision-making framework for risk management.

KEYWORDS *Input-Output model, natural capital, industrial sectors, inoperability*

---

* To whom all correspondence should be addressed

Introduction

Sustainability assessment of industrial systems poses several formidable challenges for both engineers and policy makers alike. These challenges stem from the highly complex, hierarchical, and interconnected nature of the industrial systems. Such complex systems are observed to have high efficiency, performance, and robustness but are often fragile and vulnerable to unanticipated perturbations (Carlson and Doyle, 1999; Doyle, 2007). Proper understanding of such disruptive scenarios and their impact is pivotal for guiding design of complex systems and their long term sustainability. There is a need for holistic system modeling tools for understanding the resilience of complex industrial systems, quantifying risks, and guiding a decision making framework for effective resource allocation. Among the several available tools and techniques, Input-Output analysis offers a good basis for understanding the complexity of industrial systems.

Input-Output (IO) models have been used for various purposes including economic, environment and energy modeling to determine the total monetary or physical transactions required to deliver a final demand of a given product. Input-output analysis was originally developed by the Nobel laureate Wassily Leontief to study the monetary transactions between industries in an economy. Economic Input-Output (EIO) model divides the whole economy into industries or sectors and tracks monetary transactions between each pair of industries. As an example, an IO model for a three sector economy is shown in figure 1. The core of an IO model is the transaction table with z_{ij} in the table representing the monetary value sector j pays to sector i for that much worth goods and services sector i provides to sector j. Besides inter-industry transactions, each sector 'i' also sells f_i worth of goods to the consumers called as final demand; and sector 'j' gets contribution as value added (v_j), which includes employee compensation, profit of business owners, and government tax. The column at the right and the row at the bottom are total output and total input of each sector in monetary terms, x_i. Direct Requirement Matrix (A) is a normalized matrix of Z containing direct coefficient $a_{ij} = z_{ij}/x_j$. For a new final demand F_{new}, $A \times F_{new}$ is the direct economic activity, while the total economic activity is calculated by

$$X = (I - A)^{-1} F_{new} \qquad (1)$$

where $(I - A)^{-1}$ back tracks the supply side information, and is called as Leontief inverse. The IO model represented by (1) is called as demand driven.

Ghosh matrix is another way to define transaction coefficients: $b_{ij} = z_{ij}/x_i$, then similarly, the total economic activity generated by new value added V_{new} is given as

$$X = (I - BT)^{-1} V_{new} \qquad (2)$$

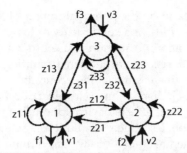

		Purchasing sector (j)			Final	Total
		1	2	3	demand	output
Selling	1	z_{11}	z_{12}	z_{13}	f_1	x_1
sectors	2	z_{21}	z_{22}	z_{23}	f_2	x_2
(i)	3	z_{31}	z_{32}	z_{33}	f_3	x_3
Value-added		v_1	v_2	v_3		
Total input		x_1	x_2	x_3		

FIGURE 1
Input-Output table for a 3 sector economy.

Here $(I - B^T)^{-1}$ is called Ghosh inverse and the model represented by (2) is called as supply driven model.

IO model provides a snapshot of an economy at a specific period with readily available empirical data. One of the major uses of input-output information, in the format of an input-output model, is to assess the effect on an economy of changes in elements that are exogenous to the model of that economy such as changes in the final demand or changes in the value added (Miller and Blair, 1985). IO model has also been widely used as an economic impact and forecasting tool.

Models based on the EIO model have also been developed for material flow analysis, energy analysis, and environmental life cycle assessment by combining the EIO model with data about emissions and resource use (Bullard and Herendeen, 1975; Dincer et al., 2003; eiolca.net). The idea is that material, energy and emissions flow are analogous to monetary flow in economies. For example, Economic Input-Output life cycle assessment (EIOLCA) focuses on the impact of emissions and energy use by combining economic data with the emissions details for specific sectors and is mainly an output-side approach (eiolca.net). More recently, Ecologically based life cycle assessment (Eco-LCA) has been developed which is novel and unique in its ability to quantify the contribution of ecosystem goods and services (Ukidwe and Bakshi, 2004). Ukidwe and Bakshi, 2007).

The present work focuses on exploring the utility of IO models for studying the effect of sudden shocks and quantifying the associated risks to complex industrial systems. We are using the IO model to understand the impact of changes in the availability of natural resources including natural capital.

This includes understanding the potential impact of loss of services such as pollination, water scarcities, and soil fertility or loss of certain key resources such as fossil fuels. Such information is used to determine sectors that are likely to face maximum risk due to sudden disruptions. Since the EIO model considers a static and linear state of the economy, it is not able to simulate long-term effects of such disruptions. However, it is appropriate for gaining insight into the short-term effects of disruptions before any adaptation due to market forces or policies. Such simulation is relevant to understanding the effect of environmental changes as well as human-induced changes such as terrorism and natural disasters.

Methodology

Both the demand driven and supply driven model represented by equations 1 and 2 are restrictive as they limit our ability to observe the impact due to changes in either the final demand (F) or the value added (V). Final demand and value added in models 1 and 2 are considered exogenous while the total sector throughput 'X' is considered endogenous and given the changes in exogenous variables, the impact on endogenous variables is estimated. This approach is restrictive for impact studies such as studying the impact of supply shortages of certain key resources or loss of ecosystem services on complex industrial systems. There is a need for a mixed model that considers exogenous specification of certain variables and endogenous specification of others. In the present case, the problem reduces to the specification of final demand for some sectors and endogenous specification of the output 'X' for some other sectors that experience supply constraint.

Miller and Blair (1985) discuss such an approach by modifications over the demand driven model represented by equation 1 for a 'n' sector economy by exogenous specification of the final demand for the first 'k' sectors (exogenous specification of $F_1, \ldots \ldots, F_k$) and endogenous specification of the total output $(X_{k+1}, \ldots \ldots, X_n)$ of the remaining 'n-k' sectors that experience supply constraint. By mathematical manipulation and rearrangement they obtain equation (3) as below.

$$\begin{bmatrix} P & 0 \\ R & -1 \end{bmatrix} \begin{bmatrix} X \\ F \end{bmatrix} = \begin{bmatrix} I & 0 \\ 0 & S \end{bmatrix} \begin{bmatrix} \bar{F} \\ R \end{bmatrix} \tag{3}$$

$\bar{F}$ is the k-element column vector of elements F_1 through F_k, whose values are exogenously determined

$\bar{X}$ is the (n-k) element column vector of elements X_{k-1} through X_n, whose values are exogenously determined

X is the k-element column vector of elements X_1 through X_k, whose values are to be determined

F is the $(n\text{-}k)$ element column vector of elements $F_{k\text{-}1}$ through F_n, whose values are to be determined

$P,R,Q,$ and S are the matrices that are a function of the original direct requirements matrix 'A'

From (3) the total output 'X' can be determined as

$$X = F^{-1}Q\bar{X} \qquad (4)$$

Additionally we define a new quantity 'q_i' called as 'normalized degraded production' or a measure of the sector 'inoperability' defined as

$$q_i = (X_i - \bar{X}_i)/X_i \qquad (5)$$

Here inoperability is similar to the concept introduced by Anderson et. al. (2007) describing the extent to which a system deviates from its planned performance under no disturbance. Inoperability 'q_i' takes on values that fall between 0 and 1, with 0 corresponding to the no production disturbance and 1 corresponding to a completely inoperable system.

Besides quantifying the economic loss incurred by individual sectors, the advantage of this approach originally proposed by Miller and Blair (1985) lies in its ability to identify the top sectors with greatest sensitivity in terms of operability to a given perturbation (sudden shock or shortage or a key resource). The model provides complementary perspective of identifying the critical sectors under a perturbation by quantifying the economic loss and estimating the extent of inoperability experienced by the different sectors. The utility of the approach described above is illustrated with the help of two case studies using a customized and aggregated 31 sector model of the U.S. economy for the year 2002 (Choi et al. 2009).

Results

The first case study involves studying the effect of sudden oil shock or disruption on the oil and gas supply on the total output of the other sectors. A 10 percent reduction in the total output of the Oil and Gas Extraction sector is considered and its effect on the output of the other sectors is simulated. The item of interest here is not so much the reason for the origin of such an oil shortage but studying the impact of such a supply disruption on the complex industrial networks. Figure 2(a) shows the top 10 sectors that experience the highest percentage degraded production or the percentage inoperability under such an oil shock or disruption in oil and gas supply. It is observed that support activities for mining, coal mining, and natural gas distribution are the top three sectors experiencing the highest inoperability because of their direct reliance on the Oil and Gas sector. The other sectors in the top 10 are the different electricity generation sectors.

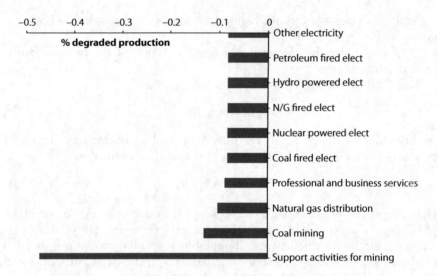

FIGURE 2(A)
Top 10 industrial sectors with the highest inoperability under a 10 percent reduction in the output of 0il and gas extraction sector.

Figure 2(b) shows the top 10 industrial sectors that suffer the highest magnitude of annual economic loss. It is interesting to note from figure 10 that the top 10 industrial sectors that suffer the highest annual production loss are not the same as those experiencing the highest inoperability shown in

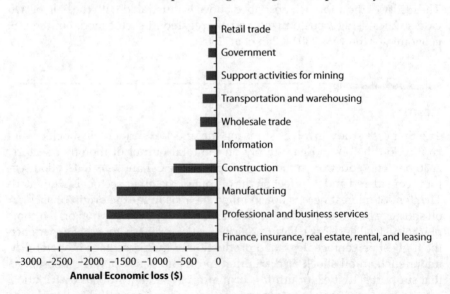

FIGURE 2(B)
Top 10 industrial sectors with the highest annual economic loss under a 10 percent reduction in the output of oil and gas extraction sector.

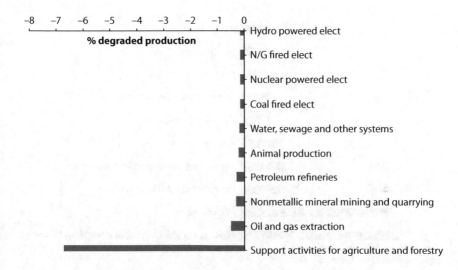

FIGURE 3(A)
Top 10 industrial sectors with the highest inoperability under a 10 percent reduction in the output of grop production sector.

figure 2(a). Finance, insurance, real estate, rental and leasing suffer the highest annual economic loss followed by professional and business services and manufacturing sectors. The reason for this can be attributed to the large economic size of these sectors that translate into big annual economic losses despite low inoperability experienced by these sectors. This is important as potential risk management options should not only focus on reducing the inoperability experienced by industrial sectors under a given perturbation but should also address the mitigation of the magnitude of the economic loss.

A second case study involves simulating the effect of a 10 percent reduction in the total output of the crop production sector on the other industrial sectors. Such a reduction in the crop production could result from a number of reasons such as loss of certain key ecosystem service such as pollination or water scarcity.

Figure 3(a) highlights the top 10 industrial sectors that experience the highest production inoperability due to a reduction in the output of the crop production sector. It is observed that the industrial sector of Support Activities for Agriculture and Forestry suffers the highest inoperability followed by Oil and Gas Extraction sector. Electricity production sectors are also among the top 10 industrial sectors experiencing the production inoperability due to such a disturbance. The production inoperability experienced by the Oil and Gas Extraction sector and the Electricity Production sector is due to the reliance of the Crop Production sector on these sectors. Any production disturbance experienced by the Crop Production sector thus affects the total output of these sectors. Such an insight is intuitively not very obvious whereas IO

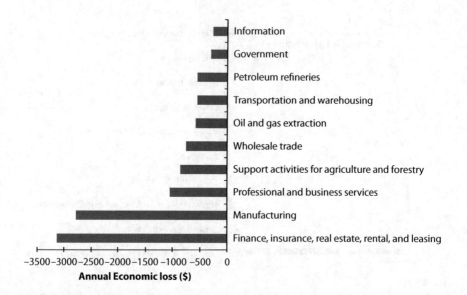

FIGURE 3(B)
Top 10 industrial sectors with the highest annual economic loss under a 10 percent reduction in the output of crop production sector.

model can capture such indirect and cascading effects in a systematic fashion. Figure 3(b) shows the top 10 industrial sectors experiencing the highest annual economic loss.

Conclusions

An IO based approach is presented and discussed for modeling the effect of sudden shocks and disturbances on the complex industrial networks. The utility of the approach is illustrated using two case studies involving reduction in the production of Oil and Gas Extraction and Crop Production sectors. The approach is suitable for modeling the effect of perturbations such as resource shortage and catastrophes on the industrial sectors. Besides quantifying the economic impact, the methodology can identify industrial sectors with greatest sensitivity under a given perturbation. It is concluded that sectors that suffer the highest economic loss under a given perturbation are not always the ones with the highest production degradation. The current approach has focused on using only the economic model. The model can however, be extended further by incorporating environmental life cycle information to quantify the impact on emissions and resource consumption. In addition to the resource quantity, there is a need to include the price dimension in the model and hence the resulting feedback effects on the final demand and the total sector output. Finally, integration of the coarser IO model with the fine scale process model can help understand the impact of sudden shocks on the process operation. This can lead to the identification of alternate process operating conditions and possible development of heuristics for guiding process designs that are robust to the impact of sudden shock and disturbances.

References

Anderson, C. W., Santos, J. R., and Haimes, Y. Y. A risk based input-output methodology for measuring the effects of the august 2003 northeast blackout. *Eco. Sys. Res.* 19(2): 183–204, 2007.

Bullard, C.W. and Herendeen, T. A. The energy cost of goods and services. *Energy Policy*, 3(4):268–278, 1975.

Carlson, J. M. and Doyle, J. Highly optimized tolerance: A mechanism for power laws in designed systems. *Phy. Rev. E.* 60(2):1412–1427, 1999.

Choi, J.-K., Bakshi, B.R., and Haab, T. A Framework for Analyzing the Biocomplexity of Material Use: Input-Output Approach. *In prep.* 2009.

Dincer, I., Hussain, M. M., and Al-Zaharnah, I.. Energy and exergy use in the industrial sector of Saudi Arabia. Proceedings of the Institution of Mechanical Engineers, Part A: *J. of Power & Energy*, 217(5):481–492, 2003

Doyle,J. Rules of engagement. *Nature.* 446(7138):860, 2007.

EIOLCA. www.eiolca.net, accessed on December 10, 2008.

Miller, R.E. and P.D. Blair, Input-Output Analysis: Foundations and Extensions. 1985, NJ: Prentice-Hall.

Ukidwe N. U. and Bakshi, B. R. Industrial and ecological cumulative exergy consumption of the united states via the 1997 input-output benchmark model. *Energy*, 32(9):1560–1592, 2007.

Ukidwe N. U. and Bakshi, B. R. Thermodynamic Accounting of Ecosystem Contribution to Economic Sectors with Application to 1992 U.S. Economy. *Env. Sci. & Tech.* 38(18):4810–4827, 2004.

53

Analyzing Complex Chemical and Polymer Manfacturing Plants: A Macroscopic Approach

Russell F. Dunn[*]

Polymer and Chemical Technologies, LLC
Cantonment, FL 32533

Ian Bowling

Texas A&M University
College Station, TX 77843-3122

CONTENTS

ABSTRACT Advances in software and hardware for personal computers, improved printing and scanning hardware and numerous other technology advances allow for new approaches to be used in analyzing large chemical and polymer manufacturing plants. Specifically, these advances allow for complex process flow sheet development, resolution and analysis of complex material and energy balances, and identification of retrofit design opportunities. A brief description of macroscopic design approaches that use PC software and hardware, along with other technology, to allow complex analysis of large chemical and polymer plants is provided. Examples of these approaches are illustrated using several polycarbonate monomer

---

[*] To whom all correspondence should be addressed

and resin manufacturing plants and polyvinyl chloride monomer and resin manufacturing plants that are currently operating in the United States.

KEYWORDS *Process integration, pinch technology, polyvinyl chloride, polycarbonate*

Introduction

Advances in software and hardware for personal computers, improved printing and scanning hardware and numerous other technology advances allow for new approaches to be used in analyzing large chemical and polymer manufacturing plants. Specifically, these advances allow for complex process flow sheet development, resolution and analysis of complex material and energy balances, and identification of retrofit design opportunities.

Macroscopic Design Approaches

The past decade has seen significant industrial and academic efforts devoted to the development of holistic process design methodologies that target energy conservation and waste reduction from a systems perspective. Implicit in the holistic approach, however, is the need to realize that changes in a unit or a stream often propagate throughout the process and can have significant effects on the operability and profitability of the process. Furthermore, the various process objectives (e.g., technical, economic, environmental, and safety) must be integrated and reconciled. These challenges call for the development and application of a systematic approach that transcends the specific circumstances of a process and views the environmental, energy, and resource-conservation problems from a holistic perspective. This approach is called *process integration*. It emphasizes the unity of the process and it is broadly divisible into the categories of *mass integration* and *energy integration*. A review of process integration methodologies was published by Dunn and El-Halwagi in 2003.

One of the most significant advancements to aid in the application of these methodologies has been the development of proprietary plant flow diagrams by Polymer and Chemical Technologies, LLC. The diagrams provide details to the unit operation level. Industrial experience has demonstrated that it is crucial that these detailed flow diagrams are developed prior to applying process integration methodologies so that the effect of all design modifications can be fully evaluated.

Example: PVC Monomer and Polymer Plant

A Process Integration Study was conducted for a PVC monomer and polymer chemical complex. The objective of this study was to identify site energy conservation projects (process designs), water conservation projects and productivity enhancement projects that would result in reduced operating costs. These projects were identified using pinch technology, source-sink analysis, process modeling and optimization and other design tools. In addition, a site material and energy balance model was developed.

Objectives

The study objectives were to identify retrofit process design projects that would:

- reduce site utility consumption, thus reducing site operating costs
- reduce site water consumption and/or wastewater generation
- reduce raw material costs and/or increase productivity
- develop a site material and energy balance model

This study covered the following operating units:

- Chlor-Alkali Plant
- Ethylene Plant
- Ethylene Dichloride Plant
- Vinyl chloride Monomer Plant
- Utilities
- Environmental Operations
- PVC Plant

Results

The following were identified:

- 39 energy conservation designs (subsets of these designs are also feasible)
- 6 water conservation and/or waster reduction designs (energy conservation projects also affect water usage and water discharge reductions)
- 3 raw material cost and/or productivity improvement opportunities

The designs identified can result in excess of $10,000,000/yr in reduced operating costs and/or additional manufacture of products for sale.

Key Findings

- This study generated valuable tools in the way of flow diagrams, pinch diagrams, source-sink diagrams and the site material/energy balance.

- Targets from composite pinch curves clearly indicated heat integration potential.

- Heat integration and additional steam generation opportunities were identified. The site can accommodate approximately 50MM BTU/hr less steam generation via multiple boilers by operating the boilers at their lowest sustainable rates; thus, a minimum of 50 MM BTU/hr of steam savings through integration and/or additional steam generation can be realized.

- Water projects provide opportunities for wastewater reduction having the potential of:
  - Up to 40,000,000 gallons per year reduction
  - Evaporation capacity increase of nearly 5%
  - 5% reduction in steam usage

- Other projects identified also offer production increases in excess of 2000 ECU's per year.

- Projects identified indicate a potential annual savings as low as $300,000 to in excess of $10,000,000 with additional capital investment.

Example: Polycarbonate Monomer and Polymer Plant

An Energy Conservation Study was conducted for a polycarbonate polymer and chemical complex. The objective of this study was to identify site energy conservation projects (process designs) that would result in reduced operating costs. These projects were identified using pinch technology, process modeling and optimization and other design tools.

This study covered the following operating units:

- Chlor-Alkali Plant
- Phosgene Plant

- BPA Plant
- Utilities
- Environmental Operations
- Polycarbonate Plant

Results

Detailed flow diagrams of each plant were generated and site utility usage was collected and/or calculated and then compiled. Heat pinch diagrams were generated to aid in the development of feasible energy conservation designs.

Thirteen energy conservation designs were identified (subsets of these designs are also feasible). Five of these designs are recommended for implementation. The recommended designs can result in a range of $500,000/yr to $4,000,000/yr in reduced energy consumption costs and up to 57 MMBTU/hr energy savings in both steam consumption and cooling water usage. The five recommended designs allow utility consumption savings equivalent to as much as 67,000 lb/hr of reduced steam consumption.

Conclusions

The chemical process industry is facing increased pressure to develop processes and products that are energy efficient, environment-friendly and less expensive. Besides, the globalization of the world economy has led to increased competition. Another trend is an increased need to minimize environmental discharges resulting in "greener" processes. These drivers have resulted in significant focus being placed on the concepts of "life cycle analysis", "sustainability" and "industrial ecology". The key to tackling all these challenges lies in process integration, which involves leveraging all process resources in an optimal fashion to reduce overall cost and increase productivity while simultaneously minimizing energy use and lowering adverse environmental impact. However, previous attempts at process integration were arbitrary or based on experience. There is a need for accurate and systematic ways to reduce costs and improve overall process and utility performance in the chemical process industries. In response to this demand, the area of process systems engineering (which involves the application of computer based optimization and design techniques) has blossomed with several applications. A key addition to these approaches has been the development of proprietary detailed process flow diagrams.

References

Dunn, R. F., and M. M. El-Halwagi. (2003). Process Integration Technology Review: Background and Applications in the Chemical Process Industry," *Journal of Chemical Technology and Biotechnology* 78(9): 1011–1021.

Dunn, R. F., and G. E. Bush. (2001). Using Process Integration Technology for Cleaner Production, *Journal of Cleaner Production* 9 (1): 1–23

Dunn, R.F., and H. Wenzel. (2001). Process integration design methods for water conservation and wastewater reduction in industry, Part 1: Design for Single Contaminant, *Clean Products and Processes* 3: 307–318

Dunn, R.F., H. Wenzel, and M. Overcash. (2001). Process integration design methods for water conservation and wastewater reduction in industry, Part 2: Design for Multiple Contaminants, *Clean Products and Processes* 3: 319–329

54

Environmentally Benign Process Design of Polygeneration Energy Systems

Pei Liu and Efstratios N. Pistikopoulos[*]

Imperial College London
London, SW7 2AZ

Zheng Li
Tsinghua University
Beijing, 100084

CONTENTS

ABSTRACT Polygeneration is a promising energy conversion technology which could provide opportunities to potentially tackle serious worldwide problems of shortage of fossil fuels and global warming caused by extensive green-house gas emissions. A polygeneration plant produces electricity and at least one type of synthetic chemical fuel with similar properties as conventional ones, offering the possibility for the energy utilization system to shift from fossil fuels to synthetic substitutions with few or minor modifications to the existing infrastructure. Other essential properties of polygeneration such as high energy conversion rate and low/zero emissions make it an ideal technology in terms of reduction of green-house gas emissions. However, designing a polygeneration energy system according to both economic and environmental objectives remains a challenging task.

In this paper, we present an energy systems engineering approach towards the design of environmentally benign polygeneration energy systems. The approach features: (i) a superstructure representation for the polygeneration process, comprising several functional blocks, where alternative technologies or

---

[*] To whom all correspondence should be addressed at e.pistikopoulos@imperial.ac.uk

equipment types are considered, (ii) the introduction of environmental impact indicators to account for the cradle-to-grave emissions in the process, and (iii) a multi-objective multi-period mixed-integer optimization model. A detailed study reveals interesting trade-offs for possible polygeneration design scenarios over a long-term operating horizon from an environmental and profitability viewpoint.

KEYWORDS *polygeneration, energy systems, environmental impact, multi-objective optimization, mixed-integer nonlinear programming (MINLP)*

Introduction

Ever-increasing oil depletion and green-house gas (GHG) emissions have become two major worldwide problems faced by all human-beings in the 21st century. According to the projection of U.S. Department of Energy (DOE), world wide consumption of petroleum and other liquid fuels will rise from 83 million barrels oil equivalent per day in 2004 to 97 million in 2015 and 118 million in 2030. Considering the world oil reserves of 1317.4 billion barrels by 2007, it leads to a reserve-to-production ratio of only 37 years. On the other hand, DOE projects the world carbon dioxide emissions will go from 26.9 billion metric tons in 2004 to 33.9 billion metric tons in 2015 and 42.9 billion metric tons in 2030. Emissions from coal combustion alone contributed 39.4 percent in 2004 and will contribute 43.1 percent by 2030 (DOE, 2007).

This serious energy and environmental situation makes it urgent to seek technologies that can reduce the currently huge pressure on oil oriented liquid fuels and carbon dioxide emissions. Polygeneration is one such potential energy conversion technology, both cost-effective and environmentally friendly, and provides opportunities to meet increasing energy demands and environmental constraints simultaneously.

A polygeneration plant is a multi-in multi-out energy system that produces electricity, chemicals, and synthetic fuels. A generic polygeneration process, as can be seen in Figure 1, starts from gasification of coal, biomass, petroleum coke, or other feedstocks, and produces electricity and synthesis fuels via integrated power generation and chemical synthesis processes.

High energy conversion efficiency and low/zero carbon dioxide emissions are two major advantages of polygeneration energy systems. Due to higher efficiency, production cost of methanol in a methanol/electricity polygeneration plant is 40 percent lower than that in a stand-alone methanol plant (Ni et al., 2000). Polygeneration also provides opportunities to realize a low/zero emission system via a pre-combustion carbon dioxide capture and sequestration (CCS) unit, together with a water-gas shift reactor in the chemical synthesis unit.

Despite of the benefits of a polygeneration process, designing such a complicated and highly-integrated system over its long-term operating horizon

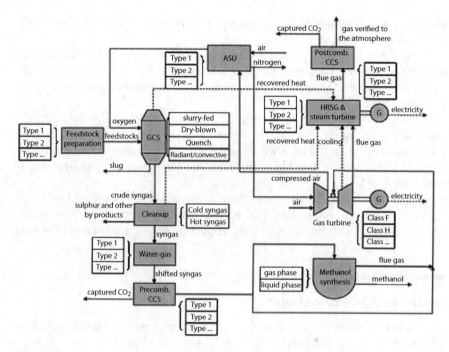

FIGURE 1

A superstructure representation of a polygeneration process.

considering both economic and environmental performances remains a challenging problem, with the following scientific issues to be solved:

- Many technical alternatives for each functional unit.
- High degree of integration among functional units.
- Trade-off between profitability and environmental impacts.
- Unpredictable behavior of the chemical synthesis unit due to different composition of inlet gas.

Some of these issues have been addressed in our previous work (Liu et al., 2007; Liu et al., 2008). In this proposed work, we present a superstructure based modelling and optimization strategy to represent all these challenges via its application on a methanol/electricity polygeneration plant.

Superstructure Representation

To capture all potential technologies and types of equipment and all possible combinations among them, a superstructure of a polygeneration plant is built, shown in Figure 1. In this superstructure, a polygeneration plant is divided into ten functional blocks, each having its own candidate technologies and types of equipment.

A plant represented by such a superstructure will be in operation over a long-term horizon. The whole operating horizon is divided into several time intervals. Time-variant parameters are denoted as piecewise functions over these intervals.

Mass and energy balances of input and output streams of each functional block are established. For the methanol synthesis block, chemical kinetics and phase equilibrium are formulated to handle the different mole compositions of inlet syngas resulted from different technologies implemented in upstream blocks.

Net present value (NPV) of the plant over its whole operating horizon is selected as an economic objective, whilst a cradle-to-gate life cycle assessment based GHG emission indicator behaves as an environmental objective.

Mathematical Formulation

As a detailed mathematical model is presented somewhere else (Liu et al., In preparation), the major parts of the model are listed below:

- Selection of technologies and types of equipment for each function block is represented by a set of binary variables
- The whole operating horizon is divided into several time intervals to represent time-variant parameters
- Mass and energy balances are established for each functional block
- A reference variable is selected for each functional block to represent the connections between inlet and outlet streams
- A first principle sub-model is built for the methanol synthesis block, based on chemical kinetics and phase equilibrium
- NPV is selected as the economic objective, accounting investment costs and discounted cash flows over all time intervals
- A Life Cycel Assessment (LCA) based indicator of GHG emissions over the whole life time of a plant is selected as the environmental impact objective.

The LCA based GHG emissions come from three categories. The first is the cradle-to-gate emissions during feedstock production, including extraction and transportation to site (Hugo et al., 2005), as follows:

$$ghg_{e,fds} = \sum_t \sum_i m_i \cdot \gamma_{e,i} \cdot \tau_t \tag{1}$$

where $\gamma_{e,i}$ is the cradle-to-gate inventory coefficient of GHG emission e for feedstock i.

The second part represents the emissions produced during equipment production, installation and plant construction via Economic Input-Output Life Cycle Assessment (Carnegie Mellon University Green Design Institute, 2008), as follows:

$$ghg_{e,eqp} = \sum_j inv_j \cdot \alpha_{e,j} \qquad (2)$$

The third part represents the emissions produced throughout the plant operating period, denoted as $ghg_{e,opt}$, which can be obtained directly from the model.

Then a single indicator of GHG emissions is obtained by weighing each kind emission e via their impact factor σ_e (Intergovernmental Panel on Climate Change, 2007), as follows:

$$ghg = \sum_e (ghg_{e,fds} + ghg_{e,eqp} + ghg_{e,opt}) \cdot \sigma_e \qquad (3)$$

After obtaining the environmental objective ghg, together with the economic objective NPV, a multi-objective optimization problem can be solved to generate optimal process designs according to different designing principles in terms of profitability expectation and GHG emissions allowance.

Case Study

A case study has been performed using the proposed methodology on a methanol/electricity polygeneration plant over a 15-year horizon. Three technologies for the gasification block, three technologies for the syngas cleanup block, and two technologies for the methanol synthesis block have been suggested, generating 18 possible technology combinations.

The model for the case study involes 9 binary variables, 1642 continuous variables, and 1781 equality and inequality constraints. It is solved in GAMS (GAMS Development Corporation, 2008), using BARON solver to obtain a feasible starting point and DICOPT solver to continue. After properly scaled, it takes 182.76 seconds of CPU time on a Pentium 4 platform to solve and produce 20 pairs of optimal solutions, the so called Pareto curve, shown in Figure 2.

The Pareto curve consists of very different process structures and operating mode. Out of the 18 possible technology combinations or process structures considered in the model, only four appear in the Pareto curve, as labeled in Figure 2, where 'H' represents hot clue gas clean up, 'RC' represents radiative

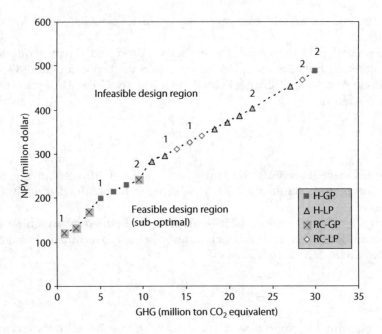

FIGURE 2
Pareto curve for polygeneration energy systems design.

and convective heat exchange gasifier, 'GP' represents gas phase methanol synthesis, and 'LP' represents liquid phase methanol synthesis. Different types of equipment and technologies are needed to meet specific design targets. For instance, a gasifier with a radiative/convective heat exchanger and a gas phase methanol synthesis reactor are selected in a process optimized with very low emissions. On the other hand, a gasifier with hot syngas cleanup and a liquid phase methanol synthesis reactor are preferable for a high profitability design strategy.

Table 1 summarizes key plant-wide operating variables for two sets of the four different structures labeled in the Pareto curve.

The length of the operating horizon also plays a key role. Figures 3 highlights this via two additional scenario analyses, one over a 10-year operating horizon and one over a 20-year horizon. As can be seen, only three types of process structures appear on the Pareto curve for the 20-year horizon scenario. Although four structures also appear on the Pareto curve for the 10-year horizon scenario, their sequence is quite different from that on the Pareto curve for the 15-year horizon scenario.

It indicates the process structures and operating modes can be very sensitive to the length of operating horizon. In a long-term operating horizon, technical parameters of a technology, such as thermal efficiency and heat recovery rate, have dominant significance, as they have great impacts on

TABLE 1

Process Operating Mode at Each Optimal Design Point (Point Number
Corresponds to its Sequence in Figure 2, from Left To Right)

Structure	No.	Coal Con. Rate (kg/s)	Pre-Comb. CCS (kg/s)	Post-Comb. CCS (kg/s)	Chem./ Power Ratio	Shift Ratio
■	1	43.5	12.8	76.7	0.23	0.47
(H-GP)	2	36.4	9.7	0.0	0.45	0.11
▲	1	41.6	0.0	66.5	0.51	0.00
(H-LP)	2	39.1	0.0	34.9	0.54	0.00
×	1	46.6	106.2	1.0	1.00	1.00
(RC-GP)	2	41.7	12.8	61.0	0.25	0.46
◆	1	40.1	0.0	55.4	0.52	0.00
(RC-LP)	2	37.2	15.4	0.0	0.36	0.41

variables which hold through out the whole horizon, such as purchase of
feedstocks and emissions. Other the other hand, in a short-term operating
horizon, parameters like investment costs and emissions from equipment
production dominant others, as they only influence variables at the plant
installation and construction stage.

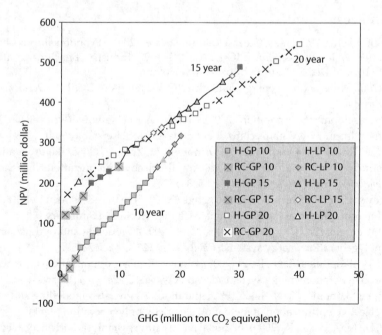

FIGURE 3

A comparison of process designs over operating horizons of different length.

Conclusions

A superstructure based multi-objective mixed-integer optimization methodology is proposed for the design of polygeneration energy systems where both profitability and environmental impacts are taken into account, based on which Pareto trade-off curves can be obtained for different operating horizon time to guide the design process.

Acknowledgements

The authors gratefully acknowledge the financial support from BP and its contribution in the inception, progress, and completion of this research study. Pei Liu would also like to thank Kwoks' Foundation for providing scholarship.

References

Carnegie Mellon University Green Design Institute. (2008). Economic Input-Output Life Cycle Assessment (EIO-LCA), US 1997 Industry Benchmark model. Retrieved Dec, 2008, from http://www.eiolca.net.

DOE (2007). International energy outlook 2007. Washington, DC, *U.S. Department of Energy*.

GAMS Development Corporation. (2008). GAMS—A user's guide. Retrieved Jan, 2008, from http://www.gams.com/docs/gams/GAMSUsersGuide.pdf.

Hugo, A., Rutter, P., Pistikopoulos, S., Amorelli, A., Zoia, G. (2005). Hydrogen infrastructure strategic planning using multi-objective optimization. *International Journal of Hydrogen Energy* **30**(15): 1523–1534.

Intergovernmental Panel on Climate Change (2007). Fourth Assessment Report of the Intergovernmental Panel on Climate Change, Technical Summary.

Liu, P., Gerogiorgis, D. I., Pistikopoulos, E. N. (2007). Modeling and optimization of polygeneration energy systems. *Catal. Today* **127**(1–4): 347–359.

Liu, P., Pistikopoulos, E. N., Li, Z. (2008). A mixed-Integer optimization approach for polygeneration energy systems design. *Comput. Chem. Eng.* (In press).

Liu, P., Pistikopoulos, E. N., Li, Z. (In preparation). Aggregated modelling and multi-objective optimization of methanol/electricity polygeneration plants.

Ni, W. D., Li, Z., Yuan, X. (2000). National energy futures analysis and energy security perspectives in China—strategic thinking on the energy issue in the 10 th Five-Year Plan (FYP). *Workshop on East Asia Energy Futures*, Beijing.

55

Determination of Optimal Design and Control Decisions for Reactor-Separator Systems with Recycle

Mohd Kamaruddin Abd Hamid, Gürkan Sin and Rafiqul Gani*

Computer Aided Process-Product Engineering Center (CAPEC),
Department of Chemical and Biochemical Engineering,
Technical University of Denmark, DK-2800, Kgs. Lyngby, Denmark

CONTENTS

ABSTRACT Two simple yet powerful techniques used within a new model-based methodology for integrated process design and control (*IPDC*) to determine optimal design decisions are presented. These are attainable region (*AR*) and driving force (*DF*) techniques whose concepts are used to find the optimal design targets as an alternative to the use of optimization/search algorithms. Accordingly, the optimal solution to the design problem is to be found by locating the maximum value of *AR* and *DF* for reactor and separator units respectively. For control problem, the minimum value of the derivative (concentration in *AR* or *DF*) with respect to manipulative variables provides an optimal solution, which ensures process controllability and resiliency as well as determines controller structure selection. While other optimization algorithms may or may not able to find the optimal solution, depending on the performance of their search algorithm and computational demand, the

---

* Corresponding author. Tel.: (+45) 4525 2882, Fax: (+45) 4593 2906, e-mail: rag@kt.dtu.dk

use of *AR* or *DF* concept is simple and able to find at least near-optimal design (if not optimal) to integrated design and control problems. In this paper, we demonstrate successfully the potential use of AR technique in finding the optimal solution for the integrated design and control of a single reactor for the synthesis of ethylene glycol.

KEYWORDS *Model-based methodology, integrated process design and control, attainable region, ethylene glycol*

Introduction

The need to understand how process design decisions influence the control performance of the system is becoming increasingly accepted in both academia and industry (Seferlis and Georgiadis, 2004). To assure that design decisions give the optimum economic and best control performances, the control aspects should be considered in the early stage of process design. By considering control aspects together with the economic issues in the process design stage, the integration of process design and control (*IPDC*) can be achieved. In practice, the systematic tools that can assist in deciding the optimal solution for IPDC problems are needed.

Various attempts to integrate control considerations at the design stage have been proposed by many researchers. Sakizlis et al. (2004) presented an overview of the state of the art of optimization-based methods of *IPDC*. They classified optimization-based methods into two categories, multi-objective approach and dynamic optimization approach. In the former, the control cost and the design cost are accounted for by different objective functions requiring the solution of complex optimization problems. Whereas, latter considers a single economic objective but the system operation is represented with dynamic models. Thus, this approach can be formulated as a mixed-integer dynamic optimization (*MIDO*) problem. However, their drawback is the *MIDO* problem will give rise to a highly nonlinear optimization formulation, which requires an effective global search algorithm to find the optimal solution.

The use of model-based analysis for solving *IPDC* problems is an alternative approach, which emphasizes process knowledge (Ramirez and Gani, 2007; Kiss et al., 2007). In this approach, a first-principles model is used to analyze *IPDC* problems without involving rigorous optimization algorithms. Russel et al. (2002) propose a systematic analysis of the model equations as pre-solution step for *IPDC* problems. By using model analysis the relationships between the design and control variables can be identified, which helps to understand, define and address issues related to *IPDC* problems.

Problem Formulation

IPDC in this work is treated as a *MIDO* problem, where the economic performance is optimized in order to design a cost effective and highly controllable
process. A general formulation for *IPDC* problem can be presented below:

$$\min J(x, u, y) \tag{1}$$

s.t.

$$h_1(x, u, y) = dx/dt \tag{2}$$

$$h_2(x, u, y) = 0 \tag{3}$$

$$g(x, u, y) \leq 0 \tag{4}$$

$$x_L \leq x \leq x_U \tag{5}$$

$$u_L \leq u \leq u_U \tag{6}$$

$$y \in \{0, 1\} \tag{7}$$

where x is the vector of design variables and u the vector of control variables.
In the objective function, Eq. (1), J, represents the expected total annualized
cost (*TAC*) of a system. The system dynamics is described by a set of differential equations given in Eq. (2). The steady-state system is described by a function given in Eq. (3). In Eq. (4), the possible inequality constraint is expressed.
The bounds on design and control variables are represented in Eqs. (5)–(6),
respectively. In Eq. (7), y comprises the binary variables for the process and
the control structure (corresponding to whether a manipulated variable is
paired with a particular controlled variables or not).

Once the *MIDO* model has been defined, the next step is solving this model
by applying a decomposition-based approach, where the *MIDO* model is
decomposed into a number of sub-problems.

Decomposition-based Solution Strategy

In *IPDC* problem, combinatorial in nature, can be solved in many ways.
However finding a solution may become cumbersome especially when the
constraints are nonlinear and/or the number of variables are large, which
causes difficulties in convergence and high computational cost. Since typically
a number of constraints is involved, the feasible region can be very small
compared to the search space. All of the feasible solutions to the problem
may lie in a relatively small portion of the search space. The ability to solve

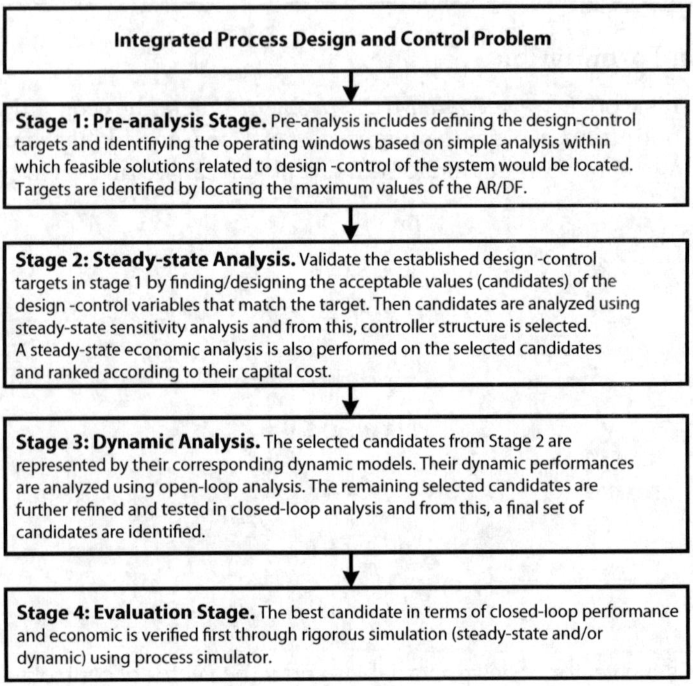

Integrated Process Design and Control Problem

Stage 1: Pre-analysis Stage. Pre-analysis includes defining the design-control targets and identifiying the operating windows based on simple analysis within which feasible solutions related to design -control of the system would be located. Targets are identified by locating the maximum values of the AR/DF.

Stage 2: Steady-state Analysis. Validate the established design -control targets in stage 1 by finding/designing the acceptable values (candidates) of the design -control variables that match the target. Then candidates are analyzed using steady-state sensitivity analysis and from this, controller structure is selected. A steady-state economic analysis is also performed on the selected candidates and ranked according to their capital cost.

Stage 3: Dynamic Analysis. The selected candidates from Stage 2 are represented by their corresponding dynamic models. Their dynamic performances are analyzed using open-loop analysis. The remaining selected candidates are further refined and tested in closed-loop analysis and from this, a final set of candidates are identified.

Stage 4: Evaluation Stage. The best candidate in terms of closed-loop performance and economic is verified first through rigorous simulation (steady-state and/or dynamic) using process simulator.

FIGURE 1
Decomposition method for IPDC problems. (Hamid and Gani, 2008).

such problems depends on the ability to identify and avoid the infeasible portion of the search space. Hence, one approach to solve this *IPDC* problem is possible by applying decomposition method, in which the problem is decomposed into sub-problems, which are relatively easy to solve (See Figure 1).

Figure 1 shows a decomposition method for *IPDC* problem (Hamid and Gani, 2008). Accordingly the problem is decomposed into four sequential hierarchical stages: (1) pre-analysis, (2) steady-state analysis, (3) dynamic analysis, and (4) evaluation stage.

Figure 2 shows how the general *IPDC* problem (*MIDO*) corresponds to the four hierarchical stages of the model-based methodology. It shows how the model-based methodology is equivalent to the general *IPDC* problem formulation in Eqs. (1)–(7). However with a single exception that the problem in decomposition methodology is rearranged in the reverse order compared to the general *IPDC* problem (Eq. 1 to 7). As each sub-problem is being solved, a large portion of the infeasible part of the search space is identified and hence eliminated, thereby leading to a final sub-problem that is significantly smaller *MIDO* or *DO* problem, which can be solved more easily.

In this contribution, we demonstrate the concept of *AR* (and *DF*) to find the optimal design decisions as an alternative to the use of search algorithms.

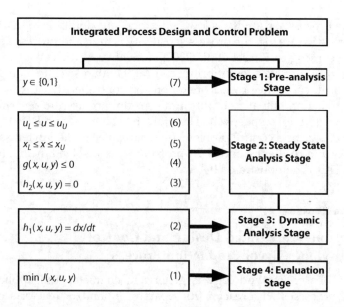

FIGURE 2
A solution methodology for *IPDC: MIDO* type *IPDC* problem (left) and its correspondence to the four sequential hierarchical stages of the decomposed *IPDC* problem (right).

Accordingly, the *AR* (and *DF*) are used in the first stage of the model-based methodology to select the optimal design and control solution. Since decision of design and control are representing by y (binary variable) in Eq. (7), the use of *AR* and *DF* concepts enumerate the decision (binary) variables is stage 1, in which assisting the optimal design decisions.

Attainable Region and Driving Force Concepts

The *AR* concept is used for dimensioning reaction units, while *DF* concept is applied for separation units in chemical systems. For the reactor design problem, the idea is to locate the maximum value of *AR* for design feasibility, and from there the operating conditions in terms of residence time, temperature, volume, etc. are identified. The same idea is used to solve separator design problem using *DF* concept. *DF* is defined as a measure of the relative ease of separation (Gani and Bek-Pedersen, 2000). When *DF* is zero, no separation is possible. When *DF* is large, separation becomes easy. By employing this technique, one can determine values of design variables for separation systems at the largest value of *DF*.

For control problem, the value of the derivative of *AR* or *DF* with respect to manipulative variables will determine process sensitivity and flexibility as well as the controller structure selection. If values of the derivative are small, the process sensitivity is low and process flexibility is high. If values of the derivative are big, the process sensitivity is high and process flexibility is low. According to Russel et al. (2002), derivative of the constitutive variables (e.g. reaction rate or equilibrium constant) with respect to manipulative variables influences the process operation and controller structure selection. Basically, reaction rate and equilibrium constant are used to determine the value of *AR* and *DF*, respectively. Therefore, by using *AR* and *DF* concepts to solve the design and control problem, insights can be gained in terms of controllability and resiliency, and *IPDC* problems can be solved in an integrated manner. This is demonstrated below.

Application—Integrated Design and Control of an Ethylene Glycol Production Process

The model-based methodology is generic in character and applicable to chemical processes with reactor (*R*), separator (*S*) and/or reactor-separator-recycle (*RSR*) systems. A reaction (*R*) system has been selected to evaluate the *AR* concept. It is important to note that for *S* and *RSR* systems, the procedure would be exactly the same.

Case Study: Reaction (R) System

The objective is to determine the optimal design (with respect to controllability and economic) that gives the highest concentration of the desired product at given feed flow rate and concentrations. The process reactions are given as:

$$EO + W \xrightarrow{r_1} EG \tag{8}$$

$$EO + EG \xrightarrow{r_2} DEG \tag{9}$$

$$EO + DEG \xrightarrow{r_3} TEG \tag{10}$$

In Eq. (8), ethylene oxide (*EO*) and water (*W*) react to produce ethylene glycol (*EG*). Eqs. (9)–(10) are side reactions where the excess *EO* react with *EG* and *DEG* to produce diethylene glycol (*DEG*) and triethylene glycol (*TEG*), respectively. The reaction rates for the above reacting system are:

$$r_1 = k_1 C_{EO} C_W, \quad r_2 = k_2 C_{EO} C_{EG}, \quad r_3 = k_3 C_{EO} C_{DEG}$$

where kinetic parameters; $k_1 = 5.238 \exp^{30.163 - 10583/T}$ [m³/kmol-h], $k_2 = 2.1k_1$ and $k_3 = 2.2k_1$ are taken from Parker and Prados (1964).

Stage 1: Pre-analysis. The *AR* analysis was carried out by evaluating the concentration of the desired product, *EG* by solving the rate equations

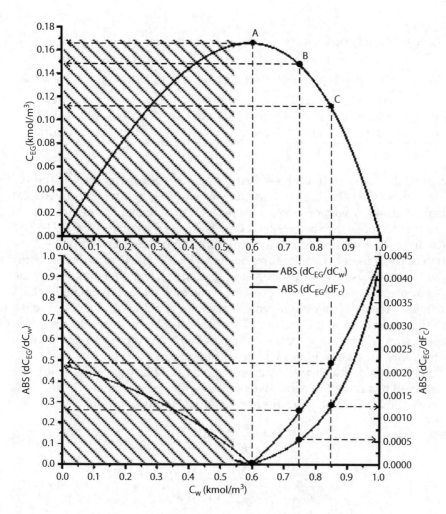

FIGURE 3

Top: *AR* diagram for concentration of ethylene glycol and water; Bottom: Corresponding derivatives of the C_{EG} with respect to C_W and F_c.

for the *EG* with respect to the limiting reactant *W*. Figure 3(top) shows the *AR* diagram for the EG with respect to *W*. The shaded area represents the infeasible region where at $C_W = 0.54$ kmol/m³, the concentration of *EO* is exhausted, thereby, turning-off the operation. From the figure, the optimum value of the attainable concentration of *EG* is selected at point A (at the maximum value of *AR*) to be set as a design target. Another alternative designs (point B and C) that are arbitrarily chosen, are also selected as targets. The reason for this is to compare the design selected by the *AR* concept to be optimal, to arbitrarily chosen alternative designs. In this stage, three alternative designs are selected representing three binary variables of *y* (y_1, y_2, y_3).

TABLE 1
Value of Design-Control Variables at Different Targets (Points A, B and C).

	Design Variable				Control Variable	
Point	V (m³)	F_o (m³/h)	T (K)	F_c (m³/h)	C_W (kmol/m³)	C_{EG} (kmol/m³)
A	11.89	1000	402.2	1341	0.6000	0.1667
B	11.89	1000	378.2	688	0.7500	0.1471
C	11.89	1000	366.9	366	0.8453	0.1118

Stage 2: Steady-state analysis. In this stage, the selected targets (A, B and C) are validated by finding the feasible values of the design-control variables (temperature, residence time, volume, etc.) that match those targets. In order to simplify the search space, the volume of reactor is fixed at 11.89 m³ (by fixing the volume of reactor represents an existing system). Then, the remaining variable, temperature, needs to be identified in order to match the target. By using their corresponding steady-state models, values of temperature and its corresponding cooling water flow rate are calculated. Table 1 tabulates the value of design-control variables at points A, B and C. From a design point of view, points B and C are not feasible since they do not satisfy the design criteria (lower concentration of EG). Thus, from an *IPDC* point of view, only the control issues will be highlighted below.

Stage 3: Dynamic analysis. Sensitivity and flexibility at points A, B, C are analyzed by taking the derivative of the C_{EG} with respective to C_W and F_c as derived in Eqs. (11)–(12). Results are shown in Figure 3(bottom).

$$\frac{dC_{EG}}{dC_W} = 0 \tag{11}$$

$$\frac{dC_{EG}}{dF_c} = \left(\frac{dC_{EG}}{dC_W}\right)\left(\frac{dC_W}{dT}\right)\left(\frac{dT}{dF_c}\right) = 0 \tag{12}$$

From Figure 3 (bottom), values of dC_{EG}/dC_W and dC_{EG}/dF_c at point A -that corresponds to zero value of the derivative- are compared to the points B and C. According to Russel et al. (2002), at the minimum value of derivatives, the process sensitivity is low and process flexibility is high. Since the derivative at point A is smaller than points B and C, process sensitivity of point A is low. From a control point of view, any changes in C_W will give smaller changes in C_{EG} at point A than points B and C. Therefore, by maintaining C_W at point A, the desired C_{EG} can more easily be obtained (controlled) than other points. In order to maintain/controlled C_W at it desired value, F_c will be manipulated. Suppose that a disturbance moves C_W (±10%) away from its setpoints (points A, B, C). In order to maintain the setpoints, F_c will be adjusted and the control cost (calculated as percentage changes in F_c) is evaluated.

From Figure 3, at points B and C, any changes to the C_W will easily move C_{EG} away from its steady state value in a big scale compared to point A.

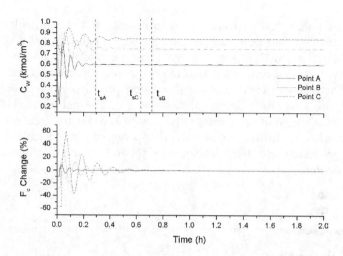

FIGURE 4
Closed loop dynamic analysis—responses of C_W and F_c change (%).

On the other hand, negative changes in C_W at points B and C means increased the volume and the operating time while positive changes means decreased the volume and the operating time. Since at point A process sensitivity is low, any changes to the C_W will require smaller changes in the manipulated variable than other points, and therefore, should easily bring the system back to the set-point (steady state).

Figure 4 shows closed loop responses of C_W and its corresponding change of F_c at points A, B and C. It is clearly shown that the time required for the C_W to come back to its setpoint, settling time (t_s), at point A is shorter than points B and C. This is because at point A, the process requires smaller changes in F_c in order to maintain its setpoint.

The same analysis is also applicable for the DF-based separation design and control analysis since the graphical concept of AR and DF regions are similar. Design of the separation system should be performed to maximize the total DF. From control point of view, this design means that smaller changes in the manipulative variables are needed for the control action. If the separation becomes more difficult (minimum DF), the operation, and control action becomes more expensive.

Conclusions and Future Work

In this paper, the use of AR concept within a model-based methodology of $IPDC$ for chemical process is successfully implemented. The results demonstrate the potential use of the AR concept to assist in selection of the optimal design of a single reactor that can ensure stable and controllable process. It was confirmed that designing a reactor at the maximum value of the *AR* leads to a process with lower sensitivity, higher flexibility and better controllability. It is also confirmed that closed loop performances of a process designed at the maximum value of *AR* are more promising than other points. All in all, this case study demonstrates how a simple technique such as *AR* is valuable for the solution of integrated design and control problems in process industries. As future perspective, this promising methodology will be further applied to solve *IPDC* problems for separation and reactor-separator-recycle systems.

Acknowledgement

The financial support from the Ministry of Higher Education (*MoHE*) of Malaysia and Universiti Teknologi Malaysia (*UTM*) is gratefully acknowledged.

References

Gani, R., Bek-Pedersen, E. (2000). A simple new algorithm for distillation column design. *AIChE J.*, 46(6), 1271–1274.

Hamid, M. K. A., Gani, R. (2008). A model-based methodology for simultaneous process design and control for chemical processes. In *Proceedings of the FOCAPO 2008*, Massachusetts, USA, 205–208.

Kiss, A. A., Bildea, C. S., Dimian, A. C. (2007). Design and control of recycle system by non-linear analysis. *Comput. Chem. Eng.*, 31, 601–611.

Parker, W. A., Prados, J. W. (1964). Analog computer design of an ethylene glycol system. *Chem. Eng. Prog.*, 60(6), 74–78.

Ramirez, E., Gani, R. (2007). Methodology for the design and analysis of reaction-separation systems with recycle. 1. The design perspective. *Ind. Eng. Chem. Res.*, 46, 8066–8083.

Russel, B. M., Henriksen, J. P., Jørgensen, S. B., Gani, R. (2002). Integration of design and control through model analysis. *Comput. Chem. Eng.*, 26, 213–225.

Sakizlis, V., Perkins, J. D., & Pistikopoulos, E. N. (2004). Recent advances in optimization-based simultaneous process and control design. *Comput. Chem. Eng.*, 28, 2069–2086.

Seferlis, P., Georgiadis, M. C. (2004). *The integration of process design and control.* Amsterdam: Elsevier B. V.

56

Study on Near-Zero Flaring for Chemical Plant Turnaround Operation

Qiang Xu*, Chaowei Liu, Xiongtao Yang, Kuyen Li, Helen H. Lou and John L. Gossage
Department of Chemical Engineering, Lamar University
Beaumont, TX 77710

CONTENTS

ABSTRACT Flaring emission during chemical plant turnaround operations generates huge amounts of air pollutants and also results in tremendous raw material and energy loss. This paper addresses near-zero flaring studies for chemical plant turnaround operation to advance current endeavors on flare minimization. Since off-spec products are inevitable during plant turnaround operations, they must be either recycled to the upstream process for online reuse or stored somewhere temporarily for future reprocessing when the plant manufacturing becomes stable, such that the off-spec products can be saved instead of being flared. A dynamic simulation based general methodology framework has been developed. The efficacies of the development are demonstrated by a virtual plant startup test.

KEYWORDS *Plant-wide dynamic simulation, near-zero flaring, emission source reduction, turnaround operation*

---

* To whom all correspondence should be addressed. E-mail: Qiang.xu@lamar.edu, Phone: (409) 880-7818

Introduction

During a chemical plant turnaround operation (plant startup, shutdown, or process upset management), off-spec product streams will be generated. These off-spec products usually have to be sent to flaring system for destruction. Flaring is crucial to the safety of chemical plant personnel and equipments. However, excessive flaring emits huge amounts of volatile organic compounds (VOC), NOx, and etc. The flaring emissions cause highly localized and transient air pollution events, which is harmful to the public health. Flaring also results in tremendous raw material and energy loss that could generate much needed products from the industry (Pohl *et al.*, 1986). It has been estimated that an ethylene plant with a capacity of 1.2 billion pounds of ethylene production per year will easily flare about 5.0 million pounds of ethylene during one single startup (Xu and Li, 2008). Assume 98% flaring efficiency (destruction efficiency), the resultant air emission will include at least 40.0K lbs CO, 7.5K lbs NO_x, 15.1K lbs hydrocarbons, and more than 100.0K lbs VOCs. Thus, as the increasingly strict environmental regulations and economic competitions, flare minimization has become one of the major concerns for chemical process industry (CPI).

Current practice of flare minimization in CPI plants is not standardized due to the complexity of the different process and operating procedures. This causes flare minimization depending almost exclusively on the experienced and well trained operators, engineers, and administrators. However, pure industrial-experience based methods are often limited, when they have to confront complex plant-wide turnaround operations with critical control and safety issues.

Quantitatively, a cost-effective way to find flare minimization opportunities is to use plant-wide dynamic simulation (DS) to virtually test plant operating strategies that will be taken by the plant operation personnel (Li *et al.*, 2007). It critically examines the potential process operational risks and infeasibilities. Based on the DS results, the feedback will help the plant improve the startup operating strategies and thus reduce flare emissions.

This paper addresses near-zero flaring studies for chemical plant turnaround operation to advance current endeavors in flare minimization. Since off-spec products are inevitable during plant turnaround operations, they must be either recycled to the upstream process for online reuse or stored somewhere temporarily for future reprocessing when the plant manufacturing becomes stable, such that the off-spec products can be saved instead of being flared. To reach the near-zero flaring target, effective recycle design and aggressive operation strategy have to be simultaneously considered to reduce the turnaround time. In this paper, a DS based general methodology framework has been developed. The efficacies of this methodology (help accomplish environmental and economic win-win situations) are demonstrated by a virtual test.

General Methodology

The developed methodology framework integrates the activities of process design (modification), dynamic simulation, and the utilization of industrial expertise. The first step for flaring emission reduction is to identify an efficient and economic design for recycling and storage of all the possible off-spec product streams. To address the conceptual design at the plant level, a design superstructure with all the possible solutions for recycling and storage of off-spec products needs to be created. Such a superstructure is shown in Fig. 1. Since off-spec products are usually generated in separation stages, the design superstructure mainly focuses on the recycles from separation sections to the other sections of a plant.

As illustrated in Fig. 1, an off-spec product stream may possibly be recycled back to any upper stream sections. A recycling stream can be either transferred through some pipes designated for each individual separation section, or mixed with other recycling streams in advance, then jointly transferred

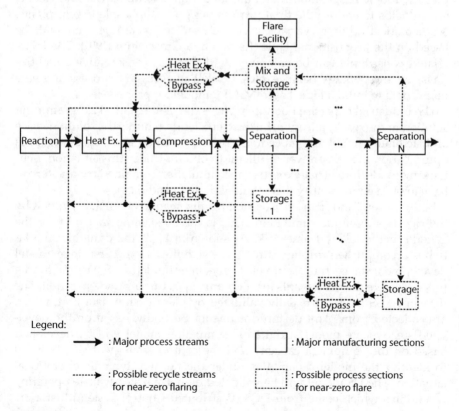

FIGURE 1
Superstructure for near-zero flaring design.

back to some upper-stream sections, accordingly. During the recycling, off-spec product steams can be stored temporarily, or reused directly. According to the processing requirements, each recycled stream might be preheated or cooled down before reuse. Thus, the design superstructure shown in Fig. 1 suggests all the possible scenarios listed above. Note that the dashed lines and boxes in Fig. 1 are possible pipes and process units that might be selected for near-zero flaring. Possible energy recovery opportunities are not included.

Based on the design superstructure, rigorous models could be used to select the design solution candidates. It should be note that the integration of industrial expertise is one of the important keys to ensure the success for near-zero flaring. The industrial expertise can be presented in propositional logic, which means a set of literals are connected by logic operator such as "OR", "AND", "IMPLICATION", "EQUIVALENCE". For instance, "if an off-spec product stream occurs, which mainly consists of light components such as CH_4 AND C_2H_4, then (IMPLICATION) it is better to preheat them first, AND recycle them to the entrance of compressor section." Note that such logic propositions contain lots of valuable industrial knowledge, which helps to endure the final process design is applicable to real plants. Mathematically, binary variables will be used to create linear inequalities based on the propositional logic (Raman and Grossmann, 1991). The logic-related constraints can be combined with process conservative equalities (mass balance, energy balance, etc.), as well as necessary process and economic data to build a logic-based MINLP model for process design.

The industrial expertise not only bridges the mathematical programming solution to the real application, but also efficiently reduces the solution space and facilitates the solving procedure. Also note that for many cases, rigorous optimization models are very difficult to obtain. Under this situation, heuristic methods (industrial-expertise and simulation combined try-and- error) have to be used to identify the potential design candidates.

Both rigorous and heuristic solution results need to be virtually tested by the rigorous dynamic simulation. This is the most important part for the general methodology framework. As shown in Fig.2, the dynamic simulation is accomplished through three-stage activities. In the first stage, model development starts with a setup of steady-state simulation (SS) model for the process where the plant wants to study during its turnaround operation. The developed SS model is usually validated by plant design data first, which are collected from plant design documents. Secondly, based on the developed steady-state model, a dynamic simulation model will be developed based on the equipment capacity and the control strategy. It will be used to identify the model capabilities under the normal dynamic operational situation. The DS model will be validated by normal steady-state operating conditions, which come from DCS (Distributed Control System) historian. Quite often the real plant data are not in mass or energy balance. Under such

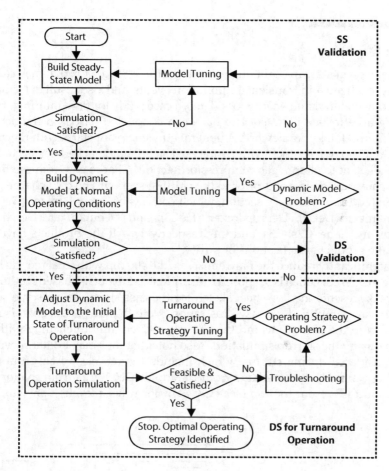

FIGURE 2
Dynamic simulation framework (Xu *et al.*, 2008).

condition, data verification, data reconciliation, and industrial expertise support become necessary.

At last, when the dynamic simulation model is proved, it will be used to examine the feasibility of the specified designs and operational strategies. If the feasibility test is passed, it indicates the selected design and operational strategy has been identified. Otherwise, troubleshooting on simulation models, various input data as well as the proposed designs and operation strategies will be conducted. Such troubleshooting efforts will eventually lead to the most desirable design and operational strategy towards near-zero flaring. Due to the inherent complexity of plant-wide dynamic simulation and the possible data incompleteness, industrial expertise is required at every stage of activity.

Case Study

Ethylene plant startups emit huge amounts of VOC and NOx that may cause highly localized and transient air pollution events, and also result in tremendous raw material and energy loss. Thus, a case study for flare minimization during an ethylene plant startup has been conducted based on the developed methodology framework. A general ethylene plant starts with thermal cracking of raw feedstock with furnaces. The effluent cracked gas is cooled and then sent to charge gas compression section. Charge gas from the final stage compression is dried prior to the chilling section. The cold charge gas is then fed to the recovery section, which consists of Demethanizer (DeC1), Deethanizer (DeC2), Depropanizer (DeC3), and Debutanizer (DeC4) to recover methane, C2s, C3s, and C4s respectively. All the products have to meet the product specifications or purity.

To significantly reduce the flare emission, the startup with total recycle has been considered, which is proposed from industrial-expertise based methods. Figure 3 presents the flowsheet sketch for the case study, where some subsystems are simplified for easy illustration. Note that furnace section is not included in the case study and the charge gas feed after quenching will be given during the startup simulation. Also note that four major outer recycles are considered during startup, which include the recycles of H2 from the chilling train; H2 and methane from the top of DeC1, C2s from the top of DeC2, and C3s from the top of DeC3. A major inner recycle is the stream

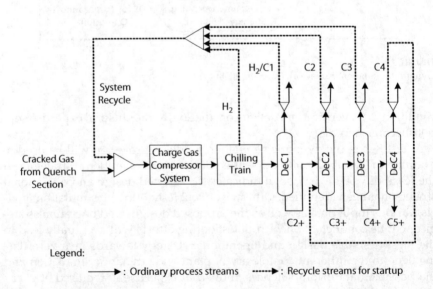

FIGURE 3
Flowsheet of the case study.

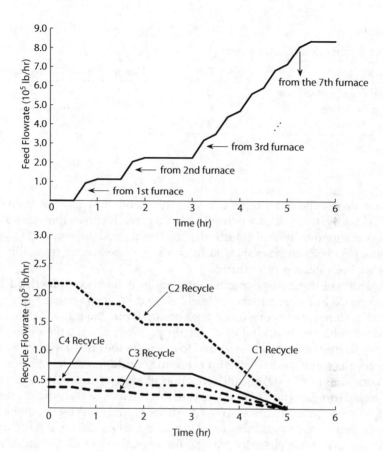

FIGURE 4
Startup procedure 1.

of C4s from the top of DeC4 fed back to the bottom of DeC2. With these recycles, the startup emission are expected to be greatly reduced due to two reasons: one is that the huge amount of components from H_2 though C_4 will be reused instead of being flared during startup; the other is that with the help of these recycled streams, the key units of DeC1 through DeC4 can gear towards their normal operation conditions quickly, which means the startup time will be reduced compared with the startup without recycles.

Based on the designs shown in Fig. 3, two startup operational strategies have been proposed, which needs rigorous dynamic simulation to virtually test. The first startup procedure is shown in Fig. 4, where the cracked gas from seven furnaces is brought up to the system sequentially. Meanwhile, the flowrates for various recycles also changes correspondingly. Note that all the recycle flowrate will finally becomes zero when the whole plant gears to its normal operation status. The second startup procedure is similar to the

TABLE 1

Startup Flaring Result Comparison

Scenarios	Startup Time (hr)	Flaring Reduction Percentage (%)
Shortest Startup in the Past	25	-----
Startup Procedure 1	6	~ 91
Startup Procedure 2	14	~ 62

first one, except the duration of every startup action in Fig. 4 has been doubled. Certainly, the first procedure is more aggressive than the second one, because it supposedly will reduce the startup time significantly. However, rigorous DS needs to be employed to check the operational feasibility and safety for both startup procedures.

Based on the rigorous dynamic simulation, it has been identified both startup procedures are virtually feasible based on the proposed design in Fig. 3. The flare emission results are summarized in Table 1, which are also compared with the historian best startup performance for the plant. Note that 98% flaring efficiency is assumed to calculate the emission results.

The results show the first startup procedure could reduce as much as 91% emissions due to the significant reduction of startup time. It suggests the near-zero flaring for the CPI plants is possible if effective and efficient design and operation strategies are employed. Although the best case study has not been applied in reality and the detailed economic benefits need more investigation, it is at least virtually feasible for operation, which provides great opportunities for future applications.

Conclusions

Near-zero flaring for CPI plant turnaround operation benefits environmental and industrial sustainability, but also presents tremendous challenges. Since off-spec products are inevitable during the plant turnaround operation, they must be either recycled to the upstream process for online reuse or stored somewhere temporarily such that they can be saved instead of being flared. This paper presents a cost-effective methodology framework for study on the near-zero flaring, which integrates the activities of process design (modification), dynamic simulation, and industrial expertise. The virtual test has demonstrated the efficacy of the proposed methodology framework. Future studies are still needed to advance current developments.

Acknowledgments

This work was in part supported by Texas Commission on Environmental Quality (TCEQ), Texas Air Research Center (TARC), and Texas Hazardous Waste Research Center (THWRC).

References

Li, K., Xu, Q., Lou, H. H., Gossage, J. L., Singh, A., Vragolic, S. Kelly, T. (2007). Flare Minimization during Plant Startup via Dynamic Simulation, *PSE ASIA 2007*, Xi'an, China.

Pohl, J., Lee, J., Payne R. (1986). Combustion Efficiency of Flares, *Combustion Science and Technology*, 50, 217.

Raman, R. Grossmann, I. E. (1991). Integration of Logic and Heuristic Knowledge in MINLP Optimization for Process Synthesis, *Comput. Chem. Eng.*, 16, 155.

Xu, Q., Li, K. (2008). Dynamic Simulation for Chemical Plant Turnaround Operation. *Integrated Environmental Management Consortium Meeting, Houston*, Texas, June 25.

Xu Q., Li K., Yang X, Liu C, Romero R., Mekala U., Lou H. H., Gossage J. L.(2008). Flare Minimization for Chemical Plant Turnaround Operation via Plant-wide Dynamic Simulation, *Proceedings of FOCAPO 2008*, 247.

57

Optimal Synthesis and Scheduling Strategies for Batch Azeotropic Distillation Processes

Vincentius Surya Kurnia Adi and Chuei-Tin Chang[*]

Department of Chemical Engineering, National Cheng Kung University, Tainan,
Taiwan, Republic of China

CONTENTS

ABSTRACT By addressing both flow sheet generation and scheduling issues, a sequential procedure has been developed in this work for designing batch azeotropic distillation systems. The proposed strategies are applied in two stages. Firstly, an integer program (IP) is solved to produce the optimal flow sheet (or state-task network). A mixed integer linear programming Model (MILP) is then constructed accordingly for generating the optimal short-term and cyclic schedules. This approach is applied to a heterogeneous ternary system in the present paper as an example. Satisfactory process configurations and production schedules can both be produced in all the cases we have studied so far.

KEYWORDS *Batch scheduling, azeotropic distillation, state-task network, mathematical programming model*

Introduction

The design of azeotropic distillation processes has always been an important issue for process engineers. For example, Feng et al. (2000) used a graphical technique to identify all possible operations in an azeotropic distillation system by resorting to the first principles and by logically sequencing such units. However, since the available studies are all concerned with the continuous processes, it is desirable to modify the existing methods for the synthesis and scheduling of batch processes.

To solve the general batch scheduling problems, a large number of methods have been developed. The STN-based MILP model is considered to be one of the most effective tools for this purpose (Ierapetritou and Floudas, 1998). Janak et al. (2006) recently developed an updated version with the unit-specific event-based continuous-time formulation to produce the production schedule for any large-scale multipurpose batch plant. This formulation incorporates several important features including various storage policies, variable batch sizes and processing times, batch mixing and splitting, sequence-dependent changeover times, intermediate due dates, products used as raw materials, and several modes of operation. It should be noted that a specific state-task network must be synthesized in advance before the construction of schedule-generating model for each application. However, there are in fact an extremely large number of alternative means to break up an azeotrope. In this study, any given azeotrope-entrainer system is divided into a finite number of lumped materials according to the classification approach suggested by Feng et al. (2000). All possible processes are then identified on the basis of this classification system. More specifically, it is assumed that there are only three types of feasible batch operations, i.e., distillation, decanting and mixing. A modified version of the logic-oriented integer program (IP) suggested by Raman and Grossmann (1991) has been adopted in the present work for synthesizing the optimal STN.

In short, a two-stage synthesis and scheduling strategy is followed in the present work to design the batch azeotropic distillation processes. An appropriate STN is first constructed with an IP model, and then the optimal schedule can be generated on the basis of an event-point based MILP model. Details on system classification, STN identification and schedule synthesis are discussed sequentially in later sections. An example is provided to illustrate the model construction procedures and also to demonstrate the benefits of the proposed approach.

System Classification

In this study, the residue curve maps (RCMs), the phase envelopes with the corresponding tie lines, and the iso-volatility curves, i.e., the so-called distillation boundaries, are all assumed to be available for the system under study. As mentioned previously, since an infinite number of mixtures can be identified in a multi-component system, there is a need to divide the corresponding RCM into a finite set of areas, lines and points, and treat each of them as a "lumped" material. The partition approach suggested by

Feng et al. (2000) is simplified for the homogeneous systems in this study to reduce the implementation effort, while a slightly modified version is used for the heterogeneous systems. The general partitioning procedure is summarized below:

1. RCM partition according to the critical curves and lines, i.e., the distillation boundaries and pseudo-boundaries, and the two-phase envelopes.

2. Carry out further partition on the basis of the uniqueness of intermediate products obtained from separation.

3. Identify the plausible operation units and their concomitant materials.

4. Identify the indispensable operation units.

5. Identify the operation units for generating the feeds to the indispensable operation units selected above.

6. Identifying additional operation units which can be used to facilitate separation.

The partitioned RCM for a heterogeneous ternary system, i.e., ethanol, water and toluene, is shown in Figure 1 as an example.

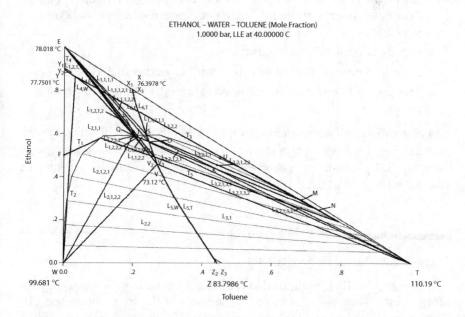

FIGURE 1
Complete partition of the ethanol (E)-water (W)-toluene (T) system with operating lines: F, feed; X, Y, Z, binary azeotropes; V, ternary azeotrope.

Plausible Operations

The plausible operations in the aforementioned systems can be defined according to a modified version of the conventional approach (Feng et al., 2000). Only two different types of batch operations, i.e., mixing and distillation, are considered for processing the homogeneous mixtures, while decantation is adopted as the third alternative in the heterogeneous systems. Since the conventional approach was developed primarily for the continuous processes, it is necessary to provide additional options in the present study to facilitate selection of column type, i.e., rectifier or stripper, and the cut number for every batch distillation operation.

To limit the search space, the candidate operating units are chosen to satisfy the following criteria:

1. The number and compositions of distillation products are dependent upon the selected column type and cut number. Each product must be located at a singular point, on a distillation boundary, or on a bounding surface.

2. The lumped materials in the same region (bounded by the same set of distillation boundaries and/or liquid-liquid equilibrium envelopes) are not allowed to be mixed.

3. No mixer yields any lumped material occupying a singular point, distillation boundary, liquid-liquid equilibrium envelope, or bounding surface.

4. Exactly two materials are mixable.

5. A desired product cannot be mixed with any other material.

6. The product from any mixer is not allowed to be fed to another mixer.

Every operating unit is represented by the symbol $(\{\cdots\}, \{\cdots\})$, in which the two curly brackets represent sets of input and output materials respectively. Since there are 522 plausible operations identified for the ethanol-water-toluene system, only a partial list is shown below.

State-Task Network Synthesis

In this work, the STN configurations are identified with an integer programming model. This model is built on the basis of the formulation used by Raman and Grossmann (1991) for logic inference with modification to characterize the intrinsic nature of batch operations. As mentioned before, there can be only three types of operations in the azeotropic distillation system,

TABLE 1

Plausible Operating Units of the Ethanol (E)-water (W)-toluene (T) System

Index	Operating unit	Type
1	$(\{L_{1,1,1,1}\}, \{V_1, EX\})$	Top Product Distillation
2	$(\{L_{1,1,1,1}\}, \{E, L_{6,E,1}\})$	Bot. Product Distillation
3	$(\{L_{1,1,1,1}\}, \{V_1, X_1, E\})$	Total Distillation
91	$(\{L_{1,1,2,1,1}\}, \{QR, MN\})$	Decantation
92	$(\{L_{1,1,2,1,1}\}, \{RJ, MN\})$	Decantation
93	$(\{L_{1,1,2,1,1}\}, \{JS, MN\})$	Decantation
123	$(\{F, L_{4,E,1}\}, \{L_{2,1,1}\})$	Mixing
124	$(\{F, L_{4,E,2}\}, \{L_{2,1,1}\})$	Mixing
125	$(\{F, L_{4,E,2}\}, \{L_{2,1,2,1}\})$	Mixing
...etc	...etc	...etc

i.e., mixing, distillation and two-phase separation (decantation). For example, let us consider a lumped material (say A) and its two downstream units, i.e., a separator $(\{A\}, \{B, C\})$ and a mixer $(\{A, D\}, \{E\})$.

In an IP model, the distillation operation can be described with two inequality constraints:

$$(1 - y_A) + (1 - z_j) + y_B \geq 1 \tag{1}$$

$$(1 - y_A) + (1 - z_j) + y_C \geq 1 \tag{2}$$

Also, the corresponding constraint for mixer should be written as

$$(1 - y_A) + (1 - y_D) + (1 - z_k) + y_E \geq 1 \tag{3}$$

where, y_A, y_B, y_C, y_D and y_E are binary variables denoting the presence (1) or absence (0) of material A, B, C, D, and E respectively.

Constraints (1) and (2) imply that the lumped materials B and C can be produced only when material A is present and, also, the j th (distillation/decantation) operation is chosen, i.e., $z_j = 1$. On the other hand, constraint (3) can be used to impose the requirements that both feeds (A and D) should be present and the k th (mixing) operation must also be selected in order to produce material E. Finally, in making a selection to process A alone or to consume both A and D, the following constraints could be adopted:

$$(1 - y_A) + z_j + z_k \geq 1 \tag{4}$$

$$(1 - y_A) + (1 - y_D) + z_k \geq 1 \tag{5}$$

Notice that the above formulation approach is general.

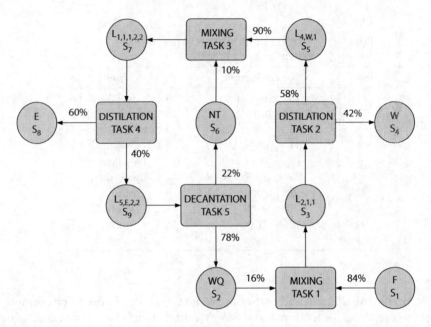

FIGURE 2
State-task network of the ethanol-water-toluene system.

One of the design objectives used in this work is to minimize the number of operation units, i.e.,

$$\min\left(\sum_{j=1}^{N_{opj}} z_j + \sum_{k=1}^{N_{opk}} z_k\right) \tag{6}$$

where N_{opj} and N_{opk} is the total number of separation and mixing operation respectively.

The STN for ethanol-water-toluene system can be generated with the afore-mentioned approach and it is shown below.

Scheduling Strategies

As indicated previously, the continuous-time representation is adopted in the present work to construct the scheduling models. The basic models and its implementation strategies are presented below.

Short-Term Scheduling Model

After obtaining the optimal STN, a mixed-integer linear program can be constructed accordingly for scheduling purpose. The formulation proposed by Ierapetritou and Floudas (1998) is directly adopted in this work. To create

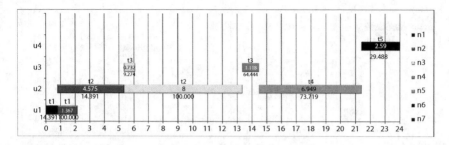

FIGURE 3
Gantt chart of ethanol-water-toluene system (short-term schedule).

such a model for a particular application, it is necessary to first postulate an enough number of *event points* corresponding to either the initiation of a task and/or the beginning of unit utilization. The locations of these points on time axis are unknown. A simple trial-and-error procedure has been used for determining the appropriate number of event points needed.

Another key idea in this formulation is the decoupling of *task events* (*i*) and *unit events* (*j*). This is achieved by using two separate sets of variables to respectively represent the *task events* (i.e., the beginning of the task), denoted as $wv(i,n)$, and the unit events (i.e., the beginning of unit utilization), denoted as $yv(j,n)$. If task event *i* starts at event point *n*, then $wv(i,n) = 1$; otherwise, it is zero. Similarly, if unit event *j* takes place at event point *n*, then $yv(j,n) = 1$; otherwise, it is zero. Processing time of each task is assumed to vary with the amount of material being handled.

A typical short-term schedule for the ethanol-water-toluene system can be found in Figure 3.

Cyclic Scheduling Model

The above model can be reformulated to generate cyclic schedules (Wu and Ierapetritou, 2004). A periodic scheduling approach resides primarily on the following assumption: For the case that the time horizon is long compared with the duration of individual tasks, a proper time period which is much smaller than the whole time horizon exists, within which, some maximum capacities or crucial criteria have been reached so that the periodic execution of such schedule will achieve results very close to the optimal one by solving the original problem without any periodicity assumption. As a result, the problem size can be significantly reduced. Besides the obvious advantage in computation, the solution is more convenient and easier to implement in practice since the same schedule is repeated many times.

In this approach, the model variables should include the time length of a cycle as well as the detailed schedule within this period. Unlike short-term scheduling where all intermediates other than those provided initially have to be produced before the beginning of the tasks, each unit schedule can

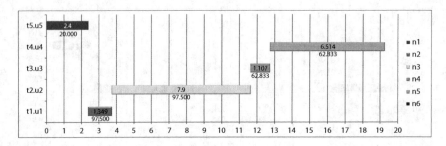

FIGURE 4
Gantt chart of ethanol-water-toluene system (cyclic schedule).

start with certain amounts of intermediates as long as storage capacity constraints are not violated. The initial amounts of intermediates are equal to the inventories accumulated at the end of every unit period, so as to preserve the material balance across the cycle boundaries.

An example of the cyclic schedule for the ethanol-water-toluene system can be found in Figure 4.

Conclusion

An effective and systematic procedure has been developed to synthesize the STNs and also production schedules for batch azeotropic distillation systems. The feasibility of this procedure is demonstrated with a heterogeneous ternary system in this paper. The proposed STN synthesis method has been rendered possible by resorting to a systematic approach to classify the entire space of a RCM into a finite number of areas, lines and points and by constructing an integer programming model for logic inference. The continuous-time formulation has also been adopted for short-term and cyclic scheduling of the batch distillation processes. The former task can be accomplished with a MILP model, while the later with a MINLP model. The effectiveness of this approach has been verified with extensive case studies.

References

Brooke, A., Kendrick, D., Meeraus, A., & Raman, R. (1998). GAMS: A user's guide. *GAMS Development Corporation*.

Feng, G., Fan, L. T., Friedler F., & Seib, P. A. (2000). Identifying operating units for the design and synthesis of azeotropic-distillation systems. *Industrial & Engineering Chemistry Research*, 39, 175–184.

Ierapetritou, M. G. & Floudas, C. A. (1998). Effective continuous-time formulation for short-term scheduling. 1. Multipurpose batch processes. *Industrial & Engineering Chemical Research*, 37, 4341–4359.

Janak, S. L., Floudas, C. A., Kallrath, J., & Vormbrock, N. (2006). Production scheduling of a large-scale industrial batch plant. I. Short-term and medium-term scheduling. *Industrial & Engineering Chemistry Research*, 45, 8234–8252.

Raman, R. & Grossmann, I. E. (1991). Relation between MILP modeling and logical inference for chemical process synthesis. *Computers & Chemical Engineering*, 15, 73–84.

Wu, D. & Ierapetritou, M. (2004). Cyclic short-term scheduling of multiproduct batch plants using continuous-time representation. *Computers & Chemical Engineering*, 28, 2271–2286.

58

A Backoff-based Strategy to Improve
Robustness in Model-based Experiment
Design Under Parametric Uncertainty

Federico Galvanin[a,*], **Massimiliano Barolo**[a],
Fabrizio Bezzo[a] **and Sandro Macchietto**[b]

[a] *DIPIC—Dipartimento di Principi e Impianti di Ingegneria Chimica
Università di Padova, via Marzolo 9, I-35131, Padova (Italy)*
[b] *Department of Chemical Engineering, Imperial College London
South Kensington Campus, SW7 2AZ London (U.K.)*

CONTENTS

ABSTRACT   Model-based experiment design techniques are a valid tool
for the rapid development and assessment of dynamic deterministic models. Once a set of constraints on the experiment design variables and on the
predicted responses is superimposed, uncertainty in the model parameters
can lead the constrained design procedure to predict sub-optimal and/or
unfeasible experiments. In this paper a backoff methodology is proposed to
formulate and solve the experiment design problem by explicitly taking into
account the presence of parametric uncertainty, to ensure the feasibility of
the planned experiment. The effectiveness of the proposed methodology is
discussed through an illustrative case study concerning parameter identification in a physiological model of type 1 diabetes mellitus.

KEYWORDS   *Experiment design, parameter estimation, backoff problem*

---

* To whom all correspondence should be addressed: federico.galvanin@unipd.it

Introduction

Mathematical models are essential to the simulation, design and optimisation of chemical and biological processes. A large class of deterministic models can be represented by systems of differential and algebraic equations (DAEs). The goal of every model building procedure is to identify both the model structure and the model parameters in order to represent in the most reliable and accurate way the undergoing phenomena. Modern model-based design of experiment (MBDoE) techniques allow for the rapid development and assessment of dynamic models, yielding the most informative set of experimental data in order to estimate precisely the parametric set of a given model. Their effectiveness has been demonstrated in several topics and a recent and exhaustive review can be found in Franceschini and Macchietto (2008). The MBDoE problem for improving parameter estimation is a particular form of dynamic optimisation involving the maximisation of a measure of the expected information, usually computed as a particular metric of the Fisher information matrix (Pukelsheim, 1993). The technique is usually embodied in a sequence of three key activities (experiment design, experiment execution, parameter estimation) and it allows the definition of a set of active constraints on both state and design variables during the design optimisation (Bruwer and MacGregor, 2002) to perform a constrained MBDoE. The goal of a constrained MBDoE is to achieve optimality (maximisation of the expected information) and feasibility (no constraints violation) during the experimental trials. Since the methodology is model-based, parametric mismatch can affect the consistency of the whole design procedure. Despite the importance of the problem of ensuring optimally informative and feasible experiments, there has been relatively little work in the area of model-based experiment design about the development of techniques to overcome the above issues. Robust techniques for optimal experiment design have been proposed in the literature focusing on the problem of optimality of the design in the presence of parametric uncertainty, solving a max-min optimisation problem ("worst case approach") or performing a dynamic optimisation over all the predicted uncertainty region ("expected value approach") of model parameters (Asprey and Macchietto, 2002; Rojas et al., 2007).

Within process systems design, constrained optimisation under uncertainty, seen as a trade-off between feasibility and optimality, has long been recognised as a key issue (Grossmann and Sargent, 1978; Halemane and Grossmann, 1983). Starting from the study of steady-state processes and moving towards dynamic systems, several approaches have been proposed to solve the process design optimisation in the presence of parametric uncertainty. In general, uncertain parameters are described by probability distribution functions and the design problem is formulated using probabilistic decision criterions, such as the maximization of the expected value of a given process performance metric. In this way the design solution (formally an "overdesign") represents the best decision starting from the actual knowledge available about the process. A way to ensure feasibility (often of greater importance than optimality) in the presence of parametric uncertainty is to introduce some backoffs from active constraints. In the backoff problem (Bahri et al., 1995) the operating point is moved away from the nominal conditions in order to ensure the process feasibility and to compensate for the effect of the disturbances. In this paper, a general methodology is proposed and discussed to address in a systematic way the problem of the constrained optimal experiment design under parametric uncertainty. The adoption of a backoff policy allows to guarantee the feasibility of the optimally designed experiment in the presence of parametric uncertainty. The technique is particularly suitable for planning the experimentation in such systems (e.g., physiological systems, reactive systems) where there is a limited knowledge about the parametric system and the operability is strictly reduced by the presence of active constraints on state variables. The proposed technique is illustrated and discussed through a case study concerning parameter identification of a physiological model of type 1 diabetes mellitus.

Problem Statement: MBDoE with Backoff

A general dynamic deterministic model can be described by a set of DAEs in the form

$$\begin{cases} \mathbf{f}\left(\dot{\mathbf{x}}(t), \mathbf{x}(t), \mathbf{u}(t), \mathbf{w}, \theta, t\right) = 0 \\ \hat{\mathbf{y}}(t) = \mathbf{g}(\mathbf{x}(t)) \end{cases} \tag{1}$$

where $\mathbf{x}(t)$ is the N_x-dimensional vector of time-dependent state variables, $u(t)$ and $\mathbf{w}$ are the N_u and N_w-dimensional time-dependent and time-invariant control variables (manipulated inputs) respectively, θ is the N_θ-dimensional set of unknown model parameters to be estimated, and t is the time. The symbol $\wedge$ is used to indicate the estimate of a variable (or a set of variables).

Model-based experiment design procedures aim at decreasing the model parameter uncertainty region by acting on the n_φ-dimensional vector of design vector variables φ:

$$\varphi = \{\mathbf{y}_0, \mathbf{u}(t), \mathbf{w}, \mathbf{t}^{sp}, \tau\} \tag{2}$$

where $\mathbf{y}_0$ is the set of initial conditions of the measured variables, and τ is the duration of an experiment. The set of time instants at which the output variables are sampled is a design variable itself, and is expressed through the n_{sp}-dimensional vector $\mathbf{t}^{sp}$ sampling times. If $\mathbf{G}(t)$ is a N_c-dimensional set of (time-varying) active constraints on the state variables, the optimal design under constraints solution has to satisfy, at the same time, the optimality condition

$$\varphi = \arg\min \psi(\mathbf{V}_\theta(\hat{\theta}, \varphi)) \tag{3}$$

and the feasibility condition

$$\tilde{C} = x(u, w, \hat{\theta}, t) - G(t) \leq 0 \tag{4}$$

where ψ is a metric of the variance-covariance matrix of model parameters V^θ expressing the chosen design criteria and $\tilde{C}$ is a N_c-dimensional set of constraint functions. In addition to (3) and (4), an n_φ-dimensional set of constraints on design variables may be present, too, usually expressed as

$$\varphi_i^l \leq \varphi_i \leq \varphi_i^u \quad i = 1 \dots n_\varphi \tag{5}$$

with lower (superscript l) and upper (superscript u) bounds on φ. For a single experiment the variance-covariance matrix of model parameters is

$$V_\theta(\hat{\theta}, \varphi) = \left[\Sigma_\theta^{-1} + H_\theta(\hat{\theta}, \varphi) \right]^{-1} \tag{6}$$

where H_θ is the N_θ-dimensional Fisher information matrix and prior information on the model parameter uncertainty region in terms of statistical distribution (for instance, a uniform or Gaussian distribution) can be included through matrix Σ_θ.

The solution to the constrained MBDoE optimisation problem is the optimal design vector φ that through the model (1) satisfies the design optimality condition (3), the feasibility constraints on the state variables (4) and the constraints on the design variables (5).

In the presence of parametric mismatch the real system is defined by the parametric set θ, which is different from the current (estimated) one. If φ' is an optimal design vector in the form (4) satisfying the optimality condition (3), the feasibility condition can be always guaranteed only if

$$\tilde{C} = x(u', w', \theta, t) - G(t) + \beta(t) \leq 0 \tag{7}$$

where $\beta(t)$ is a N_c-dimensional set of time-dependent backoff functions, taking into account the effect of the parametric mismatch on the state variables at the newly designed experimental conditions. In order to solve the constrained MBDoE optimization problem with backoff the standard optimization procedure was coupled with a stochastic simulation over the expected uncertainty region of model parameters (Figure 1) providing the proper set of backoff functions. The stochastic simulation is articulated into four key steps:

1. definition of the expected uncertainty region of model parameters through a statistical distribution $p_{\hat{\theta}}$;
2. parallel execution of N' simulations with $\hat{\theta}_i \in p_{\hat{\theta}}$ ($i = 1, \dots, N'$);
3. mapping of the uncertainty of system responses through a proper statistical distribution $p_x \mid x_i \in p_x$ ($i = 1, \dots, N'$);
4. definition of the backoff β as a function of p_x.

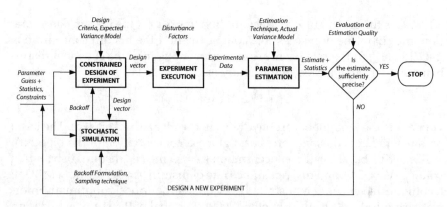

FIGURE 1
MBDoE with backoff: scheme.

Globally, the backoff vector β is a function of the design vector, of the probability distributions of x, and of the sampling technique, given a prior information on the parametric set. The experimenter can control the amount of backoff in order to keep the feasibility constraints (7) within a prescribed level of confidence. The considerable computational challenge required for solving the MBDoE optimisation problem with backoff is mainly concentrated in the stochastic simulation block, where the simulation is carried out over the entire predicted dominion of uncertainty of model parameters. However, prior information on the parametric set, sensitivity analysis and the selection of a proper sampling setting can drastically reduce the global computational effort (which, in any case, is executed off-line).

Case Study

The case study considered is a model of glucose homeostasis in the form proposed by Lynch and Bequette (2002). The model is represented by the following set of differential and algebraic equations:

$$\frac{dG}{dt} = -\theta_1 G - X(G + G_b) + D(t) \tag{8}$$

$$\frac{dX}{dt} = -\theta_2 X + \theta_3 I \tag{9}$$

$$\frac{dI}{dt} = -n(I + I_b) + \frac{u(t)}{V_I} \tag{10}$$

where G is the blood glucose concentration (mg/dL), X the insulin concentration (mU/L) in the non accessible compartment, I the insulin concentration (mU/L) and $u(t)$ the rate of infusion of exogenous insulin. The meal disturbance model $D(t)$ adopted in this study is:

$$D(t) = 2.5At\exp(-0.05t) \tag{11}$$

where A is the amount of carbohydrates in the meal (fixed at 60 g). The basal parameters kept constants are: basal glucose concentration in the blood (G_b = 81 mg/dL), basal insulin concentration (I_b = 15 mU/L), insulin distribution volume (V_I = 12 L), and disappearance rate of insulin (n = 5/54 min^{-1}). MBDoE techniques have been demonstrated as a valuable tool to design optimally informative clinical tests (Galvanin et al., 2009). The goal of the study is to design the optimal test conditions in terms of the insulin infusion rate u in order to achieve the maximum possible information from the experiment to estimate the set of model parameters precisely. The test has to be optimally informative and safe for the subject. A constrained MBDoE with backoff is designed, where the optimised design variables are the profile of the insulin infusion rate $u(t)$ (approximated as piecewise constant with n_{sw} = 7 switching times and n_z = 8 switching levels) and the vector of sampling times (n_{sp} = 10 samples with a minimum time between consecutive measurements of 10 min). The measured variable is the blood glucose concentration G, with a 3% expected relative error on the measurements. The constraints on the system are related to normoglycaemia attainment, and are the upper (G_1 = 150 mg/dL) and lower (G_2 = 60 mg/dL) thresholds on G, which is the only state variable being constrained and the only measured variable (i.e. $x_1 = y = G$). In fact, the lower bound only is a hard constraint not to be violated. However, here as a matter of example both constraints will be treated as hard ones. The constrained MBDoE is performed on the parametric set $\hat{\theta} = [0.0287\ 0.0283\ 1.30E{-}5]^T$ describing a healthy subject, while the real subject is assumed to be affected by diabetes with $\theta = [0.0155\ 0.0250\ 1.20E{-}5]^T$.

Two different designs have been compared:

1. standard constrained design (MBDoE);
2. constrained design with backoff (MBDoE-B).

The E-optimal criterion is chosen for both configurations. The backoff functions are evaluated through a stochastic simulation (N' = 500) where the parameters belong to a family of independent normal distributions ($p_{\hat{\theta}}$) whose means and standard deviations are given by vectors $\hat{\theta}$ and $\sigma = [0.0065\ \ 0.0013\ \ 0.0033E{-}5]^T$.

The uncertainty region of G and the backoff vector were evaluated considering the maximum and the minimum profile of the measured variable over the expected uncertainty region of model parameters. As can be seen from Figure 2, the test provided by a standard constrained MBDoE violates

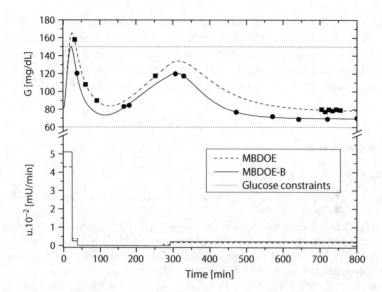

FIGURE 2
Estimated glucose concentration profiles, test samples and manipulated input configuration.

the hyperglycaemic threshold when applied to a subject with diabetes, while the backoff approach allows to design a safe test for the subject preserving at the same time the quality of the estimation in terms of statistical indexes (Table 1).

Conclusions

In this paper a new methodology for the constrained MBDoE in presence of parametric uncertainty exploiting a backoff-based approach was proposed and discussed. The new methodology has been applied to a simple model of the glucose homeostasis to design the best insulin infusion rate profile to be

TABLE 1

Parameter Estimation from Selected Configurations and Statistics (95% t-values and Confidence Intervals, $t^{rif} = 1.898$).

Par	MBDoE Estimate	t_i	MBDoE-B Estimate	t_i
θ_1	0.0102 ± 0.0051	1.99	0.0148 ± 0.0025	5.99
θ_2	0.0244 ± 0.0035	7.02	0.0251 ± 0.0017	14.63
θ_3	$1.16\text{E-}5 \pm 1.04\text{E-}6$	11.12	$1.20\text{E-}5 \pm 1.06\text{E-}6$	11.40

adopted in order to estimate precisely the parameters of a subject affected by diabetes when no preliminary information about the subject is available. The backoff strategy allows to estimate the parametric set describing the subject with diabetes in a safe manner, while a standard design, even if informative, leads the subject to temporary hyperglycaemic conditions.

References

Asprey, S. P., Macchietto S. (2002). Designing robust optimal dynamic experiments. *J. Process Control.*, *12*, 545.

Bahri, P.A., Bandoni P.A., Barton G.W., Romagnoli J.A. (1995). Backoff calculations on optimising control: a dynamic approach. *Computers. Chem. Eng.*, *19*, S699.

Bruwer, M. J., MacGregor J. F. (2006). Robust Multi-Variable Identification: Optimal Experiment Design with Constraints. *J. Process Control*, *16*, 581.

Franceschini, G., Macchietto S. (2008). Model-based Design of Experiments for Parameter Precision: State of the Art. *Chem. Eng. Sci.*, *63*, 4846.

Galvanin, F., Barolo M., Macchietto S., Bezzo F. (2009). Optimal design of clinical tests for the identification of physiological models of type 1 diabetes mellitus. *Ind. Eng. Chem. Res.*, *48*, 1989.

Grossmann, I.E., Sargent R.W.H. (1978). Optimum Design of Chemical Plants with Uncertain Parameters, *AIChE J.*, *37*, 517.

Halemane, K.P., Grossmann I.E. (1983). Optimal Process Design under Uncertainty, *AIChE J.*, *29*(3), 425.

Lynch, S.M., Bequette B.W. (2002). Model predictive control of blood glucose in type I diabetics using subcutaneous glucose measurements. *Proc ACC*, 4039.

Pukelsheim, F. (1993). *Optimal Design of Experiments*; J. Wiley & Sons: New York, U.S.A.

Rojas C.R., Welsh J.S., Goodwin G.C., Feuer A. (2007). Robust optimal experiment design for system identification. *Automatica*, *43*, 993.

59

A Novel Perspective in the Conceptual Design Paradigm: Beyond the Steady-State Solution

Flavio Manenti[*], Davide Manca and Roberto Grana

CMIC dept. "Giulio Natta", Politecnico di Milano
Piazza Leonardo da Vinci, 32, 20133, Milano, Italy

CONTENTS

ABSTRACT This manuscript proposes a novel approach to the conceptual design of industrial processes. The conventional conceptual design paradigm can be defined as the interaction between the process simulation, characterized by continuous variables, and the discrete optimization, where some specific superstructures are selected by means of the Boolean logic or, more in general, of integer variables. The paper proposes a novel approach that goes beyond the paradigm currently conceived in the literature. This approach is not limited to identify a single set of optimal steady-state operating conditions and an optimal equipment layout, but it focuses on the selection of a series of candidate process layouts, each of them optimal for a specific situation, making more profitable the definition of production campaigns both on the short- and medium-term horizons.

KEYWORDS *Dynamic conceptual design, production campaigns, MINLP superstructures*

[*] To whom all correspondence should be addressed: phone: +39.02.23993273; fax: +39.02.70638173; Email: flavio.manenti@polimi.it

Introduction

The optimization of chemical and industrial processes is a multifaceted problem that involves several issues. Some of them are nowadays well-understood and fully implemented, whereas some others are still open issues for the scientific and industrial community. This manuscript discusses the process design optimization, in terms of conceptual design.

Conceptual design denotes an optimization problem where some specific superstructures are selected by means of the Boolean logic. A superstructure comprises the theoretical coexistence of more than one sub-processes (comprising one or more process units) to perform one or more operations.

The conventional approach to conceptual design removes the suboptimal equipment layouts and identifies the best solution according to some predefined criteria, *e.g.* environment, sustainability, economics... This gives rise to either a single- or a multi-objective optimization problem where continuous variables (process conditions and geometries) are mixed to integer and/or Boolean decisional variables.

However, why should the conceptual design stop at the single (and steady-state) aprioristic solution of optimal process design?

Actually, the conceptual design can be seen also from a dynamic point of view where the final solution may comprise the coexistence of two or more superstructures that might be dynamically activated/deactivated according to the economic convenience of some specific production periods (e.g. night and daytime, work-days and weekends, summer and winter...).

As an example, let us consider the typical dynamic electric energy market of an industrialized country. It consists of high daytime prices and low prices at night. It often happens that a specific process design is economically convenient throughout the day, whereas another process layout is preferable during the night, according to some predetermined objectives.

On this subject, the selection of more coexisting sub-structures leads to an increase in the capital investment, since more process units are needed, but at the same time this approach may successfully reduce the breakeven point while shortening the return from the investment.

The paper proposes a fully integrated approach to solve a mixed-integer nonlinear programming (MINLP) problem based on a detailed simulation of the process layout.

Dynamic Conceptual Design

Often the resulting problem in the conceptual design is a mixed-integer linear or nonlinear programming (MILP or MINLP), where continuous process variables and geometric specifications are mixed to integer and Boolean decision variables (Dakin, 1965; Floudas, 1995; Grossmann and Kravanja, 1995; Biegler and Grossmann, 2004).

The scientific literature proposes several applications of conceptual design, but they do not account for the possibility of interacting layouts. To do so, the classical mixed-integer problem should be formulated as a time-dependent structure:

$$\min_{x(t),y(t)} \Phi(x(t), y(t), t)$$

$$s.t.:$$

$$h(x(t),\ y(t)) = 0 \tag{1}$$

$$g(x(t), y(t)) \geq 0$$

where Φ is the economic objective function, involving both investment and operating costs; $x \in \mathbb{R}^n$ are the continuous (or process) variables; $y \in \{0,1\}^m$ are the discrete variables; the constraints h and g represent material and energy balances as well as process constraints. All the continuous and the discrete states are time-dependent and, at the same time, they may be part of the objective function.

The proposed novel approach to process design, the dynamic conceptual design (see also eq. (1)), goes beyond the *status quo* as it is conceived in the literature. It is not limited to the steady-state optimum search of the best process design, which allows reaching the maximum profit by considering the nominal production. Conversely, it can select more than one sub-structure that are considered economically convenient by basing the decision criterion on different production scenarios.

The plant may switch from a configuration to another one by opportunely detecting the market conditions, the energy price, the raw material costs… so to enlarge the net operating margin of the overall production that has to come. This opportunity is evaluated neither *a posteriori* nor on-line, but it is selected and maximized for the economic profit (*i.e.* investment and operating costs) off-line during the process design activity.

As an example, let us consider the electric energy market. This market (Manenti, 2008; Manenti and Manca, 2008) involves a series of fluctuations (*e.g.* daily, weekly, seasonal). It often happens that a specific process design is the best one throughout the day, whereas another process structure achieves higher profits during the night.

On this subject, the optimal design may correspond to the selection of more than a single sub-structure. This unavoidably requires an increase in the capital investment, but it is usually significantly smaller if compared to the whole investment and, in addition, the breakeven point may result shorter and, consequently, the whole return on investment may be higher.

We developed a test case consisting of a detailed nonlinear simulation of the hydrodealkylation process, which usually belongs to oil refineries and petrochemical plants, and a simple superstructure consisting of a single Boolean variable to discriminate between two process configurations.

Test-Case

The industrial case study focuses on the toluene hydrodealkylation to benzene process layout optimization (Douglas, 1988).

It consists of a reaction and a separation zones. Fresh hydrogen (H_2) and toluene (C_7H_8) are preheated and fed to a plug flow reactor to produce benzene (C_6H_6). The reactor geometry has been already optimized in order to improve the yield in benzene and to reduce biphenyl ($C_{12}H_{10}$) production (Grana *et al.*, 2009). After a quench, the flowrate exiting the reactor is fed to the separation section to sequentially separate incondensables components from the C_6H_6 and the heaviest compounds (C_7H_8 and $C_{12}H_{10}$). A stream of H_2 (and CH_4) and a liquid stream of C_7H_8 are both recycled to the reaction section.

The process simulation is performed by PRO/II, a commercial simulator developed by Simsci-Esscor, Invensys Process Systems (Simsci-Esscor, 2002). The superstructure is directly modeled within the process simulation and this is made feasible by introducing flow mixers and splitters to get N alternatives while defining the following constraints:

$$\sum_{i=1}^{N} y_i F_i = 1 \qquad (2)$$

where F_i is the normalized process flowrate exiting from the splitter unit.

The superstructure is then managed by the optimization package mode-FRONTIER developed by Esteco and Enginsoft (Esteco-EnginSoft, 2008). The multi-membered evolution strategy has been adopted to solve the mixed-integer problem. The overall problem comprises the automatic interaction of these programs.

Figure 1 shows the two process layouts selected for the test case:

- The former layout is the classical hydro-de-alkylation process, where the hot stream exiting the reactor preheats the inlet flowrate by means of a feed effluent heat exchanger.

  This configuration is denoted by the Boolean variable $y = 0$.
- When $y = 1$, the hot stream is not anymore used to preheat the fresh feed, but it produces electric (which is sold).

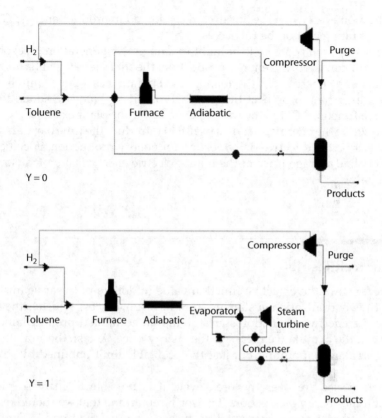

FIGURE 1
Basic (up) and expanded (down) process configurations.

By considering the third level of economic potential *EP3* (Douglas, 1988), the dynamic conceptual design can be formulated as follows:

$$\max_{y,\omega} \Phi(y,\omega) = EP3_{steady} - CI_{therm} - CV_{furn} +$$

$$+ y(P_{el}(p(\omega),\omega) - CI_{el} - CV_{therm}(\omega))$$

$$s.t.:$$

$$h(y) = 0$$

$$g(y) \geq 0$$

(3)

where $EP3_{steady}$ accounts for the end-product prices, the raw materials, and process unit installation (capital investment and operating costs). CI_{therm} is the additional capital investment required to introduce a new heat exchanger (or enlarge the existing one) in order to satisfy the reactor temperature specification even when the hot stream is adopted to produce energy. CV_{furn} is the operating cost of the furnace.

When $Y = 0$ the problem is brought to a no power generation case; otherwise, $Y = 1$ considers the energy production: the time-dependent terms such as revenues by energy sale $P_{el}(p(\omega), \omega)$ and operating cost due to purchasing the auxiliary fuel to preheat the raw materials $CV_{therm}(\omega)$ are added to the objective function. CI_{el} is the investment cost for the plant expansion.

To describe the energy price, it is useful to introduce the time partialization ω to describe the effective daily portion for energy production. Specifically, a cumulative average price of the hourly electric energy is adopted (source: GME, Italy).

Numerical Results

Figure 2 shows the objective function value in dollars per year against the daily time partialization ω starting from 10:00 am (circles), which is the time where the energy production starts giving an additional profit. Diamonds represent the cumulative mean for the energy price. At last, the straight line is the breakeven, pointing out just the profitable limit obtained by setting $y = 1$.

When circles are over the breakeven, it is possible to increase profits through the energy production. It is worth remarking that we should expect a profit higher than the breakeven at the beginning of the trend in Figure 2, since we are plotting it starting from the first advantageous time interval.

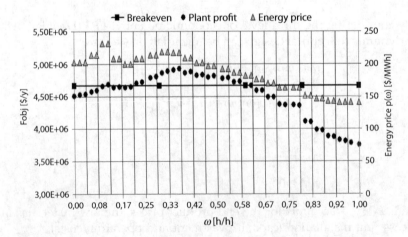

FIGURE 2
Benefit in plant profits through the dynamic conceptual design.

As a matter of facts, we are also considering the capital allowance related to the new process units for power generation. As a result, the first hours of energy production per day are necessary to cover the additional capital investment.

Moreover, if we consider the cost of the additional fuel to preheat the fresh feed flowrate entering the reactor, while we are producing electric energy, the real margin is further reduced. Actually, the first time derivative of the objective function is:

$$\left.\frac{dF}{d\omega}\right|_y = P_{el} - CV_{therm} \tag{4}$$

meaning that we can increase the net profit margin only when the derivative of the objective function is positive, since $P_{el} > CV_{therm}$.

Conversely, when the derivative is negative, even though the plant profit is beyond the breakeven point, because of the energy production and sale, the overall cost for the additional fuel required by the furnace is considerably higher, decreasing the additional margin: $P_{el} < CV_{therm}$.

It means than, the energy production is not anymore profitable. The installation of a new power generation section is economically profitable if the production campaign produces and sells the electric energy in the time interval 10:00am–06:00pm, which corresponds to $\omega = 0.33$.

In terms of net profit margins, the benefits of the dynamic conceptual design and consequently of a production campaign are reported in Figure 3. It is worth adding that, for space reasons, we considered only the typical daily fluctuations of energy market in a summer day and, also, we assumed negligible both the weekly and the seasonal variations.

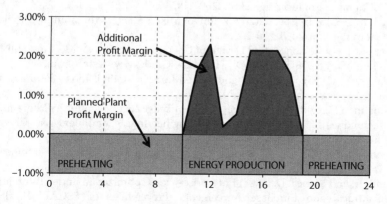

FIGURE 3

Additional net profit margin by adopting an optimal production campaign for the hydrodealkylation process.

Conclusions

The novel approach to conceptual design discussed and implemented in this paper goes beyond the traditional approach based on the optimization of the plant layout accounting for constant market, demand, and operating costs.

The dynamic conceptual design gives the opportunity of focusing on more than a single optimal configuration and provides a series of alternative layouts, each of them optimal for a specific market and production scenario. In this context, it is possible to plan an optimal production campaign that allows deactivating the current configuration while activating another process layout when profitable.

The simple superstructure adopted in this work clarifies the economical benefits of this approach, even accounting for the additional capital investment.

A possible future development may be the application of the dynamic conceptual design and the campaign production to the oil market, in order to face the strong market volatilities and crude oil price changes.

Acknowledgments

Authors acknowledge the support of Esteco and Enginsoft in setting up modeFRONTIER.

References

Biegler, L.T., and Grossmann, I.E. (2004) Retrospective on Optimization, Computers & Chemical Engineering, 28(8), 1169–1192.

Dakin, R.J. (1965) A Tree Search Algorithm for Mixed-Integer Programming Problems, Computer Journal, 8, 250–255.

Douglas, M.J. (1988) Conceptual Design of Chemical Process, NY, McGraw-Hill.

Esteco-EnginSoft (2008) modeFRONTIER User Guide, www.esteco.com.

Floudas, C.A. (1995). Nonlinear and Mixed-Integer Optimization—Fundamentals and Applications. New York, NY, USA.

Grossmann, I.E., and Kravanja, Z. (1995) Mixed-Integer Nonlinear-Programming Techniques for Process Systems-Engineering, Computers & Chemical Engineering, 19, S189-S204.

Manenti, F. (2008) The 3-D Supply Chain Management, Chemical Engineering Transactions, ISBN 0390–2358, 305–312.

Manenti, F., and Manca, D. (2008) Enterprise-wide Optimization under Tight Supply Contracts and Purchase Agreements, Proceedings of ESCAPE-18, ISBN-978/0/444/53228/2.

Simsci-Esscor (2002) PRO/II, User Guide, Lake Forest, CA, USA, www.simsci-esscor.com.

60

Stochastic Modeling of Biodiesel Production Process

Sheraz Abbasi, Urmila Diwekar
University of Illinois -Chicago
Chicago, Il 60680

CONTENTS

ABSTRACT There are inherent uncertainties in the biodiesel production process arising out of feedstock compositions, operating parameters and mechanical equipment design and can have significant impact on the product quality and process economics. The uncertainties are quantified in the form of probabilistic distribution function. Stochastic modeling capability is implemented in the ASPEN process simulator to take into consideration these uncertainties and the output is evaluated to determine impact on plant efficiency.

KEYWORDS *Uncertainties, biodiesel, feedstock, stochastic modeling*

Introduction

Biodiesel is renewable fuel derived from plant oils. It is an alternative to diesel derived from crude oil. Biodiesel burns cleaner than conventional diesel since it does not contain any sulfur content in it. It is derived from various plant and vegetable oils. Refined biodiesel consists mainly of fatty esters, free fatty acids and some triglycerides. Biodiesel feedstock consists of vegetable oils, animal fats and recycled grease. In US, biodiesel comes mainly from Soybean bean oil, tallow and some palm oil. Plant derived oils contain high percentage of triglycerides which are very large molecules. Before these oils can be used in an internal combustion engine, these triglycerides need to be broken down to reduce their viscosity. This process involves transesterification and involves breaking down triglycerides to methyl esters. One mole of triglyceride reacts with three moles of methanol to yield three moles of methyl ester and one mole of glycerol. Biodiesel feedstock is classified according to the content of: 1) triglyceride, 2) free fatty acid. The quality of the biodiesel produced depends on the fatty ester content. The triglycerides found in biodiesel are: 1) Tripalmitin, 2) Triolein, 3) Tristearin, 4) Trilinolein, 5) Trilinolenin. The amount of triglyceride contents varies with the feedstock. The type and amount of triglycerides in the feedstock varies considerably because of nature as a bio-based material.

Background

The biodiesel production process involves feedstock reaction in either plug flow reactor, batch reactor or continuous stir tank reactor (CSTR). The feedstock consists of the oil, typically soybean oil, methanol in a 6:1 molar ratio, and a catalyst (sodium hydroxide or sodium methoxide). The oil feedstock along with methanol and the catalyst is fed to the reactor. The product leaving the reactor is sent to an atmospheric tank where it settles out into two layers: 1) oil plus methanol, 2) glycerol plus methanol. The methanol and glycerol are separated in a distillation column. From the oil plus methanol layer, methanol is removed in a distillation column. The remaining liquid is consists of un-reacted triglycerides which are recycled, free fatty acids and methyl esters. Un-reacted triglycerides are decanted off by washing with hydrochloric acid and the remaining fatty acids and methyl esters which remain form purified biodiesel. This biodiesel is later on blended with gasoline in various ranges as per the requirement of the market.

There are several uncertainties in the biodiesel production process. These uncertainties are due to the nature of the biobased based feedstock. This also directly leads to different processing conditions.

Problem Statement and Approach

The overall objective of this study is to evaluate the impact of uncertainties on the production of biodiesel by evaluating the amount of biodiesel produced (methyl ester plus free fatty acid) and the quality of biodiesel produced (methyl ester content). This evaluation allows a determination of the plant efficiency which could impact process economics.

The uncertainties which are considered in this study were: 1) Uncertainties in the feedstock, 2) amount of methanol produced, 3) reactor operating temperature. The approach taken is as follows: To evaluate the impact of the above mentioned uncertainties, the approach taken is:

- Prepare ASPEN model of biodiesel production process
- Specify a fixed mass input to the reactor.
- Vary the composition of the feed by varying the amount of triglyceride feed to the reactor.
- Vary the methanol flow and operating temperatures

To achieve the above, following procedure adopted was:

- Use of stochastic simulation block in ASPEN
- Uncertainties in feed composition are assigned a probabilistic distribution function.
- Graphical analysis of the output results

Stochastic Modeling

Stochastic modeling approach involves the following procedure:

1. Specifying uncertainty in key parameters
2. Specifying the correlation structure of any independent parameters
3. Sampling the distribution of the specified parameters in an iterative fashion
4. Propagating the effect of uncertainties through the process flow
5. Applying graphical and statistical techniques to analyze the results.

The benefits of this technique are:

1. Uncertain parameters can be described
2. Impact of uncertainties can be evaluated by describing an output variable and drawing the cumulative probability distribution (CFD) graphs

Once probability distributions are assigned to the uncertain parameters, the next step is to perform sampling operations from the multi-variable uncertain parameter domain. In the stochastic modeling used in this study, Hammersly Sequence Sampling (HSS) technique (Diwekar, 2008) is used for efficient evaluations. HSS uses and optimal design scheme for placing the n points on a k-dimensional hypercube. This scheme ensures that the sample set is more respresentative of the population, showing uniformity properties in multi-dimensions, unlike Monte Carlo or other sampling techniques.

ASPEN Biodiesel Model

For the ASPEN biodiesel model, a CSTR is used as the reactor. Triolein (oleic acid triglyceride) is the triglyceride is only triglyceride which is available in the ASPEN property databank. Other triglycerides are not available in the ASPEN property databank and therefore triolein is used. The oil feed flow to the reactor is kept at a constant 10000 lb/hr. A calculator block is used so that as the triglyceride flow to the reactor increases, a corresponding decrease in free fatty acid flow rate is observed and also vice versa. The flow of triolein to the reactor is represented by a normal distribution curve. The reaction kinetics are considered to be power law and second order with respect to triolein. The value of the pre-exponential factor is assumed to be k = 1e5 and activation energy is A = 13 kcal/mol. No competing side reactions are considered because physical property data for other triglycerides commonly found the biodiesel feedstock is not available in ASPEN. The range of uncertainties in the biodiesel production that was characterized using literature data is shown in Table 1.

The probability distribution of the input parameters is shown in figure 1 to 3.

Results and Discussion

The results of stochastic modeling are shown in figure 4 to 6. The output variables of stochastic modeling are set as: 1) biodiesel produced lb/hr, 2) methyl oleate (methyl ester), 3) unreated triglyceride.

TABLE 1

Uncertainties in Input Parameters

Parameter	Distribution	Percentile	
		0.1[th]	0.9[th]
Triolein lb/hr	normal	4000	8000
Methanol lb/hr	normal	1200	1700
Reactor temp F	normal	140	200

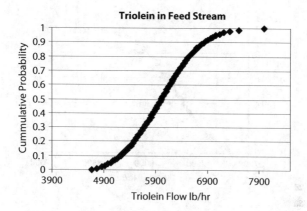

FIGURE 1
Triolein in feed stream.

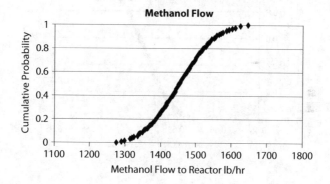

FIGURE 2
Methanol feed to reactor.

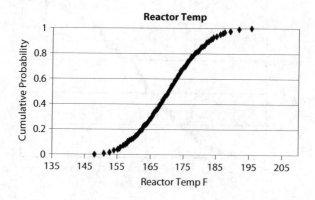

FIGURE 3
Reactor temperature.

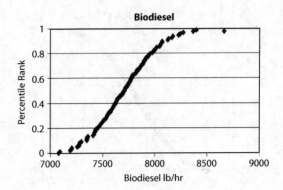

FIGURE 4
Biodiesel production lb/hr.

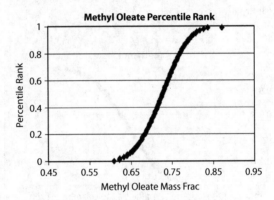

FIGURE 5
Methly-oleate mass fraction in biodiesel.

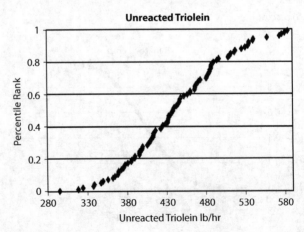

FIGURE 6
Unreacted triglyceride.

TABLE 2

Biodiesel Production

Min	Max	Range
7077.844	8658.948	1581.104
MEAN	MEDIAN	VARIANCE
7708.867	7701.329	91162.38

TABLE 3

Methlyl Oleat Mass Fraction

Min	Max	Range
0.6088414	0.8717907	0.2629493
MEAN	MEDIAN	VARIANCE
0.7268474	0.7270023	2.62E-03

TABLE 4

Unreacted Triolein

Min	Max	Range
294.6475	593.9769	299.3293
MEAN	MEDIAN	VARIANCE
443.8207	438.8737	4128.745

Plant efficiency is determined by considering the methyl ester produced per hour based on the feed flow rate of 10000 lb/hr of oil. The plant efficiency is based on a single pass transesterification and single pass conversion of triolein in the reactor and single pass separation. No recycle of unreacted triglyceride or separated methanol is considered.

The joint contribution of the three input parameters is shown on the plant efficiency. The center line is a base case value with 45% triglyceride in the feed, reactor temperature of 160 F and methanol flow of 1500 lb/hr. The base case plant efficiency is about 56%. By considering a triglyceride content range from 40% to 80%, the plant efficiency also varies from 43% to 70%. The is range of values for plant efficiency demonstrates that:

- Plant efficiency is not linearly related to the triglyceride content of the oil.
- The impact of uncertainties on the production of biodiesel is significant
- The wide range in plant efficiency highlights that uncertainties in the feedstock composition cannot be ignored because they can have a major impact on the process economics.

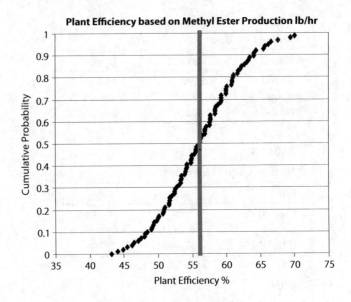

FIGURE 7
Single pass plant efficiency.

Conclusion

This study has described a process for the evaluating the uncertainties in the biodiesel production process. These uncertainties arise out of the feedstock and operating conditions. The uncertainties propagate through the flow sheet and their impact on plant efficiency is significant. The evaluation of uncertainties by using stochastic modeling can be extended to other process pathways for determining plant economics. The results of this study could be used for plant design, R&D and selection of feedstock and determining operating envelops.

Acknowledgments

We gratefully acknowledge the support of Vishwamitra Research Institute.

References

Diwekar U M. (2008), Introduction to Applied Optimization, Springer, Cambridge, MA.

Diweker, U.M. and E.S.Rubin (1990), Stochastic Modeling of Chemical Processes, Computers Chemical Engineering, Vol. 15, No. 2, pp 105–114, 1991.

Subramanayan K. and U. Diwekar (2007), User's manual for stochastic block, VRI.

Myint, L.L. and El-Halwagi, M.M (2008), Process Analysis and Optimization of Biodiesel Production from Soybean Oil, Clean Technologies and Environmental Policy, pp the 85–110, 2008

Noureddini H, Zhu D(1997) Kinetics of Transesterification of Soybean Oil. JAOCS 74:1457–1463

http://www.nrel.gov/docs/fy03osti/31460.pdf

61

Minimization of Fresh Water Consumption for Particulate Carbon (PC) Power Plants

Juan M. Salazar and Urmila M. Diwekar*

Vishwamitra Research Institute
Center for Uncertain Systems: Tools for Optimization and Management
Clarendon Hills, IL 60514

CONTENTS

ABSTRACT Coal-fired power plants are widely recognized as major water consumers whose operability has started to be affected by drought conditions across some regions of the country. Water availability also restricts the construction of new plants to satisfy the increasing power demand since these new facilities must include water-expensive carbon sequestration technologies. Therefore, national efforts to reduce water withdrawal and consumption have been intensified. Water consumption in thermoelectric generation is strongly associated to losses on cooling systems and to gas purification operations. These processes are affected by uncertain variables like atmospheric conditions. Thus, minimization of water consumption requires optimal operating conditions and parameters, while fulfilling the environmental constraints. Particulate Carbon (PC) power plants are studied in this work. Optimization under uncertainty for these large-scale complex processes with black-box models cannot be solved with conventional

---

* To whom all correspondence should be addressed

stochastic programming algorithms because of the computational expenses involved. Employment of the novel better optimization of nonlinear uncertain systems (BONUS) algorithm dramatically decreased the computational requirements of the stochastic optimization. Operation condition including boiler temperature and reactant ratios were calculated to obtain the minimum water consumption under the above mentioned uncertainties.

KEYWORDS *Particulate carbon (PC), water consumption, stochastic optimization*

Introduction

The US Department of Energy (DOE) and its national energy technology laboratory (NETL) have deeply and widely study the characteristics and conditions of coal-fired power plants to ensure the availability of clean technologies to fulfill the increasing demand of electricity (DOE/NETL, 2007c). Water consumption is one of the characteristics that need to be addressed when assessing the capabilities of clean production of coal-fired power plants (DOE/NETL, 2007a). Makeups, blowdowns, process water and cooling system are responsible for water consumption and cooling system is the largest water consumer in power plants (DOE/NETL, 2007b). R&D activities have been carried out to find alternative sources of cooling water. Important efforts have been made in advanced cooling technologies, advanced recycling techniques and advanced waste water treatment processes (Feeley et al., 2006). Additionally, comprehensive models of particulate carbon (PC) power plants have been built employing Aspen Plus® to evaluate the cost and performance of the processes (DOE/NETL, 2007c). Performance evaluation with these models includes the determination of water consumption based on a rigorous mass balance as suggested by the DOE/NETL guidelines for energy systems analysis (DOE/FE-NETL, 2004).

This paper complements those research efforts with a computational tool that can determine the conditions which minimize the water consumption in a 548 MW PC plant. It has been reported that cooling tower evaporative loses are the most significant loses within the cooling system and they are affected by environmental parameters like air humidity and also by operational parameters like the rate of power generation (DOE/NETL, 2007b). This implies that computational tools of process analysis should include the influence of uncertain parameters in the models. Process modeling under uncertainty or stochastic modeling, has been employed for process analysis in power system in the past (Diwekar and Rubin, 1991). When including the optimization techniques the resulting non-linear stochastic programming (NLSP) can become highly computationally demanding (Sahin and Diwekar, 2004). These computational expenses have been drastically decreased with the better optimization of non-linear uncertain systems (BONUS) (Sahin and

Diwekar, 2004). The BONUS algorithm has been integrated to Aspen Plus® models through the advanced engineering co-simulator (APECS) and CAPE-OPEN interface. The advantages of BONUS are exploited to calculate the conditions in which a reduction on the average consumption of water for the PC plant can be reached.

Particulate Carbon (PC) Model

The PC model employed in this work is described in detail by the report on cost and performance of fossil energy plants (DOE/NETL, 2007c). Particularly, this is a supercritical steady state model that doesn't include carbon sequestration technology and is intended to generate 548 MW. The process is divided in two main sections: Boiler section and steam generation section. The boiler is modeled as a series of two reactors (a stoichiometry reactor and a Gibbs reactor) and flue gas desulfurization (FGD) unit is attached to the boiler effluent. The steam section comprises the typical high, intermediate and low pressure turbine train and is a steam re-heating scheme with feed water heating from turbine extracted steam. Sequential modular (SM) is the Aspen Plus® standard modeling approach. This approach requires the employment of several design specifications with their corresponding nested convergences to obtain accurate modeling and demands large computational time.

The calculation of the water consumption was originally performed outside Aspen Plus® environment (DOE/NETL, 2007c). Therefore, the calculation was integrated to the original PC models including a heat exchanger and a calculator block. The complete description of the cooling tower model is reported by the literature (Hensley, 2006). The evaporation and drift losses are calculated based on the range or the difference between the hot and cold water entering and leaving the cooling tower. Hot water temperature was left fixed at 110°F but cold water temperature depends on the ambient wet-bulb temperature which is variable (originally fixed for design purposes at 51.5°F) and the approach which is a characteristic of the humidification capabilities of the tower (fixed and equal to 8.5°F). The uncertain variable chosen for the stochastic simulation and optimization is the wet-bulb temperature.

Uncertainties in PC Process

NETL reports on water consumption make several assumption on the PC models (DOE/NETL, 2007a; DOE/NETL, 2007b; DOE/NETL, 2007c). Some of these assumptions could become important sources of uncertainty since they are directly related to the water consumption and can be fairly variable. Air humidity (expressed as wet bulb temperature) and power load are two examples of such uncertain variables. Air humidity continuously changes along the US along with seasons and its influence on the water consumption was mention in a previous section. Power load changes the turbine duty and pressure profile (Cotton, 1998) and subsequently modifies the condenser and cooling tower loads. This paper focuses on the consideration of the wet bulb temperature variation and the variation of both parameters will be considered in future publications.

The main source of information for the design conditions of the cooling tower is the text by Hensley on cooling tower fundamentals (Hensley, 2006). The author suggests average measurements of the wet bulb temperature to determine the range and the approach temperature that guarantee the appropriate cold water temperature at peak demand. However, for more complex process or more accurate designs (as those required for the case in this study), the reference suggests the inclusion of wet-bulb temperature during critical months (summer) and in some case in the entire year. Therefore, if design activities require this level of accuracy, simulation purposes additionally will require detailed data for the particular location of the process. Detailed information about air humidity on different locations within the US can be obtained from the real-time weather tool of EnergyPlus energy simulation software which is available at the DOE's energy efficiency and renewable energy program (EERE) website (http://www1. eere.energy.gov). The PC models were originally established for a plant located at Midwestern US. Therefore, weather data for the years 2006 and 2007 from 8 US Midwestern urban centers (Chicago, Detroit, Indianapolis, Minneapolis, Saint Louis, Des Moines, Kansas City and Cincinnati) were requested using the software and processed to generate the probability density functions for the stochastic simulation (Diwekar and Rubin, 1991). The data was organized in four seasons, starting at September 2005 and finishing at August 2007. The original data contained dry bulb temperature and dew point as humidity information; they were converted to wet bulb temperature. Histograms to generate the corresponding probability density functions were generated for an average Midwestern urban center during each of 4 seasons of a year.

Novel Stochastic Optimization Algorithm

The excessive number of model calculations required to determine the probabilistic objective function and constraints in stochastic programming problems has been considerably reduced by BONUS algorithm. The algorithm makes use of a reweighting method using Kernel density estimation to approximate the function and its derivatives. These probabilistic approximations are fed to the non-linear optimizer at the outer loop of the stochastic programming. A brief description of the procedure is described bellow (Sahin and Diwekar, 2004):

1. Generate a set of samples for the stochastic simulation. Data for decision variables is sampled from uniform distributions and data for uncertain variables is withdrawn from the corresponding probability distributions.

2. Run the model for the number of samples previously generated.

3. Select the starting point for the to estimate the objective function with the weights calculated by Gaussian Kernel estimator for the probability density function of decision variables (Equation (1), where $f(u)$ is the probability density function of uncertain variable u, N_{samp} is the number of samples taken and h^2 is the variance of the data set)

4. Perturb the decision variables, estimate the objective function and estimate the derivative using the reweighting approach.

5. Repeat step 4 until allowed Kuhn-Tucker error is reached

$$f(u_i) = \frac{1}{N_{samp}h} \sum_{i=1}^{N_{samp}} \frac{1}{\sqrt{2\pi}} e^{-\frac{1}{2}\left(\frac{u-u_i}{h}\right)^2} \tag{1}$$

The selection of decision variables for the optimization problem was based on the parameters that could be changed in the process model. This model is described in detail by the NETL report on cost and performance baseline for fossil energy plants (DOE/NETL, 2007c). A set of variables that were described in the report as assumptions or assigned parameters were selected to be considered as decision variables. To quantify the influence of these variables in the water consumption, stochastic simulation was employed as described in the literature (Diwekar and Rubin, 1991). Partial rank correlation coefficients were used as measures of such influence since they provide a measure of the relationship between the output and input variables for a non-linear function as that represented in the PC models (Diwekar and Rubin, 1991). A stochastic simulation was run for 300 samples of 10 potential decision variables and the 5 most influential ones were selected to run the BONUS algorithm.

BONUS algorithm employs a sequential quadratic programming (SQP) method for the solution of the non-linear programming problem. The model is a non-convex function whose minima depend on the starting point for the SQP routine. Therefore, different values of the initial point were evaluated to determine local minima that could represent any improvement from the base case.

Implementation of BONUS in Aspen Plus® Models

To take full advantage of the BONUS algorithm, a simulation tool was built around the steady state simulator using the CAPE-OPEN interface. Fortran calculator blocks were included to the model to manipulate a sensitivity analysis that performs the stochastic simulation. A SQP routine was coupled with the reweighting estimation to do the optimization and added to the simulation. Under these conditions, the model is required to run the number of selected samples (600 for this particular case) and once every time that a modification to the NLP solver is required (different starting points, decision variable bounds or perturbation size).

Results and Discussion

Characterization of the Uncertainty

Figure 1 presents the characterization of the uncertain parameter wet-bult temperature as a histogram for each of the four seasons for 8 US Midwestern urban centers. It can be noticed that wet bulb temperature during fall and spring is more variable than in winter and summer. This variability will be reflected in simulation results on water consumption. These histograms are fit to lognormal distributions that are employed as probability density functions for the stochastic optimization. The fall season (Sept-Nov) was chosen for the study since the mean wet-bulb for this season (approx 10°C or 50°F) is close to the assumed by the original case (DOE/NETL, 2007c).

Optimization Under Uncertainty Using BONUS

The absolute value of the partial rank correlation coefficients of 10 variables was sorted to determine the 5 most influential parameters (Table 1). Boiler Temperature, air excess to the boiler, $CaCO_3/SO_2$ molar ratio to the FGD unit, losses at the power generator and oxygen/sulfur molar ratio at the FGD unit were chosen as decision variables.

Different starting points for the SQP routine were studied changing the initial values of boiler temperature and air excess. Figure 2 shows the BONUS results for boiler temperature ranges of 210–280°F and air excess of 15–25%. It can be noticed that a monotonic reduction of the average water consumption is observed with boiler temperature reduction. However, variation of air excess percentage for the initial points yields a minimum when started from 220°F and 15% air excess. Thus, when the boiler temperature is pushed to an assumed lower bound of 210°F the air excess

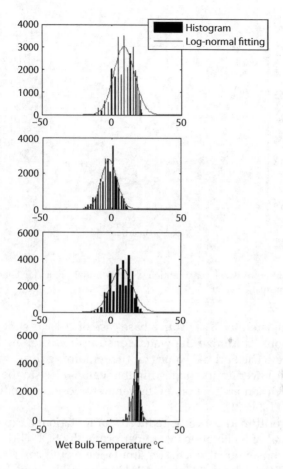

FIGURE 1
Probability density functions of 4 seasons (fall to winter from top) for wet bulb temperature on 8 US Midwestern cities.

TABLE 1

Absolute Values of Partial Rank Correlation Coefficients (PRCC)

Decision Variable	PRCC	Importance
Air excess	0.230024	2
Boiler Temperature	0.295324	1
O_2/SO_2 ratio	0.028274	5
$CaCO_3/SO_2$ ratio	0.043097	3
Generator losses	0.031358	4

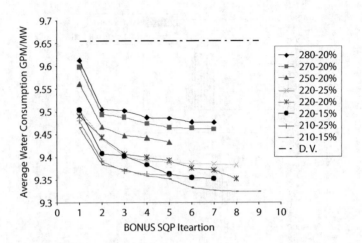

FIGURE 2
BONUS minimization of water consumption, various initial points for boiler T(°F) and air excess(D.V: Deterministic Value).

needs to be adjusted to 16 % (from a base case of 20%) as shown in Table 2. These conditions minimize the water consumption keeping constant the boiler efficiency. The optimal values for the remaining three variables were located at their lower limits suggesting that values close to the stoichiometric ratios for FGD and reduction on the generator loses can reduce the water consumption (Table 2).

BONUS algorithm drastically reduced computational expenses of this analysis. A total of 56 iterations were performed by SQP with different initial points. A conventional stochastic simulation would run 400 samples per iteration of SQP which would involve 22400 runs of the model. BONUS only required 609 runs, a reduction of 97% in computational time. Each model run requires approximately 9 to 10 minutes of CPU time.

TABLE 2

Local Optima of the Decision Variables and Objective Function to Minimize Water Consumption in a PC Plant

Variable	Base Case	Optimal
Boiler Temperature (°F)	270	210
Air excess (%)	20	16.2
$CaCO_3/SO_2$ molar ratio	1.04	1.01
Generator losses	0.015	0.005
O_2/SO_2 molar ratio	1.1	1.0
Obj. function BONUS (GPM/MW)	9.60	9.32
Obj. function Stoc. Sim.(GPM/MW)	9.62	9.21

Conclusion

Air humidity affects the water evaporation in the cooling tower of PC plants. Reduction in the boiler temperature, molar ratios of oxygen to the FGD, calcium to the FGD and generator loses; require an adjustment of the air excess to the boiler to minimize water consumption. Water consumption can be reduced in 4.26%, equivalent to 0.41GPM/MW or 4.5×10^5 tons per year for a 548 MW plant. BONUS algorithm reduced the calculation expenses of the stochastic programming in 97% by approximating the objective function with a reweighting scheme.

Acknowledgments

NETL's office of R&D provided the base case model.

References

Cotton, K. C. (1998). Evaluating and improving steam turbine performance (2nd ed.). *Cotton Fact Inc.* Rexford, NY.

Diwekar, U., Rubin, E. S. (1991). Stochastic modeling of chemical processes. *Computers & Chemical Engineering, 15*(2), 105.

Feeley, T. J.,III, McNemar, A., Pletcher, S., Carney, B., Hoffmann, J. (2006). Department of Energy/Office of fossil energy's power plant water management R&D program. *Proc.Int.Tech.Conf.Coal Util.Fuel Syst., 31st* (Vol. 1) 447.

Hensley, J. C. (Ed.). (2006). Cooling tower fundamentals (2nd ed.). *SPX Cooling Technologies, Inc.* Overland Park, KS.

DOE/FE-NETL. (2004). Quality guidelines for energy systems studies

Sahin, K., Diwekar, U. (2004). Better optimization of nonlinear uncertain systems (BONUS): A new algorithm for stochastic programming using reweighting through kernel density estimation. *Annals of Operations Research, 132*(1–4), 47.

DOE/NETL. (2007a). Estimating freshwater needs to meet future thermoelectric generation requirements.

DOE/NETL. (2007b). Power plant water usage and loss study.

DOE/NETL. (2007c). Cost and performance baseline for fossil energy plants.

62

Integrated Design and Control under Uncertainty—Algorithms and Applications

Jeonghwa Moon, Seonbyeong Kim, Gerardo J. Ruiz, and Andreas A. Linninger

Department of Chemical Engineering, University of Illinois at Chicago Chicago, IL 60607

CONTENTS

ABSTRACT High performance processes require design that operates close to design boundaries and specifications while still guaranteeing robust performance without design constraint violations. In order to safely approach tighter boundaries of process performance, much attention has been devoted to integrating design and control in which dynamics as well the design decisions are taken simultaneously in optimal fashion. However rigorous methods solving design and control simultaneously lead to challenging mathematical formulations which easily become intractable numerically and computationally. This paper introduces a new mathematical formulation to reduce this combinatorial complexity of integrating design and control. We will show substantial reduction in the problem size can be achieved by embedded control decisions within specific designs. This embedded control decisions avoid combinatorial explosion of control configuration by using a full state space model that does not require pairing of control variables and loops. The current capabilities of the methodology will be demonstrated using a realistic reactor-column flowsheet.

KEYWORDS   *Design and control integration, process and operational uncertainty, design and analysis of dynamic flexibility*

Introduction

In classical process development, design and control decisions are made separately in spite of the common aim of ensuring robust plant operations. Thus, controller tuning and optimization is limited by process dynamics already fixed in the design phase. Also it does not guarantee robustness under uncertainty and often fails to describe actual process dynamics. In order to achieve the best overall performance of the system under realistic implementable control and operational uncertainty, it is necessary to consider process design and control decision simultaneously.

During the last 30 years, several methodologies have been developed for the integration of design and control. Details of these methodologies are identified and discussed elsewhere (Vassilis et al., 2004; Ramirez & Gani, 2007). In spite of their progress in integration of design and control field, the available methodologies still offer insufficient insight to the problems. Recently, we proposed a new method, *entitled embedded control optimization approach* (Malcolm et al, 2007). This integrated design method can operate satisfactorily under adverse input conditions, while delivering products within desired quality specifications. In this paper, this embedded control optimization methodology is reviewed and enhanced to incorporate full state space estimator for better dynamic process performance. Also reactor-column flowsheet is considered as a more realistic example for this methodology.

Methodology

Problem Decomposition

The conceptual problem of the integration of process design and control is a stochastic infinite dimensional mixed integer dynamic optimization problem which is extremely challenging for existing mathematical programming techniques. To overcome the intractability of original problem, Pistikopoulos and co-workers proposed problem decomposition algorithms (Vassilis et al., 2004). In their problem decomposition, the main design optimization problem is solved in a discretized sampling space; rigorous flexibility tests in a second stage ensure the dynamic feasibility over the entire uncertain space. If current candidate designs are not feasible in critical scenarios, these critical scenarios are incorporated into the sampling space and the main problem is solved again with additional critical scenarios. By repeating this iterative

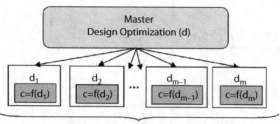

FIGURE 1
Proposed embedded control optimization structure. It optimizes control choice with given design.

process, optimal designs can be achieved. This framework is also used in our algorithm.

Embedded Control Optimization

Even though, the problem decomposition substantially reduces the problem size, it still remains a challenge due to combinatorial complexity of the NP-hard search space. Specifically, introduction of control decisions such as the insertion of feedback loops, or pairing of manipulated and control variables causes combinatorial explosion in the possible integrated design and control realizations. We therefore propose to separate the design decisions from the control decisions as shown Figure 1. At the master level, we fix design decisions such as reactor sizes, and residence time that govern dynamic process performance. No control decisions are made at this level. Once the main design decisions are specified, we assess the dynamic performance of process by using a simplified, yet reasonably competitive control schemes based on full state space identification and least square regulation. Optimal control action is calculated with relative ease based on linear state space models which are obtained dynamically in each time step. In order to keep the model consistent with the nonlinear process dynamics, the process identification is repeated in each time step. The linear state space model has the advantage that the optimal control trajectories are relatively easy to calculate. The detailed description is found elsewhere (Malcolm et al., 2007).

Flexible Design of Reactor-column Flowsheet

This section demonstrates the effectiveness of embedded control optimization for integrated control and design by designing a reactor-column flowsheet process with uncertain reaction coefficients. This is similar to processes

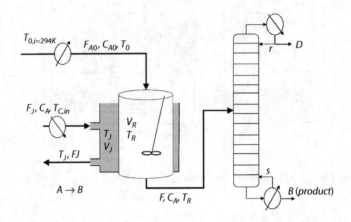

FIGURE 2
Reactor-column flowsheet. This is composed of a reactor and a column. The reactor is surrounded by a cool jacket.

introduced and studied by Luyben (2007). The aim of this case study is to determine optimal design specifications with reasonable control for dynamically flexible operations. This task of design and control integration should be done simultaneously with reasonable computational effort.

Process description. This flowsheet has one reaction and one distillative separation step, as shown in Figure 2. The precursor A enters the reactor to be converted into the product B with an exothermic irreversible first order reaction.

$$A \xrightarrow{k_0} B \qquad (1)$$

The reaction cannot be driven to full conversion, because this operation would require too high temperature causing the product B to be destroyed. In addition, this high temperature might violate safety constraints and lead to reactor explosion. To control the reactor temperature, it is equipped with a cooling jacket. Its effluent F is directed to the continuous distillation column in order to separate the product B from unreacted raw material A.

Design & manipulated variables. Reactor diameter (D_R) and length (L_R), and heat transfer area (A_J) are treated as design variables. Also the number of stages (N_s), feed stage, and column diameter (N_f) are design variables of the column,. For manipulated variables, we select input feed temperature (T_0), coolant temperature (T_J), reflux ratio (r), and reboil ratio (s). The values of the design variables and manipulated variables need to be determined to insure safe operation within desired quality standards, under any operating conditions and in the presence of uncertain reaction conditions.

Operational constraints. For safe operation, this process needs to satisfy three constraints at all times.

For safety, reactor temperature (T_R) should never exceed 385K,

$$T_R \leq 385 \tag{2}$$

For the product quality, the mole fraction of component A (x_A) of final product B should be less than 0.05.

$$x_A \leq 0.05 \tag{3}$$

For minimum process productivity, the conversion ratio (χ) should be greater than 0.7.

$$\chi \geq 0.7, \quad \chi = 1 - \frac{C_A}{C_{A0}} \tag{4}$$

Thus, the control variables are T_R, x_A, and χ. Among these constraints, the safety constraint needs to be enforced dynamically for all time periods of possible scenarios, as opposed to merely the steady state which is the sole concern in classical flexibility analysis. Some design specifications may not satisfy safety constraints dynamically, even though they do not fail in the steady state as shown Figure 3. Therefore, we need to consider the dynamic flexibility for the optimal dynamic process performance. Productivity and quality constraints are soft constraints, which lead to performance losses that cost money, but need not be enforced rigorously for all time intervals.

Uncertainty scenarios. We wish to investigate the impact of two main uncertain parameters, associated with chemical reactions. The first parameter is the preexponental factor k_0, the second is heat of reaction λ. Their nominal values and variance are illustrated in Table 1.

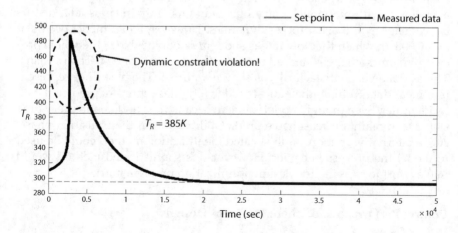

FIGURE 3
Example of safety constraint violation in dynamic performance in spite of flexibility in the steady state.

TABLE 1

Nominal Values and Expected Deviation of Uncertain Parameters

	θ^N	$\Delta\theta^+$	$\Delta\theta^-$	var
k_0	20.75e6	2.07e6	2.07e6	10%
λ	−17.43e3	−1.74e3	−1.74e3	10%

As a first attempt to perform the stochastic optimization, we chose 10 samples in the uncertain space of reaction conditions using Latin hypercube method, and evaluated the probabilities of each parameter set to calculate expected cost.

We now wish to rigorously determine the design and manipulated variables such that the process does not violate the constraints and produces product B in desired purity limit in every realization of the reaction conditions and design variability in the dynamic performance due to uncertainty.

Identification Test

In order to employ the embedded control optimization, our methodology dynamically performs adaptive state space identification. This repeated identification is to convert a highly nonlinear process model into a simple linear state representation. With this simplified linear model, the linear regulator can find optimal control moves. The performance of process in response to step change in feed rate as well as the impact of sinusoidal disturbance is show in Figure 4. In each step, there is a little change because the process is operated with moderately varying conditions, only in sinusoidal disturbances slightly impacts process dynamics. However, a big impact occurs at $t = 40,000$, in which the flow rate is suddenly doubled. Now, the whole process dynamics changes and adjustment for the new state space are made. This is done very satisfactorily, as shown Figure 4. The blue lines represent nonlinear data from mathematical model; red lines represent predicted data by identifier. Even though model is linear, because it is updated in every time step, it can maintain reasonable predictability even in the transition phase. As an alternative, a more sophisticated identification method could be used to model the process behavior. However this sophisticated method would make effort for solving the design optimization problem harder.

Control Performance of There Different Designs

Next, we tested controllability of several designs under specific disturbance scenarios. Dynamic controllability of three reactors with different volumes of 12, 23, 143 m^3, corresponding to different resident times were investigated

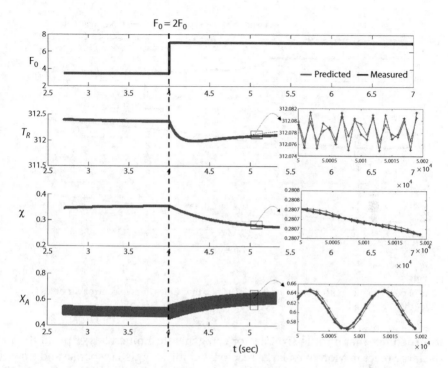

FIGURE 4
Performance test of identifier. This figure shows sequential least identification method predicts dynamic system behavior well, even in dynamic transition phase.

in response to dramatic increase of the feed rate. The doubling feed occurs at $t = 40{,}000$.

Case 1: Small reactor, $V_R = 12\ m^3$. When doubling the input feed, the small reactor looses the controllability- the control is not capable of handling this upset to process conditions as shown in Figure 5. The reactor temperature drops and reaction conversion declines, and product quality constraint is violated.

Case 2: Middle reactor, $V_R = 23\ m^3$. We tested the performance of a middle size reactor. With this design specification, the controller rejected the disturbance well. The quality constraint was violated briefly in the instance of the disturbance inception, but the controller quickly removed this quality issue and kept values of all process variables under the set points, as shown Figure 6. This design handles disturbance very well with a simple control scheme.

Case 3: Large reactor, $V_R = 143\ m^3$. Finally we tested a very large reactor. Our model demonstrated that the feed flow disturbances do not much affect the process dynamics in this over-dimensioned design. The large reactor is very robust against the disturbance, but they are not competitive because of cost.

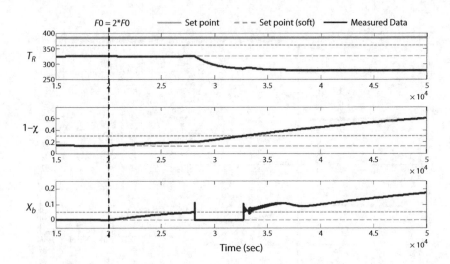

FIGURE 5
Control performance with $V_R = 12 \ m^3$. it fails to keep the values of two control variables (χ, x_A) under the set points.

It turned out that middle size reactor can generate a much larger profit than the large reactor. Moreover, large reactors exhibit sluggish response and slow servo performance when set points are changed to adjust to different product specifications. This example supports the notion that arbitrary overdesign of equipment is not a solution to ensuring dynamic process flexibility of high performance processes.

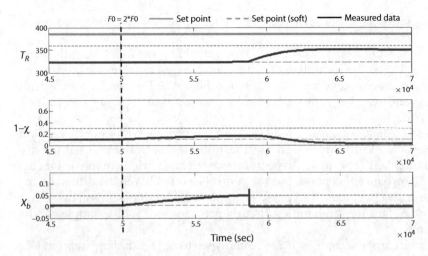

FIGURE 6
Control performance with reactor of $V_R = 23 \ m^3$. This figure shows that the controller handles disturbance without any difficulty.

TABLE 2

Best Result of Each Iteration for the Dynamic
Integrated Design and Control

No	Design Variables				Profit(k$)
	D_R(m)	L_R(m)	N_s	N_f	
0	5.00	10.00	20	10	1,599
1	5.00	10.00	21	10	1,651
9	2.48	5.05	23	11	5,922
21	2.43	4.96	23	11	6,099
34	2.44	4.98	22	11	6,225

Optimal Integrated Design with Control

The previous analysis of different design demonstrated the trade off between controllability and servo performance. When also considering costs and profits what would be the best integrated design and control? For maximizing the performance, while at the same time planning flexible operation, we performed the design optimization under uncertainty as follows. For capital cost, we consider the reactor, the column and the heat exchangers. Operating cost considered energy consumption of the input feed temperature, cooling temperature, the reboiler and condenser. Also product prices are considered in the total annual profit. The master level of this problem is to maximize total profit. To solve optimal design problem, a *Nelder-Mead simplex method* is used in master level of our methodology, embedded control was used to adjust control decisions. We found the best optimal design after 34 iterations. Table 2 shows optimal designs and profits at each iteration.

Conclusions & Future Work

This paper describes a conceptual framework for design and control integration developed by our group. Our methodology is to recast integrated design and control problem into a solvable mathematical programming formulation. The case study described control and design integration for a simple flowsheet. In the future, more challenging flowsheets will be examined. Also we wish to improve the quality of identification for highly nonlinear processes using more advanced identification such as subspace identification method or nonlinear model predictive control. However these advanced algorithms are computationally expensive; hence the trade off between accuracy and performance of algorithms need to be considered.

Acknowledgements

Financial support from NSF Grant CBET-0626162 is gratefully acknowledged.

References

Luyben, W. L. (2007). Chemical reactor design and control. *Wiley-Interscience.*

Malcolm, A., Polan, J., Zhang, L., Ogunnaike, B. A., & Linninger, A. A. (2007). Integrating systems design and control using dynamic flexibility analysis. *AIChE Journal*, 53, 2048–2061.

Ramirez, E., & Gani, R. (2007). Methodology for the design and analysis of reaction-separation systems with recycle. 2. Design and control integration. *IECR*, 46, 8084–8100.

Vassilis S., Perkins J. D., & Pistikopoulos, E. N. (2004). Recent advances in optimization-based simultaneous process and control design. *Comp. & Chem. Engr.*, 28(10): 2069–2086.

63

Optimal Design of Cryogenic Air Separation Columns Under Uncertainty

Yu Zhu and Carl D. Laird*

Artie McFerrin Department of Chemical Engineering,
Texas A&M University
College Station, TX 77843

CONTENTS

ABSTRACT   Cryogenic air separation, while widely used in industry, is an energy intensive process. Effective design can improve efficiency and reduce energy consumption, however, uncertainties can make determination of the optimal design difficult. This paper addresses the conceptual design of air separation considering two types of unknown information: uncertain physical properties and variable product demands. A rigorous, highly nonlinear model including three columns with recycle is built to capture the coupled nature of air separation systems. Using a multi-scenario approach to discretize the uncertainty space gives rise to a large-scale, structured nonlinear programming formulation. IPOPT, a rigorous interior-point implementation, is used to efficiently solve this difficult nonlinear optimization problem. The optimal value of the design variables found with and without considering uncertainties are compared in detail.

---

* Corresponding author: Tel.:+1 979 458 4514; fax:+1 979 845 6446 E-mail address: carl.laird@ tamu.edu

KEYWORDS *Cryogenic air separation, conceptual design, uncertainty, multi-scenario, energy consumption*

Introduction

Cryogenic air separation columns (ASC) are widely used in many industries to produce significant quantities of high purity industrial gas, and these processes consume a large amount of electrical energy. The industrial gas industry consumed approximately 31,460 million kilowatt hours (Over $700 million/y) in the USA in 1998, which accounts 3.5% of the total electricity purchased by the manufacturing industry (Karwan et al. 2007). Therefore, it is necessary to suitably design air separation columns and reduce energy consumption. However, it is important to consider potential uncertainties during the design phase. Adopting deterministic values of operating parameters without considering the impact from unknown information can produce a design that does not perform as expected. Unknown information can be classified into two categories (Rooney et al. 2003). *Process uncertainty* includes values that are unknown at the design stage and the operation stage. These include, for example, unmeasured disturbances, and unknown model parameters. *Process variability* includes values that are not known at the design stage, but can be measured during operation. This variation may be compensated by control variables.

As for the ASC system, process uncertainty can arise from unknown physical properties. For example, activity coefficient models for N_2-Ar-O_2 systems contain binary interaction parameters that are sensitive to argon purities and pressures (Harmens, 1970). Process variability, on the other hand, can arise because of changing product demands. In order to satisfy variable product demands, the ASC system may be required to switch among different operating conditions. The argon product variability is often ignored for it causes less uncertainty than other products. However, it effects the decision of design variables significantly, which can be demonstrated in this paper.

This paper addresses the problem of determining optimal design variables for air separation columns while considering both unknown activity coefficients and variable argon product demands using a multi-scenario programming approach. This is challenging for two reasons. First, an extremely complex and highly nonlinear rigorous model has to be built to catch the coupled nature of air separation systems. Second, the multi-scenario approach results in very large scale nonlinear programming problems and requires efficient solution strategies.

Multi-scenario Programming Approach

The multi-scenario formulation can be expressed in general form as:

$$\min_{d,z,y} P = f_0(d) + \sum_{k \in K} \sum_{q \in Q} \omega_{ik} f_{ik}\left(d, u_k, l_{qk}, \theta_k^v, \theta_q^u\right) \tag{1}$$

$$\text{s.t.} \quad \left.\begin{aligned} h_{qk}\left(d, u_k, l_{qk}, \theta_k^v, \theta_q^u\right) &= 0 \\ g_{qk}\left(d, u_k, l_{qk}, \theta_k^v, \theta_q^u\right) &\leq 0 \end{aligned}\right\} k \in K, q \in Q$$

Where the design variables are given by d, control variables are given by u, and the state variables are given by l. Inequality and equality constraints are given by g and h respectively. In the multi-scenario formulation, the uncertainty space is separated into discrete points. The index set K is defined for discrete values of variable parameters, θ^v, and the index set Q is defined for discrete values of unknown parameters, θ^u. The objective function includes fixed costs related to the design variables and a weighted sum arising from a quadrature representation of the expected value of the objective over the uncertainty space. Discretization points are selected for this quadrature, however realizations can be added to enforce feasibility at additional points. This gives a large-scale nonlinear multi-scenario problem with significant coupling or interaction induced by both the control and design variables.

We assume that the control variables u can be used to compensate for measured variable parameters, θ^v, but not the uncertainty associated with unknown parameters, θ^u. Thus, the control variables are indexed over k in the multi-scenario design problem, while the state variables, determined by the equality constraints, are indexed over q and k.

We have developed a package, SCHUR-IPOPT, that uses an internal decomposition approach for the parallel solution of structured nonlinear programming problems. This package is built upon on the existing primal-dual interior-point NLP solver, IPOPT (Wächter, 2006), where solution of the overall NLP problem is obtained by the approximate solution of a sequence of barrier sub-problems with a barrier parameter μ that approaches zero. Global convergence is promoted by a filter-based line-search strategy (Wächter, 2006).

The dominant computational expense of this algorithm is the solution of the augmented linear system resulting from a Newton iteration of the primal-dual equations. Given a problem with a particular structure, a decomposition approach can be devised to exploit the structure of this augmented system and produce efficient solutions in parallel.

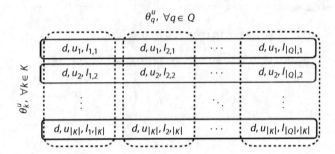

FIGURE 1
Potential decomposition layouts.

We focus on a parallel algorithm for block structured problems with complicating variables. In this formulation, each block is otherwise independent, except for a relatively small number of complicating variables that couple the blocks. This structure is suitable for formulation of multi-stage dynamic batch optimization, large-scale parameter estimation, and multi-scenario optimization under uncertainty problems, where the complicating variables are common variables that couple the different scenarios.

The structured linear system arising here can be solved efficiently in parallel using a Schur-complement decomposition. The Schur-complement is formed in parallel by repeated backsolves for each of the common variables. With this approach, scalability in the number of blocks is exceptional when there is an additional processor available for each new block. Previous results on a large distributed cluster have demonstrated that the solution time is almost flat as blocks and processors are added (Laird, 2008, Zavala et al., 2008). While the computational effort scales well with additional blocks, this straightforward approach scales at best linearly and at worst cubic with the number of common variables.

Considering formulation (1), the discretized scenarios can be represented as shown in Figure 1.

If the problem is decomposed with a single scenario for each block (and hence each processor), then the common variables in the parallel decomposition include both the control variables and the design variables. However, there is no restriction that each individual block needs to consider only a single scenario. If the problem is decomposed according to the dashed lines, where each scenario within the dashed lines is grouped into a single block, the number of common variables considered in the parallel decomposition still includes both the control variables and the design variables. However, if the problem is decomposed according to the solid lines, then the number of common variables considered by the parallel decomposition includes only the design variables. With this scheme, the coupling induced by the control variables is handled internally by the serial linear solver.

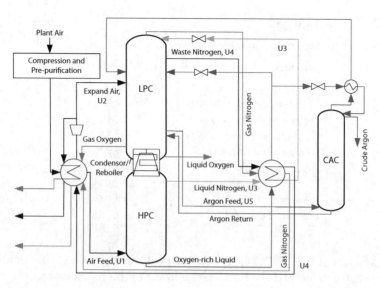

FIGURE 2
Cryogenic air separation flowsheet.

Process Description

The ASC plant studied includes a double-effect heat integrated distillation column with a side crude argon column (CAC). Addition of the CAC introduces two mass and energy integration structures into the double-effect distillation column and makes modeling and operation of the ASC significantly more difficult. Figure 2 shows the simplified structure of this ASC system. The air feed stream is first compressed and pre-purified. After being cooled by a primary heat exchanger, a portion of the air feed steam is introduced into the low pressure distillation column (LPC) with 70 theoretical stages. The remaining feed enters the bottom of the high pressure distillation column (HPC) with 36 theoretical stages. A side vapor stream is withdrawn at the 28th tray of LPC and distilled in the CAC, while the liquid from the bottom of this CAC is returned to LPC at the location of vapor stream withdrawal. Therefore, liquid oxygen product is directly taken from combined condenser/reboiler and gas oxen product is taken from the bottom of LPC. Liquid nitrogen product is directly taken from the top of HPC while gas nitrogen product is taken from the top of LPC.

Based on process dynamics of the ASC system, five main control variables, $u = [U_1\ U_2\ U_3\ U_4\ U_5]$, are selected to compensate for variability of argon product demands: the feed air stream of HPC (U_1), the feed air stream of LPC (U_2), the reflux flow from HPC to LPC (U_3), the waste nitrogen stream (U_4) and the side withdrawal from LPC to CAC (U_5). The five main design variables are

TABLE 1

Operating Conditions

Variables	Value (Units)
Gas oxygen product, mol/s	2.44
Liquid oxygen product, mol/s	0.64
Oxygen product purity	$\geq 98\%$
Gas nitrogen product, mol/s	13.13
Nitrogen product purity	$\geq 99.99\%$
Argon product purity	$\geq 96\%$
Pressure of LPC, MPa	0.13–0.14
Pressure of HPC, MPa	0.68–0.69
Pressure of CAC, MPa	0.12–0.13

the diameters of LPC, HPC, and CAC, the heat transfer area of the combined condenser/reboiler, and the brake horsepower of the compressor.

Distillation Column Model

The distillation column (LPC, HPC, and CAC) models are derived from the mass and energy balances, coupled with the equilibrium relationships,

$$y_{ij} = \gamma_{ij} K_{ij} x_{ij} \tag{2}$$

$$K_{ij} = P_j^s(T_j)/P_j \tag{3}$$

$$\log \gamma_{1,j} = \left(\alpha_{12} x_{2j}^2 + \alpha_{31} x_{3j}^2 + (\alpha_{12} + \alpha_{31} - \alpha_{23}) x_{2j} x_{3j} \right)/T_j \tag{4}$$

$$\log \gamma_{2j} = \left(\alpha_{12} x_{1j}^2 + \alpha_{23} x_{3j}^2 + (\alpha_{12} + \alpha_{23} - \alpha_{13}) x_{1j} x_{3j} \right)/T_j \tag{5}$$

$$\log \gamma_{3j} = \left(\alpha_{23} x_{2j}^2 + \alpha_{13} x_{1j}^2 + (\alpha_{23} + \alpha_{13} - \alpha_{23}) x_{2j} x_{1j} \right)/T_j \tag{6}$$

Where the liquid and vapor compositions are given by $x_{i,j}$ and $y_{i,j}$ respectively. The activity coefficients, γ, are calculated using Margule's equation, and ideal vapor-liquid equilibrium constants K_{ij} are calculated using Antoine's equation with saturation pressure P_i^s. The variables α are the binary interaction parameters of activity coefficients ($1-N_2$, $2-Ar$, $3-O_2$).

Conceptual Design Formulation

The following formulations (Douglas 1988, Peters et.al 2002) are given to describe conceptual design of cryogenic air separation columns.

The diameter of the distillation columns are calculated using,

$$D_{m,j} = 0.0164 V_{m,j}^{0.5} \left[378 M_g^2 \left(\frac{T_{m,j}}{520} \right) \frac{14.7}{\rho_{m,j}} \right]^{1/4} \tag{7}$$

$$D_m = \max(D_{m,j}), m \in (LPC, HPC, CAC) \tag{8}$$

The height of the distillation column is calculated as

$$H = 2.4n \tag{9}$$

And the heat transfer area in the combined condenser/reboiler can be calculated by

$$A = Q_l / (U \Delta T) \tag{10}$$

Where ΔT is the temperature driving force. The capital costs of column shells and trays are estimated with the following equations:

$$CSC_m = \left(\frac{M\&S}{280} \right) 102 D_m^{1.066} H_m^{0.802} (c_{in} + c_m c_p) \tag{11}$$

$$CTC_m = \left(\frac{M\&S}{280} \right) 4.7 D_m^{1.55} H_m (c_s + c_t + c_m) \tag{12}$$

Where c_p, c_m and c_{in} are the pressure range, construction material and installation cost coefficients. c_s and c_t are the tray spacing, and design cost coefficients.

The capital cost of heat exchanger in combined condenser/reboiler can be calculated by

$$HEC = \left(\frac{M\&S}{280} \right) 102 A^{0.65} (c_{in} + c_m (c_t + c_p)) \tag{13}$$

The capital cost of a main compressor can be calculated by

$$BHP = \left(\frac{(U_1 + U_2)}{1 - \Delta F_l} \frac{k}{k-1} RT_{in} \left(\left(\frac{P_{out}}{P_{in}} \right)^{\frac{k-1}{k}} - 1 \right) \right) \tag{14}$$

$$CPC = \left(\frac{M\&S}{280} \right) 518 \, (BHP)^{0.82} (c_{in} + c_t) \tag{15}$$

The major operating cost of cryogenic air separation columns is electricity. We assume the other operating costs can be ignored and that a liquefier is not installed in the system.

The electricity cost is given by

$$EC = C_{ele}BHP/\eta \tag{16}$$

Where C_{ele} is electricity price and η is the efficiency of the compressor. The total annual cost (TAC) of our air separation process is given by the following form,

$$TAC = \left(\sum_m (CSC_m + CTC_m) + HEC + CPC \right)/t_p$$

$$+ \sum_{k \in K} \sum_{q \in Q} (\omega_{qk} EC_{qk}) \tag{17}$$

Where ω_{qk} are weights which can be calculated according to the above distribution assumptions, t_p is the payback time, which is assumed to be 3-7 years. The other costs such as pipelines and valves are not included in this study.

Optimal Results

The objective function of the above problem focuses on how to minimize the total annual cost under the above inequality and equality constraints. We assume the argon product demands change among (0.135 ± 0.0081) *mol/s* and select twelve discrete points from this range with uniform distribution. The value α_{12} are assumed change from 7.0 to 9.5 according to (Harmens, 1970). We select eight discrete points from this range with uniform distribution. The total number of variables in the multi-scenario formulations is over 220,000.

In order to obtain comprehensive design information, the design variable results with and without considering uncertain parameters are compared and listed in the Table 2. In addition, as the payback time increases from 3 years to 7 years, the corresponding changes in the design variables are listed in Table 3.

TABLE 2

Design Results with and Without Considering Uncertainties

Variables	Nominal	Uncertain	Difference
Dia. of LPC, m	0.66	0.72	9.1%
Dia. of HPC, m	0.88	0.91	3.4%
Dia. of CAC, m	0.44	0.54	22.7%
BHP, Kw	90	100.1	11.2%
Heat exch. area, m²	24	28	16.7%
TAC, $10^5	1.412	1.463	3.6%

TABLE 3

Changes of Design Results Under Uncertainty as Payback Time Increases

Variables	Payback				
	3 year	4 year	5 year	6 year	7 year
Dia. of LPC, m	0.72	0.72	0.73	0.74	0.76
Dia. of HPC, m	0.91	0.91	0.92	0.93	0.95
Dia. of CAC, m	0.54	0.6	0.63	0.66	0.69
BHP, Kw	100.1	100	99.6	99	98.5
Area, A, m^2	28	28.5	29.1	29.9	30.9
TAC, $10^5	1.463	1.205	1.152	1.116	1.090

Table 2 shows the effect of uncertain parameters on HPC is the smallest and the difference between nominal and uncertain cases is only 3.4%. On the other hand, the diameter of CAC is affected significantly by uncertainties.

Table 3 shows that increasing the payback time significantly affects the diameter of CAC and the heat exchanger area, while other design variables are not as sensitive to the increase. The total annual cost correspondingly reduces as the percentage of install cost decreases.

Conclusion

This work uses multi-scenario programming method to determine optimal design variables for cryogenic air separation columns under two different kinds of uncertainties: unknown physical properties and variable product demands. Based on an extremely complex and highly nonlinear model, we adopt nonlinear solver, IPOPT to obtain optimal solution for such large scale problem.

As expected, the optimal values of design variables are more conservative when uncertainties are considered. The diameter of crude argon column is the most sensitive to such uncertainties. Impacts of different payback time on design variables are also discussed. Heat exchanger area and the diameter of the crude argon column increase significantly while the brake horsepower and total annual cost decrease as the payback time increases.

This research further demonstrates that rigorous, large-scale nonlinear optimization problems can be solved efficiently with modern tools.

References

Rooney, W. and Biegler, L. (2003) Optimal Process Design with Model Parameter Uncertainty and Process Variability, Nonlinear Confidedence Regions for Design Under Uncertainty, AIChE Journal, 49, 2.

Karwan, MH., Keblis, M. (2007) Operations planning with real time pricing of a primary input. *Computers & Operations Research.*, 34, 848.

Harmens, A. (1970) Vapour-liquid equilibrium N_2-Ar-O_2 for lower argon concentrations. *Cryogenics*, 6, 406.

Fourer, R., Gay, D.M., and Kernighan, B.W. (1992) AMPL: A modeling language for mathematical programming. Belmont, CA: *Duxbury Press*.

Wächter A., Biegler LT. (2006). On the implementation of an interior-point filter line-search algorithm for large-scale nonlinear programming. *Math. Programm.*, 106, 25

Laird, C.D., and L. T. Biegler, L.T. (2008), Large-Scale Nonlinear Programming for Multi-scenario Optimization, pp. 323–336, in Modeling, Simulation and Optimization of Complex Processes, H. G. Bock, E. Kostina, H-X Phu, R. Ranacher (eds.), Springer

Zavala, V.M., Laird, C.D., Biegler, L.T. (2008) Interior-Point Decomposition Approaches for Parallel Solution of Large-Scale Nonlinear Parameter Estimation Problems, Chem. Eng. Sci. 63, 19.

Douglas, JM. (1988) Conceptual design of chemical processes. New York, *McGraw-Hill*.

Peters, M., Timmerhaus,K., West, R. (2002) Plant design and economics for chemical engineers. New York, *McGraw-Hill*.

64

Sensitivity Assessment of Flotation Circuit to Uncertainty Using Monte Carlo Simulation

Edelmira Galvez[1], Felipe Sepulveda[1], Marcelo Montenegro[2] and Luis Cisternas[2*]

[1]CICITEM, Dept. Met. Engn.,Universidad Catolica del Norte, Antofagasta, Chile
[2]CICITEM, Chem. Engn. Dept.,Universidad de Antofagasta, Antofagasta, Chile

CONTENTS

ABSTRACT In the treatment of several minerals it is common that using flotation as a concentration step, before others treatment. Other applications of flotation include pretreatment in desalination, separation of oil from wastewaters and recovery of plastics from waste. Flotation circuit of several stages is usually used because generally it is not possible to obtain the target results in one stage. The design of these network process, however, assumes that process data are fixed and well-defined, whereas the actual operating conditions such as stages recoveries and feed grade may fluctuate over time. Those fluctuations can affect the design objectives, for example recovery and product quality. This paper applies the Monte Carlo simulation for the analysis of different types of flotation circuits, represented through a superstructure, under uncertainty in grade feed and recovery of each flotation stage. This work demonstrates that the use of Monte Carlo simulation can be an appropriate tool to evaluate and select flotation circuits. The circuits were evaluated, with the help of radial/spider graphic, using metric indices of process efficiency, capacity and stability.

KEYWORDS *Flotation circuit, uncertainty, Monte Carlo*

---

* To whom all correspondence should be addressed.

Introduction

Froth flotation of minerals is one of the most widely accepted industrial practices for separation of valuable components from associated gangue materials in minerals. The aim of the flotation process is to achieve maximum recovery and highest concentrate grade. The flotation circuit consists of several stages, including rougher, cleaner, and scavenger operations. Each of them, can include one or more mechanical or column banks. Also, other operations such as grinding and dewatering are included. The final product from the circuit is the concentrate, from the cleaner operation, and the tails from the scavenger operation.

The methodologies presented in the literature for the design of flotation circuit use deterministic values for mineralogy and grade of the feed. Also, the flotation models used in the circuit design are based on empirical models, usually considering that the flotation process is similar to a pseudo-first order reaction between particles and bubbles (constant bubbles concentration) and the particles can be classified in classes with the same floatability (Mendez et al. 2008). However, the feed to flotation process has variability in the grade and mineralogy, because of the heterogeneous nature of the mining deposits or fraction mineral (Gy, 1999, 2004a, 2004b; Tutmez, 2006). Therefore, the feed grade and mineralogy change with time. On the other hand, the individual recoveries for each flotation stage changes because of the complex behavior of the mineral and the interaction with operational variables as type and amount of reagents. The objective of this work is to analyze how these uncertainties affect the flotation circuits.

There are few works in the literature on uncertainty analysis applied to mineral processing. Xiao (2004) compared the propagation of variance or the law of propagation of error (LPE) method with the Monte Carlo Method, finding that LPE has several limitations. However, in his work, only one flotation circuit was analyzed as example. In this work Monte Carlo Simulation is utilized to analyze several flotation circuits. In this approach a probability distribution function for each parameter is sampled to provide parameter values. Usually, a computer-generated random number is used to obtain a parameter value based on the probability distribution function. When a value for each parameter has been calculated, the model is quantified to obtain an answer. This answer is then placed into a frequency table. The entire process is then repeated until a desired number of iterations is reached. Monte Carlo analysis has been applied to several processes, including moving bed systems (Kurup et al., 2008), water adsorption (Neves et al., 2008), vacuum membrane distillation (Imdakm et al., 2007), flow of fluids through porous media (Jain et al., 2003), and water networks (Raymond et al., 2007).

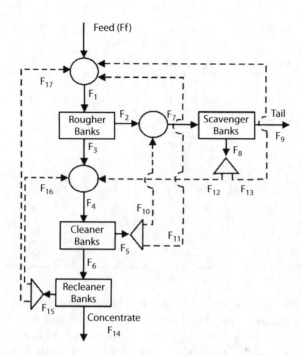

FIGURE 1
Superstructure for flotation circuits.

Procedure

Figure 1 shows a superstructure that represents twelve flotation circuits, depending on the distribution of streams. For example, the tail from the cleaner stage can be recycled to the scavenger stage or the rougher stage. The alternatives included in this study are indicated in Table 1 where 1 represents the existence of the stream and 0 is nonexistence. A general model, based on mass balances was developed for the superstructure. For the flotation stage modeling the concentrate to feed ratio for class i (T_i) was used. The concentrate to feed ratio can be determined with different models, according to the equipment type and kinetics model, or can be determined from plant data (see figure 2).

Several studies were considered, including uncertainty in the T_i values for rougher, cleaner, recleaner and scavenger, considering cases that only one T_i has uncertainty and cases with uncertainties in all T_i. Also, the uncertainty in the feed grade was analyzed. Then the T_i for the rougher, cleaner, recleaner, and scavenger stages, were assigned a normal distribution, with fixed mean and standard deviation. For the study of the feed grade a normal distribution was also assigned, with a fixed value of mean, but with several values of standard deviation.

TABLE 1
Flotation Circuits Studied Based on Figure 1

Alternatives	F_{10}	F_{11}	F_{12}	F_{13}	F_{16}	F_{17}
A	0	1	0	1	0	1
B	0	1	1	0	0	1
C	1	0	0	1	0	1
D	1	0	1	0	0	1
E	0	1	0	1	1	0
F	0	1	1	0	1	0
G	1	0	0	1	1	0
H	1	0	1	0	1	0
I	0	1	0	1	0	0
J	0	1	1	0	0	0
K	1	0	0	1	0	0
L	1	0	1	0	0	0

For each study, a value for each parameter was sampled from the normal distribution function, and the flotation circuit was simulated. Then new parameters were sampled and the circuit was simulated again. Based on a previous analysis, 2000 simulations were carried out for each study. Several metric indices were determined to evaluate each circuit, which were normalized with the objective to be able to compare the circuits using all indices. Finally, the results were represented in radial graphics.

Result

For analysis of the results metric indices were used to represent the efficiency, capacity and stability of the process. Among the indices are 1) global recovery of the species of value, a measure of the efficiency of the process in recovering the species of value, 2) concentrate grade, a measure of the selectivity of the process, 3) global separation factor, a measure of the quantity of

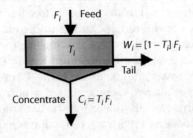

FIGURE 2
Flotation stage representation.

TABLE 2

Recovery Mean Values

Flotation Circuit	All	Rougher	Cleaner	Recleaner	Scavenger
A	86.55	86.60	87.31	87.31	87.34
B	87.78	87.83	88.38	88.38	88.41
C	67.89	67.92	68.37	68.36	68.44
D	70.17	70.22	70.49	70.48	70.56
E	89.29	89.34	89.94	89.95	89.97
F	90.30	90.35	90.81	90.82	90.83
G	68.56	69.61	69.61	69.96	70.04
H	71.76	71.77	72.02	72.02	72.09
I	90.96	90.97	91.49		91.52
J	91.83	91.84	92.24		92.26
K	76.79	76.79	77.14		77.21
L	78.64	78.66	78.96		78.92

necessary mineral to produce the concentrate, 4) enrichment ratio, a measure of the increase of the grade from the feed to the concentrate. Each one of these indices is evaluated using their mean value, standard deviation, kurtosis and skewness. Kurtosis is a measure of the "peakedness" of the probability distribution of a valued random variable. Skewness is a measure of the asymmetry of the probability distribution of a valued random variable.

The results of the simulations were represented by histograms and tabulated. For example, the tables 2 and 3 show the results in the global recovery

TABLE 3

Recovery Standard Deviation

Flotation Circuit	All	Rougher	Cleaner	Recleaner	Scavenger
A	3.14	2.74	0.52	0.52	0.95
B	2.70	2.27	0.48	0.48	0.93
C	4.12	1.69	2.24	1.01	1.87
D	3.84	1.12	2.16	0.98	1.89
E	2.56	2.25	0.56	0.20	0.78
F	2.17	1.86	0.51	0.18	0.76
G	3.92	1.37	1.37	0.86	1.82
H	3.67	0.82	2.25	0.82	1.83
I	2.12	1.94	0.37		0.67
J	1.79	1.59	0.34		0.65
K	2.77	1.38	1.83		1.52
L	2.49	0.90	1.73		1.51
Mean	2.94	1.66	1.20		1.27

TABLE 4

Concentrate Grade Mean Values

Flotation Circuit	All	Rougher	Cleaner	Recleaner	Scavenger
A	33.47	33.47	33.47	33.47	33.47
B	33.13	33.13	33.13	33.13	33.13
C	33.51	33.51	33.51	33.51	33.51
D	33.20	33.20	33.20	33.20	33.20
E	33.48	33.48	33.48	33.48	33.48
F	33.15	33.15	33.15	33.15	33.15
G	33.51	33.52	33.52	33.52	33.52
H	33.22	33.22	33.22	33.22	33.22
I	19.37	19.37	19.32		19.31
J	16.94	16.93	16.90		16.89
K	20.28	20.28	20.24		20.24
L	17.98	17.98	17.95		17.95

of the species of value for the 12 analyzed circuits, when uncertainty is included in the T_i values. Table 2 gives the mean values of the recovery, and table 3 gives the standard deviation. The alternative J shows better recoveries, independently if the uncertainty exists in rougher, cleaner, scavenger or all values of T_i, that doesn't mean that alternative J is the best option, because at the same time, this option, has the smallest value of concentrate grade (Table 4).

If the uncertainty is represented by normal distribution functions, the results usually have the same distribution type. However, in some cases the distribution presents some degree of skewness; for example, figure 3 shows that the histogram of the concentrate grade of circuit A, when the grade feed presents a normal distribution with standard deviation of 0.1, has negative skew.

The uncertainty in the recovery of the species of value depends on the origin of the uncertainty. For example, in table 3 is observed that the circuit A is more sensitive to the uncertainty in the values of T_i of the rougher stage than of the cleaner stage. On the other hand, the circuit D shows more dependence on the uncertainty in the cleaner stage than in the rougher stage.

The obtained values of the metric indices were normalized using the maximum and minimum values, so that the best flotation circuit performance was evaluated with the value 1, while the worst flotation circuit was evaluated with the value 0. Figure 4 and 5 show the radial graphs for all the flotation circuit studied.

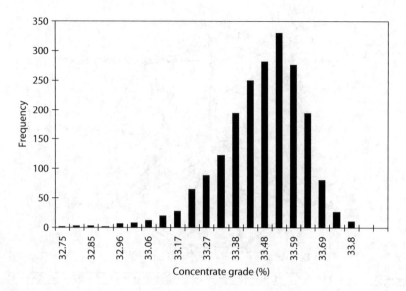

FIGURE 3
Histogram for concentrate grade of circuit A.

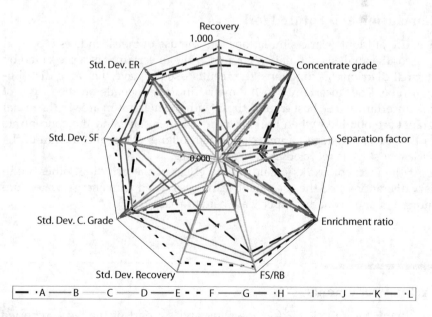

FIGURE 4
Radial graph for uncertainty in T_i values.

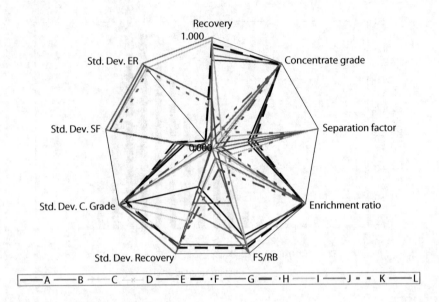

FIGURE 5
Radial graph for uncertainty in feed grade.

Conclusion and Future Work

The use of Monte Carlo simulations and the use of metric indices is a good tool for the analysis of flotation circuits. If the uncertainty is represented by normal distribution functions, the results usually have the same distribution type. The uncertainty in the metric indices depends on the origin of the uncertainty. The best results, those with the maximum area in the radial graph, are obtained when the tail of the cleaner stage and the concentrate of the scavenger stage are recycled to the rougher stage, and the tail of the recleaner stage is recycled to the cleaner stage.

At the moment works including the effect of the grinding, other auxiliary operations, and the use of metric indices related with energy, water consumption and economic aspects are being carried on.

Acknowledgments

We thank the CONICYT for financing and support of the work reported in this manuscript (FONDECYT 1060342). MM thanks to the University of Antofagasta for financing support.

References

Gy, P. (2004a). Sampling of discrete materials-a new introduction to the theory of sampling.Chemometrics and Intelligent Laboratory system 74.7–24.

Gy, P. (2004b). Sampling of discrete materials. Quantitative-sampling of zero-dimensional objects. Chemometrics and intelligent systems. 74,25–38.

Gy, P. (1999). Optimizing the operational strategy of a mine-metallurgy quarry-cement works complex,Canadian metallurgical Quartely 38,157–163.

Imdakm, AO., M. Khayet, T. Matsuura (2007). A Monte Carlo simulation model for vacuum membrane distillation process, Journal of Membrane Science, 306 (1–2), 341–348.

Jain, S., M. Acharya, S. Gupta, AN. Bhaskarwar (2003). Monte Carlo simulation of flow of fluids through porous media, Computers & Chemical Engineering, 27(3), 385–400.

Kurup, A.S., HJ. Subramani, MT. Harris (2008). A Monte Carlo-based error propagation analysis of Simulated Moving Bed systems, Separation and Purification Technology, 62 (3) 582–589.

Mendez, D.A., ED. Gálvez, LA. Cisternas (2008). State of the art in the conceptual design of flotation circuits, International Journal of Mineral Processing, In Press.

Neves, RS., AJ. Motheo, RPS. Fartaria, FMS. Silva Fernandes (2008). Modelling water adsorption on Au(210) surfaces: II. Monte Carlo simulations, Journal of Electroanalytical Chemistry, 612 (2), 179–185.

Tutmez, B. (2006). Uncertainty oriented fuzzy methodology for grade estimation, computer. Geosciences 33, 280–288.

Tan, RR., DCY. Foo, ZA. Mana (2007). Assessing the sensitivity of water networks to noisy mass loads using Monte Carlo simulation, Computers & Chemical Engineering, 31 (10), 1355–1363.

65

Incorporating Sustainability and Environmental Impact Assessment into Capstone Design Projects

Jeffrey R. Seay[1*] and Mario R. Eden[2]

[1]*Department of Chemical and Materials Engineering, University of Kentucky, Paducah, Kentucky 42002*

[2]*Department of Chemical Engineering, Auburn University, Auburn, Alabama 36849*

CONTENTS

ABSTRACT Due to recent interest in global climate change, the traditional design heuristics that have served as the backbone of capstone design must now be updated to include the topics of sustainability and environmental impact assessment. A capstone design program that includes sustainability requires careful selection of the design problems. Although sustainable design principles can be incorporated into any process, the challenge is to choose processes that clearly illustrate these principles to students. In this contribution, two projects based on the manufacture of biodiesel will be highlighted. Through these sustainable design projects, the students have the chance to experience how the design choices made at the conceptual level

---

* To whom all correspondence should be addressed

impact not only the economic performance of a process, but also the environmental performance and overall sustainability of a chemical process.

KEYWORDS   *Sustainability, capstone design, biodiesel, glycerol*

Introduction

The ultimate goal of any capstone design course is to encourage the students to systematically apply the methods they have learned in unit operations, reaction engineering, process optimization, process controls and engineering economics to the conceptual design of a complete chemical manufacturing process. To these traditional methods, it is now important to add sustainability, environmental impact assessment and product lifecycle assessment. By incorporating these additional elements, students have the opportunity to learn, at the undergraduate level, how the choices made during conceptual design impact the environmental as well as the economic performance of chemical processes.

Because of the addition of these two elements, the choice of the process to be considered by the students is of paramount importance. In addition to being sufficiently well defined enough for the students to handle, the process must also lend itself to the application of alternatives that enhance its sustainability. In other words, the students must have the opportunity to analyze a more traditional as well as sustainable versions of a process for the production of a chemical product to truly appreciate the impacts of their design decisions. Collaboration with an industrial partner can help to identify processes that meet these criteria. Collaboration not only gives the students access to real processes and research, but also provides them with an insight into the challenges facing industry in implementing sustainable and environmentally benign processes.

To quantify the potential environmental impacts of conceptual design choices, the students must utilize an assessment methodology. In the capstone design course, the students can apply the Waste Reduction (WAR) Algorithm, a calculation tool developed by the U.S. Environmental Protection Agency for estimating the potential environmental impact of a chemical process, to the conceptual design of alternative processes for the manufacture of an industrial chemical product (Young *et al.*, 2000).

The WAR Algorithm is used to calculate the potential environmental impacts of a given quantity of material if it were directly emitted to the environment (Young *et al.*, 2000). By incorporating the environmental impact assessment, the students can include sustainability and potential environmental impacts in their evaluation of their proposed conceptual processes, while not neglecting the important economic considerations typically used to guide the selection of the preferred process option.

In addition to the environmental impacts of waste product generation within the process, the inclusion of heat integration using thermal pinch analysis will

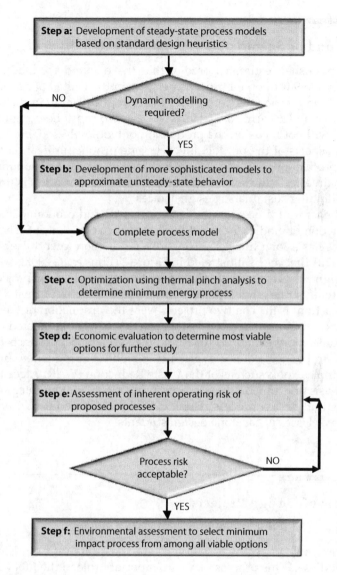

FIGURE 1
Flowchart of process integration methodology (Seay and Eden, 2009).

also be considered. Typically, the justification for including heat exchanger networks as part of a process is based on the potential operating cost savings due to lower energy consumption. However, by also including environmental considerations during heat integration activities, the true optimum process based on energy, economics and environmental considerations can be discovered. A flowchart illustrating how these items are incorporated into conceptual process design is illustrated in Figure 1 (Seay and Eden, 2009).

Background of Sustainable Design Projects

Recently published estimates predict that the demand for biodiesel will grow from 6 to 9 million metric tons per year in the United States and from 5 to 14 million metric tons per year in the European Union in the next few years (Blume and Hearn, 2007). However, the two capstone design projects presented will not focus on the production of biodiesel itself, but rather on other key aspects of the product lifecycle–side product utilization and raw material production. Understanding how each step in the manufacturing supply chain affects the sustainability of the biodiesel process is important to understanding sustainable design choices.

Like all chemical processes, the environmental and economic impacts of the production of biodiesel are not limited to the process itself. The biodiesel process has a long lifecycle, from crop production to end use in motor vehicles. This lifecycle is illustrated in Figure 2 (Elms *et al.*, 2007).

The design projects that were chosen for the capstone design projects were intended to illustrate other aspects of the product lifecycle, not just the esterification reaction itself. The two projects were the production of the sodium methylate catalyst used in esterification reaction of a common biodiesel production route and the manufacture of industrial chemical products from the crude glycerol generated as a byproduct of biodiesel production. In both of these examples, tools such as of the Waste Reduction (WAR) Algorithm, life cycle analysis and inherently safe design principles will be integrated into the conceptual design process. These two projects were introduced in two consecutive classes of capstone design students.

Overview of Design Projects

Project 1 – Production of Sodium Methylate Catalyst

The first example project is based on producing a catalyst for the manufacture of biodiesel. This process plays an important role in the lifecycle of the biodiesel process, since the energy consumed in manufacturing the catalyst, and other raw materials, must be factored into the overall environmental impact assessment.

To ensure that the students were considering a realistic industrial process, the students were given as their design basis, a recently published patent of a process designed to manufacture sodium methylate catalyst (Guth, *et al.*, 2004). Numerous pathways are available for the production of biodiesel; however the pathway utilizing sodium methylate is advantageous because it eliminates the production of water from the transesterification reaction that produces biodiesel. However, the production of the sodium methylate

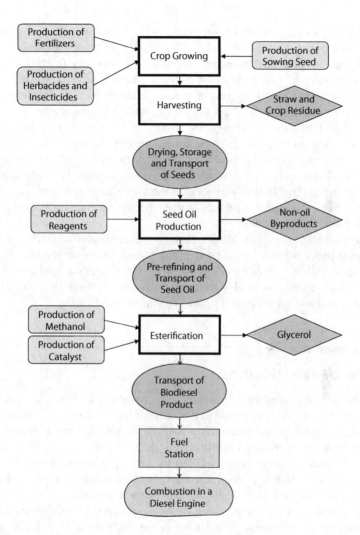

FIGURE 2
Biodiesel product lifecycle (Elms *et al.*, 2007).

catalyst is energy intensive, so it makes a good classroom example of the importance of looking at the total product lifecycle.

In addition to the process itself, the students also considered the economic impacts on the biodiesel process of sodium methylate manufacture. This was done to illustrate how one manufacturing process fits into an overall product lifecycle.

Project 2 – Production of Acrolein from Crude Glycerol

The second project is based on developing processes for the manufacture of products from the glycerol generated as a byproduct of biodiesel manufacture.

For every 9 kilograms of biodiesel produced, 1 kilogram of crude glycerol is formed as a byproduct (Chiu *et al.*, 2006). Identification of industrial uses for this glycerol is important to the economic viability of biodiesel.

It is well known that many industrially importance chemical products can be manufactured from glycerol. These products include acrolein, acrylic acid, 1,2-propanediol and 1,3-propanediol, among others (Neher *et al.* 1995). Although each of these processes is similar in nature, the production of acrolein was chosen as the process to be considered by the students.

Although a commercial process for the manufacture of acrolein from glycerol has yet to be realized, the students had the opportunity to consider how laboratory research can be incorporated into the design process. A search of the current literature provided ample information for the students regarding reactor conversion and yield.

In addition to the sustainable process, the students also considered a more traditional process for acrolein production based on the catalytic partial oxidation of propylene. By comparing the sustainable process to the traditional process, the students were able to identify the cost targets that would make switching to the glycerol based process economically viable.

Student Design Objectives

Capstone design projects are typically team oriented, with students working in groups to meet their objectives. For each of the processes describe above, the students were given conversion and yield data for the reaction, along with the reactor operating conditions, and an overall required production rate. Each student group was required to develop a model for both a traditional and a sustainable production case. However, how the students chose to model each case was left open to the group's discretion.

Since the projects are open-ended there is not one single solution, which is often difficult for the students to grasp in the beginning. The challenge is to identify the aspects of the project that should be addressed at a certain time and argue why others should not. This requires not only strong technical abilities, but also excellent time management skills. Although the projects are open-ended, each group is required to achieve certain objectives regarding integration, and economic assessment.

Sustainability Objectives

The primary purpose for selecting these processes as the basis for the capstone design projects was to incorporate sustainable production processes into the curriculum. To meet this objective, the students were required to consider the environmental impacts of their design choices. Using standard design heuristics for both processes, optimized conceptual designs were generated.

The learning goal of this aspect of the project was for the students to understand which variables, feedstock choice, energy integration, production rate, etc., had the greatest influence on the sustainability and environmental impacts. The ability of students to understand and apply principles of environmental impacts and sustainability was included as part of the project assessment. An Ethics, Safety, Society, Environment Assessment Rubric was included in the instructor evaluation of the design projects.

Since economics are also considered at each step, the students get rapid feedback on the effects of their design choices. For cases where the sustainable process is not economically viable, the students can calculate the raw material cost targets for which it would be.

Process Integration Objectives

An important aspect of sustainable process design is optimizing the energy utilization. For the capstone design project, students utilize thermal pinch analysis to determine the heat integration potential of their conceptual designs. Increased energy integration enables higher utilization of raw materials and minimizes the use of external utilities (El-Halwagi and Spriggs, 1998). This minimizes the environmental impacts of a process. Incorporating this technique early in the conceptual design process will guide the students in making environmentally friendly, yet economically viable choices.

For both the glycerol dehydration process and sodium methylate process, heat integration is especially important. This is due to the fact that both processes are quite energy intensive. Therefore, the students must carefully consider impacts of heat integration on the sustainability and environmental impact of the process, as it pertains to the economic viability of their conceptual designs.

Incorporating Environmental Impact Assessment

As previously discussed the WAR Algorithm is a calculation procedure developed by the U.S. Environmental Protection Agency for quantifying the potential environmental impacts (PEI) for a given process (Young *et al.*, 2000). As part of the capstone design course, the students are instructed to incorporate the results of PEI calculation along with economic considerations when evaluating process options. Adding the PEI to the standard design heuristics ensures that the most environmentally friendly of the economically viable process options is selected. Calculation of the PEI is straightforward and provides important insight into the effect of process changes on the environmental impact.

The WAR Algorithm can also be used to consider the impacts of the energy usage of a process. As part of the design project, students are encouraged to explore the influence of their heat integration activities on the PEI of the process. This adds another dimension to the evaluation of heat exchanger networks. Again, by integrating all these elements; sustainability, environmental impacts and economics, the students gain a more holistic understanding of their conceptual design and how the design variables interact with each other.

Product Life Cycle Considerations

When assessing sustainability, a holistic view of the entire product lifecycle is required. Choosing design projects that are part of a larger product life cycle give the students perspective as to the impacts–negative and positive–of the choices made when designing the process under consideration. By requiring the students to consider how the manufacturing part of the process fits into the entire product lifecycle is an important part of the learning process. This objective is implemented as an open ended part of the capstone design project assignment.

Industrial Collaboration

Through collaboration with an industrial partner, the experience of the design project for the students has been greatly enhanced. Both of the capstone projects presented were selected, in part, because of their importance to industry. A series of lectures from industrial researchers currently developing processes similar to the capstone design projects was included in the course schedule.

A site visit to the production facilities involved was also arranged for the students. This visit was arranged during the second half of the semester so that the students would have the opportunity to ask questions about the processes they were trying to model. Finally, reviewers from industry were included in evaluating the final reports and presentations of each of the student groups.

Results and Conclusions

In conclusion, these capstone design projects have allowed students to directly experience how design decisions influence the sustainability of a process. This project also highlighted the need to include sustainability discussions in previous core chemical engineering courses. Given the recent concerns regarding global climate change, it is important for students to understand how design decisions effect the impacts a process can have on the environment. The impacts of including industrially relevant research and the inclusion of industry experts during the course lectures and evaluation of the student projects was immediately reflected positively in the student course evaluations.

Acknowledgements

The authors would like to thank Evonik Degussa Corporation for their support of the development of these capstone design courses. Furthermore, the assistance and constructive criticism by Mr. Robert D'Alessandro, Mr. Lee Daniel and Mr. Robert Klein during the student presentations is very much appreciated.

References

Blume, A.M., & A.K. Hearns. (2007). The evolution of biodiesel, Biofuels, 20–23.

Chiu, C.-W., M.A. Dasari & G.J. Suppes. (2006). "Dehydration of glycerol to acetol via catalytic reactive distillation", AIChE Journal, 52.

El-Halwagi, M.M. and H.D. Spriggs. (1998). "Solve Design Puzzles with Mass Integration", Chemical Engineering Progress, Vol 94, No. 8.

Elms, Rene', B. Shaw, G. Nworie and M. M. El-Halwagi. (2007). "Techno-Economic Analysis of Integrated Biodiesel Production Plants", 2007 AIChE Annual Meeting, Salt Lake City, Utah.

Guth, J., Fredrich, H., Sterzel, H., Karbel, G., Burkart, K., Hoffmann, E. (2004). Method for Producing Alkali Methacrylates. European Patent EP 1242345 B1.

Neher, A., T. Haas, D. Arntz, H. Klenk and W. Girke. (1995), "Process for the production of acrolein", United States Patent 5, 387,720.

Seay, J.R. and M.R. Eden. (2009). "Incorporating Environmantal Impact Assessment into Conceptual Process Design: A Case Study Example" Journal of Environmental Progress and Sustainable Energy, Published Online: Dec 10 2008, DOI: 10.1002/ep.10328.

Young, D., R. Scharp and H. Cabezas, (2000). "The waste reduction (WAR) algorithm: environmental impacts, energy consumption, and engineering economics", Waste Management, Vol. 20.

66

Process Synthesis Targets: A New Approach to Teaching Design

Bilal Patel, Diane Hildebrandt* and David Glasser
University of the Witwatersrand, Johannesburg, South Africa
Private Bag X3, WITS, 2050

CONTENTS

ABSTRACT The design course is an integral part of chemical engineering education. A novel approach to the design course was recently introduced at the University of the Witwatersrand, Johannesburg. The course aimed to introduce students to systematic tools and techniques for setting and evaluating performance targets for processes as well as gaining insight into how these targets can be achieved. The main objectives were efficient use of raw materials, energy and improved environmental performance (reducing CO_2 emissions). The approach is to use fundamental principles–mass, energy and entropy. The course was well received by students and allowed the students to gain a better understanding of developing processes which are efficient and have less impact on the environment.

KEYWORDS *Targeting, process synthesis, process design, process integration, thermodynamics*

---

* To whom all correspondence should be addressed; E-mail: Diane.Hildebrandt@wits.ac.za

Introduction

"What is the minimum amount of carbon dioxide that a process can produce?" This may seem like a trivial question but it is not a question usually
asked when processes are being designed. In many cases, there is a lack of
a quantitative description of what is the highest efficiency, least amount of
energy or lowest amount of carbon dioxide that can be achieved for a particular process i.e. what is the theoretical achievable target. Without being
able to answer such simple questions it is hard to make good decisions with
regard to the design of processes.

In this regard, a novel approach to the chemical process design course was
recently introduced at the University of the Witwatersrand, Johannesburg.
The course aimed to introduce students to systematic tools and techniques for
setting and evaluating performance targets for processes as well as gaining
insight into how these targets can be achieved. The main objectives, in terms
of the targets set for the process design, were efficient use of raw materials,
energy and improved environmental performance (reducing CO_2 emissions).

Philosophy

The decisions taken in the early stage of the design process or the conceptual
phase is of vital importance as the economics of the process is usually set at
this stage. Biegler et al. (1997) estimates that the decisions made during the
conceptual design phase fix about 80% of the total cost of the process. Once
the process structure has been fixed, only minor cost improvements can be
achieved. Thus, the success of the process is largely determined by the conceptual design (Meeuse, 2002).There is therefore a need for systematic procedures
to generate as well as identify the most promising alternatives. Without such
procedures, even an experienced designer might not be able to uncover the
best process structure and will be stuck with a very poor process structure.
Ideally, these procedures should be applied in the early stages of the design
and should require minimum information since the use of rigorous design
methods to evaluate alternatives can be time and capital intensive.

The philosophy underlying the course is to look at the process holistically.
The design of a flowsheet is approached with this overall analysis as its foundation. We address the overall process by tools and techniques developed
within the framework of process synthesis and integration, which provides a
holistic approach to process design i.e. considering "the big picture first, and
the details later" (Srivinas, 1997).We aim to introduce a method of providing
insights and setting targets for the *overall process* based on fundamental concepts, as well as developing systematic procedures to attain these targets.

Targeting allows one to identify a benchmark for the performance of a system before the actual design of the system is carried out (El-Halwagi, 2006).

These benchmarks are the ideal or ultimate performance of such a system and provide useful insight into the process. These targets are usually based on fundamental engineering principles, for example, thermodynamic principles but can be based on heuristics or cost estimates. Targets are usually independent of the structure of the process i.e. the ultimate performance of the system can be determined without identifying how it can be reached (El-Halwagi, 2006). Thus, these targets reduce the dimensionality of the problem to a manageable size (El-Halwagi, 2006). These targets are also useful in evaluating existing systems as one can easily compare the current performance of the system to the ideal performance of the system.

Every chemical process can be considered in terms of a number of inputs and outputs. These inputs or outputs can be classified into three variables: mass, heat and work. In quantifying these variables, the concept of mass and energy conservation has been extensively used. Mass and energy balances have traditionally been used in the analysis of individual units and flowsheets. They, however, can also be used for synthesizing chemical processes. Another tool, the second law of thermodynamics (or the entropy balance) is also useful for synthesizing or analyzing chemical processes, especially since it can quantitatively assess the efficiency and sustainability of processes. The law of mass conservation (mass balance), the first law of thermodynamics (energy balance) as well as the second law of thermodynamics will be employed as the basis of the approach.

A back-to-front synthesis approach based on determining the target overall mass balance for a process is proposed. The overall mass balance can be determined by applying atomic species balances based on the inputs and outputs of the process, as well as various other criteria (such as environmental aspects, cost etc). The energy requirements are then accounted for by applying the energy balance to the overall mass balance. The work requirements are also determined for each overall mass balance based on the entropy balance. An understanding of which of the three variables is the limiting target is also very important, in that it gives insight into what is the important or limiting parameter in the design and operation of the process.

These targets also give insights, at a very early stage, into integration of the process. It can be shown that targets can be achieved by considering mass, heat and work integration of a process (Patel et al, 2007). This is useful not only for the design of new processes but for retrofitting as well.

Tools for Determining Process Target

Mass Balance for Synthesizing Processes

We begin by first introducing the law of mass conservation as it usually precedes any thermodynamic analysis. The law of mass conservation or mass (material) balance is a central concept in chemical engineering and is usually

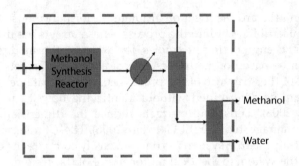

FIGURE 1
Simplified methanol synthesis flowsheet.

the first principle a chemical engineer is taught as an undergraduate. Simply stated, the law of mass conservation states that the mass flows entering and leaving a system must be equal at steady state. Mass balances are used in the design of new processes or the *analysis* of existing processes. It can be applied to individual units, for example, a reactor or a distillation column but it can also be applied to an entire flowsheet.

Since chemical processes may involve reactions, recycles, bypasses or purges, the mass balances can become quite complex. To overcome this complexity, atomic species balances are usually employed since the mass balance simply reduces to *input = output*. The atomic species balance can be very useful from a synthesis perspective as well. It allows one to perform a mass balance without having to consider each individual unit or the recycle flows; instead one can focus on the entire process. To illustrate these points, we will consider the following example.

Consider the following simplified methanol synthesis flowsheet (Figure 1), which was given to the students.

The students were told that the water and methanol product were formed in a ratio of 0.5:1. They were asked to determine the overall balance for the process, which satisfies this ratio. This is a simple task, but a systematic solution would require performing an atomic species balance for carbon, hydrogen and oxygen. We are not given the feed material to the process, thus this has to be determined. If one considers the possible feed materials, which is usually synthesis gas in the case of methanol synthesis, one can write the mass balance as follows:

$$a\ CO + b\ H_2 + dCO_2 - 1\ CH_3OH - 0.5\ H_2O = 0$$

One can write the species balance as follows:
C Balance: $a + d = 1$
H Balance: $2b = 4 + 1$
O Balance: $a + 2d = 1 + 0.5$

Note that there are three equations with three unknowns (a, b, d), thus there exists a unique solution. Solving gives the mass balance

$$\tfrac{1}{2}\,CO + 1\tfrac{1}{2}\,H_2 + \tfrac{1}{2}\,CO_2 - 1CH_3OH - 0.5H_2O = 0$$

Note that this is the mass balance for the process and does not have to correspond to any reaction(s) that may be taking place in the process. We note that in this process CO_2 is a feed material to the system. So, if one were to consider this process, one would consider it to have an environmental benefit.

However, one has to note that, the synthesis gas feed needs to be produced and one has to account for this i.e. consider the entire process. If one writes an overall balance for the process, starting with methane as a feedstock

$$(1 + r/2)\,CH_4 + (1/2 + r)\,O_2 - CH_3OH - r/2\,CO_2 - r\,H_2O = 0$$

where r represents the amount of water produced.

Thus, it is clear when one considers the overall process, it is clear that the more water that is emitted from the process, the more methane feedstock (and oxygen) will be required and the more carbon dioxide will be produced. This means higher usage of natural resources as well as higher emissions.

Producing water in the synthesis section has an effect on the overall process – even though it did not appear to be an issue for the environment when just considering that section by itself.

In order to remedy this situation, one could recycle the water back to the process; this reduces the overall mass balance from (if one considers the ratio of CH_3OH to H_2O as 1:0.5)

$$5/4\,CH_4 + O_2 \rightarrow CH_3OH + \tfrac{1}{4}\,CO_2 + \tfrac{1}{2}\,H_2O \quad \text{to:}$$
$$1CH_4 + \tfrac{1}{2}\,O_2 \rightarrow CH_3OH$$

In this way, one reduces the methane and oxygen required and the carbon dioxide emissions as shown in Figure 2.

In particular, the water could be recycled back to the synthesis section reactor. The flowsheet is shown in Figure 3. Students were asked to simulate this process in ASPEN PLUS to show that this is possible.

The above example illustrates the power of an overall balance as a tool for improving processes. The mass balance can also be used to set targets

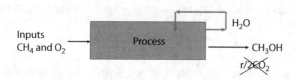

FIGURE 2
The effect of recycling water on the overall process.

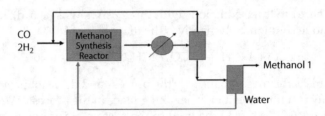

FIGURE 3
Recycling water to the synthesis section reactor.

for processes. The mass balance represents the minimum amount of inputs required to produce a specified product. Also, the mass balance can also give insights into by-products or wastes that may be formed. The mass balance can also be a target for process integration, thus (providing insight into recycling or reuse of materials). Consider another example, the synthesis of one kmol of methanol from methane and oxygen. The mass balance can be represented as follows:

$$aCH_4 + b\,O_2 - CH_3OH = 0$$

By applying the atomic species balance, one can determine the minimum amount of methane and oxygen required.

C Balance: a = 1
H balance: 4a = 4 or a = 1
O balance: 2b = 1 or b = 0.5

Thus, the mass balance is given as follows:

$$1CH_4 + \tfrac{1}{2}\,O_2 - CH_3OH = 0$$

In order to produce 1 kmol of methanol, a minimum of 1 kmol of methane and 0.5 kmol of oxygen is required. This is a *target* for the process since it represents the minimum quantities of the various components in the system.

If one, for example, fed excess oxygen (1 kmol instead of 0.5 kmol), the mass balance will need to change and *by-products or wastes will have to be formed*. In this particular case, the by-products/waste that could be formed are water and carbon dioxide. The mass balance can be written as follows:

$$a\,CH_4 + 1\,O_2 + b\,CO_2 + c\,H_2O - 1\,CH_3OH = 0$$

The atomic species balance can be used to determine the values of a, b and c.

C balance: a + b = 1
H balance: 4a + 2c = 4
O balance: 2 + 2b + c = 1

Notice that a negative value for a, b or c would mean that the component occurs as a product since all components were initially specified as feed material. The resulting mass balance is given as follows:

$$1.25\,CH_4 + 1\,O_2 + 0.25\,CO_2 + 0.5\,H_2O - 1\,CH_3OH = 0$$

By adding more oxygen than required, more methane has to be added to the process and both carbon dioxide and water are produced.

The implications of by-products/waste formation are clear from the example given above. Thus, any process that puts in higher mass flows than the target values implies:

- The formation of waste/by-products which may not be desirable
- Results in the wastage of feed material, which implies more handling i.e. larger equipment and more separation steps. It also results in increased operating cost since more feed material will be required, additional energy will be required to process feed material and for separation.

Energy Balance for Synthesizing Processes

The first law of thermodynamics or the energy balance states that the energy flows entering and leaving a system must be equal at steady state. Energy flows can be in the form of heat or work. The first law (energy balance) is the most common method of assessing or analyzing the energy requirements and energy efficiency of processes. The energy balance can be applied to individual units as well as *entire processes*.

Consider a flow process at steady state as shown in Figure 4

The energy balance is given as follows:

$$\Delta H + \tfrac{1}{2} \Delta u^2 + g\Delta z = \sum Q + \sum W_s \qquad (1)$$

where

ΔH is the difference in enthalpy of the output and input streams
$\quad (\Delta H = m_e H_e - m_i H_i)$
Δu is the difference in the velocity of the output and input streams (kinetic energy)
Δz is the difference in height of the output and input streams relative to a reference plane (potential energy)
g is the gravitational constant

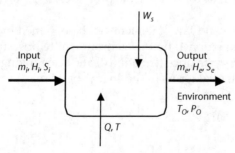

FIGURE 4
General flow process at steady state.

Q refers to the heat flow in or out of the process, (a positive value would mean that heat is required whereas a negative value indicates that work has to be released from the process)

W_s refers to the shaft work entering or leaving the system

The energy balance can also be applied to a process without the details of the flowsheet or units into consideration. Thus, the energy balance gives an indication of the quantity of energy flowing in the process. It can also be used to obtain a target for the minimum amount of energy required or produced by the overall process. Further details can are given in Patel et al (2007).

Second Law of Thermodynamics

Although the first law can provide one with a target for the minimum energy requirement (in terms of heat and work), it has its limitations in that it does not allow one to differentiate between different forms of energy and does not allow one to determine if the energy is being used efficiently. The use of the second law of thermodynamics, in conjunction with the first law is more useful for *analyzing* chemical processes. It has been successfully applied in analyzing and improving existing flowsheets (Denbigh, 1956; Riekert, 1974).

Consider again the system (flow process at steady state) shown in Figure 4. The entropy balance for the system can be written as follows:

$$\Delta S = \sum \frac{Q}{T} + S_{gen} \tag{2}$$

where

ΔS is the difference in entropy of the output and input streams ($\Delta S = m_e S_e - m_i S_i$)

Q refers to the heat flow in or out of the process, (a positive value would mean that heat is required whereas a negative value indicates that work has to be released from the process)

T is the temperature at which the heat is transferred to the system

S_{gen} refers to the entropy generated in the system (for a reversible process $S_{gen} = 0$, whereas for irreversible processes, $S_{gen} > 0$)

A combined first-second law statement can be obtained by multiplying the entropy balance (equation 1) by the environment temperature, T_O, subtracting it from the energy balance (equation 2) and rearranging. We will neglect the kinetic and potential terms in the energy balance.

$$\Delta H - T_O \Delta S = Q\left(1 - \frac{T_O}{T}\right) + W_S - T_O S_{gen} \tag{3}$$

We define

$$\Delta B = \Delta H - T_O \Delta S$$

where
  ΔB is the difference in the availability function ($B = H - T_O S$) or exergy of the output and input streams. Availability or Exergy can be defined as the maximum amount of work that can be achieved from a system with respect to its environment. Note also that the availability function can also be related to the Gibbs free energy, $G = H - TS$ when $T = T_o$.

  $Q\left(1 - \dfrac{T_O}{T}\right)$ can be considered as the work equivalent of heat

  $T_o S_{gen}$ is referred to as lost work or irreversibility and is equal to zero for reversible processes

If the process is reversible, i.e. $S_{gen} = 0$, then we will define the temperature at which this occurs at to be the Carnot Temperature, T_{carnot}. Substituting these relations into equation 3 gives:

$$\Delta G_{process} = \Delta H_{process}\left(1 - \frac{T_o}{T_{carnot}}\right) \tag{4}$$

Thus, in order for a process to be reversible the temperature, T_{carnot}, of the heat that we add to the process must satisfy the above equation as $\Delta G_{process}$ (minimum work load) and $\Delta H_{process}$ (minimum heat load) are fixed. We call this temperature T_{carnot} because equation 4 is essentially the equation of a heat engine that supplies the work ΔG reversibly from the heat ΔH. One can imagine the process as a heat pump, since heat is supplied from the surrounding at a low temperature and "pumped" up to a higher temperature by the addition of work.

The Carnot temperature is very instructive in terms of giving insights into the work flows of the process. It allows one to determine whether processes are feasible to operate as a single stage, reversible process. If no feasible temperature exists, one can determine whether other methods of supplying work are feasible (Patel et al, 2005).

Structure of Course

These synthesis techniques are offered as part of a senior-level design course. These techniques are taught over half a semester. Students are required to apply these tools to projects chosen from literature. Recent projects include the synthesis of ammonia and Fischer-Tropsch synthesis. A three-day course covering these techniques is also given to industry.

Student's feedback on these techniques was very positive. Some comments made by students include:

- "Very helpful in understanding design concepts; quite different from the normal way of teaching design, interesting"

- "I really enjoyed the course. I felt it was valuable to develop the process from the very basics (mass, energy and work integration) since it helps with the overall understanding of the process and I have learnt a great deal from that"
- "It made me aware of the impact that a trivial decision can have on an overall process. I learnt the true meaning of 'efficiency' and the necessity as an engineer to be responsible in my carrying out of any design process"

Conclusion

A new design approach was introduced which presents a unique and systematic approach to the conceptual design of chemical processes. The approach focuses on the synthesis aspects of chemical engineering design and provides a comprehensive analysis of mass, energy and work flows in a process. The approach allows students to develop a better understanding of developing processes which are efficient and environmentally friendly. The responses from students towards the course content and structure were very favourable.

References

Biegler, L.T., Grossmann, I. E. and Westerberg, A. W. (1997). Systematic Methods of Chemical Process Design, Prentice Hall, New Jersey.

Denbigh, K.G. (1956). "The second law efficiency of chemical processes", Chem Eng Sci., 6 (1), 1–9.

El-Halwagi, M. M., (2006). Process Integration, Academic Press, Amsterdam.

Meeuse, F. M. (2002). On the design of chemical processes with improved controllability characteristics, PhD Thesis, Delft University of Technology, The Netherlands.

Patel B, Hildebrandt D, Glasser D and Hausberger B. (2005). Thermodynamic analysis of processes. 1. Implications of work integration. Ind. Eng. Chem. Res., 44, 3529–3527.

Patel B, Hildebrandt D, Glasser D and Hausberger B, Synthesis of processes from a mass, energy and entropy perspective. (2007). Ind. Eng. Chem. Res., 46, 8756 –8766.

Srivinas B. K. (1996). "An overview of Mass Integration and its application to Process development", Technical Information Series, General Electric Company.

Riekert, L. (1974) The efficiency of energy-utilization in chemical processes. Chem. Eng. Sci., 29 (7), 1613–120.

67

Design and Optimization of the Substrate Geometry for Zinc Sulfide Deposition

Yousef Sharifi and Luke E. K. Achenie[*]
Virginia Polytechnic Institute & State University (Virginia Tech)
Blacksburg, VA 24060

CONTENTS

ABSTRACT Advances in technology demand high volume production of semiconductors. Chemical vapor deposition (CVD) is a standard technique for producing semiconductors when quality and cost matter the most. The substrate geometry has a large effect on the deposition uniformity or rate, which in turn plays a critical role in assessing the performance of the CVD reactor. Unfortunately, in the open literature, not much attention has been paid to the substrate geometry. Instead, standard geometries and those produced via heuristics have been employed. This paper proposes a deposition model with flexible substrate geometry. The optimal geometry is determined through shape optimization, such that a performance criterion (such as deposition uniformity or deposition rate of zinc sulfide) is achieved. Overall the model results compared favorably with experimental data from the open literature for conventional substrate geometries, namely vertical and horizontal slab-shape.

KEYWORDS *Design, optimization, substrate geometry, CVD, zinc sulfide, deposition rate*

---

[*] To whom all correspondence should be addressed

Introduction

Chemical Vapor Deposition (CVD) is the method of choice for large-scale production of semiconductors (Jensen, 1994). Many factors influence the efficiency of a CVD reactor; some of these are operating conditions (Moffat and Jensen, 1988), substrate position, deposition on reactor walls and mixing methods (Goela and Taylor, 1988).

Unfortunately there has not been much research interest in the effect of substrate geometry on the deposition process inside a CVD reactor. This paper explores the effect of substrate geometry on deposition rate, uniformity of deposited films and deposition on both sides of substrate.

Modeling

We consider a two-dimensional (2-D) steady state model with gas phase reaction for the horizontal CVD reactor illustrated in Figure 1. Various substrate geometries are created using 6-ponit Bezier curves (Farin, 1997). We have adopted a zinc sulfide chemistry primarily because of broad application of zinc sulfide (Tran, 2002), simple kinetics, and availability of experimental data to validate the model (Zhenyi and Yichao, 2002, Smith, 1989). Zinc sulfide applications demand high quality and very low impurity. To achieve this, low concentrations (2% in an inert carrier gas such as argon) of zinc and hydrogen sulfide are used in a CVD reactor. A reaction between hydrogen sulfide and zinc produces zinc sulfide particles; these are then deposited on the substrate. This process includes a gas phase reaction together with

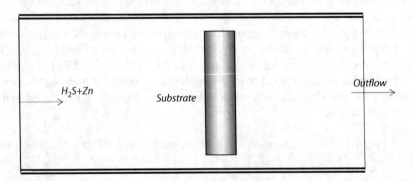

FIGURE 1
Schematic of CVD reactor.

physical surface deposition. The change of volume due to reaction for the diluted reactants is not significant and can be ignored.

The 2-D Fluid dynamics model adopted here includes momentum, mass, and heat balance (Bird et al., 2001). Thermal diffusion has significant effects on the deposition process (Coltrin et al., 1986) and is included in this model. Fluid density is estimated based on the ideal gas law. Other transport properties such as binary diffusion, viscosity, and thermal conductivity are calculated using the Chapman-Enskog correlations (Reid et al., 1987). Finally, a fully developed boundary condition is used at the reactor inlet.

Continuity

$$-(\nabla.\rho u) = 0 \tag{1}$$

Momentum

$$\nabla.\eta(\nabla u + (\nabla u)^T + \rho(u.\nabla)u + \nabla P = F \tag{2}$$

Convection and Conduction

$$\nabla.(-k\nabla T + \rho c_p T u) = 0 \tag{3}$$

Convection and Diffusion

$$\nabla.(-D_i^T \nabla(\ln(T))) - D_i \nabla c_i + c_i u) = R_i \tag{4}$$

where

$$D_i^T = \sum_{j \neq i}^{n} \frac{kc_p^2}{\rho} M_i M_j D_i \tag{5}$$

Here, ρ is the fluid density, and u the velocity vector, η dynamic viscosity, F the force term including gravity, P the pressure, c_p the heat capacity at constant pressure, k the thermal conductivity, c_i the species concentration, D_i^T thermal diffusion coefficient, D_i the binary diffusion coefficient of species i in argon, and R_i the gas phase reaction rate.

The deposition process involves two steps, formation of zinc sulfide in the gas phase (Eq. 6), and deposition of zinc sulfide on the substrate (Eq. 7). For this reaction, the equilibrium rate constant is estimated using the Gibbs free energy (Fogler, 1992).

$$Zn(g) + H_2S(g) \Leftrightarrow ZnS(s) + H_2(g) \tag{6}$$

$$ZnS(s) \Leftrightarrow ZnS(s)_s \tag{7}$$

$$K = \exp\left(\frac{\Delta G}{RT}\right) \tag{8}$$

$$\Delta G = 82.1T - 5.9T(lnT) - 0.62 \times 10^{-3}T^2$$

$$- 76400 \ kcal.mol^{-1} \tag{9}$$

Here ΔG is Gibbs free energy, R is the universal gas constant, and T is temperature in Kelvin. In a typical CVD process, the deposition quality depends on the value of the Gibbs free energy. A Gibbs free energy smaller than 30 *kcal. mol^{-1}* is needed for high quality deposition (Goela and Taylor, 1988). For this reaction, the Gibbs free energy of 36 *kcal. mol^{-1}* at 973 K is indicative of high performance deposition.

The deposition rate of *ZnS* on the substrate approximated as the product of the reaction probability and the effusive flux:

$$r_{ZnS} = \gamma \sqrt{\frac{RT}{2\pi M_{ZnS}}} c_{ZnS} \tag{10}$$

$$\gamma = 5 \times 10^{-2} \exp\left(\frac{-13.4 kcal.mol^{-1}K}{RT}\right) \tag{11}$$

Here, γ is the sticking coefficient of the *ZnS* particles on the substrate, R is the universal gas constant, M_{ZnS} is the molar mass and T is the temperature in Kelvin, c_{ZnS} is gas phase concentration of *ZnS*.

Bezier Curves

A Bezier curve is defined by Bernstein polynomials as (Farin, 1997):

$$B(t) = \sum_{i=0}^{n} B_{n,i} P_i \qquad B_{n,i} = C_n^i t^i (1-t)^{n-i} \tag{12}$$

$$C_n^i = \frac{n!}{i!(n-i)!}$$

Here, $B(t) = x(t)$ or $y(t)$ and $P_i = x_i$ or y_i. Thus the coordinates of the control points are

$$x(t) = \sum_{i=0}^{n} B_{n,i} x_i \quad y(t) = \sum_{i=0}^{n} B_{n,i} y_i \tag{13}$$

In order to obtain several feasible geometries for the substrate, we employed a 6-point Bezier curve.

Results and Discussions

We studied the effect of different substrate geometries in order to (a) study the effect on the deposition rate and (b) identify substrate geometry with uniform deposition rate along the substrate surface. To the best of our knowledge, most of the geometries are not standard geometrical shapes (i.e. not available in the open literature). The results are for the same system but with different substrate geometry. We employed an inlet velocity of 0. $m.s^{-1}$, inlet temperature of 298 K, and substrate temperature of 973 k. For all cases expansion of the gas due to change in temperature/density creates an increase in velocity around the substrate.

The Momentum balance coupled with heat and mass conservation equations are solved using COMSOL MULTIPHYSICS (http://www.comsol.com/). Substrate geometries, velocity streamlines, velocity profiles are embedded in the deposition rate figures. The calculated growth rate is in the range of $0.2 - 1.5\ \mu m.\min^{-1}$ depending on the operating conditions. For a reactor operating at 973K and 4000 Pa, predicted growth rate of $0.5\ \mu m.\min^{-1}$ is in good agreement (i.e. validated) with experimental measurements (Zhenyi and Yichao, 2002, Smith, 1989, Goela and Taylor, 1988). We have shown a few geometries (Figures 2, 3 and 4) in which the Reynolds number varies in the range 10-150 to ensure operation in the laminar flow region. For a conventional substrate, normal to the flow direction (Figure 2), the deposition rate is higher at the edge and decreases towards the center of the substrate (not shown). This difference in deposition rate is mostly due to a change in the boundary layer thickness along the substrate. The boundary layer thickness is smaller at the edge and is higher at the center, this leads to a difference in the deposition rate. With regard to the back deposition case, we observe that the deposition rate on the edge of the substrate is almost 200% higher than the deposition rate at the center of the substrate. This difference in deposition rate along the substrate affects the uniformity of the deposited materials. A bowl-shaped substrate geometry (Figure 4) led to the highest deposition rate and the most uniform deposition (Figure 5).

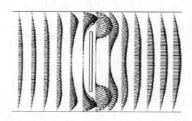

FIGURE 2
Streamlines for vertical substrate.

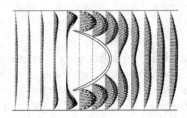

FIGURE 3
Streamlines for convex substrate.

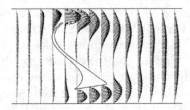

FIGURE 4
Streamlines for Bowl-shaped substrate.

Conclusions

This paper proposes a deposition model with flexible substrate geometry. The optimal geometry is determined through shape optimization, such that a performance criterion (such as deposition uniformity or deposition rate of zinc sulfide) is achieved.

The above ideas are expanded as follows. In the two-dimensional substrate model, the geometry was represented using Bézier curves. The control points on the Bézier curves formed a subset of the design variables within the

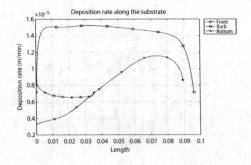

FIGURE 5
Front and Back Deposition rates for Bowl-shaped.

optimization (i.e. nonlinear programming) framework. The latter included a multi-physics model in which the two-dimensional fluid flow was considered to be symmetric along the vertical axis.

The key conclusion from the numerical simulations is that due to flow behavior inside a CVD reactor, at identical operating conditions, shape geometry plays an important role in the deposition process. Specifically, changing the shape of the substrate created a special flow pattern that significantly affected the boundary layer thickness and mass transfer around the substrate. Flow recirculation improved the mass transfer, and flow along the substrate lowered the boundary layer thickness, these together enhanced the deposition rate. Overall the model results compared favorably with experimental data from the open literature for conventional substrate geometries, namely vertical and horizontal slab-shape. In a broader context, the suggested framework can (and is being used by us) for treating other processes such as carbon nanotube formation. A full paper based on our simulations can be found in Sharifi and Achenie (2007).

Nomenclature

R	the universal gas constant
u	velocity vector
T	temperature in Kelvin
ΔG	Gibbs free energy
F	the force term including gravity
P	the pressure
c_p	heat capacity at constant pressure
k	thermal conductivity
D_i^T	thermal diffusion coefficient
c_i	species concentration
R_i	gas phase reaction rate
$B(t)$	Bernstein polynomials
C_n^i	Bernstein coefficient
$B_{n,i}$	Bernstein coefficient
P_i	control points for Bernstein polynomials
n	Bezier curve order
x_i, y_i	Bezier curve coordinates
η	dynamic viscosity
ρ	fluid density
γ	sticking coefficient of the ZnS particles on the substrate
D_i	binary diffusion coefficient of species i in argon
K	equilibrium rate constant
M_{Zns}	molecular weight

c_{Zns} gas phase concentration of ZnS
r_{Zns} deposition rate of Zns
M_i molecular weight of species i

References

Bird, R.B., Stewart, W.E. and Lightfoot, E. N. (2001). Transport Phenomena. *Wiley*.

Coltrin, M. E., Kee, R.J. and Miller, A. (1986), A Mathematical Model of Silicon Chemical Vapor Deposition. *J. Electrochemical Soc, 133, 1206–1213.*

Farin, G. (1997), Curves and Surfaces for Computer-Aided Geometric Design. *Academic Press Inc.*

Fogler, H. S. (1992). Elements of Chemical Reaction Engineering, *Prentice-Hall*, NJ.

Goela, J .S. and Taylor, R. L. (1988), Monolithic Material Fabrication by Chemical Vapour Deposition, *J. Mater. Sci., 23, 4331–4339.*

Jensen, K. F. (1994). Transport phenomena in epitaxy systems, handbook of crystal growth. *[ed.] D Hurle. Amsterdam : Elsevier, Vol. 36.*

Moffat, H. K. and Jensen, K. F. (1988). Three-Dimensional Flow Effects in Silicon CVD in Horizontal Reactors., *J. of Electrochemical Society, 135, 459–471.*

Reid, R. C., Prausnitz, J. M. and Poling, B. E. (1987). The properties of gas and liquids, *McGraw Hill* , NY.

Sharifi, Y. and Achenie, L.E.K. (2007). Effect of substrate geometry on deposition rate in CVD, *Journal of Crystal Growth, 304, 520–525.*

Smith, P. B. (1989). Preparation and characterization of ZnS thin films produced by metalorganic chemical vapor deposition. *Journal of Vacuum Science & Technology A: Vacuum, Surfaces, and Films , 7, 1451–1455.*

Tran, P. T. (2002), Use of Luminescent CdSe-ZnS Nanocrystal Bioconjugates in Quantum Dot-Based Nanosensors. *Naval Research Lab Washington DC center for Biomolecular Science and Engineering, A637744.*

Zhenyi, F and Yichao, C. (2002). CVD growth of bulk polycrystalline ZnS and its optical properties. *Journal of Crystal Growth, 237, 1707–1710.*

68

An Integrated Approach to Chemical Engineering Undergraduate Curriculum Reform

Lale Yurttas[1], Larissa Pchenitchnaia[1], Jeffrey Froyd[1],
Mahmoud El-Halwagi[1], Charles Glover[1], Irvin
Osborne-Lee[2], Patrick Mills[3], Ali Pelehvari[3]

[1]Artie McFerrin Department of Chemical Engineering, Texas A&M University (TAMU)
[2]Department of Chemical Engineering, Prairie View A&M University (PVAMU)
[3]Chemical and Natural Gas Engineering Department,
Texas A&M University Kingsville (TAMUK)

CONTENTS

ABSTRACT Sponsored by the National Science Foundation (NSF), three chemical engineering departments within the Texas A&M University System have made significant progress in their efforts to renew their curricula to address pressures of multi-disciplinary technological developments and the growing breadth of abilities and knowledge areas expected for competitive chemical engineering graduates. Three broad strategies to accomplish project goals have been implemented: (1) curriculum content reform and development; (2) student assessment activities, and (3) faculty development initiatives.

The three strategies are being implemented through six key mechanisms:

(i) Identifying and organizing curriculum development activities around four course strings to improve integration of learning outcomes and activities.

(ii) Developing interlinked curriculum components to organize and reinforce core ideas in chemical engineering curricula.

(iii) Using service learning in required chemical engineering courses.

(iv) Integrating comprehensive assessment plans and processes throughout the chemical engineering curriculum and using the data to make and evaluate changes.

(v) Offering faculty development activities to offer knowledge and development opportunities for chemical engineering faculty members.

(vi) Sharing our experiences with audiences beyond the Texas A&M University System.

In the course of the project, faculty incorporated innovative pedagogies, such as teams, cooperative learning, and project based learning. Faculty in a community-centered effort have developed outcomes and expectations for incoming students in the courses that comprise the four course strings. Web-based modules that span courses and application areas have been developed and are being integrated into core courses. Consistent conversations about continuous improvement and assessment plan engaged faculty in a collective effort for sustained change. This paper will discuss the outcomes and experiences of the three-year, NSF-sponsored project to reform chemical engineering undergraduate curricula.

KEYWORDS: *Curriculum, reform, education, chemical engineering*

Introduction

In a time of rapid change, academic programs must experiment and evolve in order to keep pace with advances in knowledge, changes in professional practice, and shifting conditions in society. The need for responsive academic programs is particularly a concern in scientific and technological fields where the growth of knowledge is exponential. Three chemical engineering departments at TAMU, PVAMU, and TAMUK are continuing their efforts to restructure their four-year undergraduate curricula to overcome the following limitations:

a) Gap between fundamentals and applications; especially in the emerging areas of chemical engineering

b) Low emphasis on solving open-ended synthesis tasks; especially for complex problems

c) Lack of a formal framework for effective integration of the various courses to avoid unnecessary redundancy and to link the outcomes of one course with the required concepts, knowledge, and skills for another

d) Ineffective usage of assessment tools and outcomes

e) Reluctant participation of the faculty as a whole in concerted discussions that cover the broad aspects of the curriculum

The project is designed and implemented to achieve four objectives. Students will be able to:

a) Apply fundamental ideas in chemical engineering over a greatly expanded range of time and length scales;

b) Apply ChE fundamental ideas to emerging application areas;

c) Construct solutions for more complex, more open-ended synthesis tasks; and

d) Transfer fundamentals and knowledge to novel challenges.

Three major strategies for project implementation include (1) curriculum content reform and development; (2) student assessment activities; and (3) faculty development initiatives. The three strategies are being implemented through six key mechanisms:

(i) Identifying and organizing curriculum development activities around four course strings to improve integration of learning outcomes and activities

(ii) Developing interlinked curriculum components to organize and reinforce core ideas in chemical engineering curricula

(iii) Using service learning in required chemical engineering courses

(iv) Integrating assessment plans and processes throughout the chemical engineering curriculum

(v) Offering faculty development activities to offer knowledge and development opportunities for chemical engineering faculty members

(vi) Implement dissemination to share our experiences with an audience beyond Texas A&M University

Curriculum Reform

Beginning with major curriculum reform initiatives in the late 1980s (e.g., Froyd & Rogers, 1997), engineering curriculum renovation has generated significant conversations. Major contributions to these conversations were made during the lifetimes of the NSF Engineering Education Coalitions (Borrego, 2007; Froyd, 2005). Some of the more lasting of the innovations to emerge from these and other curriculum reforms efforts include first-year engineering courses that emphasize engineering design processes (e.g., Mara et al., 2000), interdisciplinary capstone design courses (Ollis, 2004), and student-active pedagogies such as cooperative learning e.g, Smith et al., 2005) and inquiry-guided learning (Lewis and Lewis, 2008; McDermott, 1996). Armstrong (2006) proposed a vision for future chemical engineering education which is structured around three core principles: molecular transformation, multi-scale description, and systems analysis and synthesis. Other promising approaches include student learning communities (e.g., Froyd and Ohland, 2005; Taylor, Moore, MacGregor, & Lindblad, 2003) and service learning (Eyler & Giles, 1999). As demonstrated across the NSF Engineering Education Coalitions, a major challenge is not initiating curricular reform, but institutionalizing the reform for the majority of the students on a sustainable basis (Clark, Froyd, Merton, & Richardson, 2004; Colbeck, 2002). With the end of the engineering coalitions, the National Science Foundation moved to funding departments to support major redesigns of their curricula. The department-level reform project in the Department of Chemical Engineering at Texas A&M University is building on practices generated by previous curriculum reform initiatives and working to involve the entire department in contributing to and applying the ideas generated in the project (Yurttas, Christensen et al., 2007; Yurttas, Kraus et al., 2007). Details on each of the three strategies are provided in the following sections.

Course Strings

The first key strategy for curriculum reform and development involves organizing undergraduate ChE courses into four course strings: thermodynamics and kinetics; emerging fundamentals and applications; transport phenomena; and systems design. Course string faculty committees were developed to address the following key issues: (1) what must undergraduate engineers learn/accomplish in the course string to be successful throughout their academic career and in the next generation professional settings; (2) what obstacles exist to providing the necessary educational experiences, and (3) how can we effect change and what changes (integration) need to be made to an existing curriculum. Course string faculty committees continue to hold regular meetings every semester to address these questions. Syllabi analysis provided invaluable information to enhance the alignment of the courses. As a result of course string faculty committees' working sessions, the department faculty had an opportunity to discuss undergraduate curriculum in depth and to affect the following changes:

1. **Integrated outcomes for all chemical engineering courses:** All undergraduate course syllabi were updated to include revised course outcomes. These outcomes were developed to insure continuity among the courses so as to deliver the overall curriculum outcomes.

2. **New assessment techniques:** The department implemented end-of-semester student and faculty course evaluations to assess students' achievement of course outcomes.

3. **Course portfolios:** One of the main curriculum management tools established as a result of course string committees' work is use of course portfolios. In addition to helping individual faculty members analyze achievement of student learning outcomes, course portfolios provide helpful reference documents for other faculty teaching the same course or related courses in a sequence by providing a detailed record of approaches and outcomes.

Interlinked Curriculum Components

Interlinked curriculum components (ICCs) are web-based learning sites for students that may address new technologies, non-traditional applications, and even common foundations that span all courses. ICCs can be used by students to review concepts and applications, to learn new applications, and to develop an appreciation and understanding of the common threads and methods of the various courses. Thus, the ICCs are envisioned as an integrating tool that will help students see the collection of courses in their program as a unified curriculum. The ICCs also allow faculty to see presentations of the topics and work towards better unification to their discussions. Currently, eleven faculty members from the three departments along with their students and associates are working on interlinked curriculum components implementation. An ICC coordination committee was formed to coordinate the progress reports from ICC coordinators.

The ICCs that are being developed address the following topics: conservation principles; materials; system synthesis and integration; microchemical systems; molecular modeling; and environment and sustainability. ICC development may be viewed at http://che.tamu.edu/orgs/NSFCR/ or at the new site http://ALChemE.tamu.edu/.

Service Learning Activities

Service learning increases retention and student interest in engineering through projects that apply engineering principles to real problems and engages students in social responsibility. To enrich student experiences, service learning has been implemented in the first ChE sophomore-level course by a collaborative student and faculty effort. The project was given several times to the introductory level material and energy balances classes during fall 2006, spring 2007, fall 2007. The classes were presented with a project agreed upon through collaboration with the department and Habitat for Humanity. The projects were to design a "green" home focusing on conservation aspects such as energy, water and waste. The overall process included four major stages: formulation of the project, project promotion, designing and project completion, and project reflection.

Adapting the project to past classes' suggestions, the fall 2008 class was assigned a more focused project to investigate the environmental impacts of compact fluorescence light bulbs and incandescent light bulbs through life cycle assessment for the City of College Station. The current project is smaller in scope; however, the project still meets all the objectives and goals espoused above (Christensen & Yurttas, 2009).

Assessment Plan at College Station

The project assessment plan includes the following components:

- Project Objectives
- Project Outcomes
- Implementation Strategies
- Implementation Activities
- Assessment Methods
- Assessment Metrics

The plan focused on three categories: students, curriculum, and faculty.

Project Outcomes

The project outcomes are organized by students, curriculum, and faculty. They provide more detail about the aims the project hopes to achieve in these three areas.

1. ChE Students Will be Able to

1.1. Apply conservation laws framework to spatial and temporal scales ranging from micro to macro

1.2. Design a system, component, or process in emerging applications (bio, nano, semiconductor)

1.3. Identify, formulate, and solve open-ended problems

1.4. Apply chemical engineering education to address global and societal problems

1.5. Integrate multiple components of a system

1.6. Synthesize flowsheets to address technical, economic, environmental and energy objectives

1.7. Use information technologies to learn and apply chemical engineering concepts

2. Our New Curriculum will be Designed to

2.1. Accomplish student learning outcomes 1 through 6

2.2. Eliminate unnecessary repetition and less applicable subject matter

2.3. Integrate content and ensure continuity

2.4. Be adoptable/adaptable by other ChE programs

2.5. Use an approach that revisits topics at successively higher conceptual levels to help students develop and reinforce a conceptual framework and fundamental concepts

3. ChE faculty will be Able to

3.1. Use information technologies and science of learning principles to provide structure, conceptual materials, applications, and reinforcement

3.2. Assimilate new research on assessment, learning and teaching that will build on and enhance their knowledge of new technologies, curriculum content and reform

3.3. Integrate their courses content and outcomes with preceding and subsequent courses

3.4. Eliminate unnecessarily repetitious materials

3.5. Maintain consistency and continuity in courses they teach and continuously improve the teaching quality

3.6. Develop assessment strategies for curriculum reform and development components that align with learning outcomes and current research on the science of learning

Conclusions

Renewing an entire chemical engineering curriculum has proven to be a daunting undertaking for several reasons. First, the project team has found that there is a lack of applicable assessment tools and processes with which it can evaluate the abilities of chemical engineering undergraduates with respect to the learning outcomes. Although the project team has found some relevant assessment tools, there are many learning outcomes for which the project team, together with the entire department, has had to create assessment tools and processes, instead of using proven or promising alternatives. Second, engaging faculty members across the three chemical engineering departments has been more challenging than expected. Finding time and support to engage groups of faculty members in productive conversations about desirable learning outcomes, evaluating the extent to which students are achieving this outcomes, and constructing instructional materials, e.g., ICCs, to support instruction, especially in non-traditional subject areas, has taken longer than anticipated. Faculty members, in spite of their commitment to undergraduate education, are juggling multiple, competing responsibilities. In spite of these two challenges, the chemical engineering department has created an operational assessment plan that is being applied to analyze the learning and development of chemical engineering students with respect to a forward-looking set of learning outcomes. Further, the project has developed ICCs that will be useful in supporting the move to a renewed curriculum. These resources will be available to other chemical engineering departments. Hopefully, these will be lasting contributions to chemical engineering education.

Acknowledgments

The authors would like to acknowledge support from the National Science Foundation under Grant No. EEC-0530638. Any opinions, findings, conclusions or recommendations expressed in this material are those of the authors and do not necessarily reflect the views of the National Science Foundation.

References

Armstrong, R.C. (2006) 'A vision of the curriculum of the future', Chemical Engineering. Education, Vol. 40, No. 3, Spring, pp.104–109.
Borrego, M. (2007). Development of Engineering Education as a Rigorous Discipline: A Study of the Publication Patterns of Four Coalitions. *Journal of Engineering Education*, 96(1), 5–18.

Christensen, J., & Yurttas, L. (2009). *Service-learning and Sustainability: Striving for a Better Future.* To be presented in ASEE Annual Conference and Exposition.

Clark, M., Froyd, J., Merton, P., & Richardson, J. (2004). The evolution of curricular change models within the Foundation Coalition. *Journal of Engineering Education,* 93(1), 37–47.

Colbeck, C. L. (2002). Assessing Institutionalization of Curricular and Pedagogical Reforms. *Research in Higher Education,* 43(4), 397–421.

Eyler, J., & Giles, D. E., Jr. (1999). *Where's the learning in service-learning?* San Francisco, CA: Jossey-Bass.

Froyd, J. E. (2005). The Engineering Education Coalitions Program. In *Educating the Engineer of 2020: Adapting Engineering Education to the New Century.* Washington, DC: National Academies Press.

Froyd, J. E., & Ohland, M. (2005). Integrated engineering curricula. *Journal of Engineering Education,* 94(1), 147–164.

Froyd, J. E., & Rogers, G. J. (1997). *Evolution and Evalution of an Integrated, First-Year Curriculum.* Paper presented at the Frontiers in Education Conference. Retrieved December 1, 2008, from http://fie-conference.org/fie97/papers/1102.pdf

Lewis, S. E., & Lewis, J. E. (2008). Seeking Effectiveness and Equity in a Large College Chemistry Course: An HLM Investigation of Peer-Led Guided Inquiry. *Journal of Research in Science Teaching,* 45(7), 794–811.

Marra, R. M., Palmer, B., & Litzinger, T. A. (2000). The Effects of a First-Year Engineering Design Course on Student Intellectual Development as Measured by the Perry Scheme. *Journal of Engineering Education,* 89(1), 39–45.

McDermott, L. C. (1996). *Physics By Inquiry.* New York: John Wiley & Sons.

Ollis, D. F. (2004). Basic Elements of Multidisciplinary Design Courses and Projects. *International Journal of Engineering Education,* 20(3), 391–397.

Smith, K. A., Sheppard, S. D., Johnson, D. W., & Johnson, R. T. (2005). Pedagogies of engagement: classroom-based practices. *Journal of Engineering Education,* 94(1), 1–16.

Taylor, K., Moore, W. S., MacGregor, J., & Lindblad, J. (Eds.). (2003). *Learning Community Research and Assessment: What We Know Now.* Olympia, WA: The Evergreen State College, Washington Center for Improving the Quality of Undergraduate Education, in cooperation with the American Association for Higher Education.

Yurttas, L., Christensen, J., Haney, J. S., El-Halwagi, M., Froyd, J. E., & Glover, C. (2007). *Enhancement of Chemical Engineering Introductory Curriculum through Service-Learning Implementation.* Paper presented at the ASEE Annual Conference & Exposition. Retrieved December 5, 2008, from http://papers.asee.org/conferences/paper-view.cfm?id=4246

Yurttas, L., Kraus, Z., Froyd, J. E., Layne, J., El-Halwagi, M., & Glover, C. (2007). *A Web-based Complement to Teaching Conservation of Mass in a Chemical Engineering Curriculum.* Paper presented at the ASEE Annual Conference & Exposition. Retrieved December 5, 2008, from http://papers.asee.org/conferences/paper-view.cfm?id=3510

69

Interval Constraint Satisfaction Model for Energy and Energy-Based Emissions Targeting under Varying Process Conditions

*Mahmoud Bahy M. Noureldin, Mana M. Al-Owaidh and Majid M. Al-Gwaiz

mahmoudbahy.noureldin@aramco.com, mana.owaidh@ aramco.com and majid.gwaiz@aramco.com
Process and Control Systems Department, Saudi Aramco Dhahran, Saudi Arabia

CONTENTS

ABSTRACT In this paper a new interval based constraint satisfaction model for simultaneous energy and energy-based emissions targeting under varying process conditions is introduced. The model calculates minima and maxima energy utility targets and energy-based GHG emissions under all possible combinations of process modifications without exhaustive enumeration.

KEYWORDS Systematic targeting, heat integration, constraint satisfaction, energy-based GHG emissions

Introduction

Industrial communities worldwide are ravenous to produce more-with-less through cost effective but benign new products and processes. The chemical process industry has a reputation of being one of the most energy-intensive manufacturing sectors in the industrial community. Many techniques emanated in the early seventies for systematic heat integration have been in use ever since in the industrial community and are now gaining more momentum. Even small improvements to the design of new processes that can reduce the energy consumption and GHG emissions are being adopted. The process design and operation alternatives for chemical processes are cumbersome and the problem of optimizing the different process structures, design parameters and operating conditions is combinatorial.

Constraint Satisfaction (CS), known to be a little younger than mathematical programming, is an optimization field that is consistently gaining ground in reducing the computational effort needed to solve combinatorial optimization problems. A crucial component in the Constraint Satisfaction approach is the Constraint Propagation.

Constraint propagation is an efficient method for reducing the search space of combinatorial search and optimization problems and has become more popular in the last two decades. The basic idea of constraint propagation methods is to detect and remove inconsistent variable assignments that do not participate in any feasible solution throughout the repeated analysis and evaluation of the variables, domains and constraints describing a specific problem instance.

Constraint Logic Programming (CLP), sometimes called Propagation, is a technique used in Constraints Satisfaction applications. CLP is a unique problem-solving paradigm that establishes a clear distinction between two pivotal aspects of any problem; a precise definition of the constraints that define the problem, and the algorithms and heuristics used to solve it. CLP is increasingly being used as a problem-solving tool for many engineering problems. It was originally developed for solving CS feasibility problems but has recently been used in optimization problems too.

The HEN synthesis problem under all possible combinations of process parameters modifications is a relatively large problem. Upon decomposition to targeting and synthesis sub-problems, the HEN synthesis problem can be easily tackled using CS to find the heating and cooling utilities minima and maxima targets under all possible combinations of process changes.

In this paper, a new interval based CS model and a case study for simultaneous energy and energy-based emissions targeting under varying process conditions are introduced. The model calculates minima and maxima energy utility targets and energy-based GHG emissions under all possible combinations of fuzzy process conditions.

Interval Constraint Propagation

Interval arithmetic deals with processing intervals that bound real numbers. Consider a real variable, x, bounded by two other numbers, $xl \leq x \leq xu$. One can define an interval X such that $x \in X$ where $X = [xl, xu]$. Similarly, an interval Y can be defined to include a real variable y such that $y \in Y$. An interval arithmetic operation,*, (for example, addition, subtraction, multiplication, and division) is defined by:

$$X^*Y = \{x^*y: x \in X, y \in Y\} \tag{1}$$

A particularly useful property of interval arithmetic operations is closure under the * operation:

$$x^*y \in X^*Y, \tag{2}$$

which means that the sum, difference, product, and quotient of two real numbers belongs to the original intervals. Rules for interval operations are defined by:

$$X^*Y = [xl, xu] * [yl, yu] = [\min\{x^*y: x \in X \& y \in Y\},$$
$$\max\{x^*y: x \in X \& y \in Y\}] \tag{3}$$

For our specific arithmetic operations,

$$X + Y = [xl, xu] + [yl, yu] = [xl + yl, xu + yu], \tag{4}$$

$$X - Y = [xl, xu] - [yl, yu] = [xl - yu, xu - yl], \tag{5}$$

$$X\,Y = [xl, xu]\,[yl, yu] = [\min(xl\,yl, xu\,yu, xl\,yu, xu\,yl),$$
$$\max(xl\,yl, xu\,yu, xl\,yu, xu\,yl)], \text{ and} \tag{6}$$

$$X / Y = [xl, xu] / [yl, yu] = [xl, xu]\,[1/yu, 1/yl] \quad \text{if } 0 \notin \text{ to } [yl, yu]. \tag{7}$$

For each continuous function, $f(x)$, where x is an n-dimensional vector, X is an n-dimensional vector space, and $x \in X$, one can use interval arithmetic to identify bounds on the range of the function. Consider a function $f(x)$ whose range over interval X is defined as $f(X)$, i.e. $f(X) = \{f(x): x \in X\}$.

An interval function F is called an inclusion function for f over interval X if

$$f(X) \subseteq F(X). \tag{8}$$

Inclusion functions are extremely important in interval analysis as they provide bounds on ranges without exhaustive enumeration. There are two common methods for constructing inclusion functions: natural interval extensions and centered forms. A natural interval extension is an expression where each x in the various terms of $f(x)$ is replaced with its domain, X, and the mathematical operators are replaced with interval operations. Centered forms are inclusion functions that represent a generalization of the algebraic centered forms for real variables. A particularly useful centered form is based on the natural interval extension of Taylor's expansion of the function, such as Ratscheck and Rokne (1984) and Moore (1988).

Constraint Satisfaction Using CLP

Constraint logic propagation known as CLP is a problem-solving paradigm that establishes a clear distinction between two pivotal aspects of a problem: a precise definition of the constraints that define the problem, and the algorithms and heuristics that enable the selection of the decisions to solve the problem. Because of these capabilities, constraint programming is becoming a popular tool for solving many engineering problems [4,5]. Constraint programming was originally developed for solving feasibility problems but has recently been used in optimization problems (ILOG Inc., 1999). Although constraint programming methods are efficient in solving feasibility/targeting problems, optimization entirely depends on building the correct constrained model. To show how constraint logic propagation with interval labels works, this paper gives numerical examples, the first of which is the "Waltz" algorithm, first introduced in the mid seventies (Waltz, 1975).

Suppose the following relations model/define our problem:

$$x + y = z, \quad y \leq x$$

and we can start with the following bounds:

$$x \in [1,10], \quad y \in [3,8], \quad z \in [2,7],$$

This model can be implemented in a data structure given by the following constrained network:

The famous "Waltz" algorithm proceeds as follows:

- The constraint queue begins with both constraints (CON1, CON2).
- CON1 $(x + y = z)$ is popped from the queue.
- Since $x \geq 1$ and $y \geq 3$, CON1 gives $z \geq 4$; therefore reset the bounds of z to [4, 7].
- Since $z \geq 7$ and $y \geq 3$, CON1 gives $x \leq 4$; therefore reset the bounds of x to [1, 4].

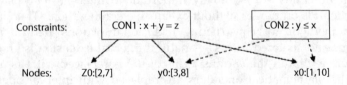

FIGURE 1
Network constraints.

- Since x and z have been changed, add CON2 to the queue.
- CON2 (y ≤ x) is popped from the queue.
- Since x ≤ y, CON2 gives y ≤ 4; therefore reset the bounds of y to [3, 4].
- Since y ≥ 3, CON2 gives x ≥ 3; therefore reset the bounds of x to [3, 4].
- Since x and y have changed, add CON1 to the queue.
- CON1 (x + y = z) is popped from the queue.
- Since x ≥ 3, y ≥ 3, CON1 gives z ≥ 6; therefore reset the bounds of z to [6, 7].
- Since only z has changed and z has no other constraints beside CON1, nothing is added to the queue.
- Since the queue is empty, the algorithm terminates.

This simple numerical example demonstrates how to use constraint logic propagation techniques in solving engineering problems that can be modeled using constraints and interval labels consisting of deterministic bounds (Davis, 1987).

A system that includes equality constraints (such as mass, heat balances and performance equations) and inequalities constraints (such as the ones including operating windows and design parameters and logical expressions constraints of superstructures and/or operating philosophies) can be easily modeled in the CS environment as shown below (Russian Institute of Artificial Intelligence, 1999):

$$X^2 + 6.0x = y - 2^k$$
$$kx + 7.7y = 2.4$$
$$(k-1)^2 < 4$$
$$(\ln(y + 2x + 12) < (k + 5)) \quad \text{or} \quad (y > k^2) \rightarrow (x < 0.0) \text{ and } (y < 1)$$

Where k is integer; x, y are real and → stands for implication.

The problem above can exhibit several solutions that all lie in the following intervals.

$$k = [0, 2],$$
$$x = [-6, -1e-10] \text{ and}$$
$$y = [0.311688, 1].$$

The simultaneous HEN synthesis and process operations optimization problem can be represented as a three-level optimization problem; minimize area cost subject to minimum number of units subject to minimum utility cost, or in two levels (BLPP) minimize capital cost subject to minimum energy and GHG emissions cost.

Model Description and CS Formulation

Hot stream temperatures are shifted down sequentially based on the first set of desired minimum temperature differences, $\Delta T_{min}{}^i$, between the hot and cold resource streams to form a set of possible discrete temperature values for a continuum of possible values for $\Delta T_{min}{}^i$. $\Delta T_{min}{}^i$ represents the ΔT_{min} of hot stream (i) which refers to the minimum temperature approach between a specific hot stream and all other cold streams. In our examples, we will define $\Delta T_{min}{}^i = 1°F$ for all streams i. The shifted supply and target temperatures of resource hot streams, and the actual supply and the target cold stream temperatures are then sorted in a descending order, with duplicates removed. In this formulation, each successive temperature pair represents the boundaries of a temperature step and defines a new temperature step "s". Note that each supply and target temperature input is in the form of intervals (e.g., a pair of range boundaries) rather than single discrete numbers.

The total number of temperature steps is "$N + 1$", where $s \in \{0,1,2,...,N\}$, and the temperature step number "0" represents the external energy utility temperature step. In this temperature step, known as the external energy utility step, we initialize the energy output to zero by setting $Q_{s=0}{}^{low\_output} = Q_{s=0}{}^{high\_output} = 0.0$ "energy units".

Each temperature step, "s" greater than 0, where $s \in \{1,2,...,N\}$, has energy surplus, $Q_s{}^{surplus}$, and energy output, $Q_s{}^{output}$. These two variables also have lower and upper limits, which are defined as follows (for $s \in \{1,2,...,N\}$):

$$Q_s^{low\_surplus} = \left(\sum_{k=1}^{n_S} FCp_k^{low} - \sum_{j=1}^{m_S} FCp_j^{high} \right)(Th_s - Tc_s), \tag{9}$$

$$Q_s^{high\_surplus} = \left(\sum_{k=1}^{n_S} FCp_k^{high} - \sum_{j=1}^{m_S} FCp_j^{low} \right)(Th_s - Tc_s), \tag{10}$$

$$Q_s^{low\_output} = Q_{s-1}^{low\_output} + Q_s^{low\_Surplus} \quad \text{and} \tag{11}$$

$$Q_s^{high\_output} = Q_{s-1}^{high\_output} + Q_s^{high\_Surplus} \tag{12}$$

Where n_s and m_s are the number of the resource hot and cold streams represented in the sth temperature step respectively, Th_s is the higher shifted temperature for the hot streams, and Tc_s is the lower shifted temperature for the cold streams. Th_s and Tc_s represent the temperature boundaries. Logic propositions such as "or, and,$\rightarrow$" are used as logical operators for the model.

$[FCp_k]$ and $[FCp_j]$ are heat capacity flowrate intervals with the following definitions:

FCp^{low}_k : is the lower bound of the Heat Capacity Flowrate term resulting from the multiplication of the value of the flow's lower bound by the specific heat value Cp of the hot stream number k in flow-specific heat units.

FCp^{high}_k : is the higher bound of the Heat Capacity Flowrate term resulting from the multiplication of the value of the flow's upper bound by the specific heat value Cp of the hot stream number k in flow-specific heat units.

FCp^{low}_j : is the lower bound of the Heat Capacity Flowrate term resulting from the multiplication of the value of the flow's lower bound by the specific heat value Cp of the cold stream number j in flow-specific heat units.

FCp^{high}_j : is the higher bound of the Heat Capacity Flowrate term resulting from the multiplication of the value of the flow's upper bound by the specific heat value Cp of the cold stream number j in flow-specific heat units

The energy and energy-based emissions targeting model takes the following general form:

$$\sum_{k \in Hs}[FCp_k] - \sum_{j \in Cs}[FCp_j] + [Qh_s] - [Qc_s] - [Q^{output}_s] = 0 \quad \forall\ S \in \{0,1,2,...,N\} \quad (13)$$

$$[M(ng)] = [Q\_consumed] \div \{Hv(ng) \times \eta\} \quad (14)$$

$$[M\_CO_2] = \frac{[M(ng)] \times (wt\%\_C) \times (MW\_CO_2)}{(MW\_C)} \quad (15)$$

$$[Q^{surplus}_0] = 0.0; [Q^{output}_0] = 0.0 \quad (16)$$

$$\text{Non-negativity interval constraints} \quad (17)$$

where
  $[M(ng)]$ = Natural gas consumption interval (ton/y)
  $[Q\_consumed]$ = Thermal load consumed for both heating and cooling (MW)
  $[\eta]$ = Heater/Electricity generation efficiency interval
  $[Hv(ng)]$ = Heating value of fuel
  $[M\_CO_2]$ = CO_2 emissions interval (ton/year)
  $[M(ng)]$ = Natural gas consumption interval (ton/year)
  $wt\%\_C$ = Carbon content in natural gas
  MW_CO_2 = Molecular weight of carbon dioxide
  MW_C = Molecular weight of carbon

TABLE 1

Heat Integration Problem Data

Stream	Ts (°C)	Tt (°C)	FCp(kW/°C)
H1	400	320	[2.0 : 3.0]
H2	370	320	[3.5 : 4.0]
C1	300	420	[1.8 : 2.2]
C2	300	370	[1.7 : 2.2]

Case Study

The problem data for the simultaneous energy and GHG emissions targeting study is shown in the table above.

A furnace is used to deliver the required process heating with an efficiency of about 65%, of-course an efficiency range could be easily inserted in the model, and a cooling water system is available to deliver the cooling duty required. Natural gas is used as fuel to the furnace with assumed carbon percentage of 71% and a heating value of 11500 MM cal/ton. In this case study only FCps are given in interval form. In industrial application not only the streams FCps are in interval form but also both supply and target temperatures will be in interval forms too.

After formulating the targeting problem as a constraint satisfaction problem, we solve the problem in the ILOG environment or using any other constraint satisfaction-based software. We obtain the minima heating and cooling energy consumption, the energy-based GHG emissions maxima heating and cooling energy consumption, and the GHG emissions without enumeration. In this case study we are assuming negligible GHG emissions due to energy consumed in cooling water pumps. Refer to (Noureldin et al., 1999–2008) and (Russian Institute of Artificial Intelligence, 1999).

The global minima heating and cooling duties upon applying heat integration concept among the problem hot and cold streams using 10°C as a minimum approach temperature are 54 and 11 kW respectively. Another important result from the problem constraints satisfaction under the given conditions is the identifications of the global maxima heating and cooling duties. These values are 94 and 159 kW respectively. The global minimum and global maximum of energy-based GHG emissions, again upon neglecting GHG emissions due to energy consumed in cooling water pumps are 52.16 and 90.8 ton/y respectively.

It is instructive to mention here that the global minimum heating, cooling and energy-based GHG emissions targets are rarely realized simultaneously and hence, the optimal process conditions that result in global minimum heating utility consumption and consequently global minimum energy-based GHG emissions, neglecting GHG due to operation of cooling water pumps are different than the ones that render global minimum cooling utility consumption.

Conclusion

Energy systems utilities consumption and emissions targeting under all possible combinations of process modifications can be obtained without enumeration using CS technique. The CS model used in this paper for targeting energy systems consumption and GHG emissions can result in new insights and rigorous tight bounds on the energy system variables during synthesis phase. Therefore, CS can be used as a preparatory step for large energy systems syntheses to reduce the search for the optimal solution under fuzzy process conditions using mathematical programming. In future papers, interval collapsing algorithm will be introduced to determine the exact real number values for optimal process conditions that render the desired global minimum heating and cooling utilities and the global minimum energy-based GHG emissions.

References

Davis, E. (1987), Constraint propagation with interval labels, *Artificial Intelligence, 32,* 99–118.

Moore, R.E. (1988). Reliability in computing: The role of interval methods in scientific computing. *Academic Press.* San Diego, USA.

Noureldin, M. B., EL-Halwagi, M. M. (1999). Interval-based targeting for pollution prevention via mass integration. *Computers & Chemical Engineering, 23,* 1527–1543.

Noureldin, M. B. (2001). Systematic approach to chemical kinetics analysis using constraint logic propagation. *Poster session: Kinetics, Catalysis and Reaction Engineering, AIChE annual meeting.* Nevada, USA.

Noureldin, M. B. (2003). Improved system and computer software for modeling energy consumption, *Patent No. 527,244,* New Zealand.

Noureldin, M.B., Hasan, A. K. (2006). Global energy targets and optimal operating conditions for waste-energy recovery in Bisphenol-A plant, *Applied Thermal Engineering, 26,* 374–381.

Noureldin, M. B., Al-Gwaiz, M. M. (2008). A hybrid interval constraint propagation & MP method for energy systems synthesis, *proceedings of 100th AIChE annual meeting,* Pennsylvania, USA.

Ratschek, H., Rokne, J. (1984). New computer methods for global optimization. *Ellis Horwood/John Wiley and Sons.* New York, USA.

Russian Institute of Artificial Intelligence (1999). UniCalc Solver for Mathematical Problems user guide (version 3.41).

Waltz, D. (1975). Understanding line drawings of scenes with shadows. *The Psychology of Computer Vision, McGraw-Hill,* N.Y., USA.

70

Design of Inter-Plant Water Networks on Mathematical Approach

Cheng-Liang Chen[*], Szu-Wen Hung and Jui-Yuan Lee

Department of Chemical Engineering, National Taiwan University Taipei 10617, Taiwan

CONTENTS

ABSTRACT   This paper aims to develop a general mathematical formulation for integrated design of inter-plant water networks. Based on the superstructures of all possible design options for processing facilities, e.g. water-using units, water mains, and receiving tanks, the design problem is formulated as a mixed-integer nonlinear program (MINLP), from the bilinear terms in balance equations and binary variables in operating constraints. The formulation can be used to apply various integration schemes by imposing the additional constraints, and five of them are considered in the present study with the objective to minimize the freshwater consumption. A representative example is provided to illustrate the application of the proposed MINLP formulation.

KEYWORDS   *Inter-plant integration, water network, mixed-integer nonlinear program (MINLP)*

---

[*] To whom all correspondence should be addressed. Tel.: +886 2 33663039. Fax: +886 2 23623040.
  E-mail: CCL@ntu.edu.tw.

Introduction

The rising costs of water supply and effluent treatment coupled with the ever-tightening environmental regulations have stimulated the development of effective approaches to reduce freshwater consumption and wastewater production in the recent past. Process integration regarding water minimization is believed as a helpful strategy for industries to enhance the competitiveness.

The methodologies on water minimization can be generally classified into two broad categories: the graphical based pinch analysis (Wang and Smith, 1994a,b; Kuo and Smith, 1998) and the mathematical-model based optimization technique (Galan and Grossmann, 1998; Huang et al., 1999; Gunaratnam et al., 2005; Karuppiah and Grossmann, 2006). In addition, a new design manner with internal water mains to simplify the network complexity and to improve the overall practicability was proposed (Feng and Seider, 2001; Wang et al., 2003; Zheng et al., 2006; Ma et al., 2007). More recently, Liu et al. (2008) presented a hybrid structure for the design of water networks, in which work the compromise between network complexity and freshwater requirement has been discussed. All of the above-mentioned works have focused on water integration within single plants. In order to further enhance water recovery potential, inter-plant water integration should be taken into account (Olesen and Polley, 1996; Liao et al., 2007; Chew et al., 2008).

This paper presents a general formulation for water integration within a production site, e.g. an industrial park of different plants, or a process plant with operations grouped into different geographical locations according to the plant layout and/or processing tasks. Opportunities for the transfer of water between plants (or, zones) are considered, and design options are fully incorporated in the formulation, so it serves as a base to apply various integration schemes. An illustrative example is provided to demonstrate the application of the proposed formulation.

Problem Statement

Given is a total site with a set of plants. Each plant includes a set of water-using units where a set of transferable contaminants with fixed mass loads are required to remove. A set of freshwater supplies with known concentrations are available for service; a set of inner-plant and/or inter-plant water mains would be used to increase the network flexibility. The objective is to determine an optimal operating strategy for the inner-plant and inter-plant water integration scheme, which targets the minimum freshwater consumption and wastewater production.

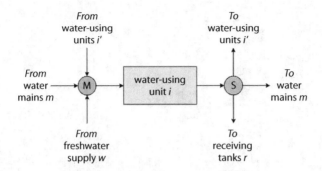

FIGURE 1
Superstructure for a water-using unit.

Mathematical Formulation

Superstructures for processing equipments are built and shown in Figures 1-3, on which the mathematical formulation is based. In addition to the constraints to ensure design specifications and feasibility, the problem consists of water and contaminant balance equations. Following this, a general formulation is presented first, and then various integration schemes can be represented by imposing supplemental constraints appropriately.

Figure 1 depicts the superstructure of all possible flow connections for a water-using unit i, where the input water would come from other water-using units i', water mains m, and freshwater supplies w. Similarly, the output water may be sent to other water-using units i', water mains m, or receiving tanks r. According to the superstructure, the water and contaminant balances for each water-using unit can be derived as given in Eqs. (1)-(3) and Eqs. (4)-(7), respectively. Note that c is the index for contaminants.

$$f_i^{in} = \sum_{i' \in I} f_{i'i} + \sum_{m \in M} f_{mi} + \sum_{w \in W} f_{wi} \quad \forall i \in I \tag{1}$$

$$f_i^{out} = \sum_{i' \in I} f_{ii'} + \sum_{m \in M} f_{im} + \sum_{r \in R} f_{ir} \quad \forall i \in I \tag{2}$$

$$f_i^{in} = f_i^{out} \quad \forall i \in I \tag{3}$$

$$f_i^{in} c_{ic}^{in} = \sum_{i' \in I} f_{i'i} c_{i'c}^{out} + \sum_{m \in M} f_{mi} c_{mc} + \sum_{w \in W} f_{wi} C_{wc} \quad \forall c \in C, i \in I \tag{4}$$

$$f_i^{in} c_{ic}^{in} + M_{ic}^{load} = f_i^{out} c_{ic}^{out} \quad \forall c \in C, i \in I \tag{5}$$

$$c_{ic}^{in} \leq C_{ic,max}^{in} \quad \forall c \in C, i \in I \tag{6}$$

$$c_{ic}^{out} \leq C_{ic,max}^{out} \quad \forall c \in C, i \in I \tag{7}$$

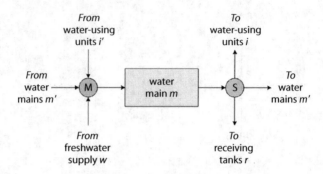

FIGURE 2
Superstructure for a water main.

Figure 2 depicts the superstructure of all possible flow connections for a water main m, where the input water may come from water-using units i, other water mains m', and freshwater supplies w. Meanwhile, the output water may be sent to water-using units i, other water mains m', or receiving tanks r. Water mains would be distributed to individual plants (zones), or centralized, whose purpose is to simplify the overall network complexity. In accordance with this superstructure, the water and contaminant balances are derived in Eqs. (8)-(10) and Eq. (11), respectively.

$$f_m^{in} = \sum_{i \in I} f_{im} + \sum_{m' \in M} f_{m'm} + \sum_{w \in W} f_{wm} \quad \forall m \in M \tag{8}$$

$$f_m^{out} = \sum_{i \in I} f_{mi} + \sum_{m' \in M} f_{mm'} + \sum_{r \in R} f_{mr} \quad \forall m \in M \tag{9}$$

$$f_m^{in} = f_m^{out} \quad \forall m \in M \tag{10}$$

$$f_m^{in} c_{mc} = \sum_{i \in I} f_{im} c_{ic}^{out} + \sum_{m' \in M} f_{m'm} c_{m'c} + \sum_{w \in W} f_{wm} C_{wc} \quad \forall c \in C, m \in M \tag{11}$$

Figure 3 shows the superstructure for a receiving tank, like a buffer where the wastewater is held before being sent to treatment systems. The water balance for each receiving tank is given in Eq. (12). It is worthy of mention

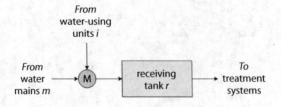

FIGURE 3
Superstructure for a receiving tank.

that effluent treatment is quite important for environmental concern, but this part is omitted and kept for future work.

$$f_r = \sum_{i \in I} f_{ir} + \sum_{m \in M} f_{mr} \qquad \forall r \in R \tag{12}$$

Equation (13) defines the lower and upper limits to the connecting flow rates with the use of binary variables. $y^* = 1$ denotes the existence of the connection between*.

$$F_*^L y_* \le f_* \le F_*^U y_* \tag{13}$$

$$* \in \{ii', im, ir, mi, mm', mr, wi, wm; \forall i, i' \in I, m, m' \in M, r \in R\}$$

So far, the aforementioned equations constitute a basis for inter-plant water integration, as a general case of hybid design options. For convenience, the equations are grouped to a set $\Omega_0 \equiv \{\text{Equations (1)-(13)}\}$. Five integration schemes are considered in this work, as described below: (Note: IPI = inter-plant integration; p = index for plants; M^{cen} and M^{cen} are sets of centralized and inner-plant mains).

Case 1: No IPI without using water mains:

$$\Omega_1 \equiv \Omega_0 \cap \left\{ x \left| \begin{array}{l} \sum_{m \in M} f_m^{in} = 0 \\[2mm] \sum_{i \in I_p} \sum_{i' \in I - I_p} f_{ii'} = 0 \qquad \forall p \in P \end{array} \right. \right\} \tag{14}$$

Case 2: No IPI with inner-plant water mains

$$\Omega_2 \equiv \Omega_0 \cap \left\{ x \left| \begin{array}{l} \sum_{m \in M^{cen}} f_m^{in} = 0 \\[2mm] \sum_{i \in I} \sum_{i' \in I} f_{ii'} = 0 \\[2mm] \sum_{i \in I_p} \sum_{m \in M - M_p} (f_{im} + f_{mi}) = 0 \qquad \forall p \in P \\[2mm] \sum_{m \in M_p} \sum_{m \in M - M_p} f_{mm'} = 0 \qquad \forall p \in P \end{array} \right. \right\} \tag{15}$$

Case 3: direct IPI (Chew et al., 2008)

$$\Omega_3 \equiv \Omega_0 \cap \left\{ x \left| \sum_{m \in M} f_m^{in} = 0 \right. \right\} \tag{16}$$

Case 4: indirect IPI (Chew et al., 2008)

$$\Omega_4 \equiv \Omega_0 \cap \left\{ \mathbf{x} \left| \begin{array}{l} \sum_{m \in M^{inn}} f_m^{in} = 0 \\ \sum_{i \in I_p} \sum_{i' \in I - I_p} f_{ii'} = 0 \quad \forall p \in P \end{array} \right. \right\} \tag{17}$$

Case 5: IPI with centralized/inner-plant water mains, without the use of cross-plant pipeline

$$\Omega_5 \equiv \Omega_0 \cap \left\{ \mathbf{x} \left| \begin{array}{l} \sum_{i \in I} \sum_{i' \in I} f_{ii'} = 0 \\ \sum_{i \in I_p} \sum_{m \in M - M^{cen} - M_p} (f_{im} + f_{mi}) = 0 \quad \forall p \in P \\ \sum_{m \in M_p} \sum_{m \in M - M^{cen} - M_p} f_{mm'} = 0 \quad \forall p \in P \end{array} \right. \right\} \tag{18}$$

Finally, the objective shown in Eq. (19) is to minimize the freshwater consumption. Besides, the design problem is a mixed-integer nonlinear program (MINLP) for the binary variables and bilinear terms.

$$\min_{\text{s.t. } \Omega_t} obj = \sum_{w \in W} \sum_{i \in I} f_{wi} + \sum_{w \in W} \sum_{m \in M} f_{wm} \quad t \in \{1, 2, 3, 4, 5\} \tag{19}$$

Illustrative Example

The example with fifteen water-using units divided into three individual plants (each adapted from Wang et al., 2003, Example 2; Zheng et al., 2006, Case study 2; and Gunaratnam et al., 2005, Case study 1) is provided to illustrate the application of the proposed formulation. The solution tool is the General Algebraic Modeling System (GAMS, Brooke et al., 2003), and the solver for MINLP is BARON.

In case 1, without any interplant integration, the respective freshwater consumptions of individual plants are 111.81 ton/h, 111.83 ton/h and 183.59 ton/h, with a total amount of 407.24 ton/h. In case 2, inner-plant water mains are put in each plant to enhance the operability, but the mixing effect increases the overall freshwater consumption from 407.24 ton/h to 421.82 ton/h. In case 3, direct IPI is carried out and the overall freshwater consumption is reduced from 407.24 ton/h to 354.46 ton/h in comparison with case 1. However, direct IPI would be hard to operate for control problems and the network complexity. Therefore, indirect IPI is performed in case 4. Although the overall freshwater consumption is slightly increased to 355.54 ton/h because of the mixing

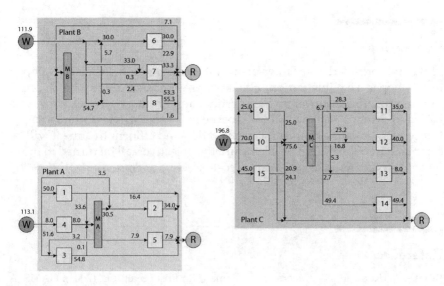

FIGURE 4
Resultant network configuration for case 2.

effect, the network structure will be simpler. For some larger scale problems, centralized water mains would be employed in addition to inner-plant mains for practical needs so case 5 is to be examined. The freshwater consumption is slightly increased to 362.1, but the operability of the whole plant can be improved when compared to cases 3 and 4. The resultant network configurations for cases 2 and 5 are shown in Figs. 4 and 5, respectively.

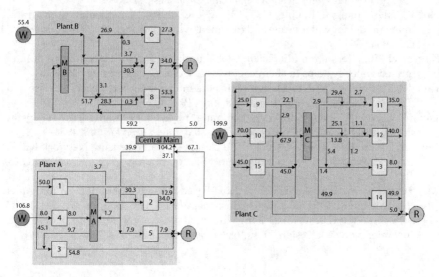

FIGURE 5
Resultant network configuration for case 5.

Conclusion

A general formulation for inter-plant water integration has been developed in this paper. Five integration schemes were considered with the objective to minimize the freshwater consumption, and an illustrative example was solved to demonstrate the validness of the proposed MINLP formulation. In future work, treatment units for regeneration and effluent treatment will be combined into the design problem, and the objective will be turned to minimize the total annual cost (TAC) for cost-effective operation.

References

Brooke, A., Kendrick, D., Meeraus, A., Raman, R., Rosenthal, R.E. (2003). GAMS: A User's Guide. The Scientific Press. Redwood City, CA.

Chew, I.M.L., Raymond, T., Ng, D.K.S., Foo, D.C.Y., Majozi, T., Gouws, J. (2008). Synthesis of Direct and Indirect Interplant Water Network. *Ind. Eng. Chem. Res.*, *47*, 9485.

Feng, X., Seider, W.D. (2001). New Structure and Design Methodology for Water Networks. *Ind. Eng. Chem. Res.*, *40*, 6140.

Galan, B., Grossmann, I.E. (1998). Optimal design of distributed wastewater treatment networks. *Ind. Eng. Chem. Res.*, *37*, 4036.

Gunaratnam, M., Alva-Argáez, A., Kokossis, A., Kim, J.K., Smith, R. (2005). Automated Design of Total Water Systems. *Ind. Eng. Chem. Res.*, *44*, 588.

Huang, C.H., Chang, C.T., Ling, H.C., Chang, C.C. (1999). A mathematical programming model for water usage and treatment network design. *Ind. Eng. Chem. Res.*, *38*, 2666.

Karuppiah, R., Grossmann I.E. (2006). Global optimization for the synthesis of integrated water systems in chemical processes. *Comput. Chem. Eng.*, *30*, 650.

Kuo, W.C.J., Smith, R. (1998). Designing for the Interactions between Water-Use and Effluent Treatment. *Trans. Inst. Chem. Eng.*, *Part A 76*, 287.

Liao, Z.W., Wu, J.T., Jiang, B.B., Wang J.D., Yang, Y.R. (2007). Design Methodology for Flexible Multiple Plant Water Networks. *Ind. Eng. Chem. Res.*, *46*, 4954.

Liu, Y.Z., Duan, H.T., Feng, X. (2008). The Design of Water-Reusing Network with a Hybrid Structure through Mathematical Programming. *Chin. J. Chem. Eng.*, *16*, 1.

Ma, H., Feng, X., Cao, K. (2007). A Rule-Based Design Methodology for Water Networks with Internal Water Mains. *Trans. Inst. Chem. Eng.*, *Part A 85*, 431.

Olesen, S.G., Polley, G.T. (1996). Dealing with Plant Geography and Piping Constraints in Water Network Design. *Trans. Inst. Chem. Eng.*, *Part B 74*, 273.

Wang, B., Feng, X., Zhang Z.X. (2003). A Design Methodology for Multiple-Contaminant Water Networks with Single Internal Water Main. *Comput. Chem. Eng., 27,* 903.

Wang, Y.P., Smith, R. (1994a). Wastewater Minimization. *Chem. Eng. Sci., 49,* 981.

Wang, Y.P., Smith, R. (1994b). Design of distributed effluent treatment systems. *Chem. Eng. Sci., 49,* 3127.

Zheng, X.S., Feng, X., Shen, R.J., Seider, W.D. (2006). Design of Optimal Water-Using Networks with Internal Water Mains. *Ind. Eng. Chem. Res., 45,* 8413.

71

Design and Optimization of Energy Efficient Complex Separation Networks

Gerardo J. Ruiz, Seonbyeong Kim, Jeonghwa Moon,
Libin Zhang and Andreas A. Linninger[*]
Laboratory for Product and Process Design
Departments of Chemical Engineering & Bioengineering
University of Illinois, Chicago, IL 60607, USA

CONTENTS

ABSTRACT Separation processes account for about 50% percent of capital
and operating costs. Separations also require the highest energy demand in
the chemical industry. Rising energy consumption combined with the environmental impact increases the need for energy saving separation processes.
Fortunately complex column networks have the potential for major energy
savings estimated to range between 30 to 70% over simple column configurations. In this paper, a computer-aided synthesis method is introduced to create
optimal complex column arrangements which encode the cost and states of
global solutions with minimum user input. A robust feasibility test helps to
ensure realizable operation of systematically generated networks. This method

---

[*] To whom all correspondence should be addressed

builds on thermodynamic transformations entitled *Temperature Collocation* which provides crucial advantages to determine the operating conditions, structure, and size of the separation network for achieving the desired product cuts. The computational approach guarantees realizable column profiles validated with industrially accepted simulation software such as *Aspen* HYSYS.

KEYWORDS *Difference point equations, complex column, temperature collocation*

Introduction

Chemical distillation is one of the most predominantly used and versatile method of separation, in the petrochemical and commodity industry. It continues to occupy 40–70% of capital and operating costs of chemical manufacturing. Distillation processes account for more than 60% of the total process energy for the manufacture of commodity chemicals (DOE 2005).

Previously energy consumption in separations was a minor important factor, because of low energy prices and less stringent environmental standards. Thus, mainly simple distillation column configurations were built. However the rising energy cost and concerns over atmospheric carbon emissions redefine the design objectives for industrial separations with a new focus on energy conservation and the emission reduction associated with it. In the past ten years, several research groups have advanced systematic methods for identifying optimal distillation network.

Recently, computer-aids separation synthesis methods for energy savings in distillation have been revisited critically. Specifically it was demonstrated that substantial energy savings can be achieved by designing the network simultaneously. Other options with high potential include complex column configurations. These complex column configurations have the potential of achieving up to 70% energy savings over simple column networks (Hilde K. Engelien 2005). The well considered Petlyuk complex column configuration is an excellent example to realize energy efficient separations. With full thermal coupling, this complex column configuration earned energy savings up to 40% while also reducing the numbers of reboilers and condensers.

Since complex separation networks entail tight structural and physical interaction between internal flows with multiple feed and product streams, its numerous possible arrangements and internal connections increase the design problem's combinatorial complexity. Another complicating factor stems from numerical difficulties with complex column profile computations, in which small changes in the design parameters like product purity or reflux ratio often produce large changes in the resulting column profiles. Thus, our research aims at developing computer-aided systematic design procedures to avoid intractable computational cost for identifying optimal network configurations and to prevent numerical failures associated with the extraordinary sensitivity of column profile calculations to even the smallest design variable perturbations. We also wish to synthesize separation networks with realistic column profiles- often referred to as sloppy slits- rather than producing theoretical idealized results, which cannot be validated by practicing engineers with the help of industrial standard flowsheet simulator software packages like *Aspen* HYSYS or *Aspen* Plus.

This paper shows our progress to address complex column synthesis. A crucial feature of our approach stems from massive size reductions enabled by a new column profile computation algorithm called *Temperature Collocation*. Two case studies show the robustness for finding optimal solution of certain classes of complex synthesis problems. We also highlight the ability to validate our solution with rigorous flowsheet simulators. Finding design parameters that will lead to a feasible flowsheet for a novel problem is often a time consuming task in its own right. This task might be made much easier in the future with the help of our computation methods.

Towards Generalized Separation Synthesis

Our approach towards a general solution of the separation synthesis problem has three major components: (i) systematic network generation, (ii) thermodynamic problem transformation to tame numerical challenges and reduce size, and (iii) advanced algorithm engineering to seek simultaneously in the structural and parametric design space.

Generating Complex Column Configurations

Complex distillation configurations have been studied previously by Agrawal and Fidkowski (Agrawal and Fidkowski 1998; Agrawal 2003). They systematically evaluated the energy consumption of basic feasible configurations of complex networks and advocate the energy savings that can potentially be achieved through feasible thermal coupling. Another important advancement is Agrawal's procedure for the exhaustive enumeration of complex flowsheets, which always contains the network structure of the most energy efficient solution. This complete synthesis of candidate flowsheet structures is an important element in a fully automatic separation synthesis methodology. To complete the synthesis, the optimal operating parameters like reflux, intermediate product purities, column height and diameter for all the candidate networks in the superstructure would have to be determined.

We propose to combine network generation methods like Agrawal's with robust global search to systematically identify optimum distillation schemes for given product purity targets. The solution seeks to determine both structural as well as parametric design decisions. Unfortunately, the resulting synthesis problem is NP-hard and suffers from challenges associated with the thermodynamic topology of the feed mixture which cannot be anticipated a priori for multi-component streams with different vapor-liquid interactions. At present, our evolving design method comprises two stages:

Step 1. Generic structure synthesis: Generate all possible basic network configurations using only structural information thus yielding a superstructure containing all potentially optimal networks for the separation of a given feed. The superstructure could be enumerated exhaustively or incorporate equipment constraints when addressing retrofit problems with existing column inventory.

Step 2. Network Task Optimization: Rigorous optimization of all candidate solutions in their respective parametric design spaces to find the globally optimal structure. Network optimization task should take into account the specific feed composition and product targets as well as the specific thermodynamic properties of the chemical species to be separated. In a brute force approach to the synthesis problem, one could advantageously solve independent optimization sub-problems for each network in parallel on separate processors. More elegant would be deployment of thermodynamically motivated implicit pruning criteria, which would allow us to discard inferior designs without explicitly performing network optimization. Such an *admissible heuristics*, however, have not been developed for complex column networks.

Continuous Complex Column Model Synthesis

Complex columns have multiple feed and product steams requiring some modifications to profile computations. Fortunately, Hildebrandt and Glasser worked out a continuous model called difference point equations model, valid for complex column sections in any possible network configuration (Tapp, Holland et al. 2004). A complex column section is defined as a column segment between feed or product trays. Eq. (1) expresses the liquid composition profiles of a complex section in terms of the generalized reflux R_Δ and the difference point composition $X_{\Delta i}$. It also applies to simple sections as a special case, thus rendering a single mathematical expression for all sections in any separation network composed of complex or simple columns. Columns–simple or complex–are said to be feasible if and only if every pair of composition profiles belonging to adjacent column sections intersect. A network is feasible, if all its columns are feasible.

$$\frac{dx_i}{dn} = \left(1 + \frac{1}{R_\Delta}\right)(x_i - y_i) + \frac{1}{R_\Delta}(X_{\Delta i} - x_i) \tag{1}$$

$$R_\Delta = L/\Delta \quad X_{\Delta i} = (Vy_i - Lx_i)/\Delta$$

Where x_i is the liquid composition, y_i is the vapor composition, n is the number of stages, and $\Delta = V$-L is the flow rate difference point. Unfortunately, in higher dimensional space of multicomponent mixtures, it is not a simple task to ascertain the intersection of column section profiles with modest computational effort. We discovered a thermodynamics transformation of the column stage number, n, into the equilibrium bubble point temperature, T. This thermo-dynamically motivated transformation offers massive size reductions in the column profile computations and is a key feature in a rigorous feasibility criterion based on minimum bubble point distance functions which eases the search. We will show in the next section, how dimensionless bubble point space together with the generalized profile map or equations can be used to advantageously synthesize the energy efficient separation networks.

Bubble Point Temperature Collocation and Feasibility Test for Complex Separation Networks

We eliminate the tray number in favor of the bubble point temperature as independent variable. Theoretically, this transformation requires the implicit function theorem for a one-to-one relationship between column height and temperature; this assumption may not hold in highly non-ideal mixtures, but has been tacitly assumed for well-behaved separation tasks we have so far investigated. Applying the thermodynamic transformation called *temperature collocation* changes the continuous column profile equations in eq. (1) to a new composition profiles with temperature instead of height as new independent integration variable like shown in eq. (2).

$$\frac{\partial x_i}{\partial T} = -\left(\left(1 + \frac{1}{R_\Delta}\right)(x_i - y_i) + \frac{1}{R_\Delta}(X_{\Delta i} - x_i)\right) \times$$

$$\frac{\displaystyle\sum_{i=1}^{c}\left(\frac{\partial K_i}{\partial T} x_i\right)}{\displaystyle\sum_{i=1}^{c}\left[\left(\left(1 + \frac{1}{R_\Delta}\right)(x_i - y_i) + \frac{1}{R_\Delta}(X_{\Delta i} - x_i)\right) K_i\right]} \tag{2}$$

This transformation has important advantages including massive problem size reduction avoiding infinite tray numbers near pinch or stationary points. A detailed derivation can be found elsewhere (Zhang and Linninger 2004).

Once compositions profiles have been calculated as functions of temperature, the column height can easily be recovered with the help of eq. (3). The calculation of the column height from bubble point temperatures has been used in the case studies to estimate the capital cost.

$$\frac{dn}{dT} = \frac{\partial n}{\partial x_i}\frac{\partial x_i}{\partial T} = -\frac{\displaystyle\sum_{i=1}^{c}\left(\frac{\partial K_i}{\partial T} x_i\right)}{\displaystyle\sum_{i=1}^{c}\left[\left(\left(1 + \frac{1}{R_\Delta}\right)(x_i - y_i) + \frac{1}{R_\Delta}(X_{\Delta i} - x_i)\right) K_i\right]} \tag{3}$$

To test the feasibility of adjacent sections i and $i + 1$ in complex column k, we generalize the feasibility criterion in terms of dimensionless bubble point distances (BPD), initially introduced only for simple columns (Zhang and Linninger 2004). According to this generalization of the minimum bubble point distance approach, a complex column k is feasible if the sum of all profile distances of all adjacent sections i and $i + 1$ is within a small tolerance of zero (ε) as in eq. (4). It worth pointing out that the aggregated bubble point distances for a complex column is still a scalar delineating the vicinity to a realizable design. The ability to judge the closeness of a design to realizable specifications is of key importance in the genetic search deployed successfully for the synthesis of simple configurations (Zhang and Linninger 2006).

The entire network is feasible if all its columns are feasible as in eq. (5). Where N is the total number of column sections in each column k.

$$Z(k) = \sum_{i=1}^{N-1} BPD(i, i+1) < \varepsilon_1 \tag{4}$$

$$\Psi(k) = \sum_{k=1}^{K} Z(k) < \varepsilon_2 \tag{5}$$

Methods and Case Studies

In mixtures with four or more components, the design problem involves the search for one or more compositional degree of freedom of any product stream. This multidimensional global optimization problem is solved using a hybrid method that combines stochastic genetic algorithm (i.e. product specification) with rigorous finite element collocation of column profiles (Zhang and Linninger 2006).

Separation of Quaternary Mixture

First example illustrates the potential energy saving of complex column configuration over simple column sequences for the separation of a quaternary mixture of methanol, ethanol, 1-propanol, and acetic acid. This complex network uses two simple columns and one complex column as shown in *Figure* 1(a). The composition profiles diagram for column I and column III are showed in *Figure* 1(b) and (c). The dimension (height) and minimum vapor rate (V_{min}) for all columns are collected on *Table* 1. We also estimated total cost of the complex network. The complex configuration analyzed uses half total vapor rate and saves 20% of total costs compared to the simple column configuration. This case study illustrates a feasible solution to the synthesis problem using a complex network, and potential energy savings realizable with complex columns.

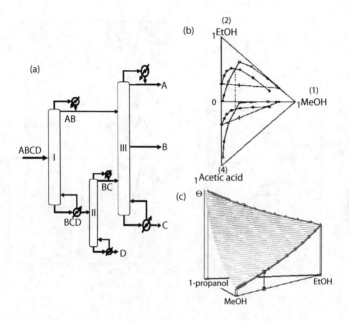

FIGURE 1
(a) Complex column configuration, (b) the quaternary simple column profile, and (c) temperature collocation profile for ternary complex column.

Initialization of Complex Distillation Networks with Aspen HYSYS

Figure 2(a) shows the complex network used to separate the same quaternary mixture of the first case study. The composition profiles found with *Aspen* HYSYS are indicated in *Figure* 2(b). *Figure* 2(c) shows the composition profiles using temperature collocation transformation. These profiles as well as the equivalent number of equilibrium stages were computed using eq. (3).

We initialize the *Aspen* HYSYS flowsheet with our column profiles as initial guesses. The *Aspen* simulation converged with only one or two iterations. This result confirms the accuracy of our methodology is acceptable for a rigorous simulator. In a similar approach we anticipate all separation networks synthesized with our methodology will be easily verified using automatic interface with *Aspen*, this will be a very useful tool for implementing this design idea in the industrial practice.

TABLE 1
Height and V_{min} for each Column in the Network

	Col. I	Col. II	Col. III
Height (m)	10.9	26.6	30.5
Vmin (kmol/h)	13.2	58.9	71.3

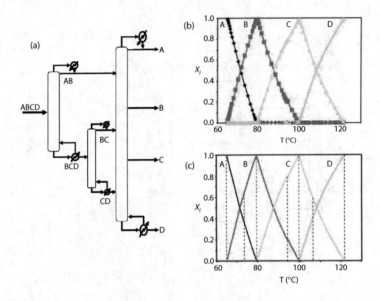

FIGURE 2
Complex column configuration (a), Aspen HYSYS simulation profiles for tray temperature (b),
and Temperature Collocation profiles (c).

Conclusions

In this article, we advocated the recent progress for synthesizing energy efficient
complex distillation networks. An approach towards a generalized solution of
the separation synthesis problem is presented. Difference point equation with
temperature as an independent variable instead of tray number was employed
to find the composition profiles of complex column sections. Temperature collocation and minimum bubble point distance (MIDI) algorithm were effective to find a feasible separation by intercepting profiles. The first case study
demonstrates the potential to 50% of energy savings using a complex column
network compared to the simple column configuration. The second case study
demonstrates current state of the art of separation synthesis in conjunction
with computer simulations to fully integrate complex separation networks.
Finally, the seamless integration of rigorous flowsheet simulators to validate
the predictive results of our scientific method was demonstrated.

Acknowledgments

Financial support by DOE Grant: DE-FG36-06GO16104 is gratefully acknowledged. We also like to thank Dr. Agrawal's helpful discussion on complex column. We acknowledge Dr. Chau-Chyun Chen for providing an Aspen software
research license.

References

Agrawal, R. (2003). "Synthesis of multicomponent distillation column configurations." *AIChE J* **49**(2): 379–401.

Agrawal, R. and Z. T. Fidkowski (1998). "Are thermally coupled distillation columns always thermodynamically more efficient for ternary distillations?" *I&EC Research* **37**(8): 3444–3454.

DOE. (2005). "Technical Topic Description." from www.doe.gov.

Hilde K. Engelien, S. S. (2005). "Minimum energy diagrams for multieffect distillation arrangements." *AIChE J* **51**(6): 1714–1725.

Tapp, M., S. T. Holland, et al. (2004). "Column Profile Maps. 1. Derivation and Interpretation." *I&EC Research* **43**(2): 364–374.

Zhang, L. and A. A. Linninger (2004). "Temperature collocation algorithm for fast and robust distillation design." *I&EC Research* **43**(12): 3163–3182.

Zhang, L. and A. A. Linninger (2006). "Towards computer-aided separation synthesis." *AIChE J* **52**(4): 1392–1409.

72

A Systematic Methodology for Molecular Synthesis Using Combined Property Clustering and GC⁺ Methods

Nishanth G. Chemmangattuvalappil, Charles C. Solvason, Susilpa Bommareddy and Mario R. Eden*
Department of Chemical Engineering, Auburn University
Auburn, AL 36849

CONTENTS

ABSTRACT Property integration techniques have enabled a systematic procedure for the identification of suitable molecules to meet certain process performance. Reverse problem techniques have been used to identify the property targets corresponding to the optimum process performance. Algorithms exist for identifying molecules with the identified properties by combining property clustering and group contribution methods (GCM). Yet, there are situations when the property contributions of some of the molecular groups of interest are unavailable in literature. To address this limitation, an algorithm has been developed to include the property contributions predicted by combined GCM and connectivity index (GC⁺) methods into the cluster space.

KEYWORDS *Property clustering, molecular design, GC⁺ method*

---

* To whom all correspondence should be addressed

Introduction

The integrated property clustering and group contribution (GC) techniques have provided tools to design molecules corresponding to optimum process performance (Eljack et al., 2007). However the property contributions of all the candidate molecular groups may not be available in the literature. Recently, techniques have been developed to predict the property contributions of molecular groups using connectivity indices (CI) (Gani et al., 2005). So, an algorithm to include the property contributions of the CI groups into the clustering framework is needed.

Combined GC-CI Method for Property Estimation

In GCM, the property function $f(Y)$ of a compound is estimated as the sum of property contributions of all the molecular groups present in the structure (Marrero and Gani, 2001):

$$f(Y) = \sum_i N_i C_i + \sum_s N_s C_s + \sum_t N_t C_t \tag{1}$$

N_i, N_s and N_t are the numbers of first, second and third order groups and C_i, C_s, C_t are their respective property contributions. If the contribution of any group is unavailable, a connectivity index (CI) based expression can be used (Gani et al., 2005):

$$f(Y^*) = \sum_i a_i A_i + b(^v\chi^0) + 2c(^v\chi^1) + d \tag{2}$$

Here $^v\chi^0$ and $^v\chi^1$ are the zero and first order CIs, A_i is the number of atom i, a_i is the estimated contribution of atom i, while b, c and d are adjustable parameters. The values of adjustable parameters are available in literature (Gani et al., 2005).

Molecular Design using Property Clustering

Methods for the application of property clustering for molecular design have been developed (Eljack et al., 2007, Shelley and El-Halwagi, 2000). If P_{jg} is the contribution of property j from group g, n_g is the total number of that group in the molecule, molecular property operator ψ^M_{j}, normalized molecular property operator Ω^M, Augmented property index AUP and molecular property cluster $C^M{}_j$ are defined as:

$$\psi_j^M(P_j) = \sum_{g=1}^{N_g} n_g P_{jg} \tag{3}$$

$$\Omega_j^M = \frac{\psi_j(P_{ji})}{\psi_j^{ref}(P_{ji})} \tag{4}$$

$$AUP = \sum_{j=1}^{N_P} \Omega_j^M \tag{5}$$

$$C_j^M = \frac{\Omega_j^M}{AUP} \tag{6}$$

For the groups whose property contributions are not available, the property operator is defined by Eq. 7. Note that there will be different values for the same group corresponding to different bonds it can make with other groups.

$$\psi_{jk}(P_j) = \sum_i a_{m,i} A_{m,i} + b({}^v\chi^0)_m + 2c({}^v\chi^1)_{mk} \tag{7}$$

Visual Solution

The original algorithm for molecular design using molecular property clusters (Eljack et al., 2007) has been extended to include CI groups as explained below:

1. The property targets must be converted to property clusters using the Eqs. 3 to 6 and form a target region on a ternary diagram (simplex) according to the algorithm developed by Eden et al. (2004). The feasibility region boundaries can be represented by six unique points as shown in fig. 1.

2. Generate the first order molecular property operators. For non-GC groups, identify the possible types of atoms that can form bonds with it and estimate the possible zero order and first order CI values. Generate the molecular property operators based on CI. Form a locus of points in the simplex with CI groups. Calculate AUP and molecular cluster values of all the groups and plot all the molecular groups on the simplex.

3. Since molecular clusters obey linear mixing rules, they can be mixed on a simplex using lever arm rules (Eljack et al. 2007). When mixing a CI group with a GC group, the number of hydrogen atoms bonded

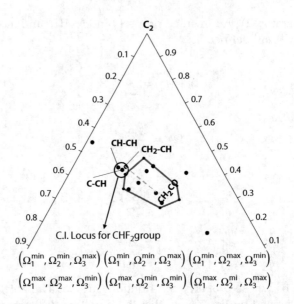

$$\left(\Omega_1^{min},\Omega_2^{min},\Omega_3^{max}\right)\left(\Omega_1^{min},\Omega_2^{min},\Omega_3^{max}\right)\left(\Omega_1^{min},\Omega_2^{max},\Omega_3^{min}\right)$$

$$\left(\Omega_1^{max},\Omega_2^{max},\Omega_3^{min}\right)\left(\Omega_1^{max},\Omega_2^{min},\Omega_3^{min}\right)\left(\Omega_1^{max},\Omega_2^{mi},\Omega_3^{max}\right)$$

FIGURE 1
CI + GC group mixing example.

to the GC group will define the corresponding group in the CI locus. The CI group corresponding to the same number of hydrogen atoms in the GC group must be chosen for mixing. The example in fig.1, explains the mixing of a CH_2CO group with a CHF_2 (CI group).

4. Formulations with zero free bonds, *AUP* value inside the *AUP* range of the sink and cluster location inside the feasibility region are possible solutions.

Algebraic Approach

The aim is to account for the effects of higher order groups while designing molecules. The property target can be represented as a function of molecular groups from GCM models (Chemmangattuvalappil et al., 2009, Eljack et al., 2007). The bounds on each property are:

$$P_{ij}^{lower} \le P_{ij} \le P_{ij}^{upper} \quad j = 1, 2, \ldots N_j; \quad i = 1, 2, \ldots \tag{8}$$

$$\Omega_j^{min} \le \Omega_{ij} \le \Omega_j^{max} \tag{9}$$

For CI groups, Ω_{ij} values of all potential bonds are estimated and the smallest value is selected to ensure that no feasible solution is left out.

$$\Omega_{iif} = \sum_{g=1}^{N_g} n_g \Omega_{jg1} + \sum_{m=1}^{N_m} n_m \Omega_{CI} \tag{10}$$

Where Ω_{jg1} is the normalized property operator of first order group g, n_m is the number of missing groups and Ω_{CI} is its minimum contribution. Let the number of first order groups given in the set $(n_{gk}: n_{gn})$ form the second order group s. η is the number of occurrences of one specific first order group in a selected second order group. Suppose, some of the second order groups are completely overlapped by a bigger second order group and some of the former groups are not overlapped. In that case, if $(n_{gk}: n_{gn})$ has subsets of smaller second order groups $(n_{gl}: n_{gm})$ with some of the first order components of $(n_{gk}: n_{gn})$ then, the normalized property operator for the second order group contribution Ω_{ijs} is:

$$\Omega_{ijs} = \left(\sum_{s=1}^{N_s} \min \left(\frac{n_{gk}}{\eta_k} : \frac{n_{gn}}{\eta_n} \right) \Omega_{jg2} \right)$$

$$+ \sum_{s=1}^{N_s} \left(\min \left(\frac{n_{gl}}{\eta_l} : \frac{n_{gm}}{\eta_m} \right) - \min \left(\frac{n_{gk}}{\eta_k} : \frac{n_{gn}}{\eta_n} \right) \right) \Omega^*_{jg2}$$

(11)

Here Ω_{jg2} and Ω^*_{jg2} are the property contributions of the group s and the smaller unoverlapped second order group respectively. If t is the index for third order groups, the normalized property operator for molecule i is:

$$\Omega_{ij} = \Omega_{ijf} + \Omega_{ijs} + \Omega_{ijt}$$

(12)

It can be seen that the property operator of the target molecule has been defined in terms of the number of first order groups. A few structural constraints must also be satisfied along with the property constraints. The number of each group should be non-negative. The minimum number of molecular fragments forming a ring must be three and for the design of aromatic compounds, there must be multiples of six aromatic carbon atoms. For the fused ring compounds, the number of carbon atoms that form the ring can be defined after the first order level estimation based on the expected number of rings.

$$n_g \geq 0$$

(13)

$$\sum n_{gr} \geq 3 \text{ or } 0$$

(14)

$$\sum n_{ac} = 0, 6, 12....$$

(15)

Where n_{gr} is the number of groups forming ring compounds and n_{ac} is the number of aromatic carbon atoms. If N_r is the number of rings in the final molecule and FBN_g is the number of free bonds in each group then:

$$\sum_{g=1}^{N_g} n_g FBN_g - 2 \left(\sum_{g=1}^{N_g} n_g FBN_g - 1 \right) - 2N_r = 0$$

(16)

TABLE 1

Property and Operators

Property	ψ_j	GC$^+$ expression
H_v (kJ/mol)	$H_v - h_{v0}$	$\Sigma n_g h_{v1} + f(Y^*)$
T_b (K)	$Exp(T_b/T_{b0})$	$\Sigma n_g t_{b1}$
T_m (K)	$Exp(T_m/T_{m0})$	$\Sigma n_g t_{m1}$

Case Study

A case study on a metal degreasing process has been revisited (Eljack et al., 2007, Eden et al., 2004). Here, we are designing the molecules that satisfy the property targets shown in Table 1 and Table 2 identified during the process design stage. The groups being considered for designing the molecule are shown in fig. 2. The Δh_v value of the SO group is not available in literature. To estimate the Δh_v value, the CI method can be used. The SO group can form bonds with two other groups as it has two free bonds in the structure. The possible values for connectivity indices and Δh_v for the SO group with all possible substituents are estimated and given in Table 3.

Visual Solution

The target properties, their bounds, property operators, their reference values, and the normalized property operator are given in Table 1 and 2.

The property targets are converted into corresponding cluster values and the boundaries of the feasibility region is determined. These points are plotted on a simplex to obtain the feasibility region corresponding to the target properties. Now, the property contributions of all the groups are obtained and converted into normalized property operators. Eq. 2 is used to estimate the contribution of the SO group for h_v for the different possible bonds. The estimated values of χ^1 and h_v corresponding to SO groups with different substituents are shown in Table 3. The normalized operators and the clusters for all groups are calculated. Now, the cluster values of different combinations of molecular groups are plotted on the simplex by satisfying structural constraints. When combining SO groups with other groups make sure that the

TABLE 2

Property Targets

Property	ψ_j^{ref}	LB	UB	Ω_{min}	Ω_{max}
H_v	20	50	100	1.53	3.53
T_b	4	480	540	2.16	2.83
T_m	4	280	350	1.67	2.68

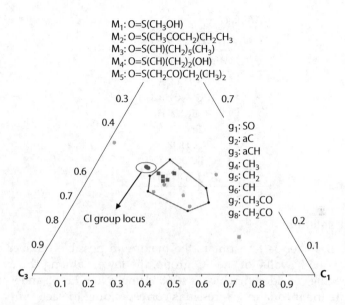

FIGURE 2
Molecular design using GC+ groups.

SO group corresponding to the proper valence delta is used. For instance, if one CH_2CO and one CH_3 are combined with the SO group, the SO group corresponding to carbon atoms with three hydrogen and two hydrogen are to be used. The molecular group formulations which fall inside the feasibility region and satisfy the AUP constraint of the sink are potential solutions, a part of which is shown in fig. 2.

TABLE 3
CIs and Group Contributions

Group	χ^0	χ^1	h_v
C_3C_3	1.02	0.56	17.72
C_3C_2	1.02	0.47	17.26
C_3C_1	1.02	0.43	17.06
C_3aC	1.02	0.33	16.58
C_2C_2	1.02	0.40	16.94
C_2C_1	1.02	0.38	16.80
C_2aC	1.02	0.31	16.46
C_1C_1	1.02	0.35	16.68
C_1aC	1.02	0.30	16.40
$aCaC$	1.02	0.21	16.28

TABLE 4

Potential Higher Order Groups

Higher Order Groups
$(CH_3)_2CH$
$CH(CH_3)CH(CH_3)$
CH_3COCH_2
CH_3COCH
$CHOH$
CH_3COCH_nOH
$CH_m(OH)CH_n$

Algebraic solution

Here, the first step is to estimate the maximum possible number of each group. For the h_v value of the SO group, the lowest among the estimated values is being considered, which is 16.68 kJ/mol. Now, Eqs. 8 and 9 are used to generate the inequality expressions corresponding to each property. All first order groups are maximized subject to the structural constraints in Eqs. 13 to 16:

SO: 1 aC: 1 aCH: 5 CH_3: 6 CH_2: 8 CH: 4 OH: 1 CH_3CO: 1 CH_2CO: 2

Higher order groups possible from these first order groups are listed in Table 4. All possible combinations of first order groups are generated and the property contributions from higher order groups have been estimated using Eq. 12. The molecular property operators are generated subjected to the constraints in Eqs. 13–16 and *AUP* values for each combination are calculated. The combinations whose *AUP* values are within the limits are potential solutions. The property values of those combinations are back calculated to confirm they are real solutions. A part of the final solution is shown in Table 5.

TABLE 5

Part of Final Solution

Final Solution
$(CH_3CO)CH_2(SO)CH_3$
$CH_3CH_2(SO)CH_2COCH_3$
$CH_3CO(SO)CH_3$
$CH_3(SO)CH(CH_3)COCH_3$
$CH_3(SO)CH(OH)CH_3$
$CH_3(CH_2CO)SOCH(CH_3)_2$
$(CH_3)SO(CH_2)_3CH_3$
$CH_3(CH_2)_2SO(CH_2)_2CH_3$
$CH_3(CH_2CO)SOCH_2CH_3$

Conclusions

In this work, a modified algorithm has been developed for molecular design using the GC+ technique when the property contributions of some of the candidate groups are not available in literature. A visual approach along with an algebraic approach has been developed for molecular design using the new algorithm. The expressions for molecular property operators developed in our previous paper have been modified using the third order group contributions to make the design more accurate and to extend its application range. The future work need to be focused on incorporating more structural information during molecular design.

Acknowledgments

The funding for this work was provided by the National Science Foundation CAREER program.

References

Chemmangattuvalappil N.G, Solvason C.C., Eljack F.T., Eden M.R. (2009). A Novel Algorithm for Molecular Synthesis Using Enhanced Property Operators. Computers & Chemical Engineering, 33(3), 636–643.

Eden M.R., Jørgensen S.B., Gani R., El-Halwagi M.M. (2004). A Novel Framework for Simultaneous Separation Process and Product Design. Chemical Engineering & Processing, 43, 595–608.

Eljack F.T., Eden M.R., Kazantzi V., El-Halwagi M.M. (2007a). Simultaneous Process and Molecular Design—A Property Based Approach. AIChE Journal, 53(5), 1232–1239.

Eljack F.T., Solvason C.C., Eden M.R. (2007b). An Algebraic Property Clustering Technique for Molecular Design. Computer Aided Chemical Engineering 24, T2–326

Gani R., Harper P.M., Hostrup M. (2005). Automatic Creation of Missing Groups through Connectivity Index for Pure-Component Property Prediction. Ind.Eng. Chem.Res. 2005, 44, 7262–7269

Marrero J., Gani R. (2001). Group Contribution Based Estimation of Pure Component Properties. Fluid Phase Equilibria, 183–184, 183–208.

Shelley M.D., El-Halwagi M.M. (2000). Component-less Design of Recovery and Allocation systems: A Functionality–based Clustering Approach. Computers & Chemical Engineering, 24, 2081–2091

73

Process Optimization Using a Hybrid Disjunctive-Genetic Programming Approach

Wei Yuan[1], Andrew Odjo[2], Norman E. Sammons Jr.[1], Jose Caballero[2] and Mario R. Eden[1*]

[1]Department of Chemical Engineering, Auburn University, Auburn, AL 36849
[2]Institute of Chemical Process Engineering, University of Alicante, Alicante 03690

CONTENTS

ABSTRACT Discrete optimization problems, which give rise to the conditional modeling of equations through representations as logic based disjunctions, are very important and often appear in all scales of chemical engineering process network design and synthesis. With the increase in the problem scale, dealing with such alternating routes becomes difficult due to increased computational load and possible entanglement of the results in sub-optimal solutions due to infeasibilities in the MILP space. In this work, Disjunctive-Genetic Programming (D-GP), based on the integration of Genetic Algorithm (GA) with the disjunctive formulations of the Generalized Disjunctive Programming (GDP) for the optimization of process networks, is presented. The genetic algorithm is used as a jumping operator to the different terms of the discrete search space and for the generation of different feasible fixed configurations. This proposed approach eliminates the need for the reformulation of the discrete/discontinous optimization problems into direct MINLP problems, thus allowing for the solution of the original problem as a continuous optimization problem but only at each individual discrete and reduced search space.

KEYWORDS *Process optimization, disjunctive programming, genetic algorithm*

---

[*] To whom all correspondence should be addressed

Introduction

Processes with nonlinear functions and discontinuities in the objective and/ or constraint space are found in a large number of synthesis problems. It has been shown that using disjunctions for expressing the discrete decisions, which conditions the selection of process units among various alternatives, can be very beneficial in handling of discontinuities in chemical process systems (Vecchietti et al., 2003; Turkay and Grossmann, 1996). The Generalized Disjunctive Programming (GDP) is an efficient method for Mixed Integer Non-Linear Programming (MINLP) for discrete/continuous optimization problems (Turkay and Grossman, 1996; Lee and Grossman, 2000). In GDP, problems are modeled with Boolean and continuous variables for the optimization of a given objective function subject to different types of constraints. Generally, GDP represents discrete decisions in the continuous space with disjunctions, and constraints in the discrete space with logic propositions.

Evolutionary search methods, of which the most widely used is the Genetic Algorithm, originally initiated by Holland (1975), has been known to be less susceptible to the existence of local optimal solution in optimization problems, yielding very good solutions even with discontinuous objective and/ or constraint functions. The evolutionary search method solutions have been found in some cases, to outperform those obtained through traditional deterministic approaches, by exhibiting robustness through the use of the objective function information and not derivatives, as well as by the ease at which they handle discrete and integer variables and non-smooth and non-continuous functions (Androulakis and Venkatasubramanian, 1991). Yet, the overall applicability of these algorithms to constrained problems remains an active research focus, especially taking into consideration the fact that they may exhibit slow convergence and may have difficulties finding the optimal solution to a problem with very small feasibility space. However, promising results are being obtained in joint heuristic and evolutionary (deterministic and stochastic) approaches to the efficient solution of engineering processes (Leboreiro and Acevedo, 2004).

This work is based on the integration of GA with the disjunctive representation of discrete/continuous optimization problems. The strategy involves the decoupling of the disjunctions and the propositional logics from the overall disjunctive formulation to the GA space where these constraints are treated before being returned as active terms with satisfied logics to the resulting NLP space. The population of disjunctive terms, corresponding to the different process superstructures or configurations are manipulated by the genetic operators in an evolutionary manner in order to obtain the best NLP solution. The implication of this is the final solution of an optimization problem with smooth nonlinear constraints and objectives confined within a reduced search space corresponding to specific active terms of a set of disjunctions, as determined by the GA. An advantage of this approach is that, apart from the fact that the dimensionality of the optimization problem can be considerably reduced due to the reduction in the total number of constraint equations (only common and active disjunctive terms equations are analyzed at each NLP call), only feasible superstructures are generated.

Disjunctive-Genetic Programming Method

The basic idea behind the D-GP approach involves the decoupling of the disjunctive terms represented by Boolean variables as well as the propositional logic constraints these variables form, to the genetic algorithmic space where they are encoded into chromosomes with specific structures corresponding to potential superstructure solution alternatives. In a typical solution framework such as the Big M - OA approach (Williams, 1999), the reformulation of optimization problems with discontinuous functions modeled with disjunctions involves conversion of the Boolean variables to binary and the proposition logics into simple linear equations, leading to MINLP problems, which in turn are solved using an iterative MILP-NLP scheme (Figure 1a).

An important feature of our proposed approach is that no MINLP reformulation of the disjunctive representation is needed. Terms of the disjunctions only need to be identified, well defined and coded into strings or chromosomes. For the adaptation of GA to the efficient handling of the generated chromosomes, we propose a segment-based crossover and mutation strategy. Furthermore, we apply the all-feasible populations approach, consisting of the creation and use of chromosomes that always satisfy the logic constraints (Odjo et al., 2008). The basic steps of the D-GP framework are depicted in Figure 2.

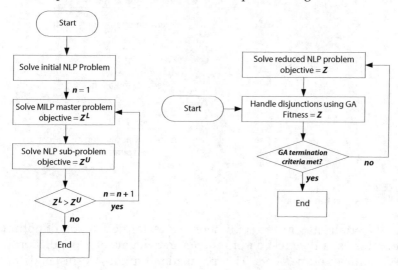

FIGURE 1
(Left) MILP-NLP iterative solution approach, and (right) the GA-NLP framework.

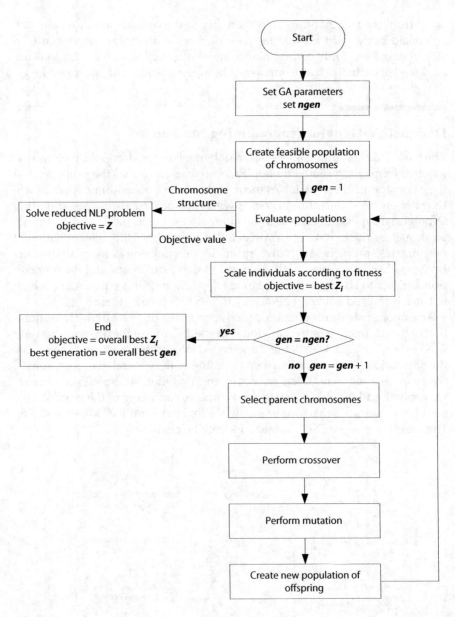

FIGURE 2
The D-GP solution framework.

An important step in the application of GA to the solution of optimization problems is the creation of random individuals of a population with a fixed number of members. This population typically corresponds to the search space of the optimization problem, for example randomly generated temperature or pressure values between a specified lower and upper limit.

However, in large synthesis problems with large number of variables giving rise to huge combinatorial problems such as possible different flowsheet configurations of a process with large number of units and process streams, and coupled with the existence of a narrow feasibility region due to different specifications and restrictions, the traditional approach of the generation of individuals within the optimization search space might lead to slow convergence and even sub-optimal solutions. In this work, a population of feasible chromosomes approach is adopted. The important aspect of this approach is that infeasible chromosomes have been eliminated from the initial population before the GA fitness function evaluation stage, as opposed to the traditional approach where feasibility constraints defined by the logic propositions and embedded as linear algebraic equations were satisfied during the GA fitness function evaluation.

Segment-Based Cross-over and Mutation Strategy

In order to conserve parts of the feasible configuration from the parents to the offspring, and also to achieve less disruptive chromosomes, the cross-over operation strategy employed in this work involved the selection of two strings (parent chromosomes) with the subsequent identification and exchange of identical segments of the parents. The number of identical segments in a parent string or chromosome is independent of the size of the optimization problem, but depends on the number of disjunctions in the synthesis problem. The genes in each segment represent the terms of the disjunctions. The rules governing the cross-over of identical segments, involve the selection and direct inter-exchange of the bulk of those segments, but not individual genes in them.

Mutation, carried out at random on any bit of a string (gene) of the population, often leads to infeasible configurations. This can be improved, however, if the operator is modified by incorporating specific information on

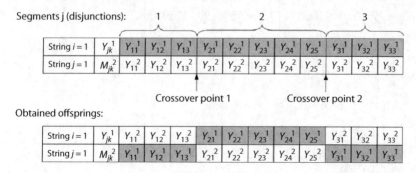

FIGURE 3

Adopted crossover operation strategy for the D-GP approach.

desired mutation routes. Rather than carry out mutations on individual genes, mutations are performed on segments of specific size and position in the string, corresponding to individual disjunctions and the order in which they are placed, respectively. It is worth noting, that the length of each segment (the number of genes in it) corresponds to the number of terms in each disjunction. In this work, this leads to the creation of mutant genes, which basically can be considered as sets of disjunctions of the current problem with different configurations of the propositional logics. In this way, diversity of the next population is guaranteed without loss of feasibility.

Case Study-Heat Exchanger Network Design

The heat exchanger network synthesis problem developed by Turkay and Grossman (1996) consists of a tubular heat exchanger, a heater, and cooler. Specifically, the problem involves the optimization of process models with discontinuous cost functions and fixed charges defined over different area regions of the heat exchangers. More details on this problem can be found in Turkay and Grossman (1996a) and Caballero et al. (2007). The problem was expressed in disjunctive form as given below, and it was sought to minimize the total cost and select the corresponding heat exchangers based the optimized values of the areas of these.

$$\min Z = \sum_{j \in J} CU_j + \sum_{k \in K} IC_k$$

s.t.

$$
\begin{bmatrix}
Y_{k,1} \\
IC_k = 2750 A_k^{0.6} + 3000 \\
0 \le A_k \le 10
\end{bmatrix}
\vee
\begin{bmatrix}
Y_{k,2} \\
IC_k = 1500 A_k^{0.6} + 15000 \\
10 \le A_k \le 25
\end{bmatrix}
\vee
\begin{bmatrix}
Y_{k,3} \\
IC_k = 600 A_k^{0.6} + 46500 \\
25 \le A_k \le 50
\end{bmatrix}
$$

$$A_k \ge 0 \quad \forall i \in I = \{1..3\}$$

$$k \in K = \{k \mid heat\ exchangers\}$$

$$i \in I = \{i \mid heat\ exchanger\ areas\}$$

$$j \in J = \{j \mid Service\ streams\}$$

$$Y_{i,k} = \{TRUE, FALSE\}$$

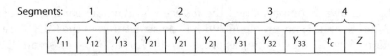

FIGURE 4
Chromosome structure for the 3 heat exchanger network problem.

In the formulations, *CU* and *IC* represent the utility and the investment costs of the whole process respectively. The solution of this problem using the D-GP as proposed in this work involves the decoupling of the logical propositional constraints from the general optimization model, and consequent generation, in a genetic algorithmic framework, of feasible configurations which satisfy these logics. A chromosome was defined, consisting of 11 genes with four distinct segments:

The first three segments correspond to the disjunctions and the 9 genes they contained correspond to the terms of these disjunctions. The last segment, common to all chromosomes, contains two genes which stores information on the performance of the particular chromosome, namely the NLP termination criteria (values −1 and +1 for infeasible and feasible NLP solution respectively), and objective function respectively. Logics were satisfied through the comparison of each randomly generated segment with a set of predefined conditions in form of simple linear equations defining the logics. To create a chromosome, an empty string is chosen, and the genes were assigned randomly to it with real values 0 and 1. If the created chromosome satisfies the logical prepositional constraints, it is chosen, else it is dropped (the chromosome is illegal) and a new combination chosen until the logics are satisfied. No penalization is applied for illegal chromosomes. By this method, a finite number of always feasible populations were generated.

TABLE 1

Model Parameters

Model Parameters		D-GP	GDP-OA
Objective value, 10^3 \$/yr		125.60	150.32
Total investment, 10^3 \$/yr		75.99	100.36
Total utility cost, 10^3 \$/yr		49.61	50.16
Optimal Areas, m^2	Heat exchanger	25	24.95
	Heater	26.18	21.94
	Cooler	30.28	28.07
Number of variables		4	15
Total number of equations		12	39
Constraints:	Nonlinear	6	36
	Linear	6	3
Number of bonds		6	15

Table 1 shows a comparison of the reduced NLP model parameters of the D-GP and the NLP sub-problems (with fixed structures) of the MINLP reformulation of the GDP technique as presented by Turkay and Grossman (1996a) of the same heat exchanger problem using the Outer Approximation coupled with the Big M algorithm. The NLP sub-problem of the GDP formulation found an objective function which was almost 20% worse than that found by the D-GP. A clear reduction in the total number of constraints equations as well as variables of the NLP problem can be observed using our proposed approach. This is because at each GA-NLP iteration, only those constraints of the active terms of the disjunctions which correspond to the configuration of each chromosome were solved, leading to a reduced search space and dimensionality of the NLP problem.

Conclusions

In this paper a joint genetic algorithm-disjunctive representation approach to synthesis of process networks involving discrete/discontinuous functions has been presented. Special modified GA operators were presented to better handle the proposed solution approach, and it was found that the generation of feasible populations yield better results as compared to the case when the population is composed of randomly generated individuals that underwent feasibility test at the fitness function evaluation stage. An interesting feature of the D-GP approach is that it eliminates the need for the reformulation of the GDP problems into a direct MINLP problem, thus allowing the solution of the original problem as a continuous optimization problem but only at each individual discrete and reduced search space. The effectiveness of the G-DP has been illustrated using a benchmark case study of heat exchanger network design. An important aspect of this work is the use of an evolutionary algorithm for the efficient handling of discontinuities in optimization problems, while simultaneously applying deterministic approaches to handle the continuous functions.

References

Androulakis, I. P., Venkatasubramanian, V. A. (1991). Genetic algorithmic framework for process design and optimization. Computers and Chemical Engineering, 15 (4), 217–228.

Caballero, J. A., Odjo, A. O., Grossmann, I. E. (2007), Flowsheet Optimization with Complex Cost and Size Functions Using Process Simulators. *AIChE Journal*, *53*, 2351.

Holland, J. H. (1975), Adaptation in Natural and Artificial Systems. Ann Arbor: *University of Michigan Press.*

Leboreiro, J., Acevedo, J. (2004). Process synthesis and design of distillation sequences using modular simulators: a genetic algorithm framework. *Computers and Chemical Engineering*, 28, 1223–1236.

Lee, S., Grossmann, I.E. (2003), Global Optimization of Nonlinear Generalized Disjunctive Programming with Bilinear Equality Constraints: Applications to Process Networks. *Computers and Chemical Engineering*, 27, 1557.

Odjo A.O., Sammons Jr. N.E., Marcilla A., Eden M.R., Caballero J. (2008), A Disjunctive-Genetic Programming Approach to Synthesis of Process Networks, Proceedings of 18th International Congress of Chemical and Process Engineering (CHISA).

Raman, R., Grossmann, I. E. (1994). Modeling and Computational Techniques for Logic Based Integer Programming. *Computers and Chemical Engineering*, 18, 563.

Turkay, M., Grossmann, I.E. (1996), Logic-Based MINLP Algorithms for the Optimal Synthesis of Process Networks. *Computers and Chemical Engineering*, 20, 959.

Vecchietti, A., Lee, S., Grossmann, E. I. (2003). Modeling of discrete/continuous optimization problems: characterization and formulation of disjunctions and their relaxations. *Computers and Chemical Engineering*, 27(3), 433–448.

Williams, H.P. (1999). Model Building in Mathematical Programming, John Wiley and Sons, Ltd, Chichester.

74

A Graphical Approach to Process Synthesis Based on the Heat Engine Concept

Baraka Celestin Sempuga, Bilal Patel, Brendon Hausberger, Diane Hildebrandt, * **David Glasser,**

Centre of Material and Process Synthesis, School of Chemical and Metallurgical Engineering, University of the Witwatersrand, Private Bag 3, WITS 2050, South Africa

CONTENTS

ABSTRACT   The issues surrounding energy utilisation and its environmental impact have lead to the need to create more efficient processes which produce fewer waste products and use less energy. In this regard, innovative ways of designing processes are required that minimize both use of raw materials and energy.

In this paper, the notion of heat engines as applied to chemical processes is shown. The paper shows that one can look at chemical processes holistically, where only the inlet and outlet streams are considered, and represent them in the ΔH-ΔG space. The ΔH gives an indication of the energy requirements of the process while the ΔG gives an indication of the work requirements of the process.

Processes can be classified in different thermodynamic regions as defined by the ΔH-ΔG space, and their feasibility in terms of their heat and work requirement can be determined. This approach can be used as a starting point in developing more reversible process flow sheets; it can be used to determine the best method of supplying or recovering energy from a process. In particular the approach can be used to analyse processes and determines if the use of heat to supply energy is advisable in terms of process

---

* To whom correspondence should be addressed: Tel: +27 11 717 7527. Fax: +27 11 717 7557. E-mail: diane.hildebrandt@wits.ac.za.

reversibility. The approach is also used to investigate and discuss the possibility of combining chemical processes classified in different thermodynamic regions in the ΔH-ΔG space, with the purpose to make infeasible processes possible or, minimize or even nullify the work requirement for the combined process.

KEYWORDS *Second law, heat engine, Carnot temperature, process synthesis, process reversibility*

Introduction

The second law of thermodynamics has been found to be a tangible basis on which process efficiency could be assessed. K.G Denbigh (1956) developed equations for calculating the thermodynamic efficiency of chemical processes, using as basis the second law of thermodynamics, which sets a certain limit on the conversion of heat into work. He used the notion of heat engine efficiency, calculated relative to a reversible heat engine, to assess the efficiency of chemical processes. Patel et al. (2004) has shown that in some cases chemical processes can be analysed in terms of their heat and work requirement and in particular that when chemical processes are looked at as 'simple processes', there exist a temperature which he called 'Carnot temperature' at which one can satisfy the work requirement using the heat that needs to be added or removed from the process.

This paper uses the notion of heat engine and Carnot temperature as applied to chemical processes. The paper shows that one can look at chemical processes holistically, where only the inlet and outlet streams are considered, and represent them in the ΔH-ΔG space. Processes can be classified in different thermodynamic regions as defined by the ΔH-ΔG space, and their feasibility in terms of their heat and work requirement can be determined. This approach can be used as a starting point in developing more reversible process.

Chemical Process and Heat Engine

Consider a simple representation of a chemical process as shown in Figure 1. The process could comprise several unit operations such as reactors, separation units, heat exchangers etc. Let us however consider only the inlet and the outlet streams of the process which we assume to be at ambient conditions T_O and P_O. The energy balance for the process is given by:

$$Q = \Delta H_{process} \tag{1}$$

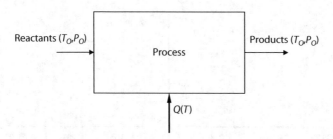

FIGURE 1
A simple chemical process.

Suppose that a reversible heat engine is used to supply the process heat requirement as shown in Figure 2; the work that the heat engine would require to pump a quantity of heat Q_O from the environment at T_O, and supply a quantity of heat Q_H to the process at the process temperature T_H, is given by (Smith, J.M et al. 2001):

Patel et al. (2004) has shown that for a reversible simple chemical process (Figure 1) the energy and entropy balance give:

$$W = Q_H\left(1 - \frac{T_o}{T_H}\right) \tag{2}$$

$$\Delta G_p(T_o, P_o) = \Delta H_p(T_o, P_o)\left(1 - \frac{T_o}{T_{Carnot}}\right) \tag{3}$$

By comparing equations (1), (2) and (3), it is clear that the process work requirement, given by the change in the Gibbs free energy across the process $\Delta G(T_o, P_o)$, matches exactly the heat engine work requirement W. We could then say that:

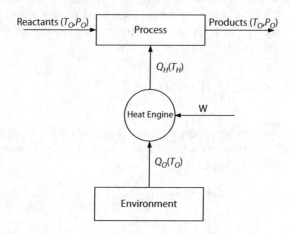

FIGURE 2
Heat engine as a heat source for a chemical process.

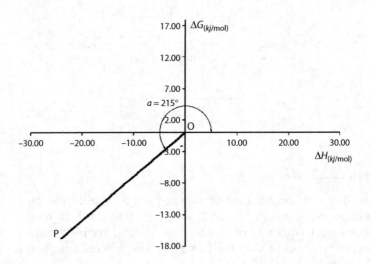

FIGURE 3
Process representation in ΔH-ΔG domain.

The Gibbs Energy change of a reversible chemical process as described in Figure 1, is equivalent to the work required by a heat engine to pump heat, at ambient temperature, from the environment to the process at the process temperature.

If we assume that the major heat load in the process is the reactor then equation (3) for the process can be represented by equation (4), where ΔG_{rxn} (T_o, P_o), and ΔH_{rxn} (T_o, P_o), are the Gibbs energy and enthalpy change of the process nett reaction, respectively.

A combination of equation (4) and a plot of ΔG versus of ΔH can provide us with insights and information regarding the process. To demonstrate this let us look at the dehydration of methanol to form dimethyl-ether (DME). The overall process reaction, the molar enthalpy and the molar Gibbs-free energy change of reaction are given in equation (5).

$$\Delta G_{rxn}(T_o, P_o) = \Delta H_{rxn}(T_o, P_o)\left(1 - \frac{T_o}{T_{Carnot}}\right) \tag{4}$$

If ε is the extent of reaction then $\Delta H_{rxn} = \varepsilon\Delta H_{rxn}$ and $\Delta G_{rxn} = \varepsilon\Delta G_{rxn}$. A linear relationship can be obtained between ΔH_{rxn} and ΔG_{rxn} by varying the extent of reaction as shown in Figure 4.

$$2CH_3OH \rightarrow CH_3OCH_3 + H_2O$$

$$\left\{\begin{array}{l} \overline{\Delta H}_{rxn} = -24.05\,\frac{kJ}{mol} \\ \\ \overline{\Delta G}_{rxn} = -16.05\,\frac{kJ}{mol} \end{array}\right\} \tag{5}$$

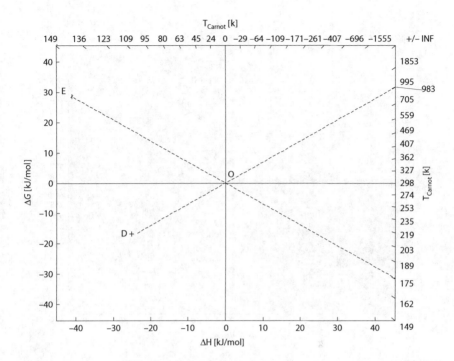

FIGURE 4
Temperature scale in ΔH-ΔG domain.

In Figure 4, the process feed is located at the origin (O) where the extent of reaction is zero; any point along the line represents the process output whose overall mass balance is an incomplete reaction, until the point at the other end of the line (P) where the process reactor goes to completion. The end point would also represent a process whose overall mass balance is a complete reaction even if the reaction in the reactor does not go to completion but has a recycle system.

Equation (4) can be rearranged to obtain:

$$T_{Carnot} = \frac{T_o}{1 - \frac{\Delta G_{rxn}}{\Delta H_{rxn}}} \tag{6}$$

The Carnot temperature (T_{Carnot}) for a process can be represented on a temperature scale in the ΔH-ΔG domain as shown in Figure 4, where a chemical process is represented by a point such as D or E. The Carnot temperature is determined by the intersection of the temperature scale with a line linking the origin and the point that represent the process. Thus if D represents the DME process as described by equation (6), then the point of intersection of the line DO with the temperature scale gives the Carnot temperature of the DME process which is 983 K.

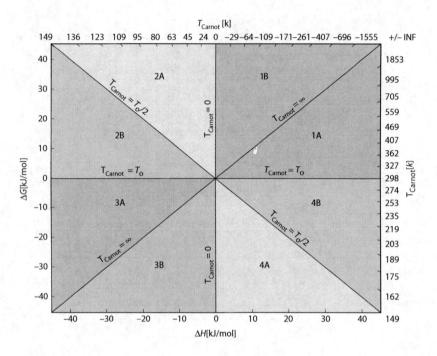

FIGURE 5
Thermodynamic regions in ΔH-ΔG domain.

T $_{Carnot}$ and the Process as a Heat Engine

In light of the variation of T_{Carnot} in the ΔH-ΔG domain, a number of thermo-dynamic regions, numbered 1 through to 4, can be identified as shown in Figure 5. Each region is identified by the magnitude and direction of heat and work flows to the process.

Processes in region 1 require heat and work input while those in region 3 reject heat and work. In these regions both the heat and work flows have the same direction, however, we can see in Figure 5 that the Carnot tempera-ture for processes in region 1A and 3A varies from ambient to infinity. This implies that in region 1A and 3A addition or removal of the process heat requirement at the process Carnot temperature is sufficient to reversibly sup-ply or recover the process work requirement. On the other hand processes in region 1B and 3B require that the process heat requirement be supplied or removed at negative Carnot temperature (which has no physical meaning) for them to be reversible; this is because $\Delta H < G$; hence, in region 1B and 3B, the heat requirement cannot satisfy the work requirement irrespective of the temperature at which the heat is added or removed.

Processes in region 2 (Figure 5) require heat output and work input and their Carnot temperatures are below ambient. This corresponds to refrigeration processes where work must be supplied to the process to remove heat. If the refrigeration process is run reversibly then the cold Carnot temperature will be reached.

Processes in region 4 (Figure 5) require heat input and work output at temperatures below ambient. This means that there is opportunity to recover work from processes in region 4 by absorbing heat at ambient temperature and reject heat to the process at a temperature below ambient.

Applications

The ΔH-ΔG space is quite useful since any number of processes can be placed on the diagram once their Gibbs energy and enthalpy is known. This approach also allows determining whether heat at an appropriate temperature is sufficient to meet the work requirement of a chemical process or other means should be considered. For example, one can determine whether steam can supply the heat requirements for a process reversibly. The approach can also be used to investigate and discuss the possibility of combining chemical processes classified in different thermodynamic regions in the ΔH-ΔG space, with the purpose to make infeasible processes possible or, minimize or even nullify the work requirement of the combined process. Due to space limitations, these applications cannot be discussed in detail and will be the focus of a subsequent paper.

Conclusion

In this paper we have used heat engine concepts to analyse the reversibility of chemical processes. This can easily be demonstrated in the ΔH-ΔG space. When the overall reaction of a chemical process is determined, one can consider only the inlet and outlet stream properties and represent the process in the ΔH-ΔG space. With this representation we can be able to do the following:

- Determine the reversible temperature of the process. This is the temperature at which heat must be added or removed from the process for it to be reversible.
- Classify the process in one of the thermodynamic regions defined in ΔH-ΔG space, to determine whether it is possible to reversibly satisfy the process work requirement by adding or removing the process heat requirement at an appropriate temperature.

Further works will reveal that one can use the ΔH-ΔG domain to combine processes for optimal heat and work integration.

References

Denbigh K.G.(1956) "The second law efficiency of chemical processes", Chem. Eng. Sci, 6.

Patel B., Hildebrandt D.,and Glasser D.(2004), Overcoming the Overall Positive Free Energy of a Process: Using the Second Law to understand how this is Achieved, Presented at the AIChE Annual Meeting, Austin, Texas, November 2004

Smith, J.M., Van Ness, H. C., Abbott; M. M. (2001) Introduction to Chemical Engineering Thermodynamics, Sixth Edition, McGraw-Hill, New York

75

CFD-based Shape Optimization
of Pressure-driven Microchannels

Osamu Tonomura[*], Manabu Kano and Shinji Hasebe

Dept. of Chem. Eng., Kyoto University
Nishikyo-ku, Kyoto 615-8510, Japan

CONTENTS

ABSTRACT In the design of micro chemical devices, the shape of microchannels is an important design factor for achieving high performance. Computational fluid dynamics (CFD) is often used to rigorously examine the influence of the shape of microchannels on heat and mass transport phenomena in the flow field. However, the rash combination of CFD and the optimization technique based on evaluating gradients of the cost function requires enormous computation time when the number of design variables is large. Recently, the adjoint variable method has attracted the attention as an efficient sensitivity analysis method, particularly for aeronautical shape design, since it allows one to successfully obtain the shape gradient functions independently of the number of design variables. In this research, an automatic shape optimization system based on the adjoint variable method is developed. To validate the effectiveness of the developed system, pressure drop minimization problems of a 180° curved microchannel are solved. These design examples illustrate that the pressure drop of the optimally designed microchannels is decreased by 20 % ~ 40 % as compared with that of the initial shape.

---

[*] Tel.: +81 75 383 2637; Fax: +81 75 383 2657. E-mail address: tonomura@cheme.kyoto-u.ac.jp.

KEYWORDS   *Optimal shape design, CFD simulation, adjoint variable method, microchannel, pressure drop*

Introduction

In recent years, micro chemical process technology has attracted considerable industrial and academic attention in various fields (Ehrfeld et al., 2000b; Hessel et al., 2005; Kockmann, 2006; Kano et al., 2007). The main characteristic of micro chemical processes is the small diameter of the channels ensuring short radial diffusion time. This leads to a narrow residence time distribution, high heat and mass transfer. In addition, micro chemical processes have a high surface to volume ratio allowing efficient heat removal and high molar fluxes.

Microfabricated chip devices, usually called "μ-TAS", have emerged as powerful tools for carrying out chemical and biomedical analysis. Such chip devices can achieve a significantly shorter separation time as well as a more accurate analysis compared to conventional laboratory scale systems. In addition, especially in the chemical industry, R&D on micro chemical processes has been energetically conducted for realizing the production of specialty products that have been difficult to produce in conventional chemical plants. To further accelerate industrial applications of micro chemical process technology, several projects have been launched throughout the world. Through the R&D activities, the necessity of developing a systematic design method of micro devices has been recognized. Micro device design problems are different from those of conventional devices. In a conventional design problem, the unit operations are modeled by using terms such as perfect mixing, piston flow, and overall heat transfer coefficient. In other words, each unit operation is modeled as a lumped parameter system. However, the desired performance of micro devices can be achieved by the precise control of the temperature, the residence time distribution and/or the degree of mixing. Hence, each micro device is modeled as a distributed parameter system. And then, the shape of the device must be included in the design variables in addition to the size of the device. The existent papers (Ehrfeld et al., 2000a; Tonomura et al, 2002) convey the fact that the manifold shape of a plate-fin micro device affects the flow distribution in the parallelized microchannels, through computational fluid dynamics (CFD) simulation. Therefore, design problem of micro devices is regarded as a shape optimization problem, in which a cost function defined on a flow domain and/or on its boundary is minimized or maximized under several constraints.

With the advances in computational resources and algorithms, CFD-based optimal shape design is an interesting field for industrial applications such as aerospace, car, train, and shipbuilding. In such design, the computation of the cost function gradient, namely the sensitivity of some performance measure, is the heart of the optimization. Recently, the adjoint variable method (Soto et al., 2004) has attracted the attention as an efficient sensitivity analysis method, since it allows successfully obtaining the shape gradient functions independently of the number of design variables. In this research, an automatic shape optimization system based on the adjoint variable method is developed using C language on a Windows platform. In addition, in order to validate the effectiveness of the developed system, the optimal shape design problems of the pressure-driven microchannels are solved.

General Formulation of the Adjoint-Based Shape Optimization

The optimization technique based on evaluating gradients of the cost function is the easiest way. For each design variable, its value is varied by a small amount, the cost function is recomputed, and the gradients with respect to it are measured. In this case, the number of CFD solutions required for N design variables is $N + 1$. Consequently, the gradient-based method requires enormous computation time when the number of design variables is large. In this study, the adjoint variable method is adopted to obtain gradients in a more expeditious manner.

In a fluid dynamic design optimization problem, the cost function depends on design parameters and the changes in flow variables due to them. The cost function can be written as

$$I = I(\mathbf{W}(\beta), \beta) \tag{1}$$

where I is the cost function, W is the flow variable vector, and β is the design variable vector that represents the surface shape of channels. The cost function I is minimized or maximized subject to partial differential equation (PDE) constraints, geometric constraints, and physical constraints. Examples for the cost function I are drag or pressure drop, for PDE constraints $R(W, \beta)$ the Euler/Navier-Stokes equations, for geometric constraints $g(\beta)$ the volume or cross sectional area, and for physical constraints $h(W)$ a minimal pressure to prevent cavitation.

The principles of the evaluation of gradients based on adjoint variables are given here (Kim, S., et al., 2004; Soto et al., 2004). A total differential in the cost function I and the PDE constraint R results in:

$$dI = \left(\frac{\partial I}{\partial \mathbf{W}}\right) d\mathbf{W} + \left(\frac{\partial I}{\partial \beta}\right) d\beta, \tag{2}$$

$$dR = \left(\frac{\partial R}{\partial \mathbf{W}}\right) d\mathbf{W} + \left(\frac{\partial R}{\partial \beta}\right) d\beta = 0. \tag{3}$$

Next, a Lagrange multiplier λ is introduced to add the flow equation to the cost function:

$$dI = \left(\frac{\partial I}{\partial W}\right)dW + \left(\frac{\partial I}{\partial \beta}\right)d\beta - \lambda^T\left\{\left(\frac{\partial R}{\partial W}\right)dW + \left(\frac{\partial R}{\partial \beta}\right)d\beta\right\}.$$

$$= \left\{\left(\frac{\partial I}{\partial W}\right) - \lambda^T\left(\frac{\partial R}{\partial W}\right)\right\}dW + \left\{\left(\frac{\partial I}{\partial \beta}\right) - \lambda^T\left(\frac{\partial R}{\partial \beta}\right)\right\}d\beta \qquad (4)$$

This implies that if we can solve:

$$\lambda^T\left(\frac{\partial R}{\partial W}\right) = \left(\frac{\partial I}{\partial W}\right), \qquad (5)$$

the variation of I is given by:

$$dI = \left\{\left(\frac{\partial I}{\partial \beta}\right) - \lambda^T\left(\frac{\partial R}{\partial \beta}\right)\right\}d\beta = Gd\beta. \qquad (6)$$

Equation (6) means that the variation of I exhibits only derivatives with respect to β, and that the shape gradient function G is independent of the number of design variables. In the case that R is PDE, the adjoint equation (5) is also PDE, and the appropriate boundary conditions must be determined. The effectiveness of the adjoint-based shape optimization is emphasized along with the increase in design variables.

A Shape Optimization System Development

In this research, an automatic shape optimization system based on the adjoint variable method is developed using C language on a Windows platform. The procedures for building the system are shown in Fig. 1. In principle, after a new shape is obtained, a new grid is generated, and the solution is restarted. For every design cycle, the following steps are required:

1) assume an initial shape,
2) generate computational grids,
3) solve the flow equations, viz. the Navier-Stokes equation and the continuity equation, for deriving the flow velocity and the pressure,
4) solve the adjoint equation to obtain the set of Lagrange multipliers,
5) calculate the shape gradient functions,

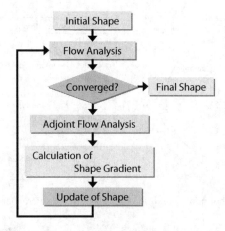

FIGURE 1
Flow chart of shape optimization.

6) obtain a new shape by moving each point on the boundary,
7) go to step 2 unless the change in the cost function is smaller than a desired convergence parameter.

Each design cycle requires a numerical solution of both the flow and the adjoint equations, whose computational time is roughly twice that required to obtain the flow solution.

Case Studies

Flow in microchannels is driven by the pressure difference or the electric potential between inlet and outlet. For pressure-driven flow, an important issue is how to reduce the pressure drop required to realize a desired flow rate in a microchannel. Curved microchannels are often used to provide long flow passage in a compact device. Modification of the shape of curved channels may decrease the pressure drop. In this study, design examples are presented to demonstrate the effectiveness of the developed system for microchannel shape optimization problems. The adjoint variable method is applied to all gradient computations. For convenience, the physical coordinates system is transformed to computational coordinates in the flow and adjoint flow analysis. The two-dimensional computations in the case studies are performed on Windows Intel® 3.0 GHz Pentium 4 processors.

180° Curved Microchannel

The design example is a shape optimization problem of 180° curved microchannels in incompressible flows. The goal of the optimization is to

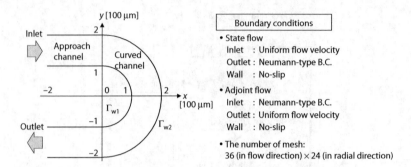

FIGURE 2
Analytical domain and boundary condition for a 180° curved microchannel.

minimize pressure drop for various inlet Reynolds numbers: Re = 0.1, 1, 10, and 100. The cost function is written as

$$I = -\int_{\Gamma_i + \Gamma_o} p u_i n_i ds,$$ (7)

which means the pressure drop between inlet and outlet. The initial shape of the 180° curved microchannel and the main design conditions are shown in Fig. 2. The width of the initial shape is 100 μm. The curved channel is connected with inlet and outlet straight channels. The total number of mesh is 864. The design boundaries are assigned to Γ_{w1} and Γ_{w2}, which correspond to the curved channel, and the design variables are associated with the grid points on both design boundaries. For pressure-driven liquid flow in a microchannel, the no-slip boundary condition is usually valid. The streamwise velocity component at the entrance is specified, and the transverse velocity component at the entrance is assumed to be zero. The prescribed pressure $p = 0$ is assumed at the exit boundary.

Design results at Re = 10 are presented here. Figure 3 shows the initial shape, the final shape under no volume constraint, and the converged shape under

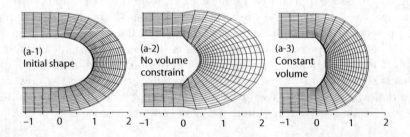

FIGURE 3
Design results: (*a*-1) initial shape, (*a*-2) final shape under no volume constraint at *Re* = 10, (*a*-3) optimal shape under a constant volume constraint at *Re* = 10. Reference frame is prepared below each shape.

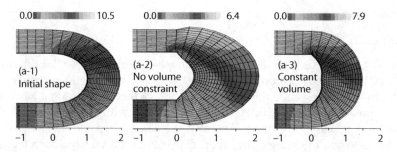

FIGURE 4
Results of pressure distribution for initial and converged shape. Yellow and blue represents high and low pressure, respectively.

a constant volume constraint. Figure 4 illustrates the pressure distributions respectively. As seen in Fig. 4, under no volume constraint, the width of the curved channel is widened, and the shape of the channel is significantly modified. The wider channel makes the flow velocity lower, and a large reduction of pressure drop can be achieved. On the other hand, under a constant volume constraint, both the inside and outside surfaces of the curved channel are moved toward the inlet and outlet, and the flow passage is shortened.

Under a constant volume constraint, the cost function is converged in 92 design iterations and pressure drop is reduced by 27.6%, as compared with that of the initial curved shape. Each design iteration requires approximately 10 seconds. On the other hand, if not under volume constraint, the design cycles are stopped in 30 design iterations due to the fluctuation of the cost function, and pressure drop is decreased by 39.3%, as compared with that of initial curved shape.

In addition, the influence of Re on the result of shape optimization is investigated. Figure 5 shows the optimal shapes at Re = 0.1, 1, and 100 under the constant volume constraint. On the basis of these results, the corresponding reductions in pressure drop are 29.0%, 28.9%, and 19.6%, respectively, as compared with that of initial curved shape. The optimal shapes at Re = 0.1 and 1 are almost the same as that at Re = 10. The optimal shape at Re =100 is different from the others, and its curvature is small due to large flow inertia.

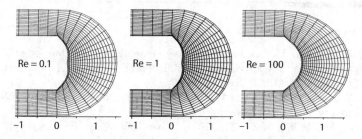

FIGURE 5
Optimal shape at Re = 0.1 (left), 1 (middle), and 100 (right). Reference frame is prepared below each shape.

Conclusions

Most of the previous work on numerical shape optimization for fluid flow can be found in the field of aerodynamics. The nonlinear optimization problems are solved by using descent algorithms based on gradient information. However, the computation of the gradient of the cost function with respect to the design variables is usually very heavy. Recently, the adjoint variable method, which enables us to obtain gradients in a more expeditious manner, has been focused on. In this work, an automatic shape optimization system based on the adjoint variable method is developed by using C language on a Windows platform.

Since the pressure drop in microchannels is an important characteristic related to the energy demand for process optimization, the developed system is applied to the pressure drop minimization problems of microchannels. The last section demonstrates by representative examples that the adjoint variable method can be used to formulate computationally feasible procedures for the shape design of pressure-driven microchannels. The computational time of each design cycle is of the same order as two flow solutions, since the adjoint equation is of comparable complexity to the flow equation. The developed system is quite general and is not limited to particular choice of cost function. Our future work will focus on the extension of the developed system to shape optimization problems of thermo-fluidic microdevices.

Acknowledgments

This research was partially supported by the New Energy and Industrial Technology Development Organization (NEDO), Project of Development of Microspace and Nanospace Reaction Environment Technology for Functional Materials.

References

Ehrfeld, W., Hessel, V., Loewe, H. (2000a). Extending the Knowledge Base in Microfabrication Towards Chemical Engineering and Fluiddynamic Simulation. *In Proceedings of IMRET 4*, 3–20.

Ehrfeld, W., Hessel, V., Lowe, H. (2000b). Microreactors: New Technology for Modern Chemistry. Wiley, VCH, Weinheim.

Hessel, V., Hardt, S., Lowe, H. (2005). Chemical Micro Process Engineering: Fundamentals, Modelling and Reactions. Wiley, VCH, Weinheim.

Kim, S., J. Alonso and Jameson, A. (2004). Multi-element High-lift Configuration Design Optimization Using Viscous Continuous Adjoint Method. *Journal of Aircraft*, 41, 1082–1097.

Kockmann, N. (2006). Micro Process Engineering: Fundamentals, Devices, Fabrication, and Applications. Wiley, VCH, Weinheim.

Kano, M., Fujioka, T., Tonomura, O., Hasebe, S., Noda, M. (2007). Data-based and Model-based Blockage Diagnosis for Stacked Microchemical Processes. *Chem. Eng. J.*, 62, 1073–1080.

Soto, O., Löhner R., Yang, C. (2004). An Adjoint-based Design Methodology for CFD problems. *International Journal of Numerical Methods for Heat & Fluid Flow*, 14, 6, 734–559.

Tonomura, O., Tanaka, S., Noda, M., Kano, M., Hasebe, S., Hashimoto, I. (2003). CFD-based Optimal Design of manifold in Plate-fin Microdevices. *In Book of Abstracts of IMRET 7*, 334–336.

76

Process Superstructure Optimization Using Surrogate Models

Carlos A. Henao, Christos T. Maravelias*

Department of Chemical and Biological Engineering University of Wisconsin
1415 Engineering Drive, Madison—WI 53706

CONTENTS

ABSTRACT Process synthesis through superstructure optimization techniques are generally regarded as theoretically powerful; however, they have not been widely used in practice since they typically result in large-scale nonconvex mixed-integer nonlinear programming (MINLP) models which can not be solved effectively. To address this limitation, we propose a framework leading to substantially simpler formulations which can be solved to optimality. This new approach includes the replacement of complex *first-principle* unit models by compact and yet accurate surrogate models. We show how all the relevant variable relationships established by a unit model, can be expressed in terms of a subset of the original model variable set. We discuss how this subset of variables can be identified, and we present a method to develop high quality surrogate models through artificial neural networks. Finally, a superstructure optimization problem is presented to illustrate the proposed framework.

KEYWORDS *Process synthesis, superstructure optimization, surrogate models, artificial neural networks*

---

* To whom all correspondence should be addressed

Introduction

Current methodologies for the synthesis of chemical processes mainly belong to one of two very different groups: the more traditional sequential and conceptual methods, and the more recent and systematic superstructure optimization techniques. In the traditional approach, engineering decisions are considered to have a natural hierarchy defining a sequential order in which they have to be made (Douglas, 1988). This leads to a design procedure in which the main subsystems of the plant are designed one at a time, disregarding the two-way interaction between decisions made at different stages. On the other hand, superstructure-based approaches initially consider a network composed by all potentially useful unit operations and all the relevant interconnections between them (Grossmann, 1985). This "superstructure" can be formulated as an optimization model with binary selection variables for the *activation* of a subset of units. The optimization model includes the reformulated unit models, interconnection equations and other constraints such as thermodynamic properties' calculation equations. The solution of the optimization models indicates which of the initial units and interconnections are to be kept, as well as the optimal values of the operational conditions.

In theory, the second approach is more powerful because it considers a large number of process alternatives and it involves the simultaneous determination of the optimum structure and operational conditions. However, this power comes at a price: the mathematical complexity of the resulting optimization model, typically, a large-scale non-convex mixed-integer nonlinear program (MINLP).

To address this limitation we present a framework focused on the reduction of the mathematical complexity in superstructure MINLPs. First, we present how the number of equations and variables can be greatly reduced by replacing detailed first-principle models of complex process units by surrogate models. To this end, we propose a novel unit model variable analysis to determine the minimal set of variables which have to be included in a surrogate model. Second, we illustrate how surrogate models can be generated using existing process simulators. Finally, we comment on the formulation of computationally tractable MINLP superstructure optimization models using the proposed surrogate models.

Approaches to engineering design using surrogate models (Won and Ray, 2005) have been motivated by the existence of highly accurate although computationally expensive computer programs which can simulate the behavior of particular engineering systems. In the case of chemical process engineering, a commercial process simulator can be used to generate a set of simulation cases whose values are then fitted using a general purpose multivariable mapping.

Among the most popular techniques to generate such surrogates we find Kriging and Artificial Neural Networks (ANN) (Haykin 1999). Caballero and Grossmann (2008), and Davis and Ierapetritou (2007) have applied

Kriging-based models to the solution of special classes of MINLP's in the synthesis-optimization of chemical process. The literature also presents several works applying ANNs to process modeling, optimization and control problems (e.g. Fernandez 2006, Mujtaba et al. 2006). However, surrogates in general, have not yet been used to reduce the complexity of superstructure models. Furthermore, no thorough and systematic treatment of aspects such as the selection of independent-dependent surrogate variables has been presented. These aspects, as well as others such as sampling will be considered here.

Unit and Surrogate Model Variable Analysis

In an attempt to reduce the mathematical complexity of superstructure MINLPs, we propose to replace original first-principle unit models with surrogate models. However, for this replacement to be helpful, such surrogates have to include as few of the original model variables as possible, making them compact and hence simplifying both their generation and later use. To achieve this goal, it is necessary to analyze the original unit models, the interconnections between the set of process units included in the superstructure, and the scope of the optimization problem.

In essence, a unit model is a set of equations establishing relationships between the unit variables. However, only a subset of these variables, henceforth denoted by I_{UM}, can be used to fully determine the values of the remaining variables, henceforth denoted D_{UM}. In other words, there is an implicit independent-dependent relationship between the variables in I_{UM} (selected to close the degrees of freedom of the model) and its complement D_{UM}.

From a modeling perspective, a detailed unit model can be replaced by a surrogate only if the latter enforces the same relationships established by the former among the variables connecting the unit equations to the rest of the MINLP. In other words, the surrogate has to capture only the relationships among the unit variables that appear in either other unit models (e.g. inlet and outlet stream variables), other constraints, or the objective function (e.g. unit capacities involved in capital cost calculations). We denote this set of connecting variables as C_{UM}.

Set C_{UM} is uniquely determined by the superstructure topology, the form of the objective function, and other design constraints. However, set I_{UM} is not unique (i.e. many different sets of variables can be chosen to close the degrees of freedom of a particular unit model), and its selection has to be based on the model structure. Particularly, a selection of I_{UM} is valid if and only if fixing such variables transforms the unit model into a structurally non-singular system of equations; that is, a square system with no over-determined subsystems. We propose the following mixed integer programming (MIP) model (M) as a way to select the variables in I_{UM} in a systematic way:

$$\max \quad \sum_{j \in J} w_j \cdot y_j \tag{1}$$

$$s.t. \quad \sum_{j:(j,k)\in A} x_{jk} = 1 \quad \forall k \in \mathbf{K} \tag{2}$$

$$\sum_{k:(j,k)\in A} x_{jk} = 1 - y_j \quad \forall j \in \mathbf{J} \tag{3}$$

$$x_{jk}, y_j \in \{0,1\} \tag{4}$$

Here $\mathbf{J}$ and $\mathbf{K}$ are, respectively, the set of variables and equations in the unit model. $\mathbf{A}$ is the set of edges of the bipartite graph describing the variable-equation incidence structure of the unit model (i.e. $\mathbf{A} = \{(j,k): \text{variable } j \in \mathbf{J}$ appears in equation $k \in \mathbf{K}$), x_{jk} is a selection binary variable matching variables and equations, y_j is a selection binary variable defining $\mathbf{I}_{UM}$ as $\{j : y_j = 1\}$, and w_j is a preference coefficient which can be adjusted to favor the inclusion in $\mathbf{I}_{UM}$ of some variables over others. Constraint (3) forbids any matching involving variables selected as part of $\mathbf{I}_{UM}$, while both (2) and (3) impose a perfect matching between the rest of the variables in the model and its equations. This perfect matching is what guarantees the aforementioned non-singularity (Bunus and Fritzson 2004).

Finally, a couple of important aspects have to be considered. First, some of the variables in the optimization problem might be fixed in the form of parameters or operational specifications (e.g. a production level in a reactor). For a particular unit, this set of fixed variables, here denoted as $\mathbf{F}_{UM}$, affects the selection of $\mathbf{I}_{UM}$. In fact, the inclusion of variables $\mathbf{F}_{UM}$ in $\mathbf{I}_{UM}$ has to be enforced by making $y_j = 1 \, \forall \, j \in \mathbf{F}_{UM}$. If model (M) becomes infeasible in the presence of these additional constraints, then the original optimization problem specification scheme is inconsistent and has to be changed. Second, we are interested in building mappings characterizing only relevant features, and since the variability of every variables in a unit model is completely explained by the variability of the variables in $\mathbf{I}_{UM} \backslash \mathbf{F}_{UM}$, the final selection of the dependent and independent variable sets for a surrogate mapping, henceforth denoted as $\mathbf{I}_S$ and $\mathbf{D}_S$, is given by:

$$\mathbf{I}_S = \mathbf{I}_{UM} \backslash \mathbf{F}_{UM} \tag{5}$$

$$\mathbf{D}_S = \mathbf{C}_{UM} \backslash (\mathbf{F}_{UM} \cup \mathbf{I}_{UM}) = \mathbf{C}_{UM} \backslash \mathbf{I}_{UM} \tag{6}$$

Also (5) and (6) lead to:

$$\left| \mathbf{I}_S \right| = \left| \mathbf{I}_{UM} \right| - \left| \mathbf{I}_{UM} \cap \mathbf{F}_{UM} \right| = \left| \mathbf{I}_{UM} \right| - \left| \mathbf{F}_{UM} \right| \tag{7}$$

$$\left| \mathbf{D}_S \right| = \left| \mathbf{C}_{UM} \right| - \left| \mathbf{C}_{UM} \cap \mathbf{I}_{UM} \right| \tag{8}$$

Here, the second equalities in (6) and (7) consider $\mathbf{F}_{UM} \subset \mathbf{I}_{UM}$. Now, since $\mathbf{F}_{UM}$ and $\mathbf{C}_{UM}$ are fixed by the original optimization problem and the unit model,

(8) suggests that the number of variables in the surrogate model, particularly $|\mathbf{D_S}|$, can be reduced by including in $\mathbf{I_{UM}}$ as many of the variables in $\mathbf{C_{UM}}$ as possible. Finally, if data is to be obtained from commercial process simulators, it is convenient to select a $\mathbf{I_{UM}}$ set similar to the set of variables used by the simulator to close the degrees of freedom of a particular unit model, henceforth denoted as $\mathbf{N_{UM}}$. This reduces the use of external specification loops (e.g. HYSYS adjust functions, ASPEN PLUS process specifications, etc.) and the associated computational burden. These requirements can be satisfied by selecting large values for $w_j \; \forall j \in \mathbf{C_{UM}} \cup \mathbf{N_{UM}}$ compared to the values for the rest of the variables.

Surrogate Model Generation

The variable analysis presented in the previous section is general and applicable to any kind of detailed process unit and surrogate model. In this work, we use ANNs due to their excellent fitting characteristics, low complexity and because they can be easily reformulated using selection binary variables and incorporated in superstructure models.

One of the most common ANN configurations is the so called multi-layer perceptron (MLP) (Haykin 1999). An MLP is a highly parallel computational structure composed by a series of connected "layers", each one composed itself by a set of elemental computational units called "neurons". Mathematically, every layer k in a network acts as a transformation between and input vector u^{k-1} and an output vector u^k according to:

$$u^k = f(W^k \cdot u^{k-1} + b^k), \qquad k = 1,..., K \qquad (9)$$

The set of all relations (9) defines the mapping between the network input u^0 and output u^K. Here, parameters b^k and W^k of layer k (composed by n_k neurons) are known as the layer bias vector (n_k-dimensional) and weight matrix ($n_k \times n_{k-1}$ dimensional) respectively. Scalar function f(·), known as activation function, (generally tanh(·) or sigmoid(·)) acts on every component of its argument vector to generate the output vector u^k. The last layer ($k = K$) in an ANN is called the output layer, while the rest are called hidden layers.

In this work, once $\mathbf{I_S}$ and $\mathbf{D_S}$ have been determined for a particular unit, the input space is sampled using a variance reduction technique, such as Latin Hypercube. The sample points are used to specify different simulation cases. The obtained results are then used to train a MLP. Scaling and principal component analysis were used to pre-process the data, in order to reduce the training burden. Also, Bayesian regularization and early stopping where implemented within the training procedure in order to avoid over-fitting, enhancing the generalization capabilities of the network. This is important when dealing with noisy data.

Example 1: Reactor Optimization

The production of Maleic Anhydride (MA) from Benzene (B) involves the following vapor phase catalytic reactions:

$$C_6H_6 + (9/2) \cdot O_2 \rightarrow C_4H_2O_3 + 2 \cdot CO_2 + 2 \cdot H_2O$$

$$C_4H_2O_3 + 3 \cdot O_2 \rightarrow 4 \cdot CO_2 + H_2O$$

$$C_6H_6 + (15/2) \cdot O_2 \rightarrow 6 \cdot CO_2 + 3 \cdot H_2O$$

A CSTR is fed with 966 kmol/h of air and 34 kmol/h of benzene at 300 K, 1013 kPa. Assuming no pressure drop, the goal is to find the operating temperature and reactor volume maximizing the annualized profit. Reaction kinetics are from Westerlink and Westerterp (1988). Annualized profit accounts for revenue from Maleic Anhydride at current price (www.icis.com), capital cost (Guthrie, 1969) and utility cost.

The idea is to use a MLP to replace the original CSTR model. Here, $\mathbf{N_{UM}}$ includes the inlet stream variables (temperature T_I, pressure P_I and component molar flows $\mathbf{F_I}$), reactor volume (V_R), pressure drop (ΔP_R) and outlet temperature (T_O). $\mathbf{C_{UM}}$ includes the Maleic Anhydride production ($F_{MA,O}$) (involved in the revenue term of the objective function), V_R (involved in the capital cost term of the objective function) and the reactor heating (Q_R) (involved in the operational cost term of the objective function). Finally, $\mathbf{F_{UM}}$ includes T_I, P_I, $\mathbf{F_I}$, ΔP_R.

Since, the CSTR model is relatively simple and $\mathbf{F_{UM}} \subset \mathbf{N_{UM}}$, we can determine $\mathbf{I_{UM}}$ without solving problem (M). In fact selecting $\mathbf{I_{UM}} = \mathbf{N_{UM}}$ satisfies all the desirable characteristics of $\mathbf{I_{UM}}$. In this way $\mathbf{I_S} = \{V_R, T_O\}$ and $\mathbf{D_S} = \{Q_R, F_{MA,O}\}$.

Figure 1 presents the values of $F_{MA,O}$ obtained from the original and surrogate models. Here, the surrogate is a MLP with a linear output layer, 2 hidden layers with 5 neurons each, and using a tanh($\cdot$) as activation function. The maximum deviation obtained for the dependent variables was only 0.4% of the mapping range. The optimization problem was solved using CONOPT in GAMS v22.5. The optimal reactor temperature is 670 K and the optimal reactor volume is 40m^3.

Superstructure Model Formulation

Since a superstructure consists of all potentially useful process units, the corresponding optimization model should account for the selection/non-selection of these units. This implies that a unit model (or its surrogate) should be enforced only if the corresponding unit is selected. This is typically achieved using disjunctive programming methods employing binary

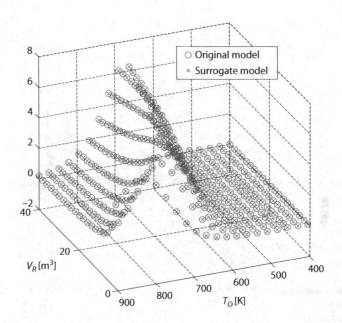

FIGURE 1

$F_{MA,O}$ [kmol/hr] Vs. V_R[m³] - T_O [K].

selection variables (Lee and Grossmann, 2000). The resulting models however are often computationally challenging.

Despite being nonlinear, multilayer perceptron models can be reformulated into models that do not contain additional nonlinear terms. Thus, the resulting superstructure optimization models are computationally tractable. The following example illustrates how such a reformulation was used to address a superstructure optimization problem.

Example 2: Superstructure Optimization

In the context of the production on Maleic Anhydride from Benzene, consider the simple process superstructure presented in Figure 2, which consists of five units: a CSTR (R_1), a PFR (R_2), a flash tank (F), a splitter (S), and a mixer (M). Our goal is to find the reactor type (CSTR or PFR), reactor temperature and volume, flash temperature and recycle fraction, that maximizes the annualized profit. The system feed, kinetics and annualized profit calculations are as in the previous example.

The formulation involves the reformulated surrogate models for the mixer ($|I_S|$=7, $|D_S|$=7), CSTR ($|I_S|$=9, $|D_S|$=8), PFR, ($|I_S|$=9, $|D_S|$=8) and flash tank ($|I_S|$=8,$|D_S|$=15). The optimization was performed using GAMS 22.5 – DICOPT. The optimal solution involves a 40 m³ PFR at 680K, a flash unit at 333 K, and recycle stream that is equal to 46% of the flash vapor stream.

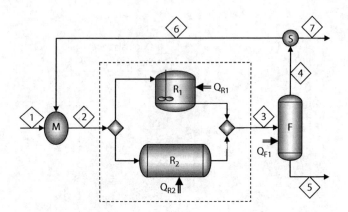

FIGURE 2
MA process superstructure.

Conclusions

In this paper, we presented a framework for the formulation of computationally tractable superstructure optimization models. The key idea is the replacement of process unit models by ANNs. In achieving this, we presented a systematic method for the selection of connecting, independent and dependent variables; we discussed how surrogate models can be generated using process simulators; and illustrated the applicability of our framework using a small superstructure example.

References

Bunus, P,. Fritzson, P. (2004). Automated static analysis of equation-based components. *Sim. Tran. Soc. Mod. & Sim. Int.* 80, 321.

Caballero, J.A., Grossmann, I. E. (2008). An Algorithm for the Use of Surrogate Models in Modular Flowsheet Optimization. *AICHE J.* 54, 2633.

Davis, E., Ierapetritou, M. (2007). A Kriging Method for the Solution of Nonlinear Programs with Black-Box Functions. *AICHE J.* 53, 2001

Douglas, J.M. (1988). *Conceptual Design of Chemical Processes.* McGraw-Hill, New York.

Fernandes, FAN. (2006). Optimization of Fischer-Tropsch Synthesis Using Neural Networks. *Chem. Eng. & Tech.* 29, 449.

Grossmann, I. E. (1985). Mixed-integer programming approach for the synthesis of integrated process flowsheets. *Comput. Chem.. Eng.,* 9, 463.

Guthrie, K. M. (1969). Data and Techniques for Preliminary Capital Cost Estimating. *Chem. Eng.* 76,114.

Haykin, S. (1999). Artificial Neural Networks: A comprehensive foundation. *Prentice Hall.* Upper Saddle River, NJ.

Lee, S., Grossmann, I. E., (2000). New algorithms for nonlinear generalized disjunctive programming. *Comp. Chem. Eng.* 24,2125.

Mujtaba, IM; Aziz, N; Hussain, MA. 2006. Neural network based modelling and control in batch reactor. *Chem. Eng. Res. Des.* 84,635.

Westerlink, E. J., Westerterp, K. R. (1988). Safe Design of Cooled Tubular Reactors for Exothermic Multiple Reactions: Multiple reactions networks. *Chem. Eng. Sci.* 43, 1051.

Won, KS., Ray, T. (2005). A Framework for Design Optimization Using Surrogates. *Eng. Optim.* 37,685.

77

Mathematical Modeling—Knowledge Acquisition about Brain Physics

Brian J. Sweetman, Sukhraaj S. Basati, Madhu S. Iyer and Andreas A. Linninger[*]
Laboratory for Product and Process Design
Departments of Chemical Engineering & Bioengineering
University of Illinois at Chicago
Chicago IL, 60607

CONTENTS

ABSTRACT In this article, three biomedical applications for the advantageous use of mathematical modeling techniques are explored. Significant progress toward improved understanding of human intracranial dynamics via medical imaging and advanced mathematical tools is demonstrated. Such tools are used to reconstruct cerebrospinal fluid flow fields and tissue stresses and strains from incomplete experimental data. Second, a novel volume sensor for the treatment of hydrocephalus is reported. Medical imaging and rigorous mathematical analysis lead to optimal sensor design and optimal placement for highest sensitivity. Third, mathematical modeling is applied to improve current methods of invasive drug delivery into the human brain. A case study of chemotoxin delivery to a glioblastoma is introduced as a distributed optimization problem. The model is intended to rationally design drug delivery techniques, which to date are mainly developed by intuition. The application of these systems techniques advance an understanding of brain physics and may lead to improved therapeutic options for patients suffering from brain diseases.

---

[*] To whom all correspondence should be addressed

805

KEYWORDS   *Brain physics, medical imaging, convection enhanced drug delivery*

Introduction

In process systems engineering, mathematical modeling is widely used to increase knowledge about dynamics and steady state performance of chemical reactions, transport phenomena, and separation processes. In biological sciences and medicine, mathematical models are often viewed with suspicion because necessary simplifications to represent the complex interactions between biochemistry, genetics, and transport are unavoidable. However, in our group we have successfully introduced mathematical modeling techniques for quantifying dynamic fluid flow phenomena in the brain (Linninger et al., 2008a). We advocate the use of rigorous conservation balances, chemical reaction, and transport phenomena in systems engineering for knowledge discovery in biological systems analysis. In this article, it is demonstrated that mathematical analysis of distributed biological systems aids in quantitative interpretation of incomplete experimental data and helps to better correlate intricately linked phenomena in biological systems.

This article describes several recent developments and examples of the successful application of rigorous mathematical modeling principles for the discovery of transport phenomena in porous brain tissue. The first example illustrates the use of fluid mechanics to understand cerebrospinal fluid (CSF) flow in the human brain. The second example demonstrates the potential of mathematical programming for medical device design. Methods developed in process systems engineering are now advantageously applied to optimize clinical devices. Specifically, systems engineering methods are shown to be effective for the optimal design of an impedance catheter to measure the ventricular volume in the brain. Finally, the concept of rational therapy design is introduced in which distributed mathematical programming techniques with partial differential equation constraints help develop patient-specific treatment plans. Mathematical programming solutions will render optimal therapy parameters in a systematic and mathematically concise fashion.

Mathematical Modeling Case Studies

Case Study A—Intracranial Dynamics

The first example for deploying basic transport mechanisms to explain phenomena in biological systems is intracranial dynamics. Intracranial dynamics describes the interactions of blood flow, CSF flow, and the deformation of porous brain tissue. While specific data about viscosity, density, and porosity about the brain and its fluids are known, there currently exists no comprehensive mathematical description of blood flow or its interaction with CSF and the flow field that results from the pulsating vasculature. While detailed aspects of normal pulsatile deformations of brain tissue can be quantified using existing medical imaging techniques, the complete picture of cause and effect is missing. Quantification of force interactions between blood, CSF, and the brain is necessary to understand normal dynamics as well as pathological conditions affecting the human brain.

In spite of these challenges, progress in medical imaging combined with better software for reconstruction makes it possible to render complex geometries of human organs on a computer. Even for an entire patient-specific human brain, one can transform medical images into computer representations containing precise coordinates as well as the connectivity of surfaces and three dimensional spaces for a specific patient. Meshing reconstructed surfaces and volumes yields a computational domain composed of tetrahedral balance envelopes which allow the numerical solution of basic conservation balances for mass, species, and momentum. A detailed description of our patient-specific image reconstruction methodology is discussed elsewhere (Linninger et al., 2007).

We compare the computational domain and special fluid and solid boundary conditions with CINE-phase contrast MRI measurements of CSF velocities to predict the entire CSF flow field inside the patient's central nervous system (CNS). Specifically, first principle conservation balances such as the Navier and Navier-Stokes equations are solved over the computational domain to obtain blood and CSF flow fields, pressure fields, and volumetric flow rates. In addition, tissue deformations along with material stresses and strains are computed. The purpose of the work is to combine actual measurements of the normal CSF dynamics to produce a three dimensional, first principle explanation of the pulsatile dynamics of blood and CSF in the human CNS.

Typically, medical imaging modalities render measurements at a few specific areas of interest, so that the entire flow field cannot be acquired experimentally. In addition even accurate measurements in specific areas of interest are not sufficient to comprehend the driving forces as well as the cause and effect relationships that produce complex CSF and blood flow patterns in the brain. However, the solution of three dimensional conservation balances with experimental measurements provides a complete picture of brain dynamics. Specifically, we accurately reproduce and predict the complex flow patterns in the subarachnoidal spaces and brain ventricles. Furthermore, dependent states that cannot be easily measured such as intracranial pressures, deformations, strains, and displacements of brain tissue are quantified. Therefore, more insight is gained from experimental measurements with the help of mathematical modeling.

First principles models from reconstructed images explains subtle changes in the flow fields and quantifies other symptoms associated with pathologies

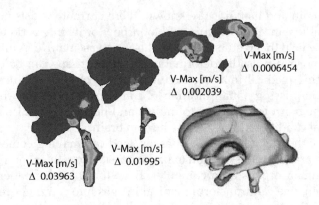

FIGURE 1

CSF velocity field in the cerebral ventricular system of patient with communicating hydrocephalus. The velocity field is displayed in sagittal cross-sections of 5 mm increments from the mid-sagittal plane. In hydrocephalus, aqueductal flow increases.

of intracranial dynamics. Specific indicators we discovered include the decrease in the prepontine stroke volume and the increase in the aqueductal flow as highlighted in Fig. 1.

Future work will refine the model in the spirit of hierarchal decomposition to accurately capture the cerebral capillary network, which in the current models is grossly simplified. In a step by step process, it is believed that mathematical modeling will be able to uncover the functionality of the physical brain which to date is still elusive. More detail regarding the modeling approach and transport mechanisms are explained in Somayaji et al. (2008).

Case Study B—Computer Aided Design of Volume Sensor

The second application shows opportunity for mathematical programming techniques of distributed systems and engineering design optimization in clinical research. In classical biological research, there is a tendency to perform design optimization directly on animal models. This practice often leads to very expensive experiments on animals such as rats, pigs or primates in an iterative trial and error process. Design parameters are improved by several stages from small animals to large animals until a device or treatment for humans has been generated. It is no surprise that the scaling procedure is very cumbersome and often entails a substantial amount of time and cost, not withstanding the suffering of the test animals. Scaling is a very well known task in chemical process design, which, thanks to system engineering methods, has been largely improved with the help of flow sheet simulations and other mathematical modeling techniques. Experimental and computational methods accelerate and intensify the knowledge gain and design decisions for clinical device development guided by animal experimentation. We point out critical advantages that come from distributed systems design and simulation efforts to accelerate the scale-up procedure with minimal animal suffering.

As an example, we present the development of a medical device to measure the ventricular volume of the human brain. The enlargement of ventricles, also known as ventriculomegaly, is associated with a disease known as hydrocephalus causing patients to suffer from nausea, severe headaches, and even life-threatening intracranial pressure rises. Our research points to an improved treatment option based on directly measuring the ventricular volume coupled with feedback process control well known in our community, but rarely used in medicine.

For making this feedback control system possible, a sensor to detect the actual ventricular volume is necessary. Our group has been working on developing such a sensor based on an impedance measurement principle. The sensor is capable of rendering a strong correlation between the ventricular volume and a voltage potential drop acquired by a series of electrodes allocated in a fixed distance inside the lateral human ventricle. The measured voltage drop is due to a small electromagnetic field induced by two excitatory electrodes. Experiments demonstrated that the voltage drop is a function of the ventricular size according to the distribution of current within the cerebrospinal fluid enclosed by the surrounding brain tissue (Linninger 2008c). Due to the contrast in electrical conductance of brain tissue and CSF, the expansion of the cavity directly affects the voltage drop detected on the measurement electrodes. However, the determination of optimal volume sensor parameters requires intensive animal testing and experimentation by trial and error. Sensor parameters such as electromagnetic field strength, electrode quantity, electrode spacing, catheter positioning, instrumentation frequency and gain, and biocompatibility are part of a large design problem. Instead of simultaneously optimizing all these parameters for each animal simultaneously in realistic conditions, the development process can be accelerated using computational methods including image reconstruction as described before.

The computer simulation allows accurate calculation of the three dimensional electromagnetic field as a function of ventricular expansion. In addition it is possible to place a catheter into the computer simulation and solve the Maxwell equations to predict what the sensor would measure if it had been placed in a patient with enlarged ventricles. It would be impossible to experiment with voltage potential changes in a human suffering from hydrocephalus. This computer simulation serves as a *virtual laboratory* to explore and experiment with different catheter parameters such as spacing, and positioning in order to minimize noise and maximize the sensitivity of the catheter and its position with respect to the enlarging ventricles. Physiological parameters based on animal experimentations and accurate

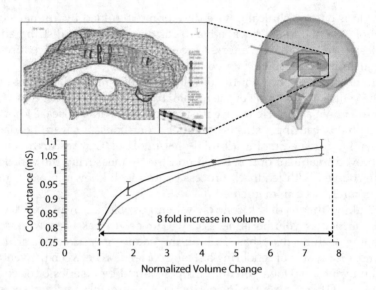

FIGURE 2

Maxwell's equations are solved in a three dimensional, unstructured, computational mesh derived from a patient's MR images to predict the electric vector field resulting from a sensor. The bottom graph shows simulated volume-conductance relationships predicted over a range of an eight fold increase in ventricular volume.

three dimensional patient-specific reconstructions of a human brain were used to optimize the device *in silico* as shown in Fig. 2.

A challenge previously not addressed is where to place the catheter so that the particular shape of the ventricular enlargement is captured with minimal accuracy loss. Current experiments are underway to optimize the sensor parameters using computational simulations which would be impossible to conduct in humans. At the same time similar computational models for rats have been developed, guiding the necessary animal experimentation. With simultaneous computer design procedures, the animal testing will follow a systematic process, since experimental data will be supported with rigorous mathematical techniques.

We hope that the computational methods accelerating the scale-up processes in chemical industry over the last fifty years will also be successful in introducing a scientific method in the development of clinical devices.

Case Study C—Catheter Placement for Drug Delivery

Currently, biomedical therapy devices are usually optimized by trial and error as pointed out previously. The next case demonstrates the large potential of mathematical programming for rational therapy design in biomedical

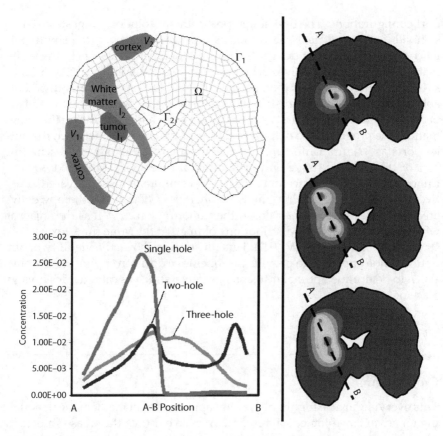

FIGURE 3

(Left) The conceptual treatment plan for glioblastoma targets the tumor (I1) and adjacent white matter (I2), without exceeding toxicity limits in the cortex (V1, V2). (Right) The optimal drug delivery plans with single, two- and three-hole catheters.

research. The objective is to maximize chemo-toxic agents in tumors. High concentrations of the chemotherapeutic agents in the brain can destroy the tumor and prevent reoccurrence. At the same time, the drug concentration should be low in other areas to spare critical regions in the cortex that would otherwise lead to detrimental side effects. A conceptual therapy plan given by the physician is depicted in Fig. 3. The problem we wish to address with rigorous optimization is to precisely locate the optimal catheter position and design for administering the cytotoxin. Three catheter designs were available to choose from: single port, two port, and a three port catheter (Linninger et al., 2008b).

The distributed optimal drug delivery problem was solved with mathematical programming methods. Fig. 3 compares achievable concentration profiles and distribution volumes in different brain regions with various port configurations. The best single-port catheter solution suggests an infusion site positioned near the tumor as expected. However, the single outlet configuration cannot deliver the desired therapeutic dosage to the white matter. The treatment only reached 46% of targeted volumes, I1, and I2, in sufficiently high concentration so that the risk for recurrence is unacceptable. A two-hole catheter distributes the drug over a wider range with two peak concentrations, the first near the tumor and the second inside the white matter. The program also determined the optimal distance between the outlet ports as 39 mm. Unfortunately, therapeutic thresholds were reached in 72% of the targeted tissue only. The optimal three-port catheter doses the tumor at therapeutic concentrations, I1, and simultaneously delivered effective drug concentrations into the white matter, I2. The chemotherapy delivered by a three-hole catheter dosed the tumor region as well as the adjacent white matter tracts (100% distribution volume). At the same time, toxic levels stayed safe in the cortex (V1, V2). This illustration shows the viability of our optimization approach to provide pre-operative decision support to the physician to determine optimal infusion parameters such as catheter design and placement.

Conclusions

This overview article highlights several applications of mathematical modeling for advancing biomedical research. We have shown three case studies in which rigorous mathematical modeling has led to insight into pathological conditions, design of medical devices, and design of rational therapy parameters. These examples and the underlying systems methodologies provide a scientific approach for discovery about complex brain functions and is a contribution to a true interdisciplinary research effort. We hope that this overview provides a motivation for process systems engineers to venture into the field of biomedical research with methods that have been proven successful in the chemical industry and hold enormous potential to advance knowledge in biomedical research.

Acknowledgments

The authors would like to gratefully acknowledge partial financial support of this project from NIH Grant 5R21EB004956, a grant from the Stars Kids Foundation, as well as NSF Grant CBET-0756154.

References

Linninger, A.A., Xenos, M., Zhu, D.C., Somayaji, M.B., Penn, R. (2007). Cerebrospinal fluid flow in the normal and hydrocephalic brain. *IEEE Trans. Biomed. Eng., 54,* 291.

Linninger, A.A., Somayaji, M.R., Zhang, L., Hariharan, M.S., Penn, R. (2008a). Rigorous mathematical modeling techniques for optimal delivery of macromolecules to the brain. *IEEE Trans. Biomed. Eng., 55(9),* 2303.

Linninger, A.A., Somayaji, M.R., Mekarski, M., Zhang, L. (2008b). Prediction of convection-enhanced drug delivery to the human brain. *J. Theor. Biol., 250, 125.*

Linninger, A. (2008c). Monitoring and controlling hydrocephalus. US Patent 005440. 10 Jan. 2008.

Somayaji, M.R., Xenos, M., Zhang, L., Mekarski, M., Linninger, A.A. (2008). Systematic design of drug delivery therapies. *Comput. Chem. Eng., 32, 89.*

78

Automated Targeting for Total Property Network with Bilateral Constraints

Denny Kok Sum Ng*[1], Dominic Chwan Yee Foo[1] and Raymond R. Tan[2]

[1]*Department of Chemical and Environmental Engineering*
University of Nottingham Malaysia, Broga Road, 43500 Semenyih,
Selangor Malaysia
[2]*Department of Chemical Engineering, De La Salle University-Manila,*
2401 Taft Avenue, 1004 Manila, Philippines

CONTENTS

ABSTRACT This paper presents an optimization-based procedure known as *automated targeting technique* to determine the minimum flowrate/cost for a *total property network* (TPN), prior to detailed design. A TPN consists of the individual elements of material reuse/recycle, waste interception for further recovery (reuse/recycle) or for final discharge. Due to the close interaction among these individual elements, they should be considered simultaneously during network synthesis. In addition, the targeting procedure address the bilateral problem where the property operator values of the process sinks and sources exist in between the property operator values of external fresh resources. This concept is important whenever a nominal desired or targeted sink property operator value exists that is neither the highest (superior) nor the lowest (inferior) value of the process sources and sinks, as in the conventional cases. An industry case study is solved to illustrate the proposed approach.

KEYWORDS *Bilateral property integration, process integration, resource conservation, targeting, waste minimization, optimization*

---

* To whom all correspondence should be addressed

Introduction

Resource conservation is an effective way to reduce operational cost and to enhance sustainability. El-Halwagi (1997) defined process integration as *a holistic approach to process design, retrofitting and operation which emphasises the unity of the process*. Over the past decades, extensive works have been reported for the synthesis of water (Wang and Smith, 1994; El-Halwagi et al., 2003; Manan et al., 2004) and utility gas (Alves and Towler, 2002; El-Halwagi et al., 2003) networks as the special cases of mass integration. The net result of resource conservation activities is the simultaneous reduction of fresh material consumption and waste generation.

Noted that however, most previous works have been restricted to "chemo-centric" or concentration-based systems where the quality of streams are described in terms of the concentration of contaminants. However, there are many applications in which stream quality is characterised by physical or chemical properties rather than contaminant concentration. This led to the concept of property integration (Shelley and El-Halwagi, 2000).

Most recently, Ng et al. (2008) introduced a new variant known as *bilateral property integration problem* in which the property operator values of the process sinks and sources exist in between the property operator values of external fresh resources (Ng et al., 2008). This concept is important whenever a nominal desired or targeted sink property operator value exists that is neither the highest (superior) nor the lowest (inferior) of all operator values of the process sources and sinks, as in the conventional cases (Kazantzi and El-Halwagi, 2005; Foo et al. 2006).

In this work, a recently developed automated targeting approach (Ng et al., 2008) is extended to the TPN wherein material reuse/recycle, waste interception for further recovery and/or for final discharge are considered simultaneously. In addition, the targeting procedure address the bilateral problem where the property operator values of the process sinks and sources exist in between the property operator values of external fresh resources. Based on the concept of insight-based targeting, the automated targeting technique is formulated as a linear programming (LP) model, which ensures that a global optimum can be found if one exists. The main limitation of this work is that it is restricted to single property problems.

Problem Statement

The problem of a TPN is stated as follows:

Given a set of *sources i* that may be reused/recycled and/or intercepted to be sent to the process sinks or for final discharge. Each source has a flowrate, F_i and is characterised by a single constant property, p_i. A set of process sinks *j* are specified. Each sink requires a flowrate of F_j and is restricted to complies with the predetermined allowable property constraints as follows:

$$p_j^{\min} \le p_j \le p_j^{\max} \tag{1}$$

where $p_j^{\min}$ and $p_j^{\max}$ are the specified lower and upper bounds of the admissible properties for sink *j*. External sources are readily available to supplement the flowrate required by the sinks ($F_{FW,j}$). For final discharge, the environment legislation restricts the quality of waste ($p^{discharge}$) to comply with the property limit ($p^{environment}$):

$$p^{\text{discharge}} \le p^{\text{environment}} \tag{2}$$

A linearized property mixing rule is needed to define all possible mixing patterns among the individual properties (Shelley and El-Halwagi, 2000):

$$\psi(\bar{p}) = \sum_i x_i \psi(p_i) \tag{3}$$

where $\psi(p_i)$ and $\psi(\bar{p})$ are linearizing operators on source property p_i and mixture property $\bar{p}$, respectively; while x_i is the fractional contribution of stream *i* in the total mixture flowrate. The optimization objective is to minimize flowrate/cost for a TPN.

Automated Targeting Technique

The automated targeting technique was originally developed for mass exchange network synthesis (El-Halwagi and Manousiothakis, 1990). It was extended by Ng et al. (2008) for property network based on water cascade analysis technique (Manan et al., 2004). However, the previous work did not consider waste treatment. This aspect is now included.

Following the previous approach, the first step is to construct a revised *property interval diagram* (Ng et al., 2008). The sinks and sources are arranged in ascending order based on the property operator (ψ_k). In cases where the property operator levels for fresh resource(s) and zero property operator level that do not exist within the process sinks and sources, an additional level is added. An arbitrary value is also added at the final level (highest among all property operators) of the property interval diagram to allow the calculation of residue property load. Next the flowrate and property load cascades are performed across all property operator levels based on Equations 4 and 5 respectively.

$$\delta_k = \delta_{k-1} + (\Sigma_i F_{\text{SR}i} - \Sigma_j F_{\text{SK}j})_k \tag{4}$$

$$\varepsilon_k = \varepsilon_{k-1} + \delta_k (\psi_{k+1} - \psi_k) \tag{5}$$

As shown in Equation 4, the sum of the *net material flowrate* cascaded from the earlier property operator level $k - 1$ (δ_{k-1}) with the flowrate balance ($\Sigma_i F_{SRi} - \Sigma_j F_{SKj}$) at property operator level k form the net material flowrate of each k-th level (δ_k). Meanwhile, the property load at each property operator interval is given by the product of δ_k and the difference between two adjacent property operator levels ($\psi_{k+1} - \psi_k$). The residue of the property load of each property operator level k (ε_k) is to be cascaded down to the next property operator level. The residual property load, ε must take a non-negative value:

$$\varepsilon_k \geq 0 \tag{6}$$

For the bilateral problem, two reverse sets of cascades are carried out simultaneously for the reuse/recycle network (RRN, Ng et al., 2008). Note that each set of cascades consists of both material flowrate and property load cascades. In the first set of cascades, the property operators are arranged in ascending order where fresh resource is located at the lowest operator level. This set of cascades is termed as the *superior cascade* (referring to the fresh resource that is found at the superior operator level). In contrast, property operators are arranged in descending order in the second set of cascades (*inferior cascade*), where a second fresh resource is located at the highest operator level (Ng et al., 2008). Thus, Equation 5 is modified as Equation 7, so that the difference between the property levels k and $k + 1$ is set as positive. This ensures a feasible property load cascade to be generated.

$$\varepsilon_k = \varepsilon_{k-1} + \delta_k \left(\psi_k - \psi_{k+1} \right) \tag{7}$$

Since the superior and inferior cascades are optimized simultaneously and the property load is cascaded in two directions, all the process sinks receive property load up to their maximum limits. The accumulation of flowrate for wastewater discharge at the final property operator level is not allowed. Thus, in each level where a source exists, a new sink is added to allow for waste discharge (F_T) to avoid accumulation of waste when the net material flowrate (δ_k) is cascaded. Thus, Equation 4 is modified for both superior and inferior cascades of property operator level as follow:

$$\delta_k = \delta_{k-1} + \left(\Sigma_i F_{SRi} - \Sigma_j F_{SKj} - F_T \right)_k \tag{8}$$

To incorporate waste interception into the automated targeting technique, modification is required on the proposed model. To enhance material recovery via reuse/recycle and to comply with the environmental limits, the waste sources are sent to the waste interception network (WIN) to improve its quality. To determine the minimum treatment flowrate, flowrate and property load cascades for WIN is added. Besides, the discharge flowrate is added as a new sink at the property level of the final discharge limit. Since different sets of cascades (RRN and WIN) are included; different key properties can be used for RRN and WIN syntheses. Note that the presented automated targeting model (Equations 4–8) is linear; thus, a global optimum can be found if one exists.

TABLE 1

Limiting Data for Case Study

Process	Flowrate (kg/day)	Density, ρ (kg/m³)	Operator, $\psi(\rho) \times 10^{-4}$ (m³/kg)	TSS, C (ppm)	Operator, $\psi(C)$ (ppm)
(Sink)					
Clay solution (SK1)	10	1120	8.93	-	-
(Source)					
Over flow (SR1)	4	1018	9.82	800	800
Bottom flow (SR2)	6	1200	8.33	2000	2000
Reject stream (RR)	∞	1600	6.25	7472.5	7472.5
Purified stream (RP)	∞	1100	9.09	10	10
Fresh water (FW)	∞	1000	10		
Clay	∞	2600	3.85		

Case Study

Table 1 shows the limiting data for a water recovery problem in palm oil production, adapted from Ng et al. (2008). Clay bath flotation technique (Figure 1) is used to separate kernel from shell and un-cracked nut. It consists of slurry water whose proportions are chosen to achieve the desired density. However, the quality of the slurry degrades during use through inadvertent separation of the clay particles from water. Besides, the cracked mixture consists of impurities that affect the density of the solution in the clay bath. Therefore, make-up water and fresh clay are fed into the system to compensate for losses and to maintain the separation efficiency.

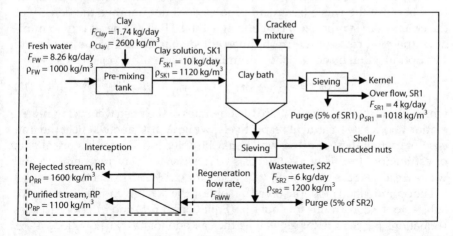

FIGURE 1
Schematic diagram of clay bath.

In order to achieve high separation efficiency of the cracked mixture, the slurry in the clay bath should posses a density of 1120 kg/m³. Thus, density is identified as the most critical property for wastewater recovery. Shelley and El-Halwagi (2000) reported the general mixing rule for density as follow:

$$\frac{1}{\bar{\rho}} = \sum_i \frac{x_i}{\rho_i} \tag{9}$$

where $\bar{\rho}$ is the mean density of the mixture; ρ_i is the density of i stream.

As shown in Table 1, two water sources (SR1 and SR2) are considered for water recovery. In this work, a filtration separation system is introduced to purify the clay bath bottom stream for further recovery. Ng et al. (2008) termed a regeneration process with two outlet streams as *partitioning regeneration* system, where a feed stream is separated into two outlet streams of different quality, i.e. a higher quality product stream and a lower quality reject stream. Both purified (RP) and reject streams (RR) from the filtration unit are allowed to be recycled to the clay bath, but not simultaneously. Five percent of the wastewater is purged from both the over flow and bottom flow streams to prevent accumulation of impurities in the clay bath. Besides, to avoid remixing of the purified and reject streams, Equation 10 is added:

$$F_{RP} \times F_{RR} = 0 \tag{10}$$

where F_{RP} and F_{RR} refer to the flowrate of the purified and reject streams. However, including this bilinear constraint converts the problem into a non-linear model. Meanwhile, concentration of total suspended solid (TSS) is chosen as the main property for final discharge and is given as 250 ppm; a fixed outlet concentration of 20 ppm is in assumed for the treatment process. Besides, the treated effluent from WIN can not be reused/recycled to the RRN because the waste treatment process does not improve the density of the effluent. To synthesize a cost effective total TPN, it is necessary to minimize the total costs of the fresh, intercepted sources and waste treatment. The optimization objective is to minimize total operating cost (TOC):

$$TOC = AF_{FW} + BF_{Clay} + CF_{RWW} + DF_{WW} \tag{11}$$

where F_{FW}, F_{Clay}, F_{RWW} and F_{WW} are the flowrates of the fresh water, clay, regeneration and waste treatment respectively; while A, B, C and D reflect the unit cost for each of these sources, which may be determined from historical data or estimation. In this work, the costs of fresh water, clay, regeneration and waste treatment are given as \$1/t, \$2/t, \$0.5/t and \$0.5/t respectively.

The optimization model is solved by solving the TOC given as in Equation 11, subject to the constraints for superior (Equations 5, 6, 8) and inferior cascades (Equations 6 – 8) of RRN; as well as the cascade for WIN (Equations 4 – 6, 10), which yields the result as shown in Figure 2. The minimum flowrate targets of fresh water and clay are found as 1.01 kg/day respectively; with

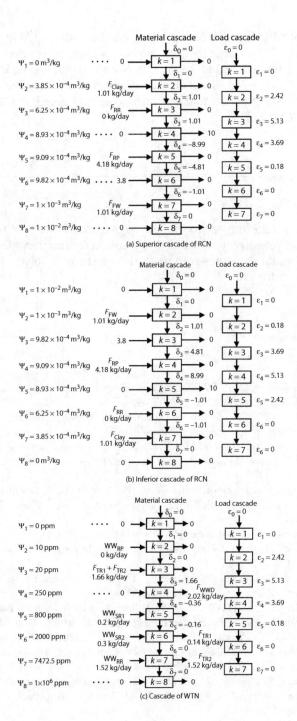

FIGURE 2
(a) Superior cascade (b) Inferior cascade of RRN (c) cascade of WIN.

5.7 kg/day of SR2 being intercepted to produce 4.18 kg/day of F_{RP} and 1.52 kg/day of WW_{RR}. Note that the purified stream is recycled to the sink and the reject stream is sent for treatment. Note also that 1.66 kg/day of waste $(F_{TR1} + F_{TR2})$ is treated and a total of 2.02 kg/day of waste (F_{WWD}) is discharged from the WIN. The total annual operating cost is targeted as \$1240 (annual operation of 365 days).

Conclusion

The automated targeting approach is extended for TPN where the individual elements of reuse/recycle, waste interception for recovery/discharge are considered simultaneously. An industrial case study is solved to illustrate the proposed approach.

References

Alves, J. J. and Towler, G. P. (2002). Analysis of refinery hydrogen distribution systems. *Ind. Eng. Chem. Res.*, 41, 5759–5769.

El-Halwagi, M. M. (1997). Pollution prevention through process integration: Systematic design tools. *Academic Press*. San Diego.

El-Halwagi, M. M., Gabriel F. and Harell, D. (2003). Rigorous Graphical Targeting for Resource Conservation via Material Recycle/Reuse Networks. *Ind. Eng. Chem. Res.*, 42, 4319–4328.

El-Halwagi, M. M. and Manousiothakis V. (1990). Automatic Synthesis of Mass-Exchange Networks with Single Component Targets. *Chem. Eng. Sci.*, 9, 2813–2831.

Foo, D. C. Y., Kazantzi, V., El-Halwagi, M. M. and Manan, Z. A. (2006). Surplus Diagram and Cascade Analysis Techniques for Targeting Property-Based Material Reuse Network. *Chem. Eng. Sci.*, 61, 2626–2642.

Kazantzi, V. and El-Halwagi, M. M. (2005). Targeting Material Reuse via Property Integration. *Chem. Eng. Prog.*, 101 (8), 28–37.

Manan, Z. A., Tan, Y. L. and Foo, D.C.Y. (2004). Targeting the Minimum Water Flowrate using Water Cascade Analysis Technique. *AIChE J.*, 50 (12), 3169–3183.

Ng, D. K. S., Foo, D. C. Y., Tan, R. R., Tan Y. L. and Pau, C. H. (2008). Automated Targeting for Conventional and Bilateral Property-Based Resource Conservation Network. *Chem. Eng. J.*, (doi: 10.1016/j.cej.2008.10.003).

Shelley, M. D. and El-Halwagi, M. M. (2000). Componentless Design of Recovery and Allocation Systems: A Functionality-Based Clustering Approach. *Comput. Chem. Eng.*, 24, 2081–2091.

Wang, Y. P. and Smith, R. (1994). Wastewater Minimisation. *Chem. Eng. Sci.*, 49 (7), 981–1006.

79

Multi-level Synthesis of Chemical Processes

Daniel Montolio-Rodriguez[1,2], David Linke[3] and Patrick Linke[1,4*]
[1]*Centre for Process & Information Systems Engineering, School of Engineering, University of Surrey, Guildford, Surrey, GU2 7XH, U.K.*
[2]*Jacobs Engineering, Process & Technology Division, Jacobs House, 427 London Road, Reading, Berkshire, RG6 1BL, United Kingdom*
[3]*Leibniz Institute for Catalysis, Richard-Willstätter-Strasse 12, 12489 Berlin, Germany*
[4]*Department of Chemical Engineering, Texas A&M University at Qatar, Education City, PO Box 23078, Qatar.*

CONTENTS

ABSTRACT   We will present a multi-level optimization approach that utilizes superstructure formulations to determine and analyse optimal process designs for heterogeneously catalyzed gas-phase reaction systems. The approach relies on a compact and practical process design representation that is used both in a high-level design-decision making framework aimed at the development of design performance targets and the identification of interactions between design performance and design complexity, and in a more detailed synthesis exercise where non-idealities for heterogeneously catalyzed gas-phase reaction systems are taken into account. The overall synthesis strategy allows to explore the major design trends at each decision level whilst effectively balancing the numerical and combinatorial complexities along the synthesis exercise. The developments are illustrated with an applications in styrene production.

---

* To whom all correspondence should be addressed

KEYWORDS *Process synthesis, optimisation, superstructure networks, multi-level design evolution*

Introduction

Systematic process synthesis methods capable of identifying optimal designs have been highlighted as key technologies to deliver major improvements in process efficiencies in the future (Tsoka et al., 2004). A number of such methods have been proposed that allow the identification of optimal conceptual designs for a given chemistry using optimization techniques coupled with process network superstructure representations (e.g., Papalexandri and Pistikopoulos, 1996; Linke and Kokossis, 2003a,b). The generic conceptual design focus of the methods lead to limitations to handle practical process constraints, complex kinetic models and to support the design-decision making processes needed to evolve designs throughout the design cycle. Truly integrated design would additionally require integration with kinetic model development activities to enable the simultaneous exploitation of synergies between chemistries and process designs. This paper focuses on addressing the shortcomings of conceptual synthesis methods by proposing with a multi-level process synthesis decision-support framework capable of supporting process design decisions at different levels of detail that can handle complex kinetics and be used to integrate process design activities into the kinetic model development stage in the future.

Synthesis Requirements and Approach

Current process design approaches follow a largely sequential information flow in terms of chemistry investigations, conceptual process design and detailed process design. This leaves many synergies that exist at the interfaces of these activities unexploited and results in designs that can substantially be improved. Ideally, the design activity should exploit the synergies between activities in the design cycle from chemistry investigation through to detailed process design. This would require an integrated process design framework, which is able to generate insights about optimal conceptual and detailed process design options on the basis of results from chemistry investigation and catalyst design. Insights from process design could then feed back to the chemistry investigation stage to ensure integrated decision making throughout the design cycle. Figure 1 illustrates the envisioned integration.

To achieve this integration, robust synthesis methods are required as enabling technologies that can reasonably quickly develop design candidates at conceptual levels as well as at detailed levels on the basis of complex

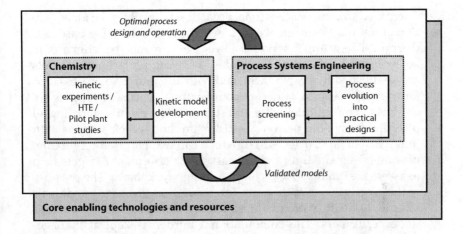

FIGURE 1
Integration and cordination of design activities at the process engineering and kinetic investigation on interface.

kinetic information so as to enable integration with the catalyst development activity. We have developed such an integrated synthesis method. At high-level, the approach employs compact and practical process design representations to screen vast numbers of possible process configurations in order to identify promising conceptual design candidates. The framework allows the development of design performance targets and the identification of interactions between design performance and design complexity to identify those promising designs that should be considered for more detailed analysis at the following levels and eventually evolved into detailed designs. As the synthesis exercise progresses, the knowledge that has emerged about the process structure is employed to reduce the number of structural alternatives to be investigated whilst more detailed process models are employed.

The developments capitalize on our previous work in the development of integrated process synthesis methods (Linke and Kokossis, 2003a,b). Due to their importance in industrial practice, we have focused our efforts on process representations specifically tailored to heterogeneously catalyzed gas-phase reacting systems.

Multi-level Process Synthesis Method

The process synthesis is performed in multiple stages to generate maximum design insight and understanding whilst identifying the most promising conceptual design candidates and these converging to the most promising detailed design candidates:

- *Conceptual screening stages:* Conceptual design involves the identification of optimal performances achieved by conventional or base case designs. These limits serve as lower performance benchmarks to be exceeded by the synthesized designs. Next, performance targets are established that can be achieved with an innovative process configuration regardless of design complexity. Performance targets are established by optimizing the full superstructure whilst allowing all possible interactions between feed streams and recycle streams with multiple reaction zones each of which can exhibit different mixing, heat management and catalyst mass. The performance targets help to assess the quality of less complex design candidates. The purpose of these early stages is to screen large numbers of design candidates in order to identify promising solutions whilst eliminating obvious underperformers. The computational burden is kept at manageable levels by employing simple process models to reduce the time required to assess large numbers of possible candidates.

- *Conceptual design stages:* In these stages, the importance of design features identified in the conceptual screening stages is analyzed. This is achieved by assessing the effect of individual structural options identified during screening on design performance. This enables assessment of complexity-performance relationships that will help the designer identify solutions that strike a good balance between design complexity and design performance.

- *Design evolution stages:* The designs developed in the conceptual design stages are analyzed and key design features are identified for which targeted superstructures are optimized to study the impact of individual design features as well as the effect of non-ideal behavior of the reaction system by including more detailed process models. The computational burden is kept at manageable levels as the combinatorial burden is greatly reduced in these stages so that the additional burden from more detailed process models does not result in increased overall computational demands. As the synthesis exercise progresses, the knowledge that has emerged about the process structure is employed to eliminate suboptimal structures and more complexity in terms of the process models can be afforded. By employing more detailed and computational demanding models in later design stages, the few optimal conceptual designs resulting from the multilevel approach can be explored thoroughly. The addition of modelling detail enables evolving the optimal conceptual designs into more realistic options. The evolution takes the form of an iterative process that is performed in multiple stages. The solution of each stage becomes the starting point of the next. The number of stages is not limited and the evolution can progress in as many stages as deemed necessary for the kind of reactive systems at hand.

Illustrative Example

A case study in the production of styrene is used to illustrate the methodology. Gas-phase heterogeneously catalysed dehydrogenation of ethylbenzene (Elnashaie et al., 2001) is the selected production route. Unreacted and produced components (ethylbenzene, styrene, toluene, benzene, ethylene, methane, steam, carbon monoxide, carbon dioxide and hydrogen) are cooled before entering the condenser. The condenser separates the condensable components (ethylbenzene, styrene, toluene and benzene) from the rest, which are purged. Condensable components are separated in a distillation sequence and benzene, toluene and styrene are the products of the process, whereas unreacted ethylbenzene is recycled to the first reactor. The reactor used in industry for the production of styrene by dehydrogenation of ethylbenzene is typically the adiabatic reactor (Elnashaie *et al.*, 2001). The objective of this application is to identify trends and key features of the system of reactors that enhance the overall process performance in terms of economic potential.

Conceptual Screening

The superstructures employed in the screening stage contained four reactor synthesis units (reaction zones) with full stream connectivity. The separation network contained a flash drum and three distillation columns to recover unreacted feed, product and byproduct. A total of 30 optimisation experiments are performed, all starting from different initial feasible points. In the screening stage, the superstructure is optimised without imposing structural constraints (i.e. reactor units can be added/deleted in the superstructure, bypasses and recycles between reactors can be identified, feeds can be distributed to any reactor zone present in the superstructure). PFRs (fixed-bed and multi-tubular) and CSTRs are considered alternatives by the reactor synthesis units. The optimization yields a maximum economic potential of 11.37 M$/yr for the optimal system, which represents an 18 % improvement as compared to an optimized classical design with an adiabatic reactor.

Conceptual Design

The analysis of the different process structures obtained in the conceptual screening stage yields the following observations in terms of design features:

- All designs identified have multiple reaction zones, most of which are multi-tubular PFRs except for about one in ten structures, where a multi-tubular reactor is followed by a CSTR.

- Ethylbenzene feed is distributed amongst reaction zones. Superheated steam is not distributed but fed to the first reaction zone.
- Recycle streams between reaction zones are present in about one third of the process design candidates. However, the recycled fractions are very close to the lower limit (5 %) imposed in the superstructure formulation.
- No bypasses are present in any of the structures.

These individual features were then studied in detail in further optimizations of reduced superstructures to establish their effect on the systems performance. Multiple reaction zones of multi-tubular PFRs and ethylbenzene side feeding were observed to have a significant impact on process performance. Recycles, bypasses and backmixed reaction zones towards the outlet of the reaction network did not significantly affect the performance. As a result of the conclusions extracted in conceptual, the proposed optimal conceptual process design candidate to be studied in detail is the structure formed by three PFRs in series with ethylbenzene feed distribution (Figure 2).

Design Evolution

In conceptual design, simple one-dimensional reactor models with temperature profiles are employed. Real reactors may not be able to attain the identified optimal temperature profiles and radial temperature effects may influence performance. In the design evolution stages, the optimal design identified in conceptual design stage is evolved in three stages to consider the implications of more detailed models on design:

- **Stage 1:** In order to suggest practical solutions to the heat management inside reactors, one-dimensional models are employed which ensure that temperature profiles for PFRs are developed can be attained with common cooling strategies.

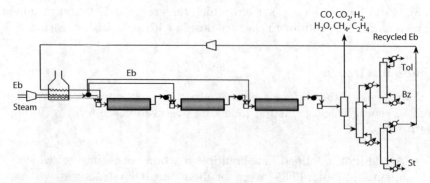

FIGURE 2
Optimal conceptual design candidate proposed as a result of the multi-level approach.

- **Stage 2:** Even more realistic models are employed for the plug-flow zones to determine the additional impact of radial temperature distributions, which is often significant in gas phase reaction systems.

- **Stage 3:** In all previous optimizations, plug flow reactors are approximated by cell models to ensure the optimization formulations include algebraic equations only in order to keep computational times at reasonable levels. This stage assesses the impact of more realistic differential algebraic PFR models with radial temperature approximation. In addition, well mixed units are replaced by more realistic fluidized bed reactors representations based on the Bubble Assemblage Model (Kato and Wen, 1969).

The evolution of the optimal conceptual design, the three multi-tubular PFRs structure with EB side feeding, reveals after design evolution stage 3 an enhancement in economic potential of 11 % as compared to optimized processes based on the adiabatic reactor design typically found in practice.

Conclusions

The developed multi-level synthesis approach aims to be an enabling technology to achieve the effective integration of process design and kinetic investigation activities in the near future. The approach relies on a multi-level strategy that identifies the maximum performance of a system regardless its complexity and helps the engineer in understanding the relationships between design features and performance at different levels of detail. The approach highlights design bottlenecks and suggests different high-performance conceptual designs for the engineer to base their design decisions around. It further allows to explore the effect of nonideal phenomena on process performance which are not usually captured by conceptual design methods. This helps in evolving conceptual designs into more realistic representations of the real system in order to ensure practicality of the developments and support validation.

References

Elnashaie, S. S. E. H., Abdallah, B. K., Elshishini, S. S., Alkhowaiter, S., Noureldeen, M. B., Alsoudani, T., 2001. On the link between intrinsic catalytic reactions kinetics and the development of catalytic process. Catalytic dehydrogenation of ethylbenzene to styrene, Catalysis Today, 64(3–4), 151–162.

Kato, K., Wen, C. Y., 1969. Bubble assemblage model for fluidized-bed catalytic reactors. Chemical Engineering Science, 24(8), 1351–1369.

Linke, P., Kokossis, A. C., 2003a. Attainable designs for reaction and separation processes from a superstructure-based approach. AIChE Journal 49(6), 1451–1470.

Linke, P., Kokossis, A. C., 2003b. On the robust application of stochastic optimisation technology for the synthesis of reaction/separation systems. Computers & Chemical Engineering 27(5), 733–758.

Papalexandri K. P., Pistikopoulos, E. N., 1996. Generalized modular representation framework for process synthesis. AIChE Journal, 42(4), 1010–1032.

Tsoka, C., Johns, W. R., Linke, P., Kokossis, A. C., 2004. Towards sustainability and green chemical engineering: tools and technology requirements. Green Chemistry, 8, 401–406.

80

On the Development of Optimal Process Design Knowledge Using Semantic Models

Claudia Labrador-Darder[1], Antonis Kokossis[1] and Patrick Linke[1,2*]

[1]*Centre for Process & Information Systems Engineering, School of Engineering, University of Surrey, Guildford, Surrey, GU2 7XH, U.K.*

[2]*Department of Chemical Engineering, Texas A&M University at Qatar, Education City, PO Box 23078, Qatar.*

CONTENTS

ABSTRACT The paper presents a systematic approach to develop process synthesis knowledge that combines optimisation, semantic models (ontologies) and analytical tools. The work addresses the representation and extraction of process synthesis knowledge during the optimisation process with the purpose to simplify and interpret design results. The simplification is achieved with a gradual evolution of the design solutions and corresponding adjustments of the optimisation search. The interpretation is accomplished with the use of analytical tools to translate data into descriptive terms understood by users. Means of analysis include dynamic ontologies populated by computer experiments and continuously upgraded in the course of optimisation.

---

* To whom all correspondence should be addressed

KEYWORDS   *Optimisation, ontology, superstructure, reactor networks*

Introduction

The paper presents a systematic approach for the extraction, interpretation and exploitation of design knowledge in process synthesis that combines optimization, semantic models (ontologies) and analytical tools. The work addresses the representation and extraction of process synthesis knowledge during the optimisation process. The approach systematically extracts information with the purpose to simplify and interpret design results to enable monitoring the search. The work is presented with application to reactor networks synthesis by means of superstructure optimisation (see e.g. Kokossis & Floudas, 1990; Marcoulaki & Kokossis, 1999) against stochastic optimisation techniques.

This paper outlines the components of the synthesis framework with emphasis on the knowledge representation and extraction aspects of the method. The approach is applied to single phase reactor networks and extended to multiphase reactor systems. The method is illustrated with an isothermal homogeneous reaction system and an isothermal multiphase reactor system example from the literature (Marcoulaki and Kokossis, 1999).

Synthesis Approach

The approach is based on the gradual accumulation of design knowledge and its deployment in the course of synthesis experiments. The method attains knowledge to reduce the synthesis structure with the use of an ontology employed in parallel to the optimisation search. The latter takes the form of a gradual process, the initial stage of which is an exhaustive superstructure. The superstructure is optimised and updated at different stages. The transition from one stage to another represents different layers of abstraction. Each stage is assigned a knowledge model populated by features obtained from the optimisation solutions that captures the information of the superstructure. At the highest (initial) level, the method employs the largest superstructure and a general knowledge model where all feasible options are embedded. In the course of optimisation, the superstructure becomes leaner whereas the knowledge model becomes richer and is populated with solution features and relationships. The proposed synthesis strategy is illustrated in Figure 1 and is presented for the synthesis of reactor network but is not restricted to any particular type of application. Different stages correspond to different superstructure optimisation exercises. The link between individual optimisation stages

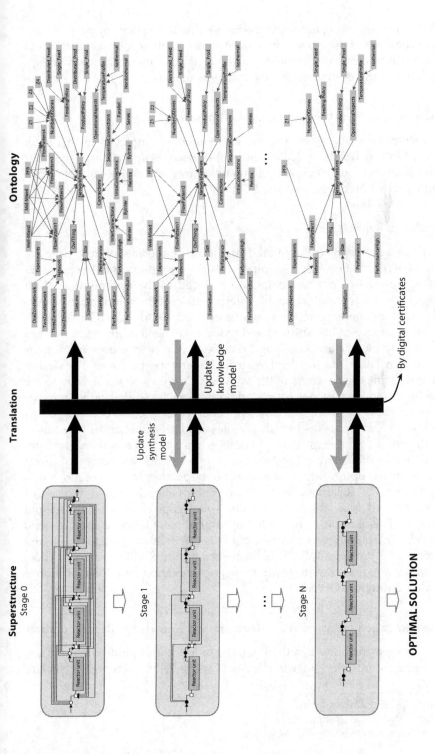

FIGURE 1

Synthesis strategy and components.

and the corresponding ontology is addressed with the development of digital certificates and the employment of clustering methods. Digital certificates embody the information of ontology in digital form. They update the ontology, which is, in turn, used to update the synthesis model.

Superstructure Representation and Optimisation

The reactor network superstructure representation employed in this work follows Linke & Kokossis (2003a) and is optimised using stochastic methods in the form of Tabu Search and Simulated Annealing. following the implementation by Linke & Kokossis (2003b).

Knowledge Representation

The knowledge representation takes the form of a formal ontology for the domain of reactor networks that represents the design information captured by the superstructure representation employed in the optimisation. With such structured representation, the interpretation of the solution can be possible and thus monitoring the search is enabled. Ontologies (Gruber, 1993) allow knowledge to be captured and made available to both machines and humans. By the use of an ontology, data related to reactor design solutions can be translated into descriptive terms understood by users. The design features define the concepts of the ontology. Concepts relate directly to the structural and operational process design variables and include: number of reactive zones, mixing per reactive zone, feeding, connections in terms of recycles and bypasses (which are classified as intra- and interconnections), product source (structural components) and size and temperature profile along the network (operational components). They represent direct links with the optimisation stage and are populated by the solutions of a particular stage. Relationships between concepts are described in terms of taxonomic and associative relationships. Knowledge is extracted from input concepts. This is used to upgrade the synthesis model or support general analysis. The ontology is encoded in OWL (Web Ontology Language) and edited using Protégé-OWL[1]. It is supported by the RACER[2] reasoner functioning as a logical classifier and ontology verification tool. Visualisation is enabled through GrOWL[3].

Communication between Superstructure and Knowledge Representation

Knowledge availability has the potential to assist and guide the transition from one optimisation stage to the next. Design solutions represent specific

---

[1] http://protege.stanford.edu/
[2] http://www.racer-systems.com/
[3] http://www.uvm.edu/~skrivov/growl/

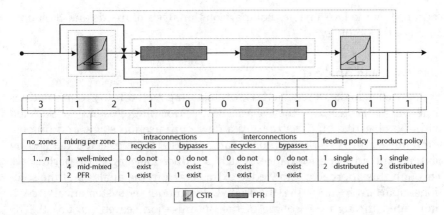

no_zones	mixing per zone	intraconnections		interconnections		feeding policy	product policy
		recycles	bypasses	recycles	bypasses		
1... n	1 well-mixed 4 mid-mixed 2 PFR	0 do not exist 1 exist	0 do not exist 1 exist	0 do not exist 1 exist	0 do not exist 1 exist	1 single 2 distributed	1 single 2 distributed

FIGURE 2
Syntheis representation and DC for a solution.

cases of the superstructure and are associated with a specific ontology. The design solutions are in numerical format unlike the ontology, which is expressed in semantic terms. This demands a translation mechanism which we establish with the introduction of digital certificates.

Digital Certificates

Digital certificates (DCs) gather specific information about design candidates in the form of a vector. They represent the information of the ontology in a numerical format to enable the analysis of solutions. DCs are issued for the solutions generated in the optimisation searches. DCs are autogenerated with the use of CLIPS[4]. CLIPS enables the construction of a rule-based system. The set of rules constitutes a knowledge base that draws on domain knowledge. In this case, rules represent "rules of thumb" which specify a set of actions to be performed for given conditions (i.e. relationships between the design variables of the superstructure). The rules relate to active reactive units and connections between them in the superstructure. The active reactive units are combined into reactive zones depending on the mixing pattern favoured and connections in the form of recycles and bypasses that relate to these reactive zones are extracted to be represented by the DCs. The rules are used as an input to the inference engine (CLIPS) which automatically matches facts (or data) against patterns (or conditions) and determines which rules are applicable. Valid rules are executed until no applicable rules remain. As a result of executing the rules, DCs are issued for each solution representing a simplified equivalent superstructure in terms of main features. For the purposes of the application, the DCs represent for each solution the number of reactive zones, the mixing pattern, the connections between reactive zones and the feeding and product policy (Figure 2). The information that is

---

[4] http://clipsrules.sourceforge.net/

captured by the DCs enables comparisons, analysis of trends and the population of the ontology.

Clustering and Analysis

The approach makes a repeated use of clustering methods to group solutions in terms of the features they share. Best performing clusters are selected with the objective of: (i) setting up a new optimisation stage, and (ii) customizing features of the optimal search. Clusters are selected depending on their spread in performance, which is reduced as stages are performed. Through this approach, knowledge extracted from the analysis of the solutions, apart from updating the knowledge model, is systematically communicated through the course of the optimisation search (Figure 1). The acquisition of knowledge subsequently guides the search towards high performance regions branching off those superstructure features that are of limited importance. In such a way, the synthesis model is updated and adapted throughout the optimisation process as stages are performed, making the optimisation more effective and robust.

Expansion to Multiphase Systems

The synthesis strategy (Figure 1) is also applied to multiphase reactor systems, namely gas-liquid and liquid-liquid systems. Modifications on the main components of the approach are introduced in order to handle the complexity involved in multiphase reactor networks optimisation problems.

The superstructure representation adopted here follows Linke & Kokossis (2003a,b). The superstructure consists of two phases: a reactive phase and a non-reactive phase. The superstructure is optimised using stochastic methods in the form of Simulated Annealing (Kirkpatrick et al., 1983).

Both the ontology and the DCs developed for single phase networks are extended to account for the new features that appear when moving to multiphase systems. Apart from the operational and structural information about the reactive phase, that for its contacting phase also needs to be incorporated along with the features resulting from the possible contacting and mixing patterns that exist between the phases. Concepts specific to multiphase systems include the definition of: the reactive/non-reactive nature of the phases, the physical state of the phases, additional features of the network such as combined mixing patterns, overall flow arrangement and mass transfer, features specific to each phase such as loops (necessary sequential connections between two non-consecutive phase compartments) and recycles (for the non-reactive phase).

For multiphase systems, DCs consist of two parts: *a phase description part*, where the information about each phase (reactive and non-reactive phase) in terms of the mixing pattern of each of the compartments belonging to the phase and the connectivity between them is included; and *a general network description part*, where information such as the number of synthesis units forming the network and the link between the contacting phases is given.

The *phase description part* is built upon the structure of the digital certificates presented for single phase applications. Here, single phase categories are duplicated due to the addition of an additional phase to include information specific to each phase.

In the *general network description part*, the connections between contacting phases are expressed in terms of the existence of mass transfer in each of the synthesis units and the flow pattern of each synthesis unit, which includes co-current and counter-current options when only PFRs are involved. The number of synthesis units forming the network is also included in this category. DCs gather information regarding: *features specific to each phase*: i) mixing pattern of each compartment, ii) connections between compartments which include sequential connections (existence of streams between two consecutive compartments) that are only included for the non-reactive phase and loops, iii) feeding policy and iv) product policy; *general features*: i) number of synthesis units, ii) existence of mass tranfer in each synthesis unit, iii) flow pattern of each synthesis unit (co-current or counter-current). Compact digital certificates (CDCs) are specifically generated to attain a clear conceptualisation of the multiphase problem which finds to be hampered by the big amount of features included in the extended DCs. Developed from the information represented by the DCs, in the CDCs only the most relevant features are considered to represent a more conceptual representation of the main design features for such systems.

Both vectors are created in parallel for each optimisation stage. On one side, CDCs contain enough information to discriminate between solutions making them a conceptual tool to be employed to monitor and guide the optimisation search. On the other hand, DCs contain much more detailed information and they are intended for a detailed analysis of the features that appear throughout the optimisation. Therefore, unless specific detail is required for a given solution, the CDCs are employed throughout the methodology.

Numerical Example

Ten initial solutions are considered as starting point for the optimisation search. The reduction of the superstructure is attained as DCs emerge with common features. Clustering identifies promising features, which are taken into account in next stages, whereas irrelevant features are gradually

TABLE 1

Results of the Optimisation Stages for Denbigh

Stage	Selection Criteria (%)	Clusters Selected				Clusters Generated	Maximum Objective
		One Units	Two Units	Three Units	Total		
1	50	3	20	11	34	46	11.4472
2	10	3	19	3	25	342	11.8905
3	5	3	15	3	21	318	12.2965
4	2	3	3	3	9	228	12.4265
5	1	3	3	3	9	63	12.7823
6	1	3	3	3	9	10	12.7823
7	1	3	3	3	9	9	12.7852
8	1	3	3	3	9	9	12.7906
9	1	3	3	3	9	9	12.8034
10	1	3	3	3	9	9	12.8034

excluded. The best 50%, 10%, 5%, 2% and 1% of the clusters generated are selected in each stage respectively.

The example presented is a multiphase version of the Denbigh reaction (Mehta & Kokossis, 2000). The example consists of a reactive gas-liquid system. The objective is to maximise the yield of component B. The superstructure includes up to three RMX units. The reactors considered includes CSTRs and PFRs. Ten initial solutions are considered as starting point for the optimisation search. The reduction of the superstructure is attained as DCs emerge with common features. Clustering identifies promising features, which are taken into account in next stages, whereas irrelevant features are gradually excluded. The best 50%, 10%, 5%, 2% and 1% of the clusters generated are selected in each stage respectively. Computer experiments are performed for a Markov chain of ten. The initial annealing temperature is set to 1000 and is decreased by an order of magnitude as stages are performed. The cooling schedule by Aarts and Van Larhoven (1985) is used ($\gamma = 0.05$). A minimum of three clusters for each one unit, two unit and three unit networks are selected amongst the clusters generated at each stage (Table 1) in order to make sure that layouts consisting of any number of units have the option to attain higher objectives in later stages of the optimisation which is not always possible due to the higher number of possible combinations of optimisation variables which increase with the number of units considered.

The optimal solution suggested by Simulated Annealing (SA) agrees with those obtained in this work (Table 2). The pair of loops found in the optimal solution for SA corresponds to the total loop identified in this work. The flow arrangement from the last unit is equivalent in both cases (the counter-current/co-current concept loses significance with no mass transfer). Computational efforts are reduced by 62% when using the proposed method,

TABLE 2

Comparison of Results for Denbigh

Denbigh case A	Stats[2]	Function evaluations	Objective (kmol/h)	Total volume (m³)	Optimal structure[1]		
This work	total	14821	12.8034	506.96			
					217.20 m³	182.67 m³	107.09 m³
Simulated Annealing	ave	3880	12.8023	506.64			
	max	5306	12.8036	513.63			
	min	3069	12.8006	499.70			
					214.54 m³	186.37 m³	107.33 m³
Simulated Annealing (Mehta,1998)	ave	–	12.2840	–			
	max		12.7800				
	min		11.7700		67.40 m³	439.90 m³	

if compared with a classical SA approach. The presented results outperform those found in the literature both in quality and maximum objective identified.

Conclusions

The paper presents a systematic synthesis approach that combines superstructure-based optimization, knowledge models and analytical tools. The work addresses the representation and extraction of process synthesis knowledge during the optimisation process and is also applicable to other types of models, applications and optimization techniques. The results show how important features and patterns are retrieved at very early stages of process design. In the initial stage, all possible solutions in terms of their features are included in the initial knowledge representation and an exhaustive superstructure where all feasible options are embedded is considered. As the optimisation goes on, relevant features emerge and others are eliminated delivering a reduced set of solutions (reduced synthesis model). This is represented in parallel in an enriched knowledge model that defines the set of optimal solutions with semantic terms understood by the user. The systematic interpretation of solutions yields to an understanding of the solution space and to a systematic reduction of the representation employed. experiments to be performed.

References

Glover, F. (1993). *Ann. Oper. Res.*, 41(3).

Gruber, T.R. (1993). *Knowl. Acquis.*, 5, 199.

Kirkpatrick, S., Gellat, C.D., Vecchi, M. P. Jr. (1983). *Sci.* 220 (4598), 671.

Kokossis, A.C., Floudas, C.A. (1990). *Chem. Eng. Sci.*, 45, 595.

Linke, P., Kokossis, A.C. (2003a). *Comp. Chem. Eng.*, 27, 733.

Linke, P., Kokossis, A.C., 2003a, *AIChE J*, 49, 1451.

Marcoulaki, E. C., Kokossis, A. C. (1999). *AIChE J*, 45, 1977.

Mehta, V. L., Kokossis, A. C. (2000). *AIChE J*, 46(11), 2256.

81

Fuzzy Logic Based System Modification
for Industrial Sustainability Enhancement

Zheng Liu[1], Yinlun Huang[1]* and Cristina Piluso[2]
[1]Department of Chemical Engineering and Materials Science
Wayne State University, Detroit, MI 48202
[2]BASF Corporation
Florham Park, NJ 07932

CONTENTS

ABSTRACT Industrial sustainability is pursued to achieve the long-term sustainable development of industrial systems. It is experienced that industrial sustainability problems are always difficult to handle, mainly because of system complexity and inherent uncertainties. In usual practice, the uncertainties encountered in Design for Sustainability (DfS) are processed mostly through deriving an agreeable solution by the parties involved. In this paper, we introduce a systematic, fuzzy-logic-based sustainability enhancement methodology that can be used readily for identifying and implementing a sustainability-bearing material-design solution for a given industrial system problem under uncertainties. The methodological efficacy is demonstrated through studying an automotive manufacturing-centered industrial regional problem for its sustainable development.

KEYWORDS Sustainability assessment, fuzzy logic, decision analysis, system improvement

---

* To whom all correspondence should be addressed. E-mail: yhuang@wayne.edu.

Introduction

Industrial sustainability is pursued to achieve the long-term sustainable development (SD) of industrial systems that can be defined as a geographic area comprising a network of manufacturing sectors, each having a number of plants. In practice, decisions and strategies for an industrial region's development must be made, reviewed, and assessed regularly by the industrial planners, business leaders, and neighboring communities.

Industrial sustainability problems are always difficult to be investigated because of their dimension and scope that carry high complexity and uncertainties contained in data, information, and possessed knowledge. In practice, the uncertainties appeared in Design for Sustainability (DfS) are handled mainly through deriving an agreeable solution by the parties involved. It is highly desirable that solutions can be identified in a holistic way. A solution approach should be capable of assessing the state of sustainability of an industrial region and the identification of superior solutions for improving system's sustainability. In this paper, a systematic, fuzzy-logic-based sustainability enhancement methodology is introduced, which contains two methods: one for system sustainability analysis under uncertainty, and the other for system design improvement for sustainable development.

Fuzzy Sustainability Assessment

One of the most challenging problems in sustainability assessment is how to deal with uncertain information. In this regard, a fuzzy logic based method is constructed, where uncertainties are expressed as fuzzy numbers and intervals. The assessment can be conducted by utilizing a sustainability assessment/enhancement oriented knowledge base, which is composed of a large number of fuzzy rules. The rules are structured in two layers. The lower layer contains the rules for assessing each of the specific types of sustainability, e.g., economic, environmental, and social sustainability, using the selected sustainability indicators. The upper layer has a number of rules designed for evaluating the overall industrial sustainability based on the triple-bottom-line information obtained from the lower layer.

Rule structure. All the rules in the knowledge base have a uniform IF-THEN structure. The rules in the lower and the upper layer of the knowledge base are, respectively, defined as follows:

$$R^{L,i}: \quad IF \left\{ x_j \text{ is } A_j^i \middle| j \in R^m \right\}$$

$$THEN \left\{ y_{k,p}^i = a_{k,0}^i + \sum_{j=1}^{m} a_{k,j}^i x_j \middle| k \in R^n, p \in R^3 \right\} \tag{1}$$

$$R^{U,i}: \quad IF \left\{ y^i_{k,p} \text{ is } B^i_{k,p} \,\middle|\, k \in R^n, p \in R^3 \right\}$$

$$THEN \left\{ z^i_S = b^i_{k,0} + \sum_{j=1}^{m} b^i_{k,j} \, y^i_{k,p} \right\} \tag{2}$$

where x_j is the j-th variable used in a sustainability indicator; $y^i_{k,p}$ is the k-th variable in the p-th sustainability type (i.e., economic, environmental, and social types); z^i_S is the overall industrial sustainability variable; A^i_j, $B^i_{k,p}$ are the fuzzy sets; $a^i_{j,k}$, $b^i_{j,k}$ are the coefficients in the rules that are derived based on the available data.

Fuzzy assessment. The assessment is conducted based on the selected rules in the two layers of the knowledge base. Thus, the rule selection process is critical in achieving a superior assessment quality. In this work, the MAX-MIN algorithm (Zimmermann (1991) is adopted for systematically activating the most suitable rules one by one. The algorithm can be mathematically expressed as:

$$\mu_j(x) = \max_{j \in J} \{ \min_{i \in I} \{ \mu_{i1}(x_1), \cdots, \mu_{ik}(x_k), \cdots, \mu_{iN}(x_N) \} \} \tag{3}$$

where $\mu_{ik}(x)$ is the membership function of variable x in fuzzy set k, representing the truth value of the i-th rule in the j-th level; $\mu_j(x)$ is the membership function of variable x in the fuzzy set that pertains to the rule selected at the j-th level.

Note that the MIN operation in Eq. (3) gives the truth value of the condition part of each rule in a rule set, while the MAX operation determines the selection of a rule from the rule set. This algorithm needs to be applied to each of the two layers in the knowledge base. That is, for an assessment task, after the data and information is input to the fuzzy assessment method, the MAX-MIN algorithm will be used to select the most appropriate rule from each of the three sustainability types in the lower layer of the knowledge base. This will result in three rules activated from each type of sustainability. The assessment in the three types will generate the information needed for the upper layer. Then, the MAX-MIN algorithm will be used again to identify the most suitable rule in the upper layer of the knowledge base and use that rule to determine the level of industrial sustainability of the industrial system.

Fuzzy Assessment Guided System Modification

System modification for sustainability improvement can be guided by the fuzzy assessment method. As the economic, environmental and social sustainability indicators as well as the composite indicator for the industrial

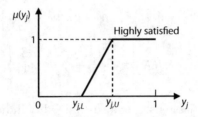

FIGURE 1
Satisfaction fuzzy set definition.

sustainability are evaluated, the satisfaction of the system's sustainability level can be determined using three fuzzy sets, each of which can be defined in the way shown in Fig. 1.

As shown in the figure above, if the sustainability indicator, y_j, has a value less than $y_{j,L}$, then that type of sustainability is completely unsatisfactory as $\mu(y_j)$ is 0; if y_j is greater then $y_{j,U}$, then $\mu(y_j)$ is 1, indicating a complete satisfaction; if y_j is between $y_{j,L}$ and $y_{j,U}$, then $\mu(y_j)$ has a value between 0 and 1, showing a partial satisfaction. A decision maker can decide if the satisfaction level acceptable or not based on the value of $\mu(y_j)$ for a specific sustainability type.

If the system's sustainability level is not fully satisfied, then the following five-step procedure can be implemented for system improvement.

Step 1. Identify the target area(s) of the system that is not fully satisfied for its sustainable development.

Step 2. For each identified area, select a feasible system modification strategy from the system modification option base provided by experts or design optimization techniques (how to develop an option base is out of the scope in this paper).

Step 3. Use the fuzzy sustainability assessment method to re-evaluate the sustainability indicators for the modified system.

Step 4. Re-evaluate the satisfaction degree of the system sustainability using the user-defined satisfaction fuzzy sets.

Step 5. If the system sustainability is highly satisfied, the system modification is accomplished; otherwise, return to *Step 2*. This process will be repeated until all the triple-bottom-line indicators for the new system have satisfied the decision maker's expectation or preset goal(s).

Case Study

An automotive manufacturing centered industrial regional problem is selected for sustainability study. The industrial region is composed of three manufacturing sectors: a chemical supply sector of two chemical solvent plants, a surface finishing sector of two electroplating plants, and an

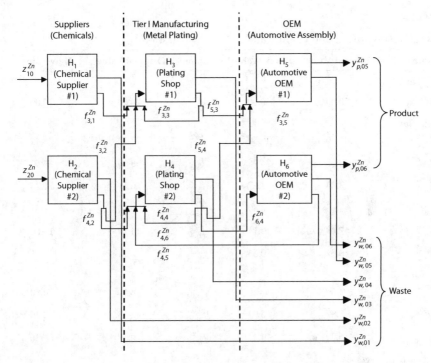

FIGURE 2
Schematic of an existing automotive manufacturing centered industrial region.

automotive sector of two OEM plants (see Fig. 2). The system's key material flow data of interest is given in the 2nd and 5th columns of Table 1. As shown, the plants in the region have already had a number of material recycles among them after implementing a number of technologies and improving their synergetic efforts.

For the purpose of illustrating the methodological applicability, only three indicators for each type of sustainability are selected for sustainability assessment.

Economic sustainability. The indicators are more profitability related. They are: (i) the value added ($x_{e,1}$), (ii) the value added per direct employee ($x_{e,2}$), and (iii) the taxes paid ($x_{e,3}$).

Environmental sustainability. The selected indicators are material, recycle, and waste focused. They are: (i) the total raw materials used per lb of product ($x_{v,1}$), (ii) the materials recycled ($x_{v,2}$), and (iii) the hazardous waste per unit value added ($x_{v,3}$).

Social sustainability. The chosen indicators are related to safety, customer satisfaction, and community involvement, which are: (i) the lost time accident frequency ($x_{s,1}$), (ii) the number of stakeholder meetings per unit value added ($x_{s,2}$), and (iii) the number of complaints per unit value added ($x_{s,3}$).

TABLE 1

System Flow Information Before and After Reintegration

State	Original Value*	New Value*	State	Original Value*	New Value*
Inflow					
z_{10}^{Zn}	50.00	50.00	z_{20}^{Zn}	70.00	70.00
Interflow					
$f_{3,1}^{Zn}$	46.50	46.50	$f_{3,2}^{Zn}$	27.72	28.44
$f_{4,2}^{Zn}$	33.88	34.76	$f_{3,3}^{Zn}$	3.90	6.64
$f_{4,4}^{Zn}$	3.83	4.67	$f_{5,3}^{Zn}$	66.41	71.74
$f_{3,5}^{Zn}$	0	2.80	$f_{4,5}^{Zn}$	0	1.87
$f_{5,4}^{Zn}$	17.48	19.14	$f_{6,4}^{Zn}$	14.29	15.66
$f_{4,6}^{Zn}$	0.57	0.63			
Waste					
$y_{w,01}^{Zn}$	3.50	3.5	$y_{w,02}^{Zn}$	8.40	6.80
$y_{w,03}^{Zn}$	7.82	6.03	$y_{w,04}^{Zn}$	2.68	2.46
$y_{w,05}^{Zn}$	8.39	4.41	$y_{w,06}^{Zn}$	0.57	0.626
Product					
$y_{p,05}^{Zn}$	75.49	81.78	$y_{p,06}^{Zn}$	13.15	14.41

* Unit: × 10^3 lbs/yr

For each indicator, three linguistic quantifiers, named LOW, MOD, and HIGH, are introduced for fuzzy set definition. The knowledge base contains a total of 27 rules (involving 3 variables, 3 possibilities). By applying the fuzzy sustainability assessment method, the values of the selected sustainability indicators for the existing system are listed in Table 2.

Further, by using the satisfaction fuzzy sets defined in Fig. 3, the composite economic and social indicators, having the value of 0.81 and 0.6 respectively, are satisfied, but the composite environmental indicator, having the value of 0.46, is not acceptable. This suggests that the system needs to be improved by implementing the five-step system modification procedure as follows.

Step 1. Environmental performance improvement is targeted.

Step 2. The option base for environmental indicator improvement contains four options. They are: Option 1 - to increase recycle flow rates, Option 2 - to introduce new recycles, Option 3 - to reduce source waste, and Option 4 - to use new technologies or alternative materials to further reduce waste.

By comparing the economic and social impacts of different options, Options 1 to 3 are selected for further consideration. It is identified that the

TABLE 2

Sustainability Indicator Calculation for the Existing System

	$x_{e,i}$	$\tilde{x}_{e,i}$	α_i	$y_{e,i}$	y_e
	409	0.83	0.30	0.25	
ECON	1.36	0.94	0.35	0.33	0.81
	24.54	0.66	0.35	0.23	
	$x_{v,i}$	$\tilde{x}_{v,i}$	β_i	$y_{v,i}$	y_v
	1.35	0.74	0.30	0.15	
ENV	8.4	0.35	0.35	0.14	0.46
	0.077	0.43	0.35	0.17	
	$x_{s,i}$	$\tilde{x}_{s,i}$	γ_i	$y_{s,i}$	y_s
	11.4	0.70	0.30	0.21	
SOC	2.2	0.44	0.35	0.15	0.60
	30.6	0.65	0.35	0.23	
		y_j	δ_i	$z_{IS,i}$	z_{IS}
		0.81	0.30	0.24	
IS		0.46	0.35	0.16	0.61
		0.59	0.35	0.21	

feasible system modifications are: (i) to improve the internal recycle capabilities of Plating Shops 1 and 2 by 70% and 22%, respectively, (ii) to introduce recycle from OEM 1 to both plating plants ($f_{3,5}^{Zn}$ and $f_{4,5}^{Zn}$ in Table 1), which can decrease 45% of the waste generated by OEM 1, (iii) to reduce the waste by 20% in Chemical Supplier 2 through internal improvement.

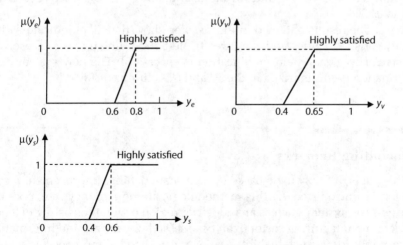

FIGURE 3
Definition of the satisfaction fuzzy sets for the case study problem.

TABLE 3

Sustainability Indicator Calculation for the Modified System

	$x_{e,i}$	$\tilde{x}_{e,i}$	α_i	$y_{e,i}$	y_e
	452.7	0.92	0.30	0.28	
ECON	1.41	0.98	0.35	0.34	0.88
	27.2	0.74	0.35	0.26	
	$x_{v,i}$	$\tilde{x}_{v,i}$	β_i	$y_{v,i}$	y_v
	1.25	0.80	0.30	0.16	
ENV	16.61	0.69	0.35	0.28	0.69
	0.05	0.62	0.35	0.25	
	$x_{s,i}$	$\tilde{x}_{s,i}$	γ_i	$y_{s,i}$	y_s
	10	0.80	0.30	0.24	
SOC	2.5	0.50	0.35	0.18	0.70
	25	0.80	0.35	0.28	
	y_j		δ_i	$z_{IS,i}$	z_{IS}
	0.88		0.30	0.26	
IS	0.69		0.35	0.24	0.75
	0.7		0.35	0.25	

Step 3. Evaluate the sustainability level of the modified system by the fuzzy sustainability assessment method. The result is shown in Table 3.

Step 4. Check the satisfaction level of the indicators also using the same satisfaction fuzzy sets shown Fig. 3. It is shown that the composite economic, environmental and social indicators are, respectively, of the values of 0.88, 0.69, and 0.7. The overall industrial sustainability value is increased from 0.61 to 0.75.

Step 5. The results obtained in the last step show that all of indicators give highly satisfied values and the three-triple-bottom-line issues are resolved satisfactorily. The system modification is successful. The new system flow information is summarized in the 3rd and 6th columns of Table 1.

Concluding Remarks

In this paper, a fuzzy-logic-based system sustainability improvement methodology is introduced. By this methodology, an industrial system, possibly a large-scale system, such as an industrial region or zone consisting of a network of manufacturing sectors, can be assessed under data and information uncertainty for its sustainability status. It can then be improved for achieving a higher level of sustainability through system modification. The case study has demonstrated the methodological efficacy and generality.

Acknowledgment

This work is in part supported by NSF (CBET 0730383 and DUE 0736739).

References

Andriantiatsaholiniaina, L. A., Kouikoglou, V. S., and Phillis.,Y. A. (2004). Evaluating Strategies for Sustainable Development: Fuzzy Logic Reasoning and Sensitivity Analysis. *Ecol. Econ.*, 48, 149.

Liu, Z. P., Huang, Y. L. (1998). Fuzzy Model-Based Optimal Dispatching for NO_x Reduction in Power Plants. *Int. J. Electrical Power and Energy Systems.* 20(3), 169.

Piluso, C. (2008). Industrial Sustainability Analysis and Decision Making under Uncertainty: A System Approach. Ph.D. Dissertation. Wayne State University, Detroit, Michgan.

Zimmermann, H. J. (1991). *Fuzzy Set Theory and Its Applications, 2nd Edition*, Kluwer Academic: Boston, MA.

82

Simultaneous Consideration of Process and Product Design Problems Using an Algebraic Approach

Susilpa Bommareddy, Nishanth G. Chemmangattuvalappil, Charles C. Solvason and Mario R. Eden[*]
Department of Chemical Engineering, Auburn University
Auburn, AL 36849

CONTENTS

ABSTRACT Introduction of the property integration framework has allowed for simultaneous representation of processes and products and established a link between molecular and process design from a properties perspective. In this contribution, an algebraic technique has been developed for solving process and molecular design problems simultaneously. The developed algorithm merges the existing algebraic techniques for process and molecular design problems and estimates the best molecule for a given process. The solution of the process design problem is achieved by solving a set of linear algebraic equality and inequality equations as a result of a constraint reduction approach. Group contribution methods are used to form molecular property clusters that are used to track properties. As both process and molecular property clusters target the optimum process performance, the inequality expressions can be solved simultaneously to identify the molecules that meet the desired process performance. Since the approach

---

[*] To whom all correspondence should be addressed

is based on an algebraic algorithm, any number of properties can be tracked simultaneously.

KEYWORDS *Property integration, algebraic approach, property operators*

Introduction

The identification of optimal molecule(s) corresponding to optimum process performance is a challenging issue. To achieve this goal, it is necessary to consider the aspects of both process and product design simultaneously. The concept of reverse problem formulation (RPF) has helped to formulate the integrated process-product design problems without leading to MINLP formulations (Eden et al., 2003). Techniques have been developed by Eden et al. (2003a, 2003b) for the identification of property targets corresponding to the optimum process performance using a visual approach. The algorithms developed by Eljack et al. (2008) and Chemmangattuvalappil et al. (2008) are useful for identifying the molecules that meet the property targets identified in the process design step. However, there is a need for a simultaneous algorithm that can be used to identify the suitable molecules corresponding to specific process performance. Since the process performance may depend upon several properties, the algorithm should be able to solve for any number of properties.

Property Clustering

Property clustering is a novel concept used to represent and track physical properties. Clusters are formed from property operator functions which are functions of the original properties tailored to obey linear mixing rules (Shelley and El-Halwagi, 2000).

$$\psi_j(P_{jm}) = \sum_{s=1}^{N_s} x_s \psi_j(P_{js})$$

(1)

Where $\psi_j(P_{js})$ is the property operator of the jth property P_{js} of stream s, x_s is the fractional contribution, N_s is the total number of streams. To make the property operator dimensionless divide it by an appropriately chosen property reference. The Augmented Property Index, AUP is defined as the sum of all the dimensionless property operators present in the system and the property cluster C_{js} is the fraction of its normalized operator to AUP (Shelley and El-Halwagi, 2000).

$$\Omega_{js} = \frac{\psi_j(P_{js})}{\psi_j^{ref}(P_{js})} \tag{2}$$

$$AUP_s = \sum_{j=1}^{NP} \Omega_{js} \tag{3}$$

$$C_{js} = \frac{\Omega_{js}}{AUP_s} \tag{4}$$

Since the normalized property operators obey linear mixing rules, they can be used to add the property contributions from different streams to obtain the property operator corresponding to the product stream. The dimensionless nature of this parameter will allow us to compare different property values on a single platform.

According to the reverse problem formulation developed by Eden et al. (2004), the first step in an integrated process-product design problem is to identify the property targets corresponding to optimum performance. The property targets to be satisfied by the sink can be stated as:

$$P_j^{lower} \le P_j \le P_j^{upper} \tag{5}$$

The normalized property operators corresponding to these targets can be written as:

$$\Omega_j^{lower} \le \Omega_j \le \Omega_j^{upper} \tag{6}$$

Qin et al. (2004) developed a constraint reduction algorithm to identify the property targets to be satisfied by different process streams corresponding to the constraints on the sink. The general dimensionless mixing rule corresponding to a property can be estimated as follows:

$$\Omega_j = \sum_{i=1}^{N_s} x_i \Omega_{j,i} \tag{7}$$

If a solvent is to be mixed with a stream of known flow rate and properties, Eq. 6 can be re-written as follows:

$$\Omega_j^{lower} \le \left(1 - \sum x_s\right)\left(\Omega_f - \sum \Omega_s\right) \le \Omega_j^{upper} \tag{8}$$

Here, x_s is the mass fraction of each individual recycle stream and Ω_f and Ω_s are the normalized property operators of the fresh solvent and the recycle streams respectively. Since the only unknown in Eq. 8 is the normalized property operators of fresh solvent properties, the maximum and minimum values corresponding to each property can be solved. The identified property ranges will then form the input to the molecular design problem.

Group Contribution Method for Property Estimation

The group contribution method (GCM) provides a convenient way to relate the molecular structure to their properties. In GCM, the property function $f(Y)$ of a compound is estimated as the sum of property contributions of all the molecular groups present in the structure (Marrero and Gani, 2001):

$$f(Y) = \sum_i N_i C_i + \sum_s N_s C_s + \sum_t N_t C_t \tag{9}$$

N_i, N_s and N_t are the numbers of first, second and third order groups and C_i, C_s, C_t are their respective property contributions.

Molecular Design Using Property Clustering

The property models in GCM can be used for molecular design using property clustering (Eljack et al., 2007). If P_{jg} is the contribution of property j from group g, n_g is the total number of that group in the molecule; molecular property operator ψ^M is defined as:

$$\psi_j^M(P_j) = \sum_{g=1}^{N_g} n_g P_{jg} \tag{10}$$

The molecular property cluster can be obtained following the same approach used to obtain property clusters.

Algebraic Approach

The property operators defined as above can only account for first order group contributions. In order to increase the accuracy of group contribution models, higher order group effects are to be considered while designing molecules. Chemmangattuvalappil et al. (2008) has developed an algorithm that can be used to identify the molecules corresponding to the property targets identified in process design step which is explained below. The property target identified by the process design is represented as:

$$\Omega_{ij}^{lower} \le \Omega_{ij} \le \Omega_{ij}^{upper} \tag{11}$$

Here, Ω_{ij} is the normalized property operator of molecule i. First calculate the normalized property operator based on first order estimation:

$$\Omega_{ijf} = \sum_{g=1}^{N_g} n_g \Omega_{jg1} \tag{12}$$

Where, Ω_{jg1} is the normalized property operator of first order group, g. Utilize the following rules to estimate the contributions of higher order groups:

Rule 1. Higher order groups can only be formed from complete molecular fragments.

Rule 2. If any of the higher order groups completely overlap some other higher order group, only the larger group must be chosen in order to prevent redundant description of the same molecular fragment.

So, if $(k{:}n)$ is the set of first order groups that are the building blocks of one second order group, s, $(n_{gk}{:}n_{gn})$ is the number of those first order groups present in the molecule, η is the number of occurrences of one particular first order group in a selected second order group, n_{gs} is the number of second order groups which can be generated from those first order groups, then:

$$n_{gs} = Min\left(\frac{n_{gk}}{\eta_k} : \frac{n_{gn}}{\eta_n} \right) \tag{13}$$

n_{gs} must be rounded down to the nearest integer number according to Rule 1. If Ω_{jg2} is the property contribution from the second order groups, the normalized property operator for the property contributions from second order groups, Ω_{ijs} is calculated as:

$$\Omega_{ijs} = \sum_{s=1}^{N_s} n_{gs} \Omega_{jg2} \tag{14}$$

Rule 2 applies when some of the second order groups are completely overlapped by some other groups.

If $(n_{gk}{:}n_{gn})$ has subsets of smaller second order groups $(n_{gl}{:}n_{gm})$ with some of the first order components of $(n_{gk}{:}n_{gn})$ and n_{gs}^* is the number of those second order groups, then:

$$n_{gs}^* = \left[Min\left(\frac{n_{gl}}{\eta_l} : \frac{n_{gm}}{\eta_m} \right) - Min\left(\frac{n_{gk}}{\eta_k} : \frac{n_{gn}}{\eta_n} \right) \right] \tag{15}$$

n^*_{gs} will be the number of those groups which are not part of any bigger second order groups. According to Rule 1, this must be rounded down to the nearest integer. If Ω_{jg2} is the contribution from the smaller second order groups, then the normalized property operator for the property contributions from smaller second order groups, Ω^*_{ijs} can be calculated as:

$$\Omega^*_{ijs} = \sum_{s=1}^{N_s} n^*_{gs}\Omega^*_{jg2} \qquad (16)$$

The third order effects can be calculated using the same approach. The normalized property operator for molecule 'i' can now be estimated as:

$$\Omega_{ij} = \Omega_{ijf} + \Omega_{ijs} + \Omega^*_{ijs} + \Omega_{ijt} + \Omega^*_{ijt} \qquad (17)$$

It is evident from Eq. (11) that, for each property, there will be two inequality expressions in the cluster space; one for the minimum value and the other for the maximum value (Qin et al., 2004). To solve for n_g, combine Eq.s (8) and (17) and split it into two equations for each property. Then, calculate the minimum and maximum values of *AUP* for the given property constraints.

To identify the possible cyclic compounds, the decision on the groups to be the part of the ring should be made ahead of design. The mathematical expression for 'Free Bond Number' (*FBN*), which is the number of free bonds in each molecular string, is (Eljack et al., 2007):

$$\sum_{g=1}^{N_g} n_g FBN_g - 2\left(\sum_{g=1}^{N_g} n_g FBN_g - 1\right) - 2N_r = 0 \qquad (18)$$

Where, N_r is the number of rings in the final molecule and FBN_g is the number of free bonds in each group. Based on these, the following expressions can be developed to ensure the existence of a meaningful molecule.

$$n_g \geq 0, \sum n_{gr} \geq 3 \text{ or } 0, FBN = 0 \qquad (19)$$

Now all possible combinations of first order groups and the possible higher order groups can be generated. This method enables identification of isomers as possible non-existence of each high order is considered. The generated molecules are screened by making sure each combination satisfies Eqs. 11 and 18.

Case Study

A current gas treatment process uses fresh methyl diethanol amine, MDEA, (HO-$(CH_2)_4$-CH_3N-OH) and two other recycled process sources (S1, and S2)

TABLE 1

Property Data for Gas Purification

P_j	Property Bounds	S1	S2	S3
V_c	530–610	754	730	790
H_v	100–115	113	125	70
H_{fus}	20–40	15	15	20
Flowrate	300	50	70	30

as a feed into the acid gas removal unit (AGRU). Another process stream, $S3$, currently a waste stream could be recycled as a feed if mixed with a fresh source to allow the mixed stream properties to match the sink (Kazantzi, 2006, Kazantzi et al., 2007). The property and flow rate data for all streams ($S1$, $S2$ and $S3$) and the sink are summarized in Table 1.

Design objectives and requirements are to find a solvent that will replace MDEA as a fresh source and that will maximize the flow rate of all available sources ($S1$, $S2$ and $S3$). The following three properties are considered: critical Volume (V_c), heat of vaporization (H_v) and heat of fusion (H_{fus}). Additionally, two thermal constraints are imposed on the synthesized molecules. The purpose of these additional constraints is to make sure that any designed molecule will remain in liquid state at the process conditions and to prevent excessive solvent losses via evaporation.

The first step is to identify the property targets for the fresh solvent. Since, the total flow rate of the recycle streams is less than the sink capacity, the complete recycle of all streams can be considered. The property operators corresponding to the target properties are given below:

$$V_{cM} = \sum_{s=1}^{Ns} x_s \cdot V_{cs} \tag{20}$$

$$H_{vM} = \sum_{s=1}^{Ns} x_s \cdot H_{vs} \tag{21}$$

$$H_{fusM} = \sum_{s=1}^{Ns} x_s \cdot H_{fuss} \tag{22}$$

Now Eqs. 6–8 can be used to obtain the property targets for the fresh stream. The identified targets along with the thermal constraints are given in table 2.

Fourteen first order groups have been considered for molecular design. The property operators for the target properties are given in table 2 and the property contributions of all the candidate groups are obtained from Marrero and Gani (2001). To insure water solubility and to reduce vapor pressure, the

TABLE 2

Property Targets

P_j	ψ_j	LB	UB
T_b	$\exp\left(\dfrac{T}{t_{b0}}\right)$	532	547
T_m	$\exp\left(\dfrac{T}{t_{m0}}\right)$	-	380
V_c	$V_c - V_{c0}$	310	610
H_v	$H_v - h_{v0}$	86.6	121.59
H_{fus}	$H_{fus} - h_{fus0}$	20	64

amine must have two or more –OH groups. To limit the extent of corrosion, only one amino group is allowed to be in the amine (N in the amino group either connects to H or C). Finally, to limit detrimental effects of direct exposure to the solvent, tertiary amines are ruled out in this case study. Eqs. 10–12 are used to obtain the maximum number of each first order groups, which are given below:

$$CH_3{:}6 \ CH_2{:}8 \ CH{:}5 \ C{:}3 \ OH{:}3 \ CH_3O{:}4 \ CH_2O{:}6 \ CHO{:}4$$
$$CH_2NH_2{:}1 \ CHNH_2{:}1 \ CH_3NH{:}1 \ CH_2NH{:}1 \ CHNH{:}1 \ CO{:}2$$

Now, different combinations of these first order groups can be made and Eqs. 13–17 are used to include the effects of higher order groups possible in those combinations. Eqs. 18 and 19 are used to ensure that the designed molecules are structurally possible. The final solution is given in table 3.

TABLE 3

Final Solution

Final Solution
2-(aminomethoxy)ethane-1,1,1-triol
((2-hydroxy-1-methoxypropan-2-yl)amino)methanediol
4-(aminomethoxy)-2,3-dimethylpentane-2,3-diol

Conclusions

In this work, an algorithm has been developed for the simultaneous consideration of both process and product design targets. Unlike database searches, which can lead to suboptimal results if the true optimal molecule is not contained in the database, the developed approach generates all the candidate molecules and thus ensures identification of the optimal structure. The algebraic approach also ensures that any number of properties can be tracked. In future, the algorithm will be extended to identify the best molecule considering the cost function and environmental impact while also concentrating on optimizing the recycling fractions of different feeds.

Acknowledgments

The funding for this work was provided by National Science Foundation CAREER program.

References

Chemmangattuvalappil N.G, Solvason C.C., Eljack F.T., Eden M.R. (2009). A Novel Algorithm for Molecular Synthesis Using Enhanced Property Operators. Computers & Chemical Engineering 33(3), 636–643.

Eden M.R., Jørgensen S.B., Gani R., El-Halwagi M.M. (2003a). Reverse Problem Formulation based Techniques for Process and Product Design. Computer Aided Chemical Engineering 15A, 451–456.

Eden M.R., Jørgensen S.B., Gani R., El-Halwagi M.M. (2003b). Property Cluster Based Visual Technique for Synthesis and Design of Formulations. Computer Aided Chemical Engineering 15B, 1175–1180.

Eljack F.T., Eden M.R., Kazantzi V., El-Halwagi M.M. (2007). Simultaneous Process and Molecular Design—A Property Based Approach. AIChE Journal, 53(5), 1232–1239.

Kazantzi V., Qin X., El-Halwagi M.M., Eljack F.T., Eden M.R. (2007): "Simultaneous Process and Molecular Design through Property Clustering Techniques", Industrial & Engineering Chemistry 46(10), 3400–3409

Marrero J., Gani R. (2001). Group Contribution Based Estimation of Pure Component Properties. Fluid Phase Equilibria, 183–184, 183–208.

Qin X., Gabriel F., Harell D., El-Halwagi M.M. (2004). Algebraic Techniques for Property Integration via Componentless Design. Ind. Eng. Chem. Res., 43, 3792–3798

Shelley M.D., El-Halwagi M.M. (2000). Component-less Design of Recovery and Allocation systems: A Functionality–based Clustering Approach. Computers & Chemical Engineering, 24, 2081–2091

83

Multi-Scale Product Design Using Property Clustering and Decomposition Techniques

Charles C. Solvason, Nishanth G. Chemmangattuvalappil and Mario R. Eden*

Department of Chemical Engineering, Auburn University
Auburn, AL 36849

CONTENTS

ABSTRACT Recent developments in the area of integrated process and product design have shown that products can be designed in terms of their properties without committing to any specific components a priori. Although current techniques make use of group contribution method (GCM) to design molecules, there are many properties, atomic arrangements, and structures that cannot be represented using GCM. One approach to expand the capability of GCM to handle a more diverse range of solutions is to combine property clustering with decomposition techniques in a reverse problem formulation. This approach first utilizes multivariate characterization techniques to describe a set of representative samples, and then uses decomposition techniques such as principal component analysis (PCA) and partial least squares (PLS), to find the underlying latent variable models that describe the molecule's properties.

KEYWORDS *Reverse problem formulation, principal component analysis, multiscale product design, property clusters*

---

* To whom all correspondence should be addressed

Introduction

One of the most common methods for designing molecules for a specific end use while minimizing computational time has been group contribution (Gani *et al.*, 2005, Marrero and Gani, 2001). However, in product design, the associated properties of concern are most often consumer attributes which do not have group contribution parameters (Hill, 2005). An effective manner to address this concern is to map the consumer attribute data from the meta-scale into a set of properties on the macro scale that can be described by group contribution (Solvason *et al.*, 2009a). This step is often performed via chemometrics which defines an empirical relationship through the use of design-of-experiments (DOE) and multivariate-linear-regression (MLR). However, the uncertainty in the relationship between the attributes and properties may lead to large target regions in the property domain for the molecular design which then must be validated in the attribute domain (Solvason *et al.*, 2009b). To improve this relationship it is beneficial to map the attribute information down to a domain subspace which exhibits a stronger attribute-property relationship. The constraints on this new domain are that it must be linear in the constituent space and it must have the ability to be described by a molecular combinatorial technique. A key difference in this method is that the domain subspace is not required to be one of the known properties described by group contribution.

Methods

The objective of this paper is to enumerate all possible molecules that meet a set of target attributes using a domain subspace mapping function combined with molecular group theory applied to chemometric data. To achieve this objective, a reverse problem formulation is applied where the attributes are mapped down to a domain subspace comprised of properties that have better attribute predictive power than conventional GCM described properties. Several tools, such as characterization and decomposition can be used to find the domain subspace. Molecular combinatorial techniques and property clustering are then applied to find and interpret the solution to the reverse problem formulation. Fig. 1 illustrates the method.

Characterization Techniques

In this method the constraints placed on finding the subspace domain align with chemometric techniques associated with characterization. Characterization is a class of tools associated with the determination of not only chemical constituency or molecular structure, but also of larger structural characteristics describing the orientation and alignment of these molecules often called microstructure

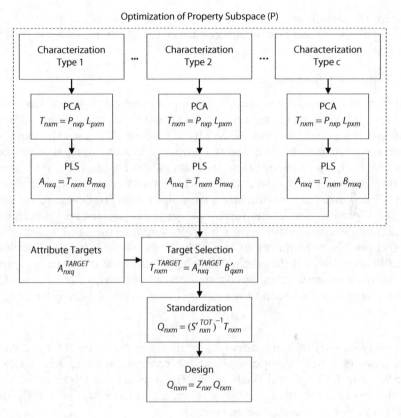

FIGURE 1
Method flowchart.

at the meso-scale. Some examples of characterization techniques include nuclear magnetic resonance (NMR), x-ray diffraction (XRD), and infrared spectroscopy (IR). The techniques are often applied to a training set of molecules defined by an experimental design used to explore the interesting facets of a set of property attributes. The added structural information available from the characterizations can be used to extend the group contribution method to higher orders as well as discern some orientation specific information.

To determine the group, structure, and orientation information that is to be used in the molecular design, a set of rules must be followed. First, the characterization must be able to completely quantify each individual molecular group used in the design. In some characterizations, progressively larger groups completely contain the information of the smaller groups, but also contain corrections for 2nd order, 3rd order, structural, and orientation effects. This hierarchal nature is handled by specifying that all groups should be combined such that the largest functional group is specified first, then the second largest, and so on. In some situations only partial overlaps may occur, for which the method will fail to specify any corrections to the first order combination at the overlap interface. To minimize this impact, it is specified that the groups be built such that the combinations occur across the C-C bond which carries the smallest amount of information (Marrero and Gani, 2001).

Decomposition Techniques

To ensure that the property subspace is appropriately orthogonal, decomposition is applied to the characterization. The most common decomposition is principal component analysis (PCA). By definition, PCA uses the variance-covariance structure to compress the property data to principal component data that contains much of the system variability. This result also improves the interpretation of the data structure by consolidating multiple property effects into single, underlying latent variables which are devoid of colinearity. The procedure for determining the eigenvector/eigenvalue pairs in PCA is susceptible to large differences in scales and variance and as such, it is general practice to mean-center and scale the data to unit variance (Johnson and Wichern, 2007). If the original property data are of the same type, then the eigenvalues can be considered measures of the contrasts, or loadings L_{pxm} of the original variables; and the eigenvectors are referred to as scores T_{nxm}.

$$P_{xxp} = T_{nxm} L'_{mxp} \tag{1}$$

In most cases the first 2 or 3 principal components can account for this variation. The remaining components can be removed without much loss of information (Johnson and Wichern, 2007). To utilize the latent variables in the property clustering algorithm, it is important to recognize that the data structure follows a linear mixing rule.

$$T_{xxm} = P_{nxp} L_{pxm} \tag{2}$$

$$\psi_{nxm}^{mix} = X_{nxp} \psi_{pxm} \tag{3}$$

The pure properties ψ_{pxm} have the same structure as the loadings L_{pxm}, thus the loadings can be thought of as the pure values of the principal components. Likewise, the response ψ_{nxm}^{mix} data are predicted mixture properties and have the same structure as the score T_{nxm} data. That means that the mixture fraction X_{nxp} in the property models is related to the multivariate data in P_{nxp}. However, there is a concerning difference between the two methods: the mixture fractions sum to 1 across the properties for each sample and the multivariate data P_{nxp} do not. In order for latent variable models to be utilized in property clustering, it is necessary to modify the latent variable structure by standardizing.

$$S_k^{TOT} = \sum_i^p X_{ik}, k \in n \tag{4}$$

$$R_{nxp} = \left(S_{nxn}'^{TOT}\right)^{-1} X_{nxp} \tag{5}$$

$$Q_{nxm} = \left(S'^{TOT}_{nxn}\right)^{-1} T_{nxm} \tag{6}$$

$$Q_{nxm} = R_{nxp} L_{pxm} \tag{7}$$

The new Q_{nxm} matrix now represents modified scores or mixtures. The loadings matrix L_{pxm} remains unchanged and the R_{nxp} matrix now represents fractions of loadings whose cumulative sum is one for each run. Unfortunately, although the components sum to one, they are sometimes negative due to the mean-centering of the multivariate property data prior to PCA. The constraint that the fractions must be between 0 and 1 is removed with no effect on the associated mathematics, only on their interpretation. At this point, the loadings L_{pxm} are the underlying latent variable domain subspace. Both full molecules and molecular group subspace properties Q_{nxm} can be found by multiplying the latent variables L_{pxm} by the associated fractions R_{nxp}. Since the molecular and group subspace property relationships in Eq. 6 were derived using a decomposition technique, the constraints imposed by decomposition should also be observed for any new molecules or mixtures created.

$$Q^{Mix}_{1xm} = Z_{1xn} Q_{nxm} \tag{8}$$

For instance, in many places in literature, NMR spectroscopy is used to characterize chemical structures. The spectra are then broken down by decomposition into principal components which assume the spectra are combinations of linearly additive subspace properties by definition. Extending this assumption to new molecule design using Eq. 8 is essentially a representation of a "linear mixture of linear mixtures" of the underlying latent variable subspace properties, all of which are linear in nature. This observation assumes that any nonlinearity in the attribute system is handled by the attribute property relationship and not the group-subspace property relationship (Muteki and MacGregor, 2006)

Property Clustering

Property clustering is a tool used to improve the interpretation of the subspace properties by deconstructing the design problem into a euclidean vector in the cluster domain and a scalar called the Augmented Property Index *AUP*. The clusters themselves are conserved surrogate properties described by property operators, which have linear mixing rules, even if the operators themselves are nonlinear. Methods for the application of group contribution method for molecular design have previously been developed using property clustering by Eljack *et al.* (2007, 2008). The modified principal components can be directly mapped in to the clustering framework, such that the loading clusters are all chosen to be positive. To ensure this constraint is met, the reference property algorithm proposed by Solvason *et al.* (2009a, 2009b) is utilized. Once the appropriate references are found, the resulting score and loading clusters can be mapped to a clustering diagram for visualization purposes. Interpretation of the clustering diagram is straightforward. The location of the loadings is indicative of the pure underlying latent variables that describe the multivariate properties. The location of the individual scores of each experimental run is also indicative of the influences of the principal components. Their proximity to the vertices indicates which principle component is influencing the properties of a particular molecule. However, since properties in the subspace can sometime be negative, then the δ_{1xn} cluster fractions can also vary beyond 0 and 1, which rules out the possibility of excluding future molecular structures from the data. Hence, all molecules and molecular groups could potentially be used in the design of the target mixture based on the first rule of clustering. However, the interpretation of similar and dissimilar groups can still be utilized and the second and third rules of clustering can be used to rule out potential groups or molecules (Eljack *et al.*, 2007, 2008).

Partial Least Squares and Molecular Design

Conducting a DOE on the important scores results in a PLS model that relates the important product attributes A_{nxq} to the underlying latent variable structure T_{nxm} from which product predictions and targets can be analyzed.

$$A_{nxq} = T_{nxm}B_{mxq} \tag{9}$$

Molecular design is then conducted by matching the targets mapped to the domain space with predicted molecules using a modified group contribution where all terms are linear, but carry combinations of 1st, 2nd, 3rd, and structural information (Chemmangattuvalappil *et al.*, 2009). An outline of the procedure is shown in Figure 1.

Case Study – Acetaminophen Tablet Design

Three attributes that are important to direct compression tablet manufacturing are disintegration time, crushing strength, and ejection force. These attributes have been notoriously difficult to analyze based on traditional mixing design because of the complex and highly nonlinear nature of pharmaceutical excipients. In order to better control these attributes, they are mapped down to a domain subspace where they can be approximated as linear combinations of molecular group parameters. The domain subspace was found to be characterized by three properties, P1, P2, and P3 using a training set of 24 excipients (Gabrielsson *et al.*, 2003).

To reduce the number of parameters in the subspace, decomposition was performed using PCA. Although the number of parameters could be

TABLE 1

Design Targets

Subspace Targets	Q1	Q2	Q3
UL	2.00	2.00	1.00
LL	1.00	0.00	0.00

reduced, it was decided to keep all of them for illustrative purposes. Using PLS models developed from the training set, consumer specific set of targets were mapped to the domain subspace as shown in Table 1.

The molecular groups identified by the characterization were identified as CH, CH_2, OH, CHO, O, CH_2-O, CHOH, CH_2OH, $CHCH_2OH$, CHCHO, Ocyc, α-pyranose, β-pyranose, and cellulose. The group specific subspace properties are shown in Fig. 2. Note the data cluster around the C_1 vertex, indicating the first principal component is dominant. Also interesting is the proximity

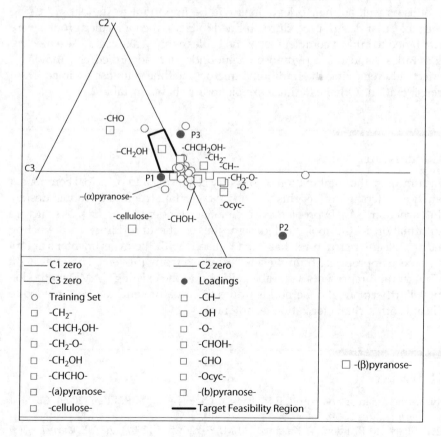

FIGURE 2
Molecular design cluster diagram.

TABLE 2

Molecular Design

Candidate Molecules	Q1	Q2	Q3
CH2OH-CH2OH	1.22	1.20	0.61
CH2OH-(α)pyranose-CH2OH	1.85	1.08	0.90
CH2OH-CH(OH)-CH2OH	1.79	1.07	0.43
CH2OH-CH2-O-CH2OH	1.75	1.06	0.15
CH2OH-O-CH2OH	1.75	1.03	0.14
CH2OH-CH(CH2OH)-CH2OH	1.80	1.45	0.57
OH-CH2-OH	1.17	0.47	0.13
OH-(α)pyranose-CH2OH	1.80	0.35	0.42
OH-CH(CH2OH)-CH2OH	1.75	0.72	0.09
OH-CH2-CH2-CH2-OH	1.75	1.29	0.23

of α-pyranose and alcohols to the target region, indicating that the alcohol groups as well as their orientations will be important to the design. Using the molecular design procedure outlined in Fig. 1, the molecular groups were combined to build a complete set of molecules as shown in Fig. 2. As expected, α-pyranose, and not β-pyranose compounds were identified as candidates, effectively removing the traditional microcrystalline cellulose excipient from consideration. Other candidate excipients are shown in Table 2.

Conclusion

In summary, the combination of property clustering and principal component analysis offers many insights and advantages for structured product design. In particular CAMD problems are no longer hindered by a lack of structure information in the molecular design. Rather, the uncertainty in predicting large molecular structures has been replaced with the experimenter's ability to choose appropriate training sets, for which many proven techniques exist. This method represents a sizeable addition to the existing CAMD methodology. Furthermore, the method is universal in nature and can be extended to include other characterization techniques.

References

N.G. Chemmangattuvalappil, F.T. Eljack, C.C. Solvason, M.R. Eden (2009), *Comp. & Chem. Eng.* 33(3), 636–643.

F.T. Eljack, M.R. Eden, V. Kazantzi, M.M. El-Halwagi (2007), *AIChE Journal*, 53(5), 1232–1239.

F.T. Eljack, M.R. Eden (2008), *Comp. & Chem. Eng.* 32(12), 3002–3010.

J. Gabrielsson, N. Lindberg, M. Palsson, F. Nicklasson, M. Sjostrom, and T. Lundstedt (2003), *Drug Development and Industrial Pharmacy*, 29(10), 1053–1075.

R. Gani, P.M. Harper, M. Hostrup (2005) *Ind.Eng.Chem.Res.* 44, 7262–7269.

R.A. Johnson and D.W. Wichern, *Applied Multivariate Statistical Analysis*. Pearson Prentice Hall, Upper Saddle River, NJ (2007).

M. Hill. (2004). *AIChE Journal*, 50(8), 1656–1661.

J. Marrero, R. Gani (2001), *Fluid Phase Equilibria*, 182–183.

K. Muteki and J.F. MacGregor (2006), *Chemom. & Intell. Lab. Sys.*, 85, 186–194.

C.C. Solvason, N.G. Chemmangattuvalappil, F.T. Eljack, M.R. Eden (2009a). *Ind. & Eng. Chem. Res.* 48(4), 2245–2256.

C.C. Solvason, N.G. Chemmangattuvalappil, F.T. Eljack, M.R. Eden (2009). *Comp. & Chem. Eng.* 33, 977–991.

84

Failure Analysis of Polymer Products and Chemical Processes to Identify Design Deficiencies

Russell F. Dunn* and **Terry Mills, III**
Polymer and Chemical Technologies, LLC
Cantonment, FL 32533

CONTENTS

ABSTRACT There have been numerous recent accidents involving motorcycles, automobiles, aircraft, piping and other engineering products. In many instances, one of the primary contributors to the accident is the failure of a thermoplastic, elastomer, thermoset or polymer fiber product. In general, these critical polymer product failures have been the result of design factors such as the improper selection of materials of construction for the intended end use, the improper processing and/or fabrication of these materials into the end product and the improper maintenance of the end product, along with other factors. In addition to the failure analysis of polymer products, failure analysis of several chemical plant accidents has also resulted in the identification of numerous process design deficiencies. Scientific methodologies and tools used for identifying the mode of failures for polymer products and chemical processes are provided. Also, issues affecting sustainability such as the public perception of the manufacturers, financial implications,

---

* To whom all correspondence should be addressed

legal ramifications, and human injuries are highlighted in several case studies. These case studies will emphasize the importance of considering these factors during polymer product and chemical process design.

KEYWORDS *Failure analysis, polymer failure, chemical release, chlorine release*

Introduction

There have been numerous recent accidents involving motorcycle, automobile and aircraft components, industrial products and other engineered parts. In many cases, the failure is related to a thermoplastic, elastomer, thermoset or polymer fiber product. In general, product failures have been largely a result of design factors such as the improper selection of materials of construction for the intended end use, the improper processing and/or fabrication of these materials into the end product and the improper maintenance of the end product, in addition to other factors. Furthermore, failure analysis of several chemical plant accidents has also resulted in the identification of numerous process design deficiencies.

Scientific methodologies and tools used for identifying the mode of failures for polymer products and chemica processes are provided in addition to case studies that highlight their usefulness.

Materials Failure Analysis: Scientific Methodologies and Tools

Standard failure analysis protocol for investigating materials failures, particularly polymers and plastics, can be used for assessing the safety and sustainability of products in their end use environment. The specific protocol used is (Parrington, 2000):

1. Information Gathering
2. Preliminary Visual Examination
3. Nondestructive Testing
4. Characterization of Material Properties through Mechanical, Chemical and Thermal Testing
5. Selection, Preservation and Cleaning of Fracture Surfaces, Secondary Cracking and Surface Condition
6. Microscopic Examination
7. Identification of Failure Mechanisms
8. Stress/Fracture Mechanics Analysis

Case Study: Polymer Product Failure

In the mid 1980's Hondo Motor Company (Honda) modified the restraint systems in their 1986–1991 Honda Accords from the use of metal seat belt buckle assemblies to the use of plastic components. Honda contracted Takata Corporation (Takata) to design and construct the plastic seat belt buckle assembly in accordance with Honda's specified criteria. Specifically, in 1986 the seat belt press release buttons were molded using ABS (acrylonitrile-butadiene-styrene) polymer. There have been numerous failures of the belt buckles containing the ABS press release buttons as a result of the fracture of these buttons. Broken pieces of the buttons can lodge in the spring mechanism inside the buckle assembly and, if this occurs, it can result in rendering the buckle assembly inoperable. Figure 1 is included as a front view of the plastic seat belt buckle assembly. Figure 2 is included as a side view of the ABS press release button indicating both an intact and broken press release button. Figure 3 is included as an example of a seat belt buckle assembly with a fractured ABS press release button. In 1992 Honda replaced the ABS press release button with a POM (polyoxymethylene or polyacetal) press release button and at the same time changed the design of the molded press release buttons, such that the POM button is not identical in design to the original ABS button. Figure 4 is included as a side view of the POM press release button indicating the added rib reinforcement.

FIGURE 1
ABS press release button in seat belt assembly.

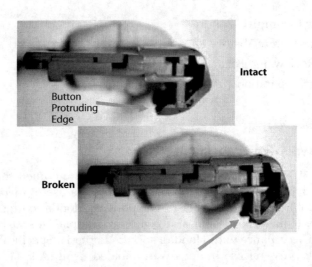

FIGURE 2
ABS polymer press release button.

The National Highway Transportation Safety Administration (NHTSA) issued Recall No. 95V-103.001 on May 24, 1995, concerning 1986–1991 Honda Accords with subject type Takata seat belt buckles after there were numerous reported failures of these buckle assemblies (NHTSA, 1997). Prior studies have reported that the plastic failures are a result of environmental degradation of the ABS plastic due to environmental exposure to UV light (Henshaw et al., 1999). However, more recent analysis has questioned the accurate identification of the root cause of the failures.

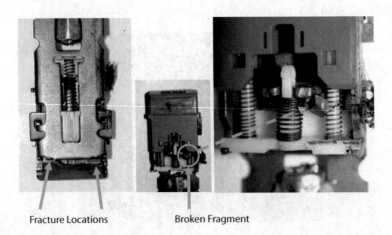

FIGURE 3
Lodged broken polymer fragment.

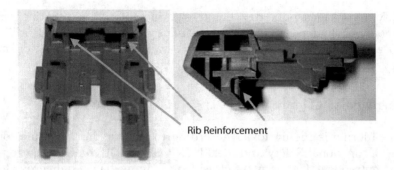

Rib Reinforcement

FIGURE 4
POM polymer press release button.

Engineering Investigation Results

- Impact resistance of notched and un-notched ABS resin is higher than the POM resin, in that the butadiene rubber content of the ABS resin provides improved impact strength and toughness to the ABS resin.

- Testing of the chemical resistance, heat resistance and UV resistance of the ABS press release button does not sufficiently explain the root cause of the button fractures.

- Honda implemented installation of a seat belt tongue guide in 1995 to reduce/eliminate button fracture. Tongue insertion into the Takata buckle assembly without the guide could be at an angle as high as 12.5° which causes a high stress on the back edge of the button surface. Finite Element Analysis (FEA) was used to show areas of high stress on the press release button during tongue insertion.

- Takata designed extra reinforcement for the POM press release buttons in 1990 and in subsequent designs. Including this reinforcement in the original ABS design would have substantially reduced the stress that caused the button fracture and fragmentation issue.

- Microscopic analysis indicates that the ABS seat belt press release buttons were fractured due to repeated impact damage combined with insufficient reinforcement strength in the button leading edge design.

Case Study: Chlorine Release

At approximately 9:20 on the morning of August 14, 2002, 48,000 pounds of chlorine was released over a 3-hour period during a railroad tank car unloading operation at DPC Enterprises, L.P., near Festus, Missouri. The facility repackages bulk dry liquid chlorine into 1-ton containers and 150-pound cylinders for commercial, industrial, and municipal use. The release involved the rupture of a chlorine transfer line and failure of emergency shutdown equipment.

Background on Chlorine

- Chlorine is highly toxic. According to the National Institute for Occupational Safety and Health (NIOSH, 2003), chlorine gas concentrations of 10 ppm are classified as "immediately dangerous to life or health" (IDLH). Liquid as well as vapor contact can cause irritation, burns and blisters.

- The current Occupational Safety and Health Administration (OSHA) permissible exposure limit (PEL) for chlorine is 1 ppm as a ceiling limit. A worker's exposure to chlorine shall at no time exceed this ceiling level.

- The National Institute for Occupational Safety and Health (NIOSH) has established a recommended exposure limit (REL) for chlorine of 0.5 ppm as a time-weighted-average (TWA) for up to a 10-hour workday and a 40-hour workweek and a short-term exposure limit (STEL) of 1 ppm.

- The American Conference of Governmental Industrial Hygienists (ACGIH) has assigned chlorine a threshold limit value (TLV) of 0.5 ppm as a TWA for a normal 8-hour workday and a 40-hour workweek and a short-term exposure limit (STEL) of 1.0 ppm for periods not to exceed 15 minutes. Exposures at the STEL concentration should not be repeated more than four times a day and should be separated by intervals of at least 60 minutes.

- The American Industrial Hygiene Association's (AIHA) Emergency Response Planning Guideline (ERPG-1) is 1 ppm for chlorine. This is the maximum airborne concentration of a chemical below which it is believed that nearly all individuals could be exposed for up to one hour.

- The American Industrial Hygiene Association's (AIHA) Emergency Response Planning Guideline (ERPG-2) is 3 ppm for chlorine. This is the maximum airborne concentration of a chemical that nearly all individuals could be exposed for up to one hour without impairing an individual's ability to take protective action.

- The American Industrial Hygiene Association's (AIHA) Emergency Response Planning Guideline (ERPG-3) is 20 ppm for chlorine. This is the maximum airborne concentration of a chemical that nearly all individuals could be exposed for up to one hour without developing life-threatening health effects.

- Chlorine, in the presence of moisture forms HCl which is corrosive to most metals, including steel, iron, brass and copper. Chlorine gas is a strong oxidizer, which may react with flammable materials. Moisture introduction into the dry chlorine gas results in corrosion of numerous types of metals, including 316 stainless steel.
- Chlorine is heavier than air and will collect in low-lying areas. The gas has a characteristic, penetrating odor, a greenish yellow color and is about two and one-half times as heavy as air. Thus, if chlorine escapes from a container or system, it will tend to seek the lowest level in the building or area in which the leak occurs. One volume of liquid chlorine, when vaporized, yields about 460 volumes of gas.

Regulations Governing Chlorine

- Chlorine is classified as a *highly hazardous chemical* in quantities greater than 1500 lbs and is regulated by the Occupational Safety and Health Administration (OSHA) under 29CFR 1910.119 *Process Safety Management of Highly Hazardous Chemicals.*
- Chlorine in quantities greater than 2500 lbs is regulated by the Environmental Protection Agency (EPA) Risk Management Plan (RMP) for accidental release prevention under 40 CFR Part 68 *Chemical Accident Prevention Provisions.*
- *Emergency action plans* for responding to chlorine releases are regulated by OSHA under 29 CFR 1910.38 *Emergency Action Plans* & 29 CFR 1910.120 *Hazardous Waste Operations and Emergency Response.*
- *Emergency response* to chlorine releases is regulated by the EPA under 40 CFR Part 68 *Chemical Accident Prevention Provisions.*

Engineering Investigation Results

- 316 Stainless Steel metal braided hose *is not* recommended for use in chlorine service. Hastelloy C-276 metal braided hose *is* recommended for use in chlorine service.
- DPC did not adequately develop a Process Safety Management (PSM) program and did not have an adequate Risk Management Plan (RMP) for chlorine.
- DPC did not have an effective *Emergency Response Plan (ERP)* for their chlorine operations.
- The perimeter alarms at the DPC Festus plant were insufficient to ensure that neighboring communities are protected from harmful exposure during an unplanned release of chlorine.
- The Level B PPE available at the DPC Festus plant was unsafe for an emergency response to an unplanned release of chlorine.

– The possibility of improper braided hose material in chlorine transfer hose was foreseeable by DPC and could have been avoided.
– The possibility of Emergency Shutdown (ESD) System failure was foreseeable and could have been avoided.

Conclusions

Sustainable product design must consider the properties of the materials of construction (ABS v. POM), end product design (press release button) and the ultimate end product use (multiple impacts and environmental exposure). This polymer case highlighted the improper end product design. The use of ABS plastic would have been sufficient if the rib reinforcement had been included in the original mold design for the press release button.

The chlorine release by DPC highlights the importance of regulations and procedures governing chemical process safety of highly hazardous materials.

Both case studies should be used to emphasize the importance of safety related issues during the design process, not after an accident has already occurred. In addition, the product and chemical process failures have resulted in diminished public perception of manufacturers, significant financial implications to the manufacturers, legal ramifications, and human injuries. Proper design can avoid these negative issues and improve the sustainability of these manufacturers.

References

Wright, David. (2001). Failure of Plastics and Rubber Products. *Rapra Technology Limited*, Shawsbury, UK, p. 92–95.

Henshaw, J. M., V. wood and A. C. Hall. (1999). Failure of Automobile Seat Belts Caused by Polymer Degradation, *Engineering Failure Analysis*, 6, 13–25.

National Highway Transportation Safety Administration (NHTSA). (1997). *Engineering Analysis EA94-036*, 1–438.

Parrington, Ronald. (2000). Fractography of Metals and Plastics, *In Proceedings of the Society of Plastics Engineers Annual Technical Conference*, Paper 0025.

Dunn et al., (2005), Failure of Plastic Press Release Buttons in Automobile Seat Belts, *Engineering Failure Analysis*, 12(1), 81–98

U. S. Chemical Safety and Hazard Investigation Board – Investigation Report Chlorine Release. (2003). DPC Enterprises, L.P., Festus, Missouri. Report No. 2002-04-I-MO. 1–99.

Figures 1–4 are Reprinted from Engineering Failure Analysis, 12(1), Dunn et al., Failure of Plastic Press Release Buttons in Automobile Seat Belts, 81–98 (2005), with permission from Elsevier.

85

Design of Secondary Refrigerants: A Combined Optimization-Enumeration Approach

Apurva Samudra, Nikolaos V. Sahinidis*

National Energy Technology Laboratory
Department of Chemical Engineering
Carnegie Mellon University
Pittsburgh, PA 15213

CONTENTS

ABSTRACT Computer-aided molecular design has emerged as a powerful technique to identify promising compounds that meet the predefined property targets. These techniques have been employed in various areas, including the design of solvents, refrigerants, and polymers.

This paper presents a mixed-integer non-linear program (MINLP) for designing a secondary refrigerant. Secondary refrigeration loops have recently been recognized as a potential solution to reduce the environmental impact of retail food refrigeration. The MINLP model adopted here includes a property prediction model (Nanda, 2001) coupled with molecular structure feasibility constraints (Sahinidis *et al.*, 2003). A large number of basic building groups are considered and no limits on the type of the molecule are imposed. To cope with the numerical challenges associated with nonlinearities of this

---

* To whom all correspondence should be addressed

model, a direct enumeration of the search space is used to solve the model for restricted molecular sizes, while a branch-and-bound algorithm is used for larger designs. Several novel solutions are obtained and analyzed in detail. We also explore the possibility of using accurate property models (Marrero and Gani, 2001) for different applications. The use of higher order property estimation, which is structure dependent, is also investigated.

KEYWORDS *Computer aided molecular design, secondary refrigerants, property prediction model*

Introduction

Retail food refrigeration accounts for 25% of hydrofuorocarbon (HFC) emissions. A typical supermarket leaks about 1 ton of refrigerant every year. Clearly, this is a major source of environmental pollution that needs to be reduced. Secondary refrigeration loops have been recognized as a potential solution to reduce the environmental impact of retail food refrigeration. The problem of "finding" a suitable refrigerant is an ideally suited problem for the solution techniques of Computer-Aided Molecular Design (CAMD). Use of CAMD in several such problems, including the design of refrigerants, solvents, polymers, pharmaceuticals, surfactants, biodegradable molecules, and compounds for nanotechnology applications have been reported.

The characteristics of secondary refrigerant that the solution molecule should posses are:

- It is nontoxic, nonflammable, noncorrosive to the equipment, and environmentally benign
- It is liquid under low temperature operating conditions (−40°C to −30°C) in order to allow single-phase operation and thus better temperature control
- It has high volumetric heat capacity to reduce the amount of refrigerant needed. The volumetric heat capacity is the product of density and liquid heat capacity
- It has low viscosity to reduce pumping costs and high thermal conductivity for high heat transfer rates in display cases

It is seen that there is a trade-off between the heat transfer characteristics and viscosity requirements of secondary refrigerants. The reason for this behavior is that high heat capacity and thermal conductivity of liquids imply higher interactions at the molecular level, resulting in high viscosity. As a result, the search for a fluid that balances these requirements is a challenging problem.

TABLE 1

Set of Descriptors for First Model

Acyclic Groups				
$-CH_3$	$-CH_2-$	$>CH-$	$>C<$	$=CH_2$
$=CH-$	$=C<$	$=C=$	$\equiv CH$	$\equiv C-$
Cyclic Groups				
$^r-CH_2-^r$	$^r_r>CH-^r$	$^r>CH-^r$	$^r_r>C<^r_r$	$^r>C<^r_r$
$^r_r>C<$	$^r=CH-^r$	$^r=C<^r_r$	$^r=C<^r$	$^r_r>C=$
Halogen Groups				
$-F$	$-Cl$	$-Br$	$-I$	
Oxygen Groups				
$-OH$	$-O-$	$>CO$	$-COOH$	$-CHO$
$=O$	$^r_r>CO$	$-COO-$	$^r-O-^r$	
Nitrogen Groups				
$-NH_2$	$>NH$	$^r_r>NH$	$>N-$	$=N-$
$^r=N-^r$	$-CN$	$-NO_2$		
Sulphur Groups				
$-SH$	$-S-$	$^r-S-^r$		

$-^r$ represents cyclic or aromatic bond

Elemental Descriptor Model

In the first model, we use elemental descriptors (shown in Table 1). The problem can be formulated as a mathematical programming problem involving continuous and integer variables. We consider N potential types of groups to be included in the compound, with n_i, $i = 1,...,N$, representing the number of times group of type i is selected in the final design. The model consists of two types of constraints

- Property prediction model
- Structural feasibility constraints

Sahinidis *et al.* (2003) discuss the structural constraints in detail. The principal properties used in the evaluation of potential alternative refrigerants include density (ρ), liquid heat capacity (C_{pla}), thermal conductivity (k), boiling (T_b) and freezing point (T_f). These estimation techniques, in turn, require

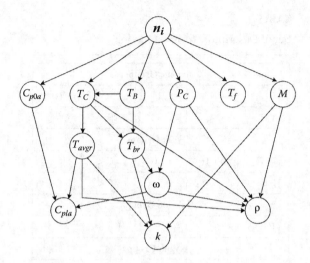

FIGURE 1

estimation of critical properties (T_c, P_c), ideal gas heat capacity (C_{p0a}), and acentric factor of molecules (w) as shown in Figure 1. We use group contribution techniques to estimate these properties. The details of the property model are discussed by Nanda (2001).

It should be mentioned that no attempt is made to include viscosity in the model, as current group contribution techniques for estimating viscosity indicated significant estimation errors. Instead, our goal is to find all compounds that satisfy all other secondary loop considerations. These compounds will be screened for viscosity considerations at a later stage.

An advantage of elemental descriptors is that they can represent a vast variety of molecules with relatively few descriptors. The use of elemental molecular descriptors, property models and the structural feasibility constraints lead to a highly nonlinear, non-convex MINLP, which presents numerical challenges to the optimization solver.

Objective Function and the Variable Bounds

In principle, a comprehensive objective function can be derived to account for all the environmental and economic factors. We use an objective function that maximizes the total volumetric heat capacity $\rho C_{pla}/M$, which is an important characteristic for evaluation of secondary refrigerants. Note, however, that the choice for the objective function is irrelevant, as we will find all feasible solutions to the above formulation. These solutions can then be sorted according to any desired selection criteria.

The operating temperature range of the refrigeration cycle provides a lower bound for the boiling point and an upper bound for the freezing temperature of the refrigerant. We introduce lower bounds for thermal conductivity, density, and heat capacity in order to restrict our search to molecules with heat transfer properties that are at least as good as those of currently used fluids.

Solution Approach

The MINLP program formed with this model was solved using a global optimization branch and reduce algorithm implemented in the BARON, a general purpose global optimization package by Sahinidis (1996). Instead of using the best solution all the feasible solution were recorded.

It is found that the highly nonlinear nature of the functions in property prediction model which involved integer variables (n_i) caused this approach to miss certain solutions.

A solution to this problem is to run direct enumeration of the search space. In enumeration, the size of the molecule was restricted by limiting the total number of groups in the molecule to 14. Enumeration of the search space is performed by evaluating all possible combinations of values of the n_i integer variables such that $\Sigma_i \, n_i \leq 14$. These combinations are then checked for structural feasibility. They were further screened by calculating the property values for these combinations and property bounds are used to eliminate the infeasible solutions. Only the feasible combinations, which represent solution molecules, are recorded. We found 784 new molecules with an objective function value greater previously found values. These molecules were then analyzed in detail.

The analysis of these solutions consists of two steps. The first step is to create the feasible molecular structure from the groups present in solution molecules. This is essential as the same set of groups in the solution can represent different compounds as well as many constitutional isomers. Most of the accurate property prediction techniques need structural information to predict the interaction effects in the compounds properties. Thus in this step the solutions and all their possible structures are recorded. The next step in the analysis is to search for these compounds in established databases and look for any experimentally determined properties for these compounds. The experimentally determined properties help us validate the use of these compounds as secondary refrigerants. The analysis of all the solutions was done using the Chemical Abstracts Service databases (CAS).

Results and Discussion

Following are the key points that the analysis of these compounds revealed.

- Most of the solutions obtained do not have any recorded existence. This means that the method provides us with many truly novel solutions. These compounds demonstrate the capability of CAMD techniques to produce completely new molecules, which are ideal for the required targets.

- This technique has shown us that the use of nonlinear property estimation techniques would always suffer the drawback of numerical
difficulties. There have been efforts in recent years to model a linear
group contribution method and we will talk about them in the next
section.
- The time required for enumeration increases exponentially and enumeration of larger molecules is not viable. However, enumeration can
be used as an effective technique for some applications of CAMD.

Linear Group Contribution Based Model

The nonlinearity and the limited range of older property model have led to the
development of a newer class of property models by Constantinou and Gani
(2001), Constantinou *et al.* (1994), Nanoolal et al. (2007), Marrero and Gani (2001)
etc. The group contribution method proposed by Marerro and Gani (MG) allows
more accurate and reliable estimations of properties for a wide range of chemical substances, including large and complex compounds. In this method, the
estimation is performed at three levels. The basic level has a large set of simple
groups that are non-overlapping and describe the entire molecule. Then, to capture the proximity effects and isomers, second-level groups are used. The second
level of estimation is intended to deal with polyfunctional, polar or nonpolar
compounds of medium size and aromatic, or cycloaliphatic compounds with
only one ring and several substituents. The third level of estimation allows estimation of complex heterocyclic and large polyfunctional acyclic compounds.

Mathematic Model of MG Group Contribution Method

The property-estimation model has the form of the following equation:

$$f(X) = \sum_{i \in F} n_i c_i + w \sum_{i \in S} n_i c_i + z \sum_{i \in T} n_i c_i$$

where F, S, and T are sets of first, second, and third order groups, respectively, n_i the number of groups of type i selected, and c_i is the contribution of
group i to property X. The weights w,z are set to 0 or 1 depending on the level
of estimation. In this work, only first-order estimation is used; therefore, w
and z are set to 0.

$f(X)$ is a function of the target property X. The selection of this function has
been based on the following criteria:

- The function has to achieve additivity in the contributions c_i
- It has to exhibit the best possible fit of the experimental data
- It should provide good extrapolating capability and, therefore, a
wide range of applicability

In any CAMD problem, we know the ranges for property X which provide us with linear constraints with integer variables using MG model.

The MG method is used to predict the critical properties of the molecule such as Critical Pressure (P_c), Critical Temperature (T_c) etc. Recently, Kolská *et al.* (2008) used the MG method descriptors and developed parameters to predict the molar heat capacity of the liquid (C_p^l). These descriptors are groups made by combining elemental descriptors to increase the accuracy of the model. There are total 108 descriptors used in the implementation. This method can be used for predicting the heat capacity at different temperatures. The model for this method is similar to the MG method but the group contribution parameters are functions of temperature.

$$C_p^l(T) = C_{p0}^l(T) + \sum_{i \in f} n_i C_{pi}^l(T) + w \sum_{i \in S} n_i C_{pi}^l(T) + z \sum_{i \in f} n_i C_{pi}^l(T)$$

$$C_{pi}^l(T) = a_i + b_i \left(\frac{T}{100} \right) + d_i \left(\frac{T}{100} \right)^2$$

Implementation

Although the correlations to predict certain properties, such as density (ρ) and thermal conductivity (k) are nonlinear, to maintain the linearity of the model, we separated these correlations from the model by using a post-processing step. Thus, the feasible solutions of the optimization model were analyzed by the post-processing step. The combined approach is very effective as the linear optimization is involved.

Apart from the property prediction constraints, the model involves the feasibility constraints, which are tailored specifically to this set of descriptors. As many descriptors are specialized subgroups (for example CH_3COO-), upper bounds on the maximum number of times a specific group can be used had to be used. This formulation differs from the previous formulation only in the functions used to evaluate the critical properties. Hence, all the bounds on the variables and objective functions are still the same.

Results and Discussion

The molecules obtained as the solution are in the stage of further analysis. The preliminary analysis of the molecules looks promising. Many molecules are chloro-fluoro carbons, which validates the use of CFCs for refrigeration. At the same time, as viscosity is not included in the formulations and as the descriptors base used is not elemental, some of the solutions turn out to be heavier molecules. The inclusion of viscosity correlations in the model is currently underway. Many of these solutions are novel compounds and have a great potential to be used as refrigerants.

Conclusion

Two different approaches are taken to solve the CAMD problem of secondary refrigerant design. The first approach is based on the elemental descriptor model. This approach results in a highly nonlinear MINLP. Use of newer, linear property model is also explored. The results are similar to the first approach and are being explored further. It is clear that use of newer property methods is necessary for larger molecules.

References

L. Constantinou and R. Gani. New group contribution method for estimating properties of pure compounds. AIChE J., 40:1697–1710, 1994.

L. Constantinou, M. L. Mavrovouniotis, and S. E. Prickett. Estimation of properties of acyclic organic compounds using conjugation operators. Ind. Eng. Chem. Res., 33:395–402, 1994.

Z. Kolská, J. Kukal, M. Z´abransk´y, and V. R˙u_zi_cka. Estimation of the heat capacity of organic liquids as a function of temperature by a three-level group contribution method. Ind. Eng. Chem. Res., 47:2075–2085, 2008.

J. Marrero and R. Gani. Group-contribution based estimation of pure component properties. Fluid Phase Equilibria, 183–184:183–208, 2001.

G. Nanda. Design of Efficient Secondary Refrigerants. Master's thesis, Department of Chemical Engineering, University of Illinois, Urbana, IL, 2001.

Y. Nannoolal, J. Rarey, and D. Ramjugernath. Estimation of pure component properties: Part 2. Estimation of critical property data by group contribution. Fluid Phase Equilibria, 252:1–27, 2007.

N.V. Sahinidis, M. Tawarmalani, and M. Yu. Design of alternative refrigerants via global optimization. AIChE J., 49:1761–1775, 2003.

N.V. Sahinidis. "BARON: A General Purpose Global Optimization Software Package," Journal of Global Optimization, 8:201–205, 1996.

86

Multidimensional Piecewise-Affine Approximations for Gas Lifting and Pooling Applications

Ruth Misener, Chrysanthos E. Gounaris and Christodoulos A. Floudas*

Department of Chemical Engineering, Princeton University
Princeton, NJ 08544-5263

CONTENTS

ABSTRACT Gas lifting requirement curves are piecewise-affine functions. In a comparative study, we exploit four algorithms, designed by Nemhauser and Wolsey (1988), Floudas (1995), Sherali (2001), and Keha et al. (2004), to solve the gas lifting problem. Each of the four algorithms is sufficient to solve the problem to global optimality and the method produced by Keha et al. (2004) has the best computational performance. The success of these four algorithms suggests the study of two- and three-dimensional piecewise-affine approximations. Using a set of grid points and function evaluations, we construct a look-up table, triangulate the grid (Chien and Kuh, 1977; Zhang and Wang, 2008), interpolate the function on discrete simplicies, formulate an MILP, and solve the MILP to optimality. We present our experience in computational performance and average error estimation. When pruning a global optimization branch-and-bound tree, the least upper bound

---

* To whom all correspondence should be addressed (floudas@titan.princeton.edu; Tel: (609) 258-4595; Fax: (609) 258-0211).

is compared to lower bounding nodes. To hasten convergence, we propose initializing a local NLP solver with the global optimum of the MILP approximation, thereby increasing the probability of reducing the upper bound.

KEYWORDS   *Gas-lifting, Pooling problems, Piecewise-linearization*

Introduction

The problem of optimizing oil well production by injecting compressed natural gas, called "lift gas," into the production tubing of an oil well has recently been addressed rigorously (Misener et al., 2009). The lift gas reduces the weight of the column of fluid, increasing the pressure gradient between the reservoir and well fluid that pushes the fluid to the surface (Bahadori et al., 2001). The fluid from the well, which includes oil, gas, and water, is treated to separate water. The oil is sold, and the remaining gas is either sold or recycled for use in another gas lifting operation (Camponogara and Nakashima, 2006).

The data sets for this problem appear as gas injection versus oil production coordinates $(q_{GAS,i,k}, q_{OIL,i,k})$ where $i \in \{1,\ldots,N_{WELLS}\}$ fixes the well index and $k \in \{1,\ldots,N_{VERTICES}\}$ represents the vertex point. Figure 1 illustrates the data given by the 6-well Buitrago et al. (1996) literature test case.

Our goal was to select the gas injection level for each well so as to maximize oil production. We chose four algorithms proposed by Nemhauser and Wolsey (1988), Floudas (1995), Sherali (2001), and Keha et al. (2004) and solved the gas

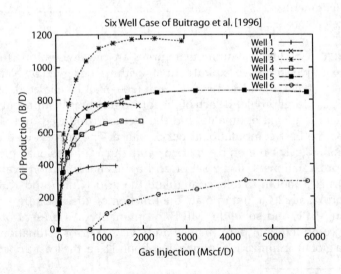

FIGURE 1
Illustration of Buitrago et al. literature test case.

lift problem using each algorithm (Misener et al., 2009). Each of the the four algorithms was sufficient to solve the problem to global optimality. However, the tests we reported in Misener et al. (2009) reveal that the special structure method from Keha et al. (2004) consistently outperforms the three other algorithms. We recommended using one of the other algorithms for challenging gas lifting problems that employ piecewise linear representations.

Motivated by the fast solution times we obtained optimizing piecewise linear approximations of functions with the form $f: \mathbb{R} \mapsto \mathbb{R}$, we explored affine approximations of functions $f: \mathbb{R}^n \mapsto \mathbb{R}$ with $n \leq 3$. The approximation, which consists of a look-up table and an interpolation algorithm, will act as a "warm start" for the upper bound of a global optimization algorithm. We solve the linear approximation of a nonlinear problem using a three dimensional interpolation scheme between grid points, using the result as an upper bound on a multidimensional function. One class of problems where this approximation scheme could be used is large-scale, highly nonlinear pooling problems.

Look-up Tables

Given a continuous function $\Omega(x): \mathbb{R}^n \mapsto \mathbb{R}$, an approximation function $\hat{\Omega}(x): \mathbb{R}^n \mapsto \mathbb{R}$ can be constructed using a look-up table and an interpolation algorithm. For the purposes of this study, a look-up table is a set of function values and associated domain points that are recorded at grid points.

Function interpolation between the look-up table grid points can be performed using a variety of algorithms, but this paper will study linear interpolation through a convex combination of the grid points (Chien and Kuh, 1977; Kosmidis et al., 2004). The linear interpolation function is uniquely defined only if each point in the domain is restricted to a single simplex, so the approximation function will interpolate function values within a tessellation of simplices.

This study restricts look-up tables to two and three dimensions because of the large number of simplices needed to triangulate dimensions greater than three (Hughes and Anderson, 1996). In other words, the algorithm developed in this study uses look-up tables to construct an approximation $\hat{\Omega}(x): \mathbb{R}^n \mapsto \mathbb{R}$ of function $\Omega(x): \mathbb{R}^n \mapsto \mathbb{R}$, when $n \leq 3$. Functions of higher dimensions can be handled using multiple look-up tables and the tight convex relaxations designed by McCormick (1976) or Meyer and Floudas (2003, 2004).

Interpolation within a Simplex

Sets of three variables are grouped (*i.e.*, $\{x_1, x_2, x_3\} = X$) for an eventual interpolation scheme. Although the result of Carathéodory proves that at most $n + 1$ grid points are needed to express each point in the domain space $X \subset \mathbb{R}^n$, there are many more than $n + 1$ grid points in any reasonable

representation of the domain space. To guarantee a deterministic interpolation outcome, only $n + 1$ grid points are activated at one time.

Assuming that the $n + 1$ appropriate grid points $x_1, \ldots, x_{n+1} \in X$ for
domain point $x \in X \subset \mathbb{R}^n$ have been activated and that we wish to approximate function $f : X \mapsto \mathbb{R}^n$, the system of equations:

$$\hat{f}(x) = w_1 \cdot f(x_1) + \cdots + w_{n+1} \cdot f(x_{n+1}) \tag{1}$$

$$x = w_1 \cdot x_1 + \cdots + w_{n+1} \cdot x_{n+1} \tag{2}$$

$$\sum_{i=1}^{n+1} w_i = 1 \tag{3}$$

$$w_i \geq 0, \quad \forall i = 1, \ldots, n+1 \tag{4}$$

uniquely determines approximates $f : X \mapsto \mathbb{R}$ in the interior of $n + 1$ grid
points.

Restriction a Simplex

To uniquely represent each point in the domain as a convex combination
of grid points, we partition the domain space into small hypercubes and
partition each of the hypercubes into non-overlapping simplices (Zhang and
Wang, 2008). This algorithm generalizes the one-dimensional piecewise-
linear approximation from Floudas (1995) and Nemhauser and Woolsey
(1988) to two and three dimensions.

Hypercube Constraints

After each variable set X is partitioned into the rows of grid points, any
point $x \in X$ is within a parallelogram (when $X \subset \mathbb{R}^2$) or parallelepiped
(when $X \subset \mathbb{R}^3$). The equations introduced in this section activate only the
grid points at the vertices of the small hypercube that contains x. Figure 2
diagrams an activated hypercube within the domain space for dimensions
two and three.

When the domain X has two dimensions, it is partitioned into $X_{i,j} \in \mathbb{R}^2$
$i = 0, \cdots, N_1, j = 0, \cdots, N_2$. Two sets of SOS1[1] binary variables, $\lambda^1 \in \{0, 1\}^{N_1}$
and $\lambda^2 \in \{0, 1\}^{N_2}$, activate each parallelogram.

$$\sum_{i=1}^{N_1} \lambda_i^1 = 1, \quad \lambda_i^1 \in \{0, 1\} \quad \forall i = 1, \ldots, N_1 \tag{5}$$

---

[1] A special Ordered Set of type I, or *SOS 1* variable, is a variable set where at most one component is nonzero.

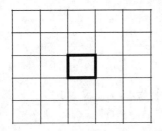
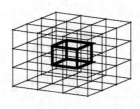

FIGURE 2

A single active hypercube in 2 and 3 dimensions.

$$\sum_{j=1}^{N_2} \lambda_j^2 = 1, \quad \lambda_j^2 \in \{0, 1\} \quad \forall j = 1,\ldots, N_2. \tag{6}$$

Only the vertices of a single activated parallelogram contribute to the inter-polation within that parallelogram, so the convex combination of continuous weights $w_{i,j} \in [0, 1] i = 0,\cdots, N_1, j = 0,\cdots, N_2$ are constrained as follows:

$$\text{Var 1} \begin{cases} \sum_{j=0}^{N_2} w_{0,j} & \leq \lambda_1^1, \\ \sum_{j=0}^{N_2} w_{i,j} & \leq \lambda_{i-1}^1 + \lambda_i^2, \quad \forall i = 1,\cdots, N_1 - 1 \\ \sum_{j=0}^{N_2} w_{N_1,j} & \leq \lambda_{N_1-1}^1 \end{cases}$$

$$\text{Var 2} \begin{cases} \sum_{i=0}^{N_1} w_{i,0} & \leq \lambda_1^2, \\ \sum_{i=0}^{N_1} w_{i,j} & \leq \lambda_{j-1}^2 + \lambda_j^2, \quad \forall j = 1,\cdots, N_2 - 1 \\ \sum_{i=0}^{N_1} w_{iN_2} & \leq \lambda_{N_2-1}^2. \end{cases}$$

When the domain X has three dimensions, it is partitioned into $X_{i,j,k} \in \mathbb{R}^3 i = 0,\ldots, N_1, j = 0,\cdots, N_2, k = 0,\cdots, N_3$. Three sets of binary variables: $\lambda^1 \in \{0, 1\}^{N_1}$, $\lambda^2 \in \{0, 1\}^{N_2}$ and $\lambda^3 \in \{0, 1\}^{N_3}$, denote the active parallelepiped. As in the two dimension case, these binary variable sets are $SOS1$:

$$\sum_{i=1}^{N_1} \lambda_i^1 = 1, \quad \lambda_i^1 \in \{0, 1\} \quad \forall i = 1,\ldots, N_1 - 1 \tag{7}$$

$$\sum_{j=1}^{N_2} \lambda_j^2 = 1, \quad \lambda_j^2 \in \{0, 1\} \quad \forall j = 1, \cdots, N_2 - 1 \tag{8}$$

$$\sum_{k=1}^{N_3} \lambda_k^3 = 1, \quad \lambda_k^3 \in \{0, 1\} \quad \forall k = 1, \ldots N_3 - 1. \tag{9}$$

Only the vertices of the activated parallelepiped contribute to the interpolation of points within that parallelepiped, so convex combination weights $w_{ijk} \in [0, 1]$ $i = 0, \ldots, N_1, j = 0, \ldots, N_2, k = 0, \ldots, N_3$ are constrained as follows:

$$\text{Var 1} \begin{cases} \displaystyle\sum_{j=0}^{N_2} \sum_{k=0}^{N_3} w_{0,j,k} \leq \lambda_1^1, \\[2em] \displaystyle\sum_{j=0}^{N_2} \sum_{k=0}^{N_3} w_{i,j,k} \leq \lambda_{i-1}^1 + \lambda_i^1, \ \forall i = 1, \ldots, N_1 - 1 \\[2em] \displaystyle\sum_{j=0}^{N_2} \sum_{k=0}^{N_3} w_{N_1,j,k} \leq \lambda_{N_1-1}^1 \end{cases}$$

$$\text{Var 2} \begin{cases} \displaystyle\sum_{i=0}^{N_1} \sum_{k=0}^{N_3} w_{i,0,k} \leq \lambda_1^2, \\[2em] \displaystyle\sum_{i=0}^{N_1} \sum_{k=0}^{N_3} w_{i,j,k} \leq \lambda_{j-1}^2 + \lambda_j^2, \ \forall j = 1, \ldots, N_2 - 1 \\[2em] \displaystyle\sum_{i=0}^{N_1} \sum_{k=0}^{N_3} w_{i,N_2,k} \leq \lambda_{N_2-1}^2 \end{cases}$$

$$\text{Var 3} \begin{cases} \displaystyle\sum_{i=0}^{N_1} \sum_{j=0}^{N_2} w_{i,j,0} \leq \lambda_1^3, \\[2em] \displaystyle\sum_{i=0}^{N_1} \sum_{j=0}^{N_2} w_{i,j,k} \leq \lambda_{k-1}^3 + \lambda_k^3, \ \forall k = 1, \ldots, N_3 - 1 \\[2em] \displaystyle\sum_{i=0}^{N_1} \sum_{j=0}^{N_2} w_{i,j,N_3} \leq \lambda_{N_3-1}^3. \end{cases}$$

These constraints restrict each point in the domain to a parallelogram (defined by 4 points) or parallelepiped (defined by 8 points). But convex combinations of points in the interior of the small two and three dimensional hypercubes will not be unique, so we partition the two and three dimensional hypercubes into non-overlapping simplices.

Triangulation Classes

In two dimensions, $X \subset \mathbb{R}^2$, there is one representative triangulation class that divides a parallelogram into non-overlapping simplices. The three

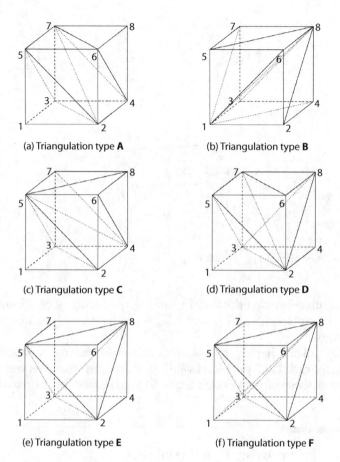

FIGURE 3
Triangulation types of the 3-cube.

dimensional case, diagrammed in Figure 3, has six representative classes that divide the parallelepiped into non-overlapping simplices (Meyer and Floudas, 2005). Each triangulation type has multiple orientations that can be chosen to minimize approximation error. To partition the domain space X, we choose a particular triangulation class. The triangulation type and orientation for each lumped variable is tessellated across the entire domain (Chien and Kuh, 1977).

Justification of Triangulation Type

The two major advantages of using triangulation B (Figure 3(b)) as the representative three dimension triangulation is that only three planes need to be introduced to isolate a point in a small hypercube into a particular simplex and each of the six simplices in triangulation type B have equal volume in

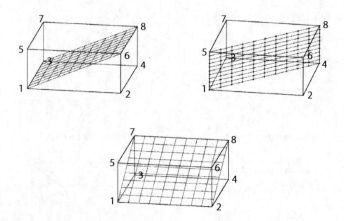

FIGURE 4
Three triangulation planes: Y_{1368}, Y_{1458} & Y_{1278}.

the case of uniform partitioning, increasing the accuracy of the interpolation. The three planes, Y_{1368}, Y_{1458} and Y_{1278}, are defined by the numbers of their vertex points.

Given an isolated hypercube of two or three dimensions, equations of the planes diagrammed in Figure 4 activate only the relevant simplex vertices. The orientation of these planes is determined so as to reduce approximation error.

Multilinear Approximation Example

This four dimension multilinear example is taken from Gounaris and Floudas (2008). This paper has only covered two and three dimension look-up tables, so the four dimension function was constructed using two look-up tables. The objective function $f : [0,1]^4 \mapsto \mathbb{R}$ is:

$$f(x) = x_1 \cdot x_2 - x_2 \cdot x_3 - x_3 \cdot x_4 + x_1 \cdot x_2 \cdot x_3 - x_1 + x_4. \tag{10}$$

TABLE 1

Partitioning Levels and Estimated Associated Errors

Partition	#of Seg	Absolute Max Err	Absolute Ave Err	Absolute Std Dev
A	4	0.040	0.0026	0.0083
B	8	0.011	0.0007	0.0021
C	16	0.003	0.0002	0.0005

TABLE 2

Optimization of the Objective Function Associated with Each of the Paritioning Schemes

	Obj	x_1	x_2	x_3	x_4	CPU (s)
$f(x)$	−1	1.00	0.00	0.00	.00	0.02
A	−1	0.00	1.00	1.00	.75	0.01
B	−1	1.00	0.00	0.75	.00	0.03
C	−1	1.00	0.00	1.00	.00	0.27

The average function value across the 1×10^7 sample points is −0.125 and there are infinitely many globally optimal solutions with $f(x) = -1$. Examples include $f(1, 0, \alpha, 0) = -1 \forall \alpha \in [0, 1]$ and $f(1, 0, 1, \beta) = -1 \ \forall \beta \in [0, 1]$. The global maximum of the function is 2. Table 1 records the partitioning scheme that was used to segment each of the 4 variables and displays the associated errors. Table 2 shows that the interpolation successfully approximated the function.

Conclusion

The success of one dimensional approximation functions in solving the gas lifting problem suggests that interpolation between predetermined grid points of a domain can be used to approximate multidimensional functions. The approximation can be used as a "warm start" for a global optimization algorithm, providing a good upper bound on the solution. A good upper bound is particularly valuable for large-scale problems such as industrial pooling problems.

Acknowledgments

The authors gratefully acknowledge support from the National Science Foundation. R.M. is thankful for her National Science Foundation Graduate Research Fellowship.

References

A. Bahadori, Sh. Ayatollahi, and M. Moshfeghian. Simulation and optimization of continuous gas lift in aghajari oil field. In *SPE Gas Technology Symposium*, Kuala Lumpur, Malaysia, 2001. Society of Petroleum Engineers.

S. Buitrago, E. Rodríguez, and D. Espin. Global optimization techniques in gas allocation for continuous flow gas lift systems. In *SPE Gas Technology Symposium, Calgary, Alberta, Canada, 1996*. Society of Petroleum Engineers. SPE 35616.

E. Camponogara and P. Nakashima. Optimizing gas-lift production of oil wells: piecewise linear formulation and computational analysis. *IIE Transactions*, 38(2):173–182, 2006.

M. Chien and E. Kuh. Solving nonlinear resistive networks using piecewise-linear analysis and simplicial subdivision. *Circuits and Systems, IEEE Transactions on*, 24(6):305–317, 1977.

C. A. Floudas. *Nonlinear and Mixed-Integer Optimization: Fundamentals and Applications.* Oxford University Press, New York, NY, 1995.

C. E. Gounaris and C.A. Floudas. Tight convex underestimators for C^2-continuous problems: II. Multivariate functions. *J. of Glob. Optim.*, 42(1):69–89, 2008.

R. B. Hughes and M. R. Anderson. Simplexity of the cube. *Discret. Math*, 158(1–3):99–150, 1996.

A. B. Keha, I. R. de Farias Jr., and G. L. Nemhauser. Models for representing piecewise linear cost functions. *Oper. Res. Letters*, 32(1):44–48, 2004.

V. D. Kosmidis, J. D. Perkins, and E. N. Pistikopoulos. Optimization of well oil rate allocations in petroleum fields. *Ind. Eng. Chem. Res.*, 43(14):3513–3527, 2004.

G. P. McCormick. Computability of global solutions to factorable nonconvex programs: Part 1-convex underestimating problems. *Math. Program.*, 10(1):147–175, 1976.

C. A. Meyer and C. A. Floudas. Trilinear monomials with positive or negative domains: Facets of the convex and concave envelopes. In C. A. Floudas and P. M. Pardalos, editors, *Frontiers in Global Optimization*, pages 327–352. Kluwer Academic Publishers, 2003.

C. A. Meyer and C. A. Floudas. Trilinear monomials with mixed sign domains: Facets of the convex and concave envelopes. *J. of Glob. Optim.*, 29(2):125–155, 2004.

C. A. Meyer and C. A. Floudas. Convex envelopes for edge-concave functions. *Math. Program.*, 103(2):207–224, 2005.

R. Misener, C. E. Gounaris, and C. A. Floudas. Global optimization of gas lifting operations: A comparative study of piecewise linear formulations. Industrial & Engineering Chemistry Research, 2009. In press.

G. L. Nemhauser and L. A. Wolsey. *Integer and Combinatorial Optimization*. J. Wiley, New York, 1988.

H. D. Sherali. On mixed-integer zero-one representations for seper-able lower-semicontinuous piecewise-linear functions. *Oper. Res. Letters*, 28(4):155–160, 2001.

H. Zhang and S. Wang. Linearly constrained global optimization via piecewise-linear approximation. *J. of Computational and Appl. Math.*,214(1):111–120, 2008.

87

Computational Design of Nanopaint: An Integrated, Multiscale Process and Product Modeling and Simulation Approach

Jie Xiao and Yinlun Huang[*]

*Department of Chemical Engineering and Materials Science,
Wayne State University, Detroit, MI 48202*

CONTENTS

ABSTRACT Nanopaint has drawn great attention in recent years, as it can offer superior properties and novel functionalities to resulting coatings. However, how to optimally design nanopaint in a cost-effective way is extremely challenging. This is mainly due to design complexity, experimental cost and, most importantly, the lack of fundamental knowledge about nanostructured material. In this paper, we introduce a computational nanopaint design methodology by resorting to multiscale modeling and simulation techniques. The developed multiscale models allow establishment of various comprehensive and quantitative correlations among material formulation, processing condition, microscopic structure, and macroscopic system performance. The correlations help enrich the fundamental knowledge base about the nanopaint material and become a key to the optimal design. On the other hand, the integrated process and product

---

[*] To whom all correspondence should be addressed

approach provides material designers with the opportunity of properly making trade-offs between product performance and process efficiency. A comprehensive study on acrylic-melamine-alumina nanoparticle contained nanopaint design demonstrates the methodological efficacy.

KEYWORDS *Computational design, nanopaint, multiscale modeling and simulation*

Introduction

Nanopaint is a type of thermoset nanocomposite (TSNC) material that contains a certain fraction of organo-modified inorganic nanoparticles in its resin. By comparing with conventional paint, nanopaint may give surface coatings very attractive functionalities and properties (Xiao and Huang, 2008a). It is experienced that nanopaint design by experiments alone requires a very long development cycle due to design complexity. The existence of vast design parameters can easily make the material development an unmanageable combinatorial problem. This could be further restricted by experimental resources.

Nanopaint development can be significantly accelerated using advanced computational methods, as they can provide great freedom and control over the investigated material parameters and product properties through virtually any number of *in silico* experiments. Moreover, computational design can help discover the fundamental knowledge and information about structure-property correlations, especially the dynamics that is difficult to observe during experiments (Xiao and Huang, 2009). However, there has been no such a computational methodology available for investigating nanopaints. The existing computational efforts using molecular dynamics (MD), lattice or off-lattice Monte Carlo (MC) techniques are exclusively limited to the study of thermoplastic nanocomposites (TPNCs) (Vacatello, 2001; Starr et al., 2002; Ozmusul et al., 2005). Note that the modeling and simulation for nanopaint design is very complicated due to the sophistication of polymer molecular weight and functional group distribution, crosslinking-bearing polymer network formation and its interaction with the nanoparticles. Monoscale molecular simulation alone cannot generate sufficient correlation information that is necessary for nanopaint design, since the effect of macroscopic processing conditions on a nanocoating structure evolution cannot be taken into account.

In this work, we will introduce a computational nanopaint design methodology by resorting to multiscale modeling and simulation techniques. The multiscale models allow establishment of various comprehensive and quantitative correlations among material formulation, macroscopic processing condition, microscopic structure, and system performance. These correlations are critical for obtaining an in-depth understanding and designing optimally nanopaint materials. A comprehensive study on the design and

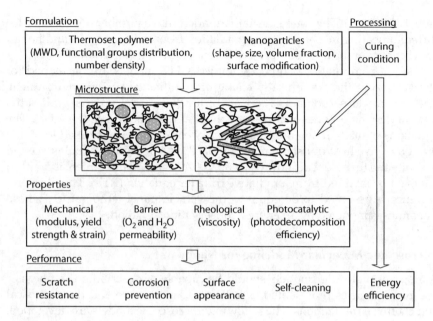

FIGURE 1
Multiscale hierarchical correlations for nanopaint design.

analysis of an acrylic-melamine-alumina nanoparticle contained nanopaint
will demonstrate the methodological efficacy.

Multiscale Modeling/Simulation for Nanopaint Design

This work focuses on the establishment of comprehensive and quantitative
correlations among nanopaint formulation, processing condition, nanocoating microstructure, property and performance (see Fig. 1). Note that as an initial step towards a thorough investigation on nanopaint design, the scratch
resistance performance of the resulting coating is chosen for study.

According to Fig. 1, the following developments are needed: (i) a macroscopic curing process modeling method, (ii) a microscopic molecular
modeling method for nanopaint, (iii) a simulation method for developing
nanocoating samples, (iv) a nanocoating sample in silico testing method,
and (v) a set of product and process performance evaluation methods. These
developments are briefly described in the following sections.

Macroscopic Curing Process Models

In this work, an automotive coating curing process is investigated. Vehicle
bodies covered by wet thin films travel one by one through a curing oven

(400~800 ft long). The oven is usually divided into a number of zones so that different heating mechanisms (i.e., radiation from the oven walls and hot air convection) can be applied under different operating conditions.

A set of computational fluid dynamics (CFD) models are developed for characterizing the dynamically changing, multistage curing environment. The oven model set contains a convection air flow model, an oven-wall radiation model and a panel heating model, all at the macroscale (Li, 2007). The equations of mass, momentum and energy conservations are used to model the convective heat transfer, and a standard k-ε turbulence model is used to describe the turbulent air flow (Li et al., 2007). The radiation intensity within the oven is obtained from a radiative transfer equation (RTE). The film temperature is assumed to be equal to the vehicle panel temperature, whose dynamics is modeled by an energy conservation equation.

Microscopic Molecular Modeling for Nanopaint

Nanopaint molecular models should be capable of describing the polymer and nanoparticles represented in a 3D simulation space and their physical and chemical interactions. The following microscopic models are developed: a polymer network model, a nanoparticle model, and a polymer-nanoparticle interaction model.

Polymer network model. The known coarse-grained bead-spring model by Kremer and Grest (1990) is extended to characterize the crosslinked polymer network. In a polymer network, each effective unit, either an effective monomer or a crosslinker molecule, is represented by a polymer bead. All the polymer beads of the same size are created through a careful selection of the degree of coarse-graining (i.e., the number of real atoms contained in each coarse-grained entity). Each bond (connecting either two adjacent effective monomers in a precursor polymer chain or an effective monomer and a crosslinker molecule) is represented by an anharmonic spring. Each pair of nonbonded polymer beads interacts via a standard Lennard-Jones (LJ) potential. The interaction potential between two bonded beads is an addition of the LJ potential and the finite extension nonlinear elastic (FENE) potential.

Nanoparticle related models. In this work, spherical shaped nanoparticles are considered, each of which is represented by a properly sized single, large bead. The modified LJ potential by Vacatello (2001) is utilized to describe the interaction between a polymer bead and a nanoparticle bead, and that between two nanoparticle beads.

Nanocoating Formation Simulation

An off-lattice MC based method is developed to describe nanocoating sample formation, which can give quantitative correlations among material formulation, curing condition and nanocoating microstructure (see Fig. 1). A nanocoating sample is developed by following a seven-step procedure, which includes those for: (i) simulation system set-up, (ii) initial configuration generation, (iii) first-stage system relaxation in an NVT ensemble, (iv) crosslinking reaction realization, (v) second-stage system relaxation in an NVT ensemble, (vi) cooling of coating samples, and (vii) third-stage system relaxation in an NPT ensemble. Due to the space limit, only the most important step (i.e., Step iv) is detailed here.

Crosslinking reaction implementation. A novel method is introduced to simulate nanocoating formation under a realistic non-isothermal curing temperature profile. The crosslinking reaction proceeds along the MC steps until a specified maximum curing time or a desired crosslinking conversion percentage is reached. A two-stage implementation approach is developed for each MC step, where the first stage is an attempted random displacement for a randomly selected entity, and the second is for conducting reactions with the selected entity. New bonds are then generated if the following three requirements are satisfied: (i) the selected entity is reactive (having un-reacted functional groups), (ii) within a pre-defined reaction distance, there exist other reactive entities, and (iii) two types of functional groups are allowed to react with each other. The two-stage approach is suitable for characterizing the continuous reaction and random movement of polymeric materials and nanoparticles in a real coating formation process. Note that a method for taking into account the effect of macroscopic curing condition on microstructure evolution will be described later.

Nanocoating in Silico Testing

The developed nanocoating is tested for property evaluation. In this work, an off-lattice MC-based tensile test method is introduced to reveal the stress-strain behavior of nanocoating samples. It is known that the mechanical properties derived from this type of behavior can indicate the nanocoating performance on scratch resistance.

An isothermal, constant strain rate, uniaxial deformation process is simulated. The simulation box is stretched in one direction (e.g., the x direction) through a series of strain increments until the desired maximum strain is reached. In each strain increment step, as an initial guess of the positions for all entities, their positions are changed affinely with the simulation box. Immediately after each strain increment, the system is relaxed for a certain number of MC cycles in an extended ensemble (Yang et al., 1997). The applied stress on the sample at any moment during the tensile test can be calculated using the Virial stress theorem (Chui and Boyce, 1999).

Product and Process Performance Evaluation

The stress-strain behavior of a nanocoating sample can be correlated to its elastic property and further to the scratch resistance performance. The elastic property (stiffness) of a nanocoating is evaluated using Young's modulus, which is quantified by the initial slope of the stress-strain curve. A qualitative correlation between Young's modulus and the scratch resistance performance is adopted. That is, an increment of the elastic modulus of a coating is an indication of the improvement of scratch resistance (Misra et al., 2004). The process performance is evaluated based on the energy efficiency for coating curing. Xiao et al. (2006) gives a detailed method for calculating energy consumption in a curing oven.

Macro-Micro Information Coupling

There exists a clear time-scale gap between the macroscopic CFD-based curing simulation and the microscopic off-lattice MC-based nanocoating structure evolution simulation. A complete curing process takes ~10^3 sec, while the characteristic time for a local diffusive motion of a coarse-grained polymer chain is only 10^{-4}~ 10^{-8} sec.

A special bi-directional coupling approach is developed to bridge the existing time-scale gap (Xiao and Huang, 2008b). The curing temperature at a specific time instant obtained from the CFD model is sent to the off-lattice MC model for evolving the nanocoating microstructure. On the other hand, the crosslinking conversion derived from the microstructure is sent back to the CFD model. Through utilizing the reaction rate information, the next time instant for sending a temperature to the off-lattice MC model can be determined (Xiao and Huang, 2009). Consequently, the microstructure evolution throughout a complete curing process under any curing condition can be revealed.

Case Study

The introduced methodology has been successfully employed to study a nanopaint material for automotive coating application, which is a mixture of hydroxyl-functional acrylic copolymer, hexamethoxy-methylmelamine crosslinker, and spherical alumina nanoparticles. The geometries of the curing oven and vehicle body, as well as the normal operating conditions are specified using practical industrial data. A total of 18 paint formulation designs are investigated in order to understand the effects of three critical material parameters, i.e., the polymer-nanoparticle interaction strength (ε^{pn}), the nanoparticle size (R^n), and the polymer number average molecular weight ($\overline{M}_n$) .

Base case. The base-case material has a reduced polymer-nanoparticle interaction strength (ε^{pn*}) of 8.0, a reduced nanoparticle radius (R^{n*}) of 7.0, and a reduced number average molecular weight ($\overline{M}_n^*$) of 8.34. Comprehensive

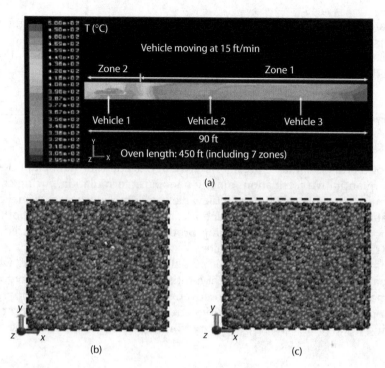

FIGURE 2
Multiscale process and product characterization.

multiscale correlations among nanopaint formulation, curing condition, coating microstructure, tensile property and scratch resistance performance have been revealed.

Figure 2(a) displays a snapshot of the CFD simulation results where three vehicles are moving through the first two zones of an oven (Li, 2007). The temperature at any specific location of any panel of each vehicle at any specific time instant during curing can be obtained. This information is utilized by the off-lattice MC simulation to evolve the nanocoating microstructure. Figures 2(b) and (c) give, respectively, the nanocoating sample microstructure and the deformed structure after the tensile test. The reduced Young's modulus quantified from the stress-strain curve is 50.94, which is 28.4% larger than the Young's modulus for a simulated pure polymeric coating system. This indicates that the nanocoating has a much better scratch resistance performance.

Integrated product and process analysis for nanopaint design. The effects of the polymer-nanoparticle interaction strength (ε^{pn}), the nanoparticle radius (R^n), and the polymer number average molecular weight ($\overline{M}_n$) on the nanocoating scratch resistance and processing efficiency are thoroughly investigated. It is found that to improve scratch resistance performance, the nanomaterial should have a larger ε^{pn} and $\overline{M}_n$, but a smaller R^n. However, such a type of material can lead to more energy consumption in manufacturing, which means a lower energy efficiency. This finding suggests nanopaint designers to consider the cost issue in product realization, e.g., the operating cost related to energy consumption.

Conclusions

One of the most valuable areas of knowledge needed for nanopaint design is the quantitative correlations among nanopaint formulation, curing condition, microstructure evolution, achievable coating properties, and final coating quality. In order to obtain such a comprehensive correlation, both the curing process and the nanocoating product need to be characterized at a wide range of length and time scales.

This work introduces a multiscale modeling and simulation method, which is capable of generating a variety of multiscale correlations necessary for an optimal design of nanopaint. Moreover, the integrated process and product approach provides material designers with the opportunity for properly making trade-offs between product performance and process efficiency.

Acknowledgments

This work is supported in part by NSF (CMMI 0700178) and the Institute of Manufacturing Research of Wayne State University.

References

Chui, C., Boyce, M.C. (1999). Monte Carlo Modeling of Amorphous Polymer Deformation: Evolution of Stress with Strain. *Macromolecules*, 32, 3795.

Kremer, K., Grest, G.S. (1990). Dynamics of Entangled Linear Polymer Melts: A Molecular Dynamics Simulation. *J. Chem. Phys.*, 92, 5057.

Li, J. (2007). *Integrated Product and Process Study: A System Approach to Multiscale Complex Systems*. Ph.D. Dissertation, Wayne State University, Detroit, MI.

Li, J., Xiao, J., Huang, Y.L., Lou, H.H. (2007). Integrated Process and Product Analysis: A Multiscale Approach to Automotive Paint Spray. *AIChE J.*, 53, 2841.

Misra, R.D.K., Hadal, R., Duncan, S.J. (2004). Surface Damage Behavior During Scratch Deformation of Mineral Reinforced Polymer Composites. *Acta Materialia*, 52, 4363.

Ozmusul, M.S., Picu, C.R., Sternstein, S.S., Kumar, S.K. (2005). Lattice Monte Carlo Simulations of Chain Conformations in Polymer Nanocomposites. *Macromolecules*, 38, 4495.

Starr, F.W., Schroder, T.B., Glotzer, S.C. (2002). Molecular Dynamics Simulation of A Polymer Melt with a Nanoscopic Particle. *Macromolecules, 35*, 4481.

Vacatello, M. (2001). Monte Carlo Simulations of Polymer Melts Filled with Solid Nanoparticles. *Macromolecules, 34*, 1946.

Xiao, J., Li, J., Lou, H.H., Xu, Q., Huang, Y.L. (2006). ACS-Based Dynamic Optimization for Curing of Polymeric Coating. *AIChE J., 52*, 1410.

Xiao, J., Huang, Y.L. (2008a). Off-lattice Monte Carlo Based Nanopaint Design for Coating Scratch Resistance Improvement. *AIChE Annual National Meeting*, paper 702d, Philadelphia, PA.

Xiao, J., Huang, Y.L. (2008b). Multiscale Model Development for Microstructure-based Product Quality Control in Polymeric Coating Curing. *AIChE Annual National Meeting*, paper 546c, Philadelphia, PA.

Xiao, J., Huang, Y.L. (2009). Microstructure-Property-Quality-Correlated Paint Design: An LMC-Based Approach. *AIChE J., 55*, 132.

Yang, L., Srolovitz, D.J., Yee, A.F. (1997). Extended Ensemble Molecular Dynamics Method for Constant Strain Rate Uniaxial Deformation of Polymer Systems. *J. Chem. Phys., 107*, 4396.

88

Product Portfolio Design for Forest Biorefinery Implementation at an Existing Pulp and Paper Mill

Virginie Chambost and Paul Stuart[*]

NSERC Environmental Design Engineering Chair in Process Integration
Department of Chemical Engineering, Ecole Polytechnique Montréal
Montréal (Québec), Canada, H3T 1J7

CONTENTS

ABSTRACT   The forest biorefinery (FBR) is being considered by many forest product companies as a solution to their stalemate situation and as an option for improving their business model. The FBR implementation implies significant challenges for mills related to key technological, economical, financial, environmental, cultural and operational risks. The definition of a biorefinery product platform needs to be systematically considered for the development of sustainable biorefinery strategies. In this article, a methodology is proposed in order to help the forestry industry identify product opportunities on the market associated with preliminary process design concepts. This

---

[*] To whom all correspondence should be addressed

methodology explores key elements for forest biorefinery implementation such as product family approach, risk definition and assessment associated with each product family, and incorporation of the most promising product family in the existing product portfolio of a given P&P company. Using a product design approach, this proposed methodology aims to reply to key challenges that the forestry industry is facing while embarking on the forest biorefinery, such as the product diversification strategy and its impact on day-to-day production, and the choice of partners for securing the biorefinery business over the longer term.

KEYWORDS *Product and process design, product portfolio, forest biorefinery implementation*

Introduction

The forest biorefinery (FBR) is considered by the forestry industry as a concrete solution for its current stalemate situation, i.e. risen costs and fiber costs as well as strengthening dollars and increasing competition. Facing increasing pulp and paper (P&P) mill closures, the industry is looking for diversifying its core business and improving its revenues at the same time. The FBR offers the potential for transforming the existing forestry industry by developing new products and maximizing existing asset utilization. Defined by Axegård (Axegård, 2005) as 'the full utilization of incoming woody biomass for the production of wood products, pulp and paper products, energy including biofuels and other organic and specialty chemicals that can be derived from wood', the FBR implies the definition of new strategic business models leading to more efficient and added-value activities. The forestry industry should evolve from being manufacturing-centric to margin-centric in order to successfully implement the FBR. The potential offered by the FBR resides in the optimal implementation of new products into the existing pulp and paper portfolio. However, how to consider FBR implementation is not obvious. With a large number of technology providers on the market looking for biomass access and with a large set of potential biorefinery products, the identification of the *right* FBR, i.e. the most successful product/process combination, is a challenge. Emerging thermochemical, biochemical and chemical technologies are speeding up global interest in the biorefinery. On the market side, a large number of replacement and substitution product opportunities are the result of e.g. new product development by chemical companies. A phased approach (Chambost, 2008) has been developed in order to guide a stage-wise implementation of the FBR considering competitive advantages and disadvantages of the forestry industry, as well as its transformation. Critical to the economic and commercial success of the FBR is the identification and management of a "biorefinery product platform". The FBR product platform definition involves the determination of product families that incorporate building blocks and value-added derivatives. The integration of the potential product family into the existing P&P portfolio, i.e. a new product portfolio, should be carefully designed in order to maximize flexibility and lead to long-term competitive advantages, as well as a sustainable business strategy. Which new product portfolio will ensure a sustainable implementation of the FBR, mitigate the main risks and maximize its benefits taking into account P&P company constraints such as limited capital spending and energy savings? The design problematic related to the FBR implementation should be based on a good understanding of the product and process design constrains. In this paper, product design fundamentals are summarized in order to support the development of a market-driven methodology that takes into account preliminary process design constrains for identifying and selecting the most promising product portfolio for a given pulp and paper mill.

Product Design

Definition of Product Design

Two schools of thinking for product design can be recognized (Hill).:

1. Industrial product design for assembled products Industrial product design is the definition of product physical forms that best meet customer needs. This design for manufacturing includes mechanical, aesthetic and ergonomic properties of assembled products (Ulrich, 2004).
2. Chemical product design for formulated products.

Chemical product design has been considered for a long time only as a tool for defining molecules to satisfy property specifications that align with customer needs (Seider). However chemical product design is now recognized as a more complicated engineering system (Stephanopoulos, 2004).

In this context, time and cost reduction to position new or enhanced products on the market are part of product design, focusing on minimal resource utilization, optimal desired product quality, cost, and efficient production and marketing (Gani, 2004). Product design is moving from being compositional specifications-centric to performance-centric (Hill). This change is supported by the cyclical evolution of the chemical industry, i.e. going from being commodity-centric and focused on market share to being specialty-centric and focused on product development, and to being customer-value centric and focused on 'solutions-to-customers channels as future growth enabler' (Stephanopoulos, 2004).

For achieving FBR implementation, the chemical product design will be considered as the predominant design paradigm in order to determine what new product functionalities or enhancements are needed on the market (Seider).

Management of Product Design

Different approaches have been considered for supporting the product/process innovation and development of new products. These methodologies, described in the literature, contain common concepts regarding product design:

- *Product and process design interaction*

The interaction between product and process design, defined as the development of a process that starts with the decision to manufacture a product and ends with the start-up and operation of this process (Janssen, 2007), has always been a major challenge to overcome in new product development. 'The basic process of innovation involves the matching of information drawn from the technological capabilities/potential and the market needs' (Allen, 2001). Both designs should interact properly. Taking the example of the Stage-Gate® approach (Cooper, 1986) sequences of product and process design concerns are revealed at diverse stages for ensuring successful and pertinent new product development.

- *Product Architecture*

Product architecture that supports the definition of product platforms and families is an enabler for product development and manufacturing efficiency via the exploitation of technical and commercial commonalities among the products offered (Chambost, 2008). Product and process design are critical for determining product architecture.

- *Product strategies define technology strategies*

Product design is considered as an enabler for the definition of an appropriate technology strategy. The concept of Funnel Development (Ulrich, 2004), that consists of six steps (planning, concept development, system level design, detailed design, test/refine and production ramp-up), underlines the selection of products based on market- and technology-based criteria, as well as the predominance of product design on technology strategy.

- *Product and process design should be iterative*

'Product design is empirical, and based on trial-and-error approaches'(Gani, 2004). Nevertheless, product design resolves multidisciplinary problems that involve multivariate trade-offs. The associated models should be flexible and

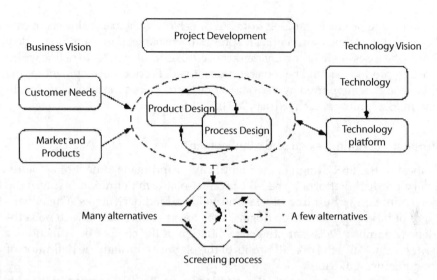

FIGURE 1
Product and process design iteration. Adapted from Stephanopoulos, 2004; Ulrich, 2004.

should consider reverse approaches that will help to reduce problem complexity. Figure 1 shows the iteration between product and process design, both supporting the product development and both managing different information.

Objective of this Paper

This paper presents a product portfolio design methodology that will help forestry companies to embark on the FBR by identifying the most promising product/process combination and related business partner options.

Application to the Forest Biorefinery

Context of the FBR: New Product Development in the P&P Industry

The FBR offers many alternatives in terms of product and process opportunities for value creation and maximization. Therefore, the design question is which product/process combination is the most promising for the implementation of the FBR over the longer term. Many technology providers have already embarked on testing and developing new or enhanced processes at pilot and demonstration scale in order to efficiently convert lignocellulosic biomass into commodity and/or specialty chemicals. The question of identifying the best product is still at an early stage. The current effort regarding product development is mainly based on technology-push, rather than market-pull, approaches. NREL defined a set of promising products, ranging from building blocks to added-value chemicals, based on key technology criteria and preliminary market understanding (Aden, 2004). The assessment of current market needs is critical in order to define which existing products should be replaced/substituted. Product development for the FBR should be supported by a strong interaction between the technology push- and market-pull approach (Ulrich, 2004).

Product Portfolio Design and Management

Critical to the FBR design is the identification and management of a "biore-finery product platform". The FBR product platform definition involves the determination of building blocks and value-added derivatives. The advan-tages of building product platforms, such as ethanol to ethylene to polyeth-ylene (Chambost, 2008), are the consolidation of the competitive position of each product, and the overall profits of the platform through the definition of manufacturing flexibility.

The design of the new product portfolio (Figure 2), i.e. integration of the FBR product platform with the existing P&P portfolio, for a given P&P mill needs to answer to two major design questions: (1) what is the best product platform that will ensure value creation and competitiveness over the long term?; (2) what is the most sustainable product portfolio alternative that will ensure success in integration and in new business model development? Conventional approaches (Sammons, 2008) have been developed for identifying the optimal biorefinery product allocation by combining process and economic modeling. Superstructures consider the FBR as an investment project, i.e. the financial return is the only target. However, the design of the new product portfolio should not only consider increased profits, but also other factors such as implied changes in existing manufacturing processes, changes in the exist-ing value chain, process risks associated with innovative processes, increased process complexity with derivatives manufacturing, yields and overall mass/ energy balances, process constraints related to supply chain flexibility, and

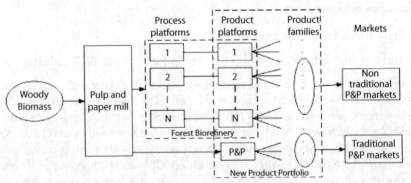

FIGURE 2
Product management for the FBR.

co-product opportunities. The FBR should support the industry transformation in order to be successful over the long term.

Methodology

Overall Methodology

A systematic approach has been defined, comprising product and process design (Chambost et al., 2008), for identifying the best product/process combination for a given company and ensuring the successful implementation of the FBR in retrofit. This product-driven approach is divided into 3 major steps which have different but consistent objectives and methodologies: step 1 is the identification of the most promising product portfolio/process/partners combinations; step 2 consists of a detailed enterprise-based analysis of each promising product portfolio and step 3 is the definition of a business plan. Step 1 is the most risky part and requires major efforts in terms of product design.

Product Portfolio Design

The sequence and types of questions shown in Figure 3 elaborate the identification of a list of promising products, the construction of a product family, the evaluation of its incorporation into a new product portfolio and the assessment of the potential of partnerships for value chain optimization. Around these questions, risks can be defined such as volatility and perturbing events on the market, or technology maturity.

The results of the methodological approach are summarized in Figure 4. This methodology proposes 3 main steps in order to identify and assess the main challenges and opportunities related to product/process options for implementing the FBR at an existing mill.

The first step aims to identify a set of chemical product families for a market region, for which 'green' replacement and/or substitution chemicals could be manufactured. This is done by 2 systematic analyses:

1. Product identification on the market

Product identification will be supported by an understanding of the existing value chains via the analysis of the current market and/or discussion with potential partners. The value chain identification takes into account all products manufactured and/or consumed on the market, as well as new products for substituting existing products.

2. Product triage and product family definition.

The large list of identified chemicals should be refined using market-based and techno-economic criteria, and process synthesis triage.

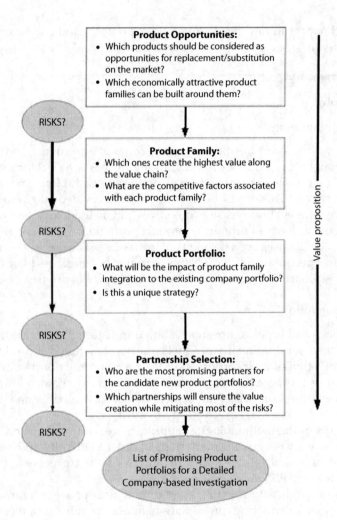

FIGURE 3
Key questions around the FBR product strategy identification.

The objective of the second step is to rank, by priority, each product family based on the identification of product family opportunity and competitiveness. These analyses, based on a SWOT matrix, aim to identify opportunities and threats as well as competitiveness of each product within a family taking into account several commercial, technical and economic criteria for determining the competitive position of each product family. This step also identifies and assesses risks based on the preliminary process and techno-economic challenges as well as market analysis and carries out a Multi-Criteria Decision-Making (MCDM) analysis for ranking each product family via panel discussions and the consideration of the analyses results. The MDCM represents the value opportunity for the P&P industry.

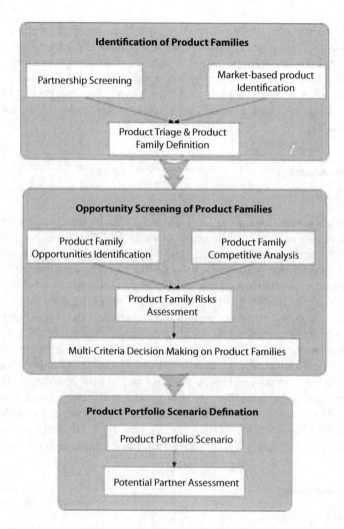

FIGURE 4
Methodology for selecting the most promising product/process scenario and partnership combination.

The last step of this methodology aims to integrate a product family into an existing product portfolio using process systems engineering tools such as life cycle analysis, supply chain management, process simulation..., and to identify strategic partnerships for the new product portfolio.

The outcomes of this methodology are the following: (a) the definition of product opportunities and their strategic potential taking into account the pulp and paper context, (b) the definition of market, process and techno-economic elements for the assessment of product family potential, (c) the identification of process/product scenarios, i.e. product families and the related technical feasibility (d) the identification of risks and the potential for their mitiga-tion, (e) the evaluation of product family integration into an existing product portfolio and (f) the value proposal regarding the most promising 'quality' partners relative to the selected product families.

This methodology serves as the basis for a more detailed product/process design analysis at an existing P&P company in order to elaborate a long-term business strategy.

Conclusion

The proposed methodology explores critical elements specific to forest prod-uct companies that consider the implementation of the biorefinery and look for product strategy development. This methodology explores the product portfolio extension of a given P&P company in order to diversify into the forest biorefinery business. This methodology also proposes a strategic approach in order to identify the set of potential products to be considered taking into account perturbing events on the market that could impact the product potential, as well as to define a product family that will provide a competitive position to a P&P company. Risk identification and assessment is a crucial activity for product family prioritization. This methodology uses a product-driven approach considering preliminary process design concepts such as technology availability, yield of production, specific to a given P&P company. This overall approach provides a strong basis for developing a long-term and profitable business plan for forestry companies that embark on the FBR, and considers a stage-wise biorefinery implementation.

Acknowledgements

This work was supported by the Natural Sciences Engineering Research Council of Canada (NSERC) Environmental Design Engineering Chair at Ecole Polytechnique in Montréal.

References

Aden, A., Werpy, T., Petersen, G., Bozell, J., Holladay, J., White, J., et al. (2004). Top Value Added Chemicals From Biomass. Volume 1-Results of Screening for Potential Candidates From Sugars and Synthesis Gas: Storming Media.

Allen, T. (2001). *Organizing for Product Development*: SSRN.

Axegård, P., (2005) "The future pulp mill–a biorefinery", Presentation at 1st International Biorefinery Workshop, July 20-21, Washington DC

Chambost, V., McNutt, J., & Stuart, P. R. (2008). Guided tour: Implementing the forest biorefinery (FBR) at existing pulp and paper mills. *Pulp and Paper Canada, 109*(7-8), 19–27.

Cooper, R. G., & Gravlin, R. (1986). *Winning at New Products*: Addison-Wesley Reading, Mass.

Gani, R. (2004). Chemical product design: challenges and opportunities. *Computers & Chemical Engineering, 28*(12), 2441–2457.

Hill, M. Chemical Product Engineering - The Third Paradigm. *Computers & Chemical Engineering, In Press, Accepted Manuscript.*

Janssen, M. (2007). *Retrofit design methodology based on process and product modeling.* Ecole Polytechnique, Montreal.

Sammons Jr, N. E., Yuan, W., Eden, M. R., Aksoy, B., & Cullinan, H. T. (2008). Optimal biorefinery product allocation by combining process and economic modeling. *Chemical Engineering Research and Design, 86*(7), 800–808.

Seider, W. D., Widagdo, S., Seader, J. D., & Lewin, D. R. Perspectives on chemical product, process design. *Computers & Chemical Engineering, In Press, Accepted Manuscript.*

Stephanopoulos, G. (2004). Invention and Innovation in a Product-Centered Chemical Industry, *WebCAST Lecture*: Massachusetts Institute of Technology.

Ulrich, K. T., & Eppinger, S. D. (2004). *Product design and development*: McGraw-Hill New York.

89

Multi-Scale Methods and Complex Processes: A Survey and Look Ahead

Angelo Lucia

Department of Chemical Engineering
University of Rhode Island
Kingston, RI 02881

CONTENTS

ABSTRACT A comprehensive overview of numerical methodologies currently available for analyzing and building understanding of complex processes is presented. Both equation-free and equation-based methods are discussed. Many multi-scale techniques, including the use of reduced order models, lifting-restricting methods, averaging strategies such as molecular dynamics, Monte Carlo simulation and the mean value theorem, stochastic sampling methods like kriging, fitting and interpolation approaches such as surrogate solution maps and gap-tooth schemes, and others are described. The importance of and ways in which information is communicated between scales is also discussed. Improvements that are needed for the next generation of multi-scale tools for analysis, visualization, simulation, and optimization of complex processes are identified and include the need to proceed with partial knowledge, techniques for handling multiplicity at various length scales, problems with many scales, and methods for predicting behavior over very long time and length scales.

KEYWORDS *Multi-scale methods, complex processes, communication between scales*

Introduction

A process is often called complex if it contains multiple time and length scales, complicated phenomenological descriptions, and feedback between various elements of the process that results in behavior that cannot be predicted by analyzing the individual components in isolation. Mathematical models of complex processes are generally described by nonlinear algebraic and/or differential equations that must be solved numerically using multi-scale methods and often require high performance computing (clusters, supercomputers, cloud computers, and so on). At fine scales, these models are often treated as equations or as input-output black boxes with and without the presence of noise. Key among the attributes of any multi-scale methodology is the communication of information between scales.

Literature Survey

Over the last decade, there has been a growing interest in the development and application of multi-scale methods in process systems engineering and many traditional processes studied by chemical engineers are complex processes. Examples include problems involving simultaneous heat and mass transfer in porous media, molecular conformation and crystal structure determination, chemical vapor deposition, and others and not all multi-scale problems are necessarily time dependent. Multi-scale modeling techniques offer valuable advantages in building better understanding and more reliable computer tools for modeling complex system behavior. Some of the more prominent methods in multi-scale modeling and applications to complex processes are surveyed in this paper.

The literature on multi-scale methods in chemical process engineering is quite extensive, covering both traditional areas such as the estimation of thermodynamic and transport properties, chemical reaction engineering, advanced polymer materials, molecular conformation and crystal structure, and others to more exotic applications in advanced materials (e.g., flow of DNA through micro-fluidic devices, defects in crystals, ionic liquids, etc.). There are many application papers as well as articles that address the development of multi-scale methodologies. Some work has been exclusively computational while others have integrated simulation and experiment.

Multi-Scale Methodologies

Any multi-scale methodology consists of models, numerical methods, and possibly experimental and/or observational data at various length/time scales as well as communication between scales. Moreover, what constitutes a small length-scale and what is large is generally problem dependent. Figure 1 gives a generic picture of the relationship between multiple length scales in problems of a physical nature, where small and large length scales imply small and large time scales respectively. This figure is also intended to show that discrete representations of geometry often gives rise to irregular elements that must be handled with care, that geometries can change from scale to scale, and that there is no requirement that the number and/or variable types at different scales be the same.

Note that the macroscopic scale is divided into rectangular control elements and that within each rectangular control element are many smaller circular elements. In addition, symmetry can often be exploited within each large or small element, if justified, to lessen computational demands. In most process engineering applications, small can range from atomistic to nano- to even millimeter length-scales while large often refers to meso- or macroscopic length scales in centimeters or meters. We also note that coarse-graining

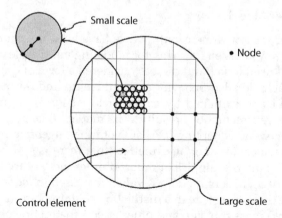

FIGURE 1
Schematic of multi-scale process.

refers to the process of constructing larger meshes within a fixed area. So for example, in Figure 1, where there are 32 control elements shown, coarse-graining at the larger length scale might consist of reducing the number of meshes (or control elements) from 32 to perhaps 8.

Traditional methodologies used for multi-scale modeling at small length scales include techniques from quantum chemistry (QC) such as density functional theory (DFT), molecular simulation methods such as molecular dynamics (MD), various Monte Carlo (MC) techniques (kinetic MC, lattice MC, etc.), transition state theory (TST), and computational fluid dynamics (CFD). These methods are well-developed and continue to evolve and find use in process engineering applications. Models at the large length scale are generally steady or unsteady-state conservation laws, Navier-Stokes equations, and so on. Pantelides (2000) gives a good early summary of the multi-scale modeling challenges and methods in process systems engineering, describing serial, parallel, and hierarchical approaches to scale integration.

In this paper, we refer to scale integration as communication between scales. We focus on general multi-scale methods that include reduced order models, equation-free methods, deterministic averaging techniques (MD, CFD, and the mean value theorem), stochastic sampling methods (MC and kriging), interpolation techniques (surrogate solution maps, gap-tooth schemes, etc.), multiple time stepping methods, and others. We also describe the means of communication between scales for these multi-scale methods. While some view multi-scale methods as methods that just perform computations and/or analyses at different length and time scales, in the author's view, true multi-scale methods alternate between computations at different scales and use two-way communication and either explicit or implicit coarse-graining to represent small scale information in some form of 'average' information at larger length and time scales.

Reduced Order Models

Reduced order (or low dimension) models have been used for a very long time in process systems engineering for modeling multi-scale chemical processes, are still used today, and can be considered a form of coarse-graining. In this approach, detailed behavior at small length and (faster) time scales is ignored and approximated by explicit (closed form) or implicit (black box) algebraic representation, resulting in models that are comprised of differential-algebraic equations (DAEs). Note that algebraic representation at small time scales does not alter the fact that there is still two-way communication. Common examples of reduced order models are the equations of state, activity coefficient models, K-values, etc. used to model equilibrium or non-equilibrium on a stage in a distillation column. Time dependent contacting, mixing, mass transfer, and other effects that define phase behavior at the molecular scale are assumed to occur infinitely fast (and reach steady-state), compared to mass and energy balance dynamics on a tray, and are approximated by algebraic equations that describe either equilibrium or non-equilibrium phase behavior. In fact, any constitutive equation can be considered a reduced order model at some scale. For example, heat capacity polynomials in temperature can be considered reduced order molecular scale models since MD at several temperatures can be used to evaluate heat capacity. Other reduced order model approaches include coupled CFD and reduced kinetic models for combustion (Androulakis, 2000; Sirdeshpande et al., 2001; Liang et al, 2007) and modal decomposition, which eliminates fast modes in dynamical systems using techniques like center manifold theory (Balakotaiah and Dommeti, 1999). Classical CSTR, PFR, and other reactor models are also reduced order models (Chakraborty and Balakotaiah, 2005) since they ignore small scale flow patterns within the reactor, use assumptions such as well mixed, no radial mixing, etc. and only model macroscopic behavior.

Equation-Free Methods

Equation-free methods for multi-scale modeling have been pioneered by Kevrekidis and co-workers (Qian and Kevrekidis, 2000; Gear et al., 2002; Siettos, et al., 2003; Kevrekidis et al., 2003). As their name suggests, these methods forego closed form model descriptions and treat macroscopic models as black boxes or experiments that result from 'wrapping' software around small scale simulation methods (MD, MC, etc.). As a result of the work of Kevrekidis, there are now equation-free, multi-scale methods in molecular dynamics (Frederix et al, 2008) and other branches of science and engineering (e.g., the heterogeneous multi-scale methods of Weinan and Engquist, 2003). The key to equation-free, multi-scale modeling is the two-way communication between small and large length scales. Kevrekidis calls this lifting/restricting. Restricting is the process of communication from small or fine scale, which is considered a high dimensional map, to low dimensional or large length scale. Lifting, on the other hand, is reverse communication, from the large to the small length scale. The result of lifting/restricting yields an implicit coarse time stepping procedure at the macroscopic scale that can be written in the form

$$M(t + \Delta t) = M(t) + R\{\mu(L[m, \Delta t])\} \tag{1}$$

where m denotes microscopic variables, M is a set of macroscopic variables, L and R are lifting and restriction operators respectively, μ is the functionality defining the microscopic simulation, and t is time.

As with any multi-scale method, the creativity lies in the way in which two-way communication is designed and implemented. In this regard, lifting is the critical issue because it can be challenging to find adequate initializations at the small scale from 'limited' low dimensional spatial and temporal information at the large scale. In addition, spatial information at the macroscopic scale must undergo dimensional reduction and is often approximated using an interpolation scheme (e.g., gap-tooth, splines, etc.). Kevrekidis and co-workers have applied their equation-free methodology to a variety of examples in process engineering including, reaction-diffusion processes, lattice gases, liquid crystals, peptide fragments, process control and others.

Deterministic Averaging Techniques

Deterministic sampling techniques like MD are frequently used at the small length scale to obtain average bulk phase thermodynamic and transport properties. Averaging in this context exploits a fundamental axiom of statistical mechanics – ergodicity (i.e., averages over time are adequately represented by ensemble averages over phase space). Averaged quantities are generally represented by the equation

$$<P> = \Sigma\, P(p^N, r^N)/K \qquad (2)$$

where P is some property, $< >$ denotes average, p^N and r^N are the momentum and position vectors for an N particle system respectively, K is the total number of integration steps, and the summation in Eq. 2 is over all integration steps. Other derived properties can often be estimated from multiple MD simulations.

Other deterministic averaging methods have also been proposed. Perhaps the most straightforward is the use of the mean value theorem within the terrain/funneling methods of global optimization recently proposed by Lucia and co-workers (Lucia et al., 2004; Gattupalli and Lucia, 2007, 2008). Geometric models at two different length scales are assumed. At the small length scale, a quadratic model with considerable noise or roughness on a surface is assumed. Average gradient and second derivative information is computed using

$$<g> = \{\textstyle\int g[z(\alpha)]d\alpha\}/\alpha \qquad (3)$$
$$<h> = \{\textstyle\int h[z(\alpha)]d\alpha\}/\alpha \qquad (4)$$

by performing terrain optimizations over some small region of the surface, where the variables α, g, and h in Eqs. 3 and 4, denote some relevant length of a path on an objective function surface or terrain, the gradient, and Hessian matrix respectively. This average information is communicated to the large length scale, which is assumed to be non-quadratic (funnel-shaped) to determine funnel parameters and an estimate of the global minimum on the surface. The estimated funnel minimum is, in turn, communicated to the small length scale in order to define the next region on the objective function surface to be explored locally. Unlike equation-free methods, the number and variable types are the same at each scale in the terrain/funneling approach.

CFD is also a deterministic method that can be used to calculate 'average' effects due to mixing (fluid flow, mass transfer and heat transfer). In this approach, momentum, energy, and continuity equations are often solved at the small scale and then averaged to give flow patterns and transfer coefficients at larger length scales.

Stochastic Sampling Methods

There are many stochastic sampling methods that are used in multi-scale modeling to obtain average information. Some are used at small length/time scales while others have been employed at large length scales. Stochastic sampling methods at small length scales are frequently based on various forms of MC simulations. Stochastic averages at the small scale are calculated using a simplified version of Eq. 2 given by

$$<P> = \Sigma \, P(r^N)/K \tag{5}$$

which does not account for kinetic energy.

Kriging is also a stochastic averaging methodology and gets its name from the geologist D. Krige (1951) who pioneered the technique in applications in hydrology. However kriging is quite different from variants of MC and often used to obtain approximate information at the large length scale. In kriging, a number of initial sampling points are placed in the feasible region and are assumed to be randomly distributed. A set of trial or test points are placed and both a kriging predictor, based on linearly weighted function values, and the expected variance associated with each test point are determined. The kriging predictor, $z(x_k)$, for a test point, x_k, is given by

$$z(x_k) = \Sigma \, f(x_i)w(x_i) \tag{6}$$

where the x_i's are initial sampling points, f is the function value, w is a weighting coefficient, and where the set of weighting coefficients is determined by solving a system of equations in covariance functions. The expected variance, $\sigma(x_k)$, is given by

$$\sigma(x_k) = \sigma_{max} - \Sigma \, w(x_i)Cov[d(x_i, x_k)] \tag{7}$$

where Cov denotes covariance and $d(x_i, x_k)$ is the distance between points x_i and x_k.

From the set of test points, predictors, and expected variances, prediction and variance maps are constructed and a set of most promising regions to perform local optimizations are identified. Davis and Ierapetritou (2007) have applied kriging to nonlinear programming problems to obtain a 'picture' of the global geometry of noisy or black box objective function surfaces. From the kriging results, local optimizations are conducted using a response surface method (RSM). While they did not alternate between kriging and RSM, due to computational expense, Davis and Ierapetritou do recognize that such an approach would produce two-way communication and a true multi-scale method.

Interpolation Methods

Interpolation is often a valuable tool in resolving spatial and temporal differences in scale. Interpolating techniques such as wavelets (Reis et al., 2008), surrogate solution maps (Kulkarni et al., 2008), and gap-tooth schemes (Kevrekidis, 2000) are among the interpolation methods that have been used to capture small and/or large scale behavior.

Gap-tooth schemes, like those proposed by Kevrekidis, are intended to provide smooth spatial descriptions at the large length scale from information inferred from the small scale through simulation.

Surrogate solution maps (e.g., Kulkarni et al., 2008) are another form of interpolation. Solutions, Y, at the small length scale for a range of explicit or implicit large scale conditions, X, are computed *a priori* and stored in the form of polynomial coefficients. Polynomial representation is used at the large length scale to lessen the computational burden using maps of the form

$$Y(X) = f(X) = \Sigma \, w_i \phi_i(X) \tag{8}$$

where $\phi_i(X)$ are polynomial (or basis) functions and w_i are weights or coefficients determined from matching values of $Y(X)$ and X at known solution points. Note that the basis functions can be orthogonal or separable, if desired. It is straightforward to couple the large and small length scale by simply using Eq. 8 to retrieve information at the small scale during large scale iterations. Two-way communication is maintained between the small and large scale in much the same way that reduced order models maintain two-way communication. In principle, maps of any functionality and dimensionality can be constructed. However, in practice, surrogate solution maps are generally restricted to low dimensionality and simple polynomial functions.

Multiple Time Step Methods

There are a number of multiple time step methods that have been developed over the years to address the accurate integration of coupled fast and slow dynamics in a variety of applications. Perhaps the best known approach of this type is a variant of the Car and Parrinello method (1985) where nuclear and electronic forces are separated. Other examples include Humphreys et al. (1994) and Barash et al. (2002). The basic idea is to integrate fast modes with a small time step, while inferring slow modes for several fast time steps, and integrate the slow modes intermittently with a larger time step. To illustrate, assume that a system of differential equations

$$y' = F(y_f, y_s, t) \tag{9}$$

has fast and slow modes, where $y = (y_f, y_s)$ is a vector of dependent variables, the symbol ' denotes the time derivative, and the subscripts f and s refer to fast and slow modes respectively. Integration proceeds as follows

$$y_f^{k+1} = y_f^k + y_f'[y_f^k, y_s^k]\Delta t_f \tag{10}$$

$$y_s^{k+1} = y_s^k + y_s'[y_f^k, y_s^k]\Delta t_s \tag{11}$$

where $\Delta t_s = k\Delta t_f$ for $k \gg 1$. Within each fast time step, extrapolation (not function evaluation) is used to update the slow variables and this is the key to reducing computational work. Multiple time stepping methods can often be developed by exploiting physics (e.g., in MD where force splitting is exploited). Nonetheless, coupling and the multi-scale nature of the problem is retained and two-way communication exists since the Eqs. 10 and 11 couple the fast and slow modes.

Multi-Scale Applications

The use of multi-scale methods in various areas of chemical engineering is reviewed because of their importance to process systems engineering.

Thermodynamic & Transport Properties

Estimation of bulk phase thermodynamic and transport properties (e.g., diffusion coefficients, viscosities, etc.) is often accomplished using quantum chemistry such as density functional theory, molecular dynamics, and/or Monte Carlo simulation at the small length scale and then averaged over many configurations to predict properties at the macroscopic length scale (usually bulk phase). Greenfield and co-workers (1998, 2001, 2004, 2007) have computed transport properties of polymers using multi-scale approaches. Angstrom-scale computations in these applications include the use of MD and TST to measure temperature-dependent diffusion coefficients, relaxation times, viscosities, and rate constants for diffusion. Larger nanometer length scale computations, on the other hand, make use of kinetic MC to determine diffusivities or infer composition changes. Pricl et al. (2006) make combined use of force field models, all-atom molecular dynamics simulations, and a continuum description of protein transport dynamics to estimate macroscopic diffusion of proteins in nano-channeled silicone membranes. Economou et al. (2008), on the other hand, use *ab initio* DFT calculations, molecular dynamics, and the tPC-SAFT equation of state to estimate structural and thermodynamic properties of imidazolium-based ionic liquids. CFD has been used to estimate flow patterns, and temperature and composition distributions on sieve trays (e.g., Rahimi et al. 2006) and in packed columns for tray and packing design. Recently Egorov et al. (2005) have made combined use of CFD and rate-based models in simulating reactive separations. There are many, many other papers and examples of the computation of thermodynamic and transport properties with multi-scale methods, which due to space limitations, are not cited in this article.

Reaction Engineering

Many multi-scale approaches have been used in reaction engineering, ranging from simple reduced order models to CFD models of convection-diffusion-reaction equations. See Chakraborty and Balakotaiah (2005) for a good survey. Reduced models like CSTR, PRF, and so on represent the simplest models and need no further discussion. At the other extreme is the computationally demanding approach of computational fluid dynamics. Here convection-diffusion-reaction equations are used as the large scale model and solved using CFD. Small scale phenomena like turbulent flows are averaged or reduced kinetics models included and incorporated in the CFD model to create a multi-scale method. Somewhere in the middle are the techniques that use spatial averaging of convection-diffusion-reaction equations. Here averaging is used at the small scale to develop low order models at the large length scale suitable for process design and control activities. The basis for one such approach is Liapunov-Schmidt bifurcation analysis and the result is a coupled or two-mode model in which both local and global behavior are described. Chakraborty and Balakotaiah (2005) call their approach a multi-mode approach and describe the application of spatial averaging to a wide variety of reactor models including isothermal tubular reactors, loop and recycle reactors, CSTR's, various models of non-isothermal reactors, and multiple reactor configurations but present very limited numerical comparisons with traditional approaches. The primary limitation of the multi-mode approach is the assumption of well defined flow fields. The authors also describe situations where multiple solutions to local models can exist and give rise to numerical difficulties.

Kulkarni et al. (2008), on the other hand, take a different approach to multi-scale modeling of catalyst pellet reactors using surrogate solution maps. Orthogonal collocation over finite elements is used to construct a set of nonlinear algebraic equations for the reactor temperature and concentration profiles at the large length scale (i.e., the bulk phase), which in this case is on the order of meters. However, these equations are coupled to the small (single catalyst pellet or millimeter) scale through the effectiveness factor. Effectiveness factor maps are generated as a function of anticipated values of the Thiele modulus, dimensionless heat of reaction and dimensionless activation energy. These effectiveness factor maps are used to determine an effectiveness factor for any temperature and concentration within the reactor. Like reduced order models, two-way communication is maintained and the method is truly multi-scale. Kulkarni et al. (2008) use their surrogate solution map approach to find solutions to the classical problem of simultaneous heat and mass transfer in a catalyst pellet described by Weisz and Hicks (1962) and present numerical results that show that surrogate solution maps can offer significant computational advantages over nested iteration or direct numerical simulation and handle multiplicity at the small length scale in an effective manner.

Molecular Modeling

MD and MC have been used to build understanding in molecular modeling for describing reaction pathways, transition states, and so on. However, it has long been known that potential energy and free energy landscapes used in molecular conformation and crystal structure determination have many, many stationary points at the small length scale. The large scale geometry is frequently more ordered and non-quadratic (e.g., funnel-shaped, helical in the case of DNA, and so on). Multiple length scales make this class of problems ideally suited for multi-scale methods. Gattupalli and Lucia (2007, 2008) have used a terrain/funneling approach to determine the molecular conformation of n-alkanes. Numerical comparisons with other molecular conformations methods that are not multi-scale methods (such as basin hopping, Wales and Doye, 1997) clearly show that exploiting the multi-scale nature of molecular conformation problems in algorithmic development leads to global optimization methodologies that have superior computational reliability and efficiency. See also Westerberg and Floudas (1999) who studied reaction pathways using global optimization and TST to determine large scale reaction rates.

Advanced Materials

There are seminal papers by de Pablo (2005), de Pablo and Curtin (2007) and Barrat and de Pablo (2007) that describe the role of multi-scale modeling techniques in advanced materials science. Models in the materials science arena typically range from continuum mechanics at the macroscopic scale to atomistic models at the quantum scale. Coarse graining generally involves a hierarchy of models from fully all-atom models to bead-spring arrangements that represent chains of molecules to primitive path analysis used for modeling entanglement in polymers. Advances in rigid materials have progressed while multi-scale modeling of soft materials remains challenging.

Dai et al. (2005, 2006) have used multi-scale methods to analyze defect evolution in crystalline systems using kinetic MC and coarse-graining in which lattice sites are grouped into cells and averaged within cells to capture void morphology. Zheng et al. (2008) describe the use of hybrid multi-scale kinetic MC method for the simulation of copper electro-deposition.

Process Systems Engineering

A decade ago, multi-scale modeling in process systems engineering was rare. As computational power has evolved (clusters, high performance computing, cloud computing), multi-scale methods have become more widespread. The previously cited papers are clear examples of this. In addition, there are papers that describe multi-scale methods in process simulation, process identification, process control, process design and optimization (Gerogiorgis and Ydstie, 2004), product/process design (Morales-Rodriguez and Gani, 2008), and others (e.g., Ingram and Cameron, 2005).

However, multi-scale modeling and multi-scale methods are not as widespread as one might expect in process systems engineering. Even as late as 2004, multi-scale modeling and multi-scale methods were still classified as new challenges in Computer Aided Process Engineering (CAPE). See Puijaner and Espuna (2004). Moreover, at the Symposium on Modeling of Complex Processes held in the spring of 2005 at Texas A & M, the subject of many of the multi-scale modeling papers in the process systems area was molecular simulation (e.g., Tunca and Ford, 2005; Faller, 2005) with few exceptions (e.g., the paper by Ge et al. (2005) concerned with scale-up and scale-down of laboratory processes to industrial operations). Nevertheless, there are papers on multi-scale methods in process systems engineering that do have a strong link to traditional macroscopic process engineering computations such as that illustrated by the work of Kevrekidis et al. (2003) and Economou et al. (2008). For example, the latter paper characterizes properties of ionic liquids and uses three length/time scales that range from the quantum scale (DFT) to the atomistic scale (MD) to the bulk phase or macroscopic scale (tPC-SAFT). Other papers on multi-scale modeling and methods that have this same strong link to traditional macroscopic or process scales include the work of Gerogiorgis and Ydstie (2004) and Kulkarni et al. (2008).

A Look to the Future

Some progress has been made in multi-scale modeling of complex processes and multi-scale methodologies and applications will continue to evolve. Needs on the horizon include (1) the use of partial information, (2) the presence of multiplicity at fine scales, and (3) integration of dynamical models over very large time and length scales. To motivate these needs, we use ocean sedimentary CO_2 sequestration (see Fig. 2) because it embodies a wide variety of issues in multi-scale modeling relevant to energy and the environment.

In ocean sedimentary storage, carbon dioxide is pumped into sediments at ocean depths of 3 km or more, where it is known that liquid carbon dioxide is slightly denser than water (i.e., the so-called neutral buoyancy zone) for sediment depths up to ~200 m. Deeper than ~200 m, water is again denser than liquid CO_2. Hydrates can also form at these depths. See Dornan et al. (2007).

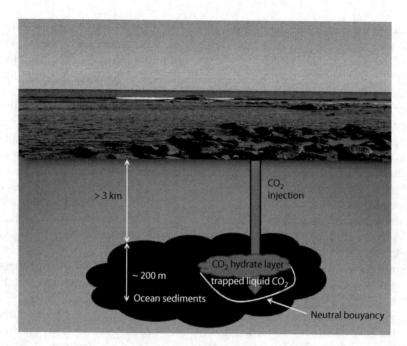

FIGURE 2
Ocean sedimentary CO_2 sequestration.

The fundamental assumption on which sequestration is based is that a CO_2 hydrate layer will form and this hydrate cap together with neutral buoyancy will trap liquid carbon dioxide (for extended periods of time on the order or hundreds to thousands of years).

The multi-scale nature of this problem is very complex. The reservoir has at least four length/time scales – the molecular scale, pore scale, sedimentary material/bulk phase scale, and reservoir scale. The associated multi-scale reservoir model is further complicated by reaction/phase equilibria, CO_2 gas hydrate formation, and multi-phase flow through porous media. CO_2 levels in the ocean, on the other hand, are driven by global vertical circulation of cold ocean waters which takes place on time scales of hundreds or thousands of years. Regional and local ocean effects must also be considered so the ocean model has several length/time scales as well.

What are the fundamental challenges and why are they multi-scale challenges?

1) Can we model the carbon exchange between ocean and reservoir?
2) Can we really store CO_2 in the ocean sediments for thousands of years?
3) If so, where do we store CO_2 to avoid disruption or leaks due to earthquakes, volcanoes, etc?

The assumption that the ocean and reservoir interact in a simple source-sink manner is flawed. In order to model true behavior, each system must drive the other and thus a multi-scale systems approach is required. However, truly coupling global ocean models with CO2 or other reservoir models is challenging.

For the reservoir, knowledge of the kinetics of hydrate formation is required at the molecular scale. However, there are at least ten or more reactions occurring simultaneously in addition to hydrate formation. Moreover, the kinetics are a function of local environment as is the formation of any number of phases. At the pore length scale, there is diffusion that affects transport and reaction so there is multi-phase flow through porous media that must be included. Next, bulk phase kinetic and equilibrium considerations in a sedimentary environment must be accounted for. This is a complex multi-phase, multi-reaction problem in which multiple solutions can exist. At the reservoir length scale, there is a multiple moving boundary problem since there is simultaneous accumulation of hydrate and liquid CO_2. Finally all of this is coupled to the ocean, in which deep ocean currents drive the flow of water through the sediments and reservoir at the concentration of CO_2 in the ocean.

Global ocean models generally consist of continuity and Navier-Stokes equations, which are simplified to include incompressible flow, constant density, and Boussinesq approximations. Nonetheless, this set of partial differential equations is extremely demanding to solve because of the large horizontal and vertical length scales involved. Spin-up time (i.e., equilibration time) for global ocean models is one to two decades, eddy diffusivities are difficult to include, and turbulence is approximated from theory since the length scale for resolving ocean turbulence is on the order of millimeters. Horizontal resolution is ten to several hundred kilometers with 20 or more vertical layers, resulting in a grid of $\sim 2 \times 10^7$ points. For the purposes of carbon sequestration, realistic bottom features must also be included. Typical integration time horizons can range from 100 to 1000 years so very coarse times steps are used. Finally, the ocean and reservoir are coupled through CO_2 concentration, salinity, and other factors. These variables can, in turn, affect ocean currents, ocean temperatures, and climate.

1) *Partial Knowledge.* Although our knowledge through observation is increasing, there is much that is not known about the ocean and the sediments (e.g., deep ocean currents cannot be measured). Thus simulation of oceans and sedimentary reservoirs requires the will and some cleverness to proceed with partial knowledge.

2) *Multiplicity at Fine Scales.* Similar to that which has been identified by Balakotaiah and Dommeti (2000) and Kulkarni et al. (2008), reacting systems in porous media can often exhibit multiplicity. The reactions that take place in the ocean during the formation of CO_2 reservoirs are no different.

3) *Integration for Very Large Time/Length Scales.* Simulation of vertical ocean transport of CO_2 requires very long time scales – on the order of 1000 years. However, behavior at much smaller length/time scales can drastically alter long term behavior. True assessment of the long term viability of carbon storage in ocean sediments must address validation, very large differences in scale, and integration over very long time scales.

Conclusions

Multi-scale modeling and multi-scale methods are still relatively new. While progress has been made on several fronts, many issues remain un-resolved. As with many disciplines, multi-scale modeling and methods provide advanced computer modeling tools but also raise new questions and concerns. In the author's opinion, the true value of multi-scale modeling is in modeling complex phenomena with some degree of quantitative confidence. However, this quantitative confidence can only be realized through model validation and the continued evolution of modeling techniques and supporting numerical methods. When designed and implemented correctly these methods make 'what if' scenarios for complex processes possible in ways that would otherwise not be possible. Multi-scale methods are the enabling technology that makes multi-scale modeling of complex processes possible and will continue to make an increasing impact in the future.

References

Androulakis, I.P. (2000). Kinetic Mechanism Reduction Based on an Integer Programming Approach, *AIChE J. 46*, 361.

Balakotaiah, V., Dommeti, S.M.S. (1999). Effective Models for Packed-Bed Catalytic Reactors. *Chem. Eng. Sci. 54*, 1621.

Barash, D., Yang, L., Qian, X., Schlick, T. (2002). Inherent Speedup Limitations in Multiple Time Step/Particle Mesh Ewald Algorithms, *J. Comput. Chem. 24*, 77.

Barrat, J.-L., de Pablo, J.J. (2007). Modeling Deformation and Flow in Disordered Materials. *MRS Bull. 32*, 941.

Car, R., Parrinello, M. (1985). Unified Approach for Molecular Dynamics and Density Functional Theory, *Phys. Rev. Lett. 55*, 2471.

Chakraborty, S., Balakotaiah, V. (2005). Spatially Averaged Multi-Scale Models for Chemical Reactors. *Adv. Chem. Eng. 30*, 205.

Dai, J. Kantes, J.M. Kapur, S.S., Seider, W.D., Sinno, T. (2005). On-Lattice Kinetic Monte Carlo Simulations of Point Defect Aggregation in Entropically-Influenced Crystalline Systems, *Phys. Rev. B 72*, 134105.

Dai, J. Seider, W.D., Sinno, T. (2006). Lattice Kinetic Monte Carlo Simulations of Defect Evolution in Crystals at Elevated Temperatures, *Molecular Simulation 32*, 305.

Davis, E., Ierapetritou, M.G. (2007). A Kriging Method for the Solution of Nonlinear Programs with Black-Box Functions, *AIChE J. 53*, 2001.

de Pablo, J.J. (2005). Molecular and Multiscale Modeling in Chemical Engineering – Current View and Future Perspectives. *AIChE J. 51*, 2371.

de Pablo, J.J., Curtin, W.A. (2007). Multiscale Modeling in Advanced Materials Research: Challenges, Novel Methods, and Emerging Applications. *MRS Bull. 32*, 905.

Dornan, P., Alavi, S., Woo, T.K. (2007). Free Energies of Carbon Dioxide Sequestration and Methane Recovery in Clathrate Hydrates, *J. Chem. Phys. 127*, 124510.

Egorov, Y., Menter, F., Kloeker, M., Kenig, E.Y. (2005). On the Combination of CFD and Rate-Based Modeling in the Simulation of Reactive Separations, *Chem. Eng. Process, 44*, 631.

Economou, I.G., Karakatsani, E.K., Logotheti, E.-G., Ramos, J., Vanin, A.A. (2008). Multi-scale Modeling of Structure, Dynamic and Thermodynamic Properties of Imidazolium-Based Ionic Liquids: Ab Initio DFT Calculations, Molecular Simulation and Equation of State Predictions. *Oil & Gas Sci. & Tech. Rev. 63*, 283.

Faller, R. (2005). Systematic Coarse-Graining of Atomistic Models for Simulation of Polymeric Systems, In El-Halwagi, M. (ed.) *Symposium on Modeling of Complex Processes*.

Frederix, Y., Samaey, G., Vanderkerchhove, C., Roose, D. (2007). Equation-Free Methods for Molecular Dynamics: A Lifting Procedure, *Proceedings in Applied Mathematics and Mechanics*.

Gattupalli, R., Lucia, A. (2007). Molecular Conformation of N-Alkanes Using Terrain/Funneling Methods, *J. Global Optim.* [*doi:10.1007/s10898-007-9206-5*].

Gattupalli, R., Lucia, A. (2008). Multi-Scale Global Optimization of All-Atom Models of N-Alkanes, *Comput. Chem. Engng.* [*doi: 10.1016/j.compchemeng.2008.11.012*].

Ge, W., Li, J., Gao, S., Song, W. (2005). Multi-Scale Approach of Multiphase Flows in Chemical Engineering, In El-Halwagi, M. (ed.) *Symposium on Modeling of Complex Processes*.

Gear, C.W., Kevrekidis, I.G., Theodoropoulis, C. (2002). Coarse Integration/Bifurcation Analysis via Microscopic Simulators: Micro-Galerkin Methods, *Comp. Chem. Engng. 26*, 941.

Gerogiorgis, D., Ydstie, B.E. (2004). Multi-Scale Modeling for Electrode Voltage Optimization in the Design of a Carbothermic Aluminum Process, In Floudas, C.A., Agrawal, R. (eds.) *Foundations of Computer-Aided Process Design CACHE Corp.*, 265.

Greenfield, M.L. (2004). Simulation of Small Molecule Diffusion Using Continuous Space Disordered Networks, *Molecular Physics 102*, 421.

Greenfield, M.L., Theodorou, D.N. (1998). Molecular Modeling of Methane Diffusion in Glassy Atactic Polypropylene via Transition State Theory. *Macromolecules 31*, 7068.

Greenfield, M.L., Theodorou, D.N. (2001). Coarse-Grained Molecular Simulation of Penetrant Diffusion in a Glassy Polymer Using Reverse and Kinetic Monte Carlo, *Macromolecules 34*, 8541.

Humphreys, D.D., Friesner, R.A., Berne, B.J. (1994). A Multiple Time-Step Algorithm for Macromolecules, *J. Phys. Chem. 98*, 6885.

Ingram, G.D., Cameron, I.T. (2005). Formulation and Comparison of Alternative Multi-Scale Models for Drum Granulation, In Puigjaner, L, Espuna, A. (eds.) *European Symposium on Computer-Aided Process Engineering (ESCAPE)–15*, Elsevier, Netherlands, 481.

Kevrekidis, I.G., Gear, C.W., Hyman, J.M., Kevrekidis, P.G., Runborg, O., Theodoropoulis, C. (2003). Equation-free, Coarse-Grained Multi-Scale Computations: Enabling Microscopic Simulators to Perform Systems Level Tasks, *Comm. Math. Sci. 1*, 715.

Krige, D. (1951). *A Statistical Approach to Some Mining Valuations and Allied Problems at the Witswatersrand*, M.S. Thesis, Univ. of Witswatrsrand, Johannesburg.

Kulkarni, K., Moon, J., Zhang, L., Lucia, A., Linninger, A.A. (2008). Multiscale Modeling and Solution Multiplicity in Catalytic Pellet Reactors, *Ind. Eng. Chem. Res . 47*, 8572.

Liang, L., Stevens, J.G., Farrell, J.T., Huynh, P.T., Androulakis, I.P., Ierapetritou, M.G. (2007). An Adaptive Approach for Coupling Detailed Chemical Kinetics and Multidimensional CFD, *Fifth U.S. Combustion Meeting*, San Diego, CA, paper # C09.

Lucia, A., DiMaggio, P.A., Depa, P. (2004). Funneling Algorithms for Multi-Scale Optimization on Rugged Terrains, *Ind. & Eng. Chem. Res., 43*, 3770.

Morales-Rodriguez, R., Gani, R. (2008). Multi-Scale Modeling Framework for Chemical Product-Process Design, In Jezowski, J., Thullie, J. (eds.) *European Symposium on Computer-Aided Process Engineering (ESCAPE) – 19*, Elsevier, Netherlands.

Pantelides, C.C. (2000). New Challenges and Opportunities for Process Modeling, In Gani, R., Jorgensen, S.B. (eds.) *European Symposium on Computer-Aided Process Engineering (ESCAPE) – 11*, Elsevier, 15.

Pricl, S., Ferrone, M., Fermeglia, M., Amato, F., Cosentino, C., Cheng, M.-C., Walczk, R., Ferrari, M. (2006). Multiscale Modeling of Protein Transport in Silicone Membrane Nanochannels. Part 1. Derivation of Molecular Parameters for Computer Simulation. *Biomedical Micro-Devices 8*, 277.

Puigjaner, L, Espuna, A. (2005). *European Symposium on Computer-Aided Process Engineering* (ESCAPE) – 15, Elsevier, Netherlands.

Qian, Y.-H., Kevrekidis, I.G. (2000). Coarse Stability and Bifurcation Analysis Using Time Steppers: A Reaction-Diffusion Example. *Proc. Natl. Acad. Sci. 97*, 9840.

Rahimi, M.-R., Rahimi, R., Shahraki, F., Zivdar, M. (2006). Prediction of Temperature and Concentration Distributions of Distillation Sieve Trays by CFD, in Sorensen, E. (ed.) *Distillation & Absorption 2006*, 220.

Reis, M.S., Saraiva, P.M., Bakshi, B.R. (2008). Multi-Scale Statistical Process Control Using Wavelets, *AIChE J. 54*, 2366.

Siettos, C.I., Armaou, A., Makeev, A.G., Kevrekidis, I.G. (2003). Microscopic/ Stochastic Timesteppers and Coarse Control: A Kinetic Monte Carlo Example, *AIChE J. 49*, 1922.

Sirdeshpande, A.R., Ierapetritou, M.G., Androulakis, I.P. (2001). Design of Flexible Reduce Kinetic Mechanisms, *AIChE J., 47*, 2461.

Tunca, C., Ford, D.M. (2005). A Hierarchical Approach to the Molecular Modeling of Permeation Through Microporous Materials, In El-Halwagi, M. (ed.) *Symposium on Modeling of Complex Processes*.

Wales, D.J., Doye, J.P.K. (1997). Global Optimization by Basin Hopping and Lowest Energy Structures of Lennard-Jones Clusters Containing Up To 110 *Atoms, J. Phys. Chem. A 101*, 5111.

Weinan, E., Engquist, B. (2003). The Heterogeneous Multi-Scale Methods, *Comm. Math. Sci. 1*, 87.

Weisz, P.B., Hicks, J.S. (1962). The Behavior of Porous Catalyst Particles in View of Internal Mass and Heat Diffusion Effects, *Chem. Eng. Sci. 17*, 265.

Westerberg, K.M., Floudas, C.A. (1999). Locating All Transition States and Studying the Reaction Pathways of Potential Energy Surfaces, *J. Chem. Phys. 110*, 9259.

Zhang, L., Greenfield, M.L. (2007). Relaxation Time, Diffusion, and Viscosity Analysis of Model Asphalt Systems Using Molecular Simulation, *J. Chem. Phys. 127*, 194502.

Zheng, Z., Stephens, R.A., Braatz, R.D., Alkire, R.C., Petzold, L.R. (2008). A Hybrid Multiscale Kinetic Monte Carlo Method for Simulation of Cooper Electrodeposition, *J. Comput. Phys. 227*, 5184.

90

Controlled Formation of Nanostructures with Desired Geometries

Earl O. P. Solis, Paul I. Barton and George Stephanopoulos

Massachusetts Institute of Technology
Cambridge, MA 02139

CONTENTS

ABSTRACT This paper provides an overview of design principles and methods associated with the controlled formation of nanostructures with desired geometries. The approach is based on a hybrid top-down and bottom-up approach: top-down formation of physical domains with externally-imposed controls and bottom-up generation of the desired structure through the self-assembly of the nanoscale particles, driven by interparticle interactions (short- and long- range) and interactions with external controls. The desired nanoscale structure must be locally stable and robust to the desired level of robustness, and it should be reachable from any random initial distribution of the nanoparticles in the physical domain. These two requirements frame the two elements of the design problem: The first defines a static optimization problem with integer (position of controls) and continuous variables (intensities of controls). The second leads to a mixed-integer, time-dependent optimization problem, akin to those encountered in optimal control. Crucial to the achievement of the design goals is the necessity to break the ergodicity of the overall system and define ergodic subsets of phase space that map the desired geometric features of the nanostructure to the features of the energy landscape within the system volume. The static and dynamic problems are solved through the formulation and solution of combinatorially-constrained mixed-integer quadratic optimization problems (QIP). The dynamics of the self-assembly process are described through multi-resolution models.

A 2-dimensional design example illustrates the design principles and methodologies involved in the controlled formation of desired nanostructures.

KEYWORDS *Stable and robust nanostructures, dynamics of self-assembly, ergodicity breaks, geometries of nanostructures*

Introduction

The theory and practice of forming closely-packed 2-dimensional films and 3-dimensional materials with desired periodic nanoscale geometries in an essentially infinite domain have advanced significantly during the last 10–15 years. For example, a large variety of self-assembled monolayers, leading to highly structured films on surfaces that provide biocompatibility, control of corrosion, friction, wetting, and adhesion, have been experimentally synthesized and theoretically analyzed (Love et al., 2005). These films are viewed as possible precursors for nanometer-scale devices for use in organic microelectronics. However, their geometries are essentially periodic and this could constitute an important limitation. The geometric features of phase-separated regions of block copolymers and blends are often of nanometer scale, periodic and dense, and can be rationally designed through the judicious selection of the monomers, length and frequency of blocks in the polymeric chain(s) (Bates and Fredrickson, 1996; Fasolka and Mayes, 2001). Furthermore, templated self-assembly techniques have been extensively developed to produce nanostructures with desired geometries through the judicious selection of the nanoparticles and the features of the environment. Examples include crystallization on template surfaces that determine the morphology of the resulting crystals (Aizenberg et al., 1999) and crystallization of colloids in optical fields (Burns et al., 1990). The templates could be physical (capillary forces, spin-coating, surface steps, and others), molecular (patterned self-assembled monolayers with specific chemical functionalization of terminal groups), or electrostatic (localized charges on the surface of a substrate). The resolution of the structural features that can be achieved by template-based approaches depends on the spatial resolution of templates. DNA-programmed placement using 2-dimensional DNA crystals as scaffolds, placement using electrophoresis, and focused placement, which uses focusing mechanisms to guide the nanoscale particles to specific locations in the physical domain that are smaller in scale than the template guiding them, are additional approaches that has been extensively reviewed by Koh (2007).

The previous methodologies for the formation of nanostructures with desired geometries, except some specific template self-assembly approaches, are not applicable for the formation of nanostructures with non-periodic features and non- close-packed constructs in finite domains. New approaches

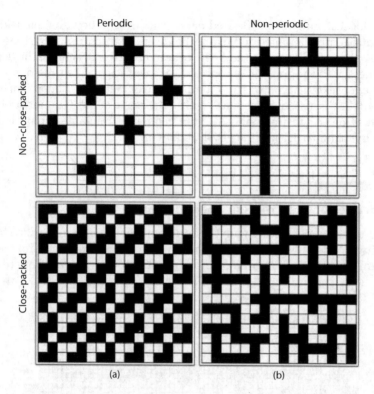

FIGURE 1

The periodic nanostructures in (a) can be achieved through self-assembly of judiciously designed nanoparticles, but the non-periodic structures in (b) require external controls for guided self-assembly.

are needed, and this need defines the scope of the present paper. For example, judiciously designed nanoparticles can self-assemble to form the periodic structures in Figure 1(a), but to form non-periodic nanostructures like those in Figure 1(b) we need external controls to guide the self-assembly process. Non-periodic structures like those in Figure 1(b) are essential for the fabrication of future electronic, magnetic and optical devices, and the construction of molecular factories and synthetic cells (Stephanopoulos et al., 2005).

The need to design and construct non-periodic and non- close-packed nanostructures systematically defines the objectives of this paper. In subsequent sections, we will present a comprehensive approach for the controlled formation of nanostructures starting from any initial distribution of nanoparticles in a 2-dimensional physical domain. Specifically, the remainder of this paper is structured as follows: Section A will introduce a 2-dimensional case study that will be used throughout this paper to define the design problem and illustrate the proposed design approaches and methodologies. In Section B we will show how the external controls are systematically placed in the physical domain to ensure local stability of the desired nanostructure and how the intensities of the controls are computed to ensure the desired structure's robustness. We will also demonstrate how to add new external controls to achieve the desired robustness, if the original set of controls was insufficient. The time-dependent placement of external controls and computation of their time-dependent intensities will be discussed in Section C within the scope of the dynamic problem, i.e., achieving the desired nanostructure geometry from any initial distribution of the nanoparticles in the physical domain.

The Scope for the Problem Formulation

Consider the nanostructure with the 2-dimensional geometric design shown in Figure 2. The physical domain is assumed to be square with dimensions (640 nm) × (640 nm), composed of an array of 64 × 64 discretized square cells

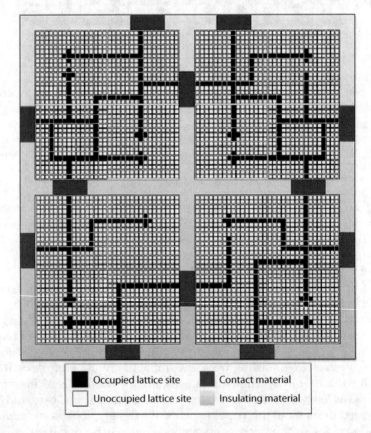

	Occupied lattice site		Contact material
	Unoccupied lattice site		Insulating material

FIGURE 2
The geometrical design of the nanostructure in the case study of this paper.

of dimensions (10 nm) × (10 nm). The number of nanoparticles in the nano-structure is equal to 493. For this example, we assume that each particle is square and takes up space equal to that of a cell. We also assume that the particles are indistinguishable and negatively charged with a charge equal to -1. In addition to long-range Coulombic repulsion, the particles interact through short-range attractive forces, e.g., hydrogen bonding. Thus, the binary interaction potential energy is given by equation (1):

$$\beta E = z_i z_j J_{ij} = z_i z_j \left[\frac{1}{r_{ij}} - \frac{1}{r_{ij}^6} \right],$$
(1)

where r_{ij} is the positional distance between particles i and j. Variables, z_i and z_j, are binary with values 0 (to represent an empty lattice site) or 1 (to represent a site occupied by a particle). This model simply represents a phenomenological model; more complex interaction energies have been used to simulate real systems. In the case studies of this paper we have used long-range Coulombic interactions and short-range Lennard-Jones attractive interactions.

It is impossible to reach the structure of Figure 2, starting from any random distribution of the particles in the domain, without external controls: the structure may not be locally of minimum energy; even if the structure is locally of minimum energy, the probability of remaining in the desired configuration (what we will refer to as its statistical robustness) may be unacceptably low for any useful purpose; dynamically we will reach different meta-stable structures, if we start from different initial distributions.

As a result, two questions arise: (a) What is the optimal placement of external controls, e.g., points at which we can externally apply charges of variable intensities, in order to render, at equilibrium, the desired structure of locally minimum potential energy and of sufficient robustness? (b) How should the placement of external charges and their intensities change with time in order to ensure that the desired structure is always reached, independently of the initial configuration of the particles in the physical domain?

Before proceeding with the methodological aspects it is important to clarify the technological limitations constraining the imposition of external controls on the physical domain of Figure 2. Clearly, photolithography (Senturia, 2000; Plummer et al., 2000) and nanoimprinting (Gates et al., 2004; Hsu et al., 2007) could produce on the substrate of the 2-dimensional domain the lines with the features of the structure in Figure 2 if they could achieve features of 10 nm size. Presently, this is not the case. But, electron beam lithography (Krapf et al., 2006; Tuukkanen et al., 2006) could generate arrays of electrodes 2 nm in diameter and about 20-30 nm from each other. For this case study we will assume that photolithography can be employed to generate controls with features of 20 nm or larger, and electron beam lithography for control points of 2 nm in size. Both will be constrained to obey distances between control points of at least 20 nm. Consequently, the contacts and insulating regions in Figure 2 can be constructed directly through photolithography, since their sizes are well above these limitations. The remaining features of the design will be achieved through controlled self-assembly of the nanoparticles with control points of 2 nm in diameter, produced by electron beam lithography. The judicious placement of these controls in the 2-dimensional domain of Figure 2 is a central design question for this paper.

The discussion in the previous paragraph implies the following fundamental design principle that we have adopted in the present paper, namely: *Construct large-size geometric features of the desired design directly through top-down technologies, e.g., photolithography, nanoimprinting. Manufacture the small-size features through controlled self-assembly of the nanoparticles.*

The controlled self-assembly of nanoparticles into the desired geometry of Figure 2 implies the need for solving the following two problems:

Static Problem: Select the positions of external controls in the 2-dimensional domain and compute their intensities in such a way that the desired nanostructure is of locally minimum potential energy. Furthermore, if the probability of the structure to remain in the desired geometry is lower than specified by the design problem, add control points and compute their intensities in order to achieve the specified robustness.

Dynamic Problem: Select the time-dependent positioning of external controls and compute their time-dependent intensities in order to ensure that the nanoparticles will always assemble into the desired structure independent of their initial distribution in the 2-dimensional domain.

For a tutorial exposition of the methodology and design principles involved in the solution of the Static and Dynamic Problems, see Solis et al. (2008).

Before proceeding with the solution of these two problems, it is important to discuss the statistical mechanical framework within which the self-assembly of the nanoparticles takes place.

The fundamental description of the controlled self-assembly process is based on the statistical mechanics of the configurations formed by the nanoscale particles. For the design under consideration in this paper, the number of particles, N, and the size of the physical domain, V, are assumed to be finite and small. Therefore, minimization of free energy cannot be used as a design objective since it is based on the assumption that $N \rightarrow \infty$ and V- with $N/V = v$ (finite). Instead, our design objective is based on the probability of achieving and retaining the desired nanostructure.

The Boltzmann probability for a system in the canonical description, i.e., constant values for the system volume, V, temperature, T, and particle number, N, is given by:

$$p(z_i) = \frac{e^{-\beta E(z_i)}}{\sum_j e^{-\beta E(z_j)}} = Z^{-1} e^{-\beta E(z_i)}. \tag{2}$$

The denominator represents the normalizing partition function. The distribution function in equation (2) represents an ensemble probability, i.e., $p(z_i)$ represents the fraction of an ensemble (or group) of systems that are in configuration z_i. It also represents the probability of finding a representative system in configuration z_i after the system has reached equilibrium. The energy parameter, β, essentially determines the "flow" of the system through phase space. The smaller the β value, the more accessible all states are to the system.

Equation (2) assumes that the "flow" through phase space allows eventual access to any other state (configuration) from any particular state. This is the *ergodic* hypothesis, and plays a very important role in the proposed formation of nanostructures with desired geometries. For the controlled self-assembling processes described in this paper, the externally imposed controls offer degrees of freedom, which are used to decrease the volume of phase space accessible to the system, i.e., decrease the number of configurations accessible from a given configuration. Such systems exhibit *non-ergodicity*, and their phase space can be decomposed into subsets that are ergodic. We will call these ergodic subsets of the complete phase space *ergodic components*. As a result, transitioning between two ergodic components of the phase space that are separated by large energetic barriers is not very probable and requires either a smaller β value or a sufficiently long period of time.

Non-ergodicity is a desirable trait for controlled self-assembly processes with a specified desired end state because it reduces the total number of undesirable "competing" states. Within each ergodic component of the phase space, one may use a variation of Boltzmann's distribution function,

$$p_\alpha\left(z_i^\alpha\right) = \frac{e^{-\beta E\left(z_i^\alpha\right)}}{\sum_k e^{-\beta E\left(z_k^\alpha\right)}} = Z_\alpha^{-1} e^{-\beta E\left(z_i^\alpha\right)}, \tag{3}$$

where Z_α is now the partition function of the ergodic component, α, and the sum over k is over all configurations in α. Configuration z_i is in the ergodic component, α.

For the solution of the Static Problem we select the external controls in such a way that they create an ergodic component in which lies the desired nanostructure. Furthermore, the external controls introduce sufficiently large energetic barriers so that the transitioning of the desired structure from its ergodic component to another one is a very low probability event.

In a similar spirit, the solution to the Dynamic Problem is based on the progressive, over time, decomposition of the initial (complete) ergodic phase space into a series of ergodic components (progressively smaller) until we reach the ergodic component defined in the Static Problem. At this point, the controls computed for the solution of the Static Problem will ensure that this structure is in *quasi-equilibrium* (of locally minimum energy) with sufficient robustness, i.e., the probability to stay at the desired geometry is larger than a prespecified threshold.

Solution Methodology for the Static Problem

The desired structure is a member of an ergodic component, which is characterized by an energy landscape that ensures that the particles are locally confined in the vicinity of their desired locations in the desired structure. This can be accomplished through the specification of the necessary degrees of freedom: the location of the point conditions (attractive or repulsive point charges) and their strength (value of charge). The qualitative features of the energy landscape are defined by the location of the point conditions and their character, i.e., attractive or repulsive, while the quantitative features are determined by the strength of the charges.

The degrees of freedom we use are isotropic and can introduce a well that attracts particles or a barrier that deflects particles. Because the presence of two wells forms a barrier and vice versa, only one of these types (wells or barriers) is needed for local stability. Hence, when given a desired structure, one should locate the well- and barrier- forming point conditions, and select the smaller number of the two. Figure 3 shows the lower-right quadrant

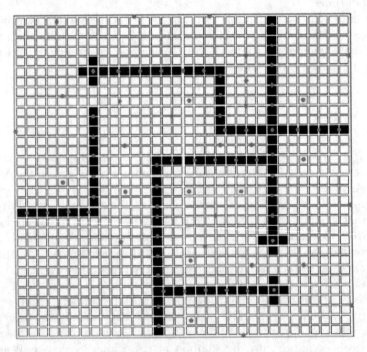

FIGURE 3
The location of point conditions (charges) as external controls in a 2-dimensional design.

Type: A   B   C

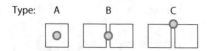

FIGURE 4
The three types of point conditions.

of the design in Figure 2. It represents a 2-dimensional desired structure in a 32 × 32 square lattice and the two types of possible point conditions: 32 repulsive (negative charges) point conditions positioned in the unoccupied regions of the domain and forming local energy barriers, and 54 attractive (positive charges) point conditions positioned in the regions occupied by particles and forming local energy wells.

To find the locations of the minimum well- and barrier- forming point conditions, we developed an algorithm which covers a 2-dimensional area with tiles of specific shapes. Specifically, for each point condition we assign a tile, which encompasses the local area that the point condition is intended to influence. Thus, the minimum number of tiles needed to cover the barrier-forming regions (i.e., the regions not occupied by particles) determines the minimum number of negative point charges, and the minimum number of tiles needed to cover the well-forming regions (i.e., the regions occupied by particles) determines the minimum number of positive point charges.

We restrict our example system to 3 types of point conditions, shown in Figure 4. It is straightforward to show that the numbers of point conditions of type A, B, and C for a square lattice of $L \times L$ cells are given by:

No. of type-A point conditions = L^2
No. of type-B point conditions = $2(L + 1)L$
No. of type-C point conditions = $(L + 1)^2$
Total No. of point conditions = $N_{PC} = 4L^2 + 4L + 1$

Though each type of point condition is isotropic the discretization of space and the different locations of each point condition type with respect to the lattice sites cause them to form different tile shapes. Figure 5 shows how each type of point condition develops different tiles of increasing sizes, i.e., possessing larger areas of influence as their strength increases; the darkest shading is the lowest strength, and the lightest shade the highest strength.

The diagram of Figure 6 shows the logical flowchart of the tiling algorithm we have used to compute the minimum number of tiles needed for barrier- or well-forming regions. The algorithm first grows all N_{PC} point conditions until an incremental growth causes the area of influence to include lattice sites of the opposing type. When the incremental growth includes only the appropriate lattice type, the point condition criteria is satisfied; when it includes the opposing lattice type, the point condition criteria is no longer satisfied and the point condition's area of influence cannot grow anymore.

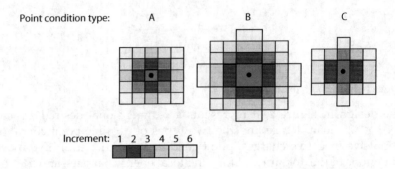

Point condition type: A   B   C

Increment: 1 2 3 4 5 6

FIGURE 5
The geometry of tiles generated from the three types point conditions in Figure 4.

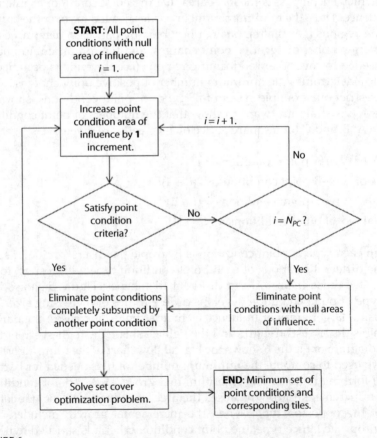

FIGURE 6
The flowchart of the algorithm generating the minimum numbers of barrier- and well-forming point conditions.

Once all N_{PC} point conditions have their designated areas of influence, we perform two rounds of elimination: (1) Remove all point conditions that have no area of influence, (2) Remove all point conditions whose areas of influence are subsumed by another point condition. This is done to reduce the number of point conditions considered in the set cover optimization problem,

$$\min_{x} \sum_{i} x_i$$

$$s,t. \sum_{i} a_{ij} x_i \geq 1 \forall j \tag{4}$$

$$x_i \in \{0,1\},$$

where x_i is a binary variable for each point condition, and a_{ij} is a binary parameter that equals 1 if lattice site j is within the area of influence of point condition i and equals 0 otherwise. The set of point conditions with $x_i = 1$ represents the minimum set that covers all the lattice sites of the appropriate type. This is an NP-hard problem with an NP-complete decision problem equivalent. If the number of point conditions considered is small enough, the set cover problem can be solved to find the minimum set. However, if the number of point conditions is too large, one must use an approximate algorithm, e.g., the greedy algorithm (Cormen et al., 2001).

For the 2D example design in Figure 3, $N_{PC} = 4,225$. Before solving the set cover problem, this number was reduced to 905 barrier- and 105 well- forming point conditions. After the optimization problem is solved, we see the minimum sets of 32 barrier- and 54 well- forming point conditions.

The minimum number of selected barrier- or well-forming point conditions generates a potential energy landscape in the 2-dimensional physical domain that ensures the local stability of the desired structure, but not its robustness. The solution of the following problem determines the degree of robustness of the desired structure against statistical fluctuations, given the positioning of the minimum number of point conditions:

$$\max \quad \delta$$

$$s \in S, \delta$$

$$s,t. \ E(s,z) - E(s, z_d) \geq \delta, \forall z \in \zeta^a \setminus z_d, \tag{5}$$

where z_d represents the desired structure, s represents the set of all point conditions strengths, and ζ_α represents the subset of configurations that defines component α. The solution to this optimization problem ensures that the difference in potential energy between any configuration in the ergodic component, in which the desired structure belongs, and the desired structure is the largest possible.

The value of δ, resulting from the solution of the optimization problem (5), is a measure of the desired structure's robustness against statistical fluctuations, a measure of the probability that the nanostructure will remain in the desired geometric configuration. If the value of δ suggests satisfactory robustness, then the Static Problem is solved, but if not, then additional point conditions are needed. Since the minimum number of point conditions, selected and used so far, is of the same type, i.e., barrier-forming (negative charges) or well-forming (positive charges), the additional point conditions will be selected from the opposite class.

The positioning of the additional point conditions is guided by the following algorithm:

(a) From the solution of the optimization problem (5), find the lattice cells where a fluctuation in the desired structure creates a configuration or a set of configurations whose energy differs the least from that of the desired structure, i.e., the difference is equal to δ. This is the *constraining feature(s)* of the desired configuration.

(b) Select a point condition from the opposite class with an energetic influence (tile) on the lattice cells of the constraining feature(s).

(c) Solve optimization problem (5) again. Return to (a) if the resulting δ remains unsatisfactory.

Applying the above methodology to the design in Figure 3 we generate the following results:

- The minimum number of barrier- and well- forming point conditions are equal to 32 and 54, respectively. We choose the minimum, i.e., 32 barrier-forming conditions. Using these point conditions to solve the optimization problem (5), we find that $\delta = -0.1$. We would like to achieve a positive δ value for robustness.

- We need 23 additional point conditions, as shown in Figure 7, to generate a $\delta = 1.2$.

Solution Methodology for the Dynamic Problem

One may consider the placement of external controls and computation of their strengths in such a way that the desired configuration corresponds to one of globally minimum energy. In this case, one may argue that any initial distribution of particles would eventually reach the desired configuration. This is similar to the 'needle-in-a-haystack' problem in protein folding (Dill and Chang, 1997) and corresponds to a formulation similar to (5) but with

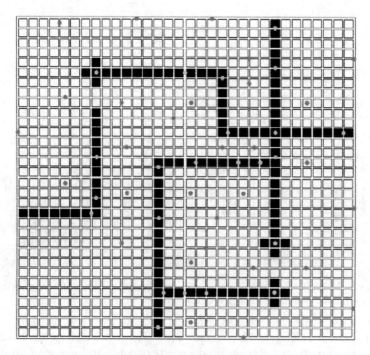

FIGURE 7
The set of 55 point conditions needed to guarantee robustness.

the search extending over the full phase space and not only the configurations of a specific ergodic component. Though it may take a long time to find the desired configuration kinetically we know that we can guarantee it with at least a certain level of robustness. Within practical time scales this is not feasible approach; the external controls create barriers between states in the phase space, thus introducing non-ergodic behavior to the system, i.e., the particles may be "locked" into a locally minimum energy configuration. To overcome trapping the system in a particular subset of configurations, we need to use time-varying control positions and strengths.

To solve the Dynamic Problem, we have adopted a multiresolution view of the evolving sets of configurations. Initially, we consider the whole phase space of possible configurations throughout the 2-dimensional physical domain. Then, we place the external controls in such a way as to render the first decomposition of the phase space into an ergodic component that ensures that the correct number of particles is in each of the four quadrants of the physical space (see Figure 8). Proceed with further decomposition of each quadrant into quadrants of half the previous scale, and continue until one has reached the scale of the features in the final design. At that point the solution to the Static Problem will ensure the stability and robustness of the desired design.

The proper placement of energy barriers at each resolution of the system volume leads to a systematic decomposition of the phase space into

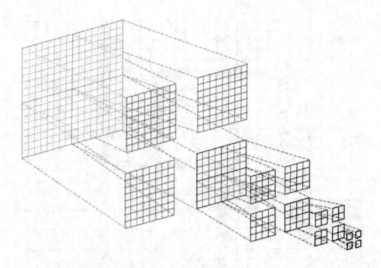

FIGURE 8
The multiresolution view of the physical domain for the specification of time-varying external controls during the solution of the dynamic problem.

progressively smaller ergodic components. Thus, external controls restrict the dynamic path that the original ergodic system follows over time and ultimately force the system to go through progressively smaller subsets of the phase space until it reaches the desired configuration through the Static Problem. Figure 9 describes the dynamic evolution for the design of Figure 3: Start with the entire phase space that contains all configurations with 119 particles at various positions of the entire physical domain ($\zeta_{\alpha 0}$ in Figure 10). The first subset of the phase space is formed so that it contains only those

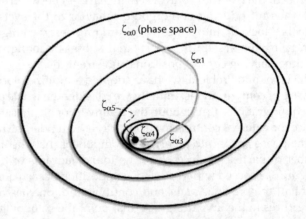

FIGURE 9
The dynamic path that restricts the system of the design in Figure 3 to progressively smaller subsets of the system's phase space.

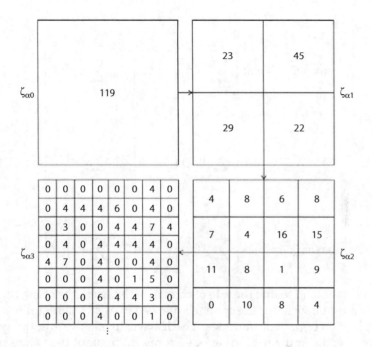

FIGURE 10
The multiresolution distribution of particles for the dynamic problem solution of the design in Figure 3.

configurations which have the following distribution of particles in the four quadrants of the physical domain: [23 (upper-left), 45 (upper-right), 29 (lower-left), 22 (lower-right)], see $\zeta_{\alpha 1}$ in Figure 10. The second subset of the phase space is smaller, i.e., it contains fewer configurations, because the allowed positions of particles are further constrained. Following the same convention as for the distribution of particles in the initial four quadrants, the distribution of particles in the second subset is given by $\zeta_{\alpha 2}$ in Figure 10. The third subset, $\zeta_{\alpha 3}$, is further smaller and contains more of the detailed geometry of the desired structure. The hierarchical constraining of phase space continues until we arrive at $\zeta_{\alpha 5}$, the desired structure in Figure 3 for the Static Problem.

The placement of controls at each of the physical subsets of the 2-dimensional domain arising from its multiresolution views is guided by the same design rules as those used for the solution of the Static Problem. However, the allowed configurations in the desired ergodic component are not described in terms of detailed geometries but are described in terms of particle densities allowed in each physical sub-domain.

For example, the following optimization problems are solved for the determination of external controls at each of the intermediate steps, until we reach the resolution of the final design:

Stage-1:

$$\max p(\zeta_{\alpha 1}) = \max \left[\sum_{z_i \in \zeta_{\alpha 1}} p(z_i) \right] = \max \left[Z_{\alpha 0}^{-1} \sum_{z_i \in \zeta_{\alpha 1}} e^{-\beta E(z_i)} \right].$$

Stage-2:

$$\max p(\zeta_{\alpha 2}) = \max \left[\sum_{z_i \in \zeta_{\alpha 2}} p(z_i) \right] = \max \left[Z_{\alpha 1}^{-1} \sum_{z_i \in \zeta_{\alpha 2}} e^{-\beta E(z_i)} \right].$$

Stage-3:

$$\max p(\zeta_{\alpha 3}) = \max \left[\sum_{z_i \in \zeta_{\alpha 3}} p(z_i) \right] = \max \left[Z_{\alpha 2}^{-1} \sum_{z_i \in \zeta_{\alpha 3}} e^{-\beta E(z_i)} \right].$$

It is important to note that the rate of change of the strengths of the external controls must be such that the system is allowed to reach local equilibrium in its distribution of particle densities at each stage. In other words, the system remains ergodic at the beginning of each resolution until the values of the external controls induce a break in its ergodicity and thus restrict the subset of phase space that the system is allowed to sample. As the values of the external controls change in time, they induce higher energy barriers and thus the rate of change of the strengths of the external controls must decrease in order to allow equilibration in the current subset of phase space. Consequently, as we proceed towards finer resolutions of the allowed configurations, the time intervals required for equilibration become progressively longer.

Conclusions

The construction of nanoscale structures with desired geometric designs can be achieved only through controlled self-assembly of nanoparticles. In this paper we have proposed a two-phase approach for the formal construction of such structures. In the first phase we solve the Static Problem that ensures the local stability and robustness of the desired structure through the judicious placement of external controls, while in the second phase we develop a time-varying placement of external controls in order to ensure that the final desired structure is reached from any initial particle distribution. In each phase a combinatorially constrained optimization problem arises and is solved in two steps: First, the minimum number of external controls is placed in the physical domain and their strengths are computed. This corresponds to a combinatorial tiling problem for covering a given area with tiles of given shapes and sizes. In the second step additional controls are introduced, if needed, in order to render the desired structure with the specified level of robustness against statistical fluctuations. The solution of the dynamic problem proceeds through a multiresolution view of the phase space of configurations. At each resolution, the desired goal is to achieve particle density specifications within sub-domains of the physical domain. Constraints imposed by technological limitations on the size and density of external controls have been integrated in the proposed approach and manifest themselves on the resolution of geometric features that can be achieved in a nanostructure.

Acknowledgments

Earl Solis and George Stephanopoulos acknowledge with gratitude the support provided for this work by Mitsubishi Chemical Holding Corporation.

References

Aizenberg, J., Black, A.J., Whitesides, G.M. (1999). Control of Crystal Nucleation by Patterned Self-Assembled Monolayers. *Nature, 398, 495–498.*

Bates, F. S., Fredrickson, G. H. (1996). Block Copolymer Thermodynamics: Theory and Experiment. *In Thermoplastic Elastomers, 2ⁿᵈ edition.* G. Holden, N.R. Legge, R. Quirk, and H.E. Schroeder, editors. Hanser Publishers, NY.

Burns, M.M., Fournier, J.M., Golovchenko, J.A. (1990). Optical Matter: Crystallization and Binding in Intense Optical Fields. *Science, 249, 749–754.*

Cormen, T.H., Leiserson, C.E., Rivest, R.L., Stein, C. (2001). *Introduction to Algorithms.* MIT Press and McGraw-Hill.

Dill, K.A., Chang, H.S. (1997). From Levinthal to pathways to funnels. *Nature Structural Biology, 4, 10–19.*

Fasolka, M. J., Mayes, A. M. (2001). Block Copolymer Thin Films: Physics and Applications. *Annual Reviews Materials Research, 31, 323–355.*

Gates, B.D., Xu, Q., Love, J.C., Wolfe, D.B., Whitesides, G.M. (2004). Unconventional Nanofabrication. *Annual Reviews Materials Research, 34, 339–372.*

Hsu, K.H., Schultz, P.L., Ferreira, P.M., Fang, N.X. (2007). Electrochemical nanoimprinting with solid-state superionic stamps. *Nano Letters, 7, 446–451.*

Koh, S.J. (2007). Strategies for Controlled Placement of Nanoscale Building Blocks. *Nanoscale Research Letters, 2, 519–545.*

Krapf, D., Wu, M., Smeets, R.M.M., Zabdbergen, H.W., Dekker, C., Lemay, S.G. (2006). Fabrication and Characterization of Nanopore-Based Electrodes with Radii down to 2 nm. *Nano Letters, 6, 105–109.*

Love, J. C., Estroff, L. A., Kriebel, J. K., Nuzzo, R. G., Whitesides, G. (2005). Self-Assembled Monolayers of Thiolates on Metals as a Form of Nanotechnology. *Chem. Rev. 105, 1103–1170.*

Plummer, J.D., Deal, M.D., Griffin, P.B. (2000). *Silicon VLSI Technology: Fundamentals, Practice and Modeling.* Prentice Hall.

Senturia, S.D. (2000). *Microsystem Design.* Kluwer Academic Publishers.

Solis, E.O.P., Barton, P.I., Stephanopoulos, G. (2008). Principles and Methodologies for the Controlled Formation of Self-Assembled Nanoscale Structures with Desired Geometries. *Molecular Systems Engineering,* C.S. Adjiman and A. Galindo, editors. Volume 6, *Process Systems Engineering Series* (2008).

Stephanopoulos, N., Solis, E.O.P., Stephanopoulos, G. (2005). Nanoscale Process Systems Engineering: Towards Molecular Factories, Synthetic Cells, and Adaptive Devices. *AIChE Journal, 51, 1858–1869.*

Tuukkanen, S., Toppari, J.J., Kuzyk, A., Hirviniemi, L., Hytonen, V.P., Ihalainen, T., Torma, P. (2006). Carbon nanotubes as electrodes for dielectrophoresis of DNA. *Nano Letters, 6, 1339–1343.*

91

Tight Energy Integration: Easier Control?

M. Baldea[1], S.S. Jogwar[2] and P. Daoutidis[2,*]

[1] *Advanced Control and Operations Research Group, Praxair Technology Center*
Tonawanda, NY 14150
[2] *Department of Chemical Engineering and Materials Science,*
University of Minnesota Minneapolis, MN 55455

CONTENTS

ABSTRACT Process integration is a key enabler to increasing efficiency in the process and energy generation industries. Efficiency improvements are obtained, however, at the cost of an increasingly complex dynamic behavior. As a result, tightly integrated designs continue to be regarded with caution owing to the dynamics and control difficulties that they pose. The present work introduces a generic class of integrated networks where significant energy flows (either arising from energy recycling or of external origin) result in dynamic models with a multi-time-scale structure. Such networks feature a clear distinction between the fast dynamics of individual units and the slow dynamics of the entire network. We draw a connection between specific (steady-state) design features and structural properties that afford the development of a framework for the derivation of low-order, non-stiff, non-linear models of the core network dynamics. Furthermore, we demonstrate that tight energy integration and the presence of significant energy flows

---

* To whom all the correspondence should be addressed. Email daoutidi@cems.umn.edu, Telephone (612) 625 8818.

facilitate, rather than hinder, control structure design and performance, and propose a cadre for hierarchical control predicated on the use of fast, distributed control for the individual units and nonlinear supervisory control for the entire network. The developed concepts are illustrated with examples and directions for future research are highlighted.

KEYWORDS *Energy integration, multi-time scale dynamics, hierarchical control*

Introduction

Efficient energy utilization is a critical need in the modern process industries and in existing and emerging energy production technologies. Process integration represents a key enabler to this end: the design and optimization of energy-integrated process networks has been an area of rich activity in process systems engineering and the significant reduction in capital and operating costs resulting from process integration is by now well-documented (Linnhoff and Hindmarsh, 1983, Linnhoff et al., 1983b, Yee et al., 1990, Annakou and Mizsei, 1996, Reyes and Luyben, 2000, Westerberg, 2004, El-Halwagi, 2006). The economic benefits associated with integrated process designs come, however, at the cost of distinct dynamics and control challenges (Silverstein and Shinnar, 1982, Georgakis, 1986, Luyben and Floudas, 1994, Malcolm et al., 2007). Integration typically results in a reduction in the number of degrees of freedom available for control, and "energy feedback" gives rise to a complex, nonlinear, overall network dynamics (Jacobsen and Berezowski, 1998, Jacobsen, 1999). Furthermore, the strong coupling among different units of an integrated network tends to limit the effectiveness of standard decentralized control strategies (Larsson et al., 2003, Kiss et al., 2005), while an ever increasing demand for frequent changes in operating conditions and targets requires going beyond regulatory control and enforcing effective transitions between different steady states.

In our recent work (Jogwar et al., 2009), we have analyzed the dynamics of integrated process networks with significant energy recycling via a Feed-Effluent Heat Exchanger (FEHE). Using singular perturbation arguments, we showed that tight energy integration induces a time scale separation at the energy balance level, with the total enthalpy of the network evolving over a longer time horizon than the enthalpies of the individual units. We demonstrated that, in the context of networks with FEHEs, controller design is facilitated, rather than hindered, by tight process integration, and proposed a two-tiered structure for energy control which was shown to exhibit superior robustness and transient performance.

In a different vein, we considered (Baldea and Daoutidis, 2008) the dynamic behavior of process networks with high energy throughput. Relying again on a singular perturbation analysis, we proved that the energy dynamics of such networks is faster than the dynamics of the variables of the material balance. Subsequently, we exploited—with excellent results—the time scale separation feature in designing a control structure with separate action at the energy balance level and at the material balance level.

In the present work, we are motivated by our prior results to define and document a generic, broad class of integrated networks where significant energy flows (either arising from energy recycling in a tightly integrated design, or of external origin) result in dynamic models with a multi-time-scale structure. Such networks feature a clear distinction between the fast dynamics of individual units and the slow dynamics of the entire network. We draw a connection between several process designs (e.g., reactors integrated with external heat exchangers or feed-effluent heat exchangers, heat integrated distillation columns, multiple effect evaporators) and structural properties that afford the development of a reduction framework for the derivation of low-order nonlinear models of the core network dynamics, which can subsequently be used for optimization and control. We propose a cadre for hierarchical control predicated on the use of fast, distributed control for individual units and nonlinear supervisory control for the entire network. Subsequently, we illustrate the application of the proposed framework on two processes (a reactor-feed effluent heat exchanger network and a double-effect distillation sequence) representative for the generic class of networks identified above. Finally, we highlight potential directions for future research.

Process Networks with Tight Energy Integration

Let us consider the network in Figure 1, consisting of N units in series and featuring a material recycle stream. Let Fo denote the feed flowrate, F_i, $i = 1,2,\ldots,N$ the outlet flowrate from the ith unit, and R the recycle flowrate (with the associated enthalpies H_i $i = 1,2,\ldots,N,R$). We assume that the means exist (*e.g.*, via direct heat exchange, transfer using a heat transfer medium, etc.) to recover energy from the output at a rate Q_{in} and recycle it to the input of the network at a rate $Q_{in} \equiv Q_{out}$.

The prototype network in Figure 1 captures the structural and dynamic properties of integrated process designs with significant energy recycling, such as processes that rely on a Feed-Effluent Heat Exchanger (FEHE) to recover energy from the products. Typically, FEHEs (Figure 2) are used for preheating feed streams using the heat of reaction carried by the network products, but are equally effective in the recovery of refrigeration (Vinson, 2006).

Likewise, the prototype network in Figure 1 can be used to represent heat integrated distillation columns that use vapor recompression to recover the latent heat of the vapor (Jogwar et al., 2009).

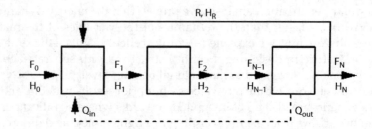

FIGURE 1
Prototype network.

The mathematical model of the prototype network in Figure 1 can be written in a generic form (Baldea and Daoutidis, 2007, Baldea and Daoutidis, 2008) as:

$$\dot{x} = f + \sum_{i=0,N} g_i F_i + \sum_{i=1}^{N-1} g_i F_i + g_r F_r$$

$$\dot{\theta} = \sum_{i=0,N} \gamma_i F_i H_i + \sum_{i=1}^{N-1} \gamma_i F_i H_i + \gamma_r F_r H_r + \quad (1)$$

$$\sum_{i=1}^{N} \gamma_d^i \delta_i + \gamma_q^{in} Q_{in} + \gamma_q^{out} Q_{out}$$

with $x \in D^x \subset \mathbb{R}^n$ being the variables in the material balance equations of units $i = 1,2,\ldots,N$, $\theta \in D^\theta \subset \mathbb{R}^p$ the variables in the energy balance (*e.g.*, temperatures, enthalpies), $\delta_i(x,\theta)$ additional energy contributions (*e.g.*, through heat of reaction, expansion work, etc.) in units $i = 1,2,\ldots,N$ and $f = f(x,\theta)$, $g_i(x,\theta)$, $g_r(x,\theta)$, $\gamma_d^i = \gamma_d^i(x,\theta)$, $\gamma_i = \gamma_i(x,\theta)$, $\gamma_q^j = \gamma_q^j(x,\theta)$ being appropriately defined vectors. For reasons that will become clear later in the manuscript, we adhere to using a separate notation for Q_{in} and Q_{out}, in spite of the fact that they represent, in effect, the same quantity.

The description in Eq. (1) assumes that individual units are modeled as lumped parameter systems and that kinetic and potential energy

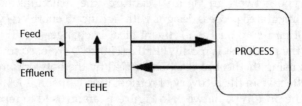

FIGURE 2
Energy flows in a network with Feed-Effluent Heat Exchanger (FEHE).

contributions are negligible. Based on the typical operation of the class of networks considered, we make the following additional assumptions:

1) At steady state, the energy flow $Q_{in} \equiv Q_{out}$ is much larger than the energy flow associated with the network material feed and output streams, respectively F_oH_o and F_NH_N. Equivalently,

$$\varepsilon = \frac{F_o H_{o,s}}{Q_{in,s}} \ll 1$$

where the subscript s denotes a steady-state value.

2) The energy flow $Q_{in} \equiv Q_{out}$ is much larger than the individual contributions in units $1,2,\ldots,N$, *i.e.* the energy flow $Q_{in} \equiv Q_{out}$ is much larger than the individual contributions in units $1,2,\ldots,N$, *i.e.*,

$$\frac{\delta_{i,s}}{Q_{in,s}} \ll 1, \forall i = 1, \ldots, N$$

3) The material flows in the network are of comparable magnitude

$$\frac{F_{i,s}}{F_{j,s}} = O(1), i, j \in 0, 1, \ldots, N, r$$

Note that assumption 2 and the overall energy balance of the network imply that the internal energy flows $F_1H_1,\ldots,F_{N-1}H_{N-1}$ are comparable in magnitude to the $Q_{in} \equiv Q_{out}$, i.e.,

$$k_i = \frac{F_{i,s} H_{i,s}}{Q_{in,s}} = O(1)$$

and, equivalently,

$$\frac{F_{o,s} H_{o,s}}{F_{i,s} H_{i,s}} = O(\varepsilon)$$

Under the above assumptions, the material and energy balance equations of the network in Figure 1 can be written as:

$$\dot{x} = f + \sum_{i=0,N} g_i F_i + \sum_{i=1}^{N-1} g_i F_i + g_r F_r$$

$$\dot{\theta} = \sum_{i=0,N} \gamma_i F_i H_i + \gamma_r F_r H_r + \sum_{i=1}^{N} \gamma_d^i \delta_i \tag{2}$$

$$+ \frac{1}{\varepsilon} H_{o,s} \left(\sum_{i=1}^{N-1} \gamma_i k_i \omega_i + \gamma_q^{in} \omega_{in} + + \gamma_q^{out} k_q \omega_{out} \right)$$

where $\omega_i = H_i/H_{i,s}$, $\omega_{in} = Q_{in}/Q_{in,s}$ and $\omega_{out} = Q_{out}/Q_{out,s}$. We additionally define $u = [u_o, \ldots, u_N, u_r]^T$, $u_i = F_i/F_{i,s}$ as a vector of scaled material flowrates, in which case the model of the network becomes:

$$\dot{x} = f + \sum_{i=0,N} g_i F_{i,s} u_i + \sum_{i=1}^{N-1} g_i F_{i,s} u_i + g_r F_{r,s} u_r$$

$$\dot{\theta} = \sum_{i=0,N} \gamma_i F_{i,s} u_i H_i + \gamma_r F_{r,s} u_r H_r + \sum_{i=1}^{N} \gamma_d^i \delta_i \qquad (3)$$

$$+ \frac{1}{\varepsilon} H_{o,s} \left(\sum_{i=1}^{N-1} \gamma_i k_i u_i \omega_i + \gamma_q^{in} \omega_{in} + + \gamma_q^{out} k_q \omega_{out} \right)$$

Thus, in what follows, we focus on systems with large energy flows of the form of Eq. (3), or more generally, systems of the form:

$$\dot{x} = f + Gu$$

$$\dot{\theta} = \varphi + \Gamma^s \omega^s + \frac{1}{\varepsilon} \Gamma^l \omega^l \qquad (4)$$

with $\varphi = \varphi(x,\theta)$ being an appropriately defined vector and $G = G(x,\theta)$, $\Gamma^l = \Gamma^l$ (x,θ), $\Gamma^s = \Gamma^s (x,\theta)$ appropriately defined matrices. $\omega^l = [\omega_{in}, u_1\omega_1, \ldots, u_{N-1}\omega_{N-1}, \omega_{out}]^T$ and $\omega^s = [u_o\omega_o, u_N\omega_N, u_r\omega_r]^T$.

The model in Eq. (4) includes a small singular perturbation parameter, ε, which indicates that such networks can potentially exhibit a two-time scale dynamic behavior. Typical closed-loop requirements—stability, output tracking, disturbance rejection—for the overall system can in such cases be achieved by designing separate (fast and slow) controllers with a coordinated control action (Kokotovic et.al., 1986, Kumar and Daoutidis, 2002). Separate reduced-order models that describe the dynamics in the fast and slow time scales are to be employed to this end. Using singular perturbation theory arguments, the following sections of the paper address this issue.

Dynamic Behavior of Networks with High Energy Recycle

In order to characterize the dynamic behavior of the class of networks under study, let us define the fast, "stretched" time scale $\tau = t/\varepsilon$. Rewriting the model in Eq. (4) in this time scale, and considering the limit case $\varepsilon \to 0$, or, equivalently, the case of an infinitely large energy recycle, we obtain:

$$\frac{dx}{d\tau} = 0$$

$$(5)$$

$$\frac{d\theta}{d\tau} = \Gamma^l \omega^l$$

which represents a description of the fast dynamics of the network, involving only the variables θ that pertain to the energy balance. Note that the fast energy dynamics described by Eq. (5) are only influenced by the energy flows $\omega_{in} \equiv \omega_{out}$ and ω_i (and implicitly by the flowrates u_i of the internal material streams, which act as the internal energy carriers).

In the case of process networks with high energy recycle such as the one in Figure 1, the total enthalpy of the network is independent of the large internal energy recycle flow $\omega_{in} \equiv \omega_{out}$. It is thus possible to verify (Jogwar et al., 2009) that only $p - 1$ of the p quasi-steady-state conditions that correspond to the fast dynamics in Eq. (5) are linearly independent. We can thus rewrite the quasi steady state conditions arising from Eq. (5) as:

$$0 = B\hat{\Gamma}^l \omega^l \tag{6}$$

with the $p \times p - 1$) matrix B having full column rank and

$$0 = \hat{\Gamma}^l \omega^l \tag{7}$$

being the linearly independent constraints that describe the equilibrium manifold corresponding to the fast dynamics.

In order to obtain a description of the slow dynamics, let us consider the limiting case $\varepsilon \to 0$ in the original time scale t, under the constraints in Eq. (7) to obtain:

$$\dot{x} = f + Gu$$

$$\dot{\theta} = \varphi + \Gamma^s \omega^s + B \lim_{\varepsilon \to 0} \frac{1}{\varepsilon} \hat{\Gamma}^l \omega^l \tag{8}$$

$$0 = B\hat{\Gamma}^l \omega^l$$

By denoting the indeterminate —yet finite— terms in Eq. (8) (which comprise of differences of large energy flows) as $z = \lim_{\varepsilon \to 0} \hat{\Gamma}^l \omega^l / \varepsilon$, we obtain a description of the slow dynamics in the form of a system of Differential Algebraic Equations (DAEs) of high index:

$$\dot{x} = f + Gu$$

$$\dot{\theta} = \varphi + \Gamma^s \omega^s + Bz \tag{9}$$

$$0 = B\hat{\Gamma}^l \omega^l$$

After specifying via control laws and energy flow correlations (e.g., in terms of temperatures and temperature gradients) the large energy flow $\omega_{in} \equiv \omega_{out}$ and, potentially, a subset $u/\hat{u}$ of the flowrates of the material streams, the algebraic constraints of Eq. (9) can be differentiated and an ODE representation of the DAE system above can be obtained. Based on the dimension of the equilibrium manifold of the fast dynamics, the order of this state space realization would be at most one.

An alternative means of describing the slow dynamics of networks with high energy throughput is the use of a coordinate change involving the total enthalpy of the network, H_T. It can be shown (Jogwar et al., 2009) that H_T evolves exclusively in the slow time scale, representing in this sense a "true" slow variable of the network.

The results derived in this subsection indicate that the structural properties of integrated networks with significant energy recycling can be exploited in the design of a control system. Specifically:

- the control objectives related to the energy balance of the individual units of networks with high energy recycle should be addressed in the fast time scale using the large internal energy flows,
- the control and, more importantly, optimization of the energy utilization at the level of the entire network should be undertaken in the slow time scale using the small energy flows of the network as manipulated inputs.

Networks with High Energy Throughput

Extending the arguments above, we notice that it is possible for the large energy flows of the network to be of *external* and distinct origin (i.e., $Q_{in} \neq Q_{out}$), while remaining of the same magnitude ($Q_{in}/Q_{out} = O(1)$). This situation is represented by the network in Figure 1 after eliminating the connection between Q_{in} and Q_{out}. This structure constitutes the prototype of numerous process unit and process network designs. For example, it can be employed to represent distillation columns (Figure 3), where a large amount of heat is required to vaporize the liquid in the reboiler and is subsequently recaptured by an external coolant in the condenser.

Multiple effect evaporators (Figure 4) also belong to the class of networks considered: the latent heat of the steam input to the first evaporator is transported through the network by the vapor streams of the subsequent evaporators and is finally released to the environment via the vapor output of the last unit.

Process networks that feature large *external* energy sources and sinks (e.g., distillation columns, multiple effect evaporators, reactors with external cooling systems, etc.) act, in effect, as an energy conduit between the source and the sink, thereby possessing a high energy throughput.

In the case of process networks with high energy throughput, it can easily be inferred that the total enthalpy of the network is dependent on the large

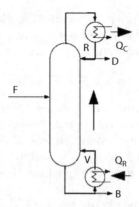

FIGURE 3
Energy flow in a distillation column.

flows w_{in} and w_{out}. By way of consequence, it can be verified (Baldea and Daoutidis, 2008) that, in this case, the steady-state conditions that correspond to Eq. (5) are linearly independent. Corroborating this observation with the developments above, we can infer that, upon setting w_{in} and w_{out} (and possibly a subset $u/\hat{u}$ of the flowrates of the material streams, with $\hat{u}$ denoting the flowrates that are not set by feedback control), by appropriate control laws, the Jacobian matrix

$$\frac{\partial}{\partial\theta}\hat{\Gamma}^l\omega^l \tag{10}$$

is nonsingular, and the equations

$$0 = \hat{\Gamma}^l\omega^l \tag{11}$$

can be solved for the quasi-steady-state values $\theta^* = k(x,\hat{u})$ of the enthalpies (or temperatures) of each unit. Substituting the solution θ^* in the model (also

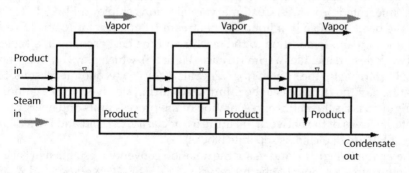

FIGURE 4
Energy flow in a multiple effect evaporator.

accounting for the fact that only $\hat{u}$ flowrates remain available after imposing the aforementioned control laws), we obtain a description of the dynamics of the network after the fast temperature "boundary layer":

$$\dot{x} = f(x,\theta^*) + \tilde{G}(x,\theta^*)\hat{u} \qquad (12)$$

which represents, in effect, the slow dynamics of the network.

In light of the above, the control of processes with high energy throughput is facilitated by the time scale separation feature of their dynamic behavior. As such, processes with high energy throughput lend themselves naturally to the use of a control structure with two tiers of control action. Specifically:

- a fast component of the control system addresses control objectives pertaining to the energy balance (e.g., temperature control), using the large energy flows ω_{in}, ω_{out} and, potentially, the flowrates of a subset of the internal material streams as manipulated inputs,

- control objectives related to the material balance, *e.g.*, the material holdups of the individual units, the total holdup of the network and the purity of the product(s), are to be addressed over a longer time horizon, employing the remaining flowrates $\hat{u}$ as manipulated inputs.

Notice that this approach contrasts the controller design proposed for networks with high energy recycle, whereby energy management at the level of the entire network is to be addressed entirely in the slow time scale.

Illustrative Examples

Analysis and Control of Reactor-Feed Effluent Heat Exchanger (FEHE) Networks

Feed-effluent heat exchangers (Figure 2) are key components in the design of heat integrated processes. Customary applications employ a FEHE to transfer the heat available in a hot effluent stream from a reactor to a cold reactor inlet stream. In addition to the reactor and the heat exchanger, a typical network design includes a trim furnace (Figure 5), which plays an important role in starting up the system (Jogwar et al., 2009),. For economic reasons, the heat load of the furnace is to be minimized during steady-state operation.

Figure 6 is a schematic representation of the process units and energy and material flows of the network in Figure 5 (for clarity, the cold and hot sides of the FEHE are depicted as separate blocks).

Referring to Figure 6 and using the notation convention established earlier, H_{tr} represents the flow of energy recycled in the FEHE, while δ_R and δ_F are, respectively, the energy flow due to generation in the reactor and the duty of the trim furnace.

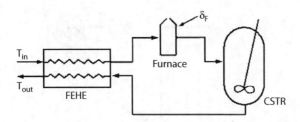

FIGURE 5
Reactor-FEHE network.

In order to minimize steady-state energy consumption, tightly integrated designs rely on a high rate of recovery and recycle of energy is in the FEHE (the heat transfer H_{tr} is much larger than the steady-state furnace duty). Moreover, the reactors employed in such integrated network designs are typically adiabatic, and the safe operation of the reactor requires that the heat effects associated the reactions, δ_R, be small.

According to the developments presented earlier, networks with FEHEs fall in the category of networks with high energy recycle, thus featuring a two-time-scale dynamic behavior. Namely, the energy dynamics of the reactor, trim furnace and the two legs of the FEHE will evolve in a fast time scale. Also based on the results established earlier, the control objectives related to the energy dynamics of the individual units should be addressed in the fast time scale by manipulating the large internal energy flows. For example, a material a bypass stream can be used to address the control of the hot side exit temperature of the FEHE and the furnace duty, a small, external energy flow, can be used to address network-level control objectives in the slow time scale (Jogwar et al., 2009).

Heat Integrated Distillation Sequences

Distillation represents one of the most energy-intensive separation processes. The design of distillation sequences with increased efficiency has been the subject of a vast body of research (King, 1980) and several energy integrated configurations have been proposed. In the case of double-effect distillation,

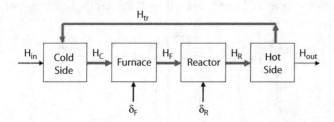

FIGURE 6
Schematic representation of the reactor-FEHE network.

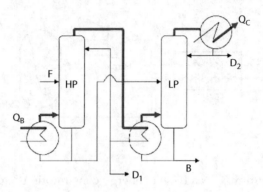

FIGURE 7
Double effect distillation.

typically used for the separation of ternary mixtures, the vapor from the top of the first, high pressure, column (denoted HP in Figure 7) is fed to a combined reboiler-condenser to provide heat for the generation of vapor in the second column (LP), which operates at a lower pressure than the first.

Resorting again to representing the condensing and boiling sides of the reboiler-condenser units as two distinct entities, Figure 8 presents the block diagram of a double-effect distillation network.

The contribution of latent heat to the enthalpy of the internal material streams of a distillation column is much larger compared to the contribution of sensible heat. The heat transfer in the reboilers and condensers of the network in Figure 8 is also dominated by latent heat effects, and the energy flows (Q_B and Q_C) are larger than the energy flows associated with the product streams. Thus, double effect distillation columns feature a high energy throughput.

Employing the results derived above, we can infer that double effect distillation processes exhibit a two time scale dynamic behavior. Specifically, the fast time scale is in this case associated with the energy dynamics, while the variables of the material balance (including the purities of the product streams B, D_1 and D_2) will evolve over a longer time horizon.

The time scale separation warrants the implementation of a hierarchical control strategy: simple, linear controllers that manipulate the energy flows

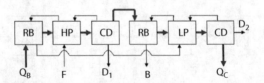

FIGURE 8
Schematic representation of double effect distillation (RB and CD denote the reboiler and condenser of a column).

Q_B and Q_D can be employed to regulate the temperatures of the reboiler and, respectively, condenser (thereby indirectly regulating the vapor boilup and reflux flows, and the temperature of the liquid and vapor on the column stages). Subsequently, the reduced-order model of the slow dynamics can be used to synthesize a (multivariable) controller relying on the flowrates of the product streams as manipulated inputs for regulating the purity of the products.

Conclusions and Future Work

In this paper, we have identified and documented the existence of a broad class of process networks where the presence of significant energy flows (either arising from energy recycling or of external origin) is reflected in dynamic models with a multi-time-scale structure. Initially, we focused on process networks with tight energy integration via high energy recycle, showing that they feature a clear distinction between the fast dynamics of individual units and the slow dynamics of the entire network. Subsequently, we demonstrated that process networks with significant energy throughput exhibit a similar (while not identical) dynamic behavior: the time scale in which network-level energy dynamics of process networks with high energy recycle evolve is much longer than in the case of networks with high energy throughput. The structural dissimilarities that are the origin of this discrepancy were identified and explained.

In both cases, we established a link between certain (steady-state) design features and structural properties that afford the development of a model reduction framework for the derivation of low-order nonlinear models of the core network dynamics. We demonstrated that the time scale multiplicity inherent to process networks with tight energy integration or high energy throughput facilitates controller design, and that such networks lend themselves naturally to the use of a hierarchical control structure. Consequently, we proposed a cadre for controller design predicated on the use of fast, distributed control for individual units and nonlinear supervisory control for the entire network, and emphasized the differences in the implementation approach that arise from the origin of the high energy flows. Finally, we highlighted the steps to be followed in the application of the proposed framework on two representative processes.

The generic class of process networks introduced and characterized in the present work is representative of numerous energy-integrated process configurations. In its present form, however, the proposed network prototype does not capture all possible energy integrated designs. For example, heat exchanger networks are high energy throughput networks that can exhibit a time scale multiplicity in the energy dynamics, owing to the simultaneous presence of several heat transfer and transport mechanisms.

Moreover, tightly integrated designs (*e.g.,* vapor recompression distillation) may feature multiple energy recycle loops and a dynamic behavior with more than two time scales. In such cases, a further classification of the control objectives and available manipulated inputs may be required, along with modifications to the proposed hierarchical controller design. The study of such networks and the appropriate expansion of the analysis and control framework developed in this paper are part of our ongoing and future work.

Acknowledgments

Partial financial support for this work by the National Science Foundation, grant CBET-0756363, is gratefully acknowledged.

References

Annakou, O., Mizsei, P. (1996) Rigorous comparative study of energy-integrated distillation schemes. *Ind. Eng. Chem. Res.,* 35:1877–1885.

Baldea, M., Daoutidis, P. (2007) Control of integrated process networks–a multi-time scale perspective. *Comp. Chem. Eng.,* 31:426–444.

Baldea, M., Daoutidis, P. (2008) Dynamics and control of process networks with high energy throughput. *Comp. Chem. Eng.,* 32:1964–1983.

El-Halwagi, M. M. (2006) Process Integration. *Academic Press.*

Georgakis, C. (1986) On the use of extensive variables in process dynamics and control. *Chem. Eng. Sci.,* 41:1471–1484.

Jacobsen, E., Berezowski, M. (1998) Chaotic dynamics in homogeneous tubular reactors with recycle. *Chem. Eng. Sci.,* 23:4023–4029.

Jacobsen, E. W. (1999) On the dynamics of integrated plants-non-minimum phase behavior. *J. Proc. Contr.,* 9:439–451.

Jogwar, S. S., Baldea, M., Daoutidis, P. (2009) Dynamics and control of process networks with large energy recycle. *Ind. Eng. Chem. Res.,* accepted for publication.

King, C. J. (1980) Separation processes. *McGraw-Hill,* New York, 2nd edition.

Kiss, A. A., Bildea, C. S., Dimian, A. C., Iedema, P. D. (2005) Design of recycle systems with parallel and consecutive reactions by nonlinear analysis. *Ind. Eng. Chem. Res.,* 44:576–587.

Kokotovic, P. V., Khalil, H. K., O'Reilly, J. (1986) Singular Perturbations in Control: Analysis and Design. *Academic Press,* London.

Kumar, A., Daoutidis, P. (2002) Dynamics and Control of Process Networks with Recycle, *J. Proc. Contr.,* 12:475–484.

Larsson, T., Govatsmark, M. S., Skogestad, S., Yu, C. C. (2003) Control structure selection for reactor, separator, and recycle processes. *Ind. Eng. Chem. Res.,* 42:1225–1234.

Linnhoff, B., Hindmarsh, E. (1983) The pinch design method for heat exchanger networks. *Chem. Eng. Sci.*, 38:745–763.

Linnhoff, B., Dunford, H., Smith, R. (1983b) Heat integration of distillation-columns into overal processes. *Chem. Eng. Sci.*, 38(8):1175–1188.

Luyben, M. L., Floudas, C. A. (1994) Analyzing the interaction of design and control -2. reactor-separator-recycle system. *Comp. Chem. Eng.*, 18(10):971–994.

Malcolm, A., Polan, J., Zhang, L, Ogunnaike, B. A., Linninger, A. A. (2007) Integrating systems design and control using dynamic flexibility analysis. *AIChE J.*, 53(8):2048–2061.

Reyes, F., Luyben, W. L. (2000) Steady-state and dynamic effects of design alternatives in heat-exchanger/furnace/reactor processes. *Ind. Eng. Chem. Res.*, 39:3335–3346.

Silverstein, J. L., Shinnar, R. (1982) Effect of design on the stability and control of fixed bed catalytic reactors with heat feedback. 1. Concepts. *Ind. Eng. Chem. Proc. Des. Dev.*, 21:241–256. Vinson, D. R. (2006) Air separation control technology. *Comp. Chem. Eng.*, 30(10–12):1436–1446.

Westerberg, A. W. (2004) A retrospective on design and process synthesis. *Comp. Chem. Eng.*, 28(4):447–458.

Yee, T. F., Grossman, I. E., Kravanja, Z. (1990) Simultaneous-optimization models for heat integration. 1. Area and energy targeting and modeling of multi-stream exchangers. *Comp. Chem. Eng.*, 14(10):1151–1164.

92

Design of Ionic Liquids via Computational Molecular Design

Samantha E. McLeese, John C. Eslick, Nicholas J. Hoffmann, Aaron M. Scurto and Kyle V. Camarda*

Department of Chemical and Petroleum Engineering
University of Kansas
Lawrence, Kansas 66045

CONTENTS

ABSTRACT Computational molecular design (CMD) is a methodology which applies optimization techniques to develop novel lead compounds for a variety of applications. In this work, a CMD method is applied to the design of ionic liquids (ILs), which are being considered for use as environmentally-benign solvents. The molecularly-tunable nature of ILs yields an extraordinary number of possible cation and anion combinations, the majority of which have never been synthesized. The product design framework developed in this work seeks to accelerate the commonly used experimental trial-and-error approach by searching through this large molecular space and providing a set of chemical structures likely to match a set of desired property targets. To predict the physical and chemical properties of an ionic liquid in a specific system, quantitative structure-property relations (QSPRs) have been developed. In this work, correlations were created for solubility, diffusivity, and melting temperature. The electronic structure of ionic liquids is quantified using molecular connectivity indices, which describe bonding environments, charge distribution, orbital hybridization and other interactions within and between ions. The resulting property prediction model is then integrated within a computational molecular design framework, which combines the

---

* To whom all correspondence should be addressed

QSPRs with structural feasibility constraints in a combinatorial optimization problem. The problem is reformulated as an MILP after exact linearization of structural constraints, and then solved using standard techniques. An example is provided for the design of ionic liquids for use within a hydrofluorocarbon (refrigerant) gas separation system. The computational efficiency and practical implementation of this product design methodology is also discussed.

KEYWORDS   *Molecular product design, ionic liquids, optimization*

Introduction

A large portion of the research currently performed today in the chemical industries is devoted to product design, that is, the search for new materials with specifically tailored properties. One new challenge within product design is in the design of environmentally-friendly refrigerants, solvents and mass separating agents. A class of compounds which are currently being studied for such applications is ionic liquids: organic salts which are liquid at and around room temperature. Ionic liquids usually possess negligible vapor pressure, and thus do not contribute to air pollution (Ren *et al.*, 2008). These compounds can be molecularly engineered to match a set of target physio-chemical properties, including solubility, diffusivity and acidity. It is estimated, however, that as many as 10^{14} unique cation/anion combinations are possible for use as ionic liquids (Holbrey and Seddon, 1999). Thus the time and expense required to perform a true search through this molecular space to find a novel ionic liquid is prohibitive. Currently, most new designs for such compounds tend to have similar structures to those previously used. In order to discover new ionic liquids which are far different from those currently used, an efficient computational screening procedure is required.

Such computational product design strategies are currently being implemented in the chemical and pharmaceutical industries; a review of industrial applications of CMD is given in Hairston (1998), and applications to other moleular systems are discussed in Venkatasubramaniam *et al.* (1994). Two major challenges arise in the development of such a computational molecular design procedure: the ability to predict the physical and chemical properties of a given molecule, and the ability to solve the large optimization problem which is derived from the search for the best molecule for a given application.

Physical Property Prediction

The simplest form of structure-property relation links physical and chemical properties of interest to the number and type of each functional group within a molecule. This group contribution approach is the basis for the well-known UNIFAC property prediction system. Other structural descriptors take into account not only the functional group types, but the topology of those groups. Bicerano (1996) correlated a large number of physical properties of polymers with topological indices, which are numerical descriptors based on the electronic structure of the atoms within a molecule, as well as on the interconnectivity of the atoms within that molecule. Gonzalez *et al.* (2007) have used these indices to fill in missing UNIFAC groups to predict physical properties of small organics. Eike *et al.* (2004) have shown the feasibility of using connectivity indices for the prediction of activity coefficients of ionic liquids. In this work, Randic's molecular connectivity indices (Randić, 1975) are used to generate structure-property correlations for ionic liquids. These descriptors provide a quantitative assessment of the degree of branching of molecules, and are based on a set of basic groups, which are defined as functional groups containing one non-hydrogen atom in a specific valence state and a given number of hydrogens. The nth order simple and valence molecular connectivity indices are given by

$$^n\chi = \sum_{(i_0,i_1...i_n)\in\varepsilon_n} \frac{1}{\sqrt{\delta_{i_0}\delta_{i_1}...\delta_{i_n}}} \tag{1}$$

$$^n\chi^v = \sum_{(i_0,i_1...i_n)\in\varepsilon_n} \frac{1}{\sqrt{\delta_{i_0}^v\delta_{i_1}^v...\delta_{i_n}^v}} \tag{2}$$

where ε_n is the edge set of n consecutive bonds between basic groups in a molecule, $i_0,i_1,...,i_n$ denote the $n + 1$ basic groups forming the n consecutive bonds, δ_{i_k}, $k = 0,...,n$ are the simple atomic connectivity indices for those basic groups (the number of bonds each group can form), and $\delta_{i_k}^v$, $k = 0,...,n$ are the atomic valency connectivity indices, which are based on the electronic structure of the basic group. Table 1 shows the basic groups employed in the example in this paper, along with their atomic connectivity indices.

TABLE 1

Basic Groups and Their Atomic and Valence
Connectivity Indices (δ and δ^v)

	δ	δ^v		δ	δ^v
$-\overset{\mid}{\underset{\mid}{C}}-$	4	4	O=	4	6
$-CH-$	3	3	$-O-$	2	6
$-CH_2-$	2	2	$F-$	1	7
CH_3-	1	1	$-\overset{\mid}{C}-$	3	5
$>C=$	3	4	$-N-$	2	1.333
$=\overset{\mid}{\underset{\mid}{S}}=$	4	2.667	$O-$	1	7

These topological indices can then be correlated with various physical properties of molecules, given a consistent set of experimental data for the properties of interest. In this work, three properties important to the use of an ionic liquid within an absorption refrigeration system are correlated: solubility of the compound in the refrigerant R-32, diffusivity between the ionic liquid and R-32, and melting point. Data for these correlations was obtained from Shiflett *et al.* (2006) and Shiflett and Yokozeki (2006). The correlations used in this work (shown below) are preliminary based on 19 common ionic liquids, and further data is currently being obtained such that the predictive capabilities of the correlations may be expanded.

In the new correlations developed in this work, zeroth- and first-order connectivity indices of both the cation and anion are applied. The correlations employed to predict the physical properties listed above are as follows:

$$D = 14.57(P) + 19.971\left(^{0}\chi_{cat}\right) + 2.212\left(^{0}\chi_{cat}^{v}\right)$$

$$+67.404\left(^{1}\chi_{cat}\right) - 98.413\left(^{1}\chi_{cat}^{v}\right) - 43.658\left(^{0}\chi_{an}\right) - 3.143\left(^{0}\chi_{an}^{v}\right) \quad (3)$$

$$+107.237\left(^{1}\chi_{an}\right) - 35.885\left(^{1}\chi_{an}^{v}\right) - 92.735$$

$$100x = 58.98(P) + 967.05\left(^{0}\chi_{cat}\right) - 66.44\left(^{0}\chi_{cat}^{v}\right)$$

$$-44.37\left(^{1}\chi_{cat}\right) + 30.01\left(^{0}\chi_{an}\right) + 15.23\left(^{0}\chi_{an}^{v}\right)$$

$$-57.36\left(^{1}\chi_{an}\right) + 29.68\left(^{1}\chi_{an}^{v}\right) - 68.73 \quad (4)$$

$$Tm = -3.934\left(^{0}\chi_{an}\right) + 76.389\left(^{0}\chi_{an}^{v}\right)$$

$$-52.389\left(^{1}\chi_{an}\right) - 23.23\left(^{1}\chi_{an}^{v}\right) - 69.001 \quad (5)$$

where D is the diffusion coefficient (m²/sec), P is the system pressure, x is the solubility in (mol/L) and T_m is the melting temperature (K). Note that the correlations include topological information about both the anion and cation, and also involve the operating pressure. These correlations should be considered preliminary, since they are based on only a few data points and a small set of unique ionic liquid structures. Nevertheless, they serve to show the capabilities of the optimization method. A larger set of data points from a set of structures which spans a larger molecular space would allow for more flexibility in the final design, but would not require a different optimization methodology.

Problem Formulation

The correlations generated in the first phase of this product design methodology are then combined with structural constraints to construct an optimization problem. The solution to this problem is a candidate ionic liquid (both basic groups and bonds) which is predicted to have properties matching a set of pre-specified target values based on the application of interest. The objective function seeks to minimize the scaled difference between the physical property values of the candidate molecule and targets.

Along with the property prediction equations and the expressions defining the molecular connectivity indices, structural constraints are also included to ensure that a stable, connected molecule is formed. These include valency and uniqueness constraints, which ensure that the valency of each atom is satisfied and two groups may only bond with one type of bond (single or double), charge constraints which insure that the cation and anion have a charge of +1 or −1, as well as connectedness constraints, which guarantee that all the basic groups within the molecule are bonded into one coherent molecule (Camarda and Maranas (1999); Lin *et al.* (2005)).

In order to store the molecular candidates computationally, a data structure is used which employs binary variables to define whether two basic groups i and j are bonded with a kth multiplicity bond. These binaries form a partitioned adjacency matrix A which is structured such that the identity of each basic group is known at each position within the molecule. This ensures that the square root terms in the connectivity index equations (1 and 2) are all constants, and thus those equations are linear functions of the unknown binary variables (Siddhaye, *et al.*, 2004). The objective function can be reformulated as a set of linear functions, and all other constraints are formulated linearly as well. Thus the problem is an MILP, which can be solved to optimality using standard techniques. The overall formulation may be posed as

$$\text{Min} \quad s = \sum_m \frac{1}{P_m^{\text{scale}}} \left| P_m - P_m^{\text{target}} \right|$$

$$\text{s.t.} \quad P_m = f_m(\chi)$$

$$\chi = g_n(a_{ijk}, w_i)$$

$$h_c(a_{ijk}, w_i) \geq 0$$

$$P_m, \chi \text{ continuous}, a_{ijk}, w_i \text{ binary}$$

where P_m is the value of the mth physical of chemical property, P_m^{target} is a preselected target value for the mth property, P_m^{scale} is a scaling factor for the mth property, $f_m(\chi)$ are the m linear property correlations which are functions of the connectivity indices χ, g_n are the defining equations for the connectivity indices which are based on the data structure a_{ijk} which is used to store bonding and w_i which stores the identity of each group i in the molecule, and $h_c(a_{ijk}, w_i)$ are structural constraints used to ensure that a stable, connected molecule is designed.

TABLE 2

Properties, Target Values and Predicted Values
for the Optimal Solution to the Example Problem

Property	Target Value	Best Solution Value
Sol.	0.80	5.34
D	20.00	20.09
T_m (K)	198.15	199.09

Results

The methodology was tested using an example set of physical property targets, as shown in Table 2.

These targets correspond to values which would be reasonable for an ionic liquid used in conjunction with R-32 within a refrigerant gas separation system. For this preliminary work, the cation was fixed to be 1-butyl-3-methylimidazolium (BMIM), which has the following structure:

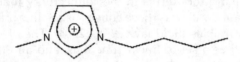

The structure of the anion was allowed to vary, but requirements were set on the number of total groups and on the existence of one sulfate group, so that an organic salt would be formed. The problem was solved using GAMS/CPLEX, and required less than 300 seconds to find an optimal solution. The optimal structure found for this example gave an objective function value of 0.90, which is the percent scaled deviation from the target property values. While it is easily possible to weight the importance of the various target properties, all three properties in this example were given even weighting. Properties can instead be given thresholds, which may be included within the constraint set and require a given property to be above or below a set value. The designed structure for this example is shown in Figure 1, which

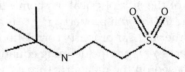

tert-buty1(2-(methylsulfonyl)ethyl)amide

FIGURE 1

Example ionic liquid designed using optimization methodology.

corresponds to an ionic liquid which could now be synthesized and tested. Further work will include integer cuts within the design formulation, to obtain a set of molecular candidates which a designer could then consider in terms of ease of synthesis prior to experimental testing.

Note the methodology in its current form does not provide information on how the new candidate structure is to be synthesized, but it does give a researcher a short list of potential compounds which are likely to match the desired physical property targets. This method may also be used to improve a given ionic liquid pair, In this case, one simply fixes certain groups within the molecule, and uses the optimization algorithm to suggest useful modifications of a given structure. The resulting optimization problems have fewer binary variables than a full design problem, and thus a larger number of modifications may be considered.

Conclusions

In this paper, a molecular design problem for the design of ionic liquids for use within environmentally-friendly refrigeration systems is considered. The product design problem has been recast as a mixed-integer linear program, and is solved to provide candidate structures for synthesis and further testing. An example is provided which designs a novel ionic liquid for use within an absorption refrigeration system in conjunction with R-32, and for which target values of three physical properties are matched. Further work will design both the anion and cation within the system, and will consider a wider range of properties.

References

Bicerano, J. (1996), Prediction of Polymer Properties, Marcel Dekker, New York.

Camarda, K.V., and C.D. Maranas (1999). "Optimization in polymer design using connectivity indices" *Ind. Eng. Chem. Res.* **38**, 1884–1892.

Eike, D. M., J. F. Brennecke, and E. J. Maginn (2004). "Predicting infinite-dilution activity coefficients of organic solutes in ionic liquids" *Ind. Eng. Chem. Res.* **43**, 1039–1048.

Hairston, D. W. (1998). "New molecules get on the fast track." *Chem. Eng.*, 30–33.

Gonzalez, H. E., J. Abildskov, and R. Gani, (2007) . "Computer-aided framework for pure component properties and phase equilibria prediction for organic systems," *Fluid Phase Equilibria*, 261, 199.

Holbrey, J. D. and K.R. Seddon, (1999). *Clean Prod. Proc.* 1, 223-236

Lin, B., S. Chavali, K. Camarda and D.C. Miller (2005). "Computer-aided molecular design using Tabu search" *Comp. Chem. Eng.*, 29, 337–347.

Ren, W. , A. M. Scurto, M. B. Shiflett and A. Yokozeki (2008). "Phase behavior and equilibria of ionic liquids and refrigerants: 1-Ethyl-3-methyl-imidazolium bis(trifluoromethylsulfonyl)imide ([EMIm][Tf2N]) and R-134a in Gas Expanded Liquids and Near-Critical Media: Green Chemistry and Engineering," Eds. K. Hutchenson, A. M. Scurto, and B. Subramaniam, ACS Symposium Series 1006: Washington, D.C., 2008; In Press.

Shiflett, M. B., M. A. Harmer, C. P. Junk and A. Yokozeki (2006). "Solubility and diffusivity of difluoromethane in room-temperature ionic liquids" *J. Chem. Eng. Data* **51**, 483-495.

Shiflett, M. B. and A. Yokozeki (2006). "Solubility and diffusivity of hydrofluorocarbons in room-temperature ionic liquids" *AIChE J.*, **52,** 1205–1219.

Siddhaye, S., K. V. Camarda, M. Southard, and E. Topp (2004). "Pharmaceutical Product Design Using Combinatorial Optimization," *Comp. Chem. Eng.*, **28**, 425–434.

Venkatasubramanian, V., Chan, K., & Caruthers, J. M. (1994). "Computer-aided molecular design using genetic algorithm" *Comp. Chem. Eng.*, **18**(9), 833–844.

93

Biodiesel Process Design through a Computer-aided Molecular Design Approach

Arunprakash T. Karunanithi[1*], Rafiqul Gani[2] and Luke E.K. Achenie[3]

[1]University of Colorado Denver, Denver, CO, USA
[2]Technical University of Denmark, DK-2800 Lyngby, Denmark
[3]Virginia Polytechnic Institute and State University, Blacksburg, VA, USA

CONTENTS

ABSTRACT This paper proposes the application of computer-aided molecular design methods for the design of biodiesel processes. Solvents are used in two stages during bio diesel production. In the first stage solvents are used for extraction of oil from seed feedstock and in the second stage solvents are used to facilitate the conversion of oil into biodiesel. This paper introduces the application of state of the art computer-aided molecular design methodology for the design/selection of solvents for the conversion of oil seed feedstock to biodiesel. A case study involving design of co-solvents for the promotion of transesterification reaction during the production of biodiesel from soybean oil is presented. The designed co-solvent is intended to increase the reaction rate by forming a single phase solution of the reactants.

KEYWORDS *Biodiesel, co-solvent, oil extraction, transesterification, CAMD*

---

* To whom all correspondence should be addressed

Introduction

Biodiesel production from feedstock such as soybean and algae involves two major steps a) extraction of oil from oil seeds and b) chemical conversion of extracted oil into biodiesel. The three common methods for extraction of oil from oil seeds are expeller/press method, solvent extraction method and supercritical fluid extraction method. There are other extraction methods such as enzymatic extraction and ultrasonic extraction that are less frequently used. Among these methods "solvent extraction" is the most frequently used approach. Solvent extraction method can be used in isolation or it can be combined with expeller/press method to extract oil. Typically hexane is used as a solvent in this approach. When combined with expeller/press method, first oil is extracted with the expeller and the residue pulp is mixed with the solvent whereby the remaining oil dissolves in the solvent. The residue pulp is filtered out and the oil is separated from the solvent by distillation. Once the oil has been extracted to the maximum extent possible, it is chemically converted to biodiesel through transesterification reaction. Chemical transesterification of oil involves reaction of a triglyceride (oil) with an alcohol (typically methanol) in the presence of a catalyst (typically a strong acid or base) to produce fatty acids alkyl esters (biodiesel) and glycerol. The transesterification is an equilibrium reaction and the transformation occurs essentially by mixing the reactants (Schuchardt et al., 1998). The presence of the catalyst accelerates the forward reaction. For an acid catalyzed process, the most commonly used catalysts are sulphuric acid and hydrochloric acid and this process is particularly effective if the vegetable oils have high free fatty acids content and more water (Yang and Xie, 2007). However the reaction time is long and a high molar ratio of methanol to oil is required.

Role of Solvents in Biodiesel Production Process

As discussed in the previous section solvents are used at various stages of the biodiesel production process. Particularly they are used for oil extraction and in certain cases they are used to facilitate the transesterification reaction. Till now promising solvents have been chosen based on bench scale experimentation and process know how. There has not been a systematic framework for design/selection of solvents for these processes. The solvents that are being used for oil extraction (e.g. cyclohexane) and esterification reaction (e.g. tetra hydro furan) have been selected on a trial and error basis. Moreover, when selecting a solvent it is imperative to consider environmental and safety issues (e.g. solvent toxicity, solvent flash point) in addition to the performance of solvents. In this paper we present a systematic computer-aided approach for solvent design/selection for biodiesel production.

Computer-aided Solvent Design

We propose an optimization based decomposition methodology (Karunanithi et al., 2005) for the design of chemical compounds (e.g. solvents) with specific properties from functional groups. The decomposition methodology involves an optimization approach in which the CAMD design problem is formulated as an optimization model where a performance requirement is posed as objective function and all other property requirements are posed as constraints. Group contribution models are used to evaluate the property constraints. Structural constraints are used to make sure that only feasible chemical compounds (solvents) are generated, i.e. compounds that satisfy valency criterion. Since both integer variables (structural) and continuous variables (e.g. mole fraction) are involved and since some of the property constraints are non-linear in nature we end up with a mixed integer non-linear programming (MINLP) optimization model. The solution to this optimization model provides the optimal molecular structure that satisfies all the property constraints with the minimal objective function value. A generic CAMD problem from Karunanithi et al. (2008) formulated as an MINLP optimization model is shown below:

$\text{Min} f_{obj} (X, Y),$

Subject to,

Structural constraints: $g_1 (Y) \leq 0,$

Pure component property constraints: $g_2 (Y) \leq 0,$

Mixture property constraints: $g_3 (X, Y) \leq 0,$

Process model constraints: $g_4 (X, Y) = 0.$

Here, Y is a vector of binary variables (integer), which are related to the identities of the building blocks (functional groups). X is a vector of continuous variables, which are related to the mixture (*e.g.*, compositions) and/or process variables (*e.g.*, flow rates, temperatures *etc.*). f_{obj} is the performance objective function, defined in terms of molecule-process (performance) characteristics and/or cost that may be minimized or maximized. g_1 and g_2 are sets of structural constraints (related to feasibility of molecular structure) and pure component property constraints (related to properties-molecular structure relationships) respectively. g_3 and g_4 are mixture property (related to properties-mixture relationships) constraints and process model (related to process-molecule/mixture relationships) constraints respectively. The solution to this generic MINLP optimization model is achieved through a decomposition based solution methodology. Detailed description about the CAMD- MINLP optimization model and the decomposition based solution methodology can be found in Karunanithi et al. (2005). The decomposition methodology utilizes two programs ProCAMD (ICAS documentation. Internal report, 2003) and OPT-CAMD+ (Karunanithi et al., 2005)

Database Search Approach

Another approach to solvent selection is to use a solvent database to select solvents by matching property requirements. Examples of well known databases that can be used for this purpose are DIPPR and CAPEC database. The property requirements are posed as solvent search criteria and the database is searched to retrieve solvent candidates that match the criteria. The drawback in the database search approach is availability of all the compounds in the database and availability of property values. Also it is rare to find property values for mixture properties in a database.

Case Study: Solvent Design for Transesterification Reaction

In this section a case study involving design of a co-solvent for the conversion of soybean oil into biodiesel through transesterification reaction is presented. As described in the first section the transesterification reaction involves the formation of vegetable oil methyl esters by the base catalysed reaction of triglycerides in the vegetable oil with methanol (Boocock, D.G.B., 2004). This reaction is known to be slow as it occurs in two phase. Low oil concentration in the methanol phase causes the slow reaction rate. The problem of the slow reaction rate can be overcome by using a co-solvent which results in the transformation of the two-phase system to one-phase system (Boocock, D.G.B., 2004). The co-solvent should satisfy the following property requirements a) should be completely miscible with both methanol and the source of fatty acid triglyceride and should produce a homogenous single phase mixture of methanol-triglyceride- co solvent. b) co-solvent should have a boiling point of less than 393 K to facilitate solvent removal after the reaction is complete c) co-solvent should have a boiling point as close as possible to that of methanol in order to facilitate removal of both methanol and co-solvent in a single stage distillation unit. In addition the solvent should satisfy other process, safety and environmental constraints, namely d) should have high solubility for fatty acid triglyceride e) should be liquid at operating conditions f) should have high flash point and g) should have low toxicity. These property requirements are translated into property constraints that can be used in the CAMD model to design optimal co-solvent for this process. Toxicity is selected as the objective function that needs to be minimized. The proposed CAMD-MINLP optimization model for the design of co-solvent is shown below

$$Min\left[-Log(LC_{50})\right] \tag{1}$$

$$\text{subject to}$$
$$\sum_i \sum_j u_{ij}(2 - v_j) = 2 \tag{2}$$

$$\sum_i \sum_j u_{ij} = N_{max} \tag{3}$$

$$\sum_j u_{ij} = 1 \tag{4}$$

$$17 < \delta < 19 \tag{5}$$

$$\frac{1}{x_2} + \frac{\partial \ln \gamma_2}{\partial x_2} \geq 0 \tag{6}$$

$$325 < T_b = 204.359 * \sum_i N_i T_{bi} + \sum_j M_j T_{bj} \geq 350 \tag{7}$$

$$T_m = 102.425 * \sum_i N_i T_{mi} + \sum_j M_j T_{mj} \leq 270 \tag{8}$$

$$T_f = 3.63 * \sum_i \sum N_i T_{fi} + 0.409 * T_b + 8843 > 200 \tag{9}$$

$$x_1 + x_2 + x_3 = 1 \tag{10}$$

u_{ij} is a binary variable (special kind of integer variable) indicating whether the ith position in a molecule has structural group j. v_j represents valence of group j. N_{max} represents maximum number of positions in a molecule. δ represents the solubility parameter of the solvent. N_i is the number of times the first order group 'i' is present in the molecule. M_j Is number of times the second order group 'j' is present in the molecule. T_f, T_b and T_m are the solvent flashpoint, boiling point and melting point respectively. T_{fi}, T_{bi} and T_{mi} are the contributions of first order groups towards flash point, melting point and boiling point respectively. T_{fj}, T_{bj} and T_{mj} are the contributions of second order groups towards flash point, melting point and boiling point respectively. LC50, a measure of solvent toxicity, represents the aqueous concentration causing 50% mortality in fathead minnow after 96 hours. γ_i is the activity coefficient of component i. H_{298} and H_{298}^m represents the heat of vaporization and molar volume of the solvent respectively. x_1, x_2 and x_3 are mole fraction of methanol, fatty acid triglyceride and solvent respectively. Equation (1) is the objective function where the toxicity of the solvent represented by $-\log$ (LC50) needs to be minimized. Equation (2), (3) and (4) represent structural constraints that are used to make sure that only chemically feasible solvents are generated. Equation (5) is the constraint on solubility parameter to achieve high solubility of fatty acid triglyceride in the solvent. Equation (6) is the Gibbs phase rule constraint to achieve a homogeneous single phase mixture of methanol, triglyceride and co-solvent. Equation (7) is the constraint on boiling point to achieve easy removal of solvent and simultaneous separation of solvent and methanol after the completion of the reaction. Equation (8) is the constraint on melting point to maintain the solvent in the liquid state at operating conditions. Equation (9) is the safety constraint represented by flash point. The group contribution model proposed by Martin and Young (2001) is used for calculating $-\log (LC50)$. The comprehensive data set from which this group contribution model was developed consisted of 397 organic compounds. This model was able to achieve a fairly good correlation of the data ($r^2 = 0.91$). The solubility parameter, boiling point, melting point and flash point are calculated using Constantinou and Gani (1994) group contribution method. The reliability of Constantinou and Gani group contribution method is extremely high and compared to other group contribution models it demonstrates significant improvements in accuracy and applicability. The absolute percentage errors for boiling point and melting point calculated using Constantinou and Gani method are 1.42% and 7.23% (Constantinou and Gani, 1994). Solubility parameter is estimated through a correlation which utilizes heat of vaporization, which in turn is estimated using Constantinou and Gani method (absolute percentage error of 2.57%). Flash point is also estimated through a correlation that relates it to boiling point. The activity coefficient is calculated using UNIFAC group contribution method (Fredunslund et al., 1975). Hence the reliability and accuracy of the utilized group contribution models are very high.

Results

The solution of the above optimization model using the decomposition methodology resulted in the optimal co-solvent structure shown in Fig. 1. The design results of the optimal solvent are shown in Table 1. Global optimality cannot be guaranteed with the decomposition methodology since the final MINLP sub-problem solved as a series of NLP problems uses a local optimization algorithm (OPT-CAMD[+]). However it is worth noting that the initial sub-problems solved using ProCAMD (ICAS documentation. Internal

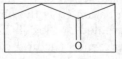

FIGURE 1
Optimal Co-Solvent- Methyl Ethyl Ketone.

TABLE 1

Design Results— Property Values of Optimal co-solvent

δ MPa $^{1/2}$	T_f K	T_m K	T_b K
18.76	262.42	187.97	343.82

report, 2003) uses an enumeration approach and hence explores the search space effectively.

Conclusions

This paper presents the application of a novel computer-aided tool for the design of biofuel processes. In particular it introduces the application of computer-aided molecular design approach for the design/selection of solvents for biodiesel process. A generic CAMD methodology that can be used for this purpose is presented and the solution method for the model is described. A case study involving the design of a co-solvent to facilitate the conversion of soybean oil to biodiesel through transesterification reaction is shown. An optimal solvent is proposed for the discussed process.

Acknowledgments

Dr. A.T.K and Dr. L.E.K.A would like to thank CAPEC, Department of chemical engineering, Technical university of Denmark, for providing the software ICAS

References

Boocock, D.G.B. (2004). Process for production of fatty acid methyl esters from fatty acid triglycerides. *US Patent number 6712867.*

Constantinou, L., Gani, R. (1994). Group-contribution method for estimating properties of pure compounds. AIChE J., 40, 1697.

Fredenslund, A., Jones, R.L., Prausnitz, J.M. (1975). Group-Contribution estimation of activity coefficients in non ideal liquid mixtures. *AIChE J., 21, 1086.*

ICAS documentation. Internal report. (2003). CAPEC, Department of Chemical Engineering, Technical University of Denmark, Lyngby, Denmark.

Karunanithi, A.T., Achenie, L.E.K., Gani, R. (2005). A new decomposition based solution methodology for the design of optimal solvents and solvent mixture. *Ind. Eng. Chem. Res., 44, 4785–4797.*

Karunanithi, A.T., Acquah, C., Achenie, L.E.K. (2008). Tuning the morphology of pharmaceutical compounds via model based solvent selection. Chinese Journal of *Chemical Engineering, 16, 465–473.*

Martin, T.M., Young, D.M. (2001). Prediction of the acute toxicity (96-h LC50) of organic compounds to the fathead minnow (Pimephales promelas) using group contribution method. *Chem. Res. Toxicol., Vol. 14, pp. 1378–1385.*

Schuchardta, U., Serchelia, S., Vargas, R.M. (1998). Transesterification of vegetable oils: a review. *J. Braz. Chem. Soc., 9, 199–210.*

94

Model Reformulation and Design of Lithium-ion Batteries

V.R. Subramanian[1,*], V. Boovaragavan[1],
V. Ramadesigan[1], K. Chen[2] and R.D. Braatz[2]

[1] *Department of Chemical Engineering, Tennessee Technological University, Cookeville, TN 38505*
[2] *Department of Chemical & Biomolecular Engineering, University of Illinois, Urbana-Champaign, Urbana, IL 61801*

CONTENTS

ABSTRACT Recently, electrochemical power sources have had significant improvements in design, economy, and operating range and are expected to play a vital role in the future in automobiles, power storage, military, mobile-station, and space applications. Lithium-ion chemistry has been identified as a good candidate for high-power/high-energy secondary batteries and commercial batteries of up to 75 Ah have been manufactured. Applications for batteries range from implantable cardiovascular defibrillators (ICDs) operating at 10 μA current to hybrid vehicles requiring pulses of 100 A. While physics-based models have been widely developed and studied for these systems, these models have not been employed for parameter estimation or dynamic optimization of operating conditions or for designing electrodes for a specific performance objective. This is an unexplored area requiring model reformulation and efficient simulation of coupled partial differential equations. This paper describes model reformulation and its application to (1) parameter estimation and prediction of capacity fade for lithium-ion batteries, and (2) optimal product design.

---

* To whom all correspondence should be addressed: Vsubramanian@tntech.edu

KEYWORDS *Lithium-ion batteries, product design, Bayesian estimation, Markov Chain Monte Carlo simulation*

Introduction

Several issues arise in the operation of lithium-ion batteries—capacity fade, underutilization, abuse caused by overcharging, and thermal run-away caused by operation outside the safe window (Newman et al, 2003). For example, the batteries used in hybrid cars operate at 50% state of charge to enhance life while compromising on utilization (energy efficiency). The capability to accurately predict capacity and internal state variables is highly desired as it can help extend the life of the battery and provide for better operational strategies. Three different approaches have been used in the literature for modeling lithium-ion batteries: (1) empirical models (Plett, 2004), (2) transport phenomena models (Doyle et al, 1993), and (3) stochastic models (Darling and Newman, 1999). Although parameter estimation, design calculations, and dynamic optimization are easiest to apply to empirical models due to their low computational cost, these models fail at many operating conditions and cannot predict the future behavior or current capacity accurately. Several recent studies have tried to understand micro- and nanoscale phenomena in batteries using stochastic methods. These models would need to be coupled with transport phenomena models to predict process behavior of batteries at the system level.

Of the three modeling approaches, transport phenomena models are currently the most promising candidates for use in design because these models can predict both internal and external behavior (system level) with reasonable accuracy. These models are based on porous electrode theory coupled with transport phenomena (Doyle et al, 1993, Botte et al, 2000) and electrochemical reaction engineering. These models are represented by coupled nonlinear partial differential equations (PDEs) in 1 or 2 dimensions and are typically solved numerically, requiring between a few minutes to hours to simulate. As an example of the form of these PDEs, the model equations for the lithium-ion battery cathode are

$$\varepsilon_p \frac{\partial c}{\partial t} = \frac{\partial}{\partial x}\left(D_{\text{eff,p}} \frac{\partial c}{\partial x}\right) + a_p\left(1 - t_+\right)j_p$$

$$I = -\sigma_{\text{eff,p}} \frac{\partial \Phi_1}{\partial x} - \kappa_{\text{eff,p}} \frac{\partial \Phi_2}{\partial x} + \frac{2\kappa_{\text{eff,p}}RT}{F}\left(1 - t_+\right)\frac{\partial \ln c}{\partial x}$$

$$\frac{\partial}{\partial x}\left(\sigma_{\text{eff,p}} \frac{\partial \Phi_1}{\partial x}\right) = a_p F j_p, \quad \frac{\partial c_s}{\partial t} = \frac{1}{r^2}\frac{\partial}{\partial r}\left(D_{sp}r^2 \frac{\partial c_s}{\partial r}\right)$$

Electrochemical modeling of a typical secondary battery involves three regions: positive porous electrode, separator, and negative porous electrode (see Fig. 1). The original model involves 10 PDEs (4 in each electrode + 2 in the separator). If 100 node points are used in the x-direction in each region and 20 node points are used in the r-direction, the original model involves 2 × 100 (separator) + 3 × 100 (macroscale in each electrode) + 1 × 20 × 100 (microscale in each electrode) = 200 + 300 × 2 + 2000 × 2 = 4800 differential-algebraic equations (DAEs) in time. This model accounts for diffusion and reaction in the electrolyte phase in the anode/separator/cathode, diffusion (intercalation) in the solid phase in the cathode and anode, ionic and electronic conductivity in the corresponding phases in the porous electrodes (cathode and anode), nonlinear ionic conductivity, and nonlinear kinetics.

While first principles-based models have been discussed in detail in the literature, attempts to rigorously estimate parameters have been minimal. As of today, literature on dynamic optimization or design of lithium-ion batteries based on physics-based models for a specific performance (e.g., minimized capacity fade) is non-existent (the linearized models analyzed in the frequency domain by Smith and Wang (2006) cannot be used for non-constant parameters). For a lithium-ion battery, the process variable could either be current or voltage (e.g., for a given load or current, the battery operates at a voltage that decays with time, or when the battery is operated at a certain voltage, current decays with time). In a hybrid environment, in which batteries operate in series-parallel combination with fuel cells or other devices, energy or power might be specified which is typically delivered by operating the battery at a particular current or voltage profile. To optimize electrochemical power sources, and to produce mini- and micro- batteries and fuel cells for the future, there is an urgent need to develop and implement effective and robust parameter estimation and optimization strategies. While existing models can be used for offline analysis and simulation purposes, these models are unsuitable for parameter estimation, dynamic optimization, or product design. Recently we have begun applying systems engineering methods with the objectives of (1) predicting underutilization and capacity fade, (2) devising optimal operating strategies, and (3) designing new materials for improved performance (e.g., extended life). This paper focuses on the following topics:

1. Development of reformulated efficient models to facilitate parameter estimation, dynamic optimization, and product design.
2. Estimation of transport and kinetic parameters as a function of a battery's cycle life.
3. Design of batteries with higher average utilization over time.

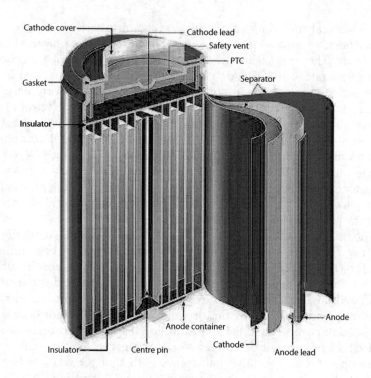

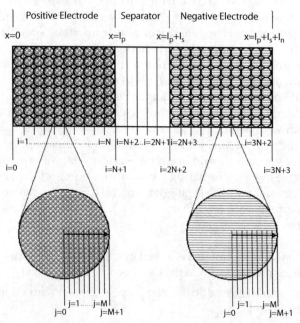

FIGURE 1
Lithium-ion battery, cross-sectional view (top) and numerical grid for modeling (bottom).

Model Reformulation

Model reduction is an active area of research for many engineering and science fields (Benner, 2008). There are standard methods available in the literature for reducing a given set of coupled partial differential equations (PDEs) to reduced order models with different levels of accuracy and detail. Proper Orthogonal Discretization (POD) uses the full numerical solution to fit a reduced set of eigenvalues and nodes to get a meaningful solution with a reduced number of equations (Cai and White, 2008). A drawback of POD is that the resulting model needs to be reconstructed when the operating current is doubled, the boundary conditions are changed, or if the parameter values are changed significantly. Orthogonal Collocation (OC) is another widely used technique to reduce the order of the models in a variety of chemical engineering problems. A drawback of the OC technique is its inability to accurately define profiles with sharp gradients and abrupt changes. Certain models might require a large number of collocation points (Wei, 1987). Orthogonal collocation was found to be only as efficient as solving the PDEs using the finite difference method for battery models and resulted in unstable codes for smaller number of points because of the model's DAE nature.

This paper describes a method for mathematical model reformulation that applies various techniques to solve for the dependent variables without losing accuracy. Specific information for the dependent variables in the x direction is provided that was not reported in past papers on the model reformulation (Subramanian et al, 2007; 2009). Volume averaging coupled with polynomial approximation for the solid-phase concentration gives high accuracy at low-to-medium rates of discharge for modeling lithium-ion batteries (Subramanian et al, 2005; Wang et al, 1998). This step converts the model from a pseudo-2D to a 1D model.

Below is a description of the step-by-step procedure to reduce the ten coupled nonlinear PDEs (in x, r, t) in the battery model to a small number of DAEs (< 50). The approach considers each dependent variable separately and finds a suitable mathematical method to minimize the computational cost associated with that particular variable. The dependency of a chosen variable with other dependent/independent variables is kept intact.

We have attempted and arrived at various possible ways of simulating this model including the finite element method (FEM), finite difference method solved using BANDJ (Newman, 1968) or DASSL (Brenan et al, 1989), and orthogonal collocation. This paper only describes the most efficient approaches we have found for solving this system of equations.

Reformulation for the Finite Difference Method

This section describes model reformulation used with finite difference method (see Fig. 2 for a flowchart of the procedure).

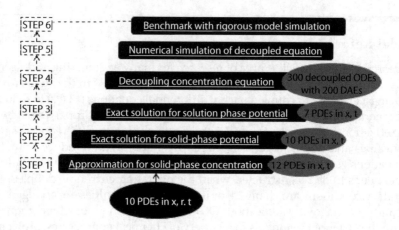

FIGURE 2
Schematic of steps involved in reformulation using the finite difference method.

The governing equation for the solid-phase potential is derived from Ohm's law for the positive and negative electrodes. If j_p was a constant, the governing equation can be solved to obtain a closed-form solution. However, j_p is a nonlinear function of the dependent variables. If finite difference method is applied in the x-direction, the solid-phase potential equation is given by

$$\sigma_{\text{eff,p}} \frac{\Phi1_{i+1} - 2\Phi1_i + \Phi1_{i-1}}{h_1^2} = a_p F j_{pi}, \quad i = 1, \dots, N, \tag{1}$$

where N is the number of interior node points. Eq. (1) can be written in matrix form as

$$A\Phi_1 = j_p + b. \tag{2}$$

The dependent variable Φ_1 can be eliminated (expressed in terms of j_p) by inverting the coefficient matrix in Eq. (2). Typically 50 node points might be needed to obtain a converged solution. Matrix methods can be used to derive and store the inverse matrix and solution *a priori* in the computer to eliminate the need for keeping Φ_1 in the model equations. A general expression can be obtained for the eigenvalues and eigenvectors as a function of N, the number of node points, so that there is no loss of accuracy. For the original equation with the boundary conditions, the exact solution for each eigenvalue is

$$\lambda_i = 2\left(1 - \cos\frac{\pi(2i-1)}{(2N+1)}\right), \quad i = 1, \dots, N. \tag{3}$$

Although similar equations can be derived for eigenvectors as a function of N, a numerical approach can be more efficient. Similar steps can be performed (although the equations are more complicated for other variables) as shown in Fig. 2 and discussed elsewhere (Subramanian, et al, 2007; 2009).

After reformulation, even if 50 node points are used in each region, the resulting 150 decoupled equations for c coupled with 100 decoupled algebraic equations (AEs) for $c_{s,surf}$ or $j_{p/n}$ and 100 decoupled ordinary differential equations (ODEs) for $c_{s,ave,p/n}$ (which occurs only in the electrode) can be solved very efficiently. After this step, the reformulated models can be run in <100 ms, which is required in a hybrid environment that may have supercapacitors with time constants <1 s.

The reformulated models have been compared with models from the literature. Both external/system (voltage-time curve, process variable) and internal variables match exactly for rates less than 2C. The computational cost is much higher for finite difference/volume/element methods that need to perform matrix algebra as a function of N, the number of node points. We could not derive any analytical expressions reported in the literature for banded matrices in mixed domains (cathode/separator/anode) with varying diffusion coefficients in each region. While it is possible that there exists an analytical solution, numerical analysis can be performed to obtain them empirically as a function of N. Our computational experience suggested that the finite difference method was not the best possible approach for the reformulation. The next section presents a more efficient method that implements model reformulation using a polynomial representation.

Reformulation for Polynomial Representation

A more efficient option than finite differences is to write j_p as a sum of polynomials or other pre-specified functions:

$$j_p = \sum_{i=0}^{N} \alpha_{pi} f_i(x). \tag{4}$$

The governing equation for the solid-phase potential can be integrated with respect to x to obtain

$$\Phi_1 = c_1 + c_2 x + \frac{a_p F}{\sigma_{eff,p}} \sum_{i=0}^{N} \alpha_{pi} \int \left(\int f_i(x) dx \right) dx. \tag{5}$$

In the literature, various kinds of polynomials such as Chebyshev polynomials (Varma and Morbidelli, 1999) have been used for model reformulation. If the pre-specified functions have exact double integrals, the solution is analytical with numerical error approaching zero as long as the functions form a complete basis and enough terms are chosen in Eq. (4). If simple polynomials are chosen, then

$$j_p = \sum_{i=0}^{N} \alpha_{pi} x^i \tag{6}$$

and

$$\Phi_1 = c_1 + c_2 x + \frac{a_p F}{\sigma_{eff,p}} \sum_{i=0}^{N} \alpha_{pi} \frac{x^{i+2}}{(i+1)(i+2)}. \tag{7}$$

The above integration constants are solved from the boundary conditions. Using a polynomial for j_p is more advantageous than using the finite difference, finite element, or finite volume methods for reformulating Φ_1 as double integration to get Φ_1 is less computationally expensive than inverting matrices. Using one of the above reformulation approaches, an analytical solution can be derived for the solid-phase potential distribution in each porous electrode. This reformulation enables a closed-form solution for the solid-phase potential distribution in each electrode as a function of other dependent variables without compromising on accuracy and without losing any physics of the battery system. Moreover, this reformulation technique reduces one PDE for the solid-phase potential to one algebraic equation. At this stage, the original model for the solid-phase potential is reduced to

$$\Phi_1(x) = 4.2 + \frac{a_p F}{\sigma_{eff,p}} \sum_{i=0}^{N} \frac{\alpha_{pi}}{i+1} \left(\frac{x^{i+2}}{i+2} - l_p^{i+1} \right). \tag{8}$$

The governing equation for the liquid-phase potential is given by the modified Ohm's law. If $\kappa_{eff,p}$ is a constant, the governing equation for electrolyte potential can be solved analytically as

$$\Phi_2(x) = k_1 - I \int \frac{1}{\kappa_{eff,p}} dx - \sigma_{eff,p} \int \frac{1}{\kappa_{eff,p}} \frac{\partial \Phi_1}{\partial x} dx$$

$$+ \frac{2RT}{F} (1 - t_+) \ln c. \tag{9}$$

$\kappa_{eff,p}$ is a nonlinear function of the dependent variable (electrolyte concentration) for various chemistries. If the function governing the variation of $1/\kappa_{eff,p}$ with respect to the other dependent variables has an exact integral, then this equation has an analytical solution. If not, simple polynomials are chosen as $\frac{1}{\kappa_{eff,p}} = \sum_{i=0}^{N} \zeta_{pi} x^i$. Then

$$\Phi_2(x) = k_1 - I \sum_{i=0}^{N} \zeta_{pi} \frac{x^{i+1}}{i+1} - \sigma_{eff,p} \Upsilon + \frac{2RT}{F} (1 - t_+) \ln c \tag{10}$$

where Υ is a product of two summations resulting from integration.

The Galerkin-collocation (GC) type weighted-average method is used to solve for the constants. For example, each constant ζ_{pi} is obtained by minimizing the residue of the governing equation with a weighting function given by the coefficient of the particular constant as

$$\frac{1}{l_p}\int_{x=0}^{l_p} w_1 Ge(\Phi_2)\Big|_{cathode}\,dx + \frac{1}{l_s}\int_{l_p}^{l_p+l_s} w_2 Ge(\Phi_2)\Big|_{separator}\,dx$$

$$+\frac{1}{l_n}\int_{l_p+l_s}^{L} w_3 Ge(\Phi_2)\Big|_{anode}\,dx = 0, \tag{11}$$

where w_1, w_2, and w_3 are the weight functions and $Ge(\Phi_2)$ denotes governing equation for Φ_2. Six of the constants in the polynomials are obtained from the boundary conditions at $x = 0$, $x = l_p$, $x = l_p + l_s$, and $x = L$. At the electrode/separator interfaces, both the electrolyte potential and its fluxes are continuous. Similarly, polynomial representations are used for the other dependent variables arising from the solid-phase concentration equations and pore-wall flux expression. The GC technique is more stable and provides accurate solutions with lesser number of terms in the polynomial representation compared to the OC technique.

The dependent variable for the electrolyte in each region is approximated with polynomial expressions and substituted in the governing equation for electrolyte concentration to get separate equations in each region and the constants are found using the weighted residual:

$$\frac{1}{l_p}\int_{x=0}^{l_p} w_1 Ge(c)\Big|_{cathode}\,dx + \frac{1}{l_s}\int_{l_p}^{l_p+l_s} w_2 Ge(c)\Big|_{separator}\,dx$$

$$+\frac{1}{l_n}\int_{l_p+l_s}^{L} w_3 Ge(c)\Big|_{anode}\,dx = 0. \tag{12}$$

Both for the electrolyte potential and concentration, six of the constants in the polynomials are found from boundary conditions. In addition, volume averaging is performed to the original set of PDEs. The electrolyte concentration can be volume-averaged over the respective region as:

$$\varepsilon_p\int_{x=0}^{l_p}\frac{\partial c}{\partial t}\,dx + \varepsilon_s\int_{x=l_p}^{l_p+l_s}\frac{\partial c}{\partial t}\,dx + \varepsilon_n\int_{x=l_p+l_s}^{L}\frac{\partial c}{\partial t}\,dx$$

$$= D_{eff,p}\left(\frac{\partial c}{\partial x}\Big|_{x=lp} - \frac{\partial c}{\partial x}\Big|_{x=0}\right) + a_p(1-t_+)\int_{x=0}^{l_p} j_p\,dx$$

$$+ D_{eff,s}\left(\frac{\partial c}{\partial x}\Big|_{x=lp+ls} - \frac{\partial c}{\partial x}\Big|_{x=lp}\right) + \tag{13}$$

$$+ D_{eff,n}\left(\frac{\partial c}{\partial x}\Big|_{x=L} - \frac{\partial c}{\partial x}\Big|_{x=lp+ls}\right) + a_n(1-t_+)\int_{x=lp+ls}^{L} j_n\,dx.$$

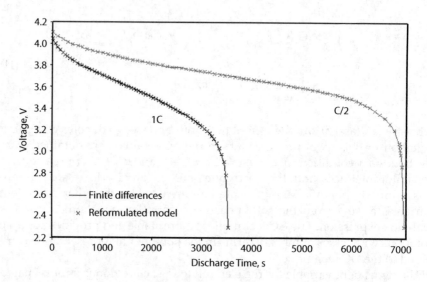

FIGURE 3
Discharge curves for 1C and 0.5C rate.

This equation can be simplified and integrated to obtain

$$C_{ave}^{Total} = \frac{\varepsilon_p C_{ave}^{Cathode} + \varepsilon_s C_{ave}^{Separator} + \varepsilon_n C_{ave}^{Anode}}{\varepsilon_p l_p + \varepsilon_s l_s + \varepsilon_n l_n} = 1000. \tag{14}$$

This equation is true for any chemistry and can also be derived from the overall mass balance of the cell. This facilitates a quicker convergence of the concentration profiles in terms of polynomials. If this condition is not used, more number of terms may be needed in the polynomial representation to maintain numerical stability.

The discharge potential is the measured variable (see Fig. 3). The discharge curves are shown for 1C (30 A/m²) and 0.5C rates of galvanostatic discharge. It can be seen that, for these rates of discharge, the reformulated model compares very well with the original numerical model. The merit of this approach is evident when comparing the number of governing equations that are solved. The reformulated model specifies 47 DAEs as opposed to 4800 DAEs for the original model. The reformulated model uses a maximum of 47 specified equations for a converged solution that compares very well even with the original finite difference code. The reformulated model predicts the intrinsic non-measurable (state) variables accurately as shown in Fig. 4, with the CPU time reduced by two orders of magnitude. The reformulated model takes only 15–50 ms to predict a discharge curve in FORTRAN environment whereas the original model can take up to a few seconds to minutes depending on the solver, environment, and the computer.

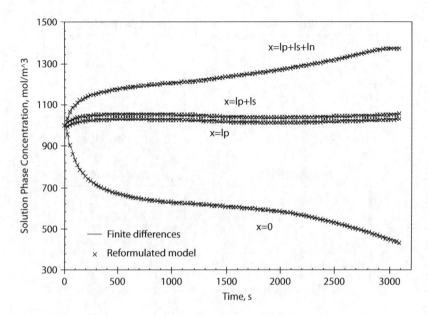

FIGURE 4
Solution phase concentration at different times for 1C rate at different interfaces.

Parameter Estimation for Capacity Fade Prediction

The protocol that is commonly used in cycling a lithium-ion battery consists of first charging at constant current then charging at constant potential until the potential reaches a uniform value across the intercalating particles, then the protocol follows discharging the battery at constant current or potential. The battery loses its capacity to hold and deliver the energy when the number of cycle increases. These losses are mainly due to the variations in the transport and kinetic parameters caused by the reduced pore volume in the porous electrodes.

In addition, researchers have observed a major loss in the discharge time period over which one can utilize the battery (Ramadass et al, 2004). This loss is an especially important prediction for batteries installed in remote applications like satellites, but is also important when using battery models to optimize operations. To enable such a level of model sophistication, a model that updates transport and kinetic parameters as a function of cycle number is developed. This information can be used to provide guidance to the development of first-principles models for capacity fade that introduce additional reactions (Ramadass et al, 2004), models for SEI layer growth, etc.

Unknown parameters that were estimated are the solid-phase diffusion coefficient D_{sn} and the reaction rate constant k_n in the negative electrode.

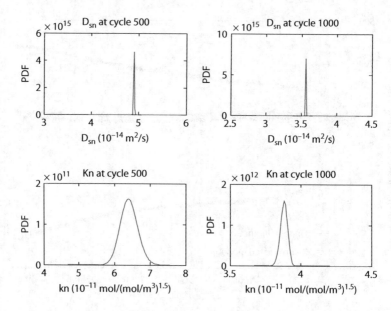

FIGURE 5
Bayesian estimation for the solid-phase diffusion coefficient D_{sn} and the reaction rate constant k_n for the negative electrode at cycle 500 and 1000.

Bayesian estimation was used to estimate the model parameters from experimental data obtained using lithium-ion batteries from Quallion® LLC. Uncertainties in the model parameter estimates were quantified by three methods: (1) estimation of hyperellipsoidal regions using linearized statistics, (2) estimation of nonlinear uncertainty regions using F-statistics, and (3) estimation of probability distributions by application of Markov Chain Monte Carlo (MCMC) simulation (Hermanto et al, 2008; Tierney, 1994). Methods 1 and 2, which are the most commonly applied, gave highly biased probability distributions in this application, whereas there is no statistical bias in method 3. Other advantages of method 3 include its explicit consideration of constraints and arbitrary non-Gaussian distributions for prior knowledge on the parameters, and that it exactly handles the full nonlinearity in the model equations. Method 3 requires many more simulation runs than methods 1 and 2, which provides further motivation for the derivation of the low-order reformulated model. Figure 5 shows the probability density functions of two model parameters obtained by MCMC simulation, with statistically significant reductions in both the solid-phase diffusion coefficient D_{sn} and the reaction rate constant k_n for the negative electrode. These model parameters reduced monotonically with cycle number, which is consistent with a monotonic decrease in the pore volume in the negative electrode.

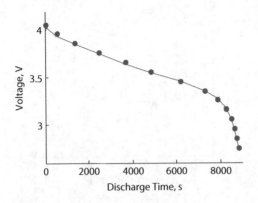

FIGURE 6

Comparison of the experimental voltage-discharge curve with the model prediction with estimated parameters for cycle 500. Each red dot is a data point, the blue line is the model prediction, and the 95% predictive intervals were computed based on the parametric uncertainties reported in Fig. 5. Similar quality fits and prediction intervals occurred for the other cycles.

The effect of the parameter uncertainties on the accuracy of the predictions of the lithium-ion battery model was then quantified by polynomial chaos expansions (Wiener, 1938). This approach avoids the high computational cost associated with applying the Monte Carlo method or parameter gridding to the simulation code by first computing a series expansion for the simulation model, followed by application of robustness analysis to the series expansion. The very low computational cost of the series expansion enables the application of the Monte Carlo method, gridding the parameter space, or the application of norm-based analytical methods (Ma and Braatz, 2001; Nagy and Braatz, 2004; 2007). 95% prediction intervals computed for each cycle provided confidence that the model can be used for predictions and design (see Fig. 6).

A battery is no longer useful once its capacity becomes too low. Even for the same manufacturing line, the lifetime varies widely from battery to battery. It would be useful for a microprocessor to provide an estimate of the number of useful cycles remaining in the battery. We explored the use of the model to predict the remaining battery life based on voltage-discharge curves measured in past cycles. To characterize the degradation in the model parameters, a power law was fit to the estimated parameter values from cycles 25 to 500 (see Figs. 7 and 8). Implicitly assuming that the changes in the parameter values are the result of the same mechanism in later cycles, the parameter values for the subsequent cycles were predicted using the power law expressions. The voltage-discharge curve predicted by this model was in very good agreement with the experimental data at cycle 1000 (see Fig. 9), indicating that the model was suitable for prediction of capacity fade.

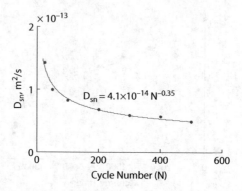

FIGURE 7
Power law fit for the solid-phase diffusion coefficient in the negative electrode based on the estimated parameter value from the first 500 cycles.

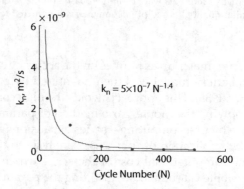

FIGURE 8
Power law fit for the reaction rate constant for the negative electrode based on the estimated parameter value from the first 500 cycles.

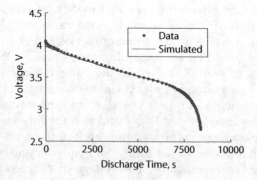

FIGURE 9
Comparison of the experimental voltage-discharge curve with the model prediction using parameter values calculated from the power law fits in Figs. 7 and 8.

The computation time required for parameter estimation using the reformulated model was between 100 and 300 ms. Compare this to the minutes to hours required for the standard finite difference method directly applied to the battery model.

Optimal Design of Lithium-ion Batteries

Prof. Newman and his group have applied macroscopic models to optimize the electrode thickness or porosity (Srinivasan and Newman, 2004). These studies have been performed by comparing the Ragone plots for different design parameters. A single curve in a Ragone plot may involve hundreds of simulations wherein the applied current is varied over a wide range of magnitude. Ragone plots for different configurations are obtained by changing the design parameters (e.g., thickness) one at a time, and by keeping the other parameters constant. This process of generating a Ragone plot is quite tedious and typically Ragone curves reported in the literature are not smooth because of computational constraints.

To our knowledge, the literature does not report the application of such first-principles models to the global optimization of multiple battery design parameters. Also, batteries are typically designed only to optimize the performance at cycle one of the battery, whereas in practice most of the battery's operation occurs under significantly degraded conditions. The reformulated model is sufficiently computationally efficient to enable the simultaneous optimal design of multiple parameters over any number of cycles by including the model for capacity fade (Figs. 7 and 8). Further, the model can be used to quantify the effects of model uncertainties and variations in the design parameters on the battery performance. As an example of such robustness analysis, the utilization averaged over 1000 cycles are reported in Fig. 10 for the battery design obtained by (1) simultaneous optimization of the applied current density (I) and thicknesses of the separator and the two electrodes (l_s, l_n, l_p) for cycle 1, and (2) variations in these four design parameters. The battery design optimized for cycle 1 does not maximize the cycle-averaged utilization.

We are also investigating the optimal design of *distributions* of properties, which cannot be reasonably handled by one-at-a-time optimization. In particular, we are optimizing porosity distributions and particle size distributions across the electrode, in addition to the standard design parameters considered in Fig. 10. Figure 11 shows some sample results in which the linear porosity distribution is optimized across the cathode thickness to maximize the utilization of the electrodes. More than 3% improvement in utilization occurred by using spatially-varying porosity within the positive electrode. The oral presentation will include additional product design results.

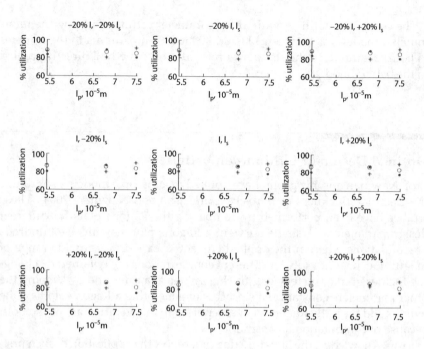

FIGURE 10
Utilization averaged over cycle 25, 500, and 1000 for a 3-level 4-factor factorial design. The title of each plot indicates the deviation in the design variables I and l_s from their values optimized for cycle 1. Circles, stars, and dots are for the l_n value optimized for cycle 1 and ±20% of that value, respectively.

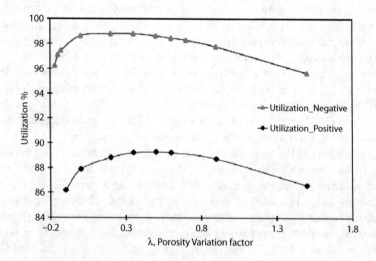

FIGURE 11
Utilization of positive and negative electrodes with linearly varying porosity as $\varepsilon_p = \varepsilon_{p0}(1 + 2\lambda(x/l_p))/(1 + \lambda)$.

Conclusions

This paper employs a systems engineering approach for the modeling and design of lithium-ion batteries. An efficient approach is presented to simulate the discharge behavior of lithium-ion battery models for galvanostatic boundary conditions with improved computational efficiency using a polynomial representation. The approach is similar to Galerkin-collocation based on weighted residuals coupled with analytic solution of the algebraic equations and volume averaging for the variables of interest. The DAE nature and structure of the model is exploited to solve for some of the dependent variables analytically. The model reformulation method is demonstrated at low-to-moderate rates of discharge. The method should be valid and applicable to other engineering models with a DAE nature such as fuel cells and monolith reactors. Also, for the use of dynamic model in feedback control algorithms with microchips, computer memory (RAM) is likely to be a concern. The proposed reduction in the number of DAEs and states would also enable the implementation of more advanced robust controller algorithms.

The reformulated model was applied with robust Bayesian estimation to predict future capacity fade in lithium-ion batteries. Predictions were demonstrated for the first 1000 cycles of a Quallion® LLC 250 mAh cell. The nonlinear nature of the models motivated the use of the MCMC approach to quantify the uncertainties in the parameter estimates. The reformulated model facilitates such a rigorous approach, as the MCMC approach requires many simulation runs to construct accurate distributions for the parameter estimates. The reformulated model should facilitate dynamic optimization for better operational strategies for improved energy efficiency and design for better electrode materials for improved performance.

Acknowledgments

The authors are thankful for the partial financial support of this work by the National Science Foundation (CBET – 0828002 and 0828123), the U.S. Army Communications-Electronics Research, Development, and Engineering Center (CERDEC) under contract number W909MY-06-C-0040, and the Oronzio de Nora Industrial Electro-chemistry Postdoctoral Fellowship of The Electrochemical Society.

References

Beck, J. V., Arnold, K. J. (1977). *Parameter Estimation in Engineering and Science*. Wiley, New York.

Benner, P. http://www.math.tu-berlin.de/~baur/D1.html (referred on December 31, 2008).

Botte, G. G., Subramanian, V. R., White, R. E. (2000). Mathematical Modeling of Secondary Lithium Batteries. Electrochim. *Acta*, 45, 2595.

Braatz, R. D., Alkire, R. C., Seebauer, E. G., Rusli, E., Gunawan, R., Drews, T. O., He, Y. (2006). Perspectives on the Design and Control of Multiscale Systems. *J. of Process Control*, 16, 193.

Brenan, K. E., Campbell, S. L., Petzold, L. R. (1989). *Numerical Solution of Initial-value Problems in Differential–Algebraic Equations*. North-Holland, New York.

Cai, L., White, R. E. (2008). Reduction of Model Order Based on Proper Orthogonal Decomposition for Lithium Ion Battery Simulations. *J. Electrochem., Soc.*, article in press.

Darling, R., Newman, J. (1999). Dynamic Monte Carlo Simulations of Diffusion in $Li_yMn_2O_4$. *J. Electrochem. Soc.*, 146, 3765.

Doyle, M., Fuller, T. F., Newman, J. (1993). Modeling the Galvanostatic Charge and Discharge of the Lithium/Polymer/Insertion Cell. *J. Electrochem. Soc.*, 140, 1526.

Hermanto, M. W., Kee, N. C., Tan, R. B. H., Chiu, M.-S., Braatz, R. D. (2008). Robust Bayesian Estimation of Kinetics for the Polymorphic Transformation of L-glutamic Acid Crystals, *AIChE J.*, 54, 3248.

Ma, D. L., Braatz, R. D. (2001). Worst-case Analysis of Finite-time Control Policies. *IEEE Trans. on Control Syst. Tech.*, 9, 766.

Nagy, Z. K., Braatz, R. D. (2004). Open-loop and Closed-loop Robust Optimal Control of Batch Processes using Distributional and Worst-case Analysis. *J. of Process Control*, 14, 411.

Nagy, Z. K., Braatz, R. D. (2007). Distributional Uncertainty Analysis Using Power Series and Polynomial Chaos Expansions. *J. Process Control*, 17, 229.

Newman, J. (1968). Numerical Solution of Coupled, Ordinary Differential Equations. *Ind. Eng. Chem. Fund.*, 7, 514.

Newman, J., Thomas, K. E., Hafezi, H., Wheeler, D. R. (2003). Modeling of Lithium-ion Batteries. *J. Electrochem. Soc.*, 150, A176.

Osaka, T., Nakade, S., Rajamäki, M., Momma, T. (2003). Influence of Capacity Fading on Commercial Lithium-ion Battery Impedance. *J. Power Sources*, 119–121, 929.

Plett, G. L. (2004). Extended Kalman Filtering for Battery Management Systems of LiPB-based HEV Battery Packs, Part 1. Background. *J. Power Sources*, 134, 252.

Ramadass, P., Gomadam, P. M., White, R. E., Popov, B. N. (2004). Development of First Principles Capacity Fade Model for Li-ion Cells. *J. Electrochem. Soc.*, 151, A196.

Smith, K., Wang, C. Y. (2006). Solid State Diffusion Limitations on Pulse Operation of a Lithium Ion Cell for Hybrid Electric Vehicles. *J. Power Sources*, 161, 628.

Subramanian, V. R., Boovaragavan, V., Ramadesigan, V., Arabandi, M. (2009). Mathematical Model Reformulation for Lithium-ion Battery Simulations— Galvanostatic Boundary Conditions. *J. Electrochem., Soc.*, 156, A260.

Subramanian, V. R., Diwakar, V. D., Tapriyal, D. (2005). Efficient Macro-Micro Scale Coupled Modeling of Batteries. *J. Electrochem. Soc.*, 152, A2002.

Subramanian, V. R., Boovaragavan, V., Diwakar, V. D. (2007). Towards Real-time (Milliseconds) Simulation of Physics Based Lithium-ion Battery Models. *Electrochem. Solid-State Lett.*, 10, A225.

Tierney, L. (1994). Markov Chains for Exploring Posterior Distributions. *Ann. Stat.*, 22, 1701.

Varma, A., Morbidelli, M. (1999). *Mathematical Methods in Chemical Engineering.* Oxford University Press, London.

Srinivasan, V, Newman, J, (2004). Design and Optimization of a Natural Graphite Iron Phosphate Lithium-Ion Cell. *J. Electrochem. Soc.*, 151, A1530.

Wang, C. Y., Gu, W. B., Liaw, B. Y. (1998). Micro-macroscopic Coupled Modeling of Batteries and Fuel Cells. Part I: Model Development. *J. Electrochem. Soc.*, 145, 3407.

Wei, J., (1987). *Advances in Chemical Engineering*, Vol. 13, Academic Press.

Wiener, N. (1938). The Homogeneous Chaos. *Am. J. Math.*, 60, 897.

95

Computer-aided Graphical Tools for Synthesizing Complex Column Configurations

Daniel A. Beneke, Ronald Abbas, Michael Vrey*, Brendon Hausberger, Simon Holland, Diane Hildebrandt and David Glasser
Centre of Material and Process Synthesis (COMPS), University of the Witwatersrand, Johannesburg, South Africa

CONTENTS

ABSTRACT There has recently been a renewed interest in the design of distillation processes due to the development of Column Profile Maps (CPMs). Using CPMs one is able to change topology within the composition space and hence many separations that have been thought of as difficult or unviable, can now be achieved. The CPM technique has also been proven to be extremely useful as a design tool as any column configuration, irrespective of complexity, can be modeled and graphically understood. This paper aims to summarize the most important and interesting results and applications obtained using the CPM technique. It shows how CPMs may be used to synthesize complex columns like a Petlyuk or Kaibel column, as well as showing how new sharp split separations can be devised.

KEYWORDS *Column profile maps, distillation design, sharp splits*

* Corresponding author email: diane.hildebrandt@wits.ac.za

Introduction

In modern chemical industries, the task of separation is a very energy consuming process, where distillation is used for about 95% of liquid separations. The energy usage from this process accounts for around 3% of the world energy consumption, as estimated by Hewitt et al. (1999).

Graphical methods for designing distillation schemes have been popular over the years. Residue Curve Maps (RCMs) are often used as a graphical method for designing multi component distillation systems. RCMs are basically a range of trajectories that track the liquid compositions of the chemical species over time in a simple distillation operation. RCMs can tell one much about the feasibility of separation and the nature of singular points, such as azeotropes and pure component vertices.

However, the RCM technique has its limitations in that it only gives information at infinite reflux, quite an impractical condition for the design engineer. Recently, in a series of papers by Tapp et al. (2004) and Holland et al. (2004 a, b) a new theory was explored in distillation: Column Profile Maps (CPMs). CPMs were derived from an adaption of ODEs proposed by Van Dongen and Doherty (1985), which take into account the net molar flows and reflux ratios in a column section. CPMs were shown to display the same topological behavior as RCMs, as well as being an extremely useful tool in distillation design by allowing the designer to set reflux ratios and net molar flows to suit the specifications of the separation.

Column Profile Maps

A CPM describes the behavior of a multicomponent system by setting appropriate parameters such as the net molar flow and the reflux ratio. The first step in constructing a CPM is to define a Column Section, which according to the definition of Tapp et al. (2004) is "a length of column between points of addition or removal of material and/or energy". A steady state material balance over a Column Section accompanied with a Taylor expansion yields:

$$\frac{dX}{dn} = \left(\frac{1}{R_\Delta} + 1\right)\left(X - Y(X)\right) + \left(\frac{1}{R_\Delta}\right)\left(X_\Delta - X\right) \tag{1}$$

$$\text{Where } X_\Delta = \left(\frac{VY^T - LX^T}{V - L}\right) \text{ and } R_\Delta = \frac{L}{V - L} = L/\Delta$$

Equation (1) is known as the Difference Point Equation (DPE). X_Δ. can be thought of as a pseudo composition vector and is valid anywhere in the composition space, even in the space outside the Mass Balance Triangle (MBT). It

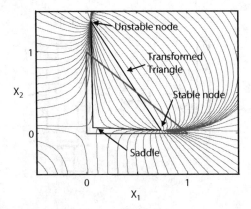

FIGURE 1
CPM for $X_\Delta = [0.3, -0.2]$ and $R_\Delta = 9$.

is however subject to the constraint that the sum of the components of X_Δ be 1. X_Δ need only be a real composition in columns sections that are terminated by a condenser or reboiler. Notice that the DPE is not bound by physically relevant initial conditions, thus one is able to perform the integration outside of the composition space. Furthermore, notice that the DPE reduces to the Residue Curve Equation at infinite reflux. Thus, for an arbitrary choice of X_Δ and R_Δ one can now begin to construct a CPM for an ideal system*, as in Figure 1.

Notice in Figure 1 how stationary points (nodes) have been shifted in the composition space, resulting in completely different profiles within the blue Mass Balance Triangle (MBT). These stationary points can be determined by algebraically solving the DPE = 0. If we connect these shifted nodes with straight lines we can see a Transformed Triangle (TT) being formed. In theory, one can now move these nodes in composition space by simply fixing the aforementioned parameters. This could lead to many new and exciting designs that have been previously thought to be unviable.

Applications

Petlyuk Design

Using the CPM design methodology, one is able to break down any column configuration into simpler Column Sections, and from there design the entire column according to the separation specifications. The famed Petlyuk

---

* *In this paper, an ideal system refers to the assumption of constant relative volatilities. Unless it is otherwise stated, $\alpha_1 = 3$, $\alpha_2 = 1$, and $\alpha_3 = 1.5$, which means that x_1 is the low boiler, x_3 is the intermediate boiler and x_3 is the high boiler.*

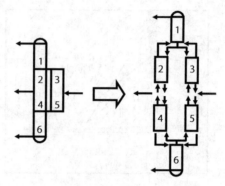

FIGURE 2

Column section breakdown for the petlyuk column.

Column, which offers significant savings in energy, can also be broken down into Column Sections (CS) as shown in Figure 2. For *simplicity*, we shall look at a the case where the Petlyuk operates at overall infinite reflux, but with CS 2-5 operating at a finite reflux, i.e. a column that draws infinitesimal product flows, but does not necessarily operate with L = V in sections 2, 3, 4 and 5. For this example, we shall set an intermediate product specification of 90% and achieving this specification will be the primary concern when deciding on a X_Δ. CS 1-6 will simply operate on Residue Curves.

It can be shown mathematically that the constraints placed on this system leads to:

- CS 2 and 4 have identical TTs
- CS 3 and 5 have identical TTs
- CS 2 and 4 and CS 3 and 5 operate on the same Difference Point, with equal magnitude but opposite signs for $R\Delta$.

The criteria for feasible column profiles is that the liquid profiles intersect twice. If one then superimposes the 2 CPMs for the coupled sections, for an appropriate selection of X_Δ and R_Δ, it can be seen that the feasible region intersects with the product specification. Hence a feasible design has been found, as shown in Figure 3.

Modeling Sharp Splits with CPMs

Invariably, the aim of any separation process is to achieve essentially pure products. Thus the sharp split constraint presents an interesting and relevant case study.

Tapp et al (2004) have shown that there are 7 regions of X_Δ placement which result in unique Pinch Point curves (see Figure 4). The boundaries of these regions correspond to the extended axes of the MBT. In terms of

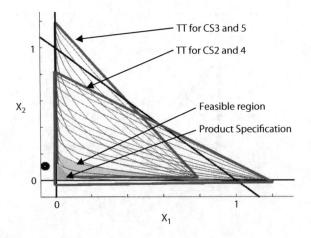

FIGURE 3
Superimposed transformed triangles for coupled column system.

CPMs, a sharp split effectively means that X_Δ is placed on the boundary of these regions. A sharp split thus displays Pinch Point Curve behavior of 2 regions.

It is interesting to note that the nodes for sharp splits are shifted in composition space in a different manner to non-sharp splits. Pinch point curves for sharp splits are linear, and appear to intersect at a point. In fact, the curves don't intersect, but merely meet at a point. The point at which this occurs is termed the "bumping point", because at this point nodes "bump" each other from their positions. For example, a saddle could be bumped from its position and be replaced by a stable node and thereby altering the topology

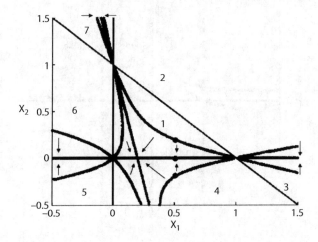

FIGURE 4
Pinch point curve behavior for different placement of X_Δ.

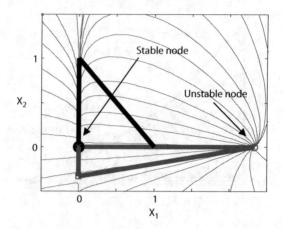

FIGURE 5
A stable node fixed on the intermediate boiler.

within the MBT drastically. This result is very useful, as one could now theoretically place a node almost anywhere in composition space to suit the separation by simply choosing R_Δ and X_Δ appropriately.

An immediate application of this is fixing X_Δ to the intermediate boiler vertex. By making use of the "node bumping" phenomenon, it is now possible to fix a stable node or an unstable node to the intermediate boiling vertex, as shown in Figure 5. This result suggests that the intermediate boiler can be completely removed in a single stripping section, and hence making the removal of the intermediate boiler significantly easier. Similarly, there

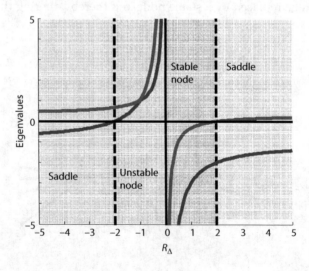

FIGURE 6
Operating regions for $X_\Delta = [0;0;1]$.

are certain choices for R_Δ and X_Δ which can fix a saddle to the high or low boiler vertex, and hence making separation much more difficult for these components.

It is of special interest to determine when and how a certain node can be fixed in composition space. For the special case where X_Δ is placed on one of the 3 pure component vertices, a node is also fixed to the same vertex. So by knowing the position of a stationary point and X_Δ, we can trace the nature of the node by varying R_Δ. For example, Figure 6 shows which values of R_Δ correspond to a specific node on the intermediate boiler vertex. The nature of the nodes are defined by the eigenvalues of the Jacobian when the DPE = 0.

Sharp Split Kaibel Column Design

This work considers the implementation of a Kaibel column, (i.e. a fully-thermally coupled column with an adiabatic wall dividing the column into two equal halves for the production of four product streams). The Kaibel Column allows for a feed mixture of four or more components from which it produces a distillate, bottoms and two product side streams. Compared to the conventional 3 column direct split sequence, the Kaibel column can be built in a single shell, making it an attractive alternative in terms of capital cost savings along with its counterpart; the Petlyuk Column. Further, the reduction in the number of reboilers and condensers' required leads to improved operating costs.

In this section of work we demonstrate the use of CPMs for the comprehensive analysis and design of Kaibel columns by applying the CPM technique for a system at sharp-split conditions. From the results of the topological analysis, it is shown that, for set product composition specifications, when using an ideal system (constant relative volatilities), there is only one set of feasible operating parameters.

The Kaibel Column Section breakdown is similar to the Petlyuk in Figure 1, but with an additional CS between CS 2 and 4, as two products are removed between these CSs. It can be shown that for the Kaibel column, CS 2 and 4's X_Δ's are placed on the intermediates pure components. A mass balance shows that the net flow through the connecting CS of the side draws is zero. As a result, this mass balance can only be satisfied completely if the difference point for component 2 (B) in this same CS ($X_{\Delta 7,\,2}$) is infinitely big. Due to the fact that this CS has a net zero flow, does not mean that the profile produced will be a residue curve, but by substituting zero net flow into the DPE the differential becomes an infinite reflux expression.

Figure 7 is the only correct CS mass balance layout in the quaternary system mass balance space. As can be seen from Figure 7 only one solution is possible as this is the only feasible mass balance that exists.

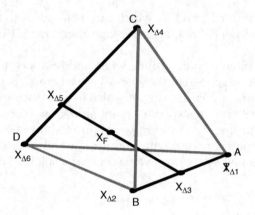

FIGURE 7
Mass balance lines between intersecting CSs.

We can represent the results on a phi space diagram as shown by Figure 8. The single zero net flow line for CS 7 of the Kaibel arrangement is the only operating line that will produce feasible results for a double shell-single reboiler system (Red line). If we shift over to a Kaibel Dividing Wall Column (DWC) we operate at a single point (black dot in Figure 8), as one cannot throttle the vapor split at the bottom of the column. This shows that there is no movement allowed to change the system by changing the liquid and vapor splits. As can be seen from Figure 8, the Petlyuk feasible region in the Phi space is much larger and thus much more operable than the single operating line for the Kaibel.

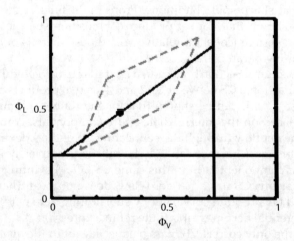

FIGURE 8
Phi space diagram for the petlyuk and kaibel.

Conclusion

In this paper it has been shown that CPMs have tremendous potential in designing and understanding simple and complex distillation systems. Nodes can almost be placed at will in composition space to suit the requirements of the separation, so much so that it is possible to place stable or unstable nodes on the intermediate boiler's vertex. The CPM technique offers a better understanding of the interaction between parameters due to its graphical nature. Furthermore, it has been shown that CPMs are extremely useful in designing complex distillation systems such as the Petlyuk or Kaibel column, and hence more efficient and creative designs can be thought of.

Acknowledgements

The authors would like to give special thanks to the NRF and the SA Research Chair Initiative.

References

Hewitt G, Quarini J, Morell M. More efficient distillation. *Chem Eng.* 1999;Oct. 21

Holland, S.T., Tapp, M.; Hildebrandt, D. and Glasser, D, Column Profile Maps. 2. Singular Points and Phase Diagram Behavior in Ideal and Nonideal Systems, *Ind. Eng. Chem. Res.*, 43(14), p3590–3603, (2004 a).

Holland, S.T., Tapp, M.; Hildebrandt, D. and Glasser, D. and Hausberger, B., Novel Separation System Design Using "Moving Triangles", *Comp. and Chem. Eng.*, 29, p181–189, (2004 b).

Tapp, M., Holland, S.T.; Hildebrandt, D., and Glasser, D., Column Profile Maps. 1. Derivation and Interpretation, *Ind. Eng. Chem. Res.*, 43(2), p364–374, (2004).

Van Dongen, D. B.; Doherty, M. F. Design and Synthesis of Homogeneous Azeotropic Distillations. 1. Problem Formulation for a Single Column. *Ind. Eng. Chem. Fundam.*, 24, p.454, 1985.

96

Integrating Product Portfolio Design and Supply Chain Design for the Forest Biorefinery

Behrang Mansoornejad, Virginie Chambost and Paul Stuart*

NSERC Environmental Design Engineering Chair in Process Integration
Department of Chemical Engineering, École Polytechnique—Montréal
Montréal H3C 3A7, Canada

CONTENTS

ABSTRACT Supply chain (SC) design involves making strategic decisions for the long term, e.g. location and capacity of facilities, flow of material between SC nodes, as well as choosing suppliers and markets. The forest biorefinery is emerging as a promising opportunity for improving the business model of forest product companies, however introduces significant challenges in terms of mitigating technology, economic and financial risks— each of which must be systematically addressed, including in SC design. In

---

* To whom all correspondence should be addressed

this regard, product portfolio definition and technology selection are two important decisions that have rarely been considered in a systematic SC evaluation. This paper presents such a methodology, in which product/process portfolio design and SC design are linked in order to build a design decision making framework. According to this methodology, "manufacturing flexibility" links product/process portfolio design to SC design, through a margins-centric SC operating policy. Techno-economic studies and advanced cost modeling along with scenario generation for price change representing market volatility are employed in the methodology.

KEYWORDS   *Forest biorefinery, supply chain design, product design*

Introduction

Pulp and Paper (P&P) companies in Canada are facing a stalemate situation (Stuart, 2006). Their business has been endangered by global low-cost competitors; therefore they are encountering declining markets and over capacity. In order to remain low-cost producers, they have cut R&D activities and spent minimum capital to modernize their mills and thus they are dealing with the lack of knowledge of product quality requirement and supply chain (SC) practices. Enterprise Transformation (ET) has been proposed by experts as a solution for rescuing Canadian P&P industry from its current situation (Chambost et al., 2008a). ET implies evolving aggressive corporate-wide initiatives designed to impact the strategies, structures and human system of the corporation – as well as to create more sustainable and profitable organizations. ET must be performed in two broad separate ways referenced as "inside-out" and "outside-in". Inside-out transformation is when the current mission/vision of the company is kept unchanged and the company is made-over in terms of its processes and manufacturing culture. Outside-in transformation involves changes in current mission/vision and the core business of the company by producing new products and providing new services. What helps P&P companies to transform their enterprise and to rescue their industry is the Forest Biorefinery (FBR). P&P companies have some competitive advantages, e.g. access to biomass and engineering know-how, established infrastructure close to forest biomass and established SC for wood, pulp and paper. Hence, taking advantage of these privileges, P&P companies can transform their enterprise by implementing the FBR, because, in order to implement the FBR, companies must produce bioproducts besides pulp and paper, which implies outside-in transformation. On the other hand, FBR implementation will change company's core business; therefore they need new management practices and manufacturing culture, which address inside-out transformation.

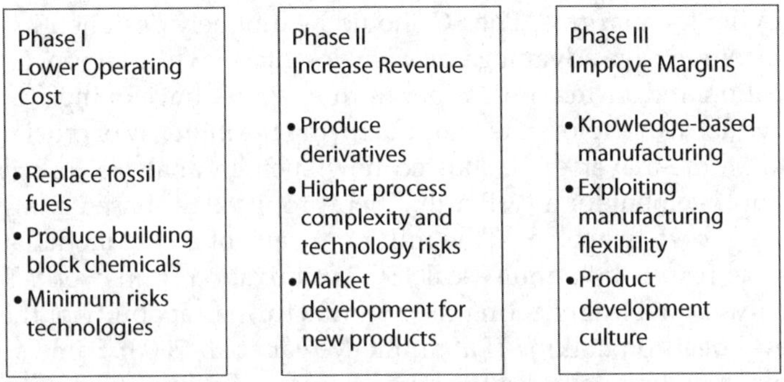

Phase I
Lower Operating
Cost
 	
• Replace fossil
  fuels
• Produce building
  block chemicals
• Minimum risks
  technologies
	Phase II
	Increase Revenue
	
	• Produce
	  derivatives
	• Higher process
	  complexity and
	  technology risks
	• Market
	  development for
	  new products
		Phase III
		Improve Margins
		
		• Knowledge-based
		  manufacturing
		• Exploiting
		  manufacturing
		  flexibility
		• Product
		  development
		  culture

FIGURE 1
Strategic implementation of the biorefinery by a P&P company.

FBR implementation can be performed based on a phased approach which can be considered by P&P companies, Figure 1 (Chambost et al., 2008 a).

In phase I, companies must lower their operating costs by producing substitute fuel products for fossil fuels such as bunker C or natural gas, or by employing new technologies with minimum risks. Such projects must compete internally for capital due to lack of capital spending budget in P&P companies. In phase II companies must increase their revenue by producing added value products. This phase includes change in the core business, which in turn implies outside-in transformation. Considering that phase II requires capital, new technology and new product delivery requirements, partnership is of crucial importance in this step. Therefore the main challenge at this phase is to select the most sustainable product/process portfolio and partner(s). In phase III companies must focus on improving margins by exploiting manufacturing flexibility through "knowledge-based manufacturing". This latter term implies advanced SC optimization techniques which identify the trade-offs between supply, demand and manufacturing capability via advanced cost accounting techniques and improve the company's margin by SC optimization given the manufacturing capability of the mills. As these activities seek improved bottom-line results via transforming the enterprise in terms of work and process steps, phase III implies an inside-out transformation (Chambost et al., 2008 a).

Producing several products implies the opportunity of taking advantage of manufacturing flexibility via the identification of product portfolios at a given mill, i.e. integration of biorefinery product families based on key building blocks and their related derivatives with existing P&P production (Chambost et al., 2008 a). Thus according to the feedstock and product price as well as supply and demand constraints the manufacturing flexibility can be exploited to produce different products with different rates in order to optimize the company's margin. Hence, the challenge in phase III is to design the SC network and to manage it given the manufacturing flexibility needed for improving the margins. The SC should be uniquely designed in order to provide a competitive advantage over the long term while supporting value chain creation and/or maximization. In this regard, developing a SC-based analysis, which can explore the manufacturing flexibility, is of crucial importance. Laflamme-mayer et al. showed how such an analysis contributes to margin improvement for a P&P mill. Their proposed SC-based analysis first identifies the cost structure of the mill by means of a cost model and then analyzes the results by a multi-scale SC optimization framework. The proposed analysis enables reflecting the manufacturing capability of the mill at the SC-level decision making (Laflamme-Mayer et al., 2008). Same approach can be employed for analyzing the manufacturing flexibility capability of the FBR in the SC-level decision making.

Given the phased approach presented above, there are two critical aspects for the FBR implementation, i.e. product/process portfolio definition related to phase II and SC network design related to phase III. What links these two aspects is "manufacturing flexibility". Manufacturing flexibility is the capability of producing different products with different rates based on the product price with the goal of mitigating risks associated with market volatility and stabilizing the margins. This type of flexibility is inherent in the process design, because the process must be designed in such a way that enables exploiting the flexibility. Therefore, product/process portfolios are defined for flexibility to stabilize the margins and to secure the return on investment. Afterwards, the range of manufacturing flexibility is established as a design target and then this established target is designed. Finally the SC network is designed so that the market requirements can be met through the designed SC network and within the designed range of manufacturing flexibility. The goal of this paper is to propose a hierarchical methodology for SC-based analysis which can integrate these three aspects, i.e. product/process portfolio design, design of manufacturing flexibility, and SC network design. The proposed methodology will be able to evaluate product/process portfolio options and the required manufacturing flexibility, and to reflect them in the SC network design.

Integration of product, process and SC design has not gained attention in the chemical engineering context. The majority of articles in the body of literature relate to assembly process environments, e.g. car manufacturers and electronics manufacturers. Huang et al. (2005) addressed the challenge of designing effective supply chain systems that integrate platform product decisions, manufacturing process decisions, and supply sourcing decisions for a product family of notebooks. Blackhurst et al. (2005) proposed a methodology, called Product Chain Decision Model (PCDM), whose objective is to model complex and dynamic systems, such as supply chains, and the decision-making processes inherent in the operation of the supply chain, product and process design decisions. An aviation electronics provider was studied as the industrial case. Fixson (2005) introduced product architecture as a tool to link product, process and SC design decisions for assembly processes. Lamothe et al. (2006) developed an optimization model for selecting a product family and designing its SC for car manufacturers.

In the context of biorefinery, Sammons et al. (2008) proposed a general systematic framework for optimizing product portfolio and process configuration in integrated biorefineries. The framework first determines the variable costs as well as fixed costs using data in terms of yield, conversion and energy usage for each process model. Then, if a given process model needs the use of solvent, molecular design techniques are used to identify alternative solvents that minimize environmental and safety concerns. Next, process integration tools, e.g. pinch analysis, are employed to optimize the models. Lastly, the optimized model will generate data for economic and environmental performance metrics. An optimization formulation enables the framework to decide whether a certain product should be sold or processed further, or which processing route to pursue if multiple production pathways exist for a special product.

The hierarchical methodology presented in this article can be differentiated from the framework proposed by Sammons et al. in the following ways: (i) It seems that their methodology does not involve market investigations before selecting the products. (ii) Also no SC metric exist in the framework. (iii) The definition of flexibility in the methodology presented in this work is different, i.e. manufacturing flexibility, which is inherent in the design and enables to produce different products with different rates, versus flexibility in choosing products and process.

Methodology

The hierarchical methodology proposed in this paper comprises three major steps each of which points out one of the three key aspects mentioned above, i.e. product/process portfolio design, design of manufacturing flexibility, and SC network design. The first step deals with product/process portfolio design through product portfolio definition and Large Block Analysis (LBA). The second step implies establishing the design target of the manufacturing flexibility and then designing the established target. The third step addresses the SC network design which involves redesigning the SC network configuration.

First Step: Product/Process Portfolio Design

At this step, the challenge is to identify the most promising product/process combination from a large range of product/process opportunities (Werpy & Peterson, 2004). Therefore this step can be divided into two consecutive parts; (1) product portfolio definition, (2) LBA for the defined product portfolios in order to generate product/process portfolios.

The selection of the most promising product portfolio includes two major concepts:

(i) First, the product identification is based on a market-driven analysis reflecting the commercial product opportunities. In this regard, products could be classified into three groups; (a) Replacement products which are identical in chemical composition to the existing products in the market, but made out of renewable feedstock, e.g. biopolyethylene. (b) Substitution products which have different chemical composition, but the same functionality, e.g. polylactic acid (PLA) instead of polyethylene terephtalate (PET). (c) Novel products like biomaterials, nanocomposites which have new functionalities and therefore no existing markets (Chambost et al., 2008 a).

(ii) The product portfolio of a mill should be the expression of value creation via the determination of key building block offering high potential for added-value products and the integration of the new identified products with traditional P&P products (Chambost et al., 2008b). Important elements should be considered while considering the definition of portfolios such as follows; (a) Manufacturing flexibility is an important criterion for product portfolio definition. It must be investigated that which set of products introduces a better potential for flexibility. (b) The defined product portfolio must be able to stabilize the margins and to secure the return on investment (ROI), thus market volatility, legislation changes and other factors must be taken into consideration. (c) The definition of product portfolio should take into account the identification of sustainable partnership models, i.e. partnering with technology providers and/ or chemical companies, in order to secure the SC and lower the risks of entering an existing/new value chain.

Product Portfolio Definition

A three-stage methodology has been developed for the definition of product portfolio, Figure 2 (Chambost et al., 2008 a). In the first stage sets of possible products must be identified. For this purpose, overall product opportunities are investigated based on market, economic and product specific information such as product functionalities, volume, market size and growth, market saturation and basic margins. The goal of this stage is to identify a list of promising bioproducts. In the second stage, based on market and competitiveness criteria and a preliminary techno-economic study, possible sets of product families can be identified. Product families comprise bioproducts that link to each other via common processing steps. For instance, ethanol, ethylene and polyethylene could form a biorefinery product family, since

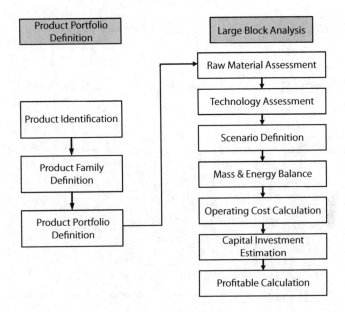

FIGURE 2
Product/Process portfolio definition.

ethanol can be converted into ethylene and further into polyethylene. At the last stage, according to a mill or company-based analysis, product portfolios will be generated. Product portfolios are the combination of existing P&P products and new biorefinery products. At this stage, mill's specifications must be taken into consideration in order to identify the opportunities for integration between P&P processes and bioprocesses in terms of feedstock, chemicals and energy. Finally a critical risks assessment is conducted for each product portfolio.

Large Block Analysis (LBA)

The objective of LBA is to provide comparable techno-economic data such as operating cost, capital investment cost and profitability, of different product/process portfolios and then to screen out non-profitable portfolios.

LBA has seven major stages which are shown schematically in Figure 2.

At the first stage, which is "raw material assessment", given the defined product portfolios, list of raw materials must be identified based on their accessibility to the mills and the maximum available volume according to their cost. The second stage is "technology assessment" in which emerging technologies for producing each product must be surveyed, taking into account the mass and energy balance, type of feedstock and technological risks of each technology. At the third stage, called "scenario definition", the combinations of raw material/process/product are generated as scenarios.

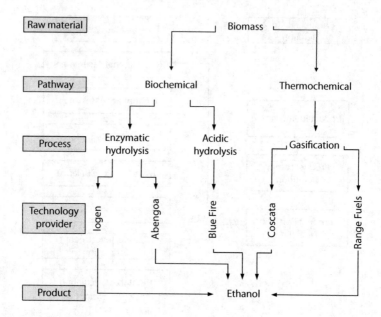

FIGURE 3
Chemical pathways from raw material to ethanol.

From raw material to product, there are different pathways and several processing routes. For example, as shown in Figure 3, for producing bioethanol from biomass, there are two pathways, i.e. biochemical and thermochemical. For each of these pathways, different types of process can be used, such as gasification for thermochemical pathway, and enzymatic or acidic hydrolysis for biochemical pathway. Finally there are many technology providers for each process. Therefore each scenario includes one type of feedstock, a specific pathway, processing route and a technology provider related to the processing route, and finally a product. Thus, to define scenarios, given the outcome of the last two stages, i.e. raw material and technology assessment, the specific technology provider, and hence its corresponding process type and pathway, and the required raw material for producing each product must be identified. At this stage, the potentials of integration of selected portfolios with the existing mill must be taken into account in terms of technological fit, integration factors and risks.

At the fourth stage, "mass and energy balance" is done for each scenario based on technology provider's and raw material specific information. The fifth stage deals with "operating cost calculation". There are two types of operating costs: variable and fixed cost. Variable costs such as costs of chemicals, fuels, etc. are calculated based on the balance sheets of the processes and price information. Fixed costs involve labor cost, maintenance, insurance and taxes, and general overhead. The sixth stage is "capital investment estimation", in which capital investment is estimated for each scenario based on published information for stand-alone bioprocesses. Then the mill's impact will be investigated in order to identify the potentials for integration with P&P processes in terms of chemicals or energy. In this regard, the existing mill system that can used for the bioprocesses must be defined and afterwards, based on the demand of the bioprocesses and the current mill specifications, the cost of modification needs of the mill system can be estimated. In the last stage, which is "profitability calculation", the profitability of each scenario according to the revenue from end products and by-products will be estimated by means of profitability measures such as NPV, IRR or ROI. After these stages, the non-profitable scenarios will be screened out and a finite number of scenarios will be selected as product/process portfolios. These portfolios will be analyzed further so that the best portfolio can be identified.

Second Step: Designing the Manufacturing Flexibility

This part of the methodology contains two steps which must be implemented for the remaining product/process portfolios; in the first step the range of flexibility is established as a design target for each process and in the second step the established target of each process is designed. In order to perform this part of the methodology, two tools are needed; (1) a SC model, (2) an operations-driven cost model. SC model aims to maximize the profitability across the entire SC by first identifying the trade-offs between the demand and production activities and then by finding the optimal alignment of manufacturing capacity and market demand. On the other hand, the operations-driven cost model is based on "operations-driven" thinking, which implies using lower-level process data and detailed process analysis in order to better reflect manufacturing capability for higher-level decision-making. Hence, the overall objective in operations-driven thinking is first to characterize the manufacturing operations (descriptive) by identifying the cost drivers and second, to provide advanced decision support (prescriptive) (Janssen et al., 2006).

SC Model

The SC model aims to calculate the optimum profitability of the whole SC. It is formulated into an optimization problem whose objective function is the sum of revenues from different main products and by-products subtracted by SC costs including feedstock cost, inventory cost, production cost, transitions and shutdowns cost. There are two types of decision variables in the mathematical formulation of the SC model. The first type is continuous variables which comprise flow of material between SC nodes, e.g. flow of feedstock from suppliers to the mill, amount of each product that must be produced, and the inventory levels for each type of feedstock and product. The second type is binary variables which imply "yes/no" type of decisions, e.g. which product must be produced or which production line must operate. Each node of the SC, i.e. suppliers, inventories, manufacturing centers, has its own constraints which must be formulated mathematically.

Operations-Driven Cost Model

The operations-driven cost model is where cost and process information are captured and systematically integrated to characterize the processes (Janssen, et al., 2006). Cost model must be made up for all remaining product/ process portfolios from previous step of the methodology. The outcome of the cost model are manufacturing costs for each design alternative (see *Designing the Established Target*), as well as profitability measures based on capital and manufacturing costs, which in this methodology will be ROI. The cost model is fed, from one side, by process information which represents the resource consumption, and on the other side, by cost information which shows the cost of each resource. Each mill is represented by a number of Process Work Centers (PWC) in which some processes are performed. The cost incurred by the processes in each PWC is calculated. Also there is an Overhead Work Center (OWC) which introduces the manufacturing overheads and non-manufacturing costs. These costs are used to calculate the final cost object, which can be the product cost for each design alternative.

Establishing the Range of Manufacturing Flexibility

For each process in each portfolio, there is a nominal production rate based on the result of "technology assessment" step of LBA. At this stage, the range of flexibility within which the manufacturing processes must operate is determined. In other words, it must be determined that, given the price volatility in the market and with the aim to maximize the profitability, to what extent each production rate must be able to vary. For this purpose, a finite number of price scenarios, representing the price volatility, are generated. It is worth mentioning that price volatility, which implies the future market situation, is modeled through aforementioned scenario generation with the aim of long-term strategic decision making. Thus product prices won't be taken in real time. Real-time analysis will be performed at tactical and operational levels which deal with the dynamic aspect of the SC. After generating price scenarios, the SC model is run for each scenario with no constraint on the rate of manufacturing processes, so that the SC model can find the optimum production rate for each process based on the product price. Figure 4 shows an example.

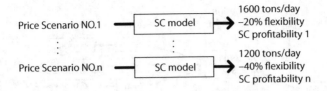

FIGURE 4
SC model input and output.

Considering that the nominal production rate is 2000 tons/day, given a decrease in product price scenario NO.1, the optimum production rate obtained by the SC model is 1600 tons/day, which represents −20% of flexibility based on the nominal rate (2000 tons/day), while the obtained result from SC model for the product price NO.n, representing a stronger decrease in price, is 1200 tons/day, which represents −40% of flexibility. This calculation must be done for each scenario, so that it is determined that to what extent each process needs to be flexible for a given price volatility. Also the SC profitability for each price scenario will be estimated and finally based on the range of flexibility and SC profitability the range of flexibility as a design target will be determined.

Designing the Established Target

At this stage, the range of flexibility will be constrained by limiting the production rates based on the result of the previous stage. For instance, given that −40% was the maximum flexibility obtained from the last stage, this percentage will be the flexibility constraint which cannot be exceeded. In other words, the SC model won't be able to go beyond this range. For this stage, a limited number of design alternatives will be generated to represent the flexibility. As it is impossible to have −40% of flexibility on a continuous process, therefore in order to have this percentage of flexibility, the production line must be divided into 2, 3 or 4 lines whose sum of production rates is equal to the nominal rate (2000 tons/day). Figure 5 shows an example.

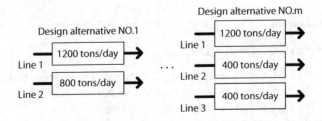

FIGURE 5
Design alternatives.

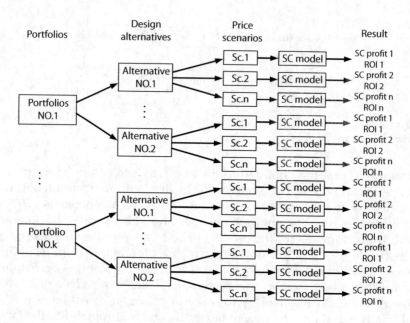

FIGURE 6
Identifying the most profitable design alternative.

As shown in Figure 5, the nominal rate (2000 tons/day) can be represented in terms of two parallel lines (1200 tons/day and 800 tons/day, or 1400 tons/day and 600 tons/day) or three parallel lines (one line for 1200 tons/day and two lines for 400 tons/day). Then, for each price scenario, the SC model will be run and the SC profitability as well as ROI will be calculated for each design alternative. Therefore, for each design alternative, a set of SC profitability and ROI will be calculated for a set of price scenarios. Based on these results, the most profitable design alternative will be identified. It must be mentioned that these steps must be performed for all remaining portfolios, so that the best design alternative for each portfolio can be delineated. Figure 6 shows this stage graphically.

Third Step: SC Network Design

The goal of this step is to design the SC network for each portfolio. For this purpose, a finite number of SC network alternatives will be generated for each portfolio. Figure 7 shows this step graphically.

SC network alternatives can be defined in terms of expansion of existing facilities or buying new facilities in different areas. Then, given the same price scenarios, SC model will be run for each SC network alternative and

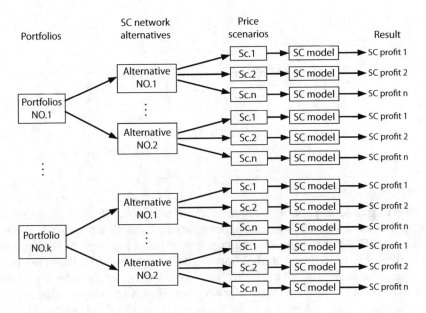

FIGURE 7
SC network design alternatives.

the SC profitability will be calculated for each of them. Based on the results, the best SC network alternative can be determined for each portfolio.

Decision Making Framework

As it was mentioned previously, the proposed methodology must be performed for all portfolios. After implementing this methodology, each portfolio can be characterized by means of several aspects, i.e. manufacturing flexibility, SC profitability and ROI. These aspects can be used as different metrics in a multi-criteria decision making framework. It is worth mentioning that all of these metrics would be SC metrics. Another type of metrics can be provided by Life Cycle Analysis (LCA) and added to the framework. Based on the result obtained by the multi-criteria decision making framework the best product/process portfolio can be determined.

Illustrative Example

This methodology will be applied in a case study at a P&P mill. This P&P mill, which produces one grade of pulp and one grade of paper, aims to implement the FBR by producing bioproducts. After product market analysis, two product portfolios are considered; the first portfolio includes the two

TABLE 1

Price Change Scenarios

		1	2	3	4	5	6
Portfolio 1	Ethanol	—	↓	↓	—	↓↓	↓↓
	Ethylene	↓	—	↓	↓	—	↓↓
					↓		
Portfolio 2	LA	—	↓	↓	—	↓↓	↓↓
	PLA	↓	—	↓	↓	—	↓↓
					↓		

grades of pulp and paper plus ethanol and ethylene, while the second port-folio contains the same pulp and paper grades plus lactic acid (LA) and poly lactic acid (PLA). Different companies provide the technology for production of such products. If, based on the characteristics of the existing P&P mill, two scenarios for each product portfolio can be considered, four product/process portfolios can be defined. After carrying out mass and energy balances for each scenario, the profitability of each scenario is obtained by calculating the operating and investment costs. The two most profitable portfolios are retained for the next step. Assuming that one portfolio involving ethanol/ethylene and one including LA/PLA are selected, in the next step the range of required flexibility must be determined and six price scenarios are generated for each portfolio, as shown in Table 1.

The first two scenarios consider a price decrease for only one of the products, while third scenario represents price decrease for both products, e.g. ethanol and ethylene. The next three scenarios follow the same rule, but consider a stronger decrease in product prices. All scenarios are defined using price decreases in order to address the worst case. For each of these scenarios, the developed SC model is run without constraint on production capacity in order to obtain the optimal production rate of each process for each scenario, and to determine the flexibility range. Given the number of retained product/process portfolios and price scenarios, i.e. two and six respectively, the SC model must be run twelve times. Table 2 shows the range of flexibility needed for each process in the case of each scenario realization.

TABLE 2

Range of Flexibility for Each Process

	1	2	3	4	5	6
Ethanol	0	−15	−20	0	−23	−30
Ethylene	−10	0	−5	−20	0	−14
LA	0	−10	−14	0	−13	−20
PLA	−8	0	−3	−10	0	−7

TABLE 3

Design Alternatives for Each Portfolio

		Design 1	Design 2
Portfolio 1	Ethanol	2 lines: 600 tons /day 1400 tons/day	3 lines: 300 tons/day 300 tons/day 1400 tons/day
	Ethylene	2 lines: 200 tons/day 800 tons/day	2 lines: 200 tons/day 800 tons/day
Portfolio 2	LA	2 lines: 400 tons /day 1600 tons/day	3 lines: 200 tons/day 200 tons/day 1600 tons/day
	PLA	2 lines: 100 tons/day 900 tons/day	2 lines: 100 tons/day 900 tons/day

Therefore, −30%, −20%, −20% and −10% of flexibility are the maximum flexibility needed for ethanol, ethylene, LA and PLA production processes, respectively. In the next step design alternatives are defined based on calculated ranges of flexibility. For each process, two alternatives have been considered. Given that the nominal production rate for ethanol/LA and ethylene/PLA is 2000 tons/day and 1000 tons/day, respectively, the design alternatives are defined as presented in Table 3.

Figure 8 illustrates the ROI of each design alternative for each price scenario.

According to the resulting ROIs, design 2 for both portfolios is a better choice. Therefore, for the next step, which deals with SC network design, two SC network alternatives are defined for each of these design alternatives. These network alternatives are presented in Table 4.

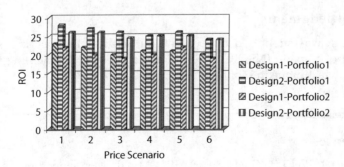

FIGURE 8
ROI of design alternatives for each price scenario.

TABLE 4

SC Network Alternatives

Alternative 1	Alternative 2
Expanding the existing warehouse	Buying a new warehouse

Again the SC model is employed (in which other step is it also used?) to calculate the profitability of these SC network alternatives for each portfolio. Eventually, the final result, SC profitability, range of flexibility and ROI, will be used as SC metrics, along with LCA metrics, in a multi-criteria decision making framework. The final result of this framework will determine the best product/process portfolio.

Conclusions

A hierarchical methodology is proposed to integrate product/process portfolio design, design of manufacturing flexibility, and SC network design. SC optimization, techno-economic study and operations-driven cost modeling are employed as tools. Scenario generation is used to address the product price volatility. Inspired by work done previously in Environmental Design Engineering Chair at Ecole Polytechnique in Montréal regarding SC-based analysis in the context of P&P industry, this methodology shows how the FBR can be implemented strategically via a step-wise approach. Through this step-wise approach, different options in terms of product/process portfolio can be studied and their potential for flexibility can be investigated. Also, the best SC network for these options can be identified. Therefore product portfolio design and process design can be reflected in the SC strategic design via this SC-based analysis.

Acknowledgments

This work was supported by the Natural Sciences Engineering Research Council of Canada (NSERC) Environmental Design Engineering Chair at Ecole Polytechnique in Montréal.

References

Blackhurst, J., Wu, T., O'Grady, P. (2005). PCDM: a decision support modeling methodology for supply chain, product and process design decisions. *Journal of Operations Management*, 23(3–4), 325.

Chambost, V., McNutt, J., Stuart, P. R. (2008a). Guided tour: Implementing the forest biorefinery (FBR) at existing pulp and paper mills. *Pulp and Paper Canada, 109*(7–8), 19.

Chambost, V., Martin, G., Stuart, P.R. (2008b). Identifying the Forest Biorefinery Product Portfolio. *Keynote at the 18th International Congress of Chemical and Process Engineering*, Prague, CZ.

Fixson, S. K. (2005). Product architecture assessment: a tool to link product, process, and supply chain design decisions. *Journal of Operations Management, 23*(3–4), 345.

Huang, G. Q., Zhang, X. Y., Liang, L. (2005). Towards integrated optimal configuration of platform products, manufacturing processes, and supply chains. *Journal of Operations Management, 23*(3–4), 267.

Janssen, M., Naliwajka, P., Stuart, P. R. (2006). Using process-based cost modeling to evaluate process modernization alternatives. *92nd Annual Meeting of the Pulp and Paper Technical Association of Canada (PAPTAC)* Pulp and Paper Technical Association of Canada, Montreal, Quebec H3C 3X6 Canada, Vol. B, pp. B61.

Laflamme-Mayer, M., Shah, N., Pistikopoulos, S., Linkewich, J., Stuart, P. (2008). Multi-scale on-line supply chain planning part a: Decision processes and framework for a high-yield pulp mill. *Submitted to:AIChE Journal.*

Lamothe, J., Hadj-Hamou, K., Aldanondo, M. (2006). An optimization model for selecting a product family and designing its supply chain. *European Journal of Operational Research, 169*(3), 1030.

Sammons Jr, N. E., Yuan, W., Eden, M. R., Aksoy, B., Cullinan, H. T. (2008). Optimal biorefinery product allocation by combining process and economic modeling. *Chemical Engineering Research and Design, 86*(7), 800.

Stuart, P. (2006). The forest biorefinery: Survival strategy for Canada's pulp and paper sector? *Pulp and Paper Canada, 107*(6), 13.

Werpy, T., Peterson, G. (2004). Top value-added chemicals from biomass feedstock—Volume I: Results of screening for potential candidates from sugars and synthesis gas. *NREL for US Department of Energy.*

97

A Novel Shortcut Method for the Design of Heteroazeotropic Distillation of Multicomponent Mixtures

Korbinian Kraemer, Andreas Harwardt and Wolfgang Marquardt*

Aachener Verfahrenstechnik, RWTH Aachen University
Templergraben 55, 52056 Aachen, Germany

CONTENTS

ABSTRACT Shortcut methods are a valuable tool in the early stages of chemical process design, where numerous flowsheet alternatives need to be evaluated to determine the most energy-efficient, feasible flowsheet. Various shortcut methods based on tray-to-tray calculation and pinch point analysis for the inspection of feasibility and the determination of the minimum energy demand for homogeneous azeotropic distillation have been published in the literature. For multicomponent heteroazeotropic distillation, however, no generally applicable shortcut methods have been presented so far. Based on the existing shortcut methods for homogeneous distillation, a shortcut method for heteroazeotropic distillation, the feed pinch method (FPM) is proposed in this work. The FPM is based on rigorous thermodynamics, applicable to heterogeneous azeotropic mixtures of any number of components and can be automated. The new method is illustrated by several

---

* To whom all correspondence should be addressed

examples of ternary as well as quaternary heteroazeotropic separations, for which feasibility and minimum energy demand is determined with paramount robustness and efficiency.

KEYWORDS　*Heteroazeotropic distillation, shortcut method, feed pinch method, rectification body method*

Introduction

When designing a sustainable distillation process for an azeotropic multi-component mixture, numerous alternative flowsheets and entrainer candidates have to be evaluated in order to determine the most energy-efficient flowsheet. In industrial practice, usually a small number of possible flowsheets are selected by heuristics and evaluated manually by repetitive simulation studies, which require detailed design specifications in the early design phase. Despite this tedious, iterative procedure, no guarantee concerning the quality of the solution can be given. On the other hand, the MINLP optimization of a flowsheet superstructure comprising all alternatives would be far too complex to solve with today's optimization algorithms. Shortcut methods for distillation column design, however, allow for an inspection of feasibility and an efficient and robust calculation of the minimum energy demand without the need for detailed column specification. Hence, these methods are perfectly suited for a fast screening and ranking of design alternatives in the conceptual design phase. Powerful and recent shortcut methods based on rigorous thermodynamics for nonideal and azeotropic mixtures are reviewed in brief in the following (see also Bausa et al., 1998). The methods are demonstrated by an example separation of the homogeneous azeotropic mixture of acetone, methanol and ethanol. Subsequently, the application to the heterogeneous case is shown. The different restrictions of the existing methods, especially for heterogeneous mixtures, are pointed out in order to motivate the development of the novel method for heteroazeotropic distillation that can be viewed as an extension of the existing methods to overcome their limitations.

Review of Existing Shortcut Methods

Levy et al. (1985) proposed the boundary value method (BVM) for the determination of the minimum energy demand for nonideal distillation. Here, column tray-by-tray profiles are calculated for each column section from the respective column ends. For a given distillate composition x_D, distillate flowrate D and condenser duty Q_D, the tray to tray profiles for the rectifying section are computed successively starting at the distillate by balancing components and energy and considering chemical equilibrium on each tray:

$$\bar{y}^*(x_{n+1}) = \frac{L_n}{L_n + D}\bar{x}_n + \frac{D}{L_n + D}\bar{x}_D, \qquad n = 1 \ldots n_f$$

$$h_{n+1}^V = \frac{L_n}{L_n + D}h_n^L + \frac{D}{L_n + D}h_D - \frac{Q_D}{L_n + D}, \quad n = 1 \ldots n_f.$$

(1)

The stripping section profile can be calculated accordingly. Separation feasibility is determined by an inspection of intersection of the column profiles. The lowest energy duty that allows an intersection of column profiles defines the minimum energy demand. However, for sharp splits, traces have to be specified for every component, since the profiles would not leave the subspace of the product components otherwise. The manifold of stripping section profiles for different trace components in the bottom product of the example separation are shown in Figure 1. The search for the minimum energy therefore requires a tedious simultaneous optimization of the energy

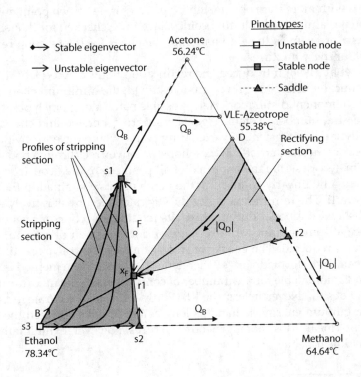

FIGURE 1
Stripping section profiles, pinch points and rectification bodies at minimum reflux for the ternary separation of acetone, methanol and ethanol.

and all trace components. Considering that the intersection of profiles needs to be checked manually, the application of the BVM is effectively limited to ternary mixtures.

In order to overcome the dependency of the shortcut results on trace components in the products, pinch based shortcut methods have been proposed by various authors. Pinch point curves for a given product and a variable reboiler/condenser duty can be calculated for each column section as solution branches of the tray-to-tray equations (1). When the energies are fixed, the pinch points are identified as fixed points of the plate-to-plate recurrence on the pinch point curves (c.f. Figure 1). These pinch points, which are insensitive towards the choice of trace components, can be classified as stable or unstable nodes or as saddles depending on the number of stable eigenvectors.[1]

The Eigenvalue criterion (EC) by Poellmann et al. (1994) can be considered as a pinch based BVM. Instead of calculating profiles from the column ends, the tray-to-tray profiles are started from points close to the saddle pinches, which are located to a small extent in the direction of the unstable eigenvectors. The minimum reflux condition is achieved at the smallest reflux ratio which makes an intersection of the profiles possible. Again, the selection of the relevant subset of active pinch points is not trivial. For multicomponent mixtures with more than one unstable eigenvector per pinch point, multidimensional manifolds of column profiles have to be checked for intersection, which is a problem as the automation of the check for intersection remains difficult for these mixtures.

Bausa et al. (1998) introduced the rectification body method (RBM) as an algorithmically accessible procedure to estimate the minimum energy duty of multicomponent distillation. Here, possible paths along pinch points with an increasing number of stable eigenvectors are generated and checked for thermodynamic consistency by excluding paths, where the entropy production does not increase strictly monotonously. Convex rectification bodies that approximately describe the manifold of all profiles are then constructed for each section by linearly connecting the pinch points contained in the paths (c.f. Figure 1). The minimum energy duty is calculated by iteratively identifying the lowest reboiler duty that results in an intersection of a set of bodies. Since all pinch points are used, no a-priori selection of relevant pinch solutions is required. The check for intersection of the convex rectification bodies can be performed very efficiently and, therefore, the method is applicable and automatable for any number of components in the mixture and for any kind of split. Nevertheless, the RBM only returns an accurate indication of the minimum energy demand as long as the profiles between the pinch points are linear. Most homogeneous mixtures, however, exhibit only very mild nonlinear behavior.

---

[1] Denotation of pinch points as in Julka and Doherty (1990): r or s for column section and number of unstable eigenvectors plus one. Second letter is for differentiation, when more than one pinch point of the same denotation occurs.

Lucia et al. (2008) proposed the shortest stripping line approach (SSL) to find minimum energy requirements in distillation. They calculate stripping profiles until a pinch occurs and then switch to the rectifying profile. The shortest stripping line which produces a feasible result, i.e. where the distillate product is reached by the rectifying profile, marks the minimum energy demand. While the SSL is based on a constant molar overflow assumption, it was automated and applied to homogeneous zeotropic and azeotropic mixtures of up to six components and multi-unit processes.

Application to Heterogeneous Mixtures

Many industrially relevant mixtures exhibit immiscibilities in the liquid phase. Moreover, the use of a heterogeneous entrainer allows for a crossing of distillation boundaries to separate azeotropic multicomponent mixtures in many industrial applications. In typical designs, a heterogeneous stream is produced at the top of the column which is then split in a decanter into an entrainer-lean distillate and an entrainer-rich reflux. Contrary to homogeneous systems, heterogeneous systems always exhibit strong nonlinearities. Shortcut methods for homogeneous systems cannot be applied without an adaption to handle the decomposition of the liquid phase in the decanter but also on the heterogeneous trays within the column. Hence, there are very few publications on shortcuts for heterogeneous systems, especially when methods are deducted that consider immiscibilities only in the decanter. A thorough analysis of the properties of heteroazeotropic distillation has been presented by Urdaneta et al. (2002). The application of the shortcut methods for homogenous distillation reviewed above is illustrated in this section with the ternary heteroazeotropic mixture of isopropanol, water and cyclohexane. A ternary feed is to be separated into a bottom product of pure isopropanol and a cyclohexane-lean product on the decanter tie-line through the minimum boiling ternary azeotrope (c.f. Figure 2). The limitations of the existing shortcut methods are demonstrated and the novel shortcut method is presented subsequently. Though it has been developed independently from the work of Lucia et al. (2008), there are some similarities.

RBM. Urdaneta et al. (2002) have extended the procedure for the calculation of the pinch points of a separation to handle heterogeneous systems. Based on these pinch point solutions, the RBM can also be in principle applied to heterogeneous systems. However, the accuracy of the RBM can be very low for these systems, as heterogeneous mixtures usually exhibit strongly curved profiles in and around the region of immiscibility. This can be attributed to the high nonlinearity of the VLLE as well as to the fact that the reflux from the decanter is of different composition than the product stream leaving the column. The profile of the rectifying section for

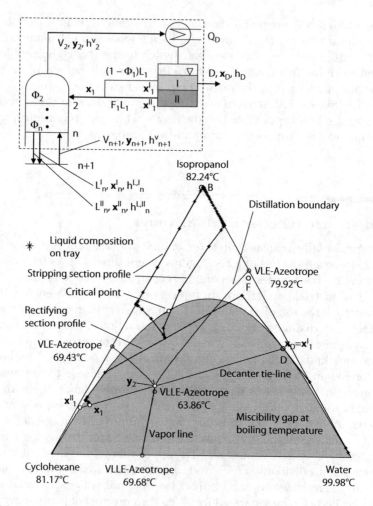

FIGURE 2
Balance envelope of the rectifying section of a heteroazeotropic distillation column and the system of isopropanol, water and cyclohexane at 1 bar including profiles at minimum reflux.

the example mixture is strongly curved towards the isopropanol vertex as shown in Figure 2. Note, that the liquid profiles of the rectifying section start with the composition of the reflux from the decanter instead of the distillate product due to the separation of the two coexisting liquid phases within the decanter. While the linear combinations of pinch points at minimum reflux approximate the stripping profiles very well, they miss the curved profiles of the rectifying section by a large margin, as shown in Figure 3. The rectification bodies can be brought to intersection at a significantly higher reflux than the minimum reflux leading to a significant overestimation of the minimum energy demand (see Table 1).

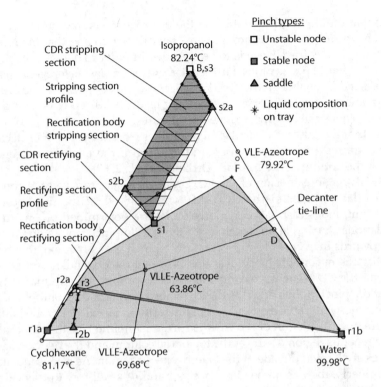

FIGURE 3
Relevant rectification bodies and CDRs at the minimum reflux as determined by the CDRM.

BVM. A method that does not rely on linearization, such as the BVM, should be well suited to handle the strongly curved profiles of heterogeneous systems. Pham et al. (1989) have extended the BVM to heteroazeotropic distillation but only consider designs with homogeneous column trays. For columns with heterogeneous behavior in the rectifying section, however, the courses of the profiles are not only dependent on the specification of trace components in the products, but also on the specification of heterogeneous trays k and the liquid phase ratio Φ_k on the last heterogeneous tray. It has been shown by Urdaneta et al.

TABLE 1

Compositions of Isopropanol, Water, Cyclohexane and Minimum Energy Demands

Molar Composition	x_F 0.66/0.32/0.02		x_D 0.4/0.57/0.03	x_B 1/0/0
$Q_{b,min}/F$ [MJ/kmol]	RBM	MAC	CDRM	FPM
	65.9	30.3	30.4	30.3

(2002) that the liquid phase ratios on the heterogeneous trays are degrees of freedom within the downwards tray-to-tray calculation and that the number of heterogeneous trays k needs to be specified in order to derive suitable values for the liquid phase ratios. Furthermore, Φ_k on the last heterogeneous tray marks an additional degree of freedom that needs to be specified. Factoring in the graphical check for intersection required by the BVM, it is obvious that this method is not suited for heterogeneous mixtures.

Note that the upwards calculation of profiles from the reboiler or decanter never requires a specification of k and Φ_k, even if there are heterogeneous trays at the bottom of the column (Urdaneta et al., 2002).

CDRM. As an extension of the EC, Urdaneta et al. proposed the continuous distillation region method (CDRM), where curved rectification bodies, so called continuous distillation regions (CDR), are determined by tray-to-tray calculations starting at an ε-vicinity of the saddle pinch points downwards and upwards in every column section. As a pinch based method, this procedure has the major advantage over the BVM that the dependency of the profiles—or the CDR—on trace components in the products is eliminated. Since the saddle pinch points describe the extreme locations of the manifold of possible profiles at the specified reflux, the full CDR is identified. For the example mixture, the calculation of the CDR is further simplified by the fact that the heterogeneous region occurs at the top of the column and the relevant saddle pinches are located outside of the heterogeneous region. As a consequence of this property, there is no need to specify k and Φ_k as all tray-to-tray calculations within the heterogeneous region are performed upwards. For all other cases, however, k and Φ_k still need to be specified properly to determine the full CDR and subsequently the minimum energy demand.

The intersecting CDRs of the stripping and rectifying section at minimum reflux are shown in Figure 3 and the minimum energy demand is given in Table 1. The graphical determination of intersection can be accomplished with little effort for this ternary example. Considering the dependence of the profiles on the specification of the ε-distance to the respective saddle pinch point and possibly on the specification of k and Φ_k, it is obvious that this manual procedure of trial and error, however, becomes very tedious when processes with several columns connected by a recycle need to be evaluated. Moreover, the construction of multi-dimensional distillation regions out of a few profiles and the check of intersection become impossible for mixtures with more than three components.

Feed Pinch Method for Heteroazeotropic Distillation

Since the published shortcut methods for the determination of the feasibility and the minimum energy demand for heteroazeotropic distillation are restricted by inaccuracies or limitations towards the number of components, it was our motivation to develop a shortcut method that can handle heterogeneous mixtures without restrictions. Specifically, the method needs to be based on rigorous thermodynamics, applicable to heterogeneous mixtures of any number of components, insensitive towards trace components as well as specifications of k and Φ_k and algorithmically accessible.

A big step towards this goal is reached in this work by a rather simple modification of the CDRM of Urdaneta et al. (2002). Instead of calculating tray-to-tray profiles from all saddle pinch points to form multiple distillation regions, the proposed feed pinch method (FPM) only requires the calculation of one tray-to-tray profile starting from the point that all possible profiles run through: the feed pinch point. This single profile is only calculated for the section that does not contain the feed pinch, i.e. when the feed pinch is the stable node pinch of the stripping section, the rectifying section profile is calculated upwards from the feed pinch. When the feed pinch is the stable node pinch of the rectifying section, the stripping section profile is calculated downwards from the feed pinch. The profile of the section to which the feed pinch belongs, in this case the stripping section profile, does not need to be calculated as the stable pinch can always be reached by a profile of the respective section. The tangential pinch as the exception to this rule can be detected by an automatic pinch reachability check (Bausa et al., 1998). The energy is increased in this case until the tangential pinch disappears.

Feasibility of the separation is detected, when the profile reaches the product composition or the unstable node pinch on the decanter tie line. The lowest reflux, for which this is possible, denotes the minimum energy demand. Note that the profile converges at the product for sharp splits but runs through the product for non-sharp splits, where the minimum energy demand is the only feasible specification for the reboiler energy. If too little reflux is specified, the profile will leave the phase diagram.

The phenomenon that the profile leaves the composition space for $Q_B < Q_{B,min}$ can be illustrated using McCabe-Thiele diagrams for binary mixtures. For minimum reflux, the operating lines of the rectifying section and the stripping section intersect at the vapor equilibrium line, resulting in an infinite number of trays for the separation (Figure 4, left). If the energy requirement (or reflux) is reduced below the minimum value, the intersection of the operating lines occurs above the vapor equilibrium line (Figure 4, right). Using the McCabe-Thiele methodology, the profile of the rectifying section, which is constructed between the operating line of the rectifying section and the vapor equilibrium line starting at the feed pinch of the stripping section, does not converge towards the distillate product, but to physically invalid values.

The only restriction of the FPM is the requirement of a feed pinch point. An example of a separation without a feed pinch has been given by Bausa et al. (1998). However, to our knowledge the vast majority of separations exhibit a feed pinch. The detection of the feed pinch, certainly, is not a trivial task. Candidate pinches are the stable node pinches of both sections.

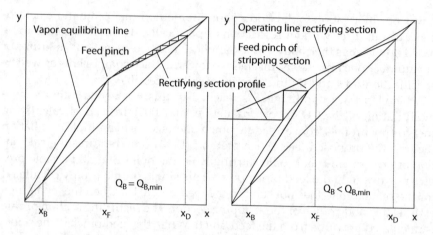

FIGURE 4

Tray-to-tray calculations for the rectifying section of a binary mixture for $Q_B = Q_{B,min}$ (left) and $Q_B < Q_{B,min}$ (right).

While there may be several stable pinch points per section, for sharp splits usually only one stable pinch point for the whole column lies on a pinch point curve that runs into the product composition of the opposite section (or into the decanter tie line when there is a decanter). This stable pinch point can then be identified as the feed pinch. If there are no pinch point curves ending in a product (for non-sharp splits) or more than one pinch curve ending in a product (for the particular case when both products are located at singular points of the composition space) profiles are calculated for all stable node pinches. The pinch point that produces a feasible profile with the lowest reflux marks the feed pinch. In this case, the minimum energy demand is still found with less tray-to-tray calculations than the CDRM would demand.

For the calculation of profiles, the BVM and the CDR method demand a slight, user specified deflection from the products or the saddle pinches, respectively. In contrast, the calculation of profiles for the FPM can be started directly at the feed pinch and, thus, the course of the profile does not depend on the specification of a deflection. An additional benefit of the FPM is the independence of the results from the specification of the design variables k and Φ_k for heterogeneous distillation. For an explanation of this property, different cases have to be considered:

- When the feed pinch occurs in the stripping section, i.e. the minimum energy demand is determined by the profile of the rectifying section (c.f. Figure 5), k and Φ_k do not need to be specified, since the calculation is carried out upwards from the feed pinch (Urdaneta et al., 2002).

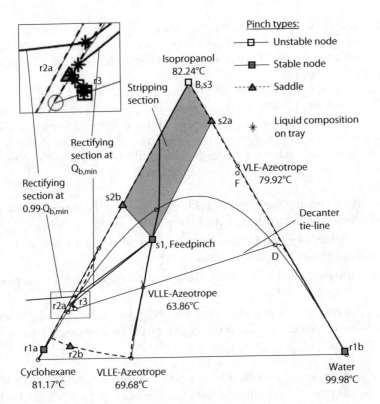

FIGURE 5
Rectifying section profiles for $Q_{B,min}$ and $Q_B = 0.99 \cdot Q_{B,min}$ calculated with FPM. The stripping section profiles, which do not need to be calculated, are indicated by the corresponding rectification body.

- When the feed pinch occurs in the rectifying section, i.e. the minimum energy demand is determined by the profile of the stripping section, two scenarios need to be distinguished:
  - The heterogeneous region is located at the top of the column: In this case, the stable node pinches of the rectifying section are always located outside of the heterogeneous region (Bausa, 2001) and no heterogeneous trays have to be considered in the stripping section.
  - The heterogeneous region is located at the bottom of the column: In this case, it can be assumed that the stripping profile is located entirely in the heterogeneous region or does not leave the heterogeneous region once it has entered it. With this assumption, there is only one viable specification for the liquid phase ratios.

Due to the independence on k, Φ_k, trace components and the distance to the saddle pinches, the profiles that have to be calculated within the FPM for the determination of feasibility and minimum energy demand are a function of the energy duty only.

An additional benefit of the FPM is the simple check for feasibility, i.e. the condition that the profile needs to reach the product composition (or the unstable node on the decanter tie line). Contrary to the BVM and the CDRM, where a multitude of possible profiles have to be checked for intersection in multi-dimensional space, the check for feasibility within the FPM offers the following advantages:

- Only one profile needs to be checked in the case of a sharp split (feed pinch can be determined a priori).
- Two points need to be checked for intersection, which is easy to carry out even in multi-dimensional space.
- The intersection occurs at a well-defined location (product composition/unstable node).
- In the case of a sharp split, only tray N_{max} has to be checked as the profile converges at the product composition if enough energy is supplied.

As a consequence of the favorable properties described above, the FPM can easily be automated by a minimization of the energy duty while checking feasibility. Note that this is possible for mixtures of any number of components as the check for feasibility is not limited to ternary mixtures. In this work, the variation of the energy duty and the check for feasibility were done manually. In our future work, however, an automation of the FPM will be implemented.

The FPM is similar to the shortest stripping line (SSL) approach by Lucia et al. (2008), although it has been developed independently. Contrary to the SSL approach, however, the FPM does not rely on a constant molar overflow assumption as it considers energy balances. In addition, balances and equilibrium for heterogeneous trays and a phase stability test (Bausa et al., 2000) are considered to handle heteroazeotropic distillation. Furthermore, a pinch point analysis, as in Urdaneta et al. (2002), is performed at first. Thus, the feed pinch does not need to be determined by tray-to-tray calculations, which would have to be carried out starting from both products when the separation topology is not known a-priori. Within the FPM, a tray-to-tray profile is therefore only calculated for the section that does not contain the feed pinch.

The application of the FPM to the ternary example mixture is shown in Figure 5. The profile of the rectifying section starts at the feed pinch, i.e. the stable pinch of the stripping section and reaches the unstable pinch of the rectifying section on the decanter tie line, which marks the composition of the reflux from the decanter. The minimum reboiler duty, for which this is possible, is determined to be $Q_{b,min} = 30.3$ MJ per kmol feed. The rectifying profile for a reboiler duty of $0.99 \cdot Q_{b,min}$ leaves the phase diagram as depicted in Figure 5.

Quaternary Heterogeneous System Water, n-Butyl Acetate, n-Butanol, Acetic Acid

At ambient pressure, the quaternary system water, n-butyl acetate, n-butanol, acetic acid exhibits three heterogeneous and four homogeneous azeotropes as well as two binary miscibility gaps between water and n-butyl acetate and between water and n-butanol at boiling temperature. As illustrated in Figure 6, the binary miscibility gaps form a coherent heterogeneous region. The specified separation, which is given in Table 2, is accomplished by a heteroazeotropic column setup with a decanter at the top of the column, where pure water is drawn off.

Bausa (2001) inspected this separation with the RBM (see Figure 7) and determined several rectification bodies for the rectifying section, one body for the stripping section and a minimum energy demand of 44.5 MJ per kmol feed. It was already noted by Bausa that the RBM with its linearized

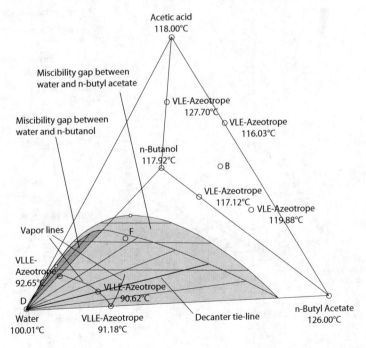

FIGURE 6
Heterogeneous system of water, n-butyl acetate, n-butanol, acetic acid at 1 bar.

TABLE 2

Compositions of Water, *n*-Butyl Acetate, *n*-Butanol, Acetic Acid and Minimum.
Energy Demands

Molar Composition	x_F 0.5/0.17/0.17/0.17	x_D 0.99/2e-3/8e-3/0	x_B 0/0.33/0.34/0.33
$Q_{b,min}/F$ [MJ/kmol]		RBM	FPM
		44.5	35.3

rectification bodies might overestimate the minimum energy demand.
Indeed, the profiles of the rectification section display a distinct curvature,
which is illustrated in Figure 7 by two profiles in the vicinity of the saddle
pinches r2 and r3. Note that these profiles pass by the stripping section with
a considerable distance to the edges of the rectification body of the stripping
section. It is therefore a very tedious, if not impossible task to determine
a CDR for the rectifying section and identify an intersection at minimum
reflux.

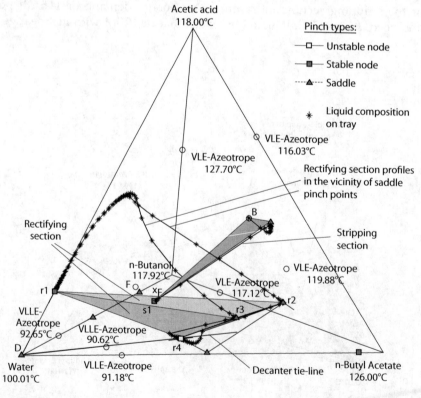

FIGURE 7
Rectification bodies and profiles at the minimum reflux as determined by the RBM.

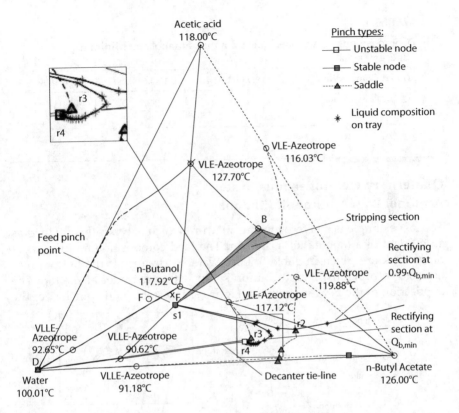

FIGURE 8

Rectifying section profiles for $Q_{B,min}$ and $Q_B = 0.99 \cdot Q_{B,min}$ calculated with the FPM. The stripping section profiles, which do not need to be calculated, are indicated by the corresponding rectification body.

The application of the FPM for this quaternary example is shown in Figure 8. The pinch point curve on which the stable pinch s1 is located terminates at the azeotrope on the decanter tie line. Since no other stable pinch point curves end at a product or on the decanter tie-line, the stable pinch s1 of the stripping section is identified as the feed pinch. If the feed pinch was not determined a-priori, a calculation of profiles from all stable pinches of the system would yield the same results with only little more effort. The upwards calculation of 50 trays for the rectifying section profile is started at the feed pinch s1. The profile, whose course is only a function of the reboiler duty, converges at the unstable node r4 on the decanter tie-line when a sufficient reboiler duty is supplied. The minimum reboiler duty, for which the profile still reaches the unstable node, was determined to be 44.5 MJ per kmol feed, about 21 % lower than the result of the RBM. The profile leaves the composition space for a reboiler duty lower than the minimum, i.e. $0.99 \cdot Q_{b,min}$, as shown in Figure 8.

TABLE 3

Compositions of Acetone, Water, Butanol, Ethanol and Minimum
Energy Demands

Molar Composition	x_F 2e-3/0.99/3e-3/8e-4	x_D 0.12/0.64/0.19/0.05	x_B 0/1/0/0
$Q_{b,min}/F$ [MJ/kmol]		RBM	FPM
		4.44	3.57

Quaternary Heterogeneous System
Acetone, Water, Butanol, Ethanol

The last example is a heterogeneous mixture which is typically obtained as an effluent of a biobutanol fermenter. The feed composition and the separation task are given in Table 3. Pure water is drawn off the bottom in a heteroazeotropic distillation column setup with a decanter at the top. The application of the FPM is shown in Figure 9. The stable node pinch r1 of the

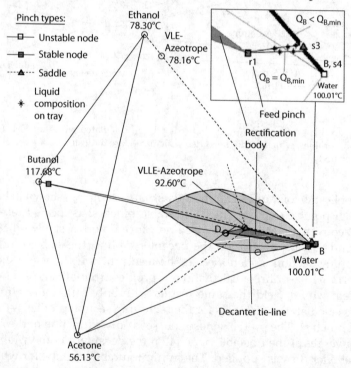

FIGURE 9

System of acetone, water, butanol, ethanol at 1 bar. Stripping section profiles for $Q_{B,min}$ and $Q_B = 0.99 \cdot Q_{B,min}$ calculated with the FPM. The rectifying section profiles, which do not need to be calculated, are indicated by the rectification body.

rectifying section is identified as feed pinch, since it lies on a pinch point curve that terminates at the bottom product. The stripping section profile is then calculated downwards from the feed pinch. The minimum energy demands calculated with the FPM and the RBM are listed in Table 3. Again, the RBM overestimates the minimum energy demand by a considerable margin (20%) due to the curvature of the stripping section profiles. With the help of the FPM, however, an accurate result for the minimum demand is obtained robustly and efficiently.

Conclusion

Various shortcut methods for homogeneous azeotropic distillation based on tray-to-tray calculations, pinch point analysis or a combination thereof are published in the literature. The published shortcut methods for heteroazeotropic distillation are either restricted by inaccuracies (RBM) due to highly curved column profiles or limited to ternary mixtures (BVM, CDRM) due to a graphical inspection of feasibility.

In this work, a shortcut method for heteroazeotropic distillation, the feed pinch method (FPM), is proposed. Like the CDRM, the FPM is based on a combination of pinch point analysis and the calculation of tray-to-tray profiles. In contrast to the CDRM, however, the FPM requires the calculation of only one tray-to-tray profile starting at the feed pinch, which offers decisive advantages: the course of this profile is a function of the reboiler duty only and the inspection of intersection with an unstable node or product composition can easily be performed algorithmically. As a consequence, the FPM can be applied to heterogeneous mixtures of any number of components. The ability of the FBM to handle multicomponent mixtures is demonstrated by two quaternary examples, where the FPM returns better results than the RBM. Furthermore, the procedure can easily be automated, since it allows an implementation of an algorithmic inspection for feasibility instead of a graphical check. The only restriction of the FPM is the requirement of a feed pinch point. To our knowledge, however, the vast majority of relevant separations exhibit a feed pinch. The a-priori knowledge of the feed pinch may further speed up the procedure but is not required. In our future work, the FPM will be extended to handle complex column setups and an automation of the FPM will be implemented.

Acknowledgments

Korbinian Kraemer would like to acknowledge financial support by the Deutsche Forschungsgemeinschaft (DFG) within project MA 1188/26-1 and the Max-Buchner-Forschungsstiftung. Andreas Harwardt would like to acknowledge financial support by the Cluster of Excellence "Tailor-Made Fuels from Biomass", which is funded by the Excellence Initiative by the German federal and state governments.

References

Bausa, J., Watzdorf, R. v., Marquardt, W. (1998). Shortcut methods for nonideal multi-component distillation: 1. Simple columns. *AIChE J.*, *44*, 2181.

Bausa, J., Marquardt, W. (2000). Quick and reliable phase stability test in VLLE flash calculations by homotopy continuation. *Comput. Chem. Eng.*, *24*, 2447.

Bausa, J. (2001). Näherungsverfahren für den konzeptionellen Entwurf und die thermodynamische Analyse von destillativen Trennprozessen. *Fortschrittsberichte VDI.*, *Reihe 3, Nr. 692*, VDI Verlag, Düsseldorf.

Julka, V., Doherty, M. F. (1990). Geometric behavior and minimum flows for nonideal multicomponent distillation. *Chem. Eng. Sci.*, *45*, 1801.

Levy, G. S., Van Dongen, D. B., Doherty, M. F. (1985) Design and synthesis of homogeneous azeotropic distillations. 2. Minimum reflux calculations for nonideal and azeotropic columns. *Ind. Eng. Chem Fundam.*, *24*, 463.

Lucia, A., Amale, A., Taylor R. (2008) Distillation pinch points and more. *Comput. Chem. Eng.*, *32*, 1350.

Pham, H. N., Ryan, P. J., Doherty, M. F. (1989). Design and minimum reflux for heterogeneous azeotropic distillation columns. *AIChE J.*, *35*, 1585.

Poellmann, P., Glanz, S. B, Blass, E. (1994). Calculating minimum reflux of nonideal multicomponent distillation using eigenvalue theory. *Comput. Chem. Eng.*, *18(Suppl.)*, 49.

Urdaneta, R. Y., Bausa, J., Brüggemann, S., Marquardt, W. (2002). Analysis and conceptual design of ternary heteroazeotropic distillation processes. *Ind. Eng. Chem. Res.*, *41*, 3849.

98

Advances in Global Optimization for Standard, Generalized, and Extended Pooling Problems with the (EPA) Complex Emissions Model Constraints

**Ruth Misener, Chrysanthos E. Gounaris and
Christodoulos A. Floudas***

*Department of Chemical Engineering, Princeton University,
Princeton, NJ 08544-5263*

CONTENTS

ABSTRACT We discuss recent advances in deterministic global optimization for (a) standard pooling problems, (b) generalized pooling problems, and (c) extended pooling problems. These three classes of pooling problems are ubiquitous in the chemical, petrochemical, manufacturing, supply chain, and wastewater treatment industries. The primary aim of pooling problems is to maximize the profit of a complex network consisting of refinery exit streams or wastewater streams, blending pools, and final products. The network is subject to constraints reflecting the material balances, and product quality restrictions.

---

* To whom all correspondence should be addressed (floudas@titan.princeton.edu;
Tel: (609) 258-4595; Fax: (609) 258–0211).

In (a), linear blending rules are employed, the nonconvexities are of the bilinear type, and the mathematical model is a nonconvex NLP. In (b), the existence of intermediate streams, the pools, and all connections are treated as discrete alternatives and the resulting model is a nonconvex MINLP. In (c), the EPA Complex Emissions Model (40CFR80.45, 2007), which legally certifies the emissions of reformulated and conventional gasoline, is introduced explicitly. Since several types of nonconvexities result from the mathematical formulation of the EPA model, we present novel theoretical and computational findings for piecewise-linear relaxations and the generation of convex envelopes using the edge-concave paradigm. We incorporate these relaxations into a global optimization algorithm and present computational results on the EPA toxics model.

KEYWORDS *Pooling problems, piecewise-linearization, EPA complex emissions model*

Introduction

In a petroleum refinery, final products are created by combining feed stocks emerging from distillation units, reformers, and catalytic crackers (DeWitt et al., 1989). Because of limited storage availability and transportation requirements, these streams are sent to common pools before being mixed into products (Visweswaran, 2009). A refinery uses the material from up to nine intermediate pools to create a plethora of final products such as three grades of gasoline, diesel fuel, aviationjet fuel and fuel oil (Rigby et al., 1995).

Optimally combining intermediate stocks into final products is of a long-standing interest in the petroleum industry. As early as the 1950s, Exxon was using linear programming to improve blending schemes (Baker and Lasdon, 1985). The objective of these early models, like the more sophisticated models that followed, was to maximize profit while meeting product-specific constraints. Blending feed stocks became more challenging in the 1970s as recognition of environmental and health hazards limited the octane-enhancing additive tetra-ethyl lead (DeWitt et al., 1989; Meyer and Floudas, 2006). Other legislation, such as the Clean Air Act of 1970, restricted the sulfur content and volatility of gasoline.

These environmental standards, coupled with limited availability of low-sulfur crude and new automobiles requiring high octane fuels (DeWitt et al., 1989), inspired extensive research interest into the pooling problem. The pooling problem involves a feed-forward network topology and a set of product quality restrictions. If intermediate storage pools were unnecessary, the problem could be expressed as a linear program, but monitoring pool composition requires nonconvex bilinear and, for large-scale problems, trilinear terms.

In this paper, we review recent advances in deterministic global optimization for the standard pooling problem, which can be modeled as a nonconvex

NLP and the generalized pooling problem, which can be modeled as a nonconvex MINLP. Finally, we introduce the extended pooling problem, based on the EPA Complex Emissions Model (40CFR80.45, 2007), and discuss novel theoretical and computational findings for generating tight convex underestimators and incorporating the relaxations into a branch-and-bound scheme.

Standard Pooling Problem

In the standard pooling problem, the flow rates on a predetermined network structure of feed stocks, pooling tanks, and final products are optimized to maximize profit subject to quality constraints on the final product composition. Although the standard pooling problem considers only fuel qualities that blend linearly, bilinear terms arise from the quality balances about the pooling tanks (Floudas et al., 1989; Floudas and Aggarwal, 1990; Floudas, 2000; Tawar-malani and Sahinidis, 2002).

Haverly (1978), who published the first algorithm to locally improve the classic pooling problem, observed that the solution to his algorithm depended on the starting point. But, in the high-throughput petroleum industry where a saving a fraction of a cent per gallon translates into large profits, even implementing local NLP solvers had a high impact. DeWitt et al. (1989) conservatively estimate that implementing the local optimizer OMEGA for gas blending improvements yielded 30 million dollars in annual revenue for Texaco.

Interest in theoretically guaranteeing global optimality for the pooling problem led to a number of advances in the field of global optimization, especially with respect to biconvex and bilinear programming. Notable contributions have been made by Floudas et al. (1989) and Floudas and Aggarwal (1990), who designed a global search algorithm based on Generalized Benders' Decomposition; Lodwick (1992), who determined implied variable bounds through preprocessing analysis; Floudas and Visweswaran (1990), Visweswaran and Floudas (1990), and Floudas and Visweswaran (1993), who developed the first rigorous deterministic global approach for biconvex and bilinear problems based on duality theory; Foulds et al. (1992), who implemented the bilinear envelopes of McCormick (1976) and Al-Khayyal and Falk (1983); Ben-Tal et al. (1994), who introduced the q-formulation; Adhya et al. (1999), who explored Lagrangian approaches; Quesada and Grossmann (1995), who used the reformulation-linearization technique of Sherali and Alameddine (1992)[1]; Audet et al. (2000), who designed a branch and cut method for quadratic programs using four classes of linearizations;

---

[1] Also see Sherali and Adams (1999) for a comprehensive study of the reformulation-linearization technique developed by Sherali and co-workers.

Tawarmalani and Sahinidis (2002), who proposed the pq-formulation; Varvarezos et al. (2008), who implemented additional refinery planning needs such as risk management into the optimization framework; Almutairi and Elhedhli (2009), who suggested a new Lagrangian relaxation for the pooling problem and demonstrated that their relaxation is often tighter than previously-developed Lagrangian relaxations; and Pham et al. (2009), who incorporated discretized pool qualities into an algorithm that quickly generates near-optimal results. To address the higher-order multilinear terms which arise in large-scale pooling problems, Meyer and Floudas (2003, 2004) derived explicit facets of the convex envelopes for trilinear monomials.

Generalized Pooling Problem

In recent years, researchers have studied a second class of pooling problems, known as generalized pooling problems. In the generalized pooling problem, inter-pool links are permitted and network components such as intermediate streams and pools are treated as discrete alternatives. The resulting non-convex disjunctive program can be modeled as a MINLP. Figure 1 indicates

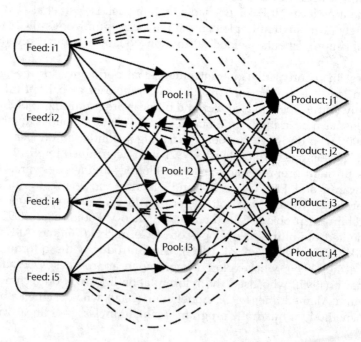

FIGURE 1
Representative superstructure for the generalized pooling problem.

the difficulty of the generalized pooling problem. Because each of the pipes depicted in Figure 1 may or may not be activated, the problem is combinatorially complex with respect to the binary decision variables and bilinear terms.

The first researchers to consider the class of generalized pooling problems, Audet et al. (2004), considered a single, predetermined network topology of three feeds, two pools, and three products. Meyer and Floudas (2006) not only allowed flow between pools, but also substantially broadened the class of generalized pooling problems to include discrete decisions, such as whether to build a pool or pipeline. Meyer and Floudas (2006) considered an industrially-relevant topological superstructure, and optimized the network configuration using disjunctive programming. Although the specific test case in Meyer and Floudas (2006) applies to wastewater treatment plants, the bilinear terms of the wastewater treatment problem match those of the pooling problem. The industrial case study presented in Meyer and Floudas (2006) optimizes a network with seven sources, ten potential plants, one sink, and three relevant qualities.

Karuppiah and Grossmann (2006) studied a variant of the generalized pooling problem by optimizing the network topology of water systems. Using disjunctive programming, Karuppiah and Grossmann (2006) demonstrated substantial objective value improvement in optimizing integrated water systems rather than sequentially optimizing freshwater and wastewater systems.

Although the combinatorial complexity of the generalized pooling problem leads to large models, both Meyer and Floudas (2006) and Karuppiah and Grossmann (2006) were able to solve industrially-relevant examples by incorporating piecewise-linear underestimators of bilinear terms into a global optimization algorithm. Based on these successes, Wicaksono and Karimi (2008) analyzed a variety of novel piecewise-linear underestimators of bilinear terms and showed that the relaxation schemes of Meyer and Floudas (2006) and Karuppiah and Grossmann (2006) can be improved using alternate mathematical representations.

Extended Pooling Problem

Environmental Protection Agency (EPA) *Title 40 Code of Federal Regulations Part 80.45: Complex Emissions Model* (40CFR80.45, 2007) codifies and legally certifies a mathematical model of reformulated gasoline (RFG) emissions based on the eleven fuel components recorded in Table 1. Final products exiting an oil refinery must comply with emissions standards, or upper bounds, on volatile organic (VOCMAX), NO$_x$ (NOXMAX) and toxics (TOXMAX) emissions (40CFR80.41, 2008).

The extended pooling problem, which was introduced by Gounaris and Floudas (2007), incorporates *Title 40 Code of Federal Regulations Part 80.45:*

TABLE 1

Fuel Components in the EPA Complex Emissions Model Bounded
by the Limits of RFG Model Accuracy

	Var	Fuel Quality	Bounds	Units
1	OXY	oxygen	0.0–4.0	wt%
2	SUL	sulfur	0.0–500.0	ppm
3	RVP	Reid Vapor Press	6.4–10.0	psi
4	E200	200°F dist. frac	30.0–70.0	vol%
5	E300	300°F dist. frac	70.0–100.0	vol%
6	ARO	aromatics	0.0–50.0	vol%
7	BEN	benzene	0.0–2.0	vol%
8	OLE	olefins	0.0–25.0	vol%
9	MTB	MTBE		wt% O_2
10	ETB	ETBE		wt% O_2
11	ETH	ethanol		wt% O_2

Complex Emissions Model (40CFR80.45, 2007) and associated legislative
bounds into the constraint set. The extended pooling problem restricts the
volatile organic, NOx, and toxics emissions of RFG by appending three sets
of emissions model equations and the following constraints to the standard
pooling problem:

$$VOC \leq VOC_{MAX} \tag{1}$$

$$NOX \leq NOX_{MAX} \tag{2}$$

$$TOX \leq TOX_{MAX} \tag{3}$$

where VOC_{MAX}, NOX_{MAX}, and TOX_{MAX} are parameters satisfying applicable
legislation.

The equations that make up the EPA Complex Emissions Model are not
only nonconvex, but also non-smooth. Figure 2 illustrates the non-smooth
nature of exhaust benzene (BENZ), a component of toxics emissions, by plotting BENZ versus the fuel quality E300 with all other fuel qualities held
constant. Additionally, the coefficients of the EPA model (40CFR80.45, 2007)
change according to the time of year, region in the country, and the type of
vehicle.

In the following sections, both a MINLP formulation and a tight MILP
relaxation of the extended pooling problem are presented. The MINLP representation of the toxics component of the EPA Complex Emissions Model is
stated without explanation because an equivalent formulation of the emissions model was recently published (Furman and Androulakis, 2008). The
MINLP representation and MILP relaxation of the extended pooling problem

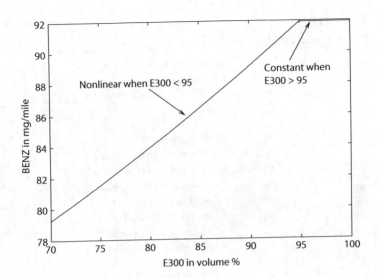

FIGURE 2
BENZ vs. E300 with other fuel qualities constant.

is then integrated into a global optimization algorithm and the extended pooling problem is solved using an example test case. Because of space constraints in this paper, only the toxics component of the emissions model is presented, and the complete the formulation, relaxation, and construction of a global optimization algorithm that includes the volatile organic and NO_X emissions will be presented elsewhere.

Formulation of the Extended Pooling Problem

Standard Portion of the Problem

Table 2 defines the indices, variables, and parameters which formulate the extended pooling problem using a representation equivalent to the standard pooling problem formulation. The objective is to maximize profit or, equivalently, minimize negative profit:

$$\min \sum_{i,l} c_i \cdot x_{i,l} - \sum_{l,j} d_j \cdot y_{l,j} - \sum_{i,j} (d_j - c_i) \cdot z_{i,j.} \tag{4}$$

Bounds limit the availability of each petroleum feed stock:

$$A_i^L \le \sum_l x_{i,l} + \sum_j z_{i,j} \le A_i^U \ \forall i \tag{5}$$

TABLE 2

Pooling Problem p-Formulation

Indices	i	feed stocks from refinery
	j	final products
	k	fuel qualities in EPA Model
	l	pools
Vars	$P_{l,k}$	quality k of pool l
	$x_{i,l}$	flow from stock i to pool l
	$y_{l,j}$	flow from pool l to product j
	$z_{i,j}$	flow from feed i to product j
	of_j	outflow rate of product j
	$u_{j,k}$	quality k of product j
	$y_{E300,j}$	binary switch for product j
	$y_{ARO,j}$	binary switch for product j
Params	c_i	cost of feed stock i
	d_j	revenue from product j
	$A_i^l - A_i^u$	availability bounds on feed i
	$C_{i,k}$	quality k of feed stock i
	$D^L - D^u$	demand bounds for product j
	$P_{j,k}^L - P_{j,k}^u$	bounds on quality k for product j
	S_l	volumetric capacity of pool l

Mass balances are defined around each pool:

$$\sum_i x_{i,l} - \sum_j y_{l,j} = 0 \quad \forall l. \tag{6}$$

Supply limits define the capacity of each pool:

$$\sum_i x_{i,l} \leq S_l \quad \forall l. \tag{7}$$

The quality balances about a pool are:

$$\sum_i C_{i,k} \cdot x_{i,l} - \sum_j P_{l,k} \cdot y_{l,j} = 0 \quad \forall l, k. \tag{8}$$

Augmenting the p-formulation, Gounaris and Floudas (2007) define outflow rates (of_j) and qualities $(u_{j,k})$. The nonconvex EPA model is integrated into the extended pooling problem by calculating the relevant emissions using the fuel qualities $(u_{j,k})$ at each product outflow. The outflow rate is:

$$of_j = \sum_l y_{l,j} + \sum_i z_{i,j} \quad \forall j \tag{9}$$

and the quality balances at the final products outflow are:

$$(u_{j,k}) \cdot (of_j) = \sum_l P_{l,k} \cdot y_{l,j} + \sum_i C_{i,k} \cdot z_{i,j} \quad \forall j, k. \tag{10}$$

Finally, hard bounds to define tight variable limits:

$$0 \le (x_{i,l}) \le \min \left\{ A_i^u, S_l, \sum_j D_j^u \right\} \tag{11}$$

$$0 \le (y_{l,j}) \le \min \left\{ S_l, D_j^u, \sum_i A_i^u \right\} \tag{12}$$

$$0 \le (z_{i,j}) \le \min \left\{ A_i^u, D_j^u \right\} \tag{13}$$

$$\min_i C_{i,k} \le (P_{l,k}) \le \max_i C_{i,k} \tag{14}$$

$$D_j^L \le (of_j) \le D_j^u \tag{15}$$

$$P_{j,k}^L \le (u_{j,k}) \le P_{j,k}^U \tag{16}$$

Equations (4)–(16) define the standard component of the extended pooling problem. Nonconvexities in this component of the extended pooling problem arise from the bilinear terms in the quality balances around the pools (Eq. (8)) and the products (Eq. (10)).

EPA Complex Emissions Portion of the Problem

The following MINLP formulation of the toxics model is presented without explanation, but a detailed discussion of constructing an equivalent MINLP representation can be found in Furman and Androulakis (2008). MINLP formulations of the volatile organic and NO_X models will be presented in a future publication.

The toxics emissions model is a function of the eleven fuel components presented in Table 1. These components are, in turn, functions of the outflow fuel qualities $u_{j,k}$. For the most part, the two are identical:

$$OXY_j = u_{j,1} \; \forall j \quad SUL_j = u_{j,2} \; \forall j$$

$$E200_j = u_{j,4} \; \forall j \quad BEN_j = u_{j,7} \; \forall j$$

$$OLE_j = u_{j,8} \; \forall j \quad MTB_j = u_{j,9} \; \forall j \tag{17}$$

$$ETB_j = u_{j,10} \; \forall j \quad ETH_j = u_{j,11} \; \forall j,$$

but the value of RVP depends on the time of year:

$$RVP_j = \begin{cases} u_{j,3} & \text{Summer} \\ 8.7 & \text{Winter} \end{cases} \quad \forall j, \tag{18}$$

the value E300$_j$ is set to 95 vol% when $u_{j,5} \geq 95$:

$$u_{j,5} - 95 \geq \left(u_{j,5}^L - 95\right) \cdot y_{E300,j} \tag{19}$$

$$u_{j,5} - 95 \leq \left(u_{j,5}^U - 95\right) \cdot (1 - y_{E300,j}) \tag{20}$$

$$E300_j - 95 \leq \left(u_{j,5}^U - 95\right) \cdot y_{E300,j} \tag{21}$$

$$E300_j - 95 \geq \left(u_{j,5}^L - 95\right) \cdot y_{E300,j} \tag{22}$$

$$E300_j - u_{j,5} \leq \left(u_{j,5}^U - u_{j,5}^L\right) \cdot (1 - y_{E300,j}) \tag{23}$$

$$E300_j - u_{j,5} \geq \left(u_{j,5}^L - u_{j,5}^U\right) \cdot (1 - y_{E300,j}), \tag{24}$$

and the value ARO$_j$ is set to 10 vol% when $u_6 \leq 10$:

$$u_{j,6} - 10 \geq \left(u_{j,6}^L - 10\right) \cdot y_{ARO,j} \tag{25}$$

$$u_{j,6} - 10 \leq \left(u_{j,6}^U - 10\right) \cdot (1 - y_{ARO,j}) \tag{26}$$

$$ARO_j - 10 \leq \left(u_{j,6}^U - 10\right) \cdot (1 - y_{ARO,j}) \tag{27}$$

$$ARO_j - 10 \geq \left(u_{j,6}^L - 10\right) \cdot (1 - y_{ARO,j}) \tag{28}$$

$$ARO_j - u_{j,6} \leq \left(u_{j,6}^U - u_{j,6}^L\right) \cdot y_{ARO,j} \tag{29}$$

$$ARO_j - u_{j,6} \geq \left(u_{j,6}^L - u_{j,6}^U\right) \cdot y_{ARO,j} \tag{30}$$

Also, MTB$_j$, ETB$_j$, and ETH$_j$ represent three of the four oxygen components, so they are constrained by OXY$_j$:

$$OXY_j \geq MTB_j + ETB_j + ETH_j \quad \forall j. \tag{31}$$

Note that $y_{E300,j}$ and $y_{ARO,j}$ are binary decision variables representing disjunctions in the EPA Complex Emissions Model (40CFR80.45, 2007) such as the one illustrated in Figure 2. These binary variables make the extended pooling problem an MINLP rather than a nonconvex NLP like the standard pooling problem.

Toxics emissions (TOX$_j$), is expressed as the sum of six components: exhaust benzene (BENZ$_j$), formaldehyde (FORM$_j$), acetaldehyde (ACET$_j$), 1,3-butadiene (BUTA$_j$), nonexhaust benzene (NEBENZ$_j$), and polycyclic organic matter (POM$_j$). The POM model is a linear multiple of the volatile organic exhaust emissions model and is excluded from this paper because of space. The toxics emissions model without the POM contribution is:

$$\mathrm{TOX}_j = \mathrm{BENZ}_j + \mathrm{FORM}_j + \mathrm{ACET}_j + \mathrm{BUTA}_j$$
$$+ 10 \cdot \mathrm{NEBENZ}_j \quad \forall j \tag{32}$$

The five components of the simplified toxics emission model are presented in Eq. (33)–(41). The model coefficients, which vary according to the time of year, region of the country, and whether the vehicle is a "high" (e = 1) or "low" (e = 2) emitter, are presented in 40CFR80.45 (2007) and Furman and Androulakis (2008).

$$\mathrm{BENZ}_j = \sum_{e=1}^{2} \frac{\mathrm{BENZ}(b) \cdot w_e^T}{e^{b_e(b)}} \times \exp\{t_{BE,e,j}\}, \tag{33}$$

$$t_{BE,e,j} = c_{e,1}^{BE}\mathrm{OXY}_j + c_{e,2}^{BE}\mathrm{SUL}_j + c_{e,3}^{BE}\mathrm{E300}_j + c_{e,4}^{BE}\mathrm{ARO}_j + c_{e,5}^{BE}\mathrm{BEN}_j \tag{33}$$

$$\mathrm{FORM}_j = \sum_{e=1}^{2} \frac{\mathrm{FORM}(b) \cdot w_e^T}{e^{f_e(b)}} \times \exp\{t_{F,e,j}\} \tag{35}$$

$$t_{F,e,j} = c_{e,1}^{F}\mathrm{E300}_j + c_{e,2}^{F}\mathrm{ARO}_j + c_{e,3}^{F}\mathrm{OLE}_j + c_{e,4}^{F}\mathrm{MTB}_j \tag{36}$$

$$\mathrm{ACET}_j = \sum_{e=1}^{2} \frac{\mathrm{ACET}(b) \cdot w_e^T}{e^{a_e(b)}} \times \exp\{t_{A,e,j}\} \tag{37}$$

$$t_{A,e,j} = c_{e,1}^{A}\mathrm{SUL}_j + c_{e,2}^{A}\mathrm{RVP}_j + c_{e,3}^{A}\mathrm{E300}_j$$
$$+ c_{e,4}^{A}\mathrm{ARO}_j + c_{e,5}^{A}\mathrm{MTB}_j$$
$$+ c_{e,6}^{A}\mathrm{ETB}_j + c_{e,7}^{A}\mathrm{ETH}_j \tag{38}$$

$$\mathrm{BUTA}_j = \sum_{e=1}^{2} \frac{\mathrm{BUTA}(b) \cdot w_e^T}{e^{d_e(b)}} \times \exp\{t_{BU,e,j}\} \tag{39}$$

$$t_{BU,e,j} = c_{e,1}^{BU}\mathrm{OXY}_j + c_{e,2}^{BU}\mathrm{SUL}_j + c_{e,3}^{BU}\mathrm{E200}_j$$
$$+ c_{e,4}^{BU}\mathrm{E300}_j + c_{e,5}^{BU}\mathrm{ARO}_j + c_{c,6}^{BU}\mathrm{OLE}_j \tag{40}$$

$$\text{NEBENZ}_j = \alpha_1 \cdot \text{BEN}_j + \alpha_2 \cdot \text{RVP}_j \cdot \text{BEN}_j +$$

$$\alpha_3 \cdot \text{BEN}_j \cdot \text{MTB}_j + \alpha_4 \cdot \text{RVP}_j^2 \cdot \text{BEN}_j +$$

$$\alpha_5 \cdot \text{RVP}_j \cdot \text{BEN}_j \cdot \text{MTB}_j +$$

$$\alpha_6 \cdot \text{RVP}_j^3 \cdot \text{BEN}_j +$$

$$\alpha_7 \cdot \text{RVP}_j^2 \cdot \text{BEN}_j \cdot \text{MTB}_j \qquad (41)$$

Relaxation of the Extended Pooling Problem

To construct a MILP relaxation of the MINLP representation of the extended pooling problem, Eq. (8), (10), (33), (35), (37), (39), and (41) must be underestimated.

In underestimating Eq. (8), we follow Meyer and Floudas (2006) and Karuppiah and Grossmann (2006) in constructing piecewise-linear underestimators for the bilinear terms. We use the recent results of Gounaris et al. (2008) which extend the work of Wicaksono and Karimi (2008) to choose the best mathematical representation of the underestimator. Specifically, we use a representation denoted "nf4r" to relax Eq. (8) (Gounaris et al., 2008). The bilinear terms $p_{l,k} \cdot y_{l,j}$ in Eq. (8) are replaced with a placeholder variable:

$$p_{l,k}^{\text{L}} \cdot y_{l,j}^{\text{L}} \le w_{l,j,k}^{p,y} \le p_{l,k}^{\text{U}} \cdot y_{l,j}^{\text{U}}. \qquad (42)$$

To construct piecewise-linear bilinear underestimators, each fuel quality $p_{l,k}$ is *ab initio* partitioned into N segments according to the variable choice of Gounaris et al. (2008):

$$p_{l,k}(n) = p_{l,k}^{\text{L}} + \frac{n}{N} \cdot \left(p_{l,k}^{\text{U}} - p_{l,k}^{\text{L}} \right) \forall l, k, \ n = 0, \ldots, N \qquad (43)$$

and a binary variable $\lambda_{l,k}(n)$ is introduced to activate one and only one domain segment:

$$\lambda_{l,k}(n) = \begin{cases} 1 & \text{if } p_{l,k}(n-1) \le p_{l,k} \le p_{l,k}(n) \\ 0 & \text{else} \end{cases} \qquad (44)$$

$$\forall l, k, n = 1, \ldots, N$$

$$\sum_{n=1}^{N} p_{l,k}(n-1) \cdot \lambda_{l,k}(n) \le p_{l,k} \le \sum_{n=1}^{n} p_{l,k}(n) \cdot \lambda_{l,k}(n) \ \forall l, k, n = 1, \ldots, N \qquad (45)$$

$$\sum_{n=1}^{N} \lambda_{l,k}(n) = 1 \forall l, k. \tag{46}$$

Continuous variable $\Delta y_{l,j,k}(n)$, $n = 1, \cdots N$ is a place holder for the flow rate $y_{l,j}$ in the bilinear relaxations:

$$y_{l,j} = y_{l,j}^{L} + \sum_{n=1}^{N} \Delta y_{l,j,k}(n) \tag{47}$$

$$0 \le \Delta y_{l,j,k}(n) \le \sum_{n=1}^{N} \left(y_{l,j}^{U} - y_{l,j}^{L} \right) \cdot \lambda_{l,k}(n). \tag{48}$$

The final relaxation of the Eq. (8) bilinear terms $\forall l, j, k$ is:

$$w_{l,j,k}^{p,y} \ge y_{l,j}^{L} \cdot p_{l,k} + \sum_{n=1}^{N} p_{l,k}(n-1). \Delta y_{l,j,k}(n) \tag{49}$$

$$w_{l,j,k}^{p,y} \ge y_{l,j}^{U} \cdot p_{l,k} + \sum_{n=1}^{N} \{ p_{l,k}(n) \cdot (\Delta y_{l,j,k}(n) - \left(y_{l,j}^{U} - y_{l,j}^{L} \right) \cdot \lambda_{l,k}(n)) \} \tag{50}$$

$$w_{l,j,k}^{p,y} \le y_{l,j}^{L} \cdot p_{l,k} + \sum_{n=1}^{N} \{ p_{l,k}(n-1) \cdot (\Delta y_{l,j,k}(n) - \left(y_{l,j}^{U} - y_{l,j}^{L} \right) \cdot \lambda_{l,k}(n)) \} \tag{51}$$

$$w_{l,j,k}^{p,y} \le y_{l,j}^{U} \cdot p_{l,k} + \sum_{n=1}^{N} p_{l,k}(n) \cdot \Delta y_{l,j,k}(n). \tag{52}$$

Because the bilinear terms in Eq. (10) are closely related to the ones in Eq. (8), we chose not to construct a piecewise-linear relaxation of the $(u_{j,k}) \cdot (of_j)$ terms. Instead, we replaced each of the bilinear terms in Eq. (10) with the continuous variable $w_{j,k}^{u,of}$ and underestimated the terms using the envelopes of McCormick (1976):

$$w_{j,k}^{u,of} \ge u_{j,k}^{L} \cdot of_j + u_{j,k} \cdot of_j^{L} - u_{j,k}^{L} \cdot of_j^{L} \tag{53}$$

$$w_{j,k}^{u,of} \ge u_{j,k}^{U} \cdot of_j + u_{j,k} \cdot of_j^{U} - u_{j,k}^{U} \cdot of_j^{U} \tag{54}$$

$$w_{j,k}^{u,of} \le u_{j,k}^{L} \cdot of_j + u_{j,k} \cdot of_j^{U} - u_{j,k}^{L} \cdot of_j^{U} \tag{55}$$

$$w_{j,k}^{u,of} \le u_{j,k}^{U} \cdot of_j + u_{j,k} \cdot of_j^{L} - u_{j,k}^{U} \cdot of_j^{L} \quad \forall j, k \tag{56}$$

We underestimated the convex Eqs. (33), (35), (37), and (39) using outer approximation. We partitioned the continuous variables $t_{BE,e,j}$, $t_{Ar,e,j}$, $t_{Fr,e,j}$, and $t_{BU,e,j}$ and constructed supporting hyperplanes at each partition point.

Equation (41), representing nonexhaust benzene emissions, is the final non-convex equation in the extended pooling problem. Although Eq. (41) is not edge-concave, we use the paradigm of edge-concavity to efficiently generate a tight lower bound on NEBENZ. Tardella (1988/89, 2003, 2008) introduced edge-concave functions, a class of functions that admit a vertex polyhedral convex envelope. Using the theoretical results of Tardella (2003), Meyer and Floudas (2005) developed an algorithm to generate the convex envelope of any three-dimensional edge-concave function.

According to Tardella (2003), a function on a box is edge-concave if and only if it is componentwise concave, that is:

$$\frac{\partial^2 \text{ NEBENZ}_j}{\partial \text{RVP}_j^2} = 2\alpha_4 \cdot \text{BEN}_j + 6\alpha_6 \cdot \text{RVP}_j \cdot \text{BEN}_j + 2\alpha_7 \cdot \text{MTB}_j \leq 0 \ \forall j \quad (57)$$

$$\frac{\partial^2 \text{ NEBENZ}_j}{\partial \text{MTB}_j^2} = 0 \leq 0 \quad \forall j \quad (58)$$

$$\frac{\partial^2 \text{ NEBENZ}_j}{\partial \text{BEN}_j^2} = 0 \leq 0 \quad \forall j \quad (59)$$

Since Eq. (58) and (59) are always true, the remaining task is to see when Eq. (57) is negative. Because $\frac{\partial^2 \text{ NEBENZ}_j}{\partial \text{RVP}_j^2} \not\leq 0$, Eq. (41) is not edge-concave. However:

$$\text{NEBENZ}_j - \alpha_4' \cdot \text{RVP}_j^2 \cdot \text{BEN}_j \quad (60)$$

is edge-concave when:

$$\alpha_4' = \alpha_4 + 3 \cdot \alpha_6 \cdot \text{RVP}_j^L + \alpha_7 \cdot \text{MTB}_j^L, \quad (61)$$

so we underestimate Eq. (60) using the algorithm of Meyer and Floudas (2006), which results in the facets of the convex envelope, and the remainder $(\alpha_4' \cdot \text{RVP}_j^2 \cdot \text{BEN}_j)$ using the recursive arithmetic techniques of Maranas and Floudas (1995) and Ryoo and Sahinidis (2001). We will not explicitly state the linear equations describing the convex envelope of Eq. (60) because they change as bounds are tightened within a global optimization algorithm, but the method to develop the convex envelope can be found in Meyer and Floudas (2006).

NEBENZ has a global minimum of 0, but the underestimate attained using only recursive arithmetic techniques is –13.56 in Region 1 and –11.49 in Region 2 (Maranas and Floudas, 1995; Ryoo and Sahinidis, 2001). Combining recursive techniques with the edge-concave algorithm of Meyer and Floudas (2005) improves the lower bound by more than 20%, to –10.61 in Region 1 and –9.01 in Region 2.

The relaxations in this section are substituted for the nonlinear components presented in the Eq. (4)–(41). The next section presents an extended pooling problem test case that integrates the MINLP formulation and MILP relaxation into a global optimization algorithm.

Example of the Extended Pooling Problem

We use the topology illustrated in Figure 3 and defined by the parameters listed in Tables 3–6 as an extended pooling problem test case. The test case has four grades of petroleum feed stocks, three fuel additives, one pool, and two final products. The four feed stocks represent the characteristics of four intermediate stocks leaving the distillation, reforming, or catalytic cracking units of a refinery. Each feed stock is estimated to have a market value based on its composition. The pool has capacity $S_1 = 300$.

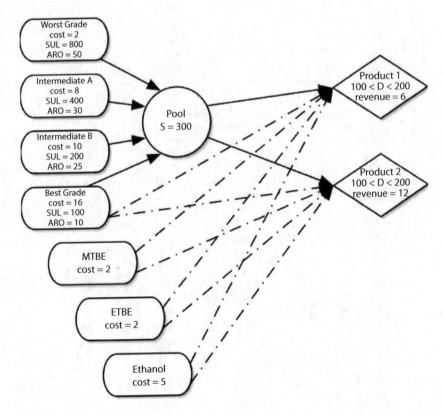

FIGURE 3
Topology for the extended pooling problem.

TABLE 3

Cost (c_i) and Availability (A^L_i & A^U_i) of Feed i

i	1	2	3	4	5	6	7
C_i	2	8	10	16	2	2	5
A^L_i	50	0	0	0	0	0	0
A^U_i	400	200	200	100	10	10	50

TABLE 4

Quality Bounds k on Product j ($P^L_{j,k}$ & $P^U_{j,k}$)

k	$P^L_{1,k}$	$P^L_{2,k}$	$P^U_{1,k}$	$P^U_{2,k}$
1	0.3	0.3	4.0	4.0
2	50.0	50.0	500.0	250.0
3	6.4	6.4	10.0	8.0
4	30.0	30.0	70.0	60.0
5	70.0	70.0	100.0	85.0
6	0.0	0.0	30.0	25.0
7	0.0	0.0	2.0	0.5
8	0.0	0.0	25.0	10.0
9	0.1	0.1	4.0	4.0
10	0.1	0.1	4.0	4.0
11	0.1	0.1	4.0	4.0

TABLE 5

Price (d_j) and Demand (D^L_j & D^U_j) of Product j

j	d_j	D^L_j	D^U_j
1	6	100	200
2	12	100	200

TABLE 6

Quality k of Raw Material i ($C_{j,k}$)

k	$C_{1,k}$	$C_{2,k}$	$C_{3,k}$	$C_{4,k}$	$C_{5,k}$	$C_{6,k}$	$C_{7,k}$
1	0.1	0.2	0.4	0.7	100	100	100
2	800	400	200	100	0	0	0
3	6.0	8.8	8.0	8.0	8.4	8.0	9.6
4	20	60	55	50	100	100	100
5	70	85	80	75	100	100	100
6	50	30	25	10	0	0	0
7	0	0.8	1.0	0.2	0	0	0
8	10	15	15	5	0	0	0
9	0	0	0	0	100	0	0
10	0	0	0	0	0	100	0
11	0	0	0	0	0	0	100

In reality, transportation considerations may require additives to be mixed into the gasoline at a distribution station (*e.g.*, ethanol is rarely transported by pipeline with other gasoline components), but the test case described in this study simplifies the problem by assuming that additives are blended into the final products at the refinery.

We developed a C++ program that interfaces with the linear solver CPLEX (ILOG, 2005) to minimize the MILP relaxations of the extended pooling problem. The upper bounds are obtained by making system calls to the local nonlinear solver MINOS (Murtagh et al., 2004) through the modeling language GAMS (Brooke et al., 2005). Note that system calls to GAMS slow the program substantially and an interface with an open-source nonlinear solver to generate upper bounds can lead to significant improvements.

Using the upper and lower bounding strategies, we designed a global optimization algorithm to reach ε-convergence. Each of the $p_{l,k} \cdot y_{l,j}$ terms in Eq. 8, are underestimated using Eq. (42)–(52) with $N = 16$ segments and the continuous variables $t_{BE,\,e,j}$, $t_{A,e,j,j}$, $t_{F,\,e,j}$, and $t_{BU,\,e,j}$ are partitioned into 32 segments to create an outer approximation for Eq. (33), (35), (37), (39).

TABLE 7
Global Solution (Within 0.49% Optimality Gap)

	$i = 1$	$i = 2$	$i = 3$	$i = 4$
$x_{i,1}$	57.24	0.00	152.62	3.93
$x_{i,1}$	1.21	3.48	0.10	0.10
$z_{i,2}$	74.23	6.52	0.38	0.20
	$j = 1$	$j = 2$		
$y_{1,j}$	95.11	118.67		
of_j	100	200		

k	$P_{1,k}$	$u_{1,k}$	$u_{1,k}$
1	0.33	4.00	4.00
2	358.80	342.49	250.00
3	7.46	7.51	7.70
4	45.54	47.60	49.13
5	77.24	78.04	77.21
6	31.42	30.00	22.35
7	0.72	0.68	0.50
8	13.47	12.88	9.85
9	0.00	3.48	3.26
10	0.00	0.10	0.19
11	0.00	0.10	0.10

After solving both the full MILP relaxation to develop a lower bound and locally solving the original MINLP representation to obtain an upper bound, we perform strong branching on the product flow rates (of_j). After each strong branching step, the intermediate flow rates $y_{l,j}$ are optimally tightened by replacing the objective function in Eq. (4) with $y_{l,j}$ $\forall_{l,j}$ and maximizing and minimizing the temporary objective function. We also optimally tighten the quality variables ($p_{l,k}$ & $u_{j,k}$) satisfying:

$$\left| \frac{\frac{\sum_i c_{i,k} x_{i,l}}{\sum_j y_{l,j}} - p_{l,k}}{p_{l,k}} \right| \geq 0.005 \tag{62}$$

$$\left| \frac{\frac{\sum_i \{z_{i,j} \cdot C_{i,k}\} + \sum_l \{p_{l,k} \cdot y_{l,j}\}}{of_j} - u_{j,k}}{u_{j,k}} \right| \geq 0.005, \tag{63}$$

i.e., the variables deviating more than 0.5% from their value in relation to the other variables. For a toxics standard TOXMAX = 61, the optimality gap reduces to 0.49% (LB = −66.98) after 187.4 seconds on a Pentium 4 running Linux. Table 7 displays the variable values at this solution.

Conclusion

This paper has reviewed the recent advances in deterministic global optimization for (a) standard pooling problems, (b) generalized pooling problems, and (c) extended pooling problems. Although small to medium-sized standard pooling problems have been successfully addressed by a number of researchers, industrially-sized pooling problems, the combinatorially-complex generalized pooling problem, and the newly-introduced extended pooling problem offer interesting areas of research. Solving these pooling problems will help improve today's energy systems in a cost-effective and environmentally conscious manner. Although these problems have proved challenging, new global optimization methods, such as the relaxation techniques presented in this paper, may lead to major improvements.

Acknowledgments

The authors thankfully acknowledge support from the National Science Foundation. R.M. is grateful for her National Science Foundation Graduate Research Fellowship.

References

40CFR80.41. Code of Federal Regulations: Standards and requirements for compliance, 2008. http://frwebgate.access.gpo.gov/cgi-bin/get-cfr.cgi.

40CFR80 45 Code of Federal Regulations: Complex emissions model, 2007. http://frwebgate.access.gpo.gov/cgi-bin/get-cfr.cgi.

N. Adhya, M. Tawarmalani, and N. V. Sahinidis. A Lagrangian approach to the pooling problem. *Ind. Eng. Chem. Res.*, 38(5):1965–1972, 1999.

F. A. Al-Khayyal and J. E. Falk. Jointly constrained biconvex programming. *Math. of Oper. Res.*, 8(2): 273–286, 1983.

H. Almutairi and S. Elhedhli. A new Lagrangean approach to the pooling problem. *J. of Glob. Optim.* ,2009. Forthcoming.

C. Audet, P. Hansen, B. Jaumard, and G. Savard. A branch and cut algorithm for nonconvex quadratically constrained quadratic programming. *Math. Program.*, 87(1):131–152, 2000.

C. Audet, J. Brimberg, P. Hansen, S. Le Digabel, and N. Mladenovic. Pooling problem: Alternate formulations and solution methods. *Manag. Sci.*, 50(6):761–776, 2004.

T. E. Baker and L. S. Lasdon. Successive linear programming at Exxon. *Manag. Sci.*, 31(3):264–274, 1985.

A. Ben-Tal, G. Eiger, and V. Gershovitz. Global minimization by reducing the duality gap. *Math. Program.*, 63(2):193–212, 1994.

A. Brooke, D. Kendrick, and A. Meeraus. GAMS: A user's guide, 2005. GAMS Development Corporation.

C. W. DeWitt, L. S. Lasdon, A. D. Waren, D. A. Brenner, and S. Melham. OMEGA: An improved gasoline blending system for Texaco. *Interfaces*, 19(1):85–101, 1989.

C. A. Floudas and A. Aggarwal. A decomposition strategy for global optimum search in the pooling problem. *ORSA J. on Comput.*, 2, 1990.

C. A. Floudas and V. Visweswaran. Primal-relaxed dual global optimization approach. *J of Optim. Theory and Appl.*, 78(2):187–225, 1993.

C. A. Floudas and V. Visweswaran. A global optimization algorithm (GOP) for certain classes of nonconvex NLPs: I. Theory. *Comput. & Chem. Eng.*, 14(12):1397–1417, 1990.

C. A. Floudas, A. Aggarwal, and A. R. Ciric. Global optimum search for nonconvex NLP and MINLP problems. *Comput. & Chem. Eng.*, 13(10):1117–1132, 1989.

C.A. Floudas. *Deterministic Global Optimization: Theory, Methods and Applications.* Nonconvex Optimization and Its Applications. Kluwer Academic Publishers, Dordrecht, Netherlands, 2000.

L. R. Foulds, D. Haughland, and K. Jornsten. A bilinear approach to the pooling problem. *Optim.*, 24:165–180, 1992.

K. C. Furman and I. P. Androulakis. A novel MINLP-based representation of the original complex model for predicting gasoline emissions. *Comput. & Chem. Eng.*, 32: 2857–2876, 2008.

C. E. Gounaris and C.A. Floudas. Formulation and relaxation of an extended pooling problem. In 2007 *AIChE Annual Meeting*, Salt Lake City, Utah, 2007. AIChE.

C. E. Gounaris, R. Misener, and C.A. Floudas. Computational comparison of piecewise-linear relaxations for pooling problems. 2008. Submitted for Publication.

C. A. Haverly. Studies of the behavior of recursion for the pooling problem. *ACM SIGMAP Bulletin*, 25:19–28, 1978.

ILOG. CPLEX. http://www.ilog.com/products/cplex/, 2005. Version 9.0.2.

R. Karuppiah and I.E. Grossmann. Global optimization for the synthesis of integrated water systems in chemical processes. *Comput. & Chem. Eng.*, 30:650–673, 2006.

W. A. Lodwick. Preprocessing nonlinear functional constraints with applications to the pooling problem. *ORSA J. on Comput.*, 4(2):119–131, 1992.

C. D. Maranas and C. A. Floudas. Finding all solutions of nonlin-early constrained systems of equations. *J. of Glob. Optim.*, 7(2):143–182, 1995.

G. P. McCormick. Computability of global solutions to factorable nonconvex programs: Part 1-convex underestimating problems. *Math. Program.*, 10(1):147–175, 1976.

C. A. Meyer and C. A. Floudas. Trilinear monomials with positive or negative domains: Facets of the convex and concave envelopes. In C. A. Floudas and P. M. Pardalos, editors, *Frontiers in Global Optimization*, pages 327–352. Kluwer Academic Publishers, 2003.

C. A. Meyer and C. A. Floudas. Trilinear monomials with mixed sign domains: Facets of the convex and concave envelopes. *J. of Glob Optim.*, 29(2):125–155, 2004.

C. A. Meyer and C. A. Floudas. Global optimization ofa combinatorially complex generalized pooling problem. *AIChE J*, 52(3):1027–1037, 2006.

C. A. Meyer and C. A. Floudas. Convex envelopes for edge-concave functions. *Math. Program.*, 103(2):207–224, 2005.

B. A. Murtagh, M. A. Saunders, W. Murray, M. A. Saunders, P. E. Gill, R. Raman, and E. Kalvelagen. MINOS. http://www.gams.com/dd/docs/solvers/minos.pdf, 2004.

V. Pham, C. Laird, and M. El-Halwagi. Convex hull discretization approach to the global optimization of pooling problems. *Ind. Eng. Chem. Res.*, 48(4): 1973–1979, 2009.

I. Quesada and I. E. Grossmann. Global optimization of bilinear process networks with multicomponent flows. *Comput. & Chem. Eng.*, 19:1219–1242, 1995.

B. Rigby, L. S. Lasdon, and A. D. Waren. The evolution of Texaco's blending systems: From OMEGA to Star Blend. *Interfaces*, 25 (5):64–83, 1995.

H. S. Ryoo and N. V. Sahinidis. Analysis ofbounds for multilinear functions. *J. of Glob. Optim.*, 19(4):403–424, 2001.

H. D. Sherali and W. P. Adams. *A Reformulation-Linearization Technique for Solving Discrete and Continuous Nonconvex Problems*. Nonconvex Optimization and Its Applications. Kluwer Academic Publishers, Dordrecht, Netherlands, 1999.

H. D. Sherali and A. Alameddine. A new reformulation-linearization technique for bilinear programming problems. *J. of Glob. Optim.*, 2:379–410, 1992.

F. Tardella. On a class of functions attaining their maximum at the vertices of a polyhedron. *Discret. Appl. Math.*, 22:191–195, 1988/89.

F. Tardella. On the existence of polyhedral convex envelopes. In C. A. Floudas and P. M. Pardalos, editors, *Frontiers in Global Optimization*, pages 563–573. Kluwer Academic Publishers, 2003.

F. Tardella. Existence and sum decomposition of vertex polyhedral convex envelopes. *Optim. Lett.*, 2:363–375, 2008.

M. Tawarmalani and N. V. Sahinidis. *Convexification and Global Optimization in Continuous and Mixed-Integer Nonlinear Programming: Theory, Applications, Software, and Applications*. Nonconvex Optimization and Its Applications. Kluwer Academic Publishers, Norwell, MA, USA, 2002.

D. K. Varvarezos, B. J. Joffe, G. E. Paules IV, and T. Kunt. New optimization paradigms for refinery planning. In *Proceedings Foundations of Computer-Aided Process Operations*, pages 441–445. FOCAPO, 2008.

V. Visweswaran. MINLP: Applications in blending and pooling. In C. A. Floudas and P. M. Pardalos, editors, *Encyclopedia of Optimization*, pages 2114–2121. Springer Science, 2 edition, 2009.

V. Visweswaran and C. A. Floudas. A global optimization algorithm (GOP) for certain classes of nonconvex NLPs: II. application of theory and test problems. *Comput. & Chem. Eng.*, 14(12):1419–1434, 1990.

D. S. Wicaksono and I. A. Karimi. Piecewise MILP under-and over-estimators for global optimization of bilinear programs. *AIChE J.*, 54(4):991–1008, 2008.

99

Constructing, Modifying and Maintaining Consistent Process Models

Heinz A Preisig

Department of Chemical Engineering Norwegian University of Science and Engineering N-7491 Trondheim, Norway

CONTENTS

ABSTRACT Network modelling of chemical-physical-biological systems is being discussed in its basic structures. A framework for the implementation of this model design and maintenance environment is presented consisting of three main components: a muli-graph editor, a semantic module for the multi-graph definition, a semantic module for the nodes and the arcs implementing the application of modelling chemical-physical-biological systems.

KEYWORDS   *Computer-aided, modelling, process systems engineering*

Mind-Mapping Process Models

Models are done for a purpose, a specific application. So as the application varies, the model varies giving rise to generate different models for possibly the same (physical) object. Models are used to mimic behaviours, to map behaviours into objects that can be manipulated, that can be experimented with without having to "fool" with the real system, not at least because the real-world may not permit such manipulations. Models give freedom to the mind, allow tampering and testing, playing with what could become real before it has real-world consequences beyond the use of resources. Models thus play a central role in anything that has to do with exploring and exploiting object's behaviours. The definition of model is subject to philosophical considerations (Rosenbluth and Wiener, 1945; Apostel, 1960; Swanson, 1966; Aris, 1978; Peschel, 1976). People also use the term model for quite different things, largely depending on the context and discipline. Thus we define first our region of operation. For the time being this will be

- Emphasise on physical-chemical-biological systems
- Macroscopic field theories, thus not particulate and stochastic systems
- The term "modelling" is used for the generation of a set of equations and the instantiation of conditions and parameters.
- The solution of the model is usually its time-dependent response to a set of time-dependent stimuli.
- Application domain mainly chemical engineering or related subjects.

Having place ourselves in a context with respect to what modelling is about, what should we want to be able to do? Or maybe first what is it that we are doing:

Modelling is a major part of the chemical engineering curriculum in which, besides basic physical concepts a range of theories are communicated that enable the engineers to describe the behaviour of phys-chem-bio systems usually in the form of a set of equations. Having the equations a mathematical problem is formulated by instantiating conditions[1] and parameters[2]. This

---

[1] The term "conditions" is used for state-dependent information, which is driving the system and considered known, usually state information related to the environment of the plant being modelled.

[2] Characteristic quantities, which either are universal constants or parameters of partial empirical models fitted to experimental data.

mathematical problem is in the continuation solved either analytically or most commonly numerically. The problem may be classified as a simulation (steady state or dynamic) if other stimuli are given and one wants to obtain the result to these stimuli, or (optimal) control problem if one has characterised the plant and wants to compute the stimuli having define a desired result, or thirdly an design (realisation, design) problem if one knows the stimuli and the response, but requires to "fit" the plant by adjusting conditions and parameters such as volumes, number of trays etc. etc. These three classes of problems have stimulated different research activities, which generated a multitude of different solutions and which are more or less tailored to specific properties of the mathematical problem. In all cases, the model is kind of given, at least the structure or the superstructure (superimposed models linked by a decision network). Thus people have worked on generating solvers to these problems and there exists a quite large zoo of solvers for most of the principle mathematical problems. Not that they are perfect and of universal nature, but they do in most cases a decent job. The solvers are quite readily available and most engineers and scientist have an array of tools right at there finger tips on their personal computers, which are generating results efficiently and rapidly.

In contrast, the job of generating a decent model in the first place is seen as artistry. Today the statement *modelling being an art* is a quite commonly accepted, so why then not classify engineering schools as creative art centres? Is it not surprising that there has been so little of a visible attempt to make modelling–being defined as mapping a real-world object's behaviour into a set of equations–a structured, well-defined process. The consequences are that we have little control over this process, which is quite visible as generating plant model is today one of the major bottle necks in engineering. Having implemented several prototypes one of which was commercialised, the direct costs benefits can be estimated as we know now that the modelling process itself can be reduced by up to 2 orders of magnitudes. This translates directly into saving of manpower. However the indirect costs of not having done the model or having erroneous models are not predictable but must be enormous indeed.

Modelling is a design process that should depend on the application of the model. Today it is probably safe to assume that one has accepted the fact that models are linked with the application, as the model is tailored to reproduce those characteristics that are essential for the application. A model is thus kind of fitted to the application and consequently an outer loop exists in which one optimises the descriptive power relevant or necessary or optimal with respect to the application of the model. The performance of the process using the model is thus the criterion making it necessary to see the model not as a fixed item, but one that is being optimised with respect to the application based on a measure reflecting the use of the model.

So what kind of capabilities would we possibly like to have available as an engineer or scientist?

- **It should provide means to design and maintain consistent models.** *Deisgn*: acknowledges the fact that models cannot be synthesised, at least not as yet. Models are being designed to meet the user's specifications, which in turn are defined by the use of the model (Apostel, 1960; Aris, 1978). The term *maintain* acknowledges the fact that we are usually not able to meet the specs in one shot but may need several ones and also the use of the model may change with time or the there may be other reasons to adjust it, such as plant modifications. The term *consistent* acknowledges the fact that certain basic rules must be obeyed when constructing the model equations.

- **It should enable us to map our mental picture of the plant into a set of equations.** Thus again the aspect of design is emphasised, opening the opportunity to use, try and modify rapidly different ideas on how to capture the essentials of the process. The tool should thus enable the user to generate results within a very short time span, meaning at least two orders of magnitude reduction in the time it takes to construct a model and generate results that serve an evaluation of the model.

Mind-mapping induces the feeling of fuzziness of information. It is this picture of the user dumping ideas and the computing environment filling in the rest. This *filling in* process requires quite the opposite, namely nearly an extreme formalism that has knowledge of all the applicable concepts. Thus underlying there is an extremely hard-defined structure. Whilst the user's mind is allowed to run freely, the processing system task is to keep the information in an extremely tightly defined framework. The user is thereby put in the position of riding a carriage on a track given the task to take decisions on which track is to be chosen, decisions that are only bound to fundamental concepts.

The challenge is thus to define a tightly constraint environment that provides an extreme wide bandwidth on applications.

The Network Modelling Framework

The framework must thus be defined tightly and wishing it to be also broadly applicable it must be fundamental in terms of science. It is thus not much of a surprise that the work done over the last years when constructing such environments has mainly been a stepping back towards the fundamentals analysing the process of constructing models in detail not at least also applying the ideas in a teaching environment. The result is a step-wise systematic approach for establishing models. We have limited ourselves to macroscopic models, though the extension to microscopic systems is falling in line quite readily. The different steps are:

- Map the process into an abstract network of communicating control volumes
- Establish base model by establishing the balances of the relevant conservation equations for each node. This defines the primary state of the plant.
- Add description for the transfer defined in the network.
- Add internal dynamics for each node (reactions, transpositions).
- The secondary state variables introduced by the transfer and transposition kinetics must be obtained as the result of a mapping of the primary state space, conditions and parameters.
- Add control by linking models of measurements being a function of the primary and the secondary state with the controller and the controller with the variable resistances in the transfer laws.

Mapping the Process into an Abstract Network

This first step entails making the main time-scale and associate length-scale decisions. The abstraction lays out the basic network of capacities and connections. First the system boundaries are to be defined. This is an iterative process in which essentially a local "universe" is defined, meaning the outermost boundary is defined in which the modelled object is embedded. These boundaries are at a first instance closed, implying that no exchange occurs across these boundaries. Next the boundary of the modelled object towards the embedding environment is defined. Subsequently the object itself is subdivided recursively into smaller and smaller bits (Preisig, 2009). This process of providing this structure is completely left to the person who models the process. It is the essential design component: A person sits down and decides on how to subdivide the plant into "digestible" pieces so that the conglomerate appropriately describes the plant's behaviour. The key lies in the term appropriately, thereby requesting closing of a loop on the most outer shell, requesting a measure of performance of the process using the model in the particular application, for example control.

The result of this operation is a network in which the nodes represent capacities, smallest control volumes, and the arcs the interactions in the form of flows of extensive quantities. The network can be very large, meaning it may consist of millions of nodes, as this is the case in discretised distributed systems, such as they are common in fluid dynamics. The connections represent flows of extensive quantities, such as component mass, heat, work, momentum. The arcs are directed with the directionality introducing a reference co-ordinate system for each connection against which the directionality of the actual flow is being measured. Because the network is very large, it is convenient to overlay this network a hierarchical structure, which enables later the handling of large and complex models.

Adding the element of control adds a new type of nodes and arcs (Preisig, 1996): Control related nodes represent dynamic information processing systems. A minimum of three different arcs are introduced, namely directed arcs that connect capacities with information systems, arcs that go the opposite direction and those that connect between information processing nodes. The nodes themselves must represent realisable systems. The connection from the capacity to the information node is an observation of the state of the capacity. The arc from the information node to the target node represents a manipulating element in a stream, such as a valve and represents a process control input. All of these signals are information signals and unidirectional.

The network representation can be further enriched by specifying distributed nodes vs. lumped nodes vs. steady state nodes. This gives rise to defining also respective boundaries and connections, thus distributed boundaries and distributed connections.

Having defined modelling phys-chem-bio systems on the macroscopic field-theory level, the fundamental capturing the behaviour are the dynamic conservation principles; thus the conservation for component mass, energy and momentum. The accumulation in the capacity is balanced with the inflow and outflow of the respective quantity and in the case of component mass, the internal transposition of one kind of mass into another. It is noteworthy that flows of one kind of extensive quantity induces the flow of others, most remarkably the flow of mass induces flow of volume, thus flow of volume work and flow of energy in the form of internal energy, potential energy and kinetic energy. This information can also be added to the network through a typing mechanism enabled for nodes and arcs. The result is a set of typed sub-networks, being superimposed over the basic network.

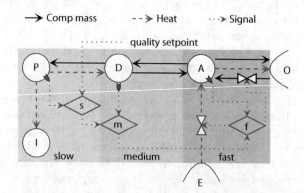

FIGURE 1

An example of abstracting a process: Model for controlling the climate in a food transport container (Verdijck, 2003). The systems (P,D,A,O) form a mass transfer network. All systems are part of a energy transfer network except the control systems (s,f,m) that form the information processing network.

Defining the Components: Interactions

Defining the interactions entails defining the transfer of extensive quantities. According to Gibbs the driving forces in the fields are defined as the partial derivatives of energy with respect to the transferred quantities. For conductive heat transfer and radiation this yields the temperature, for mass flow it is the chemical potential if the mechanism is diffusion or volumetric if it is convection, etc.

Computer-Based Modelling

The structure of the step-wise approach reflects into the architecture and nature of the tools we use to construct an environment for designing and maintaining process models. The three main components are a multi-graph editor, being the master module, a network semantics module and a node/component semantics module.

Top Module: Multi-Graph Editor

The outer shell of the environment implements a multi-graph editor implementing the handling of a large number of interacting graphs. The tool edits graphs: defines new graphs, add, deletes, moves nodes, connect nodes by arcs, imports graphs, makes nodes to represent graphs and explode nodes that contain graphs. This latter feature enables construction of networks of graphs, hierarchical graphs. The graphical user interface is controlled by an automaton that implements the semantics of the network. The set of actions is strictly limited to graph operations.

Plug-in 1: Network Semantics

The multi-graph editor has a slot to receive a module implementing the network's semantics. The network semantics are captured in an automaton, which it then provides to the multi-graph editor. On the other end, the network semantics module has a slot for receiving the semantics of the network components, namely the nodes and the arcs. The attached graphical user interface provides the user with manipulating the network semantics. Events, being generated by the network semantics GUI (Graphical User Interface) are processed internally and may generate actions that are passed on to the multi-graph editor, which in turn updates its GUI and provides the information back in sequence to the other two modules.

In terms of the model equations this part settles the rules of how the equations for the nodes and the arcs are being combined to represent the overall system. Thus for the phys-chem-bio system, these are the network

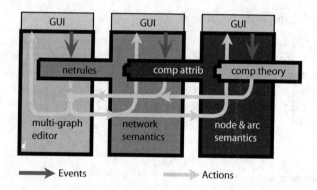

FIGURE 2
A top-down view on the environment.

of conservation equations, latter being supplied by the graph component semantics module. The automaton contains the rules on how the network is being structured, for example that the network is constructed as a strict-hierarchical graph, thus forming a tree at the end of which are primitive nodes representing the capacities. It implements also the rule that connections can only be established between leave nodes, thus primitive capacities.

An automaton is defined by a triple ▶ $\mathcal{X}$:: set of states associated with the construction of the graph i.e. deleting, inserting, creating arc, explode/group and explore. With the nodes and arcs being typed (coloured), the state set is also a function of these "colouors". ▶ $\mathcal{E}$:: a set of events generated by the GUI, the mouse and the keyboard. ▶ $\mathcal{Y}$:: a set of actions. Actions are primarily associated with handling of the graph, but may be augmented with actions associated with graph and graph component semantics.

Plug-in 2: Graph Component Semantics

This module implements the semantics of the graph components. The implementation uses equations that represent the nodes internals and the connections. This it implements the conservation for the each type of node, and a superstructure of equations representing all possible state variable transformation equations as well as geometry components.

The transfer descriptions are attached to the typed arcs. They use the graph information to construct the transfer laws and tie into the state variable transformations.

All the transformations link to the material description for which we use strictly canonical models. The canonical models are equipped with the canonical transformations and the ability to generate derivatives with respect to the canonical variables as well as all the parameters of the underlying property model (Løevfall, 2008). Thus the module provides the driving forces and sensitivities as well as all the Jacobians, Hessians or higher-order derivatives the representation requires or the solver may utilise.

Factory: Network Semantics

The network semantics are generated off-line. It requires information from the multi-editor, namely the events being generated by the GUI and the available actions. The actions are all associated with handling the components in the multi-graph. Currently the total is less than 20. The second piece of information is what kind of nodes and arcs there are so as to capture the rules in the automaton. The type information is collected from the node/arc theory module, which supplies the type information. For nodes we have (steady-state, dynamic, graph) and ([0..3]-D distributed), (external, internal) and types (physical (mass, component mass, energy, momentum), signal processing). For the connections ([0-2]-D distributed) and (mass, component mass, heat, volumetric work, signal). Further, the different components can also define their own, tailored actions, that can be activated through the multi-graph GUI. For example one wants to be able to edit the properties of a node by mouse-click on the respective node.

Factory: Graph Component Semantics

Physical-Chemical-Biological Model Network

This factory generates a super bipartite graph[3] of variable and equations that represent nodes and arcs in the network. We define a super-network or a super-graph as a graph in which some arcs may connect to several nodes. This enables us to introduce decision variables which indicated particular choices in alternative connections. Consider the following structure with all the variables being typed:

$$\text{dynamics} \quad \dot{x} = \mathbf{F}z + \mathbf{N}r \tag{1}$$

$$\text{integrator} \quad \mathbf{x} = \int_0^t \dot{x}(\tau)\, d\tau + \mathbf{x}(0) \tag{2}$$

$$\text{flow from a to b} \quad z_{a/b} := \left\{ t_{a/b}^{\langle \vartheta \rangle}(y_a - y_b) \right\} \tag{3}$$

$$\text{transposition} \quad r_s := \left\{ k_s^{\langle \kappa \rangle}(y_s) \right\} \tag{4}$$

$$\text{state-var transf} \quad 0 := \left\{ S_s^{(\sigma)}(x_s, y_s) \right\} \tag{5}$$

---

[3] Graph theory uses the term hyper graph. Since chemical engineering is used to super-structures and the here addressed problem is similar to super structures we use the term super-graph

The state of the network model for the given type is thereby given by the stack of vectors x, where the stack parameter is the system (node) index. The matrix **F** is the typed incidence matrix of the network and the z the stacked and typed transport vectors with the stack parameter being the flow index. The block matrix **N** is the block-incidence matrix of the transposition network, thus usually the reaction network and r the stack of reaction rates with the stack variable being the reaction and the system indices. The index s indicates an arbitrary system (node) s.

The second equation defines an arbitrary flow vector from a to b. Again it is of the appropriate type. Since in general several different transfer mechanism may be possible, several choices may exist. This introduces a hypergraph feature with the integer variable ϑ being the decision variable. Note that the transfer is a function of the two systems being connected, here labelled with a and b. Similar for the transposition, several alternatives may exist, with κ being the decision variable. Finally the transformation equations may also allow for a choice as for example different equation of states may be chose. Many of these transformations are though unique and must thus not be chosen. The transposition and the state-variable transformations are only a function of the local state. The representation has a number of additional properties: ▶ The dynamic equations are linear in the flows and the reaction rates. ▶ There may be alternative transport mechanisms and ▶ alternative transfer descriptions. ▶ There may be alternative transposition descriptions, reaction laws for example. ▶ The state variable transformations form mostly a lower triagonal functional system. ▶ They are often nonlinear, ▶ may have multiple solutions, and ▶ may be difficult to solve. ▶ Alternatives include equation of state and ▶ alternative geometry relations. Essential though is the **structure** of the system. The key in recent research (Westerweele, 2003; Preisig, 2004) was the recognition that the state variable transformations must be a mapping of the primary state x into the secondary state space spanned by y. ▶ The sum of the dynamic equations eliminate the production term in each control volume.

Adding Control

The control network connects through observing state-dependent information, thus directly primary state variables or secondary state variables and manipulates variables in the transfer laws. For this purpose, one has to extend the transfer laws with a manipulated variable:

$$\text{flow from a to b} \quad z_{a/b} := \left\{ t_{a/b}^{\langle \vartheta \rangle}(y_a, y_b, u_{a/b}) \right\}$$

The thus introduced arcs are of the type signal and the nodes implementing controllers are of the type information systems. Since the multi-editor is manipulating hierarchical networks, hierarchical control structures are enabled. It is certainly also possible to introduce any other artefacts such as samplers and zero-order hold for the modelling of time-discrete control systems and state discretisation or output discretisation for the implementation of discrete-event dynamic structures.

Indexing

The main structure of the algebraic representation of the model is laid out in Equation (1)- Equation (5). These equations are primarily indexed with nodes, arcs and "type" of physical quantity. The "type" if structured and one type my induce another. So, for example *component mass* implies *mass*, thus if a node, representing a capacity is connected to an arc of the type *component mass*, then the node will have to have the attribute *component mass* and *mass* too. Thus there is also a semantics associated with the "types", attributes associated with the graph components and consequently the node and arc representation. Such rules are: node inherits arc types *component mass* $\rightarrow$ *mass* , *heat* and *work* $\rightarrow$ *energy*. These types map generically into attributes attached to nodes, arcs and algebraic objects representing nodes an arcs or any parts thereof. In particular they give also rise to the definition of typed networks, which reflect into the typed incidence matrix definitions used explicitly in the model representation Equation (1). Some attributes are computed for the network according to rules, the main one being the distribution of the species in mass transfer networks. The idea of defining component mass transfer systems that build on black/white diffusivities (species does or does not diffuse) and a black/white definition of reactions (if all reactants are present, product is produced) is quite old (Preisig et al., 1990) being introduced to the chemical engineering community in (Preisig, 1994). The algorithm associated with the calculation is a recursive depth-first or width-first graph search with the additional features of implementing the "diffusivity" operator into the arc transition and the "reaction" operator into the node for additional generation of species, thus an additional attribute.

On the equation level, the diffusivities map into the definition of the attributes for the arcs representing component mass transfer and the species present in each node into attributes for the nodes. With arcs connecting with nodes, these attributes reflect into the equations. It is essential that on this level the indexing is introduced as an algebraic concept. Probably he first time the concept was presented to the public was (Westerweele et al., 1998) and later (Westerweele, 2003). The idea is to introduce a selector for sets in the form of selector matrices. Translated into index sets, a selector matrix is a binary matrix indicating selection by a binary entry. Let $\mathscr{S}$ be a set of species, called the base set and $\mathscr{S}_m$ be a set of species present in connection m then a selector matrix $S \in \mathbb{B}^{s \times n}$ can be defined in which $s := \text{card}(\mathscr{S})$ and which $n := \text{card}(\mathscr{S}_m)$. Further let the vector $c := [1,\ldots,\text{card}(\mathscr{S})]$ be the indexing vector for set $\mathscr{S}$ then the indexing vector for the set $\mathscr{S}_m$ is:

$$b := S^T c$$

This concept can be extended to higher indexed objects. For example one can build a stacked version of Equation (6). Each stack is characterised by two indices (level 1 index, level 0 index). The corresponding index sets (level 0) are then themselves indexed with level 1 sets. So in our case, the node species set, having as a base set the base species set, is indexed with the node set forming a 2-level block index set and the mass-arc species set is indexed with the mass-arc index. To be more precise in our definitions, we define index structures, which are duples each with an index and an index set, for example $s \in \mathscr{S}$. Considering mass only, we define the following index structures:

$$\text{nodes} \qquad \mathscr{N}^n := n \in \mathscr{N}$$

$$\text{species} \qquad \mathscr{S}^s := s \in \mathscr{S}$$

$$\text{mass-arcs} \qquad \mathscr{M}^m := m \in \mathscr{M}$$

Block indices can then be defined, for example:

$$\text{species in nodes} \qquad \mathscr{S}^i_n := i \in \mathscr{S}_{n \in \mathscr{N}}$$

$$\text{species in mass arcs} \qquad \mathscr{S}^j_m := j \in \mathscr{S}_{m \in \mathscr{M}}$$

Indicating the indexing as subscripts, the notation becomes precise:

$$\mathbf{z}_{\mathscr{S}^s} := S_{\mathscr{S}^s, \mathscr{S}^j_m} \overset{\mathscr{S}^j_m}{*} \mathbf{z}_{\mathscr{S}^j_m}$$

where the index set on top of the multiplication operator specifies over which axis the reduction (summation) takes place. This notation, whilst it is somewhat tedious, resolves the index broadcasting problem containing all the information required to safely code the algorithms. The component mass conservation Equation (8) without reactions, can then be written:

$$S_{\mathscr{S}^i_n, \mathscr{S}^j_m} := S_{\mathscr{S}^s, \mathscr{S}^i_n} \overset{\mathscr{S}^s}{*} S_{\mathscr{S}^s, \mathscr{S}^j_m} \tag{6}$$

$$F_{\mathscr{S}^i_n, \mathscr{S}^j_n} := F_{\mathscr{N}^n, \mathscr{M}^m} \overset{\mathscr{N}^n, \mathscr{M}^m}{} S_{\mathscr{S}^i_n, \mathscr{S}^i_m} \tag{7}$$

$$\dot{\mathbf{n}}_{\mathscr{S}^i_n} := F_{\mathscr{S}^i_n, \mathscr{S}^j_m} \overset{\mathscr{S}^j_m}{*} \mathbf{z}_{\mathscr{S}^j_m} \tag{8}$$

The first equation Equation (6) generates from two selection matrix a new one, which maps from the stacked species in mass-arcs to the stacked species in mass-nodes. The second equation Equation (7) "colours" the mass-typed incidence matrix with the generated new selection matrix through a dot product[4], which is a block-scalar matrix times a block-matrix matrix.

Implementation

For the representation of the theory a small language was defined. It has a number of base objects: ▶ A variable, which can be indexed serves as the basis for the variable definitions. A set of operations are associated with the variable object, including the base operators $(+ - * /)$ and the usual set of unitary functions. This is extended gradually to a physical variable, which has the basic SI units. The operators are also applied to the units providing the means of doing a unit check and unit generation in expressions. ▶ The set of operators is being extended by a product operator, which operates on a set of dimensions being specified as parameters to the operator. This implements a generalized tensor product. The respective dimensions are given in the form of a duple consisting of an index and an index set. The approach can be interpreted as a type of Einstein indexing. The product operator checks on the existence of the given index sets in the two objects being combined whilst the objects must not match in the other dimensions. The result is mapped into the space consisting of all the dimensions not being the operator parameters. Thus for example the product:

$$\mathbf{R}_{\mathcal{A}^i, \mathcal{D}^r} := A_{\mathcal{A}^i, \mathcal{B}^j, \mathcal{C}^k} * [\mathcal{B}^j, \mathcal{C}^k] * A_{\mathcal{B}^j, \mathcal{C}^k, \mathcal{D}^r}$$

$$:= \sum_{\mathcal{C}^k} \sum_{\mathcal{B}^j} a_{i,j,k} * b_{j,k,r}$$

The syntax for the operator is * *list* * with *list* :: [*varID*, ..., *varID*] and *list*:: a variable identifier. In addition an operator is defined, which generalises the product of a scalar with a matrix. In this case the index sets are indexed themselves as one deals with generalised block matrix.

The module is thus designed to check on consistency of units for the defined expressions and subsequently equations. It also checks on the index sets if variables are indexed and it implements those two additional operators for the purpose of enabling the handling of multi-dimensional objects such as indexed vectors, stacks, matrices, block matrices, and any of the multi-dimensional extensions. This module can be seen as an realisation of what has been defined in (Hackenberg, 2005), at least the flavour is very similar.

---

[4] The notation could be somewhat simplified by indicating only the indices on top of the dot operator, given that one uses unique indices

The associated parser is surprisingly small, less than 30 lines of code being a realisation of a Backus-Naur-type of grammar definition. The link to the multi-graph is given by a set of index sets: ▶ The node set, ▶ the arc set, ▶ the attribute set (colours, types). These sets are used to refine the knowledge representation: ▶ the attributes for each node, and ▶ for each arc.

The parser generates an abstract syntax tree with typed variables and operators. The two special operators are used in connection of the computation of the index set and their broadcasting to the various algebraic objects. The model in this generic form is written onto a file to be used by the theory plug-in as described in section on **Component Semantics**. Here again the same parser is being used, but a different abstract syntax tree is being generated. Whilst in the theory definition module, the emphasis is on checking for consistency, this is now not anymore necessary and a more number-oriented version is implemented enabling the post-computation of the index maps. The representation was chosen such that the implementation of the numerical calculations can be done with NumPy, which implements also the special multiplication operator. The special dot operator is quite readily implemented as a special method for specific variable types. The implementation uses strong typing, making extensive use of inheriting and extending objects.

Having the variables/equation hyper-network being stored as abstract syntax trees makes it quite straightforward to use a template machine for generating target code (Parr, 2004). Once one template set has been implemented it is straightforward to extend this to different target codes. Attention must be paid to making the wrappers. The splicing operation is not always trivial. Interesting target codes are C and C++ for use with standard integrators. We have lately been using BuzziFerraris library for example (Buzzi-Ferraris and Manca, 1998; Manca and Buzzi-Ferraris, 2007). MatLab® is certainly another interesting target and in the past we have also written code for Modellica® and gProms®.

Making Assumptions

Models are often modified by using insight to the process in terms of order-of-magnitude assumptions about time-scales. The three main assumptions have been discussed in (Preisig, 2008), namely small capacities, large flows and large reaction rates. All of them can be reduced to standard singular perturbation problems (Murdock, 1999). The making of assumptions may be enforced by not having enough information. Thus such assumptions may be introduced motivated mainly by the lack of information. For example if an output stream is not known, say an overflow, one may assume constant volume. Similarly for a flash tank one may assume constant volume, which gives rise for making the same assumptions in a distillation model, for example. These assumptions invariably introduce mathematical problems, mostly differential index problems (Moe, 1996; Ascher and Petzold, 1998), which however all can be resolved by trivial null-space calculations associated with the different incidence matrices. The interested reader is referred to the cited reference.

Workflow

The Normal User

For the common user, the phys-chem-bio modelling environment opens a work bench that ask for either opening an existing model for further modification or to create a new one. The multi-graph editor provides a graphical interface to the graphs each of which shows the whole model at all times. Since the model forms a strict-hierarchical tree with the leave nodes being the capacities and the control components, on one side the parent to the currently viewed node is shown whilst on the other the siblings are shown. The graph itself shows the currently viewed higher-level node being a graph on its own, with internal nodes connected by internal arcs and connected to the outside (parents or siblings) through external arcs. The arcs are typed using the plug-in interface of the network semantic module. The interface also shows the state of the automaton, which can be changed by key indicated keys or tics. The interface uses different cursors, which show possible actions displayed in textual form as well. This makes the environment largely self-documenting. Experiments with undergraduate students showed that they can use the interface within an hour even though no documentation was given except a brief demonstration.

The interface enables the definition of the hierarchical graph with typing. Next the details must be filled in. At this point, the environment is aware of the typed incidence matrices, representing the typed subnets. Thus all the information necessary to generate the structure of the balance equations in the form of Equation (1) are known. This asks for the flows and the transpositions to be specified in the next step, which is defining the different types of transfers. Since the model has multiple possible transfer descriptions this requires selecting an appropriate mechanism and possibly a choice between alternative descriptions must be taken. Once a decision is taken, a new set of variables is introduced into the model description. The respective transformation equations are automatically added as long as there is no alternatives to be chosen. One of such choices is associated with the material description, equation of states for example. Again once a choice has been made new variables are being introduced, which are filled in so as to result a unique mapping for the secondary state established through this process and the primary state as well as the parameters and specified conditions. The same process applies to the transposition/reaction part. The result is a complete model, which is always of differential index one.

The user is also given the option that some things may not be known, which then in turn requires to make some assumptions as mentioned in the previous section. The environment has the model reduction mechanisms built in, as they affect the incidence matrices.

The environment indicates where things have not yet been defined in the model on the graphical representation of the model. This makes it easy to complete the model.

The user is also given the option to save parts of the model tree defining sub-models that can be saved in a library for further use. Thus logically also the insertion of such sub-models is being enabled. As this can be done at any stage of the definition process, sub-models can be put into libraries with only the physical structure or any typing or other model detail being added. This provides a maximum of flexibility for the definition process.

Finally the user may choose a target code and a textual documentation of the model.

The Theory-Defining User

As one moves up in the hierarchy of users more knowledge is required. The interface implementing the little language is designed to catch as many as possible structural errors. However, not all can be checked. Obviously it is always possible to generate a theory that does no make any physical sense what-so-ever.

Defining New Plug-Ins

The existing plug-ins for process modelling can be replaced by other ones providing the multi-editor a complete new look and use. Obviously this requires insight into the code and its details.

Concluding Thoughts

► The abstraction of the process into a network of capacities and connections communicating extensive quantities is the most important step in modelling. It fixes all the main characteristics of the model. What comes afterwards is only adding colour, so-to-speak. ► The mapping "state variable transformations" is a centre piece in the definition framework. It is absolutely essential that this piece is in place. The author is tempted to claim that not defining these transformations properly is the most common mistake in models.

▶ There are many ways to construct such an environment. However, as far as the author is aware, this is one of a few in the zoo of PSE tools. ▶ Unsolved issues in connection with solvers, integrators in particular make little use of the model structure, fail frequently, do little to no checks on validity range of variables and their definition domain. ▶ There is no need to obscure the model with mapping it into secondary structures such as Bondgraphs or the likes. The Hamiltonian analysis, for example, can be readily done in the network modelling setting as it is only an abstraction of basic physical principles. ▶ The extension to particulate systems of stochastic nature is done readily by a proper averaging that satisfy the conservation in the respective domains and at the connecting boundaries. ▶ Potential cost savings are enormous. Direct: 2-order of magnitude savings in preparing the model. Indirect: Increased use of models. Correct models. For a start, all balances close on any hierarchical level, a feature none of to-days simulators can guarantee. ▶ Proper integration of control and phys model. ▶ Complete transparency in what is the structure, what are the partial models being used and what assumptions have been made. ▶ No errors from manipulations, no transcript errors, no coding errors, no indexing errors, complete documentation, documentation carries along to any down-stream processing. ▶ Potential of integrating different graphical representation ranging from operator interface to abstract modelling interface. Would lead to an enormous concentration of information.

References

L Apostel. *Towards the formal study of models in the non-formal sciences from the concept and the role of the model in mathematics and natural and social sciences.* D. Reidel Publishing Company, Dordrecht, The Netherlands, 1960.

R Aris. *Mathematical modelling techniques.* Pitman, London, 1978.

U M Ascher and L R Petzold. *Computer methods for ordinary differetntial equations and differential-algebraic equations.* SIAM, 1998.

G Buzzi-Ferraris and D Manca. BzzOde: a new C++ class for the solution of stiff and non-stiff ordinary differential equations systems. *Comp & Chem Eng*, 22(11):1595–1621, 1998.

Jörg Hackenberg. *Computer support for theory-based modelling of process systems.* PhD thesis, RWTH Aachen, Germany, 2005.

Børn Tore Løevfall. *Computer Realisation of thermodynamic modles using algebraic objects.* PhD thesis, NTNU, Trondheim, Norway, 2008.

D Manca and G Buzzi-Ferraris. The solution of dae systems by a nuymerically robust and efficient solver. *ESCAPE 17*, pages 93–98, 2007.

H I Moe. *Dynamic process simulation: studies on modeling and index reduction.* PhD thesis, Norwegian University of Science and Technology, Trondheim, Norway, 1996.

J A Murdock. *Perturbations, Theory and Methods.* SIAM Classics In Applied Mathematics 27, 1999.

Terence John Parr. Enforcing strict model-view separation in template engines. In *WWW'04: Proceedings of the 13th international conference on World Wide Web*, pages 224–233, New York, NY, USA, 2004. ACM. ISBN 1-58113-844-X. doi: 10.1145/988672.988703.

M Peschel. Grundprinzip der modellbildung. In F Flix, K-H Schmelowsky, M Sydow, and W Zwick, editors, *Mathematische Modellbildung in Naturwissenschaft und Technik*. Akademie Verlag, Berlin, Germany, 1976.

H A Preisig. Computer-aided modelling: Species topology. *ADCHEM 94*, pages 143–148, 1994.

H A Preisig. Computer aided modelling - two paradigms on control. *Computers & Chemical Engineering*, 20:S981–S986, 1996.

H A Preisig. A topology approach to modelling. *SIMS 45*, pages 413–420, 2004.

H A Preisig. Three principle model reductions based on time-scale considerations. *ESCAPE 18*, 2008.

H. A Preisig. A graph-theory-based approach to the analysis of large-scale plants. *Computers & Chemical Engineering*, 33:598–604, 2009. doi: 10.1016/j.compchemeng.2008.10.016.

H A Preisig, T Y Lee, and F Little. A prototype computer-aided modelling tool for life-support system models. *20th ICES*, (901269):10, 1990.

A Rosenbluth and N Wiener. The role of models in science. *Philosophy of Science*, 12, 1945.

J W Swanson. On models. *British Journal of Philosophy of Sciene*, 17(4):297–311, 1966.

G J C Verdijck. *Model-based Product Quality Control applied to climate controlled processing of agro-material*. PhD thesis, TU Eindhoven, Eindhoven, The Netherlands, 2003.

M R Westerweele. *Five Steps for Building Consistent Dynamic Process Models and Their Implementation in the Computer Tool MODELLER*. PhD thesis, TU Eindhoven, Eindhoven, The Netherlands, ISBN 90-386-2964-8 2003.

M R Westerweele, H A Preisig, and M Weiss. Concept and design of modeller, a computer-aided modelling tool. *Benelux 98*, 1998.

100

Mathematical Modeling in Process Design and Operation: Structural Assessment

Ferenc Friedler[1]* and L. T. Fan[2]

[1] University of Pannonia, Egyetem u. 10, H-8200 Veszprem, Hungary
[2] Kansas State University, Manhattan KS 66506, U.S.A.

CONTENTS

ABSTRACT Process synthesis and batch process scheduling are executed most frequently by resorting to mathematical programming. Nevertheless, its most crucial initial step, i.e., the generation of the mathematical model, has received relatively minute attention. The present contribution demonstrates that the mathematical model generated at this crucial step affects profoundly the quality of the solution and the necessary computational time. The P-graph and S-graph frameworks developed for process synthesis and batch process scheduling, respectively, provide consistent methodologies for efficaciously generating the mathematical model and its solution.

KEYWORDS *Mathematical modeling, process design, process operation, P-graph, S-graph*

---

* To whom all correspondence should be addressed (friedler@dcs.vein.hu)

Introduction

A process design or operation problem is solved customarily by a mathematical-programming method. Review of recent publications has revealed various failures in modeling process design and operation. An illegitimate mathematical model may result in a non-optimal or even infeasible solution, or it is unsolvable due to its excessive complexity. Obviously, a mathematical model should be a valid representation of the process embodying all its significant features and yet be solvable. The relationship of the mathematical model to the process being modeled as well as to the solver being deployed are usually convoluted, and therefore, the effective and valid model is extremely difficult to establish. It is worth noting that the available information on the process structure can significantly affect the procedure for modeling the process or its operation. It appears that the available literature is void of model generation; it is treated only in a limited number of publications (see, e.g., Grossmann, 1990; Kovacs *et al.*, 2000). The main emphasis of the current contribution is on model generation, especially for process synthesis and scheduling of batch processes.

Structural Difficulties in Model Generation: Process Synthesis

Let us suppose that a process synthesis problem is framed by specifying a set of potential feed streams (raw materials) and product streams together with the mathematical models of plausible processing equipment, i.e., operating units. The problem aims at generating or identifying the optimal network of operating units through optimization. It has been repeatedly demonstrated that various mathematical models of a given process synthesis problem may result in solutions with substantially different values of the cost function. Thus, it is of the utmost importance that the mathematical model generated indeed yields the optimal solution of the process synthesis problem as originally framed. Several examples illustrate that an incomplete mathematical model results in a solution far from the optimal one. For example, Kovacs *et al.* (2000) have shown that the published results (Quasada and Grossmann, 1995) can be improved as much as 30 %, thereby giving rise to modeling issue 1: *Modeling leads to an incomplete mathematical model*.

Process synthesis is initiated most commonly by constructing the so-called super-structure, which, in turn, gives rise to the mathematical model necessary for identifying the optimal solution. It is, therefore, essential that the structure or network of the optimal solution be contained in the super-structure; otherwise, the optimality of the resultant solution cannot be assured. A parametric study of a simple class of process synthesis problems illustrates that such a super-structure cannot be generated readily.

Kovacs *et al.* (1998) have analyzed the set of potentially optimal networks for separation-network synthesis problems with two three-component feed streams and three pure product streams with simple sharp separators, dividers, and mixers. This class of problems will be called the class 1 SNS problems.

As illustrated in Kovacs *et al.*, (1998), ten different networks can be optimal depending on the values of the problem's parameters. Naturally, two separators are sufficient to solve any instance of the class 1 SNS problems. The cost function is concave for any separator. It is, therefore, expected that three or more separators, i.e., redundancy, cannot give rise to the optimality. Among the ten networks that are optimal under various circumstances, however, only two networks contain two separators each; and eight networks, three separators each. Hence, eight networks contain redundant separators. In other words, for some instances of the class 1 SNS problems, the optimal solutions based on the mathematical model, which excludes redundancy, are not optimal solutions of the synthesis problems as originally posed (Kovacs *et al.*, 1998). Moreover, the inclusion of a loop in an optimal separation network of the class 1 SNS problems is unexpected (see, e.g., Floudas, 1987). On the contrary, four of the ten optimal separation networks involve looping. This implies that the optimal solution obtained from the mathematical model excluding loops for the class 1 SNS problems is not the optimal solutions for some of the SNS problems as originally posed (also see Kovacs *et al.*, 1993).

In an SNS problem for generating multicomponent product streams from multicomponent feed streams, it is often possible to bypass certain amounts of feed streams to some product streams. If the cost of bypassing is negligibly small, it is expected that the extent of bypassing would be always maximal in an optimal solution. Nevertheless, an example is given by Kovacs *et al.* (1995) to illustrate that this is not always the case: Even when the cost of bypassing is zero, it may affect the network structure, thus may resulting in an increase in the network's cost, which offsets any advantage gained from bypassing, thereby giving rise to modeling issue 2: *Modeling is based on invalid assumptions.*

These issues imply that it is essential to define a rigorous super-structure that assumedly leads to the optimality. Different mathematical models may give rise to the same optimality; nevertheless, the effort required for their solution can markedly vary. To achieve the maximum efficiency for synthesis, it is mandatory that the methods for generating the mathematical models and those for their solution be determined collectively.

Structural Difficulties in Model Generation:
Batch Process Scheduling

The solution of the batch scheduling problem of Méndez and Cerdá (2003) corresponds to a minimum makespan of 60 hours. The same problem was solved by Kim *et al.* (2000) resulting in a similar solution with a minimum makespan of 60 hours. Note that at 30 hours in the makespan of these solutions, three units exchange materials simultaneously.

It can be unequivocally demonstrated that the solution mentioned above is not optimal; in fact, it is not even a feasible solution to this problem as delineated in the following. It is impractical to have two or more units exchanging material simultaneously. Such occurrence, termed cross-transfer, cannot be achieved surely in a real setting. The mathematical formulation or model as originally proposed inherently fails to detect it. The truly optimal solution has been given by Hegyháti *et al.* (2009); its minimum makespan is 71 hours, thereby giving rise to modeling issue 3: *Modeling leads to a mathematical model whose optimal solution is unattainable or infeasible in practice.*

For the maximum throughput problem of batch processes, most of the recent mathematical formulations are based on the so-called continuous-time domain representation in which the time horizon of interest is divided into uneven time intervals using a presupposed number of time points (see Floudas and Lin, 2004 for a review). Each time point signifies the start or end of the task in a particular unit, and its location on the time horizon is determined by the optimal point. The major drawback of such formulations is that the optimality of the solution depends on the right number of time points. Unfortunately, no mathematically rigorous method is available for determining the right number of time points to ensure global optimality. Currently, an iterative procedure is adopted in which the number of time points is incremented with the solution determined at each iteration step. If the final solution does not improve during some increases in time points, it is assumed that the optimal solution for the problem at hand has been attained. This iterative technique, however, is subject to a major drawback. In some problems, even if the objective value might not improve for two or more increments in the number of time points, a further increase in the number of time points might improve the value of the objective, thus implying that iteration can terminate at a suboptimal result (Castro *et al.*, 2001), thereby giving rise to modeling issue 4: *Modeling is based on unknown information.*

P-graph Framework for Process Synthesis

The P-graph framework has been established for the effective integration of model-generation and solution in process-network synthesis (PNS). The framework includes a specific network representation and algorithms.

Conventional graphs are suitable for analyzing a process structure; however, such graphs are incapable of uniquely representing process structures in process synthesis (Friedler *et al.*, 1992). Thus, a special directed bipartite graph, P-graph, has been introduced to circumvent this difficulty. It is bipartite since its vertices are partitioned into two sets, and no two vertices in the same set are adjacent in the graph. Vertices in one of the partitions are for representing operating units, and those in the other are for representing materials.

Combinatorial Properties of Process Networks in Process Synthesis

Let M be a given finite set of all material species, or materials in short, which are to be involved in the synthesis of a process system. Suppose that a process network synthesis problem is specified by triplet (P,R,O) where P($\subset$M) is the set of products to be produced; R($\subset$M), the set of available raw materials; and O, the set of operating units.

Axioms and their Relations to the MINLP Model

It is usually assumed in the available literature that a process synthesis problem can be appropriately formulated as a MINLP problem. In this regard, the following MINLP model is deemed to describe rigorously a process synthesis problem.

$$\min \quad f(x,y)$$
$$\text{s.t.} \quad g(x,y) \leq 0,$$
$$x \in \mathbf{R}^n, \ y \in \text{integer}.$$

Naturally, if P-graph (m,o) is the structure of a feasible solution of process network synthesis problem (P,R,O), then the synthesized process must produce each product involved in set P. In the structure of this process, therefore, every final product must be represented; otherwise, the process is infeasible. The requirement for producing every product appears in the MINLP model as a constraint on the required amount to be produced. Consequently, it is embedded into the MINLP model that every product appears in a feasible process network. This can be regarded as an axiom of the combinatorially or structurally feasible networks; in fact, four other axioms have been conceived (Friedler *et al.*, 1992, 1998).

In synthesizing a process network from 35 plausible operating units, the number of possible networks is ≈34 billion (Friedler *et al.*, 1995). The five axioms reduce this search space to 3465; in other words, it is sufficient to take into account only these 3465 networks in synthesizing the process. Naturally, a question arises as to the possibility of reducing the search space further by resorting to an additional combinatorial axiom or axioms. The answer is negative: any of the 3465 networks can be optimal under certain parameters and constraints of the problem.

Rigorous Super-Structure: Maximal Structure

Suppose that S(P,R,O) denotes the set of all combinatorially feasible networks of PNS problem (P,R,O). The union of all combinatorially feasible networks, i.e., network μ(P,R,O), is a rigorous super-structure for PNS problem (P,R,O); it is called the maximal structure. It has been proved (Friedler *et al.*, 1992) that the maximal structure is also a combinatorially feasible network of the problem, i.e., μ(P,R,O)∈S(P,R,O). On the basis of this property, the maximal structure can be generated algorithmically in polynomial time (Friedler *et al.*, 1993). Algorithm SSG has been introduced for generating the elements of set S(P,R,O), i.e., the set of all combinatorially feasible networks (Friedler *et al.*, 1995). Algorithms MSG and SSG together are considered to be the first procedure for algorithmic process network synthesis. Even though this procedure is proved to be effective, it can be further improved by the accelerated branch-and-bound algorithm of PNS, i.e., algorithm ABB (Friedler *et al.*, 1996). Note that the P-graph framework has been deployed not only for PNS but also for the identification or synthesis of networks in the systems whose scales vastly different from process networks, such as catalytic pathways, metabolic pathways and supply chain networks (Feng *et al.*, 2000; Fan *et al.*, 2002; Feng *et al.*, 2003; Fan *et al.*, 2005; Lee *et al.*, 2005; Fan *et al.*, 2008). A tutorial account of the P-graph framework is given in Peters *et al.* (2002).

S-graph Framework for Modeling Process Operation

Unlike the commonly deployed MILP and MINLP approaches, the S-graph-based scheduling, developed by Sanmartí *et al.* (1998), resorts to a sophisticated graph representation that was initially applied to problems of makespan minimization. (Sanmartí *et al.*, 2002; Romero *et al.*, 2003, 2004). In an S-graph, two classes of arcs, the so-called recipe-arcs and schedule-arcs, are specified. In practice, if an arc is established from node i to node j, the task corresponding to node j cannot start its activity earlier than c(i,j) time after the task corresponding to node i started, where c(i,j) is the weight of the arc. The minimum makespan is determined via combinatorial algorithms by exploiting the specific structure of each individual problem; it is intrinsically capable of detecting and circumventing the infeasibility in optimization (avoiding modeling issue 3).

The S-graph framework has also been extended to throughput maximization (Majozi and Friedler, 2006). The solution procedure is based on guided search within a region defined by the structure of the problem. The efficiency of the search is a consequence of two main features. Firstly, redundancy is eliminated in the search. Secondly, the region comprising nodes that will never contribute to the generation of any optimal solution is identified and eliminated prior to the search, thus resulting simultaneously in the reduction of search space and CPU times. At each node of the search, a partial problem involving a fixed number of batches of the product is solved using S-graph algorithms. In generating the optimal solution of the problem with the S-graph-based method, no unknown information (e.g., number of time-points) is needed as any other known method, thereby circumventing modeling issue 4. The computation time of the new method is substantially shorter than those of others; nevertheless, its main advantage is that it assuredly gives rise to the optimality.

Conclusions

The difficulty of algorithmic model generation in process synthesis and batch process scheduling is illustrated by analyzing available modeling methodologies. A brief discourse is given to indicate that the P-graph and S-graph frameworks are consistent methodologies for effectively modeling of process synthesis and batch process scheduling, respectively.

References

Castro, P. M., Barbosa-Póvoa, A.P.F.D., Matos, H. (2001). An improved RTN continuous-time formulation for the short-term scheduling of multipurpose batch plants. *Ind. Eng. Chem. Res.*, 40, 2059–2068.

Fan, L.T., Bertok, B., Friedler, F. (2002). A Graph-Theoretic Method to Identify Candidate Mechanisms for Deriving the Rate Law of a Catalytic Reaction. *Comp. Chemistry*, 26, 265–292.

Fan, L.T., Shafie, S., Bertok, B., Friedler, F., Lee, D. Y., Seo, H., Park, S., Lee, S. Y. (2005) Graph-Theoretic Approach for Identifying Catalytic or Metabolic Pathways, *J. Chin. Inst. Eng.*, 28, 1021–1037.

Fan, L.T., Lin, Y.-C., Shafie, S., Hohn, K. L., Bertok, B., Friedler, F. (2008). Graph-theoretic and energetic exploration of catalytic pathways of the water-gas shift reaction. *J. Chin. Inst. Chem. Eng.*, 39, 467–473.

Feng, G., Fan, L.T., Friedler, F., Seib, P. A. (2000). Identifying Operating Units for the Design and Synthesis of Azeotropic-Distillation Systems. *Ind. Eng. Chem. Res.*, 39, 175–184.

Feng, G., Fan, L.T., Seib, P. A., Bertok, B., Kalotai, L., Friedler, F. (2003). A Graph-Theoretic Method for the Algorithmic Synthesis of Azeotropic-Distillation Systems. *Ind. Eng. Chem. Res.*, 42, 3602–3611.

Floudas, C.A. (1987). Separation Synthesis of Multivomponent Feed Streams into Multicomponent Product Streams. *AIChE J.*, 33, 540–550.

Floudas, C.A., Lin, X. (2004). Continuous-time versus discrete-time approaches for scheduling of chemical processes: a review. *Comp. Chem. Eng.*, **28**, 2109–2129.

Friedler, F., Tarjan, K., Huang, Y. W., Fan, L. T. (1992). Graph-Theoretic Approach to Process Synthesis: Axioms and Theorems. *Chem. Eng. Sci.*, 47, 1973–1988.

Friedler, F., Tarjan, K., Huang, Y. W., Fan, L. T. (1993). Graph-Theoretic Approach to Process Synthesis: Polynomial Algorithm for Maximal Structure Generation. *Comp. Chem. Eng.*, 17, 929–942.

Friedler, F., Varga, J. B., Fan, L. T. (1995). Decision-Mapping: A Tool for Consistent and Complete Decisions in Process Synthesis. *Chem. Eng. Sci.*, 50, 1755–1768.

Friedler, F., Varga, J. B., Feher, E., Fan, L.T. (1996), Combinatorially Accelerated Branch-and-Bound Method for Solving the MIP Model of Process Network Synthesis, Nonconvex Optimization and Its Applications. *State of the Art in Global Optimization, Computational Methods and Applications* (Eds: C. A. Floudas and P. M. Pardalos), pp. 609–626, Kluwer Academic Publishers, Dordrecht.

Friedler, F., Fan, L. T., Imreh, B. (1998). Process Network Synthesis: Problem Definition. *Networks*, 28, 119–124.

Grossmann, I.E. (1990) MINLP Optimization Strategies and Algorithms for Process Synthesis. *FOCAPD* (Eds: J.J. Siirola, I.E. Grossmann, G. Stephanopoulos) pp. 105–132, CACHE, Elsevier, Amsterdam.

Hegyhati, M., Majozi, T., Holczinger, T., Friedler F. (2009). Practical infeasibility of cross-transfer in batch plants with complex recipes: S-graph vs MILP methods. *Chem. Eng. Sci.*, 64, 605–610.

Kim, S. B., Lee, H. K., Lee, I. B., Lee, E. S., Lee, B. (2000). Scheduling of non-sequential multipurpose batch processes under finite intermediate storage policy. *Comp. Chem. Eng.*, 24,1603–1610.

Kovacs, Z., Friedler, F., Fan, L.T. (1993). Recycling in a Separation Process Structure. *AIChE J.*, 39, 1087–1089.

Kovacs, Z., Friedler, F., Fan, L.T. (1995). Parametric Study of Separation Network Synthesis: Extreme Properties of Optimal Structures. *Comp. Chem. Eng.*, 19, S107–112.

Kovacs, Z., Ercsey, Z., Friedler, F., Fan, L.T. (1998). Redundancy in a Separation-Network. *Hung. J. Ind. Chem.*, 26, 213–219.

Kovacs, Z., Ercsey, Z., Friedler, F., Fan, L.T. (2000). Separation-Network Synthesis: Global Optimum through Rigorous Super-Structure. *Comp. Chem. Eng.*, 24, 1881–1900.

Lee, D.-Y., Fan, L.T., Park, S., Lee, S. Y., Shafie, S., Bertok, B., Friedler, F. (2005). Complementary Identification of Multiple Flux Distributions and Multiple Metabolic Pathways. *Metab. Eng.*, 7, 182–200.

Majozi, T., Friedler, F. (2006). Maximization of Throughput in a Multipurpose Batch Plant under Fixed Time Horizon: S-Graph Approach. *Ind. Eng. Chem. Res.*, 45(20), 6713–6720.

Méndez, C.A., Cerdá, J. (2003). An MILP Continuous-Time Framework for Short-Term Scheduling of Multipurpose Batch Processes Under Different Operation Strategies. *Opt. Eng.*, 4, 7–22.

Peters, M. S., Timmerhaus, K. D., West, R. E. (2002). Flowsheet Synthesis and Development of Plant Design and Economics for Chemical Engineers. *McGraw-Hill*.

Quesada, I., Grossmann, I. E. (1995). Global optimization of bilinear process networks with multicomponent flows. *Comp. Chem. Engng*, 19 (12), 1219–1242.

Romero, J., Espuna, A., Friedler, F., Puigjaner, L. (2003). A New Framework for Batch Process Optimization Using the Flexible Recipe. *Ind. Eng. Chem. Res.*, 42(2), 370–379.

Romero, J., Puigjaner, L., Holczinger, T., Friedler, F. (2004). Scheduling Intermediate Storage Multipurpose Batch Plants Using the S-graph. *AIChE J.*, 50(2), 403–417

Sanmartí, E., Friedler, F., Puigjaner, L. (1998). Combinatorial Technique for Short Term Scheduling of Multipurpose Batch Plants Based on Schedule-Graph Representation. *Comp. Chem. Eng.*, 22, S847–S850.

Sanmarti, E., Holczinger, T., Puigjaner, L., Friedler, F. (2002). Combinatorial Framework for Effective Scheduling of Multipurpose Batch Plants. *AIChE Journal*, 48(11), 2557–2570.

Seo, H., Lee, D. Y., Park, S., Fan, L.T., Shafie, S., Bertok, B., Friedler, F. (2001). Graph-Theoretical Identification of Pathways for Biochemical Reactions. *Biotechnology Letters*, 23, 1551–1557.

Printed and bound by CPI Group (UK) Ltd, Croydon, CR0 4YY

23/10/2024

01777710-0001